The Ecology, Evolution, Systematics and Behaviour of Mites

The Ecology, Evolution, Systematics and Behaviour of Mites

Editors

Maciej Skoracki
Monika Fajfer

MDPI

Basel • Beijing • Wuhan • Barcelona • Belgrade • Novi Sad • Cluj • Manchester

Editors
Maciej Skoracki
Adam Mickiewicz University
Poznan
Poland

Monika Fajfer
Cardinal Stefan Wyszynski
University
Warsaw
Poland

Editorial Office
MDPI AG
Grosspeteranlage 5
4052 Basel, Switzerland

This is a reprint of articles from the Special Issue published online in the open access journal *Animals* (ISSN 2076-2615) (available at: https://www.mdpi.com/journal/animals/special_issues/1W665Y889X).

For citation purposes, cite each article independently as indicated on the article page online and as indicated below:

Lastname, A.A.; Lastname, B.B. Article Title. *Journal Name* **Year**, *Volume Number*, Page Range.

ISBN 978-3-7258-1459-6 (Hbk)
ISBN 978-3-7258-1460-2 (PDF)
doi.org/10.3390/books978-3-7258-1460-2

Contents

animals

Editorial

The Ecology, Evolution, Systematics, and Behaviour of Mites

Maciej Skoracki [1,*] and Monika Fajfer-Jakubek [2]

1 Department of Animal Morphology, Faculty of Biology, Adam Mickiewicz University, Uniwersytetu Poznańskiego 6, 61-614 Poznań, Poland
2 Department of Molecular Biology and Genetics, Institute of Biological Sciences, Cardinal Stefan Wyszynski University, Wóycickiego 1/3, 01-938 Warsaw, Poland; m.fajfer@uksw.edu.pl
* Correspondence: maciej.skoracki@amu.edu.pl

In the intricate web of biodiversity, mites serve as fundamental, yet often overlooked, architects, playing essential roles in ecosystems across the globe. Their interactions with plants, animals, and microorganisms highlight a complex array of ecological relationships that influence the distribution, diversity, and dynamics of biological communities. This Special Issue, entitled "The Ecology, Evolution, Systematics, and Behaviour of Mites", assembles a collection of studies that advances our understanding of mites through detailed examinations of their coevolutionary relationships, taxonomic diversity, molecular biology, and ecological interactions. The contributions within this Special Issue not only shed light on the multifaceted nature of mites but also emphasize the importance of interdisciplinary approaches to unravelling the mysteries of these abundant arthropods.

In this Special Issue, the coevolutionary associations between mites and their hosts are explored through several studies [1–5], providing insights into the host–parasite relationship, phylogeny, and host specificity of mites. For instance, the discovery of a new scale-mite species from Robert's Tree Iguana [1] not only enriches the taxonomic diversity of mites but also provides novel insights into the phylogenetic relationships within the Pterygosomatidae family. Similarly, the discovery of *Tinamiphilopsis temmincki* on the Tataupa Tinamou [2] contributes to the understanding of syringophilid mites' evolutionary history and host–parasite dynamics, challenging assumptions about host specificity and evolutionary pathways. Research on *Demodex* in the mouflon [3] expands our comprehension of host–parasite relationships by shedding light on the evolutionary history and ecological interactions of these skin mites in wild populations. Lastly, the description of three new feather mite species from Brazilian parrots [4] and the study on parasitic mites of African barbets [5] reveal the specificity of mite–host relationships and contribute to the broader understanding of coevolutionary dynamics between mites and birds, highlighting the role of ecological and evolutionary processes in shaping host–parasite interactions.

Significant strides in the taxonomic revision of mite groups are presented in articles [6–13], showcasing the evolving nature of mite systematics. The taxonomy of the Teneriffiidae family is clarified [6], while new species groups within the *Tenuipalpus* sensu lato are proposed [7]. A remarkable new species of phthiracaroid mites from the Peruvian Andes [8] highlights the discovery of novel taxa in underexplored regions. The revision of the genus *Neoprotereunetes* [9,10] and a comprehensive review of the *Neoseiulus* species in China [11] have improved our understanding of these groups. Furthermore, the establishment of a new subfamily, Cunaxicaudinae [12], highlights the continuous discovery of novel morphological features and their implications for understanding mite evolution and systematics. Finally, the addition of the article on three new species of *Aceria* from China [13] further enriches the contributions to the field of mite taxonomy in this Special Issue.

The incorporation of molecular techniques into mite research [14–19] has revolutionized our understanding of mite evolution, genetic diversity, and phylogenetic relationships. Studies on the biogeography of *Fuscozetes fuscipes* [14], genetic diversity in quill mites [15],

Citation: Skoracki, M.; Fajfer-Jakubek, M. The Ecology, Evolution, Systematics, and Behaviour of Mites. *Animals* **2024**, *14*, 1142. https://doi.org/10.3390/ani14081142

Received: 29 February 2024
Accepted: 8 March 2024
Published: 9 April 2024

and the ontogeny and ecology of *Nanhermannia coronata* [16] have demonstrated the utility of DNA barcoding and molecular phylogenetics in uncovering hidden diversity and clarifying taxonomic relationships. Molecular identification of Laelapidae mites [17] and description of a new *Ultratenuipalpus* species [18] further illustrate the power of molecular data in advancing our understanding of mite biology. Nonetheless, at the same time, a survey of mite contamination in public genomic databases [19] reveals the widespread presence of mite DNA in sequencing projects, offering insights into mite diversity and host associations through unintended data sources.

Ecological studies [20–22] have explored the interactions between mites, their hosts, and the environment, revealing the adaptive strategies and ecological roles of mites in various habitats. While the former research on Uropodina mites in dormouse nest boxes [20] shed light on the niche preferences and community dynamics of mites in mammalian nests, the latter explores similar ecological dynamics in bird nests, illustrating how mites adapt to and exploit these specialized niches. Research on the life-type characteristics of three spider mite pests [21] complements these insights by demonstrating the influence of host plants on mite behaviour and life history strategies, emphasizing the adaptive nature of these pests to different environmental conditions. Collectively, from examining mite communities in nests [20,22] to assessing the impact of host plants on spider mite pests 21], these studies underscore the adaptive flexibility of mites and their intricate interactions with the surrounding world. Such research is essential for understanding the complex behaviours of mites and the ecological niches they inhabit.

All the articles in this Special Issue advance our knowledge across the spectrum of mite biology, from their ecological roles to their evolutionary dynamics. By integrating taxonomic revisions, molecular analyses, and ecological studies with investigations into host–parasite interactions, this body of research illuminates the complexity of mites' life and their crucial roles within ecosystems. As we continue to explore the mysteries of the mite world, the interconnectedness becomes ever more apparent, highlighting the importance of interdisciplinary approaches in capturing the full scope of mite biodiversity and evolutionary interactions in the natural world.

Conflicts of Interest: The authors declare no conflict of interest.

References

1. Fajfer, M.; Skoracki, M. Life Stages and Phylogenetic Position of the New Scale-Mite of the Genus *Neopterygosoma* (Acariformes: Pterygosomatidae) from Robert's Tree Iguana. *Animals* **2023**, *13*, 2809. [CrossRef]
2. Skoracki, M.; Fajfer, M.; Hromada, M.; Hušek, J.; Sikora, B. *Tinamiphilopsis temmincki* sp. n., a New Quill Mite Species from Tataupa Tinamou, and the Early History of Syringophilid Mites. *Animals* **2023**, *13*, 2728. [CrossRef]
3. Kozina, P.; Izdebska, J.N.; Rolbiecki, L. A New Species of *Demodex* (Acariformes: Prostigmata) Observed in the Mouflon, *Ovis aries musimon* (Artiodactyla: Bovidae) with Data on the Parasitism and Occurrence of Other Ectoparasites. *Animals* **2023**, *13*, 2619. [CrossRef]
4. Hernandes, F.A. A Review of the Feather Mite Genus *Lopharalichus* Gaud & Atyeo, 1996 (Acariformes: Pterolichidae), with Descriptions of Three New Species from Brazilian Parrots (Psittaciformes: Psittacidae). *Animals* **2023**, *13*, 2360. [CrossRef]
5. Sikora, B.; Mahamoud-Issa, M.; Unsoeld, M.; Hromada, M.; Skoracki, M. Species Composition of Parasitic Mites of the Subfamily Picobiinae (Acariformes: Syringophilidae) Associated with African Barbets (Piciformes: Lybiidae). *Animals* **2023**, *13*, 2007. [CrossRef]
6. Mirza, J.H.; Kamran, M.; Alatawi, F.J. Taxonomy of the Family Teneriffiidae (Acari: Prostigmata: Anystoidea): Generic Synonymies with the Key to World Species of the Family. *Animals* **2023**, *13*, 3736. [CrossRef]
7. Elgoni, N.A.; Kamran, M.; Alatawi, F.J. *Tenuipalpus* Sensu Lato Donnadieu (Acari: Prostigmata: Tenuipalpidae); New Species Groups, a New Species, and Keys to the World Species. *Animals* **2023**, *13*, 3278. [CrossRef]
8. Niedbała, W.; Adamski, Z.; Laniecki, R.; Magowski, W.L. Ptyctimous Mites (Acari, Oribatida) of Peru with the Description of an Extraordinary New Phthiracaroid Mite from the Peruvian Andes. *Animals* **2023**, *13*, 2403. [CrossRef]
9. Laniecki, R.; Magowski, W.L. New Definition of *Neoprotereunetes* Fain et Camerik, Its Distribution and Description of the New Genus in Eupodidae (Acariformes: Prostigmata: Eupodoidea). *Animals* **2023**, *13*, 2213. [CrossRef]
10. Laniecki, R.; Magowski, W.L. Correction: Laniecki, R.; Magowski, W.L. New Definition of *Neoprotereunetes* Fain et Camerik, Its Distribution and Description of the New Genus in Eupodidae (Acariformes: Prostigmata: Eupodoidea). *Animals* 2023, *13*, 2213. *Animals* **2023**, *13*, 3761. [CrossRef]

11. Liu, Y.; Ren, F.-X.; Fan, Q.-H.; Ma, M. The Genus *Neoseiulus* Hughes (Acari: Phytoseiidae) in Shanxi, China. *Animals* **2023**, *13*, 1478. [CrossRef]
12. Chen, J.; Yao, M.; Guo, J.; Yi, T.; Jin, D. The Unique Cauda-Liked Structure Represents a New Subfamily of Cunaxidae: Description of New Taxa and Discussion on Functional Morphology. *Animals* **2023**, *13*, 1363. [CrossRef]
13. Tan, M.; Lian, R.; Ruan, H.; Liang, X. Three New Species of *Aceria* (Acari: Trombidiformes: Eriophyoidea) from China. *Animals* **2024**, *14*, 720. [CrossRef]
14. Seniczak, S.; Seniczak, A.; Jordal, B.H. Morphological Ontogeny, Ecology, and Biogeography of *Fuscozetes fuscipes* (Acari, Oribatida, Ceratozetidae). *Animals* **2024**, *14*, 538. [CrossRef]
15. Glowska, E.; Laniecka, I.; Ostrowska, K.; Gebhard, C.A.; Olechnowicz, J.; Dabert, M. Evaluation of Genetic Diversity in Quill Mites of the Genus *Syringophiloidus* Kethley, 1970 (Prostigmata: Syringophilidae) with Six New-to-Science Species. *Animals* **2023**, *13*, 3877. [CrossRef]
16. Seniczak, S.; Seniczak, A. Morphological Ontogeny and Ecology of a Common Peatland Mite, *Nanhermannia coronata* (Acari, Oribatida, Nanhermanniidae). *Animals* **2023**, *13*, 3590. [CrossRef]
17. Kaminskienė, E.; Radzijevskaja, J.; Griciuvienė, L.; Stanko, M.; Snegiriovaitė, J.; Mardosaitė-Busaitienė, D.; Paulauskas, A. Molecular Identification and Phylogenetic Analysis of Laelapidae Mites (Acari: Mesostigmata). *Animals* **2023**, *13*, 2185. [CrossRef]
18. Castro, E.B.; Beard, J.J.; Ochoa, R.; Bauchan, G.R.; Otero-Colina, G.; Dowling, A.P.G.; Lofego, A.C.; Feres, R.J.F. A New Species of *Ultratenuipalpus* (Acari: Tenuipalpidae) from Brazil and Re-Description of *Ultratenuipalpus meekeri* (De Leon), the Type Species of the Genus, with DNA Barcodes. *Animals* **2023**, *13*, 1838. [CrossRef]
19. Xie, J.; Zhang, Y. Diversity and Distribution of Mites (ACARI) Revealed by Contamination Survey in Public Genomic Databases. *Animals* **2023**, *13*, 3172. [CrossRef]
20. Błoszyk, J.; Hebda, G.; Kulczak, M.; Zacharyasiewicz, M.; Rutkowski, T.; Napierała, A. Communities of Uropodina (Acari: Mesostigmata) in Nest Boxes Inhabited by Dormice (*Glis glis* and *Muscardinus avellanarius*) and Differences in Percentages of Nidicoles in Nests of Various Hosts. *Animals* **2023**, *13*, 3567. [CrossRef]
21. Mushtaq, H.M.S.; Ali, H.M.S.; Kamran, M.; Alatawi, F.J. Effect of Different Host Plants on Life Type Characteristics of Three Spider Mite Pests (Acari: Prostigmata: Tetranychidae). *Animals* **2023**, *13*, 3433. [CrossRef]
22. Graczyk, R.; Indykiewicz, P.; Olszewski, A.; Tobółka, M. Mites Living in the Nests of the White Stork and Black Stork in Microhabitats of the Forest Environment and Agrocenoses. *Animals* **2023**, *13*, 3189. [CrossRef]

animals

MDPI

Article

Multi-Locus Sequence Analysis Indicates Potential Cryptic Speciation in the Chigger Mite *Neoschoengastia gallinarum* (Hatori, 1920) Parasitising Birds in Asia

Praveena Rajasegaran [1,2], Sirikamon Koosakulnirand [3,4], Kim-Kee Tan [1], Jing Jing Khoo [3], Youseuf Suliman [3], Mohammad Saiful Mansor [5], Mohd K. S. Ahmad Khusaini [6], Sazaly AbuBakar [1], Kittipong Chaisiri [7], Serge Morand [8], Zubaidah Ya'cob [1] and Benjamin L. Makepeace [3,*]

1 Tropical Infectious Diseases Research and Education Centre (TIDREC), Higher Institution Centre of Excellence (HICoE), Universiti Malaya, Kuala Lumpur 50603, Malaysia; praveenarasi94@gmail.com (P.R.); kimkee@um.edu.my (K.-K.T.); sazaly@um.edu.my (S.A.); zyacob@um.edu.my (Z.Y.)
2 Institute for Advanced Studies, Universiti Malaya, Kuala Lumpur 50603, Malaysia
3 Institute of Infection, Veterinary & Ecological Sciences, University of Liverpool, Liverpool L3 5RF, UK; s.koosakulnirand@liverpool.ac.uk (S.K.); jing.jing.khoo@liverpool.ac.uk (J.J.K.); y.m.suliman@liverpool.ac.uk (Y.S.)
4 Department of Microbiology & Immunology, Faculty of Tropical Medicine, Mahidol University, Bangkok 10400, Thailand
5 Department of Biological Sciences and Biotechnology, Faculty of Science & Technology, Universiti Kebangsaan Malaysia, Bangi 43600, Malaysia; msaifulmansor@ukm.edu.my
6 Wildlife Conservation Division, Department of Wildlife and National Parks Peninsular Malaysia, Ministry of Natural Resources, Environment and Climate Change, Kuala Lumpur 56100, Malaysia; khusainikharip@gmail.com
7 Department of Helminthology, Faculty of Tropical Medicine, Mahidol University, Bangkok 10400, Thailand; kittipong.chaisiri@gmail.com
8 IRL HealthDEEP, CNRS-Kasetsart University-Mahidol University, Bangkok 10900, Thailand; serge.morand@umontpellier.fr
* Correspondence: blm1@liverpool.ac.uk

Citation: Rajasegaran, P.; Koosakulnirand, S.; Tan, K.-K.; Khoo, J.J.; Suliman, Y.; Mansor, M.S.; Ahmad Khusaini, M.K.S.; AbuBakar, S.; Chaisiri, K.; Morand, S.; et al. Multi-Locus Sequence Analysis Indicates Potential Cryptic Speciation in the Chigger Mite *Neoschoengastia gallinarum* (Hatori, 1920) Parasitising Birds in Asia. *Animals* **2024**, *14*, 980. https://doi.org/10.3390/ani14060980

Academic Editor: Theo de Waal

Received: 18 February 2024
Revised: 9 March 2024
Accepted: 17 March 2024
Published: 21 March 2024

Simple Summary: The chigger mite *Neoschoengastia gallinarum* (Hatori, 1920) is a parasite that feeds on the skin tissue of birds across multiple countries in Southeast and East Asia. In domestic chickens, heavy infestations with this mite can lead to skin irritation and damage to the carcass, reducing economic value. In this study, we collected *N. gallinarum* samples from wild birds of conservation concern and domestic chickens in Peninsular Malaysia and Thailand. Sequence analyses of three genes from the mites were compared across four Malaysian populations, one Thai population, and previously published sequences from southeastern China. A variety of methods were applied to classify these sequences and determine the extent of interbreeding between populations. These methods agreed in identifying three clusters of sequences by country of origin, although there was partial overlap between Thailand and China. The populations from Malaysia and Thailand appear to be reproductively isolated from one another and may represent distinct species with almost identical morphological features, except for leg length. Further studies are required to determine if these genetic dissimilarities are accompanied by distinct ecological, behavioural, or pathological differences in *N. gallinarum* in different regions of Asia.

Abstract: *Neoschoengastia gallinarum* is widely distributed in Asia, preferentially parasitising birds, and heavy infestations have clinical impacts on domestic fowl. In common with other trombiculid mites, the genetic diversity and potential variation in host preferences or pathology induced by *N. gallinarum* are poorly understood. This study aimed to unravel the geographical variation and population structure of *N. gallinarum* collected from galliform birds in Peninsular Malaysia and Thailand by inference from concatenated mitochondrial-encoded cytochrome c oxidase subunit I (COI), and nuclear-encoded internal transcribed spacer 2 (ITS2) and 18S ribosomal DNA gene sequences, including a comparison with previously published data from southeastern China. Our multi-locus sequence analysis revealed three monophyletic clades comprising (A) specimens from

Peninsular Malaysia, (B) the samples from Thailand together with a minority of Chinese sequences, and (C) the majority of sequences from China. Similarly, most species delimitation approaches divided the specimens into three operational taxonomic units. Analysis of molecular variance revealed 96.41% genetic divergence between Malaysian and Thai populations, further supported by the absence of gene flow (N*m* = 0.01). In conclusion, despite the two countries sharing a land border, populations of *N. gallinarum* from Peninsular Malaysia and Thailand appear to be genetically segregated and may represent distinct cryptic species.

Keywords: Galliformes; trombiculid; molecular barcoding; trombiculiasis; chickens

1. Introduction

Trombiculid mite larvae or "chiggers" (Actinotrichida: Trombiculidae) are globally distributed etiological agents of trombiculiasis, a form of dermatitis resulting from their bites [1–3]. Trombiculiasis can affect a wide range of wild and domestic terrestrial vertebrate hosts, including humans, in which the condition is sometimes referred to as "scrub itch". During feeding, a straw-like structure called the stylostome is formed from compounds in the chiggers' saliva reacting with the host's tissues, creating a tube that extends from their mouthparts. The saliva is also thought to contain lytic enzymes and anticoagulants, which facilitate the imbibement of tissue fluid and liquified skin cells [4,5]. Hypersensitivity reactions to mite allergens may then proceed at the bite site [6,7], especially in atopic hosts, leading to potentially severe dermatitis in a variety of host species [8–10]. Crater-like pits and nodular lesions caused by chigger bites have been reported in mammals infested with chigger species from the genera *Euschoengastia* Ewing, 1938, *Gahrliepia* Oudemans, 1912, *Hyponeocula* Vercammen-Grandjean, 1960, and *Schoutedenichia* Jadin and Vercammen-Grandjean, 1954 [11–16]. Additionally, several studies have described chigger-induced skin lesions in both domestic and wild birds, sometimes accompanied by poor body condition or even mortality [2,17,18]. Importantly, some chigger species have a major clinical impact on humans as vectors of scrub typhus, a potentially fatal zoonosis caused by *Orientia* spp. bacteria [19]. However, the chigger genus primarily responsible for *Orientia* transmission to humans (*Leptotrombidium*) is not a major cause of scrub itch compared with members of the genera *Eutrombicula* Ewing, 1938, *Schoengastia* Oudemans, 1910, and *Neotrombicula* Hirst, 1925 [20].

The genus *Neoschoengastia* Ewing, 1929 has a global distribution with over 70 recorded species, most of which have a marked predilection for domestic or wild avian hosts [21–23]. While certain *Neoschoengastia* spp. have been recorded on mammalian hosts such as rodents and ungulates [24,25], they are not a recognised cause of scrub itch in humans. However, *Neoschoengastia* spp. are significant pests of domestic fowl, especially for turkeys in North America [*Neoschoengastia americana* (Hirst, 1921)], common pheasants in Japan (*Neoschoengastia shiraii* Sasa and Sato, 1953), and chickens in East and Southeast Asia (*Neoschoengastia gallinarum*) [26–28]. Recently, *N. gallinarum* was recorded for the first time in Thailand (parasitising domestic chickens), as well as being found in abundance on wild Galliformes [*Gallus gallus* (Linnaeus, 1758), *Lophura rufa* (Raffles, 1822), *Polyplectron inopinatum* (Rothschild, 1903), and *Polyplectron malacense* (Scopoli, 1786)] in Peninsular Malaysia [29]. China, Taiwan, and Vietnam are also included in this species' range [23], which is widespread and greatly reduces the economic value of poultry due to damage to the carcass [28]. However, there is a gap in knowledge concerning the genetic diversity and potential variation in host preferences or pathology induced by *N. gallinarum* across its endemic regions of East and Southeast Asia.

The use of molecular approaches for chigger species discrimination has been very limited until recently. However, the application of molecular barcoding based on the mitochondrially encoded cytochrome c oxidase subunit I (COI) gene [or occasionally the nuclear-encoded internal transcribed spacer 2 (ITS2) region] is becoming more widespread in the chigger field, with several studies from Asia and Europe using this approach for *Leptotrombidium* spp. and a number of other genera [30–35]. Although such analyses should be interpreted with caution since they are based on a single gene, they indicate that some chigger species with identical barcodes can display morphological plasticity on different hosts, whereas other species exhibit polymorphisms in the COI region without accompanying morphological variation. Notably, *N. gallinarum* is the only chigger species in which more than two genes have been applied in population genetic studies. Zhou et al. [30] used portions of the 18S and 28S rRNA genes, the complete ITS2 region, and a COI fragment to study the population structure of *N. gallinarum* in the Fujian and Guangdong provinces of southeastern China. They reported that two genotypes of COI were present, which were not linked to geographical location or morphological variation, and the relatively conserved nuclear markers did not show polymorphisms associated with the COI genotypes. They concluded that COI is useful for both interspecies and intraspecies phylogenetic analyses and the discovery of new genotypes. Meanwhile, the ITS2 and 18S rDNA genes are relatively conserved and more suitable for analysing interspecies variation and species-level identification. Here, with the aim of unravelling the geographical variation and population structure of *N. gallinarum* in Peninsular Malaysia and Thailand, we performed multi-locus sequence analyses using concatenated COI, ITS2, and 18S rRNA genes. Moreover, we applied comparative analyses with published sequences available for the Chinese populations to determine whether *N. gallinarum* displays panmixia across Asia or forms reproductively isolated populations. We present evidence suggesting that the *N. gallinarum* populations of Peninsular Malaysia, Thailand, and southeastern China constitute at least two and possibly three cryptic species.

2. Material and Methods

2.1. Study Sites and Chigger Collections

The sampling effort for the collection of *N. gallinarum* [36] from infested galliform birds was conducted at four sites in Peninsular Malaysia [Sungkai Wildlife Conservation Centre, Perak (code SWCC)—January 2021 and March 2021; Asahan Village Bestari Jaya, Selangor (BJV)—April 2021; Jemaluang Wildlife Conservation Centre, Johor (JWCC)—February 2022; Kota Tinggi Plantation, Johor (KTP)—June 2022]. Only a single site in Thailand was sampled [Saen Thong subdistrict, comprising two villages—Ban Huay Muang and Ban Santisuk—in Tha Wang Pha district, Nan province (BNAN)] in December 2022, during activities of the One Health Observatory project (ANR FutureHealthSEA) [37] (Figure 1). Details of samples collected from the five sites from each species of host are summarised in Table 1. Chigger mites were removed from predilection sites on the birds' skin (mainly breast and thigh—see Figure 2) using fine forceps. The recovered chiggers were stored in 70% ethanol at −20 °C. Chiggers from each host were counted and 10% of specimens were selected for mounting in Berlese fluid for species-level identification using an Axio Imager M2 microscope (Zeiss, Oberkochen, Germany) and ZEN 2011 imaging software [31]. These individuals were not used for DNA extraction but were retained as voucher specimens and deposited at the Tick Cell Biobank Asia Outposts Laboratory, Tropical Infectious Diseases Research & Education Centre, Universiti Malaya [29]. The remaining chiggers from each bird host were identified using the autofluorescence method [31] on a GXM-L3201 LED research fluorescence trinocular microscope (GT Vision LTD, Newmarket, UK) with reference to the voucher specimens.

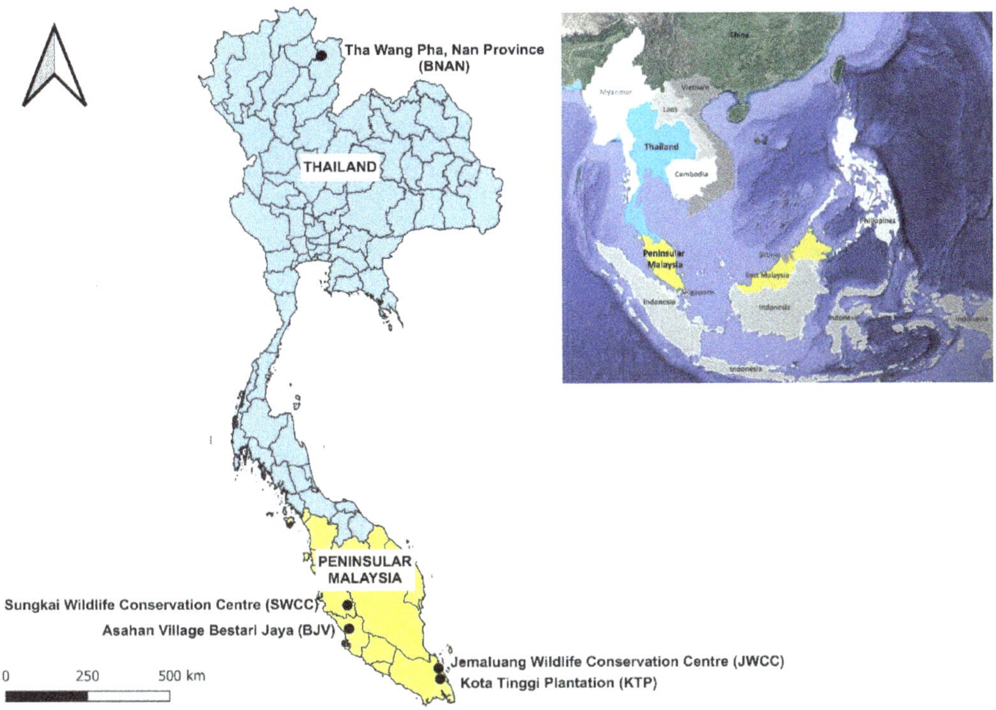

Figure 1. Map illustrating the five study sites in Peninsular Malaysia and Thailand. The inset map displays the Southeast Asian region. The main map shows the sampling localities within Malaysia [Sungkai Wildlife Conservation Centre, Perak (SWCC); Asahan Village Bestari Jaya, Selangor (BJV); Jemaluang Wildlife Conservation Centre, Johor (JWCC); Kota Tinggi Plantation, Johor (KTP)] and Saen Thong subdistrict, Tha Wang Pha district, Nan province, Thailand (BNAN).

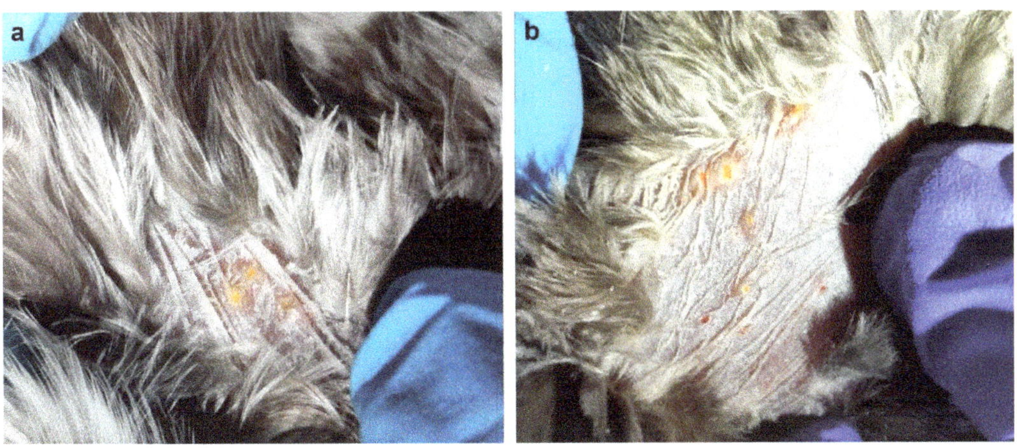

Figure 2. Multifocal coalescing pattern of chigger infestation on the dermal surface of a Malayan crested fireback (*Lophura rufa*), specifically on the (**a**) thigh and (**b**) breast areas.

Table 1. Information on geographical origin and host species of *N. gallinarum*.

Country	Population Code	Locality	Coordinates	Habitat Type	Host Species	Chigger ID	GenBank Accession No.		
							COX1	ITS2	18S
Peninsular Malaysia	SWCC	Sungkai Wildlife Conservation Centre, Perak	E101.36623, N4.06430	Sanctuary	*Lophura rufa*	SWX3	OR632279	OR636401	OR632359
					Polyplectron inopinatum	SWX10	OR632280	OR636402	OR632360
					P. inopinatum	SWX12	OR632281	OR636403	OR632361
					P. inopinatum	SWX13	OR632282	OR636404	OR632362
					P. inopinatum	SWX14	OR632283	OR636405	OR632363
					P. inopinatum	SWX15	OR632284	OR636406	OR632364
					P. inopinatum	SWX19	OR632285	OR636407	OR632365
					P. inopinatum	SWX21	OR632286	OR636408	OR632366
					P. inopinatum	SWX22	OR632287	OR636409	OR632367
					P. inopinatum	SWX23	OR632288	OR636410	OR632368
					P. inopinatum	SWX37	OR632289	OR636411	OR632369
					P. inopinatum	SWX40	OR632290	OR636412	OR632370
					L. rufa	SWX43	OR632291	OR636413	OR632371
	BJV	Bestari Jaya Village, Selangor	E101.41022, N3.37801	Village	*Gallus gallus domesticus*	KPGX5	OR632292	OR636414	OR632372
					G. gallus domesticus	KPGX6	OR632293	OR636415	OR632373
					G. gallus domesticus	KPGX9	OR632294	OR636416	OR632374
					G. gallus domesticus	KPGX10	OR632295	OR636417	OR632375
					G. gallus domesticus	KPGX11	OR632296	OR636418	OR632376
					G. gallus domesticus	KPGX18	OR632297	OR636419	OR632377
					G. gallus domesticus	KPGX19	OR632298	OR636420	OR632378
					G. gallus domesticus	KPGX20	OR632299	OR636421	OR632379
					G. gallus domesticus	KPGX21	OR632300	OR636422	OR632380
	JWCC	Jemaluang Wildlife Conservation Centre, Johor	E103.85297, N2.29136	Sanctuary	*L. rufa*	JWX3	OR632301	OR636423	OR632381
					Polyplectron malacense	JWX17	OR632302	OR636424	OR632382
					P. malacense	JWX18	OR632303	OR636425	OR632383
					P. malacense	JWX20	OR632304	OR636426	OR632384

Table 1. *Cont.*

Country	Population Code	Locality	Coordinates	Habitat Type	Host Species	Chigger ID	GenBank Accession No.		
							COX1	ITS2	18S
	KTP	Kota Tinggi Plantation, Johor	E103.86604, N2.03023	Forest	*G. gallus*	KTX7	OR632305	OR636427	OR632385
					G. gallus	KTX8	OR632306	OR636428	OR632386
					G. gallus	UMPX2	OR632307	OR636429	OR632387
					G. gallus	CHX2	OR632308	OR636430	OR632388
Thailand	BNAN	Ban Huay Muang and Ban Santisuk, Saen Thong subdistrict, Tha Wang Pha, Nan	E100.71897, N19.13999; E100.69891, N19.12957	Village	*G. gallus domesticus*	BNANX2	OR632309	OR636431	OR632389
					G. gallus domesticus	BNANX8	OR632310	OR636432	OR632390
					G. gallus domesticus	BNANX9	OR632311	OR636433	OR632391
					G. gallus domesticus	BNANX10	OR632312	OR636434	OR632392
					G. gallus domesticus	BNANX11	OR632313	OR636435	OR632393
					G. gallus domesticus	BNANX12	OR632314	OR636436	OR632394
					G. gallus domesticus	BNANX13	OR632315	OR636437	OR632395
					G. gallus domesticus	BNANX14	OR632316	OR636438	OR632396
					G. gallus domesticus	BNANX15	OR632317	OR636439	OR632397
					G. gallus domesticus	BNANX16	OR632318	OR636440	OR632398
					G. gallus domesticus	BNANX17	OR632319	OR636441	OR632399
					G. gallus domesticus	BNANX19	OR632320	OR636442	OR632400
					G. gallus domesticus	BNANX20	OR632321	OR636443	OR632401
					G. gallus domesticus	BNANX21	OR632322	OR636444	OR632402
					G. gallus domesticus	BNANX22	OR632323	OR636445	OR632403

2.2. DNA Extraction from Chiggers

Total genomic DNA was extracted from individual chigger mites of *N. gallinarum* using a QIAamp DNA Micro Kit (Qiagen, Redwood City, CA, USA) following the manufacturer's protocol. Briefly, the chiggers were washed in nuclease-free water for ethanol elimination. Next, chigger samples were digested in 180 µL tissue lysis buffer with 20 µL proteinase K and incubated at 56 °C overnight. The kit manufacturer's instructions were continued with the DNA recovered in 30 µL elution buffer and stored at −20 °C.

2.3. PCR Amplification and Sequencing of PCR Products

Amplifications of the extracted genomic DNA were performed using a universal invertebrate COI (forward–LCO1490: 5′-GGTCAACAAATCATAAAGATATTGG-3′; reverse–HCO2198: 5′-TAAACTTCAGGGTGACCAAAAAATCA-3′) primer pair [38], specific assays targeting ITS2 (forward–5.8S: 5′-CACGCCGAGCACTCGACATT-3′; reverse–28S: 5′-GATCCTTCGCTCGCCGTTACT-3′), 18S ribosomal DNA (18S) (forward–5′-GGCTCATTA AATCAGTTACGGTT-3′; reverse–5′-ATTCCTCGTTCATGGGCAAT-3′) [30], and an ND5 mitochondrial gene fragment (forward–5′-TTTCTGTATTCTGAGCCTTCT-3′; reverse–5′-ATAATAGGGGTTAGCAGAG-3′) [39] of *N. gallinarum*. Polymerase chain reaction (PCR) amplifications were conducted in 25 µL reaction volumes including 2 µL DNA template, 12.5 µL 5X Green DreamTaq Buffer, and 1 µL each primer (final concentration, 0.4 µM) in a 96-well SimpliAmp Thermal Cycler (Applied Biosystems, Inc., Foster City, CA, USA). The amplification profile was as follows: pre-denaturation at 95 °C (2 min), followed by 35 cycles of 95 °C (1 min) for denaturation; 40 °C (1 min) for annealing; 72 °C (1 min and 30 s) for extension; and a final extension at 72 °C (7 min) for COI. For ITS2 and 18S, the programme constituted 94 °C (5 min) for pre-denaturation, followed by 35 cycles of denaturation at 94 °C (30 s); annealing at 55 °C (30 s); extension at 72 °C (30 s); and a final extension at 72 °C (5 min). Lastly, for ND5, the amplification profile begins with pre-denaturation at 94 °C (5 min), followed by 35 cycles of 94 °C (30 s) for denaturation; 54 °C (30 s) for annealing; 72 °C (40 s) for extension; and a final extension at 72 °C (7 min). The amplified PCR products were electrophoresed on a 1.0% agarose gel to determine the product size before submission to Apical Scientific Laboratory Sequencing Company, Selangor, Malaysia, for further purification and Sanger sequencing.

2.4. Sequence Alignment

Both forward and reverse sequences of COI, ITS2, and 18S were analysed and edited using BioEdit v7.2.5 [40]. However, we were unable to amplify the ND5 gene fragment of *N. gallinarum* using primers from Tao et al. [39]. All successfully amplified sequences were later aligned using the ClustalX [41] program implemented in BioEdit v7.2.5 [40]. Sequences of COI, ITS2, and 18S of *N. gallinarum* were deposited in the National Center of Biotechnology Information (NCBI) GenBank database under accession numbers OR632279-OR632323, OR636401-OR636445, and OR632359-OR632403, respectively (Table 1).

The aligned COI (551 bp), ITS2 (260 bp), and 18S (729 bp) gene sequences were concatenated using Molecular Evolutionary Genetic Analysis (MEGA) software (version 11.0.11) [42], and the congruency of different partitions among these genes was calculated using a partition homogeneity test of 100 replicates implemented in PAUP 4.0a169 [43]. This generated a p-value of 0.87, indicating that the concatenated dataset was congruent between constituent genes. Thus, the 1540 bp concatenated alignment of COI, ITS2, and 18S of *N. gallinarum* was used in the present study.

2.5. Phylogenetic Reconstruction and Haplotype Network

The MEGA software (version 11.0.11) [42] was used to run Modeltest to estimate the best evolutionary model of nucleotide substitution for the concatenated sequences. Tamura 3-parameter (T92) with gamma (G) distribution rates showed the lowest Bayesian Information Criterion (BIC) and was chosen to best describe the substitution pattern in the rest of the phylogenetic analysis. Further, MEGA11 was used to compute a pairwise distance using the Kimura 2-parameter (K2P) model [44]. An initial phylogenetic tree was constructed using the Neighbour Joining (NJ) method inferred in MEGA11 with 1000 bootstrap replicates for individual genes of COI, ITS2, 18S, and concatenated datasets. Maximum Likelihood (ML) analysis was also computed on individual genes and concatenated datasets using online phylogeny software, PhyML 3.0, with an automated model selection using BIC [45]. Bayesian inference (BI) analysis was run for the concatenated dataset using MrBayes version 3.2.7 [46]. The Hasegawa–Kishono–Yano substitution model with a gamma shape parameter of 0.109 (HKY + G) was favoured as the best model by jModeltest2 [47] and implemented in the online server CIPRES Science Gateway v3.3 (https://www.phylo.org/, accessed 1 March 2024) [48]. The BI analysis was performed on two million generations of Markov Chain Monte Carlo (MCMC), and the tree was sampled every 100th generation, with the first 10% of trees discarded as burn-in. A total of 10 sequences—8 of *N. gallinarum* (COI–MK423976, MK423977, MK423978; ITS2–MK423979, MK423981, MK643333, MK643334; 18S–MK400440) from the study by Zhou et al. [30] and 3 of *Tetranychus urticae* C. L. Koch, 1836 (Acarina: Trombidiformes; COI-EU345430.1, ITS2-MH919319.1, and 18S-AB926313.1)—were obtained from GenBank and concatenated accordingly. Together with the 45 sequences from Peninsular Malaysia and Thailand, these sequences were selected to study the phylogenetic relationship with *T. urticae* as the outgroup. All trees were visualised in FigTree v1.4.4 and edited in the Interactive Tree of Life (iTOL) [49]. Minimum spanning networks (MSN) [50] among haplotypes were computed using TCS Network [51] and illustrated in PopArt v1.7 [52] to acquire a graphical representation of concatenated COI, ITS2, and 18S data.

2.6. Species Delimitation Analyses

Assemble Species by Automatic Partitioning (ASAP) [53], Automatic Barcode Gap Discovery (ABGD) [54], multi-rate Poisson Tree Processes (mPTP) [55], and Generalised Mixed Yule Coalescent (GMYC) [56] were used for species delimitation analyses. Both ASAP and ABGD were performed on a web-based server (ASAP: https://bioinfo.mnhn.fr/abi/public/asap, accessed on 8 January 2024; ABGD: https://bioinfo.mnhn.fr/abi/public/abgd/abgdweb.html, accessed on 8 January 2024) using a Kimura (K80) model with default settings, TS/TV model 2.0 [53,57]. Additionally, for ABGD entity recognition, settings were based on the suggested partition at $P = 0.01$, a relative gap width of 1 and 50 steps, $P_{min} = 0.001$, $P_{max} = 0.1$, and Nb bins for distance distribution = 20 [53]. The mPTP delimitation analysis was performed on an mPTP web service available at http://mptp.h-its.org, accessed on 1 March 2024 [55]. To initiate the GMYC species delimitation method for the concatenated dataset, an ultrametric tree was generated using BEAST v2.6.6 [58] to run on the CIPRES Science Gateway v3.3 online portal (https://www.phylo.org/, accessed 1 March 2024) [59]. Preceding this, an XML input file was created using BEAUti v2.6.6 [58] with the best-fitting model, namely (HKY + G) substitution, as determined by jModelTest2 [47]. The Markov Chain Monte Carlo (MCMC) chains were run for 30 million generations, with topologies and parameters logged every 1000 generations. The analysis was then confirmed using Tracer v1.7.1 [60] for an Effective Sampling Size (ESS) of more than 200, demonstrating that the MCMC chains had adequately converged [61]. The output tree was analysed in TreeAnnotator 2.6.6 [58], discarding the initial 10% as burn-in. The subsequent GMYC analysis for the concatenated dataset was conducted in RStudio [62] using R packages v4.3.0, including "ape" [63], "paran" [64], "rncl" [65], and "splits" [66].

2.7. Population Genetic and Demographic Analysis

Gene flow was determined by computing the level of population subdivision (F_{ST}) and the number of migrants (Nm), also using DnaSP software version 6.12.03 [67]. To resolve the interrelation between geographical distance and genetic differentiation between populations, the Mantel test was conducted in Arlequin version 3.5.2.2 [68] using 1000 permutations [69,70]. Finally, populations were divided into the broad geographical groups of Malaysia and Thailand to study the pattern of genetic structure based on the region of origin, which was examined using an analysis of molecular variance (AMOVA) by estimating the *F*-statistic (Φ_{ST}) values with 1000 permutations in Arlequin software 3.5.2.2.

3. Results

3.1. Identification Confirmation and Sequence Characteristics

The trombiculid mites collected from Galliformes were morphologically screened and measured, referring to Domrow and Nadchatram [71], which confirmed their identification as *N. gallinarum* (Figure 3) [36]. No difference in key characteristics was found for this species between Peninsular Malaysia and Thailand except for the total length of legs (Table 2). The diagnostic characters of the *N. gallinarum* mounted for brightfield microscopy were barbed galeal setae, a coxal formula of I.I.I, a palpal setal formula of BBNBB + 7B, and a scutal formula of AL > PL > AM [71], with measurements as shown in Table 2.

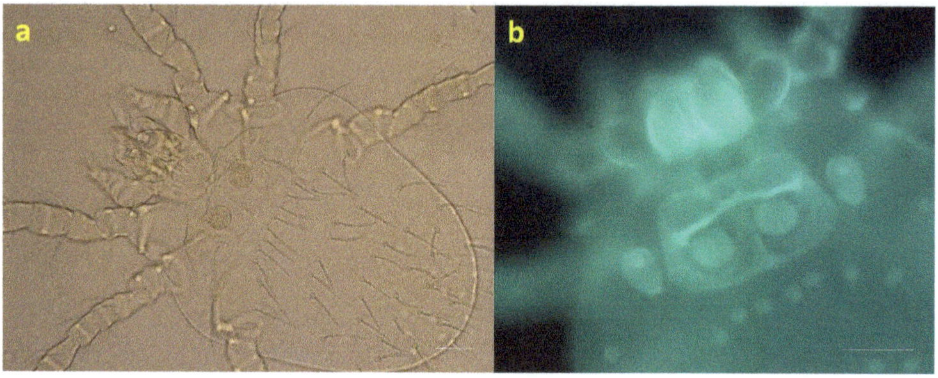

Figure 3. (**a**) Brightfield microscopic view of *N. gallinarum*; (**b**) autofluorescence (AF) imaging of *N. gallinarum* scutum (scale bars, 10 μm). Both images were obtained using a Zeiss Axio Imager M2 microscope and ZEN 2011 imaging software. The host was a Malayan peacock-pheasant (*Polyplectron malacense*).

Table 2. Diagnosis and morphometry comparisons of *N. gallinarum* voucher specimens from Peninsular Malaysia and Thailand.

	Morphometry Measurements (μm)											
	AW	**PW**	**SB**	**ASB**	**PSB**	**AP**	**AM**	**AL**	**PL**	**S**	**H**	**IP**
	Peninsular Malaysia											
$n = 11$												
Mean	52	67	42	21	25	28	30	43	39	26	43	686
Min	48	64	39	17	24	27	25	38	36	23	39	625
Max	60	74	44	25	30	31	34	48	46	31	49	704
	Thailand											
$n = 7$												
Mean	52	69	43	21	25	30	28	44	42	24	43	713
Min	49	63	41	19	24	28	25	41	38	18	39	701
Max	53	74	45	23	27	31	32	48	46	34	46	726
	Mann–Whitney U-test											
U	36.000	28.500	30.500	37.000	36.000	17.000	27.000	32.000	22.000	19.000	33.500	1.500
Z	−0.236	−0.924	−0.748	−0.139	−0.235	−1.993	−1.049	−0.596	−1.507	−1.805	−0.457	−3.361
P	0.860	0.375	0.479	0.930	0.860	0.056	0.328	0.596	0.151	0.085	0.659	<0.001 *

Note: Statistical analysis was performed with exact significance using SPSS software v. 26. AW—distance between anterolateral setae; PW—distance between posterolateral setae; SB—distance between sensilla bases; ASB—distance between sensillary bases line and anterior margin of scutum; PSB—distance between sensillary bases line and posterior margin of scutum; AP—distance between anterolateral setae and posterolateral setae; AM—length of anteromedial setae; AL—length of anterolateral setae; PL—length of posterolateral setae; S—length of scutal sensilla; H—length of humeral setae; IP—total length of leg. * Asterisk indicates the parameter with a significant statistical test ($p < 0.05$).

Segments of COI, ITS2, and 18S were successfully sequenced and concatenated from 45 individuals of *N. gallinarum* with a final alignment length of 1540 bp. Of these, 1384 were conserved sites, whereas 55 were variable sites (comprising eight singleton variable sites and 47 parsimony-informative sites).

3.2. Phylogenetic Reconstruction

The phylogenetic analysis of 45 individuals from this study was complemented by including 7 concatenated, published *N. gallinarum* sequences from Zhou et al. [30]. The topology was similar for phylogenetic trees constructed by different methods [i.e., ML or NJ (Figure 4) and BI (Figure S1)]. The tree was divided into three main clades, of which the Malaysian clade (A) was founded on the strongest evidence (100% NJ/99% ML bootstrap support). Clade B comprised the entire population from Thailand and two samples from China (NGY5 and NGFA4), whereas the remainder of the Chinese samples clustered in a third clade (C). Bootstrap support for clades B and C was moderate (>80%), while within the Peninsular Malaysia and Thailand samples, evidence for population structure within each country was variable but sometimes exceeded 80%. However, although four distinct geographic sites had been sampled in Peninsular Malaysia, these subpopulations did not cluster strictly by location (Figure 4—note distribution of sample codes from Table 1). Phylogenetic trees constructed using individual gene markers produced similar tree topologies between COI (Figure S2) and the concatenated dataset but for ITS2, sequences from Malaysia and China were not clearly separated (Figure S3). The 18S rRNA gene exhibited the highest level of conservation between the three loci as expected, with only a single polymorphic site. This comprised two alleles, one in Thailand and one in China, which were observed together in Malaysia (Figure S4, Table S4).

3.3. Pairwise Distance and Species Delimitation Analysis

Pairwise intraspecific analysis of genetic distances for concatenated sequences of *N. gallinarum* ranged from zero to 3.55% (Table S1). The highest intraspecific divergence was recorded for an individual from BJV (KPGX18) compared with four individuals from BNAN at 3.55%, whereas the lowest divergence (zero) was seen between multiple individuals within the population from Peninsular Malaysia. At the country level, the pairwise genetic distance for concatenated genes between populations from Peninsular Malaysia and Thailand was 3.36%, whereas divergences of 2.64% and 2.36% separated the populations of Peninsular Malaysia and Thailand, respectively, from the Chinese populations. Maximum pairwise distances were considerably higher for COI (9.06%—Table S2) than for ITS2 (2.7%—Table S3). The species delimitation analyses conducted using ABGD, ASAP, and mPTP consistently identified three operational taxonomic units (OTUs). Notably, the ASAP analysis produced the lowest score of 2.00, while the mPTP analysis yielded the best multi-coalescent rate score of 112.25. The OTUs comprised (1) Peninsular Malaysia only (=clade A), (2) China minority clade + Thailand (=clade B), and (3) China majority clade (=clade C), as superimposed on the tree in Figure 4. In contrast, the molecular delimitations of GMYC revealed significant discrepancies, resulting in the identification of seven OTUs: three for Peninsular Malaysia (within clade A), one for Thailand (designated within clade B), and three for China (including two within clade B and one in clade C), as illustrated in Figure 4.

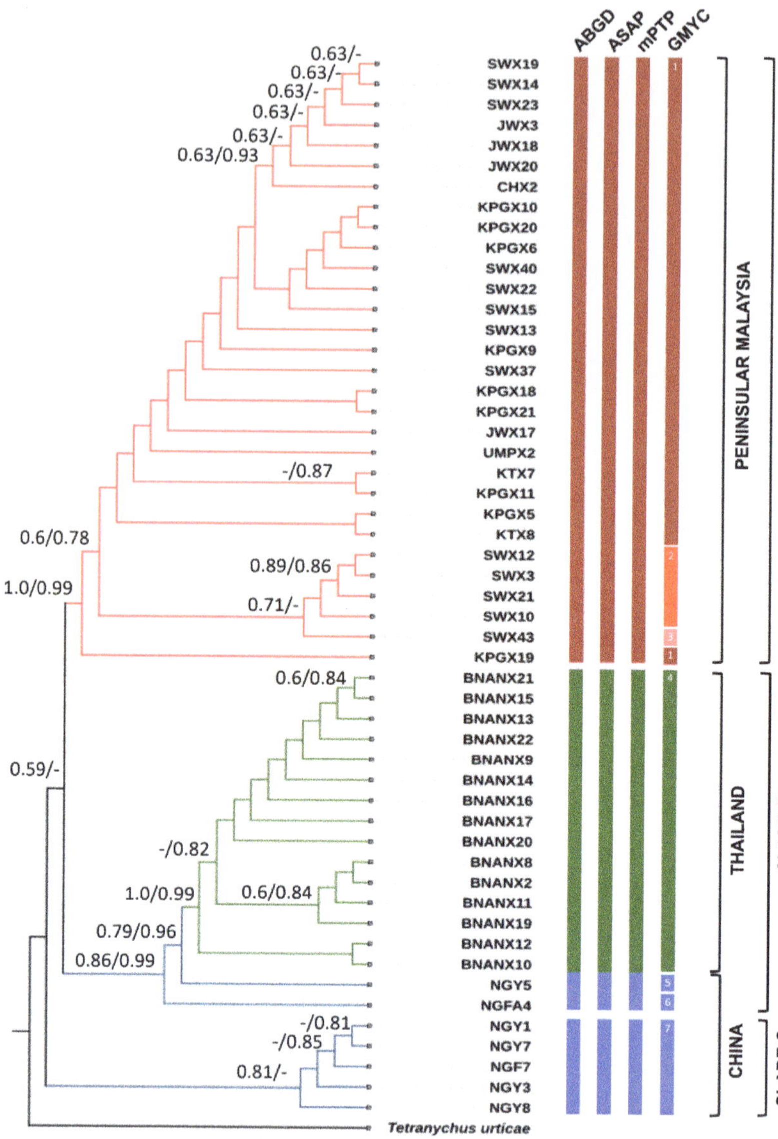

Figure 4. Phylogenetic relationships among *N. gallinarum* populations from Peninsular Malaysia (red), Thailand (green), and China (blue) inferred through Neighbour Joining (NJ) and Maximum Likelihood (ML) analysis based on the concatenated nucleotide sequences of mitochondrial cytochrome c oxidase subunit 1, second internal transcribed spacer, and 18S ribosomal DNA. Bootstrap values (NJ/ML) are shown on the branches. Vertical bars on the right are the results of species delimitation by ABGD, ASAP, mPTP, and GMYC with the population groups indicated to the right. The numbers in the vertical bars of GMYC indicate the OTUs assigned from that analysis.

3.4. Haplotype Resolution and Network Analysis

Sixteen distinct haplotypes were recognised from the MSN constructed using the concatenated *N. gallinarum* gene datasets from Peninsular Malaysia and Thailand (n = 45), with a further seven originating from the published Chinese data (Figure 5). The MSN highlighted the unambiguous separation between the populations from Peninsular Malaysia and Thailand (zero haplotypes in common), and neither were any haplotypes shared with China. However, despite the Thailand specimens originating only from two villages within the same subdistrict, they were split into 6 haplotypes compared with 10 haplotypes found across the 4 subpopulations sampled in Peninsular Malaysia. Haplotype 3 was the most prevalent, including individuals from all four Peninsular Malaysia subpopulations ($n = 13$), followed by haplotype 4 found in three subpopulations ($n = 7$). Haplotypes 2, 5–10, 14, and 16 represented singletons (Table 3). Similarly, the MSN constructed using individual gene markers revealed no shared haplotype among the three examined countries for the COI gene (Figure S5, Table S5). However, in the case of ITS2 (comprising 12 haplotypes), populations from China demonstrated evidence of haplotype sharing with both Peninsular Malaysia and Thailand (Figure S6, Table S6). Finally, the 18S rRNA gene displayed just two haplotypes: Hap 1 was the only one present in Thailand and was a rare haplotype in Malaysia (restricted to Sungkai), whereas all Chinese and most Malaysian samples belonged to Hap 2 (Figure S7, Table S7).

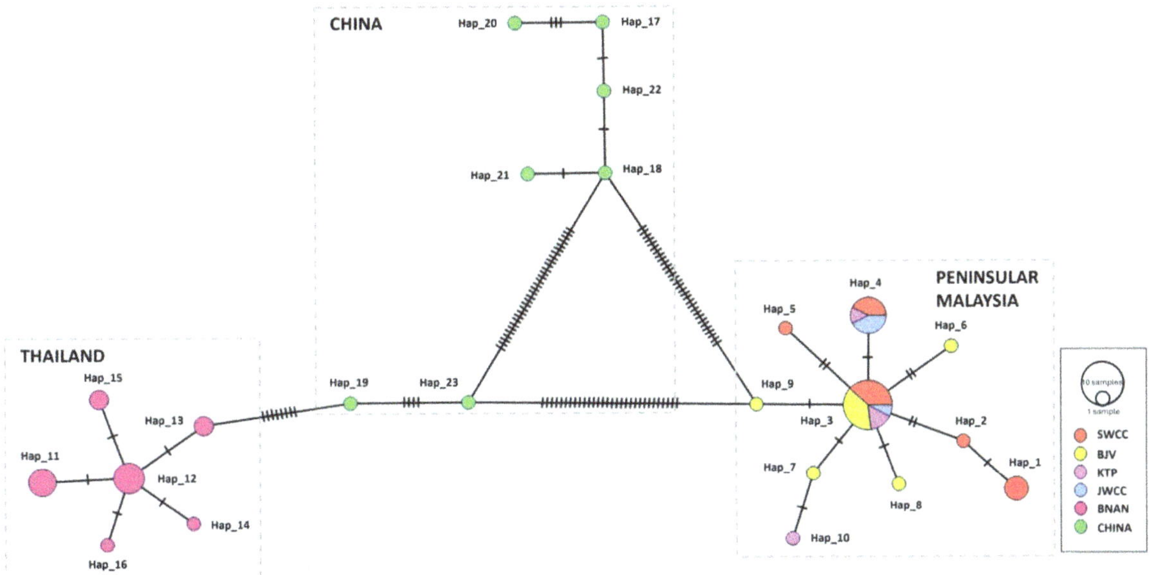

Figure 5. Minimum spanning haplotype network of *N. gallinarum* based on concatenated sequences isolated from four populations in Peninsular Malaysia, one population in Thailand, and the seven sequences from China obtained from Zhou et al. [30]. Each haplotype is represented by the coloured nodes and their relative sizes indicate haplotype frequency. Nodes of the same colour specify the haplotype from the same population. The dashed lines on each node connecting haplotypes represent polymorphisms.

3.5. Genetic Differentiation and Gene Flow

The AMOVA revealed that 96.41% of genetic variation was partitioned among groups of *N. gallinarum* from Peninsular Malaysia and Thailand (Table 4). The among-populations–within-groups variability (0.51%) was much lower than the genetic variation apparent within each population (3.08%). The variance component and fixation index were statistically significant for the among populations–within groups and within-population comparisons, but not for the among-groups analysis (Table 4).

The observed overall migrant per generation (N*m*) value of 0.02 and population subdivision (F_{ST}) value of 0.933 indicated low gene flow that led to very high genetic differentiation among most populations of *N. gallinarum* studied (Table 5). The greatest F_{ST} value was observed in comparisons between each Peninsular Malaysia subpopulation and the population from Thailand (Table 5). However, the Mantel regression analysis showed no significant relationship between net F_{ST} and geographic distance among the five subpopulations of *N. gallinarum* (r = 0.962, p = 0.109) in Peninsular Malaysia and Thailand.

Table 3. Haplotype (hap) frequency of five populations of *N. gallinarum* in Malaysia and Thailand by region.

Hap	*N. gallinarum* Individuals from Each Study Region (*n*)				
	Peninsular Malaysia				Thailand
	SWCC (13)	BJV (9)	JWCC (4)	KTP (4)	BNAN (15)
1	3	0	0	0	0
2	1	0	0	0	0
3	5	5	1	2	0
4	3	0	3	1	0
5	1	0	0	0	0
6	0	1	0	0	0
7	0	1	0	0	0
8	0	1	0	0	0
9	0	1	0	0	0
10	0	0	0	1	0
11	0	0	0	0	4
12	0	0	0	0	5
13	0	0	0	0	2
14	0	0	0	0	1
15	0	0	0	0	2
16	0	0	0	0	1
Total hap	5	5	2	3	6

Note. SWCC: Sungkai Wildlife Conservation Centre; BJV: Bestari Jaya Village; JWCC: Jemaluang Wildlife Conservation Centre; KTP: Kota Tinggi Plantation; BNAN: Tha Wang Pha, Nan Province.

Table 4. Measures of geographical population differentiation in *N. gallinarum* based on AMOVA.

Source of Variation	d.f.	Sum of Square	Variance Components	Variation (%)	Fixation Index (Φ)	Significance Test (p)
Among groups	1	442.689	22.02220	96.41	0.96407	0.197
Among populations within groups	3	4.528	0.11737	0.51	0.14299	0.031 *
Within population	40	28.138	0.70346	3.08	0.96920	0.00 *

Note: * Significant $p < 0.05$.

Table 5. Number of migrants per generation (N*m*) and population subdivision (F_{ST}) of *N. gallinarum* in relation to the geographical distance.

Populations		Distance (km)	Migrant per Generation (N*m*)	Population Subdivision (F_{ST})
SWCC	BJV	108	1.51	0.14189
SWCC	JWCC	499	0.72	0.25882
SWCC	KTP	525	2.78	0.08247
SWCC	BNAN	2081	0.01	0.96553
BJV	JWCC	421	0.46	0.35294
BJV	KTP	437	−6.38	−0.04082
BJV	BNAN	2166	0.01	0.97279
JWCC	KTP	33	2.00	0.11111
JWCC	BNAN	2552	0.00	0.98168
KTP	BNAN	2598	0.01	0.97093
Whole population			0.02	0.93312

Note. SWCC: Sungkai Wildlife Conservation Centre; BJV: Bestari Jaya Village; JWCC: Jemaluang Wildlife Conservation Centre; KTP: Kota Tinggi Plantation; BNAN: Tha Wang Pha, Nan Province.

4. Discussion

The simplest definition of cryptic species is "two or more distinct species that are erroneously classified (and hidden) under one species name" [72,73]. However, a definition that takes account of the underlying biological processes involved in cryptic speciation would add that it is a low level of phenotypic distinctiveness coupled with clear genetic differentiation that exemplifies cryptic species [73]. Evidence for cryptic speciation has been uncovered across the diversity of life and, in 2015, a review of cryptic species in Acari found that the phenomenon had been reported from 24 of the 142 acarine superfamilies, although the greatest predictor of cryptic species discovery was the research effort expended on specific taxa [74]. In the current study, populations of *N. gallinarum* from two countries (Malaysia and Thailand) exhibited similar features based on morpho-taxonomic identification, differing significantly only in the length of the legs. Minor morphological features alone are often unreliable for the accurate identification of sibling or cryptic species [75] and in *N. gallinarum*, the lack of marked morphological differences contrasted with deep splits in concatenated molecular markers between chiggers originating from Thailand and Peninsular Malaysia. Moreover, most of the published sequences from southeastern China formed a third, separated clade.

Multi-locus sequence analysis studies have increased in popularity over the years due to the reduced impact of evolutionary rates for individual genes [76,77]; for instance, several such studies have been performed in ticks, usually using concatenated mitochondrial markers [78–81]. While mitochondrial DNA undergoes a more rapid rate of mutation compared to nuclear DNA [82] and recombination in animal mitogenomes is considered rare [83], its utility in identifying distinct maternal lineages is counterbalanced by caveats when applied to the detection of reproductive isolation [84]. Hence, combining nuclear and mitochondrial loci as performed here is favourable for population genetic analyses.

Our study revealed a genetic divergence of 3.36% between Peninsular Malaysia and Thailand, revealing the potential existence of a species complex and reminiscent of recent studies in the region on *Simulium* spp. blackflies [85–87]. According to Pramual et al. [88], a divergence of >3% indicates a substantiated threshold signifying distinct separation between sister phylogroups. Notably, the ABGD, ASAP, and mPTP methods concorded in delineating the Peninsular Malaysia and Thailand specimens into two separate OTUs, and the MSN analysis showed a lack of shared concatenated haplotypes between them. However, although a proportion of the published data from southeastern China were classified in the same OTU as the Thailand specimens with most species delimitation methods, the MSN analysis demonstrated that none of the concatenated haplotypes reported from southeastern China were shared with Peninsular Malaysia or Thailand. This finding is more consistent with the GMYC analysis, but we propose that a conservative approach be taken with respect to the potential numbers of cryptic species until more data are available, especially from the Chinese populations. Despite being collected from just two villages within the same subdistrict, the Thailand specimens exhibited 6 distinct haplotypes, whereas only 10 haplotypes were found across the 4 subpopulations sampled in Peninsular Malaysia. Koopman et al. [89] proposed that the presence of shared haplotypes among different subpopulations indicates recent gene flow in the population, as seen with the specimens from Peninsular Malaysia. Haplotype 3 within the population from Peninsular Malaysia was the most prevalent and may represent the ancestral haplotype due to its representation in a significant proportion of individuals across all subpopulations and its centralised placement in the network [90]. Moreover, Hap 3 may also be a stable haplotype with diverse environmental adaptability [91].

Clearly, the COI gene provided the greatest resolution among the specimens analysed here with no haplotypes in common between countries, whereas both ITS2 and 18S exhibited shared haplotypes in two of the three countries. To the best of our knowledge, only one other analysis of the ITS2 region in chiggers has been published, and this found no evidence of intraspecific variation in the genera *Leptotrombidium*, *Neotrombicula*, and *Euschoengastia* in South Korea [32], although the geographic extent of sampling was very limited. Regarding the application of 18S rRNA sequencing in chiggers, it has been used for confirmation of species identification in studies from Brazil [92] and South Korea [93], in which the gene was found to be invariant within species. Thus, the identification here of several haplotypes for ITS2 and 18S provides corroborating evidence for cryptic speciation in *N. gallinarum* independently of the COI mitochondrial marker, even if shared haplotypes between countries are present at the nuclear level.

Greater genetic differentiation among populations can hinder gene flow [94]. This phenomenon was observed in our study, with high separation between the two sampled nations, and total interpopulation gene flow was limited (N*m* = 0.02) by the increase in geographical distance. This genetic divergence may underlie the species' adaptability to their specific geographical habitat and local environmental changes across the national border. Recently, Tao et al. [39] published a population genetic study of *N. gallinarum* in four provinces of China with a larger sample size (*n* = 192) than we achieved here. Unfortunately, a direct comparison with their study was not possible, as we were unable to amplify the ND5 locus of *N. gallinarum* used by these workers. They found that *N. gallinarum* in southern China was divided into two clades, but there was little evidence of genetic isolation between geographic sites. One exception was the population from Jiangxi, which displayed limited gene flow with *N. gallinarum* from other provinces, although it was still much greater than that between Thailand and Peninsular Malaysia. Generally, high gene flow with low to moderate genetic differentiation was observed between the subpopulations of southern China, while genetic variation within the population as a whole was higher than that among subpopulations, which is in accordance with our findings in Peninsular Malaysia. In China, trade in commercial lines of poultry between provinces may have facilitated gene flow in *N. gallinarum*, as chiggers have very limited intrinsic dispersal ability. Conversely, in Peninsular Malaysia and Thailand, traditional rearing of local chicken breeds at the village level is likely to drive reproductive isolation in parasites of poultry.

In sexual populations, increases in gene flow will lead to an increase in genetic diversity. In general, homogeneous environments contribute to reduced levels of genetic diversity, while heterogeneous environments, including variations in geography, climate, vegetation, and other factors, result in higher levels of genetic diversity [91,95]. The collection of *N. gallinarum* from both domestic and wild birds in various habitats in Peninsular Malaysia (e.g., forests, sanctuaries, and villages) may have contributed to greater genetic diversity in comparison to only one ecotype (villages) from Thailand, but broader sampling in Thailand will be required to unravel the potential impacts of environmental and host factors. In Peninsular Malaysia, *N. gallinarum* infested a wide range of bird host species, with *L. rufa* (Malayan crested fireback) and *P. inopinatum* (Mountain peacock-pheasant) noted as new host records [29] for this chigger species. Due to their decreasing population trends, *L. rufa*, *P. inopinatum*, and *P. malacense* (Malayan peacock-pheasant) are categorised as totally protected species in Malaysia [96] and classed as either "vulnerable" (*L. rufa*, *P. inopinatum*) or "endangered" (*P. malacense*) by the International Union for Conservation of Nature [97–100]. The Department of Wildlife and National Parks of Peninsular Malaysia is proactively involved in searching for these species within their native habitats. Any individuals located may be captured and subsequently placed in captivity for the specific intention of breeding [101]. This breeding program could introduce chigger mites into the captive environment, and this may explain the gene exchange between the population from KTP (forest) and those from JWCC and SWCC (captive breeding sanctuaries). Strong selection by host in *N. gallinarum* appears to be unlikely, as Hap 3 and 4 were recovered from several different bird species.

A previous study on a chigger species from Poland [*Hirsutiella zachvatkini* (Schluger, 1948)] revealed host-dependent morphological plasticity in the leg, but not scutal, characters in the absence of differentiation based on COI barcodes [102]. However, in other species from Poland and Greece, such as *Leptotrombidium europaeum* (Daniel and Brelih, 1959) and *Neotrombicula talmiensis* Schluger, 1955, respectively, high diversity in morphology was observed within a single OTU, while some congeneric specimens were morphologically similar to these two species but were assigned to different OTUs by ABGD analysis of COI sequences [35]. Beyond Europe, substantial intraspecific diversity of COI has been reported within *Leptotrombidium* spp. in Southeast Asia [31], South Korea [32], and Japan [34], and for *Walchia* spp. in Southeast Asia [31,103], sometimes even in chiggers of the same species collected from a single host. However, as chigger populations can harbour several vertically transmitted bacteria with the potential to induce reproductive manipulations [3] and cytonuclear discordance [104], it is important to investigate potential cryptic species using nuclear as well as mitochondrial loci as we have explored here.

5. Conclusions

The use of multi-locus sequence analysis of both mitochondrial-encoded and nuclear-encoded genes revealed that *N. gallinarum* populations in Peninsular Malaysia and Thailand are geographically isolated with restricted gene flow leading to unambiguous genetic differentiation. High genetic diversity was attributed to the population in Peninsular Malaysia; however, more exploration is needed to elucidate the genetic diversity of *N. gallinarum* in Thailand, which was high even in two adjacent villages within the same subdistrict. Finally, our study revealed three robustly supported genetic lineages in Asia and further denoted *N. gallinarum* as a potential species complex, although further studies are required to determine the extent of biological differences (including pathogenicity) between its members.

Supplementary Materials: The following supporting information can be downloaded at https://www.mdpi.com/article/10.3390/ani14060980/s1, Figure S1: Phylogenetic relationships among *N. gallinarum* populations from Peninsular Malaysia, Thailand, and China were inferred through Bayesian Inference analysis based on the concatenated nucleotide sequences of mitochondrial cytochrome c oxidase subunit 1, second internal transcribed spacer, and 18S ribosomal DNA. Vertical bars on the right are the population groups. Coloured branches indicate different countries: red for samples from Peninsular Malaysia, green for samples from Thailand, and blue for samples from China. Figure S2: Phylogenetic relationships among *N. gallinarum* populations from Peninsular Malaysia, Thailand, and China were inferred through Neighbour Joining (NJ) and Maximum Likelihood (ML) analysis based on the sequences of mitochondrial cytochrome c oxidase subunit 1 (COI). Bootstrap values (NJ/ML) are shown on the branches. Vertical bars on the right are the population groups. Coloured branches indicate different countries: red for samples from Peninsular Malaysia, green for samples from Thailand, and blue for samples from China. Bootstrap values less than 0.50 are not shown in the figure. Figure S3: Phylogenetic relationships among *N. gallinarum* populations from Peninsular Malaysia, Thailand, and China were inferred through Neighbour Joining (NJ) and Maximum Likelihood (ML) analysis based on the sequences of nuclear-encoded internal transcribed spacer 2 (ITS2). Bootstrap values (NJ/ML) are shown on the branches. Coloured branches indicate different countries: red for samples from Peninsular Malaysia, green for samples from Thailand, and blue for samples from China. Bootstrap values less than 0.50 are not shown in the figure. Figure S4: Phylogenetic relationships among *N. gallinarum* populations from Peninsular Malaysia, Thailand, and China were inferred through Neighbour Joining (NJ) and Maximum Likelihood (ML) analysis based on the sequences of nuclear-encoded 18S. Bootstrap values (NJ/ML) are shown on the branches. Vertical bars on the right are the population groups. Coloured branches indicate different countries, red for samples from Peninsular Malaysia, green for samples from Thailand, and blue for samples from China. Bootstrap values less than 0.50 are not shown in the figure. Figure S5: Minimum spanning haplotype network of *N. gallinarum* based on COI gene sequence isolated from four populations in Peninsular Malaysia, one population in Thailand, and the seven sequences from China obtained from Zhou et al. [30]. Each haplotype is represented by the coloured nodes and their relative sizes indicate haplotype frequency. Nodes of the same colour specify the haplotype from the same population. The dashed lines on each node connecting haplotypes represent polymorphisms. Figure S6: Minimum spanning haplotype network of *N. gallinarum* based on the ITS2 gene sequence isolated from four populations in Peninsular Malaysia, one population in Thailand, and the seven sequences from China obtained from Zhou et al. [30]. Each haplotype is represented by the coloured nodes and their relative sizes indicate haplotype frequency. Nodes of the same colour specify the haplotype from the same population. The dashed lines on each node connecting haplotypes represent polymorphisms. Figure S7: Minimum spanning haplotype network of *N. gallinarum* based on the 18S ribosomal DNA isolated from four populations in Peninsular Malaysia, one population in Thailand, and the seven sequences from China obtained from Zhou et al. [30]. Each haplotype is represented by the coloured nodes and their relative sizes indicate haplotype frequency. Nodes of the same colour specify the haplotype from the same population. The dashed lines on each node connecting haplotypes represent polymorphisms. Table S1: Pairwise genetic distance based on concatenated genes between populations from Peninsular Malaysia, Thailand, and China computed using the Kimura 2-parameter (K2P) model from MEGA11. Table S2: Pairwise genetic distance for the COI gene between populations from Peninsular Malaysia, Thailand, and China computed using the Kimura 2-parameter (K2P) model from MEGA11. Table S3: Pairwise genetic distance for the ITS2 gene between populations from Peninsular Malaysia, Thailand, and China computed using the Kimura 2-parameter (K2P) model from MEGA11. Table S4: Pairwise genetic distance for the 18S rRNA gene between populations from Peninsular Malaysia, Thailand, and China computed using the Kimura 2-parameter (K2P) model from MEGA11. Table S5: Haplotype frequency of five populations of *N. gallinarum* in Malaysia and Thailand by region based on the COI gene. Table S6. Haplotype frequency of five populations of *N. gallinarum* in Malaysia and Thailand by region based on the ITS2 gene. Table S7: Haplotype frequency of five populations of *N. gallinarum* in Malaysia and Thailand by region based on the 18S rRNA gene.

Author Contributions: Conceptualisation, P.R. and B.L.M.; Data Curation, P.R. and B.L.M.; Formal Analysis, P.R., K.-K.T., J.J.K. and Y.S.; Funding Acquisition, K.C., S.A., S.M. and B.L.M. Investigation, P.R., S.K., M.S.M., M.K.S.A.K., K.C., S.M. and Z.Y.; Methodology, P.R., S.K., M.S.M., M.K.S.A.K., S.M., Z.Y. and B.L.M.; Supervision, K.C., S.A., S.M., Z.Y. and B.L.M.; Validation, P.R., K.-K.T., J.J.K., Y.S. and B.L.M.; Visualisation, P.R., K.C. and B.L.M.; Writing—Original Draft, P.R. and B.L.M.; Writing—Review and Editing, P.R., K.C., Z.Y. and B.L.M. All authors have read and agreed to the published version of the manuscript.

Funding: This research was supported by the Royal Society International Collaboration Award (ICA\R1\191058) granted to B.L.M. and S.A. and a Mahidol-Liverpool PhD scholarship awarded to S.K., K.C. and S.M. are supported by the ANR FutureHealthSEA Project, grant number ANR-17-CE35-0003-01.

Institutional Review Board Statement: In Peninsular Malaysia, animal handling was approved by the Universiti Malaya Institutional Animal Care and Use Committee (Permission No: G8/07052020/09012020-01/R, 7 May 2020). A wildlife research permit was obtained from the Department of Wildlife and National Parks (No. W-00064-15-21, 25 February 2021). In Thailand, animal ethical clearance was obtained by the Faculty of Tropical Medicine–Animal Care and Use Committee, Mahidol University, on 28 April 2020 with the certification number FTM-ACUC 011/2020E. This study was also approved by the Animal Welfare and Ethical Review Body of the University of Liverpool with reference nos. AWC0219 (Malaysia, 20 December 2021) and AWC0179 (Thailand, 19 June 2020).

Informed Consent Statement: Not applicable as research did not include human participants.

Data Availability Statement: Sequences generated were deposited in the GenBank database (https://www.ncbi.nlm.nih.gov/GenBank, accessed 4 October 2023) under the accession numbers OR632279-OR632323 for the COI gene, OR636401-OR636445 for the ITS2 gene, and OR632359-OR632403 for the 18S rRNA gene.

Acknowledgments: We are very grateful for the support of the field team from Universiti Malaya and the Department of Wildlife and National Parks in Malaysia, especially the department staff from the Sungkai Wildlife Conservation Centre, Perak, and the Jemaluang Wildlife Conservation Centre, Johor. In Thailand, we gratefully acknowledge the Tha Wang Pha District Livestock Office as well as the village chiefs and livestock owners from the two villages (Ban Santisuk and Ban Huay Muang in Saen Thong subdistrict, Tha Wang Pha, Nan province) for facilitating fieldwork and chigger specimen collection from local poultry.

Conflicts of Interest: The authors declare no conflict of interest.

Abbreviations

ASAP: Assemble species by automatic partitioning; ABGD: Automatic barcode gap discovery; CIPRES: Cyber Infrastructure for Phylogenetic Research; COI: Cytochrome c oxidase subunit I; GMYC: Generalised Mixed Yule Coalescent; ITS2: Internal transcribed spacer 2; mPTP: multi-rate Poisson Tree Processes; PCR: Polymerase chain reaction; OTU: Operational taxonomic unit.

References

1. Stekolnikov, A.A.; Waap, H.; Gomes, J.; Antune, T. Chigger mites of the genus *Ericotrombidium* (Acariformes: Trombiculidae) attacking pets in Europe. *Vet. Parasitol.* **2016**, *221*, 60–63. [CrossRef] [PubMed]
2. Molín, J.; Asín, J.; Stekolnikov, A.A.; de Miguel, R.; Pérez, M.; Gimeno, M.; Pinczowski, P.; Luján, L. Pathology of trombiculosis caused by *Neoschoengastia simonovichi* in wild red-legged partridges (*Alectoris rufa*). *J. Comp. Pathol.* **2020**, *181*, 92–96. [CrossRef]
3. Chaisiri, K.; Linsuwanon, P.; Makepeace, B.L. The chigger microbiome: Big questions in a tiny world. *Trends Parasitol.* **2023**, *39*, 696–707. [CrossRef] [PubMed]
4. Jones, B.M. The penetration of the host tissue by the harvest mite, *Trombicula autumnalis* Shaw. *Parasitology* **1950**, *40*, 247–260. [CrossRef] [PubMed]
5. Hase, T.; Roberts, L.W.; Hildebrandt, P.K.; Cavanaugh, D.C. Stylostome formation by *Leptotrombidium* mites (Acari: Trombiculidae). *J. Parasitol.* **1978**, *64*, 712–718. [CrossRef]
6. Wright, S.M.; Wikel, S.K.; Wrenn, W.J. Host immune responsiveness to the chigger, *Eutrombicula cinnabaris*. *Ann. Trop. Med. Parasitol.* **1988**, *82*, 283–293. [CrossRef]
7. Dong, X.; Chaisiri, K.; Xia, D.; Armstrong, S.D.; Fang, Y.; Donnelly, M.J.; Kadowaki, T.; McGarry, J.W.; Darby, A.C.; Makepeace, B.L. Genomes of trombidid mites reveal novel predicted allergens and laterally transferred genes associated with secondary metabolism. *GigaScience* **2018**, *7*, giy127. [CrossRef]
8. Little, S.E.; Carmichael, K.P.; Rakich, P.M. Trombidiosis-induced dermatitis in white-tailed deer (*Odocoileus virginianus*). *Vet. Pathol.* **1997**, *34*, 350–352. [CrossRef]
9. Smith, G.A.; Shama, V.; Knapp, J.F.; Shields, B.J. The summer penile syndrome: Seasonal acute hypersensitivity reaction caused by chigger bites on the penis. *Pediatr. Emerg. Care* **1998**, *14*, 116–118. [CrossRef]
10. Leone, F.; Di Bella, A.; Vercelli, A.; Cornegliani, L. Feline trombiculosis: A retrospective study in 72 cats. *Vet. Dermatol.* **2013**, *24*, 535-e126. [CrossRef]
11. Lawrence, R.F. The larval trombiculid mites of South African vertebrates. *Ann. Natal. Mus.* **1949**, *11*, 405–486.
12. Traub, R.; Morrow, M.L. A revision of the chiggers of the subgenus *Gahrliepia* (Acarina: Trombiculidae). *Smithson. Misc. Collect.* **1955**, *128*, 1–88.
13. Farrell, C.E. Chiggers of the genus *Euschoengastia* (Acarina: Trombiculidae) in North America. *Proc. U. S. Natl. Mus.* **1956**, *106*, 85–235. [CrossRef]
14. Tanigoshi, L.K.; Loomis, R.B. Genus *Hyponeocula* (Acarina, Trombiculidae) of western North America. *Melanderia* **1974**, *17*, 1–27.
15. Wrenn, W.J. Notes on the ecology of chiggers (Acarina: Trombiculidae) from Northern Michigan and the description of a new species of *Euschoengastia*. *J. Kansas Entomol. Soc.* **1974**, *47*, 227–238.
16. Wrenn, W.J.; Baccus, J.T.; Loomis, R.B. Two new species of North American mites in the genus *Euschoengastia* (Acarina: Trombiculidae). *Southwest Nat.* **1976**, *21*, 301–309. [CrossRef]
17. Sileo, L.; Sievert, P.R.; Samuel, M.D. Causes of mortality of albatross chicks at Midway Atoll. *J. Wildl. Dis.* **1990**, *26*, 329–338. [CrossRef] [PubMed]
18. Ornelas-Almeida, M.A.; de Oliveira, F.R.B.; da Silva, A.E.; Moreira, E.L.T.; Maia, P.C.C.; Duarte, L.d.F.C.; Murphy, G.; Ayres, M.C.C. Nodular trombiculinosis caused by *Apolonia tigipioensis*, Torres and Braga (1938), in an ostrich (*Struthio camelus*) and a house sparrow (*Passer domesticus*). *Vet. Parasitol.* **2007**, *15*, 374–377. [CrossRef] [PubMed]
19. Elliott, I.; Pearson, I.; Dahal, P.; Thomas, N.V.; Roberts, T.; Newton, P.N. Scrub typhus ecology: A systematic review of *Orientia* in vectors and hosts. *Parasite Vectors* **2019**, *12*, 513. [CrossRef]
20. Womersley, H. The scrub-typhus and scrub-itch mites (Trombiculidae, Acarina) of the Asiatic-Pacific region. *Rec. S. Aust. Mus.* **1952**, *10*, 438–673.
21. Nadchatram, M.; Upham, R.W., Jr. Two new larval trombiculid mites (Acarina, Trombiculidae) collected from ground holes and birds in Malaysia. *J. Med. Entomol.* **1966**, *3*, 345–350. [CrossRef]
22. Bassini-Silva, R.; Jacinavicius, F.D.C.; Welbourn, C.; Huang-Bastos, M.; Ochoa, R.; Barros-Battesti, D.M. Taxonomic notes on *Neoschoengastia esorhina* Brennan, 1971 (Trombidiformes: Trombiculidae), a chigger species from Brazil. *Int. J. Acarol.* **2021**, *47*, 137–141. [CrossRef]
23. Stekolnikov, A.A. A checklist of chigger mites (Acariformes: Trombiculidae) of Southeast Asia. *Zootaxa* **2021**, *4913*, 1–163. [CrossRef]
24. Yeruham, I.; Rosen, S.; Hadani, A.; Braverman, Y. Arthropod parasites of Nubian ibexes (*Capra ibex nubiana*) and gazelles (*Gazella gazella*) in Israel. *Vet. Parasitol.* **1999**, *83*, 167–173. [CrossRef]
25. Barnard, K.; Krasnov, B.R.; Goff, L.; Matthee, S. Infracommunity dynamics of chiggers (Trombiculidae) parasitic on a rodent. *Parasitology* **2015**, *142*, 1605–1611. [CrossRef]
26. Kunz, S.E.; Price, M.A.; Graham, O.H. Biology and economic importance of the chigger *Neoschongastia americana* on Turkeys. *J. Econ. Entomol.* **1969**, *62*, 872–875. [CrossRef]
27. Fujisaki, K.; Taniguchi, T.; Ishizu, K.; Uchikawa, K. Fatal cases of pheasants *Phasianus colchicus var. tenebrosus*, heavily infested with *Neoschoengastia shiraii*. *Jpn. J. Sanit. Zool.* **1991**, *42*, 61–63. [CrossRef]
28. Kuo, S.C.; Ho, C.C.; Kuo, M.C.; Robbins, R.G. Case Report: Trombidiasis in native chicken (Prostigmata: Trombiculidae). *Taiwan Vet. J.* **2004**, *30*, 301–306.

29. Koosakulnirand, S.; Rajasegaran, P.; Alkathiry, H.A.; Chaisiri, K.; Round, P.D.; Eiamampai, K.; Khusaini, M.K.S.A.; Ramji, M.F.S.; Abubakar, S.; Ya'cob, Z.; et al. On the taxonomy of chigger mites (Acariformes: Trombiculidae) parasitizing birds in Thailand and Malaysia, with the description of a new species. *Acarologia* **2023**, *64*, 1109–1138. [CrossRef]
30. Zhou, Q.; Wang, Z.-X.; Tao, J.-M.; Qin, J.-P.; Lu, J.-P.; Lin, R.-Q.; Wang, L.-M.; Weng, Y.-B.; Tan, Z.-J. Characterization of *Neoschoengastia gallinarum* from subtropical China by rDNA and identification of two genotypes based on mitochondrial *cox*1. *Parasitol. Res.* **2020**, *119*, 3339–3345. [CrossRef] [PubMed]
31. Kumlert, R.; Chaisiri, K.; Anantatat, T.; Stekolnikov, A.A.; Morand, S.; Prasartvit, A.; Makepeace, B.L.; Sungvornyothin, S.; Paris, D.H. Autofluorescence microscopy for paired-matched morphological and molecular identification of individual chigger mites (Acari: Trombiculidae), the vectors of scrub typhus. *PLoS ONE* **2018**, *13*, e0193163. [CrossRef] [PubMed]
32. Lee, H.S.; Choi, K.S. Genetic variation of chigger mites in the Republic of Korea. *Entomol. Res.* **2022**, *52*, 385–393. [CrossRef]
33. Zajkowska, P.; Mąkol, J. Parasitism, seasonality, and diversity of trombiculid mites (Trombidiformes: Parasitengona, Trombiculidae) infesting bats (Chiroptera) in Poland. *Exp. Appl. Acarol.* **2022**, *86*, 1–20. [CrossRef] [PubMed]
34. Ogawa, M.; Takada, N.; Noda, S.; Takahashi, M.; Matsutani, M.; Kageyama, D.; Ebihara, H. Genetic variation of *Leptotrombidium* (Acari: Trombiculidae) mites carrying *Orientia tsutsugamushi*, the bacterial pathogen causing scrub typhus. *J. Parasitol.* **2023**, *109*, 340–348. [CrossRef] [PubMed]
35. Zajkowska, P.; Postawa, T.; Mąkol, J. Let me know your name: A study of chigger mites (Acariformes: Trombiculidae) associated with the edible dormouse (*Glis glis*) in the Carpathian–Balkan distribution gradient. *Exp. Appl. Acarol.* **2023**, *91*, 1–27. [CrossRef] [PubMed]
36. Hatori, J. "Tsutsugamushi" disease in Formosa (5th Communication). *Taiwan Igakkai Zasshi (J. Formosa Med. Soc.)* **1920**, *209*, 317–352.
37. Chaisiri, K.; Kittiyakan, A.; Kumlert, R.; Lajaunie, C.; Makaew, P.; Morand, S.; Paladsing, Y.; Tanita, M.; Thinphovong, C. A social-ecological and One Health observatory: Ten years of collaborative studies in Saen Thong (Nan, Thailand). *One Health Cases* **2023**, ohcs20230008. [CrossRef]
38. Folmer, O.; Hoeh, W.R.; Black, M.B.; Vrijenhoek, R.C. Conserved primers for PCR amplification of mitochondrial DNA from different invertebrate phyla. *Mol. Mar. Biol. Biotechnol.* **1994**, *3*, 294–299.
39. Tao, J.M.; Ashram, S.E.; Alouffi, A.; Zhang, Y.; Weng, Y.B.; Lin, R.Q. Population genetic structure of *Neoschoengastia gallinarum* in South China based on mitochondrial DNA markers. *Parasitol. Res.* **2022**, *121*, 2793–2802. [CrossRef]
40. Hall, T.A. BioEdit: A user-friendly biological sequence alignment editor and analysis program for Windows 95/98/NT. In *Nucleic Acids Symposium Series*; Information Retrieval Ltd.: London, UK, 1999; pp. 95–98.
41. Thompson, J.D.; Gibson, T.J.; Plewniak, F.; Jeanmougin, F.; Higgins, D.G. The CLUSTAL_X Windows interface: Flexible strategies for multiple sequence alignment aided by quality analysis tools. *Nucleic Acids Res.* **1997**, *25*, 4876–4882. [CrossRef]
42. Tamura, K.; Stecher, G.; Kumar, S. MEGA11: Molecular evolutionary genetics analysis version 11. *Mol. Biol. Evol.* **2021**, *38*, 3022–3027. [CrossRef] [PubMed]
43. Swoford, D.L. *PAUP: Phylogenetic Analysis Using Parsimony Version 4*; Sinauer Associates: Sunderland, MA, USA, 2002.
44. Kimura, M.A. Simple method for estimating evolutionary rates of base substitutions through comparative studies of nucleotide sequences. *J. Mol. Evol.* **1980**, *16*, 111–120. [CrossRef] [PubMed]
45. Guindon, S.; Dufayard, J.F.; Lefort, V.; Anisimova, M.; Hordijk, W.; Gascuel, O. New algorithms and methods to estimate maximum-likelihood phylogenies: Assessing the performance of PhyML 3.0. *Syst. Biol.* **2010**, *59*, 307–321. [CrossRef] [PubMed]
46. Huelsenbeck, J.P.; Ronquist, F. MRBAYES: Bayesian inference of phylogenetic trees. *Bioinformatics* **2001**, *17*, 754–755. [CrossRef]
47. Darriba, D.; Taboada, G.L.; Doallo, R.; Posada, D. JModelTest 2: More models, new heuristics and parallel computing. *Nat. Methods* **2012**, *9*, 772. [CrossRef] [PubMed]
48. Miller, M.A.; Pfeiffer, W.; Schwartz, T. The CIPRES science gateway: A community resource for phylogenetic analyses. In Proceedings of the TeraGrid 2011 Conference: Extreme Digital Discovery, Salt Lake City, UT, USA, 18–21 July 2011; Volume 41, pp. 1–8. [CrossRef]
49. Letunic, I.; Bork, P. Interactive Tree Of Life (iTOL) v5: An online tool for phylogenetic tree display and annotation. *Nucleic Acids Res.* **2021**, *49*, W293–W296. [CrossRef]
50. Bandelt, H.J.; Forster, P.; Röhl, A. Median-joining networks for inferring intraspecific phylogenies. *Mol. Biol. Evol.* **1999**, *16*, 37–48. [CrossRef]
51. Clement, M.; Snell, Q.; Walke, P.; Posada, D.; Crandall, K. TCS: Estimating gene genealogies. In Proceedings of the 16th International Parallel and Distributed Processing Symposium, Ft. Lauderdale, FL, USA, 15–19 April 2002.
52. Leigh, J.W.; Bryant, D. POPART: Full-feature software for haplotype network construction. *Methods Ecol. Evol.* **2015**, *6*, 1110–1116. [CrossRef]
53. Puillandre, N.; Brouillet, S.; Achaz, G. ASAP: Assemble species by automatic partitioning. *Mol. Ecol. Resour.* **2021**, *21*, 609–620. [CrossRef]
54. Puillandre, N.; Lambert, A.; Brouillet, S.; Achaz, G.J.M.E. ABGD, Automatic Barcode Gap Discovery for primary species delimitation. *Mol. Ecol.* **2012**, *21*, 1864–1877. [CrossRef]
55. Kapli, P.; Lutteropp, S.; Zhang, J.; Kobert, K.; Pavlidis, P.; Stamatakis, A.; Flouri, T. Multi-rate Poisson tree processes for single-locus species delimitation under maximum likelihood and Markov chain Monte Carlo. *Bioinformatics* **2017**, *33*, 1630–1638. [CrossRef]

56. Fujisawa, T.; Barraclough, T.G. Delimiting species using single-locus data and the Generalized Mixed Yule Coalescent approach: A revised method and evaluation on simulated data sets. *Syst. Biol.* **2013**, *62*, 707–724. [CrossRef] [PubMed]
57. Goulpeau, A.; Penel, B.; Maggia, M.E.; Marchán, D.F.; Steinke, D.; Hedde, M.; Decaëns, T. OTU delimitation with earthworm DNA barcodes: A comparison of methods. *Diversity* **2022**, *14*, 866. [CrossRef]
58. Bouckaert, R.; Vaughan, T.G.; Barido-Sottani, J.; Duchêne, S.; Fourment, M.; Gavryushkina, A.; Heled, J.; Jones, G.; Kühnert, D.; De Maio, N.; et al. BEAST 2.5: An advanced software platform for Bayesian evolutionary analysis. *PLoS Comput. Biol.* **2019**, *15*, e1006650. [CrossRef] [PubMed]
59. Miller, M.A.; Pfeiffer, W.; Schwartz, T. Creating the CIPRES science gateway for inference of large phylogenetic trees. In Proceedings of the 2010 Gateway Computing Environments Workshop (GCE), New Orleans, LA, USA, 14 November 2010; pp. 1–8.
60. Rambaut, A.; Drummond, A.J.; Xie, D.; Baele, G.; Suchard, M.A. Posterior summarization in Bayesian phylogenetics using Tracer 1.7. *Syst. Biol.* **2018**, *67*, 901–904. [CrossRef] [PubMed]
61. Drummond, A.J.; Rambaut, A. BEAST: Bayesian evolutionary analysis by sampling trees. *BMC Evol. Biol.* **2007**, *7*, 214. [CrossRef] [PubMed]
62. RStudio Team. *RStudio: Integrated Development for R*; RStudio: Boston, MA, USA, 2020.
63. Paradis, E.; Schliep, K. ape 5.0: An environment for modern phylogenetics and evolutionary analyses in R. *Bioinformatics* **2019**, *35*, 526–528. [CrossRef] [PubMed]
64. Dinno, A. Package 'Paran' (R package Version 1.5.2). Available online: https://cran.r-project.org/web/packages/paran/paran.pdf (accessed on 16 March 2024).
65. Michonneau, F.; Bolker, B.; Holder, M.; Lewis, P.; O'Meara, B. Package 'rncl': An Interface to the Nexus Class Library (R Package Version 0.8.3). Available online: https://cran.r-project.org/web/packages/rncl/rncl.pdf (accessed on 16 March 2024).
66. Ezard, T.; Fujisawa, T.; Barraclough, T.G. splits: SPecies' LImits by Threshold Statistics (R Package Version 1.0-14/r31). Available online: https://r-forge.r-project.org/projects/splits/ (accessed on 16 March 2024).
67. Rozas, J.; Ferrer-Mata, A.; Sánchez-DelBarrio, J.C.; Guirao-Rico, S.; Librado, P.; Ramos-Onsins, S.E.; Sánchez-Gracia, A. DnaSP 6: DNA Sequence Polymorphism Analysis of Large Data Sets. *Mol. Biol. Evol.* **2017**, *34*, 3299–3302. [CrossRef]
68. Excoffier, L.; Laval, G.; Schneider, S. Arlequin (version 3.5): An integrated software package for population genetics data analysis. *Evol. Bioinform. Online* **2007**, *23*, 47–50.
69. Slatkin, M. Inbreeding coefficients and coalescence times. *Genet Res.* **1991**, *58*, 167–175. [CrossRef]
70. Excoffier, L.; Smouse, P.E.; Quattro, J.M. Analysis of molecular variance inferred from metric distances among DNA haplotypes: Application to human mitochondrial DNA restriction data. *Genetics* **1992**, *131*, 479–491. [CrossRef]
71. Domrow, R.; Nadchatram, M. XLIII *Neoschongastia* in Malaya (Acarina, Trombiculidae). In: Macdonald, WW, editor. Malaysian Parasites XXXV-XLIX. Studies from the Institute for Medical Research. *Kuala Lumpur Gov. Fed. Malaya* **1960**, *29*, 185–193.
72. Bickford, D.; Lohman, D.J.; Sodhi, N.S.; Ng, P.K.; Meier, R.; Winker, K.; Das, I. Cryptic species as a window on diversity and conservation. *Trends Ecol. Evol.* **2007**, *22*, 148–155. [CrossRef] [PubMed]
73. Struck, T.H.; Feder, J.L.; Bendiksby, M.; Birkeland, S.; Cerca, J.; Gusarov, V.I.; Dimitrov, D. Finding evolutionary processes hidden in cryptic species. *Trends Ecol. Evol.* **2018**, *33*, 153–163. [CrossRef] [PubMed]
74. Skoracka, A.; Magalhaes, S.; Rector, B.G.; Kuczyński, L. Cryptic speciation in the Acari: A function of species lifestyles or our ability to separate species? *Exp. Appl. Acarol.* **2015**, *67*, 165–182. [CrossRef] [PubMed]
75. Yong, H.S.; Lim, P.E.; Tan, J.; Song, S.L.; Suana, I.W.; Eamsobhana, P. Multigene phylogeography of *Bactrocera caudata* (Insecta: Tephritidae): Distinct genetic lineages in Northern and Southern hemispheres. *PLoS ONE* **2015**, *10*, e0129455. [CrossRef] [PubMed]
76. Taylor, J.W.; Jacobson, D.J.; Kroken, S.; Kasuga, T.; Geiser, D.M.; Hibbett, D.S.; Fisher, M.C. Phylogenetic species recognition and species concepts in fungi. *Fungal Genet. Biol.* **2000**, *31*, 21–32. [CrossRef] [PubMed]
77. Rosenberg, N.A.; Nordborg, M. Genealogical trees, coalescent theory and the analysis of genetic polymorphisms. *Nat. Rev. Genet.* **2000**, *3*, 380–390. [CrossRef] [PubMed]
78. Al-Khafaji, A.M.; Clegg, S.R.; Pinder, A.C.; Luu, L.; Hansford, K.M.; Seelig, F.; Dinnis, R.E.; Margos, G.; Medlock, J.M.; Feil, E.J.; et al. Multi-locus sequence typing of *Ixodes ricinus* and its symbiont *Candidatus Midichloria mitochondrii* across Europe reveals evidence of local co-cladogenesis in Scotland. *Ticks Tick Borne Dis.* **2019**, *10*, 52–62. [CrossRef] [PubMed]
79. Alghamdi, S.Q.; Low, V.L.; Alkathiry, H.A.; Alagaili, A.N.; McGarry, J.W.; Makepeace, B.L. Automatic barcode gap discovery reveals diverse clades of *Rhipicephalus* spp. and *Haemaphysalis* spp. ticks from small mammals in Asir, Saudi Arabia. *Parasites Vectors* **2021**, *14*, 541. [CrossRef]
80. Bitencourth, K.; Amorim, M.; Oliveira, S.V.D.; Gazêta, G.S. *Amblyomma aureolatum* genetic diversity and population dynamics are not related to spotted fever epidemiological scenarios in Brazil. *Pathogens* **2021**, *10*, 1146. [CrossRef]
81. Páez-Triana, L.; Muñoz, M.; Herrera, G.; Moreno-Pérez, D.A.; Tafur-Gómez, G.A.; Montenegro, D.; Patarroyo, M.A.; Paniz-Mondolfi, A. Genetic diversity and population structure of *Rhipicephalus sanguineus sensu lato* across different regions of Colombia. *Parasites Vectors* **2021**, *14*, 424. [CrossRef] [PubMed]
82. Lynch, M.; Koskella, B.; Schaack, S. Mutation pressure and the evolution of organelle genomic architecture. *Science* **2006**, *311*, 1727–1730. [CrossRef] [PubMed]

83. Sharbrough, J.; Bankers, L.; Cook, E.; Fields, P.D.; Jalinsky, J.; McElroy, K.E.; Neiman, M.; Logsdon, J.M.; Boore, J.L. Single-molecule sequencing of an animal mitochondrial genome reveals chloroplast-like architecture and repeat-mediated recombination. *Mol. Biol. Evol.* **2023**, *40*, msad007. [CrossRef]
84. Skoracka, A.; Kuczyński, L.; de Mendonça, R.S.; Dabert, M.; Szydło, W.; Knihinicki, D.; Truol, G.; Navia, D. Cryptic species within the wheat curl mite *Aceria tosichella* (Keifer) (Acari: Eriophyoidea), revealed by mitochondrial, nuclear, and morphometric data. *Invert. Syst.* **2012**, *26*, 417–433. [CrossRef]
85. Rivera, J.; Currie, D.C. Identification of Nearctic black flies using DNA barcodes (Diptera: Simuliidae). *Mol. Ecol. Resour.* **2009**, *9*, 224–236. [CrossRef] [PubMed]
86. Pramual, P.; Adler, P.H. DNA barcoding of tropical black flies (Diptera: Simuliidae) of Thailand. *Mol. Ecol. Resour.* **2014**, *14*, 262–271. [CrossRef]
87. Hew, Y.X.; Ya'cob, Z.; Adler, P.H.; Chen, C.D.; Lau, K.W.; Sofian-Azirun, M.; Muhammad-Rasul, A.H.; Putt, Q.Y.; Izwan-Anas, N.; Hadi, U.K.; et al. DNA barcoding of black flies (Diptera: Simuliidae) in Indonesia. *Parasites Vectors* **2023**, *16*, 248. [CrossRef]
88. Pramual, P.; Simwisat, K.; Martin, J. Identification and reassessment of the specific status of some tropical freshwater midges (Diptera: Chironomidae) using DNA barcode data. *Zootaxa* **2016**, *4072*, 39–60. [CrossRef]
89. Koopman, W.J.M.; Li, Y.; Coart, E.; Van de Weg, W.E.; Vosman, B.; Roldán-Ruiz, I.; Smulders, M.J.M. Linked vs. unlinked markers: Multilocus microsatellite haplotype-sharing as a tool to estimate gene flow and introgression. *Mol. Ecol.* **2007**, *16*, 243–256. [CrossRef]
90. Posada, D.; Crandall, K.A. Intraspecific gene genealogies: Trees grafting into networks. *Trends Ecol. Evol.* **2001**, *16*, 37–45. [CrossRef] [PubMed]
91. Xu, Y.; Mai, J.-W.; Yu, B.-J.; Hu, H.-X.; Yuan, L.; Jashenko, R.; Ji, R. Study on the genetic differentiation of geographic populations of *Calliptamus italicus* (Orthoptera: Acrididae) in Sino-Kazakh border areas based on mitochondrial COI and COII genes. *J. Econ. Entomol.* **2019**, *112*, 1912–1919. [CrossRef] [PubMed]
92. Jacinavicius, F.C.; Bassini-Silva, R.; Muñoz-Leal, S.; Welbourn, C.; Ochoa, R.; Labruna, M.B.; Barros-Battesti, D.M. Molecular detection of *Rickettsia* genus in chigger mites (Trombidiformes: Trombiculidae) collected on small mammals in southeastern Brazil. *Rev. Bras. Parasitol. Vet.* **2019**, *28*, 563–568. [CrossRef] [PubMed]
93. Park, S.W.; Ha, N.Y.; Ryu, B.; Bang, J.H.; Song, H.; Kim, Y.; Kim, G.; Oh, M.D.; Cho, N.H.; Lee, J.K. Urbanization of scrub typhus disease in South Korea. *PLoS Negl. Trop. Dis.* **2015**, *9*, e0003814. [CrossRef]
94. Pramual, P.; Kongim, B.; Nanork, P. Phylogeography of *Simulium siamense* Takaoka and Suzuki complex (Diptera: Simuliidae) in Thailand. *Entomol. Sci.* **2011**, *14*, 428–436. [CrossRef]
95. Sun, Z.H.; Luan, F.G.; Zhang, D.M.; Chen, M.J.; Wang, B.; Li, Z.Z. Genetic differentiation of *Isaria farinosa* populations in Anhui Province of East China. *J. Appl. Ecol.* **2011**, *22*, 3039–3046.
96. Department of Wildlife and National Parks (PERHILITAN). Laws of Malaysia, Act 716, Wildlife Conservation Act 2010. Available online: https://storage.unitedwebnetwork.com/files/478/2bcd898fbf196a7cc36b99572fbc3a70.pdf (accessed on 31 October 2023).
97. BirdLife International. *Lophura rufa*. The IUCN Red List of Threatened Species 2020: E.T22727445A184588512. 2020. Available online: https://www.iucnredlist.org/species/22727445/184588512 (accessed on 31 October 2023).
98. BirdLife International. *Polyplectron inopinatum*. The IUCN Red List of Threatened Species 2023: E.T22679365A218821910. 2016. Available online: https://www.iucnredlist.org/species/22679365/218821910 (accessed on 31 October 2023).
99. BirdLife International. *Polyplectron malacense*. The IUCN Red List of Threatened Species 2022: E.T22679385A137837773. 2022. Available online: https://www.iucnredlist.org/species/22679385/137837773 (accessed on 31 October 2023).
100. Savini, T.; Namkhan, M.; Sukumal, N. Conservation status of Southeast Asian natural habitat estimated using Galliformes spatio-temporal range decline. *Glob. Ecol. Conserv.* **2021**, *29*, e01723. [CrossRef]
101. Corder, J.; Davison, G. Captive breeding challenges posed by Malaysian and Bornean Peacock-pheasants (*Polyplectron malacense* and *P. schleiermacheri*). *Zoo Biol.* **2021**, *40*, 346–351. [CrossRef]
102. Moniuszko, H.; Zaleśny, G.; Mąkol, J. Host-associated differences in morphometric traits of parasitic larvae *Hirsutiella zachvatkini* (Actinotrichida: Trombiculidae). *Exp. Appl. Acarol.* **2015**, *67*, 123–133. [CrossRef]
103. Sungvornyothin, S.; Kumlert, R.; Paris, D.H.; Prasartvit, A.; Sonthayanon, P.; Apiwathnasorn, C.; Morand, S.; Stekolnikov, A.A.; Sumruayphol, S. Geometric morphometrics of the scutum for differentiation of trombiculid mites within the genus *Walchia* (Acariformes: Prostigmata: Trombiculidae), a probable vector of scrub typhus. *Ticks Tick Borne Dis.* **2019**, *10*, 495–503. [CrossRef]
104. Cariou, M.; Duret, L.; Charlat, S. The global impact of *Wolbachia* on mitochondrial diversity and evolution. *J. Evol. Biol.* **2017**, *30*, 2204–2210. [CrossRef] [PubMed]

animals

MDPI

Article

Three New Species of *Aceria* (Acari: Trombidiformes: Eriophyoidea) from China †

Mengchao Tan [1,]*, Ranran Lian [2], Hongyan Ruan [1] and Xuhui Liang [2]

1 Key Laboratory of Beibu Gulf Environment Change and Resources Utilization of Ministry of Education, Nanning Normal University, Nanning 530001, China; rhyan@nnnu.edu.cn

2 Guangxi Key Laboratory of Agric-Environment and Agric-Products Safety, National Demonstration Center for Experimental Plant Science Education, College of Agriculture, Guangxi University, Nanning 530004, China; 18939601915@163.com (R.L.); 17776712629@163.com (X.L.)

* Correspondence: linghanxii@163.com

† urn:lsid:zoobank.org:pub:C9ADE3B4-407F-4473-ACB0-5C2CB370D90C; urn:lsid:zoobank.org:act:8BDC930C-D093-49F2-841D-0E9C91A9849E; urn:lsid:zoobank.org:act:EABCFE27-5953-45D6-8EC0-F2877C15FDF5; urn:lsid:zoobank.org:act:ED164732-7CC0-4FCF-A3F0-9347608FFA19.

Simple Summary: The superfamily Eriophyoidea includes more than 5000 species worldwide and is a group of phytophagous mites that has an important influence on the agricultural economy. Aceria is a rich genus of more than 1000 species that belongs to the family Eriophyidae, which is distributed throughout the whole world. Here, three new species, *Aceria bischofiae* **sp. nov.**, *Aceria cryptocaryae* **sp. nov.**, and *Aceria buddlejae* **sp. nov.**, from Guangxi and Chongqing Province, China (the Oriental realm), are described and illustrated.

Abstract: Three new *Aceria* species from South China are described and illustrated. *Aceria bischofiae* **sp. nov.** was collected on *Bischofia javanica* Blume (Phyllanthaceae), inducing galls on surfaces of the leaves; *Aceria cryptocaryae* **sp. nov.** was collected on *Cryptocarya metcalfiana* Allen (Lauraceae), causing the formation of erinea on the undersurface of the leaves; and *Aceria buddlejae* **sp. nov.** was collected as a vagrant on *Buddleja lindleyana* Fort. (Scrophulariaceae) leaves, and no symptoms were observed on the host plant.

Keywords: Aceriini; Eriophyidae; Guangxi; taxonomy; galls

Citation: Tan, M.; Lian, R.; Ruan, H.; Liang, X. Three New Species of *Aceria* (Acari: Trombidiformes: Eriophyoidea) from China. *Animals* **2024**, *14*, 720. https://doi.org/10.3390/ani14050720

Academic Editors: Maciej Skoracki and Monika Fajfer

Received: 16 January 2024
Revised: 20 February 2024
Accepted: 22 February 2024
Published: 25 February 2024

1. Introduction

Eriophyoidea (Acari: Prostigmata) is a large mite superfamily and among the smallest arthropods known. Until now, more than 5000 named species have been recognized, some of which are significant pests of agronomic plants [1,2], and over 80% of eriophyoid mite species are monophagous, registered on only one host plant [3,4]. Host plants supposedly played key roles in their diversification [5].

Aceria Keifer is the genus of the family Eriophyidae Nalepa with the highest number of known species. Until now, more than 1000 species names of *Aceria* have been reported around the world, of which about 81 species have been found in China [6–10]. However, some species within *Aceria* are described too simply, and their taxonomic status needs to be further clarified through more detailed morphological descriptions and comprehensive taxonomic methods.

Bischofia javanica Blume is an evergreen tree belonging to the family Phyllanthaceae, broadly distributed in China, India, Bangladesh, and Southeast Asia. The nutritional value of this plant is very high, and the leaves are widely used in the preparation of salads and condiments [11]. Until now, three eriophyoid mites have been described from the plants of the genus *Bischofia*: *Phyllocoptruta maerimae* Boczek and Chandrapatya, 2000;

Bischofius kanchanaburi Boczek and Chandrapatya, 2000; and *Diptilomiopus bischofiae* Li, Wei, and Wang, 2009 [12,13]. *Buddleja lindleyana* Fort. is a garden ornamental plant and also a commonly used medicinal plant that belongs to the family Scrophulariaceae, which is native to China and mainly distributed in most parts of southern China. It is also distributed in America, Malaysia, Africa, and so on [14]. To date, one eriophyoid mite has been described from the plants of the genus *Buddleja*: *Aculops salviifoliae* Meyer and Ueckermann, 1990 [15]. *Cryptocarya metcalfiana* Allen belongs to the family Lauraceae, which is distributed in South China. Only one eriophyoid mite has been described from the plants of the genus *Cryptocarya*: *Aceria aphanothrix* (Nalepa, 1923) [16].

This paper presents descriptions of three new *Aceria* species: *Aceria bischofiae* **sp. nov.**, *Aceria cryptocaryae* **sp. nov.**, and *Aceria buddlejae* **sp. nov.** from the subtropical zone of China (the Oriental Region).

2. Materials and Methods

Mite specimens were collected from different host plants in Guangxi and Chongqing provinces by the aid of a hand lens (80×) (brand: Binyun; model: BY2600; manufacturer: Xinxiang Optics, Hangzhou, China) in China. The mites were collected from leaf samples and stored in a 70% ethanol solution using a brush. Samples were slide-mounted in modified Berlese medium [17] without adding additional fibers [18]. All specimens were examined with an Olympus CX41 (Philippines) microscope under phase contrast (oil immersion: 100×/1.25; widefield eyepiece: 10×). Micrographs were obtained from a Nikon DS-Ri2 microscope. The morphological terminology used in the morphological description of the mites follows Lindquist [1] and Amrine et al. [19], and internal female genitalia nomenclature follows Chetverikov [20]. The generic classification follows Amrine et al. [19] in combination with descriptions of all the published genera after 2003. All morphological measurements were according to Amrine and Manson [17], as modified by de Lillo et al. [18]. Measurements refer to the length of the morphological trait unless otherwise specified and are given in micrometers (μm). The holotype female measurement precedes the corresponding range for paratypes (given in parentheses). For males, only the ranges are given. Moreover, "*" in the descriptions means there is no variation in measurements. The number of measured specimens (n) is given within parentheses in the description of each stage. Line drawings were prepared according to de Lillo et al. [18], and abbreviations used in figures follow Amrine et al. [19]. Host plant names and their synonymies are in accordance with The World Flora Online "http://www.worldfloraonline.org/" (30 May 2023).

Type materials are deposited at the Key Laboratory of Beibu Gulf Environment Change and Resources Utilization of the Ministry of Education, Nanning Normal University, Guangxi, China.

3. Results

Systematics

Family: Eriophyidae Nalepa, 1898.

Subfamily: Eriophyinae Nalepa, 1898.

Tribe: Aceriini Amrine and Stasny, 1994.

Genus: *Aceria* Keifer, 1944.

Aceria bischofiae **sp. nov. (Figures 1 and 2)**

Description: **Female** (n = 15). Body vermiform, 191 (185–202, including gnathosoma), 48 (43–48) wide, 44 (42–46) thick. **Gnathosoma** 19 (18–21), projecting obliquely downwards, pedipalp coxal setae *ep* 2 (2–3), dorsal pedipalp genual setae *d* 3 (2–4), unbranched, palp tarsus setae *v* 1 (1–2), cheliceral stylets 18 (18–20). **Prodorsal shield** 29 (27–30), including frontal lobe, 38 (33–38) wide; with a short flexible distally rounded frontal lobe, 3 (3–5), over gnathosomal base. Median lines and admedian lines are present on the posterior half of the shield; submedian lines do not reach the rear shield margin; a few short dashes medially; and some short and long dashes on the lateral margin of the shield. Tubercles of scapular setae *sc* on rear shield margin, 14 (12–15) apart, scapular setae *sc* 20 (18–21), divergently backward. **Coxae** smooth; prosternal apodeme 5 (5–6); setae *1b* 7 (7–8), tubercles *1b* 7 (6–7) apart; setae *1a* 16 (15–20), tubercles *1a* 8 (8–9) apart; setae *2a* 29 (27–33), tubercles *2a* 17 (16–18) apart. **Leg I** 24 (20–25), femur 8 (7–8), femoral setae *bv* 7 (5–7), genu 3 (3–4), genual setae *l''* 18 (16–18), tibia 4 (4–5), tibial setae *l'* 2 (2–3), tarsus 5 (5–6), tarsal setae *ft'* 6 (5–8), setae *ft''* 15 (11–16), setae *u'* 3 (2–3), solenidion *ω* 5 (5–6), curved down, distally simple, empodium simple, 4 (4–5), 5-rayed. **Leg II** 22 (20–23), femur 7 (7–8), femoral setae *bv* 6 (6–8), genu 3 (3–4), genual setae *l''* 8 (5–8), tibia 3 (3–4), tarsus 4 (4–5), tarsal setae *ft'* 5 (4–5), setae *ft''* 12 (10–12), setae *u'* 3 (3–4), solenidion *ω* 6 (5–6), curved down, distally simple, empodium simple, 4 (4–5), 5-rayed. **Opisthosoma** with 68 (67–70) dorsal semiannuli, with elongate microtubercles, and 62 (61–64) ventral semiannuli, with small elongate microtubercles on rear annulus margin; coxigenital region with 4 (3–4) semiannuli between coxae and genitalia, with fine microtubercles; last 8 (8–9) dorsal semiannuli with fine and elongated microtubercles. Setae *c2* 20 (19–21), on ventral semiannulus 11 (10–11), 40 (38–44) apart; setae *d* 31 (28–33), on ventral semiannulus 23 (22–23), 31 (30–33) apart; setae *e* 38 (36–39), on ventral semiannulus 38 (38–39), 18 (15–19) apart; setae *f* 13 (11–14), on 6th ventral semiannulus from rear, 12 (11–13) apart. Setae *h2* 38 (33–40), setae *h1* absent. **Genital coverflap** 11 (10–12), 18 (17–18) wide, coverflap with 15 (14–16) longitudinal ridges, setae *3a* 7 (5–7), 13 (12–13) apart. **Internal female genitalia**, spermathecae ovoid, oriented posterolateral; spermathecal tubes relatively short; short spermathecal tubes, directed laterad; transverse genital apodeme trapezoidal, distally folded.

Male (n = 3). Similar in shape and prodorsal shield arrangement to female. Body 175–182, 40–41 wide. **Gnathosoma** 18–19, projecting obliquely downwards, setae *ep* 2*, setae *d* 2–3, unbranched, setae *v* 1*, cheliceral stylets 18*. **Prodorsal shield** 25–26, 30* wide. Tubercles of scapular setae *sc* on rear shield margin, 13* apart, scapular setae *sc* 18–19, divergently backward. **Coxae** smooth; setae *1b* 7–8, tubercles *1b* 6* apart; setae *1a* 16–18, tubercles *1a* 8* apart; setae *2a* 30–32, tubercles *2a* 17* apart. **Leg I** 22–24, femur 7–8, femoral setae *bv* 6–7, genu 3*, genual setae *l''* 16–18, tibia 4*, tibial setae *l'* 2*, tarsus 4–5, tarsal setae *ft'* 5–6, setae *ft''* 14–16, setae *u'* 3*, solenidion *ω* 5*, curved down, distally simple, empodium simple, 5*, 4-rayed. **Leg II** 20–22, femur 7–8, femoral setae *bv* 6–7, genu 3*, genual setae *l''* 7–8, tibia 3*, tarsus 4–5, tarsal setae *ft'* 4*, setae *ft''* 12 10–11, setae *u'* 2*, solenidion *ω* 6*, curved down, distally simple, empodium simple, 5*, 4-rayed. **Opisthosoma** dorsally arched with 67–68 dorsal semiannuli, with elongate microtubercles, and 63–65 ventral semiannuli, with elongate microtubercles on the rear annulus margin; coxigenital region with 3* semiannuli between coxae and genitalia, with fine microtubercles. Setae *c2* 17–19, on ventral semiannulus 10*, 40–42 apart; setae *d* 30–34, on ventral semiannulus 22*, 30–31 apart; setae *e* 32–37, on ventral semiannulus 37*, 17* apart; setae *f* 12*, on 6th ventral semiannulus from rear, 11* apart. Setae *h2* 35–38, setae *h1* absent. **Genitalia** 9–11, 13–15 wide, setae *3a* 6–8, 11–12 apart.

Type material: Holotype, female (slide number EAA2-3.1; marked Holotype), found on *Bischofia javanica* Blume (Fam. Phyllanthaceae), Nanning Normal University, Nanning City, Guangxi, China, 23°10′55″ N, 108°17′12″ E, elevation 109 m, 23 May 2023, coll. Meng-Chao Tan. Paratypes, 14 females on 14 slides and three males on three slides (slide number EAA2-3.2~3.18; marked Paratypes), from *B. javanica*, with the same details as holotype.

Type of host plant: *Bischofia javanica* Blume (Fam. Phyllanthaceae).

Relation to the host plant: mites induce small round galls on the surfaces of the leaves (Figure 3A,B).

Etymology: the species is named after the generic name of the type of host plant, i.e., *Bischofia*, in the genitive case.

Differential diagnosis: *Aceria bischofiae* **sp. nov.** appears to be close to *Aceria varia* (Nalepa, 1892), which was originally found on *Populus tremula* L. (Salicaceae) in France and Iran [21,22]. *Aceria bischofiae* **sp. nov.** and *A. varia* have similar short median line at the basal third of shield, numerous short lines on the outer side of the shield, empodium 5-rayed, genital coverflap with 14–16 longitudinal ridges, but they differ by the number of rings of the opisthosoma (67–70 dorsal semiannuli and 61–64 ventral semiannuli in *A. bischofiae* **sp. nov.** versus 73–86 dorsal semiannuli and 64–80 ventral semiannuli in *A. varia*), setae *h1* (absent in *A. bischofiae* **sp. nov.** versus 9–10 in *A. varia*), the coxal ornamentation (smooth in *A. bischofiae* **sp. nov.** versus with distinct granules in *A. varia*), the length of the scapular setae *sc* (18–21 μm in *A. bischofiae* **sp. nov.** versus 31–35 μm in *A. varia*), the length of the scapular setae *d* and setae *e* (setae *d* 28–33 μm, setae *e* 36–39 μm in *A. bischofiae* **sp. nov.** versus with setae *d* 46–60 μm, setae *e* 15–17 μm in *A. varia*).

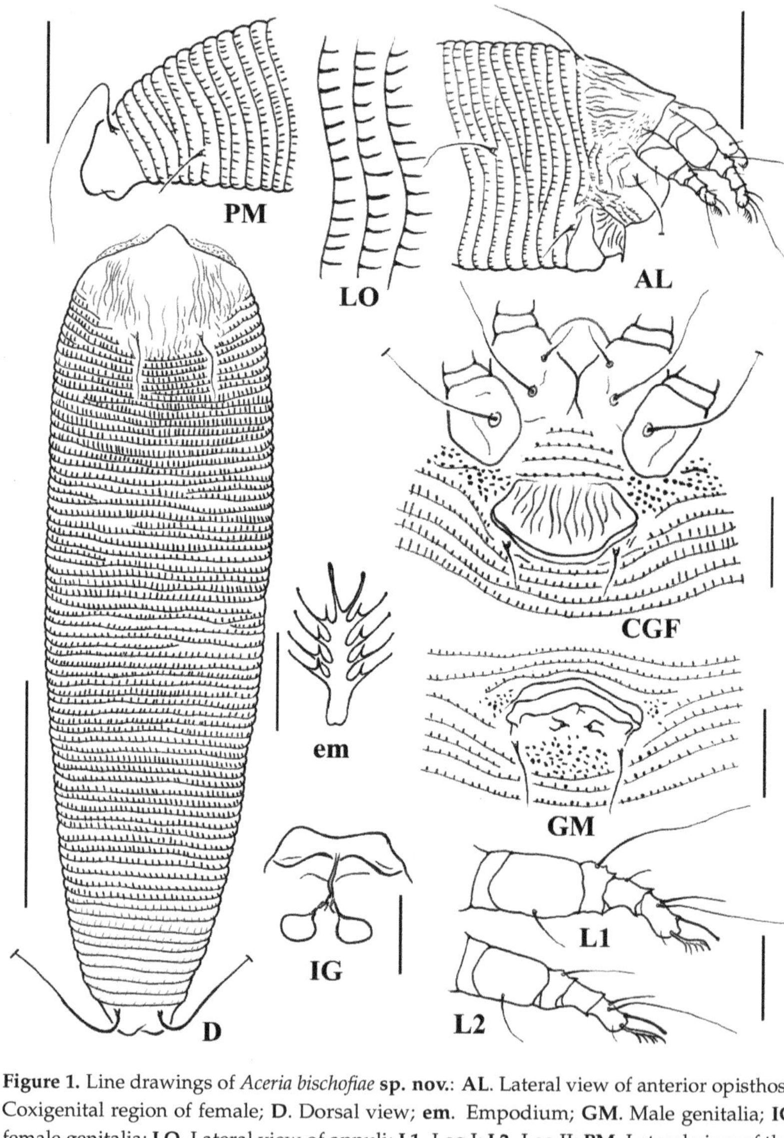

Figure 1. Line drawings of *Aceria bischofiae* **sp. nov.**: **AL**. Lateral view of anterior opisthosoma; **CGF**. Coxigenital region of female; **D**. Dorsal view; **em**. Empodium; **GM**. Male genitalia; **IG**. Internal female genitalia; **LO**. Lateral view of annuli; **L1**. Leg I; **L2**. Leg II; **PM**. Lateral view of the posterior opisthosoma. Scale bar: 50 μm for **D**; 25 μm for **AL** and **PM**; 10 μm for **CGF**, **GM**, **LO**, **L1**, **L2**, and **IG**; 2.5 μm for **em**.

This new species also has few morphological similarities to *Aceria lagerstroemiae* Kuang and Yang, 1994, collected on *Lagerstroemia indica* L. in China [23], including coxal ornamentation (with 15–18 longitudinal ridges), coxae smooth, setae *h1* absent, number of dorsal semiannuli (65–70), scapular setae *sc* length, as well as the length of ventral setae *d*, *e*, and *f*. The new species can be differentiated for prodorsal shield ornamentation (median line and admedian lines present on about posterior half of the shield, a few short dashes medially and some short and long dashes on the lateral margin of the shield in *A. bischofiae* **sp. nov.** versus shield ornamented several lines in *A. lagerstroemiae*), the number of empodium rays (5-rayed in *A. bischofiae* **sp. nov.** versus 6-rayed in *A. lagerstroemiae*), and the shape of microtubercles on dorsal semiannuli (with elongate microtubercles in *A. bischofiae* **sp. nov.** versus with semi-oval microtubercles in *A. lagerstroemiae*).

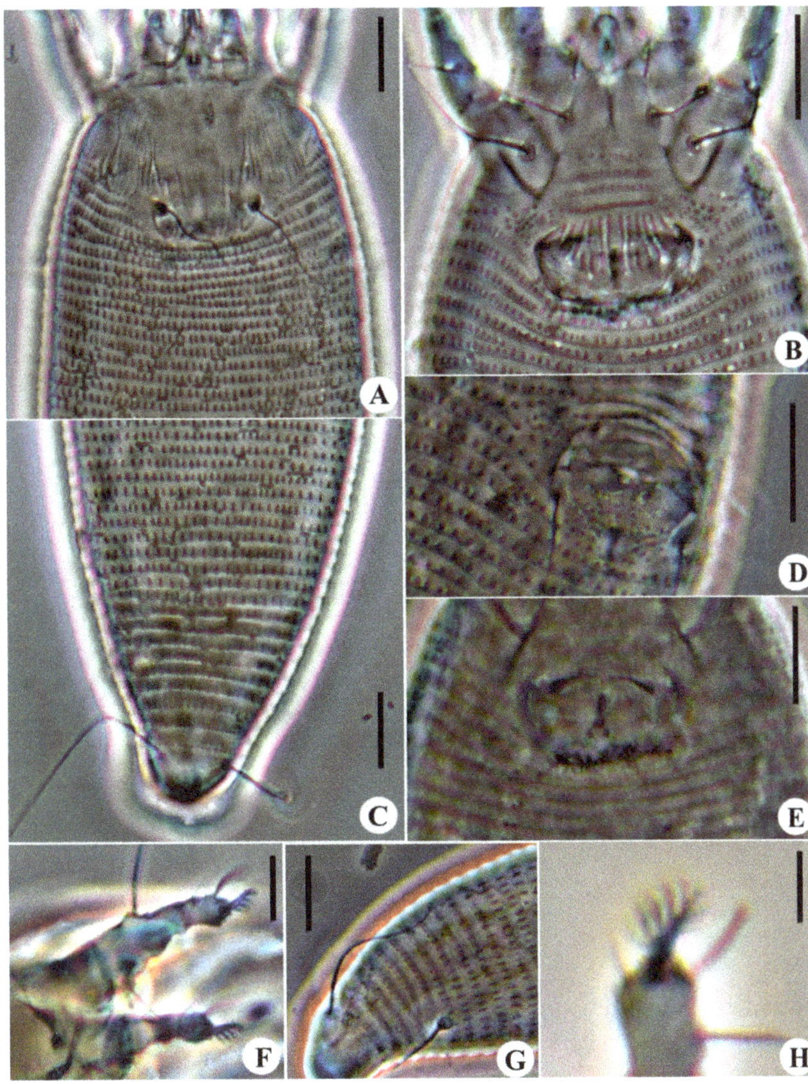

Figure 2. Images of *Aceria bischofiae* **sp. nov.**: (**A**) Prodorsal shield of female; (**B**) Female coxigenital area; (**C**) Postero-dorsal view of mite; (**D**) Male coxigenital area; (**E**) Internal genitalia; (**F**) Legs; (**G**) Postero-lateral view of mite; (**H**) Empodium. Scale bar: 10 μm for (**A**–**G**); 2.5 μm for (**H**).

Figure 3. (**A**,**B**) Damage symptoms associated with *Aceria bischofiae* **sp. nov.** on *Bischofia javanica* Blume; (**C**,**D**) Erineum caused by *Aceria cryptocaryae* **sp. nov.** on *Cryptocarya metcalfiana* Allen.

***Aceria cryptocaryae* sp. nov. (Figures 4 and 5)**

Description: **Female** (n = 15). Body vermiform, 199 (193–231, including gnathosoma), 53 (48–53) wide, 48 (45–48) thick. **Gnathosoma** 18 (18–20), projecting obliquely downwards, setae *ep* 2 (2–3), setae *d* 3 (3–4), unbranched, setae *v* 1 (1–2), cheliceral stylets 16 (15–16). **Prodorsal shield** 26 (25–26), including frontal lobe, 31 (29–33) wide. The shield pattern is distinct and composed of granules aligned and connected by lines as follows: an incomplete median line broken; two complete, sinuous subparallel admedian lines; diverging posteriorly; submedian lines incomplete, extending from the anterior margin and ending ahead of the prodorsal shield tubercle; a lateral line; and granules on each side. Tubercles of scapular setae *sc* on rear shield margin, 21 (19–21) apart, scapular setae *sc* 16 (15–16), divergently backward. **Coxae** with coarse distinct granules; prosternal apodeme 6 (5–6); setae *1b* 5 (5–6), tubercles *1b* 9 (7–9) apart; setae *1a* 17 (15–17), tubercles *1a* 8 (8–10) apart; setae *2a* 33 (30–35), tubercles *2a* 21 (20–22) apart. **Leg I** 26 (23–26), femur 8 (7–8), with fine granules, femoral setae *bv* 6 (5–6), genu 3 (3–4), genual setae *l''* 15 (13–15), tibia 5 (4–5), tibial setae *l'* 2*, tarsus 6 (5–6), tarsal setae *ft'* 14 (12–15), setae *ft''* 18 (15–18), setae *u'* 3 (2–3), solenidion ω 7 (6–7) distally slightly knobbed, empodium simple, 6 (5–6), 4-rayed. **Leg II** 24 (23–25), femur 8 (7–8), with fine granules, femoral setae *bv* 5 (5–7), genu 3 (3), genual setae *l''* 5 (5–6), tibia 4 (3–4), tarsus 5 (5–6), tarsal setae *ft'* 4 (4–6), setae *ft''* 15 (13–15), setae *u'* 3 (2–3), solenidion ω 8 (7–8) distally slightly knobbed, empodium simple, 5 (5–6), 4-rayed. **Opisthosoma** with 68 (67–69) dorsal semiannuli, with elliptical microtubercles, and 65 (65–67) ventral semiannuli, with circular microtubercles on the rear annulus margin; coxigenital region with 4* semiannuli between coxae and genitalia, with circular microtubercles; spiny microtubercles on the rear margin of the last 10 (10–11) dorsal semiannuli; elongated and linear microtubercles on the last 6 ventral semiannuli. Setae *c2* 15 (15–16), on ventral semiannulus 11 (10–11), 39 (36–42) apart; setae *d* 35 (31–35), on ventral semiannulus 22 (22–23), 31 (27–33) apart; setae *e* 41 (36–45), on ventral semiannulus 38 (38–39), 17 (17–19) apart; setae *f* 12 (11–13), on 5th ventral semiannulus from rear, 14 (13–14) apart. Setae *h2* 48 (45–52), setae *h1* 4 (3–4). **Genital coverflap** 13 (12–14), 19 (18–21) wide, with some strong granulated lines at the genital coverflap base and 8–9 longitudinal ridges distally; setae *3a* 6 (5–6), 13 (11–13) apart. **Internal female genitalia**, transverse genital apodeme trapezoidal, with thickened anterior margin; longitudinal bridge relatively long; spermathecae bulbous; both spermathecae are equal in size; spermathecal tubes short, slightly swollen, directed posterolaterad.

Male (n = 3). Similar in shape and prodorsal shield arrangement to female. Body 175–188, 44–45 wide. **Gnathosoma** 15–16, projecting obliquely downwards, chelicerae 15*, setae *ep* 2*, setae *d* 2–3, unbranched, setae *v* 1–2. **Prodorsal shield** 26–28, 32–34 wide. Tubercles of the scapular setae *sc* ahead of rear shield margin 15–16 apart, setae *sc* 14–15. **Coxae** similar to that of the female; setae *1b* 5–6, tubercles *1b* 7–8 apart; setae *1a* 14–16, tubercles *1a* 9–10 apart; setae *2a* 26–28, tubercles *2a* 17–18 apart. **Leg I** 21–24, femur 7–8, femoral setae *bv* 5–6, genu 4*, genual setae *l''* 14–16, tibia 3*, tibial setae *l'* 2–3, tarsus 5*, tarsal setae *ft'* 8–9, setae *ft''* 15–18, setae *u'* 3*, solenidion ω 6–7 slightly knobbed, empodium simple, 5–6, 4-rayed. **Leg II** 21–24, femur 7–8, femoral setae *bv* 4–5, genu 3–4, genual setae *l''* 5*, tibia 4*, tarsus 5*, tarsal setae *ft'* 4–6, setae *ft''* 14–16, setae *u'* 2*, solenidion ω 7* slightly knobbed, empodium simple, 5*, 4-rayed. **Opisthosoma** dorsally arched with 67–68 semiannuli, with elongate microtubercles on rear annular margins; 66–67 ventral semiannuli, with small circular microtubercles on rear annulus margin; 4* semiannuli between coxae and genital region. Setae *c2* 14–16 on ventral semiannulus 10*, 37–39 apart; setae *d* 30–32 on ventral semiannulus 20–21, 26–27 apart; setae *e* 33–35 on ventral semiannulus 36–38, 15–16 apart; setae *f* 10–11 on 6th ventral semiannulus from rear, 11–12 apart. Setae *h2* 40–46; setae *h1* 2*. Genitalia 12–13, 18–20 wide, setae *3a* 4*, 13–14 apart.

Type material: Holotype, female (slide number EAA2-5.1; marked Holotype), found on *Cryptocarya metcalfiana* Allen (Fam. Lauraceae), Mulun National Nature Reserve, Hechi City, Guangxi, China, 25°9′31″ N, 108°3′30″ E, elevation 306.7 m, 28 July 2021, coll. Meng-Chao Tan, An-Kang Lv. Paratypes, 14 females on 14 slides and three males on three slides (slide number EAA2-5.2~5.18; marked Paratypes), from *C. metcalfiana*, with the same details as holotype.

Type of host plant: *Cryptocarya metcalfiana* Allen (Fam. Lauraceae).

Relation to the host plant: causing the formation of erinea on the undersurface of the leaves, with slight bulges on the opposite side of the lamina. (Figure 3C,D)

Etymology: the species is named after the generic name of the type of host plant, i.e., *Cryptocarya* in the genitive case.

Differential diagnosis: *Aceria cryptocaryae* **sp. nov.** is most similar to *Aceria tribuli* (Keifer, 1974) collected from *Tribulus terrestris* L. (Zygophyllaceae) in Sudan and Egypt [24,25], in the prodorsal shield ornamentation pattern, sculpture of coxae, and coverflap. The new species is distinguishable from *A. tribuli* for the femur of legs (with fine granules in *A. cryptocaryae* **sp. nov.** versus smooth in *A. tribuli*), empodium (4-rayed in *A. cryptocaryae* **sp.**

nov. versus 6-rayed in *A. tribuli*), the number of rings of the opisthosoma (67–69 dorsal semiannuli and 65–67 ventral semiannuli in *A. cryptocaryae* **sp. nov.** versus 70–80 dorsal semiannuli and 70–75 ventral semiannuli in *A. tribuli*), the length of scapular setae *sc* (15–16 μm in *A. cryptocaryae* **sp. nov.** versus 50–55 μm in *A. tribuli*), the opisthosomal setae *c2* (15–16 μm in *A. cryptocaryae* **sp. nov.** versus 32–45 μm in *A. tribuli*), setae *d* (31–35 μm in *A. cryptocaryae* **sp. nov.** *versus* 66–75 μm in *A. tribuli*).

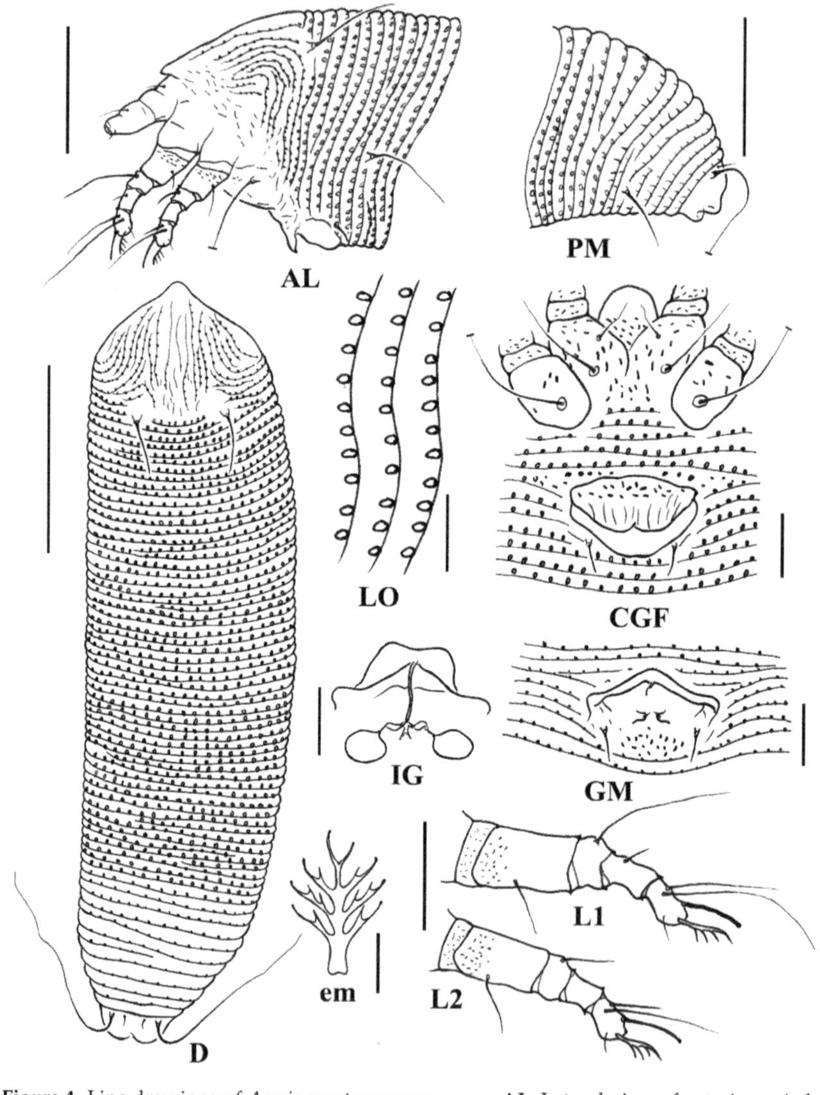

Figure 4. Line drawings of *Aceria cryptocaryae* **sp. nov.**: **AL**. Lateral view of anterior opisthosoma; **CGF**. Coxigenital region of female; **D**. Dorsal view; **em**. Empodium; **GM**. Male genitalia; **IG**. Internal female genitalia; **LO**. Lateral view of annuli; **L1**. Leg I; **L2**. Leg II; **PM**. Lateral view of the posterior opisthosoma. Scale bar: 50 μm for **D**; 25 μm for **AL** and **PM**; 10 μm for **CGF, GM, LO, L1, L2,** and **IG**; 2.5 μm for **em**.

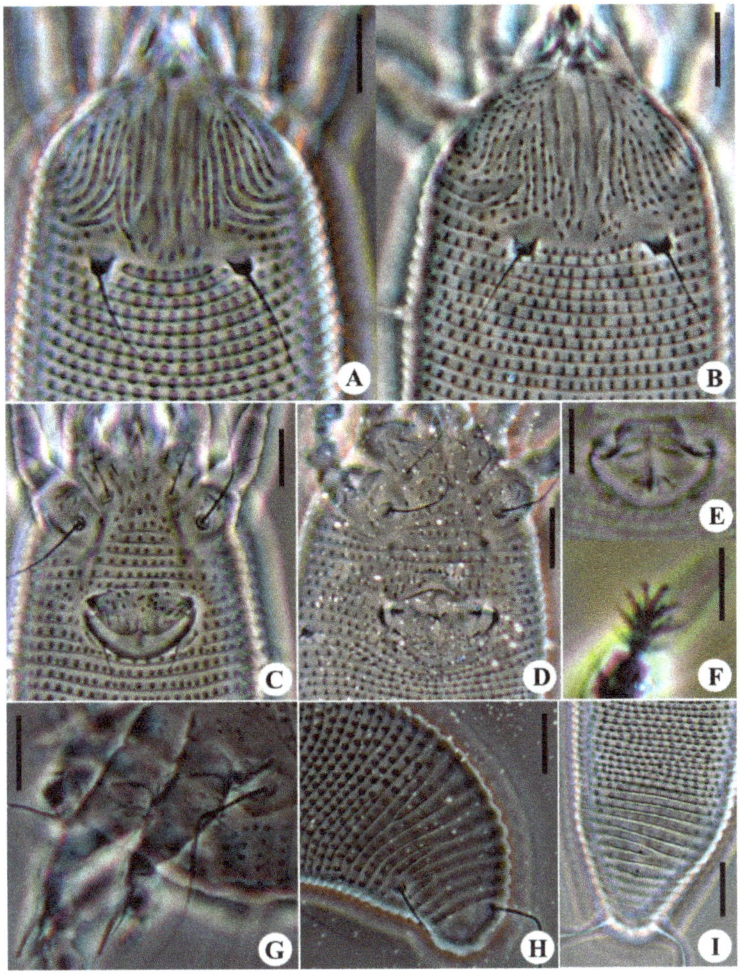

Figure 5. Images of *Aceria cryptocaryae* **sp. nov.**: (**A**) Prodorsal shield of female; (**B**) Prodorsal shield of male; (**C**) Female coxigenital area; (**D**) Male coxigenital area; (**E**) Internal genitalia; (**F**) Empodium; (**G**) Legs; (**H**) Postero-lateral view of mite; (**I**) Postero-dorsal view of mite. Scale bar: 10 μm for (**A–E**,**G–I**); 5 μm for (**F**).

***Aceria buddlejae* sp. nov. (Figures 6 and 7)**

Description: **Female** (n = 14). Body vermiform, 181 (170–195, including gnathosoma), 49 (45–49) wide, 48 (48–49) thick. **Gnathosoma** 19 (17–20), projecting obliquely downwards, setae *ep* 2 (2–3), setae *d* 5 (4–5), unbranched, setae *v* 1 (1–2), cheliceral stylets 20 (18–21). **Prodorsal shield** 29 (28–30), including frontal lobe, 39 (37–41) wide; short and rounded frontal lobe 4 (4–5) over gnathosomal base. Median lines are present on the posterior half of the shield; admedian lines are complete and sinuate; and submedian lines are present on the anterior half of the shield. Some short dashes and microtubercles are on the lateral sides of the shield. Tubercles of scapular setae *sc* on the rear shield margin, 17 (15–17) apart, scapular setae *sc* 15 (15–16), divergently backward. **Coxae** ornamented with some granules; prosternal apodeme 7 (6–7); setae *1b* 5 (5–6), tubercles *1b* 8 (7–9) apart; setae *1a* 13 (11–14), tubercles *1a* 9 (9–11) apart; setae *2a* 24 (23–26), tubercles *2a* 20 (19–22) apart. **Leg I** 25 (24–27), femur 7 (6–8), femoral setae *bv* 7 (5–8), genu 4*, genual setae *l″* 18 (17–19), tibia 5 (4–5), tibial setae *l′* 3 (3–4), tarsus 6 (6–7), tarsal setae *ft′* 8 (6–9), setae *ft″* 20 (17–20), setae u' 3 (2–3), solenidion ω 7 (6–7) distally slightly knobbed, empodium simple, 5 (4–5), 3-rayed. **Leg II** 23 (23–26), femur 7 (7–8), femoral setae *bv* 8 (7–8), genu 3 (3–4), genual setae l'' 6 (5–7), tibia 4 (4–5), tarsus 6 (6–7), tarsal setae ft' 4 (4–5), setae ft'' 19 (17–20), setae u' 3 (2–3), solenidion ω 6 (6–7) distally slightly knobbed, empodium simple, 5 (4–5), 3-rayed. **Opisthosoma** with 65 (63–67) dorsal semiannuli, with elongate microtubercles on rear annular margins, and 67 (65–69) ventral semiannuli, with spiny microtubercles on rear annulus margin; coxigenital region with 4 (3–4) semiannuli between coxae and genitalia, with fine microtubercles; spiny microtubercles on rear margin of last 7 (7–8) dorsal semiannuli; elongated and linear microtubercles on last 8 ventral semiannuli. Setae *c2* 23 (21–23), on ventral semiannulus 12 (11–12), 41 (40–43) apart; setae *d* 45 (42–45), on ventral semiannulus 23 (22–23), 33 (31–34) apart; setae *e* 13 (11–13), on ventral semiannulus 40 (40–42), 19 (18–20) apart; setae *f* 18 (16–18), on 6th–7th ventral semiannulus from rear, 17 (17–18) apart. Setae *h2* 72 (60–77), setae *h1* 6 (5–6). **Genital coverflap** 12 (11–12), 18 (15–18) wide, with some strong granulated lines at the genital coverflap base; setae *3a* 13 (10–14), 14 (11–14) apart. **Internal female genitalia**, transverse genital apodeme trapezoidal, longitudinal bridge relatively long; oblique apodeme present; short spermathecal tubes, directed laterad; spermathecae oval-shaped, relatively small.

Male (n = 1). Similar in shape and prodorsal shield arrangement to female. Body 151*, 44* wide. **Gnathosoma** 19*, projecting obliquely downwards, chelicerae 21*, setae *ep* 2*, setae *d* 3*, unbranched, setae *v* 1*. **Prodorsal shield** 21*, 32* wide. Tubercles of the scapular setae *sc* ahead of the rear shield margin are 15* apart, setae *sc* 17*. **Coxae** are similar to those of the female; setae *1b* 6*, tubercles *1b* 8* apart; setae *1a* 13*, tubercles *1a* 7* apart; setae *2a* 27*, tubercles *2a* 18* apart. **Leg I** 26*, femur 7*, femoral setae *bv* 5*, genu 4*, genual setae l'' 17*, tibia 4*, tibial setae l' 4*, tarsus 7*, tarsal setae ft' 6*, setae ft'' 16*, setae u' 3*, solenidion ω 6* slightly knobbed, empodium simple, 4*, 3-rayed. **Leg II** 25*, femur 7*, femoral setae *bv* 5*, genu 3*, genual setae l'' 4*, tibia 4*, tarsus 6*, tarsal setae ft' 4*, setae ft'' 16*, setae u' 3*, solenidion ω 7* slightly knobbed, empodium simple, 4*, 3-rayed. **Opisthosoma** dorsally arches with 61* semiannuli; 62* ventral semiannuli; 4* semiannuli between the coxae and genital region. Setae *c2* 19* on ventral semiannulus 10*, 35* apart; setae *d* 33* on ventral semiannulus 21*, 27* apart; setae *e* 13* on ventral semiannulus 35*, 15* apart; setae *f* 16* on ventral semiannulus 6th ventral semiannulus from rear, 16* apart. Setae *h2* 55*, setae *h1* 6*. Genitalia 11*, 14* wide, setae *3a* 8*, 11* apart.

Type material: Holotype, female (slide number EAA2-6.1; marked Holotype), found on *Buddleja lindleyana* Fort. (Fam. Scrophulariaceae), Chengkou County, Chongqing City, China, 32°16′29″ N, 108°46′75″ E, elevation 956.6 m, 27 August 2022, coll. Li-Mei Ren, An-Kang Lv. Paratypes, 12 females on 13 slides and one male on three slides (slide number EAA2-6.2~6.15; marked Paratypes), from *B. lindleyana*, with the same details as holotype.

Type of host plant: *Buddleja lindleyana* Fort. (Fam. Scrophulariaceae).

Relation to the host plant: vagrant on the leaves; no apparent damage was observed.

Etymology: the specific designation *buddlejae* is from the generic name of the host, *Buddleja*.

Differential diagnosis: *Aceria buddlejae* **sp. nov.** appears to be close to *Aceria noxia* Flechtmann and Tassi, 2020, that was found on *Amaranthus viridis* L. (Amaranthaceae), in the prodorsal shield ornamentation pattern, sculpture of coxae, seta *h1* present [26]. *Aceria buddlejae* **sp. nov.** can be differentiated from the above-mentioned species by the genitalia coverflap (with some strong granulated lines at the genital coverflap base in *A. buddlejae* **sp. nov.** *versus* coverflap basally with two transverse bands of coarse granules and distally with 14–16 longitudinal ridges in *A. noxia*), the frontal lobe (present in *A. buddlejae* **sp. nov.** versus absent in *A. noxia*), the number of empodium rays (3-rayed in *A. buddlejae* **sp. nov.** versus 5-rayed in *A. noxia*), the number of rings of the dorsal semiannuli (63–67 dorsal semiannuli in *A. buddlejae* **sp. nov.** versus 76–93 dorsal semiannuli in *A. noxia*), the length of the coxal seta III *2a* (23–26 μm in *A. buddlejae* **sp. nov.** versus 40–45 μm in *A. noxia*), the length of setae *c2* (21–23 μm in *A. buddlejae* **sp. nov.** versus 28–40 μm in *A. noxia*).

The new species is also similar to *Aceria hupehensis* Kuang and Hong, 1995, collected from *Castanea mollissima* Blume (Fagaceae) in China [27]. It shares the same prodorsal shield pattern, sculpture of coverflap, and number of empodium rays as *A. genistae*. However, the two species differ in: the frontal lobe (present in *A. tinctoriae* **sp. nov.** versus absent in *A. cumaniamajoris*), the number of rings of the opisthosoma (63–67 dorsal semiannuli and 65–69 ventral semiannuli in *A. buddlejae* **sp. nov.** versus 52–56 dorsal and ventral semiannuli in *A. hupehensis*), the length of setae *c2* (21–23 µm in *A. buddlejae* **sp. nov.** versus 6 µm in *A. hupehensis*), the length of setae *3a* (10–14 µm in *A. buddlejae* **sp. nov.** versus 5 µm in *A. hupehensis*).

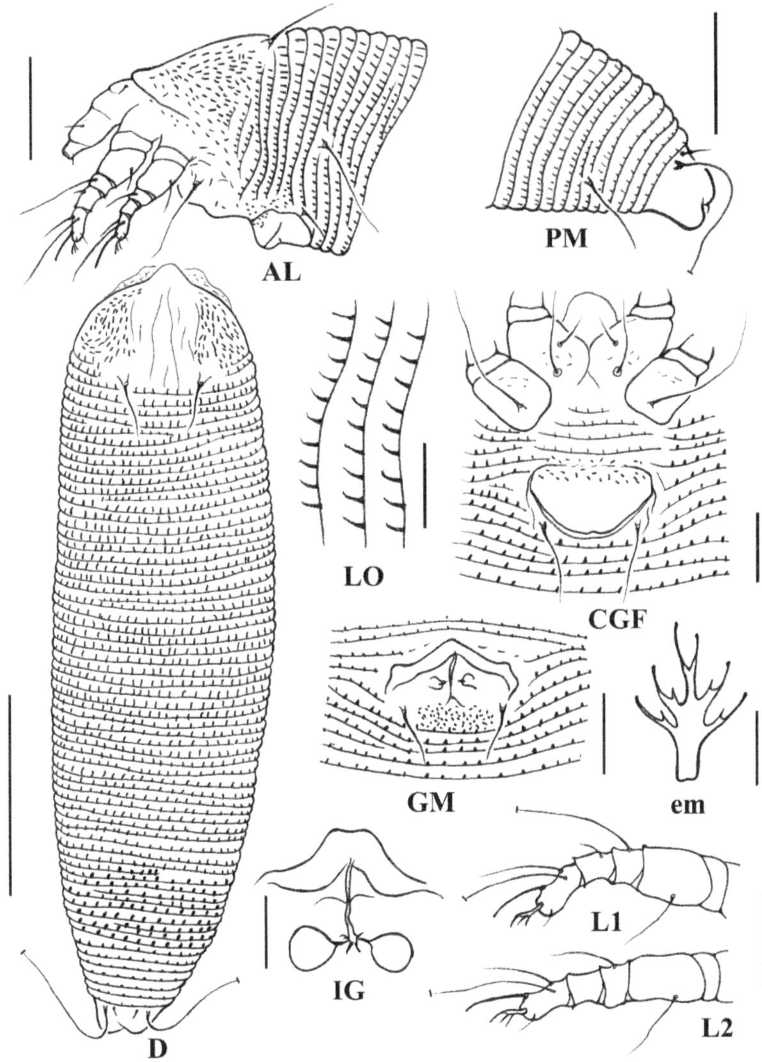

Figure 6. Line drawings of *Aceria buddlejae* **sp. nov.**: **AL**. Lateral view of anterior opisthosoma; **CGF**. Coxigenital region of female; **D**. Dorsal view; **em**. Empodium; **GM**. Male genitalia; **IG**. Internal female genitalia; **LO**. Lateral view of annuli; **L1**. Leg I; **L2**. Leg II; **PM**. Lateral view of the posterior opisthosoma. Scale bar: 50 µm for **D**; 25 µm for **AL** and **PM**; 10 µm for **CGF**, **GM**, **LO**, **L1**, **L2**, and **IG**; 2.5 µm for **em**.

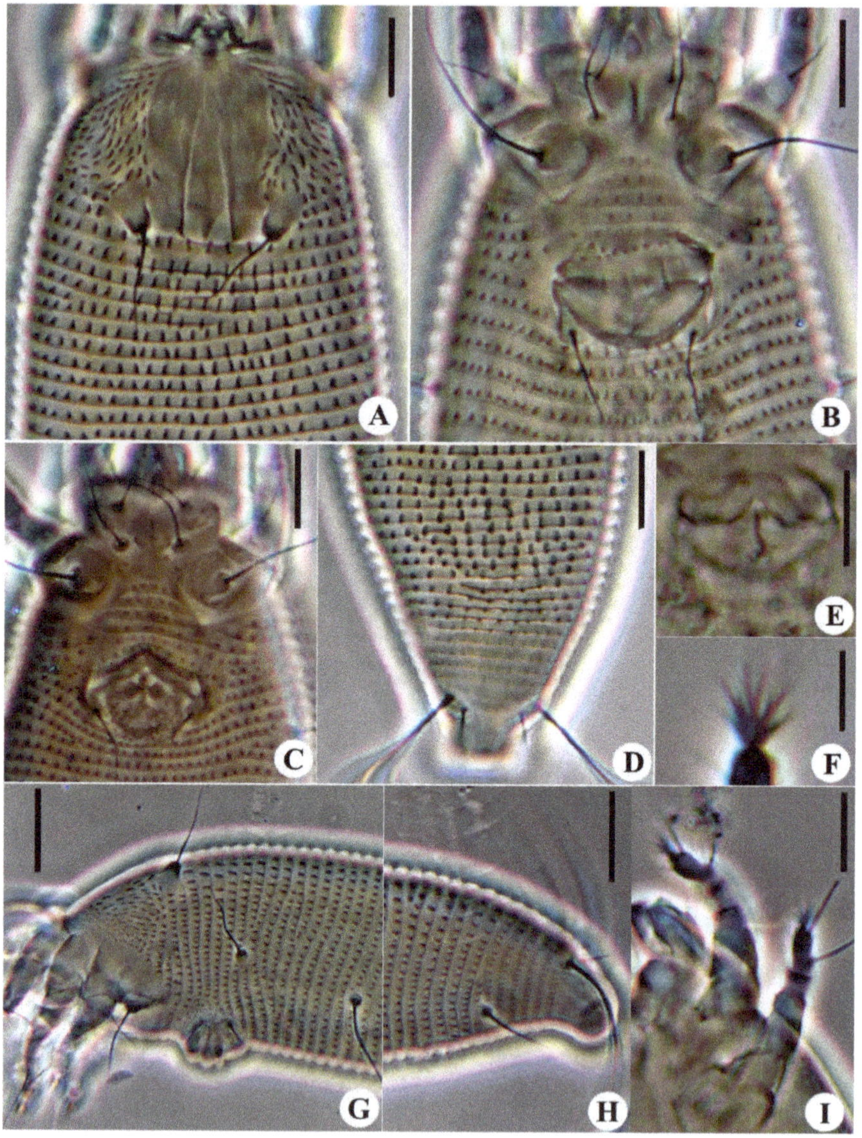

Figure 7. Images of *Aceria buddlejae* **sp. nov.**: (**A**) Prodorsal shield; (**B**) Female coxigenital area; (**C**) Male coxigenital area; (**D**) Postero-dorsal view of mite; (**E**) Internal genitalia; (**F**) Empodium; (**G**) Lateral view of anterior opisthosoma; (**H**) Postero-lateral view of mite; (**I**) Legs. Scale bar: 20 µm for G, H; 10 µm for (**A–E**), I; 5 µm for (**F**).

4. Discussion

According to the data from the published references, there are 84 species (including 3 new species in this paper) of the genus *Aceria* that have been found in China, parasitizing 34 families and 77 species of plants [9]. Among them, 7 *Aceria* species of host plants belong to the family Salicaceae, and 5 *Aceria* species of host plants, respectively, belong to the families Poaceae and Solanaceae. In relation to the host plants of *Aceria*, 41 species cause galls on leaves, 8 species cause the formation of erinea on the undersurface of the leaves, and 29 species causes vagrants on the leaves. In terms of the geographical distribution and floral distribution of *Aceria*, there are 66 species in the Oriental realm and 24 species in the Palearctic realm. *Aceria kuko* (Kishida, 1927) and *Aceria lycopersici* (Massee, 1939) are distributed in most areas of China; *Aceria dispar* (Nalepa, 1891) and *Aceria tosichella* Keifer, 1969, are widely distributed in northern China; *Aceria litchii* (Keifer, 1943) and *Aceria hupehensis* Kuang and Hong, 1995, are widely distributed in southern China; others are only distributed in local areas. The discovery of three new *Aceria* species in China indicates that the species richness of the genus is still underestimated. Undoubtedly, it is necessary to further collect and investigate the taxa of *Aceria* in the future to understand their real diversity.

However, there are still some problems in the classification of the genus *Aceria*. Due to the limitations of early microscopic techniques and the low standards for the description of new species of *Aceria*, many *Aceria* species were not described in detail when they were published. The quality of illustrations was poor or no illustrations, the host plants were not identified, the damage description of *Aceria* to the host plants was simple, and the naming was irregular, which led to the emergence of a large number of homonyms in *Aceria* in recent years. For example, the *Aceria* species studied by Nalepa were stored in the Vienna Museum of Natural History in the United States. Due to long-term poor management, the alcohol-soaked specimens had dried up, the slide specimens could not be observed, the original manuscripts were lost, and many type specimens could not be verified [18,28]. Therefore, there are still some known *Aceria* species that need to be collected again for supplementary description and revision. In addition, most *Aceria* species are mainly distinguished by similar characteristics such as dorsal shield decoration, the number of dorsal and ventral rings, the genital coverflap decoration, the number of empodium, and so on. The distinguishing features of some closely related species are very subtle, and there may be certain differences on the same host in different regions [29–31]. Therefore, a combination of molecular and morphological methods is needed to determine the taxonomic status of species.

5. Conclusions

In this research, we described three new *Aceria* species: *Aceria bischofiae* **sp. nov.**, *Aceria cryptocaryae* **sp. nov.**, and *Aceria buddlejae* **sp. nov.** We also summarized the number of species, host plants, and geographical distribution of the genus *Aceria* in China. However, because *Aceria* species in northern China have not been systematically investigated yet, it is safe to assume that many other *Aceria* species may exist and will eventually be discovered.

Author Contributions: Conceptualization, M.T.; methodology, M.T.; formal analysis, M.T., R.L., H.R. and X.L.; data curation, M.T. and R.L.; writing—original draft preparation, M.T., R.L., H.R. and X.L.; supervision, M.T.; funding acquisition, M.T. All authors have read and agreed to the published version of the manuscript.

Funding: This research was funded by the National Natural Science Foundation of China (Grant No. 32000324).

Institutional Review Board Statement: Not applicable.

Informed Consent Statement: Not applicable.

Data Availability Statement: All data are available in this paper.

Acknowledgments: We would like to thank Li-Mei Ren and An-Kang Lv (Guangxi University, China) for their help with collecting mite specimens. We also thank the reviewers for their critical review of the manuscript.

Conflicts of Interest: The authors declare no conflicts of interest.

References

1. Lindquist, E.E. External anatomy and notation of structures. In *Eriophyoid Mites—Their Biology, Natural Enemies and Control*; World Crop Pests, 6; Lindquist, E.E., Sabelis, M.W., Bruin, J., Eds.; Elsevier Science Publishers: Amsterdam, The Netherlands, 1996; pp. 3–31.
2. Zhang, Z.Q. Eriophyoidea and allies: Where do they belong? *Syst. Appl. Acarol.* **2017**, *22*, 1091–1095. [CrossRef]
3. Skoracka, A.; Smith, L.; Oldfield, G.; Cristofaro, M.; Amrine, J.W., Jr. Host-plant specificity and specialization in eriophyoid mites and their importance for the use of eriophyoid mites as biocontrol agents of weeds. *Exp. Appl. Acarol.* **2010**, *51*, 93–113. [CrossRef]
4. Yin, Y.; Yao, L.F.; Hu, Y.; Shao, Z.K.; Hong, X.Y.; Hebert, P.D.N.; Xue, X.F. DNA barcoding uncovers cryptic diversity in minute herbivorous mites (Acari, Eriophyoidea). *Mol. Ecol. Resour.* **2022**, *22*, 1986–1998. [CrossRef]
5. Xue, X.F.; Yao, L.F.; Yin, Y.; Liu, Q.; Li, N.; Hoffmann, A.A.; Sun, J.T.; Yin, Y.; Hong, X.Y. Macroevolutionary analyses point to a key role of hosts in diversification of the highly speciose eriophyoid mite superfamily. *Mol. Phylogenet. Evol.* **2023**, *179*, 107676. [CrossRef]
6. Hong, X.Y.; Xue, X.F.; Song, Z.W. Eriophyoidea of China: A review of progress, with a checklist. *Zoosymposia* **2010**, *4*, 57–93. [CrossRef]
7. Liu, L.X.; Yao, G.; Yang, J.; Wang, G.Q. One new genus, ten new species, and four new records of eriophyoid mites (Acari: Eriophyoidea) in Tibet, China. *Syst. Appl. Acarol.* **2022**, *27*, 1295–1336. [CrossRef]
8. Ruan, H.Y.; Hu, L.; Cui, X.Y.; Tan, M.C. Three new species of eriophyoid mites (Acari: Eriophyoidea) associated with *Melicope pteleifolia* from Guangxi Zhuang Autonomous Region, China. *Syst. Appl. Acarol.* **2021**, *26*, 353–366. [CrossRef]
9. Xue, X.F.; Hong, X.Y. A checklist of eriophyoid mites of China (Acariformes: Eriophyoidea). *Syst. Appl. Acarol.* **2023**, *28*, 1958–2148. [CrossRef]
10. Keifer, H.H. Eriophyid studies XIV. *Bull. Calif. Dep. Agric.* **1944**, *33*, 18–38.
11. Indra, R.; Bachheti, R.; Archana, J. Chemical composition, mineral and nutritional value of wild *Bischofia javanica* seed. *Int. Food Res. J.* **2013**, *20*, 1747–1751.
12. Boczek, J.; Chandrapatya, A. Studies on eriophyoid mites (Acari: Eriophyoidea). XLI. *Bull. Pol. Acad. Sci.-Tech.* **2000**, *48*, 348–351.
13. Wei, S.G.; Wang, G.Q.; Li, D.W.; Ou, S.S. *Eriophyoid Mites of Guangxi, China (Acari: Eriophyoidea)*; Guangxi Science and Technology Press: Guangxi, China, 2009; 329p.
14. Liu, S.S.; Feng, S.Y.; Huang, Y.Y.; An, W.L.; Yang, Z.R.; Xie, C.Z.; Zheng, X.S. Characterization of the Complete Chloroplast Genome of *Buddleja lindleyana*. *J. AOAC Inter.* **2021**, *105*, 202–210. [CrossRef]
15. Meyer, M.K.P.S.; Ueckermann, E.A. African Eriophyidae: Genus *Aculops* Keifer 1966 (Acari: Eriophyidae). *Phytophylactica* **1990**, *22*, 159–175.
16. Nalepa, A. Eriophyiden aus Java (4. Beitrag). *Treubia* **1923**, *3*, 423–432.
17. Amrine, J.W., Jr.; Manson, D.C.M. Preparation, mounting and descriptive study of Eriophyoid mites. In *Eriophyoid Mites. Their Biology, Natural Enemies and Control*; World Crop Pests, 6; Lindquist, E.E., Sabelis, M.W., Bruin, J., Eds.; Elsevier Science Publishers: Amsterdam, The Netherlands, 1996; pp. 383–396.
18. de Lillo, E.; Craemer, C.; Amrine, J.W., Jr.; Nuzzaci, G. Recommended procedures and techniques for morphological studies of Eriophyoidea (Acari: Prostigmata). *Exp. Appl. Acarol.* **2010**, *51*, 283–307. [CrossRef]
19. Amrine, J.W., Jr.; Stasny, T.A.H.; Flechtmann, C.H.W. *Revised Key to the World Genera of Eriophyoidea (Acari: Prostigmata)*; Indira Publishing House: West Bloomfield, MI, USA, 2003; pp. 1–244.
20. Chetverikov, P.E. Comparative confocal microscopy of internal genitalia of phytoptine mites (Eriophyoidea, Phytoptidae): New generic diagnoses reflecting host-plant associations. *Exp. Appl. Acarol.* **2014**, *62*, 129–160. [CrossRef]
21. Nalepa, A. Neue Arten der Gattung *Phytoptus* Duj. und *Cecidophyes* Nal. In *Denkschriften kaiser Akademie der Wissenschaften: Mathematisch-Naturwissenschaftliche Classe*; Aus der Kaiserlich-Königlichen Hof-und Staatsdruckerei: Wien, Auatria, 1892; Volume 59, pp. 525–540+4 pls.
22. Mehri-Heyran, H.; Lotfollahi, P.; de Lillo, E.; Azimi, S. Redescription of *Aceria varia* and *Tegoprionus dentatus* (Trombidiformes: Eriophyoidea: Eriophyidae) from Iran. *Pers. J. Acarol.* **2020**, *9*, 129–139.
23. Kuang, H.Y.; Yang, S.P.; Lu, Q.B. Four new species of the genus *Aceria* from Guangxi, China (Acari: Eriophyoidae). *Acta Zootaxonomica Sin.* **1994**, *19*, 296–301.
24. Keifer, H.H. *Eriophyid Studies C-9*; Agricultural Research Service, United State Department of Agriculture: Beltsville, MD, USA, 1974; pp. 1–24.
25. Elhalawany, A.S.; Mohamed, A.A.; Ueckermann, E.A. Two new species and complementary descriptions of four new records of family Eriophyidae (Acari: Trombidiformes) in Egypt. *Syst. Appl. Acarol.* **2022**, *27*, 670–696. [CrossRef]
26. Flechtmann, C.H.W.; Tassi, A.D. An *Aceria* species (Prostigmata, Eriophyidae) from *Amaranthus* in Brazil. *Pers. J. Acarol.* **2020**, *9*, 311–319.
27. Kuang, H.Y. *Economic Insect Fauna of China. Fasc. 44. (Acari: Eriophyoidea) (I)*; Science Press: Beijing, China, 1995; pp. 54–55.
28. Chetverikov, P.E.; Hörweg, C.; Kozlov, M.I.; Amrine, J.W., Jr. Reconditioning of the Nalepa collection of eriophyoid mites (Acariformes, Eriophyoidea). *Syst. Appl. Acarol.* **2016**, *21*, 583–595.
29. Lewandowski, M.; Skoracka, A.; Szydło, W.; Kozak, M.; Druciarek, T.; Griffiths, D.A. Genetic and morphological diversity of *Trisetacus* species (Eriophyoidea: Phytoptidae) associated with coniferous trees in Poland: Phylogeny, barcoding, host and habitat specialization. *Exp. Appl. Acarol.* **2014**, *63*, 497–520. [CrossRef]

30. Laska, A.; Majer, A.; Szydło, W.; Karpicka-Ignatowska, K.; Hornyák, M.; Labrzycka, A.; Skoracka, A. Cryptic diversity within grass-associated *Abacarus* species complex (Acariformes: Eriophyidae), with the description of a new species, *Abacarus plumiger* n. sp. *Exp. Appl. Acarol.* **2018**, *76*, 1–28. [CrossRef] [PubMed]
31. Li, H.S.; Xue, X.F.; Hong, X.Y. Cryptic diversity in host-associated populations of *Tetra pinnatifidae* (Acari: Eriophyoidea): What do morphometric, mitochondrial and nuclear data reveal and conceal? *Bull. Entomol. Res.* **2014**, *104*, 221–232. [CrossRef] [PubMed]

animals

MDPI

Article

Morphological Ontogeny, Ecology, and Biogeography of *Fuscozetes fuscipes* (Acari, Oribatida, Ceratozetidae)

Stanisław Seniczak [1], Anna Seniczak [2,*] and Bjarte H. Jordal [3]

1 Department of Evolutionary Biology, Faculty of Biological Sciences, Kazimierz Wielki University, 85-093 Bydgoszcz, Poland; stseni@ukw.edu.pl
2 Faculty of Applied Ecology, Agricultural Sciences and Biotechnology, Inland Norway University of Applied Sciences, 2318 Elverum, Norway
3 University Museum of Bergen, University of Bergen, 5007 Bergen, Norway; bjarte.jordal@uib.no
* Correspondence: anna.seniczak@inn.no

Simple Summary: The systematic status of *Fuscozetes* is not clear in the literature. Therefore, the morphological ontogeny of *F. fuscipes*, the type species of this genus, was investigated and compared with its congeners in this study, and a new diagnosis of *Fuscozetes* is given. The juveniles of *F. fuscipes* are light brown, with a brown prodorsum, sclerites, epimeres, and legs. In all juveniles, a humeral organ and a humeral macrosclerite are present. The gastronotum of the larva has 12 pairs of setae (h_3 is present), while the nymphs have 15 pairs. In the larva, the gastronotal shield is weakly developed, most of the gastronotal setae are short and inserted on the microsclerites, and several other macrosclerites and many microsclerites are present on the hysterosoma. In the nymphs, the gastronotal shield is well developed, with 10 pairs of setae (*d*-, *l*-, and *h*-series, and p_1), and setae p_2 and p_3 are located on a large posteroventral macrosclerite. In all the instars, femora I and II are oval in cross-section, without a large ventral carina. Mitochondrial COI sequence data revealed a deep split between the Nearctic and Palearctic populations of *F. fuscipes*, and a less, but significant, divergence within each continent. These strong geographical barriers were contrasted with multiple cases of shared haplotypes over long distances in the Palearctic, indicating high migration rates in modern times.

Abstract: The systematic status of *Fuscozetes* Sellnick, 1928, is not clear in the literature. Therefore, the morphological ontogeny of *F. fuscipes* (C.L. Koch, 1844), the type species of this genus, was investigated and compared with its congeners in this study, and a new diagnosis of *Fuscozetes* is given. The juveniles of *F. fuscipes* are light brown, with a brown prodorsum, sclerites, epimeres, and legs. In all juveniles, a humeral organ and a humeral macrosclerite are present. The gastronotum of the larva has 12 pairs of setae (h_3 is present), whereas the nymphs have 15 pairs. In the larva, the gastronotal shield is weakly developed, and most gastronotal setae are short except for a slightly longer h_2. Most of the gastronotal setae are inserted on the microsclerites except for h_3, and several other macrosclerites and many microsclerites are present on the hysterosoma. In the nymphs, the gastronotal shield is well developed, with 10 pairs of setae (*d*-, *l*-, and *h*-series, and p_1), and setae p_2 and p_3 are located on a large posteroventral macrosclerite. In all the instars, femora I and II are oval in cross-section, without a large ventral carina. Mitochondrial COI sequence data revealed a deep split between the Nearctic and Palearctic populations of *F. fuscipes*, and a less, but significant, divergence within each continent. These strong geographical barriers were contrasted with multiple cases of shared haplotypes over long distances in the Palearctic, indicating high migration rates in modern times.

Keywords: oribatid mites; Sphaerozetinae; juveniles; leg setation; stage structure; DNA barcoding

Citation: Seniczak, S.; Seniczak, A.; Jordal, B.H. Morphological Ontogeny, Ecology, and Biogeography of *Fuscozetes fuscipes* (Acari, Oribatida, Ceratozetidae). *Animals* **2024**, *14*, 538. https://doi.org/10.3390/ani14040538

Academic Editors: Monika Fajfer and Maciej Skoracki

Received: 27 December 2023
Revised: 1 February 2024
Accepted: 2 February 2024
Published: 6 February 2024

1. Introduction

Fuscozetes Sellnick, 1928, with the type species *F. fuscipes* (C.L Koch, 1844), is an average genus in terms of its number of species. It includes 15 species according to Subías [1], and two of them are considered *species inquirendae*. *Fuscozetes* belongs to the subfamily Spherozetinae sensu Shaldybina [2], which also contains the genera *Edwardzetes* Berlese, 1913; *Ghilarovizetes* Shaldybina, 1969; *Melanozetes* Hull, 1916; and *Sphaerozetes* Berlese, 1885.

The systematics of *Fuscozetes* are not clear in the literature, and its diagnosis has varied over time. Sellnick [3] paid particular attention to the skeletal characters of adults, whereas Shaldybina [2] added the number of notogastral setae, which can provide information on the phylogeny of moss mites [4–6]. Based on the morphology of the juvenile stages and adults of the *Fuscozetes* species, Shaldybina [2,7,8] limited the number of setae on the notogaster of adults to 10 or 11 pairs. Behan-Pelletier [9,10] expanded the diagnosis of *Fuscozetes* to 10–14 pairs, including c_3, but this diagnosis was problematic because it also included *Melanozetes* Hull, 1916, which has 14 pairs of notogastral setae. Based on the morphology of the juvenile stages and adults of the *Fuscozetes* species, Seniczak et al. [11] restricted the diagnosis of *Fuscozetes* to 10–13 pairs of notogastral setae, including seta c_2, and some or all the setae of the *d*-series. Next, Seniczak et al. [12] and Seniczak and Seniczak [13,14] added the length and position of solenidion ω_2 on tarsus I, which clearly separated *Fuscozetes* from *Melanozetes*, both in the nymphs and the adults. In *Fuscozetes*, solenidion ω_2 is shorter than ω_1 and is placed posterolaterally to ω_1, whereas in *Melanozetes*, solenidion ω_2 is as long as or longer than ω_1 and is placed anteriorly to ω_1.

Shaldybina [2] included *Fuscozetes* in the subfamily Sphearozetinae, which clearly differs from Ceratozetinae and Trichoribatinae in both the adult and juvenile stages. Weigmann [15] omitted this proposal and included *Fuscozetes* and related genera in Ceratozetidae, indicating that the systematics of this family need further investigation.

The juvenile instars of the *Fuscozetes* species are relatively well studied. According to the catalogue by Norton and Ermilov [16], Seniczak et al. [12], and Seniczak and Seniczak [13,14], the morphological ontogeny of seven species of *Fuscozetes* is known: *F. coulsoni* A. and S. Seniczak, 2020; *F. fuscipes*; *F. kamchatkicus* Seniczak et al., 2016; *F. pseudosetosus* Shaldybina, 1969; *F. setiger* (Trägårdh, 1910); *F. setosus* (C.L. Koch, 1839); and *F. tatricus* Seniczak, 1993. The morphological ontogeny of *F. fuscipes* has already been investigated by Seniczak [17], but this study was general and omitted the lateral aspect of the larvae, tritonymphs, and adults, as well as the leg setation, which is important for the morphological comparison of species within the genus. Here, we investigated specimens of *F. fuscipes* from Norway, which differed slightly from those studied by Seniczak [17] from Poland, illustrating some regional variability in this species. We also illustrated the morphological structures of *F. fuscipes* with SEM figures to clarify the miniscule characters of this species. In addition, we compared the molecular data (COI) of *F. fuscipes* from different locations, based on our own and public data.

2. Materials and Methods

2.1. Morphological and Biological Studies

The specimens of *F. fuscipes* used in this study were collected on 15 June 2018 by A. Seniczak from patches of *Sphagnum* mosses on the shore of lake Skomakerdiket (Bergen, Vestland, Norway, 60°23′39.7″ N, 5°21′04.7″ E, 320 m a. s. l.). The samples were extracted in Berlese funnels in the laboratory of the Department of Natural History, University Museum of Bergen (Norway), over ten days. The juveniles of *F. fuscipes* were identified using the specific characters given by Seniczak [17]. In 30 adults selected at random, the sex ratio and the number of gravid females were determined, as well as the body length and width. We also measured the morphological characters of all the instars of *F. fuscipes*, namely the total body length (in the lateral aspect, from the tip of the rostrum to the posterior edge of the notogaster), the width (in the dorsal aspect, at the widest part of the notogaster), and the length of the anal and genital openings and the setae, perpendicularly to their length. All measurements are given in μm. All light microscopy was performed using a Nicon Eclipse Ni.

We illustrated the dorsal and lateral aspects of the larvae, tritonymphs, and adults; some leg segments of these stages; the ventral regions of all instars; and the palps and chelicerae of the adults. The illustrations were prepared from individuals mounted temporarily on slides in lactic acid. In the text and figures, the following abbreviations are used: rostral (*ro*), lamellar (*le*), interlamellar (*in*) and exobothridial (*ex*) setae, lamella (*La*), translamella (*Tr*), bothridium (*bo*), bothridial seta (*bs*), notogastral or gastronotal setae (*c-*, *d-*, *l-*, *h-*, and *p*-series), porose areas (*Aa*, *A1*, *A2*, and *A3*), opisthonotal gland opening (*gla*), pteromorph (*Ptm*), cupules or lyrifissures (*ia*, *im*, *ip*, *ih*, *ips*, and *iad*), tutorium (*Tut*), pedotectum (*Pd*), circumpedal carina (*cp*), custodium (*cus*), discidium (*Dis*), humeral sclerite (*hs*), humeral organ (*oh*), subcapitular setae (*a*, *m*, and *h*), cheliceral setae (*cha* and *chb*), Trägårdh organ (*Tg*), palp setae (*sup*, *inf*, *l*, *d*, *cm*, *acm*, *it*, *vt*, *ul*, and *su*), solenidion ω, adanal and anal setae (*ad*- and *an*-series), epimeral setae (*1a–c*, *2a*, *3a–c*, and *4a–c*), genital (*g*) and aggenital (*ag*) setae, leg solenidia (σ, φ, and ω), famulus (ε), and setae (*bv*, *d*, *l*, *ft*, *tc*, *it*, *p*, *u*, *a*, *s*, *pv*, *pl*, and *v*). The terminology used follows that of Grandjean [4,5,18–21], Behan-Pelletier [9,10], and Norton and Behan-Pelletier [22]. The species nomenclature follows Subías [1] and Norton and Ermilov [16].

For scanning electron microscopy (SEM), the mites were air-dried, coated with Au/Pd in a Polaron SC502 sputter-coater, and placed on Al stubs with double-sided sticky carbon tape. The observations and micrographs were made with a Quanta™ 450 FEG scanning electron microscope.

2.2. DNA Barcoding

The specimens of *F. fuscipes* used for molecular studies were collected in Southern, Central, and Northern Norway. We also used public sequences from the BOLD database, which originated from Canada, Finland, and Germany. The outgroup sequences represented two other *Fuscozetes* species, *F. setosus* and *F. setiger*, and representatives of all the other genera of the subfamily Spherozetinae (*Edwardzetes* Berlese, 1913; *Ghilarovizetes* Shaldybina, 1969; *Melanozetes* Hull, 1916; and *Sphaerozetes* Berlese, 1885) and representatives of two other related subfamilies, Trichoribatinae (*Diapterobates* Grandjean, 1936; *Neogymnobates* Ewing, 1917; *Svalbardia* Thor, 1930; and *Trichoribates* Berlese, 1910) and Ceratozetinae (*Ceratozetes* Berlese, 1908).

The specimens of *F. fuscipes* were DNA-barcoded at the Canadian Centre for DNA Barcoding (CCDB) in Guelph, Canada. Before sending the samples to the CCDB, each specimen was photographed, and these photos are vouchers available at the Barcode of Life Data System (BOLD, http://boldsystems.org, accessed on 20 November 2023). The specimens were placed in a well containing 50 mL of 90% ethanol in a 96-well microplate and submitted to the CCDB. The mites were sequenced for the barcode region of the COI gene according to standard protocols at the CCDB (www.ccdb.ca, accessed on 20 October 2020), using either the LepF1/LepR1 [23] or the LCO1490/m HCO2198 [24] primer pairs. The DNA extracts were placed in archival storage at −80 °C, mainly at the CCDB, and some (sequencing code starting with UMNFO) at the University Museum of Bergen (ZMUB). The COI sequence chromatograms were checked for double peaks and potential NUMTs, and were blasted in GenBank to detect and exclude possible contaminants. The sequences are available in GenBank (Table 1).

Table 1. Information about sequenced specimens of *Fuscozetes fuscipes* and other oribatid species used in this study. Na—not available; these sequences are public in BOLD, but without some data.

Species	Sequence Code at BOLD	GenBank Access No.	Locality	Coordinates	Elevation m a. s. l.	Reference
Fuscozetes fuscipes (C.L Koch, 1844)	MARBN1420-23	PP215015	NO: Nordland	68.292, 14.186	60	-
	MARBN1419-23	PP214995	NO: Nordland	68.292, 14.186	60	-
	MARBN1418-23	PP215016	NO: Nordland	68.292, 14.186	60	-

Table 1. *Cont.*

Species	Sequence Code at BOLD	GenBank Access No.	Locality	Coordinates	Elevation m a. s. l.	Reference
Fuscozetes fuscipes (C.L Koch, 1844)	SOITS159-22	PP215024	NO: Finnmark	69.713, 30.871	40	-
	SOITS131-22	PP214998	NO: Finnmark	69.713, 30.871	40	-
	SOITS130-22	PP214996	NO: Finnmark	69.713, 30.871	40	-
	SOITS129-22	PP215013	NO: Finnmark	69.713, 30.871	40	-
	SOITS057-22	PP215003	NO: Finnmark	69.143, 29.240	61	-
	SOITS002-22	PP215011	NO: Finnmark	69.143, 29.240	61	-
	MARBN109-21	PP215006	NO: Trøndelag	63.489, 8.874	50	-
	UMNFO278-18	PP215005	NO: Agder	58.451, 8.705	62	-
	UMNFO439-18	PP215021	NO: Vestland	60.398, 5.351	370	-
	UMNFO438-18	PP215026	NO: Vestland	60.398, 5.351	370	-
	UMNFO437-18	PP215027	NO: Vestland	60.398, 5.351	370	-
	UMNFO277-18	MN520688	NO: Agder	58.451, 8.705	62	Seniczak et al. [25]
	ABMI236-17	MN348906	CA: Alberta	53.506, −114.960	780	Young et al. [26]
	ABMI237-17	MN355322	CA: Alberta	51.734, −113.641	904	Young et al. [26]
	AMOR043-08	MN348553	CA: Alberta	51.847, −114.764	1054	Young et al. [26]
	AMOR045-08	MN354723	CA: Alberta	51.847, −114.764	1054	Young et al. [26]
	AMOR630-10	MN354116	CA: Alberta	53.533, −113.533	780	Young et al. [26]
	AMOR631-10	MN351373	CA: Alberta	53.533, −113.533	780	Young et al. [26]
	AMOR889-10	MN350498	CA: Alberta	53.208, −115.651	780	Young et al. [26]
	MIONB042-10	HM887577	CA: Ontario	45.390, −78.906	386	Young et al. [26]
	MIONB083-10	HQ575095	CA: Ontario	45.390, −78.906	386	Young et al. [26]
	ORNA083-09	MN356812	CA: Quebec	45.610, −76.004	176	Young et al. [26]
	SSEIB5800-13	KM828928	CA: Alberta	53.567, −112.851	722	Young et al. [26]
	SSEIB8176-13	KM834684	CA: Alberta	53.567, −112.851	722	Young et al. [26]
	SSEIB8208-13	KM834777	CA: Alberta	53.567, −112.851	722	Young et al. [26]
	SSEIB8308-13	KM840220	CA: Alberta	53.567, −112.851	722	Young et al. [26]
	SSEIB8310-13	KM840191	CA: Alberta	53.567, −112.851	722	Young et al. [26]
	SSEIB8319-13	KM828000	CA: Alberta	53.567, −112.851	722	Young et al. [26]
	SSEIB8352-13	KM831118	CA: Alberta	53.567, −112.851	722	Young et al. [26]
	SSEIB8357-13	KM824343	CA: Alberta	53.567, −112.851	722	Young et al. [26]
	SSEIB8361-13	KM833473	CA: Alberta	53.567, −112.851	722	Young et al. [26]
	FINOR506-13	MZ623348	FI: Uusikaupunki	60.814, 21.216	12	Roslin et al. [27]
	FINOR507-13	MZ623712	FI: Uusikaupunki	60.814, 21.216	12	Roslin et al. [27]
	FINOR508-13	MZ626462	FI: Uusikaupunki	60.814, 21.216	12	Roslin et al. [27]
	FINOR509-13	MZ628127	FI: Uusikaupunki	60.814, 21.216	12	Roslin et al. [27]
	FINOR510-13	MZ626956	FI: Uusikaupunki	60.814, 21.216	12	Roslin et al. [27]
	TBGMI153-21	Na	GE: Thuringia	51.083, 10.426	447	-
Ghilarovizetes longisetosus (Hammer, 1952)	CHACA151-08	JX835704	CA: Manitoba	58.760, −94.069	29	Young et al. [26,28]
	CHACA976-10	HM405840	CA: Manitoba	58.786, −93.739	281	Young et al. [26,28]
	MYMCC122-11	JX834086	CA: Manitoba	58.771, −93.851	281	Young et al. [26,28]
Neogymnobates luteus Hammer, 1955	CHACB198-10	HQ558468	CA: Manitoba	58.731, −93.781	281	Young et al. [26,28]
	CHACB616-10	HQ558703	CA: Manitoba	58.623, −94.230	8	Young et al. [26,28]
	CHACB930-10	HM907357	CA: Manitoba	58.625, −93.816	38	Young et al. [26,28]

Table 1. *Cont.*

Species	Sequence Code at BOLD	GenBank Access No.	Locality	Coordinates	Elevation m a. s. l.	Reference
Ceratozetes parvulus Sellnick, 1922	MARBN125-21	PP215014	NO: Trøndelag	60.593, 7.432	1166	-
	MARBN310-21	PP215030	NO: Vestland	60.593, 7.432	1166	-
	MARBN311-21	PP215035	NO: Vestland	60.593, 7.432	1166	-
Melanozetes stagnatilis (Hull, 1914)	MARBN1344-23	PP215020	NO: Vestland	60.794, 5.055	46	-
	MARBN1345-23	PP215008	NO: Vestland	60.794, 5.055	46	-
	MARBN1347-23	PP215022	NO: Vestland	60.794, 5.055	46	-
Trichoribates berlesei (Jacot, 1929)	MARBN295-21	PP215033	NO: Trøndelag	63.405, 10.120	113	-
	MARBN327-21	PP215032	NO: Trøndelag	63.405, 10.120	113	-
	MARBN328-21	PP214999	NO: Trøndelag	63.405, 10.120	113	-
Diapterobates notatus (Thorell, 1871)	UMNFO301-18	PP215034	NO: Svalbard	78.2037, 15.319	161	-
	UMNFO302-18	PP215025	NO: Svalbard	78.2037, 15.319	161	-
	UMNFO303-18	PP215001	NO: Svalbard	78.2037, 15.319	161	-
Svalbardia lucens (L. Koch, 1879)	UMNFO340-18	PP215023	NO: Svalbard	78.209, 15.711	11	-
	UMNFO341-18	PP215028	NO: Svalbard	78.209, 15.711	11	-
	UMNFO342-18	PP215018	NO: Svalbard	78.209, 15.711	11	-
Fuscozetes setiger (Trägårdh, 1910)	UMNFO412-18	PP215004	NO: Svalbard	78.040, 13.646	161	-
	UMNFO413-18	PP215019	NO: Svalbard	78.040, 13.646	161	-
	UMNFO802-19	PP215010	NO: Vestland	60.583, 7.472	1356	-
Sphaerozetes orbicularis (C.L. Koch, 1835)	UMNFO428-18	PP215017	NO: Vestland	60.398, 5.351	370	-
	UMNFO429-18	PP215000	NO: Vestland	60.398, 5.351	370	-
	UMNFO430-18	PP215007	NO: Vestland	60.398, 5.351	370	-
Edwardzetes edwardsi (Nicolet, 1855)	UMNFO462-18	PP215031	NO: Vestland	60.398, 5.351	370	-
	UMNFO463-18	PP215012	NO: Vestland	60.398, 5.351	370	-
	UMNFO534-18	PP215009	NO: Vestland	60.584, 7.519	1356	-
Svalbardia bicuspidata (Thor, 1930)	UMNFO490-18	PP215002	NO: Vestland	60.572, 7.478	1356	-
	UMNFO491-18	PP214997	NO: Vestland	60.572, 7.478	1356	-
	UMNFO492-18	PP215029	NO: Vestland	60.572, 7.478	1356	-
Fuscozetes setosus (C.L. Koch, 1839)	UZINS275-23	Na	SK	Na	Na	-
	UZINS276-23	Na	SK	Na	Na	-

The sequences were aligned by eye in MEGA11: Molecular Evolutionary Genetics Analysis, version 11 [29]. A Bayesian inference (BI) analysis was conducted in MrBayes 3.2 using a GTR + G + I model of nucleotide substitutions [30]. Posterior probabilities were generated from 10 million generations of sampling from two independent runs using one cold (temp = 0.3) and three heated chains, excluding the first 25% of generations as burn-in. The chain convergence was assessed using a standard deviation of the split frequencies approaching 0.01 and a potential scale reduction factor (PSRF) of 1.0. The consensus tree topology was visualized in FigTree 1.4.2 (available at http://tree.bio.ed.ac.uk/software/figtree, accessed on 20 November 2023) and edited in Adobe Illustrator.

3. Results

3.1. *Diagnosis of Fuscozetes Sellnick, 1928*

Based on the morphological characters given by Seniczak et al. [11,12] and Seniczak and Seniczak [13,14] and *F. fuscipes* studied herein, the diagnosis of *Fuscozetes* is as follows: the adults are of a medium size (423–897), brown to dark brown, and with characters of

Ceratozetidae [2]. A translamella is present or absent, the lamellar cusp is rounded or with teeth, and the bothridial seta is clavate or fusiform. The notogastral setae are short to long (10–13 pairs, including c_2 and all or some setae of the *d*-series), and the porose areas (4 pairs) are of a similar size or with a larger *Aa*. Femora I and II are oval in cross-section, and solenidion ω_2 on tarsus I is shorter than ω_1 and is located posterolaterally to solenidion ω_1.

The juveniles are light brown with a brown prodorsum, sclerites, epimeres, and legs. The bothridial seta is clavate or fusiform and the gastronotal setae are short to long. In the larva, a humeral organ and a humeral macrosclerite are present or absent. The gastronotal seta c_1 is inserted on the humeral macrosclerite or microsclerite, or on the unsclerotized integument, while setae c_2 and c_3 are inserted on the microsclerites or the unsclerotized integument. The gastronotal shield is uniform, divided in two parts, structured as a pygidium, or absent (in which case most gastronotal setae are located on microsclerites); other sclerites and microsclerites are present or absent. In the nymphs, the humeral macrosclerite is present, whereas the humeral organ is present or absent. The gastronotal shield has 10 pairs of setae (*d*-, *l*-, and *h*-series, and p_1), where setae p_2 and p_3 are placed on the macrosclerite or the unsclerotized integument. In all the juveniles, the femora are oval in cross-section, without a large ventral carina. Solenidion ω_2 on tarsus I is shorter than solenidion ω_1, and is located posterolaterally to ω_1.

3.2. *Morphological Ontogeny of Fuscozetes fuscipes (C.L. Koch, 1844) (Figures 1–21)*

Oribata fuscipes C.L. Koch, 1844: Michael [31].

Fuscozetes fuscipes: Sellnick [3], Willmann [32], Shaldybina [2], Mehl [33], Karppinen and Krivolutsky [34], Golosova et al. [35], Karppinen et al. [36,37], Marshall et al. [38], Olszanowski et al. [39], Niemi et al. [40], Ryabinin and Pankov [41], Subías [1], Weigmann [15], Miko [42], Murvanidze and Mumladze [43], Schatz [44], and Murvanidze et al. [45].

3.2.1. Diagnosis

The adults are brown, of a medium size (629–897), and with the characters of *Fuscozetes* given above. The translamella is present, the lamellar cusp is long with teeth, and the bothridial seta is fusiform. The notogastral setae are long (10 pairs, including c_2), and porose area *Aa* is rounded and slightly larger than the other porose areas.

In the juveniles, the bothridial seta is fusiform and a humeral organ and humeral macrosclerite are present. In the larva, the gastronotal shield is absent; most of the gastronotal setae are short and inserted on the microsclerites; and other sclerites and microsclerites are also present. In the nymphs, the gastronotal shield is present, the setae of the *c*-series are located on the microsclerites, and setae p_2 and p_3 are placed on a large, posteroventral sclerite.

3.2.2. Morphology of Adults

The adults are brown to dark brown, oval in the dorsal and ventral view (Figures 1a, 2 and 4a,d), and of a medium size (734–897), as redescribed by Shaldybina [8] and Bayartogtokh and Weigmann [46] (but see Remarks). The mean length (and range) of females is 847.6 ± 19.7 (815–897, *n* = 23) and the maximum width is 566.2 ± 19.8 (538–587); the mean length (and range) of males is 791.7 ± 26.4 (734–815, *n* = 7) and the maximum width is 526.3 ± 20.4 (505–554). The prodorsal seta *in* is long, *le* and *ro* are of a medium size, and *ex* is short (Figures 1a, 2, 3a, 4, 5, 6a and 7c, Table 2). The notogastral setae (13 pairs, including c_2) are long, with short barbs (Figures 1b and 6a), but appear smooth under low magnification. The porose areas (four pairs) are rounded and *Aa* is slightly larger than the other porose areas (Figures 1a and 3a). The chelicerae are chelate–dentate (Figures 3b and 6c,d), with seta *cha* longer than *chb*; both setae are barbed. Most palp setae have short barbs (Figures 3c, 5a and 6c,d). The custodium (*cus*) is long and the discidium (*Dis*) is triangular. The epimeral setae are short and smooth, and the inner setae are slightly shorter than the other setae (Figures 2, 4d, 6c,d and 7a,b). The genital (*g*), aggenital (*ag*), and anal (*an*) setae are short and smooth, and the adanal setae (*ad*) are longer, with short barbs on the apical part. The medium parts of femora I and II are oval in cross-section, without a large ventral carina, but femur II has an anteroventral projection (Figures 4b, 5a, 7c and 8). Most of the leg setae have short barbs (Figures 4, 5a,b, 6c,d, 7, 8 and 9a,b). The formulae of the leg setae [trochanter to tarsus (+ solenidia)] are as follows: I—1-5-3(1)-4(2)-20(2); II—1-5-3(1)-4(1)-15(2); III—2-3-1(1)-3(1)-15; and IV—1-2-2-3(1)-12.

Remarks: The adults investigated here were larger than those studied by Willmann [32] (body size: 710 × 500), Shaldybina [8] (body size: 753–774 × 473–516), Bayartogtokh and Weigmann [46] (body size: 629–676 × 419–433), and Weigmann [15] (body length of 630–765), but the length and distribution of the notogastral setae were similar in all studies.

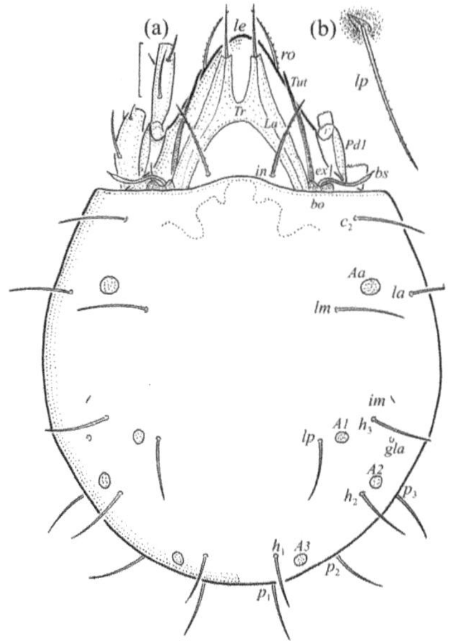

Figure 1. Adult female *Fuscozetes fuscipes*. (**a**) Dorsal aspect, with legs partially drawn; scale bar: 50 µm. (**b**) Seta *lp* (enlarged).

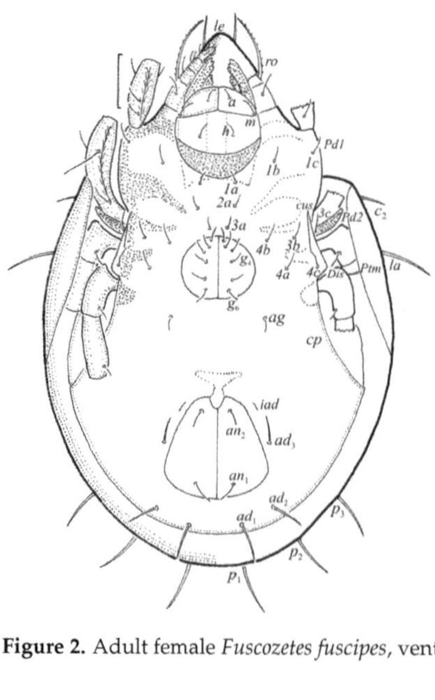

Figure 2. Adult female *Fuscozetes fuscipes*, ventral aspect, with legs partially drawn; scale bar: 50 µm.

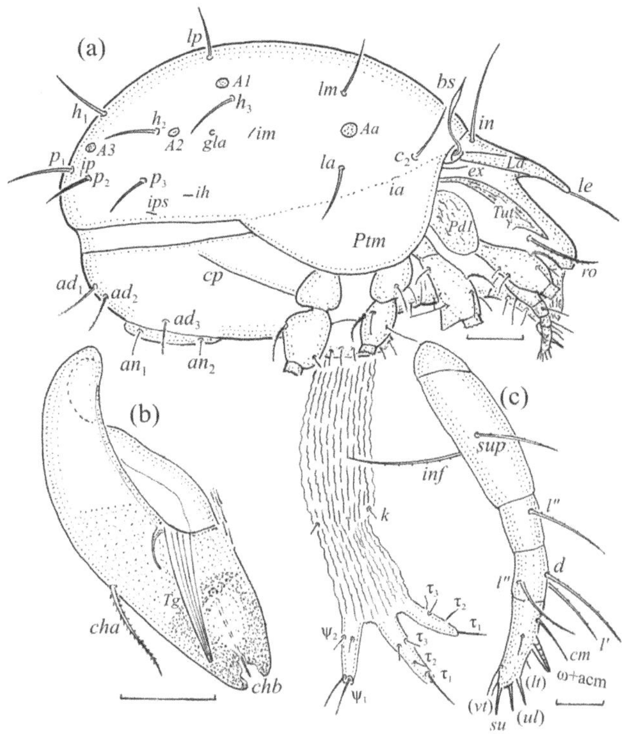

Figure 3. *Fuscozetes fuscipes.* (**a**) Female with ejected ovipositor, lateral aspect, with legs partially drawn; scale bar: 50 µm. Mouthparts of adult, right side; scale bars: 20 µm. (**b**) Chelicera, paraxial aspect. (**c**) Palp.

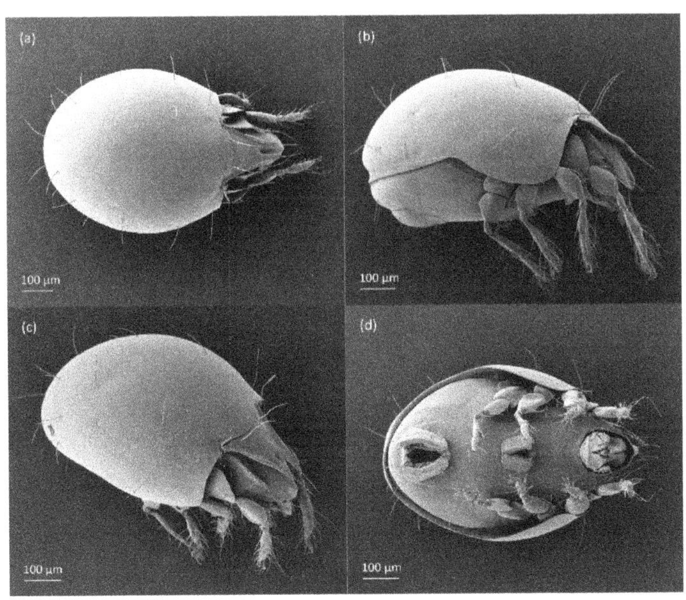

Figure 4. SEM micrographs of adult *Fuscozetes fuscipes.* (**a**) Dorsal view, (**b**) lateral view, (**c**) dorsolateral view, and (**d**) ventral view.

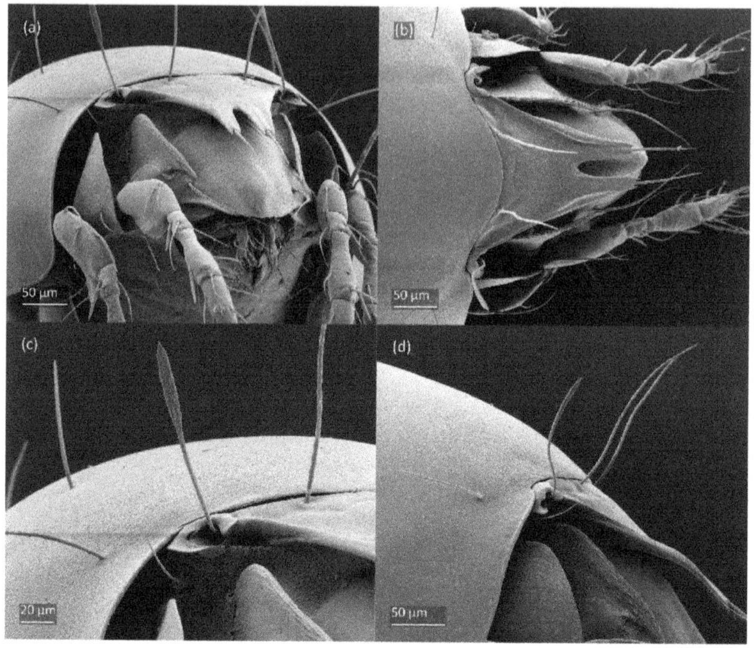

Figure 5. SEM micrographs of adult *Fuscozetes fuscipes*. Anterior part of body, (**a**) frontal view and (**b**) dorsal view; bothridial seta, (**c**) frontal view and (**d**) lateral view.

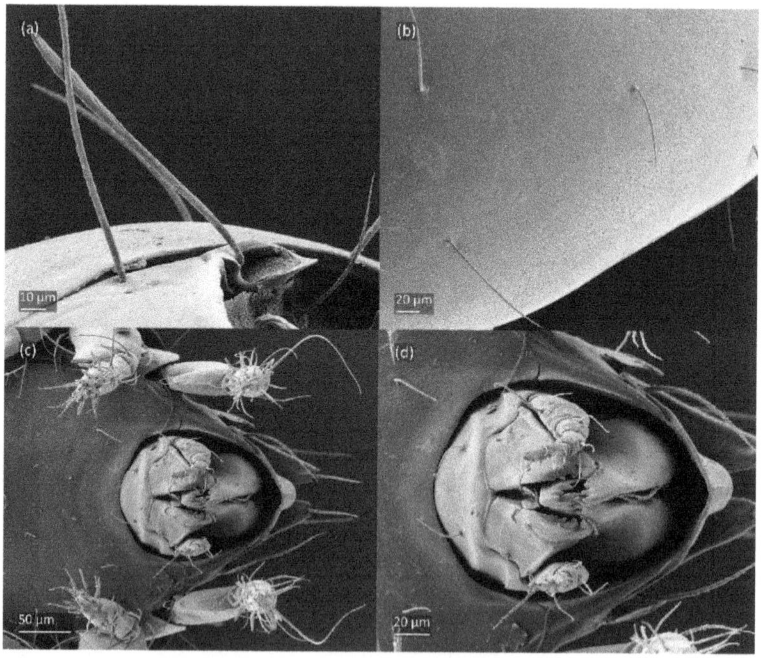

Figure 6. SEM micrographs of adult *Fuscozetes fuscipes*. (**a**) Bothridial seta, lateral view; (**b**) part of notogaster, dorsal view; and (**c**,**d**) anterior part of body, ventral view.

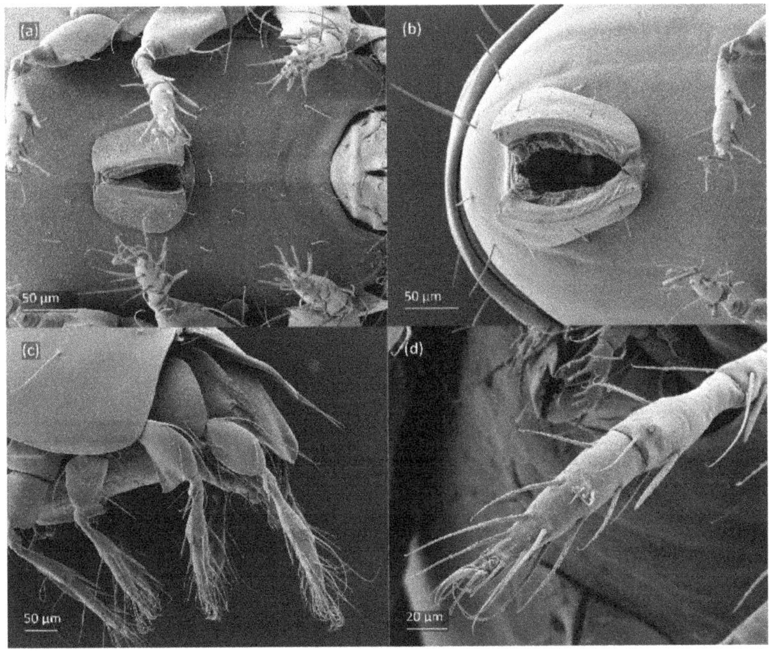

Figure 7. SEM micrographs of adult *Fuscozetes fuscipes*. Ventral view of (**a**) medial part of body and (**b**) posterior part of body; (**c**) anterior part of body, lateral view; and (**d**) part of leg I, dorsal view.

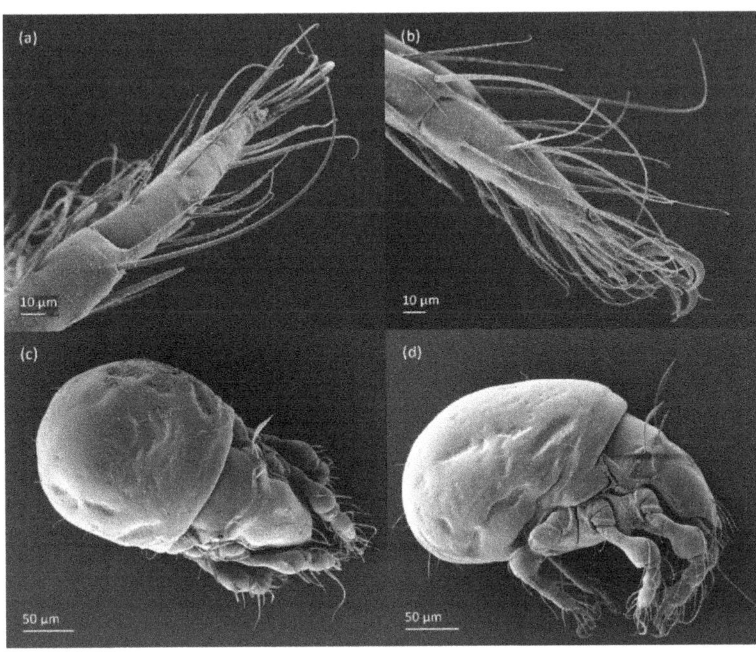

Figure 8. SEM micrographs of *Fuscozetes fuscipes*. Part of leg I of adult, (**a**) dorsal view and (**b**) lateral view; larva, (**c**) dorsal view and (**d**) lateral view.

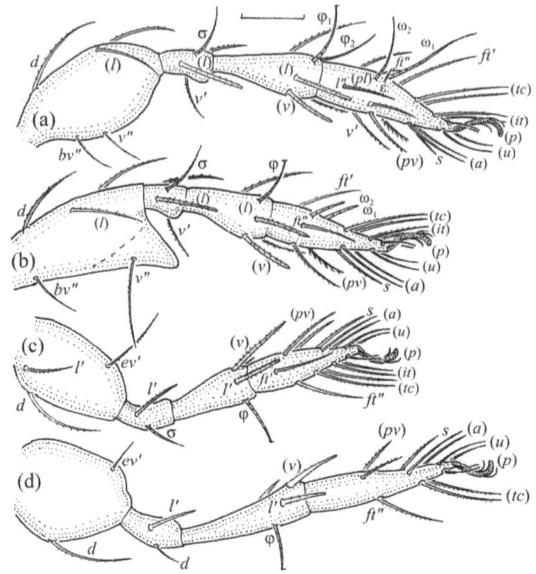

Figure 9. Leg segments of adult *Fuscozetes fuscipes* (femur to tarsus), right side; scale bar: 20 µm. (**a**) Leg I, (**b**) leg II, (**c**) leg III, and (**d**) leg IV.

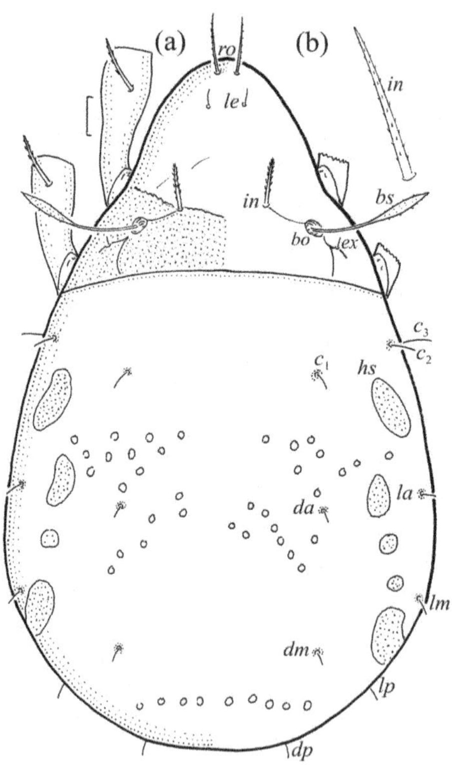

Figure 10. *Fuscozetes fuscipes* larva. (**a**) Dorsal aspect, legs partially drawn; scale bar: 20 µm. (**b**) Seta *in* (enlarged).

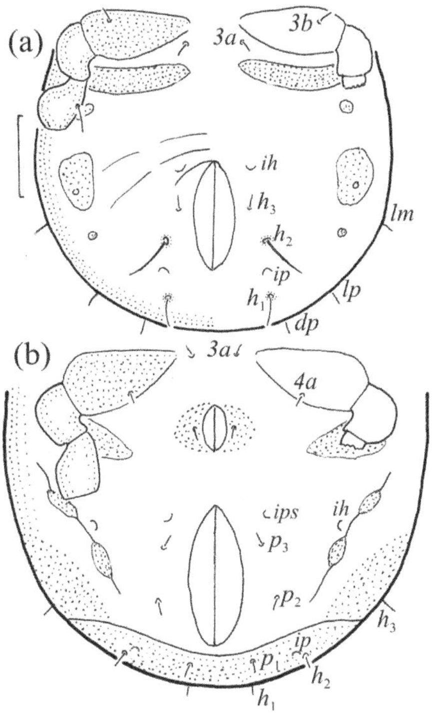

Figure 11. Posterior part of *Fuscozetes fuscipes* hysterosoma, legs III and IV partially drawn; scale bar: 20 µm. (**a**) Larva and (**b**) protonymph.

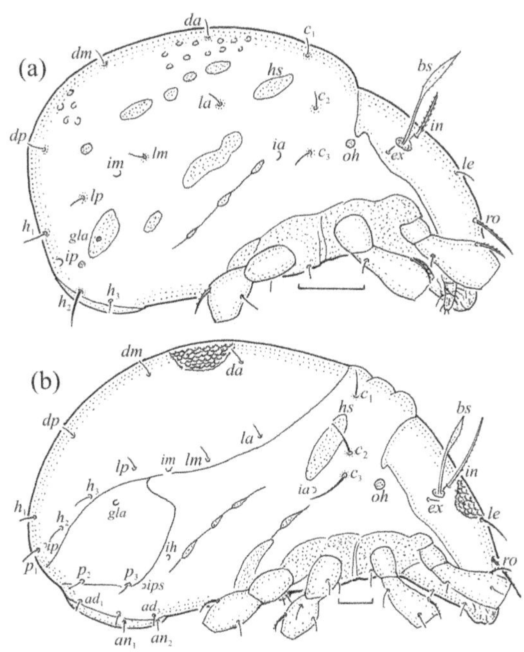

Figure 12. *Fuscozetes fuscipes*, lateral aspect, legs partially drawn; scale bars: 50 µm. (**a**) Larva and (**b**) tritonymph.

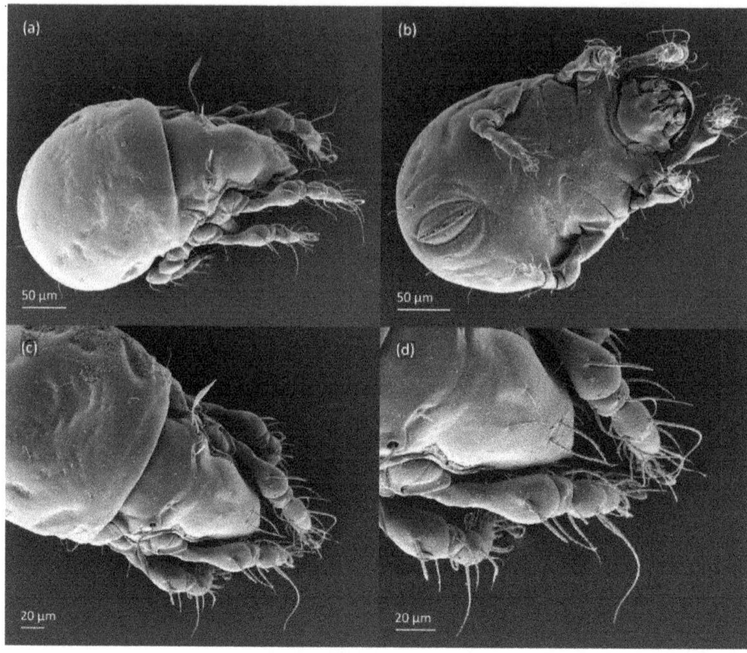

Figure 13. SEM micrographs of *Fuscozetes fuscipes* larva. (**a**) Dorsolateral view; (**b**) ventral view; and dorsal view of (**c**) anterior and medial part of body and (**d**) anterior part of body.

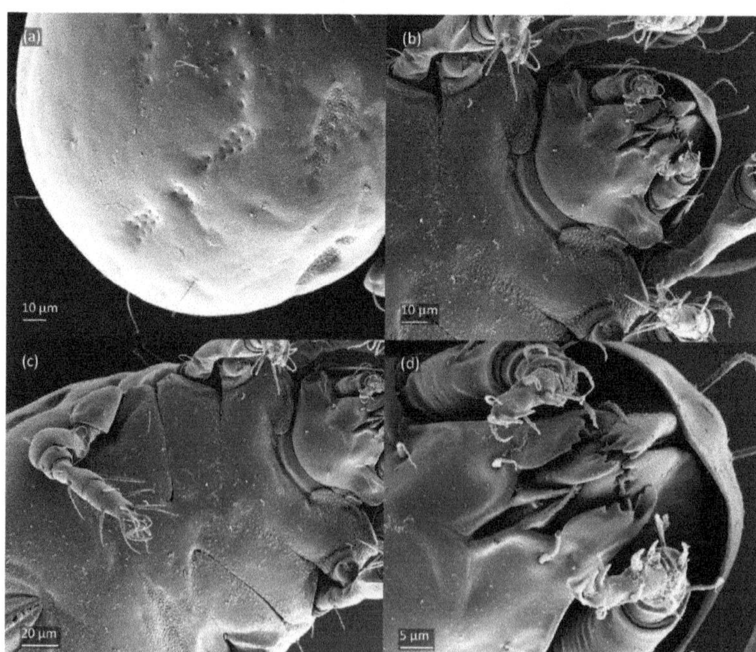

Figure 14. SEM micrographs of *Fuscozetes fuscipes* larva. (**a**) Posterior part of body, dorsolateral view; ventral view of (**b**) anterior part of body, (**c**) medial part of body, and (**d**) gnathosoma.

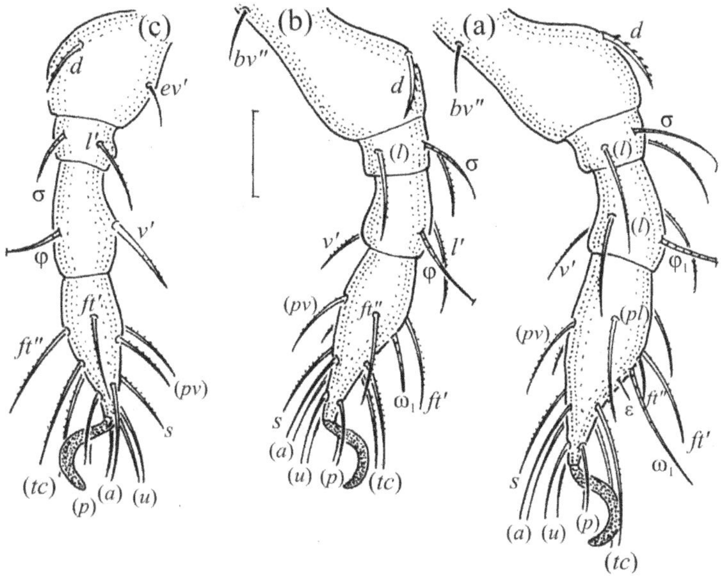

Figure 15. Leg segments of *Fuscozetes fuscipes* larva (femur to tarsus), right side; scale bar: 20 µm. (**a**) Leg I, (**b**) leg II, and (**c**) leg III.

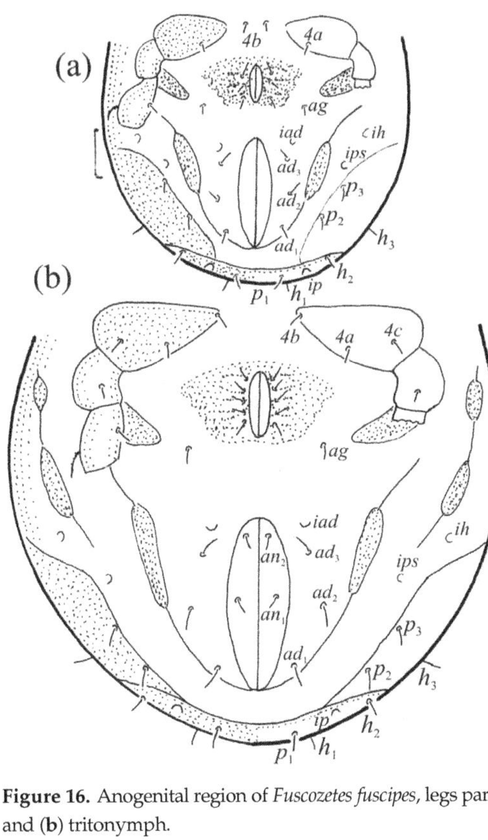

Figure 16. Anogenital region of *Fuscozetes fuscipes*, legs partially drawn; scale bar: 50 µm. (**a**) Deutonymph and (**b**) tritonymph.

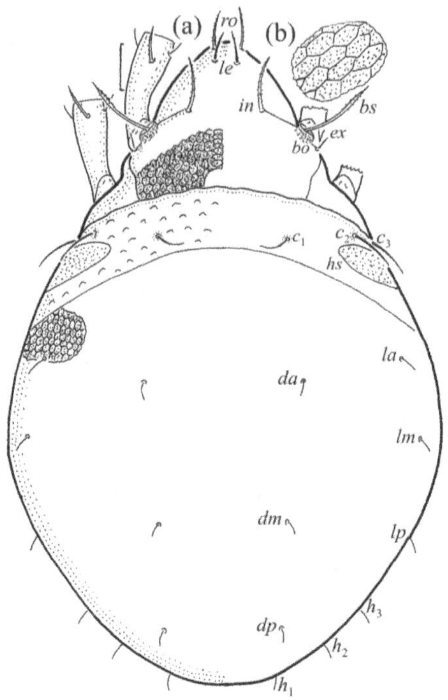

Figure 17. *Fuscozetes fuscipes* tritonymph. (**a**) Dorsal aspect, legs partially drawn; scale bar: 50 µm. (**b**) Pattern of gastronotum (enlarged).

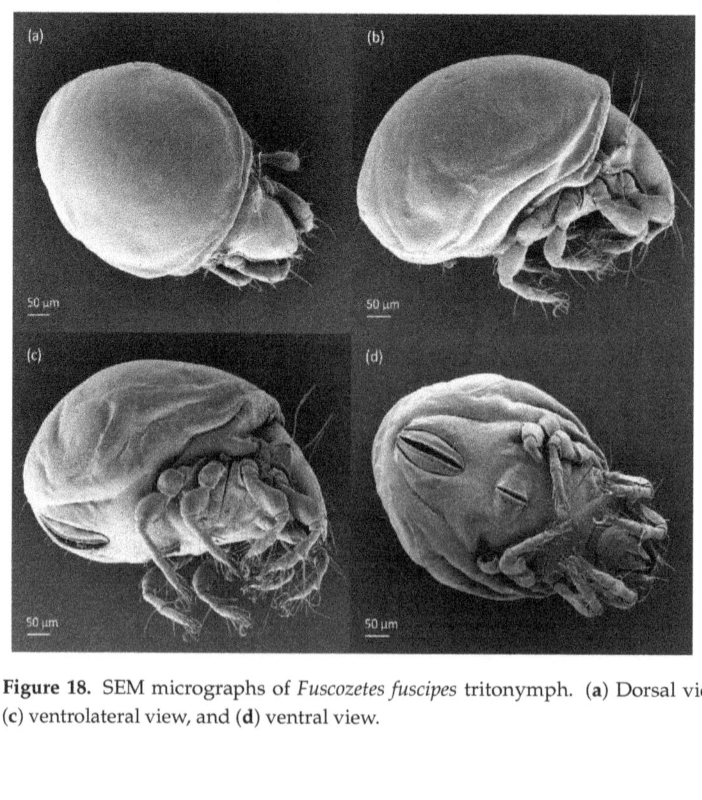

Figure 18. SEM micrographs of *Fuscozetes fuscipes* tritonymph. (**a**) Dorsal view, (**b**) lateral view, (**c**) ventrolateral view, and (**d**) ventral view.

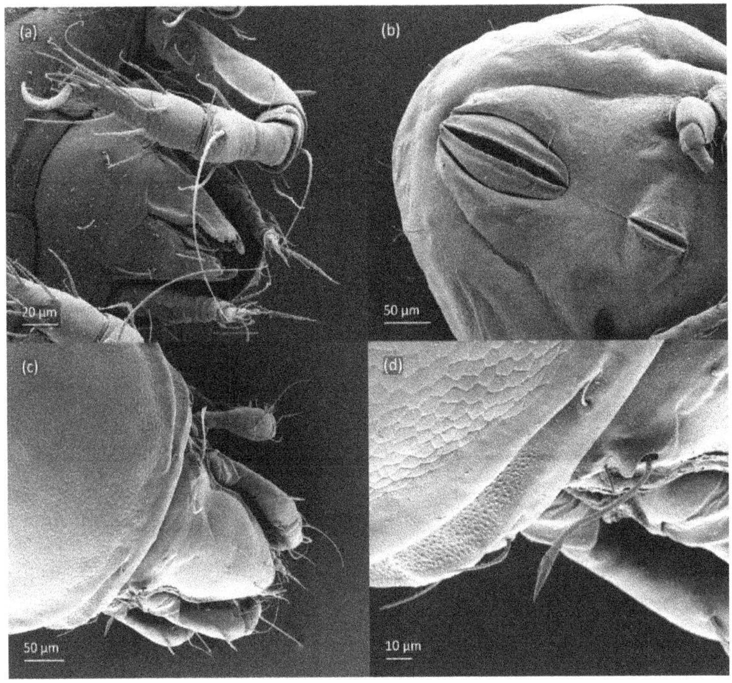

Figure 19. SEM micrographs of *Fuscozetes fuscipes* tritonymph. Ventral view of (**a**) anterior part of body and (**b**) posterior part of body; dorsal view of (**c**) anterior and medial part of body and (**d**) bothridial seta.

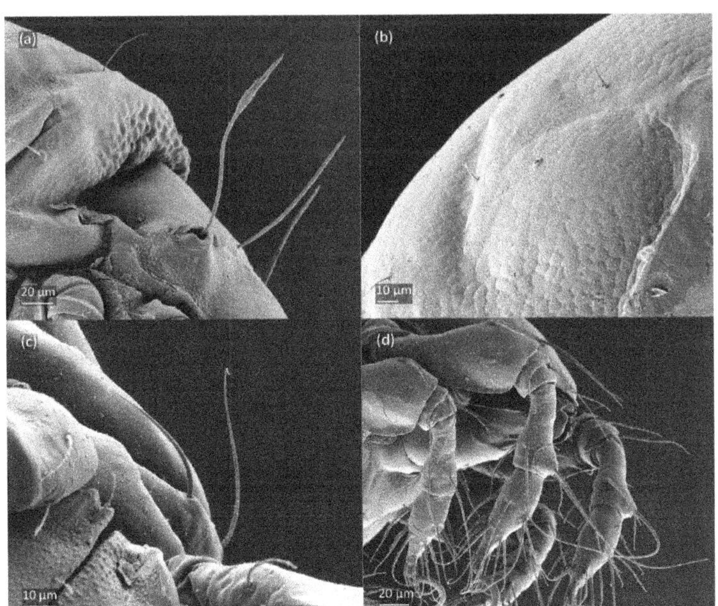

Figure 20. SEM micrographs of *Fuscozetes fuscipes* tritonymph. Lateral view of (**a**) bothridial seta and (**b**) *gla* opening; (**c**) seta c_2 and c_3, ventral view; and (**d**) legs I and II, lateral view.

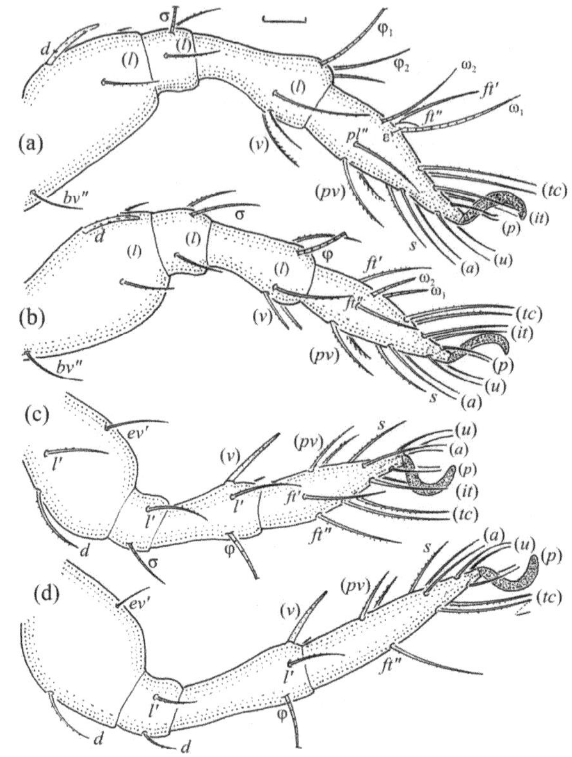

Figure 21. Leg segments of *Fuscozetes fuscipes* tritonymph (femur to tarsus), right side; scale bars: 20 μm. (**a**) Leg I, tarsus (*pl′* not illustrated); (**b**) leg II; (**c**) leg III; and (**d**) leg IV.

Table 2. Measurements of some morphological characters of juvenile stages of *Fuscozetes fuscipes* (mean measurements of 4–10 specimens in μm).

Morphological Characters	Larva	Protonymph	Deutonymph	Tritonymph	Adult
Body length	343	429	540	682	512
Body width	224	273	377	475	311
Length of prodorsum	101	138	193	258	190
Length of: seta *ro*	40	47	57	62	95
seta *le*	22	24	33	49	74
seta *in*	32	64	80	93	166
seta *bs*	77	82	99	104	85
seta c_1	16	24	32	41	Lost
seta c_2	19	28	40	51	154
seta c_3	17	29	35	44	Lost
seta *da*	14	16	28	31	Lost
seta *dp*	15	18	25	29	Lost
seta *la*	14	21	30	31	106
seta *lp*	16	23	29	29	87
seta h_1	16	18	25	31	102
seta h_2	37	18	23	29	89
seta h_3	10	21	24	30	94
seta p_1	Not developed	17	23	28	99
Genital opening	Not developed	37	49	70	114
Anal opening	87	100	121	167	163

According to Shaldybina [8] and Weigmann [15], the porose area *Aa* was found to be larger than in the adults investigated herein and as reported by Bayartogtokh [47], which may reflect regional variation in this species. Shaldybina [8] drew dark areas around the notogastral setae, which are only observed in young, light-brown adults.

3.2.3. Redescription of Juveniles

The larva is egg-shaped in its dorsal and ventral view (Figures 9c, 10a, 11a and 13b), and its body is light brown with a darker prodorsum, sclerites, epimeres, and legs. The prodorsum is subtriangular, the rostrum is rounded, setae *ro* and *in* are of a medium size and finely barbed, and the other setae are short and smooth (Figures 9c,d, 10a, 12a, 13 and Table 2). The mutual distance between setal pair *le* is almost two times longer than that between setae *ro*, and the mutual distance between pair *in* is about four times longer than that between pair *ro*. Setal pair *le* is placed approximately midway between the setal pairs *in* and *ro*. The opening of the bothridium is rounded and the bothridial seta is fusiform, with a barbed head. A ridge is present between the opening of the bothridium and the insertion of seta *in*. The prodorsum is finely porose.

The gastronotum of the larva has 12 pairs of setae, including h_3 inserted laterally to the anterior part of the anal valves (Figures 9d, 11a, 12a and 13b). The gastronotal shield is poorly developed and most of the gastronotal setae are short and smooth (Figures 9c,d, 10a, 11a, 12a, 13a–c and 14a,c), except for a slightly longer h_2. Most of the setae are located on basal microsclerites except for h_3. The humeral organ is rounded and located anteriorly to seta c_3. The humeral sclerite is oval and porose, three or four other macrosclerites are present laterally to setae *da* and *dm*, and many microsclerites are present in the central and posterior parts of the gastronotum. Three large macrosclerites are present on the lateral side of the gastronotum, including one around the *gla* opening, along with 3–4 small sclerites (Figures 10a, 11a and 12a). A large sclerite is present posteriorly to leg III. Cupule *ia* is located posteriorly to seta c_3, cupule *im* is located posteriorly to seta *lm*, *im* is located between setae h_1 and h_2, and *ih* is lateral to the anterior end of the anal opening (Figures 10a, 11a and 12a). The gland opening *gla* is located laterally to seta *lp*. The paraproctal valves (segment PS) are glabrous. The chelicera is chelate–dentate (Figure 14b,d). All the femora are oval in cross-section, a large ventral carina is absent, and most leg setae are finely barbed (Figures 9c,d, 13, 14b–d and 15).

The shape and color of the protonymph and other nymphs are the same as in the larva, but the head of the bothridial seta is slimmer, and the prodorsum and gastronotal shield have a reticulate cuticle (Figures 12b, 17, 18, 19b–d and 20a,b). The gastronotum has 15 pairs of setae because the setae of the *p*-series appear in the protonymph and are retained in subsequent nymphs; all of these setae are short and smooth except for the medium-sized *c*-series setae (Figures 11b, 12b, 16, 17, 18, 19b–d, 20a–c and Table 2). The humeral organ is located in the same location as in the larva, the humeral sclerite has seta c_1, and the other setae of the *c*-series are located on the basal microsclerites. The gastronotal shield is well developed, with 10 pairs of setae (*d*-, *l*-, and *h*-series, and p_1), and setae p_2 and p_3 are placed on a large posteroventral sclerite. Small sclerites are present laterally to the setae of the *l*-series, and a large macrosclerite is present posteriorly to leg IV (Figures 11b, 12b and 16). In the protonymph, one pair of genital setae is present on the genital valves, and two pairs are added in both the deutonymph and the tritonymph (Figures 11b, 16, 18d and 19b). The genital valves are placed on a large macrosclerite. In the deutonymph, one pair of aggenital setae and three pairs of adanal setae appear, and they remain in the other nymphs; all are short and smooth. In the tritonymph, cupules *ia* and *im* are placed in the same manner as in the larva, cupule *ip* is between setae p_1 and h_2, cupule *iad* is lateral to the anterior part of the anal valves, and cupules *ih* and *ips* are pushed laterally to cupule *iad* (Figures 11b, 12b and 16). The gland opening *gla* is placed anterolaterally to seta h_3. The chelicerae are chelate–dentate (Figures 18d and 19a). The anal valves of the protonymph and deutonymph are glabrous, and those of the tritonymph have two pairs of short and smooth setae. All the femora are oval in cross-section, a large ventral carina is absent, and most of the leg setae are finely barbed (Figures 18, 19, 20d and 21). In one deutonymph, two setae *v′* were present on trochanter III.

3.2.4. Summary of Ontogenetic Transformations

In the larva, the prodorsal setae *ro* and *in* are of a medium size, and the setae *le* and *ex* are short, whereas in the nymphs and adult, seta *in* is clearly longer than *ro*, and seta *le* is of a medium size. In all the juveniles, the bothridium is rounded, whereas in the adult, it is larger and gains scales. In all the instars, the bothridial seta is fusiform with a barbed head, but in the nymphs and adult, the head is slimmer than in the larva. The larva has 12 pairs of gastronotal setae (h_3 is present), while the nymphs have 15 pairs. The notogaster of the adult loses setae c_1, c_3, and those of the *d*-series, such that 10 pairs of setae remain on the notogaster. The formula of gastronotal setae in *F. fuscipes* is 12-15-15-15-10 (from larva to adult). The formulae of the epimeral setae are: 3-1-2 (larva), 3-1-2-1 (protonymph), 3-1-2-2 (deutonymph), and 3-1-3-3 (tritonymph and adult); the formula of the genital setae is 1-3-5-6 (protonymph to adult); the formula of the aggenital setae is 1-1-1 (deutonymph to adult); and the formula of segments PS–AN is 03333-0333-022. The ontogeny of the leg setae and solenidia of *F. fuscipes* is given in Table 3.

Table 3. Ontogeny of leg setae (Roman letters) and solenidia (Greek letters) of *Fuscozetes fuscipes*.

Leg	Trochanter	Femur	Genu	Tibia	Tarsus
Leg I					
Larva	–	*d*, *bv″*	(*l*), σ	(*l*), *v′*, φ_1	(*ft*), (*tc*), (*p*), (*u*), (*a*), *s*, (*pv*), (*pl*), ε, ω_1
Protonymph	–	–	–	–	ω_2
Deutonymph	–	(*l*)	–	φ_2	–
Tritonymph	*v*	–	–	*v″*	(*it*)
Adult	–	*v′*	*v′*	–	*l″*, *v′*
Leg II					
Larva	–	*d*, *bv″*	(*l*), σ	*l′*, *v′*, φ	(*ft*), (*tc*), (*p*), (*u*), (*a*), *s*, (*pv*), ω_1
Protonymph	–	–	–	–	–
Deutonymph	–	(*l*)	–	*l″*	ω_2
Tritonymph	*v*	–	–	*v″*	(*it*)
Adult	–	*v′*	*v′*	–	–
Leg III					
Larva	–	*d*, *ev′*	*l′*, σ	*v′*, φ	(*ft*), (*tc*), (*p*), (*u*), (*a*), *s*, (*pv*)
Protonymph	–	–	–	–	–
Deutonymph	*v*	*l′* [1]	–	*l′*	–
Tritonymph	*l*	–	–	*v″*	(*it*)
Adult	–	–	–	–	–
Leg IV					
Protonymph	–	–	–	–	*ft′*, (*p*), (*u*), (*pv*)
Deutonymph	–	*d*, *ev′*	*d*	*v′*, φ	(*tc*), (*a*), *s*
Tritonymph	*v*	–	*v′*	*l′*, *v″*	–
Adult	–	–	–	–	–

Note: structures are indicated where they are first added and are present through the rest of the ontogeny; pairs of setae are in parentheses, and a dash indicates no additions. [1] Added in some deutonymphs; if not, this seta is added in the tritonymph.

3.3. *Mitochondrial Genetic Variation*

The Bayesian analysis of the COI sequences revealed two deeply diverged clades of *F. fuscipes* corresponding to exclusively Nearctic and Palearctic populations, respectively (Figure 22), and separated by a 15.5–18.4% uncorrected p-distance. Each regional clade contained 11 haplotypes, with a maximum of 7.2% divergence in the Nearctic clade and 5.8% in the Palearctic clade. Several Palearctic haplotypes were shared between distant locations, including Finnmark and Vestland in Norway, and one haplotype was shared between Finnmark and SW Finnland. Near-identical haplotypes were found in Agder and Nordland (Norway), and in SW Finland and Germany.

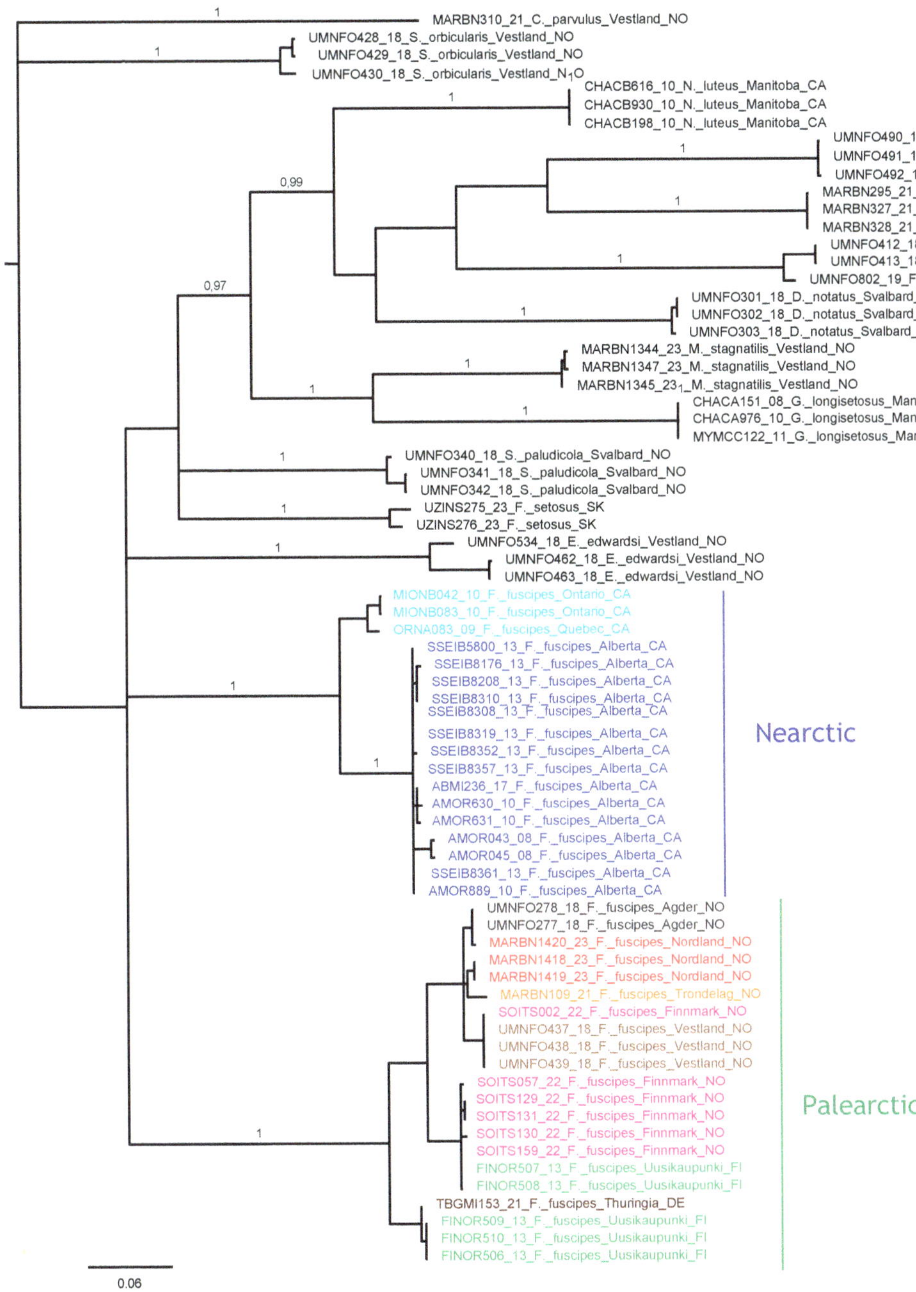

Figure 22. Bayesian tree topology based on COI nucleotide sequences (658 bp). Specimen numbers correspond to those in the BOLD database (http://boldsystems.org/, accessed on 20 November 2023). Information about barcoded specimens is presented in Table 1.

3.4. Ecology and Biology

We collected *F. fuscipes* in *Sphagnum* mosses on the shore of lake Skomakerdiket (Bergen, Norway), where this species achieved a density of 102 individuals per 500 cm^3. In this population, the juveniles made up 52% of all individuals, with the following stage structure: 8 larvae, 33 protonymphs, 4 deutonymphs, 7 tritonymphs, and 50 adults. The female-to-male sex ratio was 1:0.3, and 7% of the females were gravid and carried one or two large eggs (290 × 175), comprising 34% of the length of the females.

4. Discussion

4.1. Morphology and Development

The juvenile stages of *F. fuscipes* from Norway are generally similar to those from Russia [48] and Poland [17], except for the shape of some setae and sclerites and the number of microsclerites in the larvae. In the larvae from Norway, the prodorsal seta *in* is of a medium size and barbed, as in the larvae from Poland [17], whereas in the larvae from Russia [48], this seta is short and smooth. In all regions, the larvae of *F. fuscipes* lack a gastronotal shield, most of the gastronotal setae are located on the microsclerites and other sclerites, and microsclerites are also present on the gastronotum. However, the shape of the macrosclerites differs in these larvae, and the larvae from Norway and Russia have more microsclerites than those from Poland. In the tritonymphs from Norway and Poland, the gastronotal setae are slightly longer than in the specimens from Russia. All these differences probably illustrate regional variation in the species.

Seniczak et al. [14] compared a selection of morphological characters in several *Fuscozetes* species. In light of this comparison and this investigation, the largest is *F. fuscipes*, the smallest is *F. setiger*, and the body length of most species overlaps. These species also differ from one another in the shape of their bothridial seta, translamella, lamellar cusp, and porose area *Aa*, and in the number and shape of their notogastral setae. Most species have 10 pairs of notogastral setae (c_2 is present); two species (*F. novus* Shaldybina, 1969, and *F. tatricus*) have 11 pairs (c_2 and *dp* are present), and *F. setosus* has 10–13 pairs of setae (c_2 and some or all the setae of the *d*-series are present).

Seniczak et al. [14] also compared 23 morphological characters of the larvae and tritonymphs of *F. coulsoni*, *F. fuscipes*, *F. kamchatkicus*, *F. setiger*, *F. setosus*, and *F. tatricus*. The juveniles of *F. fuscipes* are the most similar to those of *F. setosus*, differing from them in six morphological characters, and the most dissimilar from those of *F. setiger*, differing from them in 21 morphological characters. This is not too surprising, in view of the distant phylogenetic relationship to *F. setiger* indicated by our COI data analysis (Figure 22).

The morphological ontogeny of *F. fuscipes* is generally similar to that of *F. coulsoni*, *F. kamchatkicus*, *F. pseudosetosus*, *F. setiger*, *F. setosus*, and *F. tatricus* [8,12–14,17,49–51], except for the shape of the larval gastronotum. In *F. fuscipes*, *F. pseudosetosus*, and *F. tatricus*, the gastronotal shield is absent, but macrosclerites and microsclerites can be present, whereas in the other species, the gastronotal shield and a humeral macrosclerite are present, and sometimes other macrosclerites and microsclerites. Among these species, the ontogeny of the leg setae and solenidia were investigated in detail in *F. coulsoni*, *F. kamchatkicus*, and *F. setiger* [12–14]. The ontogeny of the leg setae and solenidia of *F. fuscipes* studied herein is most similar to that of *F. setiger*, differing from it in two morphological characters, and most dissimilar to that of *F. kamchatkicus*, differing from it in five morphological characters (Table 4). Some leg characters can be diagnostic.

The morphology of the adults and juveniles of *Fuscozetes* is generally similar to those of *Melanozetes* [9–11,52–60], except for the diagnostic characters for these genera. The adults of *Fuscozetes* have 10–13 pairs of notogastral setae, including c_2 and some or all the setae of the *d*-series, whereas those of *Melanozetes* have 14 pairs of setae, including c_2 and c_3 [11,56–59].

Table 4. Comparison of some leg characters in *Fuscozetes fuscipes*, *F. coulsoni*, *F. kamchatkicus*, and *F. setiger*.

Character	*F. fuscipes*	*F. coulsoni*	*F. kamchatkicus*	*F. setiger*
Adult				
Seta v' on genu I and II	Present	Present	Absent	Present
Seta l'' and v' on tarsus I	Present	Absent	Absent	Present
Seta l' on femur III	Present	Absent	Absent	Present
Anteroventral edge on femur II	Pointed	Rounded	Rounded	Rounded
Tritonymph				
Seta v' on genu I and II	Absent	Absent	Absent	Present
Seta l' on femur III	Present	Absent	Absent	Present

The separation of the juveniles of *Fuscozetes* from those of *Melanozetes* is more difficult than the adults. The juveniles of *Fuscozetes* have a generally smaller area of sclerites on the gastronotum that those of *Melanozetes*, except for the larvae of some species. For example, the larva of *Melanozetes avachai* Seniczak et al. 2016 [58] has a weakly developed gastronotal shield, most of its gastronotal setae are located on microsclerites, femora I and II are oval in cross-section, and a large ventral carina is absent, as with that of *F. fuscipes* studied herein. The separation of the nymphs and adults of *Fuscozetes* from those of *Melanozetes* is easier than for the larva, mainly using the length and location of solenidion ω_2 on tarsus I; in *Fuscozetes*, this solenidion is shorter than ω_1 and is placed posterolaterally to ω_1, whereas in *Melanozetes*, solenidion ω_2 is as long as or longer than ω_1 and is placed anteriorly to ω_1.

The three-dimensional SEM figures of *F. fuscipes* correspond well with the line drawings of this species, which are, to some degree, subjective and depend on the technique of preparation and the author. For example, in the larva of *F. fuscipes*, some macrosclerites and microsclerites are observed in different aspects as depressions, which are rarely observed in SEM figures. In the tritonymph, the reticulation of the gastronotum and a humeral sclerite are well observed.

4.2. Ecology and Distribution

Fuscozetes fuscipes has a Holarctic distribution [1]. In Norway, it has been found in moist mosses in the north (Finnmark), west (Vestland), and south [60–63]. *Fuscozetes fuscipes* is a hygrophilous [64,65] or meso-hygrophilous species [66,67]. It inhabits wet tundra [10], *Sphagnum* mosses, and wet habitats [65,68–70] close to pools and lakes [71–74], as well as wet-to-humid forest soils and meadows [45]. In an oligotrophic bog in Norway, it was found only in the lower *Sphagnum* layer, 10–17 cm deep [63,75]. Solhøy Wunderle and Solhøy [76] consider *F. fuscipes* a true arctic and high-mountain species, and Schatz [77] and Murvanidze et al. [45] confirmed the dominance of this species in subalpine zones, while Solhøy found it in an alpine zone in Norway [75]. This species is sensitive to some pesticides [78], but it is cold-tolerant and able to survive winter temperatures of −28 °C [79].

Wallwork [80] investigated various biological aspects of *F. fuscipes* in the laboratory. Adults and nymphs were fed on the macerated leaf tissue of hemlock (*Conium maculatum* L.) and on the moist decaying leaves and petioles of yellow birch (*Betula alleghaniensis* Britt.), and the nymphs also fed on dead mites and springtails. Madge [65] observed no clear response by adults and juveniles to a higher air humidity in similar experiments, whereas in dry air, these wet-adapted mites quickly died. Nevertheless, this species survives much longer in a lower humidity than another hygrophilous species, *H. rufulus* C.L. Koch, 1835, probably due to its thicker waxy cuticle on the body [66,81].

Shaldybina [82] cultivated *F. fuscipes* under laboratory conditions at 18–20 °C; fed it on lichens, mosses, and raw potatoes; and estimated a development time of 86 days for this species. The mean time of the development of successive instars (+ immobile period between stages) was in days: egg, 12; larva, 11.5 + 4; protonymph, 12 + 4; deutonymph, 14.5 + 4; and tritonymph, 17.5 + 6.5. Among the 18 species cultivated by this author, the time of development varied between 40 and 180 days. Under natural conditions in cooler climates, the time of the development of *F. fuscipes* probably lasts longer than 86 days, which limits its population growth. In our investigation, only 7% of the females of this species were gravid, each carried one or two large eggs, and in June, the number of juveniles was approximately similar to that of the adults. In earlier studies carried out in a similar area at two lake shores, also in June, the juveniles were more abundant than the adults and made up 71% and 81% of each local population [83].

There are no specific data on the dispersal of *F. fuscipes*, but it can probably use many passive ways of spreading that are common in oribatid mites. Many oribatids can migrate over large distances with the wind (anemohydrochory) [84–87], via birds [88–92], by water currents or waves (together with the action of wind), or with objects drifting in water [84,86,93], including transport in seawater [94]. Even though Oribatida lack obvious morphological adaptations for active transport by phoresy, it has been demonstrated that they are carried on insects [95–100] and frogs [101].

The several shared or near-identical haplotypes between distant sites sampled in this study indicate high migration rates of this species in Europe, at least in modern times. A similar pattern of haplotype identity was found for *Platynothrus peltifer* (C.L. Koch, 1839) in Western Norway, Belgium, and Germany [102]; *P. punctatus* (L. Koch, 1879) in Svalbard, Western Norway, and Southern Spain [103]; and *Nanhermannia coronata* Berlese, 1913, in Northern, Central, and Southern Norway, Ireland, and Finland [104]. We may hypothesize that their migration in a latitudinal direction is largely influenced by migrating birds. In contrast, the longitudinal separation of subclades in Canada, and between Canada and Scandinavia, in two different clades of *F. fuscipes* indicates much less migration in this direction. The morphology of *F. fuscipes* from Canada has not been studied in detail, including the juvenile stages, so it is not possible to pinpoint any morphological differences between the Nearctic and Palearctic populations. To enable a firm test of species boundaries and taxon validity, more genetic data and morphological studies are needed, because a single and rapidly evolving mitochondrial marker is not sufficiently informative to conclude on such issues.

Furthermore, we note that other oribatid genera have mixed patterns of genetic differentiation across the Atlantic, e.g., *Platynothrus troendelagicus* Seniczak and Seniczak, 2022, which has identical COI haplotypes across Europe (Norway and Ireland) and Canada [105], whereas populations of *P. peltifer* have diverged significantly between Europe, USA, and Japan. It is, therefore, possible that more intensive sampling of *F. fuscipes* will support a similar cryptic-species scenario. In this context, it will be useful to study the ecological traits that may affect long-distance colonization in this species group.

5. Conclusions

1. *Fuscozetes* species are a well-formed morphological group of mites that differ clearly from closely related *Melanozetes* species, both as nymphs and adults. The adults of *Fuscozetes* have fewer notogastral setae (10–13 pairs, including c_2 and some or all the setae of the *d*-series) than those of *Melanozetes* (14 pairs, including c_2 and c_3), and the adults and nymphs have a shorter solenidion ω_2 on tarsus I than ω_1, and it is placed posterolaterally to ω_1. In *Melanozetes*, solenidion ω_2 is as long as or longer than ω_1 and is placed anteriorly to ω_1.
2. *Fuscozetes fuscipes* is a hygrophilous species and prefers wet tundra, *Sphagnum* mosses, and wet habitats close to pools and lakes.
3. Mitochondrial genetic data revealed deeply diverged populations across the Holarctic, a high local and regional genetic diversity, and several examples of haplotypes shared between distant Scandinavian localities, indicating latitudinal long-distance migration.

Author Contributions: Conceptualization, S.S. and A.S.; methodology, S.S., B.H.J. and A.S.; line drawings, S.S.; DNA barcoding, B.H.J. and A.S.; SEM imaging, A.S.; writing—original draft, S.S. and A.S.; writing—review and editing, S.S., B.H.J. and A.S. All authors have read and agreed to the published version of the manuscript.

Funding: Funding was provided by the Norwegian Taxonomy Initiative (grant No. 6-20, 70184243); some sequences were obtained with the grant to NIBIO from the Norwegian Taxonomy Initiative (grant No. 9-21, 70184244). The sequencing was financed by Norwegian Barcode of Life (NorBOL).

Institutional Review Board Statement: Not applicable.

Informed Consent Statement: Not applicable.

Data Availability Statement: The COI sequences will be publicly available in GenBank and in BOLD.

Acknowledgments: We are grateful to all the reviewers for all their suggestions, which considerably increased the scientific value of this paper. We are also grateful to Cornelya F. C. Klütsch (Norwegian Institute of Bioeconomy Research, NIBIO, Svanhovd, Norway) for making sequences of *Fusozetes fuscipes* from Finnmark available; to Steffen Roth (University Museum of Bergen, University of Bergen, Bergen, Norway) for the collection of the Svalbard specimens used in this paper; and Irene Heggstad (Department of Earth Science, University of Bergen, Bergen, Norway) for her professional help with the scanning electron microscopy.

Conflicts of Interest: The authors declare no conflict of interest.

References

1. Subías, L.S. Listado sistemático, sinonímico y biogeográfico de los Ácaros Oribátidos (Acariformes, Oribatida) del mundo (1758–2002). *Graellsia* **2004**, *60*, 3–305, updated 2023. [CrossRef]
2. Shaldybina, E.S. Family Ceratozetidae Jacot, 1925. In *Key to Soil-Inhabiting Mites–Sarcoptiformes*; Ghilarov, M.S., Ed.; Nauka Publishing: Moscow, Russia, 1975; pp. 277–303. (In Russian)
3. Sellnick, M. Formenkreis: Hornmilben, Oribatei. In *Die Tierwelt Mitteleuropas*; Brohmer, P., Ehrmann, P., Ulmer, G., Eds.; Series 3; Quelle und Meyer: Leipzig, Germany, 1928; Volume 9, pp. 1–42.
4. Grandjean, F. Les segments post-larvaires de l'hystérosoma chez les Oribates (Acariens). *Bull. Soc. Zool.* **1939**, *64*, 273–284.
5. Grandjean, F. Formules anales, gastronotiques, génitales et aggénitales du développement numériques des poils chez les Oribates. *Bull. Soc. Zool.* **1949**, *74*, 201–225.
6. Shaldybina, E.S. Some morphological features of the ceratozetidoids (Oribatei). *Notes Gorky Pedagog. Inst. Biol. Ser.* **1972**, *130*, 35–66. (In Russian)
7. Shaldybina, E.S. The postembryonic development of beetle mites of the superfamily Ceratozetoidea Balogh, 1961, and their systematics. 1. In *Acarological Conference, Lecture Theses*; Nauka: Moscow, Russia, 1966; pp. 225–226. (In Russian)
8. Shaldybina, E.S. Moss Mites of the Superfamily Ceratozetoidea (Their Morphology, Biology, System and Role in the Anoplocephalid Epizooties). Ph.D. Thesis, Gorky State Pedagogical Institute, Moscow, Russia, 1969; p. 708. (In Russian)
9. Behan-Pelletier, V.M. Ceratozetidae of the Western North American Arctic. *Canad. Entomol.* **1985**, *117*, 1287–1366. [CrossRef]
10. Behan-Pelletier, V.M. Ceratozetidae (Acari: Oribatei) of the Western North American Subarctic. *Canad. Entomol.* **1986**, *118*, 991–1057. [CrossRef]
11. Seniczak, S.; Behan-Pelletier, V.M.; Solhøy, T. Systematic value of some notogastral setae in adult Sphaerozetinae (Acari, Oribatida, Ceratozetoidea) in the light of ontogenetic studies. *Acarologia* **1990**, *31*, 385–400.
12. Seniczak, S.; Kaczmarek, S.; Seniczak, A. Morphological ontogeny of *Fuscozetes kamchatkicus* sp. nov. (Acari: Oribatida: Ceratozetidae) from Kamchatka Peninsula (Russia), with comments on *Fuscozetes*. *Syst. Appl. Acarol.* **2016**, *21*, 1017–1030. [CrossRef]
13. Seniczak, A.; Seniczak, S. Morphological ontogeny of *Fuscozetes coulsoni* sp. nov. (Acari: Oribatida: Ceratozetidae) from Svalbard, Norway. *Syst. Appl. Acarol.* **2020**, *25*, 680–696. [CrossRef]
14. Seniczak, A.; Seniczak, S. Systematic position of *Fuscozetes setiger* (Acari, Oribatida, Ceratozetidae) in the light of ontogenetic studies. *Syst. Appl. Acarol.* **2022**, *27*, 1454–1474. [CrossRef]
15. Weigmann, G. Hornmilben (Oribatida). In *Die Tierwelt Deutschlands*; Part 76; Dahl, F., Ed.; Goecke & Evers: Keltern, Germany, 2006; pp. 1–520.
16. Norton, R.A.; Ermilov, S.G. Catalogue and historical overview of juvenile instars of oribatid mites (Acari: Oribatida). *Zootaxa* **2014**, *3833*, 1–132. [CrossRef] [PubMed]
17. Seniczak, S. The morphology of juvenile stages of moss mites of the subfamily Sphaerozetinae (Acarida: Oribatida), II. *Ann. Zool.* **1989**, *42*, 237–248.
18. Grandjean, F. Étude sur le développement des Oribates. *Bull. Soc. Zool.* **1933**, *58*, 30–61.
19. Grandjean, F. Les poils des épimières chez les Oribates (Acariens). *Bull. Mus. Nat. Hist. Nat.* **1934**, *6*, 504–512.

20. Grandjean, F. La notation des poils gastronotiques et des poils doreaux du propodosoma chez les Oribates (Acariens). *Bull. Soc. Zool.* **1934**, *40*, 12–44.
21. Grandjean, F. Essai de classification des Oribates (Acariens). *Bull. Soc. Zool.* **1953**, *78*, 421–446.
22. Norton, R.A.; Behan-Pelletier, V.M. Suborder Oribatida. Chapter 15. In *A Manual of Acarology*; Krantz, G.W., Walter, D.E., Eds.; Texas Tech University Press: Lubbock, TX, USA, 2009; Chapter 15; pp. 430–564.
23. Hebert, P.D.N.; Cywinska, A.; Ball, S.L.; de Waard, J.R. Biological identifications through DNA barcodes. *Proc. R. Soc. Lond. Ser. B Biol. Sci.* **2003**, *270*, 313–321. [CrossRef]
24. Folmer, O.; Black, M.; Hoeh, W.; Lutz, R.; Vrijenhoek, R. DNA primers for amplification of mitochondrial cytochrome c oxidase subunit I from diverse metazoan invertebrates. *Mol. Mar. Biol. Biotechnol.* **1994**, 3, 294–299.
25. Seniczak, A.; Seniczak, S.; Jordal, B.H. Integrated taxonomy approach: Molecular data and ontogeny studies clarify systematic status of *Chamobates borealis* (Acari, Oribatida). *Syst. Appl. Acarol.* **2019**, *24*, 2409–2426. [CrossRef]
26. Young, M.R.; Proctor, H.C.; deWaard, J.R.; Hebert, P.D.N. DNA barcodes expose unexpected diversity in Canadian mites. *Mol. Ecol.* **2019**, *28*, 5347–5395. [CrossRef]
27. Roslin, T.; Somervuo, P.; Pentinsaari, M.; Hebert, P.N.; Agda, J.; Ahlroth, P.; Anttonen, P.; Aspi, J.; Blagoev, G.; Blanco, S.; et al. A molecular-based identification resource for the arthropods of Finland. *Mol. Ecol. Resour.* **2022**, *22*, 803–882. [CrossRef] [PubMed]
28. Young, M.R.; Behan-Pelletier, V.M.; Hebert, P.D.N. Revealing the Hyperdiverse Mite Fauna of Subarctic Canada through DNA Barcoding. *PLoS ONE* **2012**, *7*, e48755. [CrossRef]
29. Tamura, K.; Stecher, G.; Kumar, S. MEGA11: Molecular Evolutionary Genetics Analysis Version 11. *Mol. Biol. Evol.* **2021**, *38*, 3022–3027. [CrossRef]
30. Ronquist, F.; Teslenko, M.; van der Mark, P.; Ayres, D.L.; Darling, A.; Höhn, S.; Larget, B.; Liu, L.; Suchard, M.A.; Huelsenbeck, J.P. MrBayes 3.2: Efficient Bayesian phylogenetic inference and model choice across a large model space. *Syst. Biol.* **2012**, *61*, 539–542. [CrossRef]
31. Michael, A.D. *British Oribatidae*; Ray Soc: London, UK, 1884; Volume I, pp. 1–336.
32. Willmann, C. Moosmilben oder Oribatiden (Cryptostigmata). In *Die Tierwelt Deutschlands*; Dahl, F., Ed.; Gustav Fischer: Jena, Germany, 1931; Bd. 22; Volume 5, pp. 79–200.
33. Mehl, R. Checklist of Norwegian ticks and mites (Acari). *Fauna Norv.* **1979**, *26*, 1–45.
34. Karppinen, E.; Krivolutsky, D.A. List of oribatid mites (Acarina, Oribatei) of northern Palaearctic region. I. Europe. *Acta Entomol. Fenn.* **1982**, *41*, 1–18.
35. Golosova, L.D.; Karppinen, E.; Krivolutsky, D.A. List of oribatid mites (Acarina, Oribatei) of northern Palaearctic region. II. Siberia and the Far East. *Acta Entomol. Fenn.* **1983**, *43*, 1–14.
36. Karppinen, E.; Krivolutsky, D.A.; Poltavskaja, M.P. List of oribatid mites (Acarina, Oribatei) of northern Palaearctic region. III. Arid lands. *Ann. Entomol. Fenn.* **1986**, *52*, 81–94.
37. Karppinen, E.; Krivolutsky, D.A.; Tarba, Z.M.; Shtanchaeva, U.Y.; Gordeeva, E.W. List of oribatid mites (Acarina, Oribatei) of northern Palaearctic region. IV. Caucasus and Crimea. *Ann. Entomol. Fenn.* **1987**, *53*, 119–137.
38. Marshall, V.G.; Reeves, R.M.; Norton, R.A. Catalogue of the Oribatida (Acari) of Continental United States and Canada. *Mem. Entomol. Soc. Canada* **1987**, *139*, 1–418. [CrossRef]
39. Olszanowski, Z.; Rajski, A.; Niedbała, W. Roztocze Acari–Mechowce Oribatida. In *Katalog Fauny Polski–Catalogus Faunae Poloniae*; Sorus: Poznań, Poland, 1996; Volume 34, pp. 1–243.
40. Niemi, R.; Karppinen, E.; Uusitalo, M. Catalogue of the Oribatida (Acari) of Finland. *Acta Zool. Fenn.* **1997**, *207*, 1–39.
41. Ryabinin, N.A.; Pankov, A.N. *Catalogue of Oribatid Mites of the Far East of Russia. Part II. Continental Part of the Far East*; DVO: Vladivostok, Russia; Khabarovsk, Russia, 2002; p. 92. (In Russian)
42. Miko, L. Oribatid mites (Acarina, Oribatida) of the Czech Republic. Revised check-list with a proposal for Czech oribatid nomenclature. *Klapalekiana* **2016**, *52*, 1–302.
43. Murvanidze, M.; Mumladze, L. Annotated checklist of Georgian oribatid mites. *Zootaxa* **2016**, *4089*, 1–81. [CrossRef]
44. Schatz, H. Catalogue of oribatid mites (Acari: Oribatida) from Vorarlberg (Austria). *Zootaxa* **2020**, *4783*, 1–106. [CrossRef]
45. Murvanidze, M.; Todria, N.; Maraun, M.; Mumladze, L. Annotated checklist of Georgian oribatid mites–II. *Zootaxa* **2023**, *5227*, 50–62. [CrossRef] [PubMed]
46. Bayartogtokh, B.; Weigmann, G. New and little known species of oribatid mites of the genera *Arthrodamaeus* and *Fuscozetes* (Arachnida: Acari: Oribatida) from Mongolia. *Species Divers.* **2005**, *10*, 75–84. [CrossRef]
47. Bayartogtokh, B. Oribatid mites of Mongolia (Acari: Oribatida). In *Russian Academy of Sciences*; KMK Scientific Press Ltd.: Moscow, Russia, 2010; 400p.
48. Shaldybina, E.S. The postembryonal development of *Fuscozetes fuscipes* (C.L. Koch) (Oribatei, Ceratozetidae). In *Fauna, Systematics, Biology and Ecology of Helminths and Their Intermediate Hosts*; Gorky: Moskow, Russia, 1978; pp. 84–93. (In Russian)
49. Shaldybina, E.S. A new species of the genus *Fuscozetes* (Oribatei, Ceratozetidae). *Zool. J.* **1977**, *56*, 709–713. (In Russian)
50. Shaldybina, E.S. Juvenile instars of Ceratozetoidea (Oribatei). In *Fauna, Classification, Biology and Ecology of Parasitic Worms and their Intermediate Hosts*; Gorky State Pedagogy Institute: Moskow, Russia, 1977; pp. 76–89.
51. Seniczak, S. *Fuscozetes tatricus* n. sp. a New Ceratozetoid Moss Mite (Acari, Oribatida, Ceratozetidae) from Poland. *Zool. Anz.* **1993**, *230*, 169–180.
52. Shaldybina, E.S. The biology of *Melanozetes mollicomus* (Koch) (Oribatei, Ceratozetidae). *Zool. J.* **1967**, *46*, 1659–1667. (In Russian)

53. Behan-Pelletier, V.M. Ceratozetidae (Acari: Oribatida) of arboreal habitats. *Canad. Entomol.* **2000**, *132*, 153–182. [CrossRef]
54. Seniczak, S. The morphology of juvenile stages of moss mites of the subfamily Sphaerozetinae (Acarida: Oribatida), I. *Ann. Zool.* **1989**, *42*, 225–235.
55. Seniczak, S. The morphology of juvenile stages of moss mites of the subfamily Sphaerozetinae (Acari, Oribatida). III. *Zool. Anz.* **1993**, *231*, 25–38.
56. Seniczak, A.; Seniczak, S. Morphological ontogeny of *Melanozetes stagnatilis* (Acari, Oribatida, Ceratozetidae). *Syst. Appl. Acarol.* **2018**, *23*, 652–664. [CrossRef]
57. Seniczak, S.; Seniczak, A.; Kaczmarek, S. Morphological ontogeny of *Melanozetes azoricus* with comments on *Melanozetes* (Acari: Oribatida: Ceratozetidae). *Intern. J. Acarol.* **2015**, *41*, 523–536. [CrossRef]
58. Seniczak, S.; Kaczmarek, S.; Seniczak, A. Morphological ontogeny of *Melanozetes avachai* n. sp., a unique member of *Melanozetes* (Acari: Oribatida: Ceratozetidae). *Acarologia* **2016**, *56*, 463–484. [CrossRef]
59. Seniczak, S.; Seniczak, A.; Kaczmarek, S.; Marquardt, T. Morphological ontogeny of *Melanozetes interruptus* (Acari, Oribatida, Ceratozetidae), and comments on *Melanozetes* Hull. *Syst. Appl. Acarol.* **2023**, *28*, 1897–1913. [CrossRef]
60. Bayartogtokh, B.; Ermilov, S.G.; Shtanchaeva, U.Y.; Subías, L.S. Ontogenetic instars of *Melanozetes paramollicomus* sp. nov., with remarks on morphological ontogeny of Sphaerozetinae (Acari: Oribatida: Ceratozetidae). *Zootaxa* **2021**, *5086*, 69–89. [CrossRef] [PubMed]
61. Thor, S. Übersicht der norwegischen Cryptostigmata mit einzelnen Nebenbemerkungen. *Nyt Mag. Nat.* **1937**, *77*, 275–307.
62. Willmann, C. Oribatiden von der Insel Herdla. *Berg. Mus. Arbok Nat. Rekke* **1929**, *5*, 1–16.
63. Solhøy, T. Oribatids (Acari) from an oligotrophic bog in western Norway. *Fauna Norv. Ser. B.* **1979**, *26*, 91–94.
64. Madge, D.S. The behaviour of free-living mites as affected by humidity (Acarina; Oribatoidea). *Anim. Behav.* **1961**, *9*, 108. [CrossRef]
65. Madge, D.S. The humidity reactions of oribatid mites. *Acarologia* **1964**, *6*, 566–591.
66. Madge, D.S. The longevity of fasting oribatid mites. *Acarologia* **1964**, *6*, 718–729.
67. Ivan, O.; Calugar, A. The fauna of edaphic mites (Acari, Gamasida, Oribatida) in some peat bogs–Protected areas in North Moldavia (Romania). *Ann. Compl. Muz. Bucov.* **2003**, *16–17*, 127–150.
68. Borcard, D. Les Oribates des tourbières du Jura Suisse (Acari, Oribatei). Faunistique VII. Oribatuloidea (*Haplozetidae*). Ceratozetoidea. *Mitt. Schweiz. Entomol. Ges.* **1995**, *68*, 363–372.
69. Irmler, U. Long-term fluctuation of the soil fauna (Collembola and Oribatida) at groundwater-near sites in an alder wood. *Pedobiologia* **2004**, *48*, 349–363. [CrossRef]
70. Weigmann, G. Recovery of the oribatid mite community in a floodplain after decline due to long time inundation. In *Acarine Biodiversity in the Natural and Human Sphere*; Weigmann, G., Alberti, G., Wohltmann, A., Ragusa, S., Eds.; Proceedings of the V Symposium of the European Association of Acarologists, Montpellier, Berlin, 21–25 July 2004. *Phytophaga* **2004**, *14*, 201–208.
71. Seniczak, A.; Seniczak, S.; Maraun, M.; Graczyk, R.; Mistrzak, M. Oribatid mite species numbers increase, densities decline and parthenogens suffer during bog degradation. *Exp. Appl. Acarol.* **2016**, *68*, 409–428. [CrossRef] [PubMed]
72. Seniczak, A.; Seniczak, S.; Iturrondobeitia, J.C.; Solhøy, T.; Flatberg, K.I. Diverse *Sphagnum* mosses support rich moss mite communities (Acari, Oribatida) in mires of western Norway. *Wetlands* **2020**, *40*, 1339–1351. [CrossRef]
73. Seniczak, A.; Seniczak, S.; Graczyk, R.; Kaczmarek, S.; Jordal, B.H.; Kowalski, J.; Djursvoll, P.; Roth, S.; Bolger, T. A forest pool as a habitat island for mites in a limestone forest in southern Norway. *Diversity* **2021**, *13*, 578. [CrossRef]
74. Seniczak, A.; Seniczak, S.; Iturrondobeitia, J.C.; Gwiazdowicz, D.J.; Waldon-Rudzionek, B.; Flatberg, K.I.; Bolger, T. Mites (Oribatida and Mesostigmata) and vegetation as complementary bioindicators in peatlands. *Sci. Total Environ.* **2022**, *851*, 158335. [CrossRef] [PubMed]
75. Solhøy, T. Dynamics of oribatei populations on Hardangervidda. In *Fennoscandian Tundra Ecosystems, Part 2 Animals and Systems Analysis*; Wielgolaski, F.E., Ed.; Springer: Berlin/Heidelberg, Germany; New York, NY, USA, 1975; pp. 60–65.
76. Solhøy Wunderle, I.; Solhøy, T. The fossil oribatid mite fauna (Acari, Oribatida) in late glacial and early holocene sediments in Krakenes Lake, Western Norway. *J. Paleolim.* **2000**, *23*, 35–47. [CrossRef]
77. Schatz, H. Biogeography of oribatid mites (Acari, Oribatida) from the Cordillera de Talamanca, Costa Rica and Panama. In *Acarology XI: Proceedings of the International Congress*; Morales-Malacara, J.B., Behan-Pelletier, V., Ueckermann, E., Pérez, T.M., Estrada, E., Gispert, C., Badii, M., Eds.; Instituto de Biología UNAM., Facultad de Ciencias, UNAM., Sociedad Latinoamericana de Acarología: Mexico City, Mexico, 2007; pp. 151–167.
78. Voronova, L.D. The effect of some pesticides on the soil invertebrate fauna in the South Taiga Zone in the Perm Region (USSR). *Pedobiologia* **1968**, *8*, 507–525. [CrossRef]
79. Sömme, L. Overwintering ecology of alpine collembola and oribatid mites from the Austrian Alps. *Ecol. Entomol.* **1979**, *4*, 175–180. [CrossRef]
80. Wallwork, J.A. Notes on the feeding-behaviour of some forest soil Acarina. *Oikos* **1958**, *9*, 260–271. [CrossRef]
81. Madge, D.S. The effects of lethal temperatures on oribatid mites. *Acarologia* **1965**, *7*, 121–130.
82. Shaldybina, E.S. Cultivation of some oribatid mite species in laboratory conditions with the purpose of studying their life cycles. In *The First All-Union Conference on Zooculture Problems, Moscow (Extended Abstracts)*; Academy of Sciences of the USSR: Moscow, Russia, 1986; pp. 275–277. (In Russian)

83. Seniczak, A.; Solhøy, T.; Seniczak, S.; Riva-Caballero, A. Species composition and abundance of the oribatid fauna (Acari, Oribatida) at two lakes in the Fløyen area, Bergen, Norway. *Biological Lett.* **2010**, *47*, 11–19. [CrossRef]
84. Popp, E. Semiaquatile Lebensräume (Bülten) in Hoch- und Niedermooren. 2 Teil. Die Milbenfauna. *Int. Rev. Ges. Hydrobiol.* **1962**, *47*, 533–579. [CrossRef]
85. Vanschoenwinkel, B.; Gielen, S.; Seaman, M.; Brendonck, L. Any way the wind blows–Frequent wind dispersal drives species sorting in ephemeral aquatic communities. *Oikos* **2008**, *117*, 125–134. [CrossRef]
86. Vanschoenwinkel, B.; Gielen, S.; Vandewaerde, H.; Seaman, M.; Brendonck, L. Relative importance of different dispersal vectors for small aquatic invertebrates in a rock pool metacommunity. *Ecography* **2008**, *31*, 567–577. [CrossRef]
87. Vanschoenwinkel, B.; Gielen, S.; Seaman, M.; Brendonck, L. Wind mediated dispersal of freshwater invertebrates in a rock pool metacommunity: Differences in dispersal capacities and modes. *Hydrobiologia* **2009**, *635*, 363–372. [CrossRef]
88. Lebedeva, N.V.; Krivolutsky, D.A. Birds spread soil microarthropods to Arctic Islands. *Dokl. Biol. Sci.* **2003**, *391*, 329–332. [CrossRef]
89. Lebedeva, N.V.; Lebedev, V.D. Transport of oribatid mites to the polar areas by birds. In *Integrative Acarology*; Bertrand, M., Kreiter, S., McCoy, K.D., Migeon, A., Navajas, M., Tixier, M.S., Vial, L., Eds.; European Association of Acarologists: Montpellier, France, 2008; pp. 359–367.
90. Lebedeva, N.V.; Lebedev, V.D.; Melekhina, E.N. New data on the oribatid mite (Oribatei) fauna of Svalbard. *Dokl. Biol. Sci.* **2006**, *407*, 182–186. [CrossRef]
91. Lebedeva, N.V.; Melekhina, E.N.; Gwiazdowicz, D.J. New data on soil mites in the nests of *Larus hyperboreus* in the Spitsbergen archipelago. *Vestn. Juzn. Naucn. Cent. Ran* **2012**, *8*, 70–75.
92. Coulson, S.J. Association of the soil mite *Diapterobates notatus* (Thorell, 1871) (Acari, Oribatidae) with *Cynomya mortuorum* (Linnaeus, 1761) (Calliphoridae, Calliphorinae): Implications for the dispersal of oribatid mites. *Int. J. Acarol.* **2009**, *35*, 175–177. [CrossRef]
93. Schuppenhauer, M.M.; Lehmitz, R. Floating Islands: A method to detect aquatic dispersal and colonisation potential of soil microarthropods. *Soil Org.* **2017**, *89*, 119–126.
94. Coulson, S.J.; Hodkinson, I.D.; Block, W.; Webb, N.R.; Harrison, J.A. Survival of terrestrial soil-dwelling arthropods on and in seawater: Implications for trans-oceanic dispersal. *Funct. Ecol.* **2002**, *16*, 353–356. [CrossRef]
95. Norton, R.A. Observations on phoresy by oribatid mites (Acari: Oribatei). *Internat. J. Acarol.* **1980**, *6*, 121–129. [CrossRef]
96. Coulson, S.J.; Moe, B.; Monson, F.; Gabrielsen, G.W. The invertebrate fauna of High Arctic seabird nests: The microarthropod community inhabiting nests on Spitsbergen, Svalbard. *Polar Biol.* **2009**, *32*, 1041–1046. [CrossRef]
97. Waleckx, E.; Montalvo-Balam, T.J.; Pinzón-Canul, A.; Arnal, A.; Gerardo, M.; Martínez, P.A. First report of phoresy by an oribatid mite (Acari: Oribatida) on a triatomine bug (Hemiptera: Reduviidae). *Int. J. Acarol.* **2018**, *44*, 210–211. [CrossRef]
98. Ermilov, S.G. Oribatid mites (Acari: Oribatida) phoretic on passalid beetles (Coleoptera: Passalidae), with description of a new species from Indonesia. *Ecol. Montenegrina* **2019**, *22*, 90–96. [CrossRef]
99. Ermilov, S.G.; Frolov, A.V. New and interesting oribatid mites (Acari, Oribatida) phoretic on Aceraius grandis (Coleoptera, Passalidae) from Vietnam. *Syst. Appl. Acarol.* **2019**, *24*, 945–961. [CrossRef]
100. Ermilov, S.G.; Frolov, A.V. New data on oribatid mites (Acari, Oribatida) phoretic on passalid beetles (Coleoptera, Passalidae) from the Afrotropical and Oriental regions, with descriptions of three new species from Congo, Gabon and Ghana. *Syst. Appl. Acarol.* **2021**, *26*, 769–787. [CrossRef]
101. Beaty, L.E.; Esser, H.J.; Miranda, R.; Norton, R.A. First report of phoresy by an oribatid mite (Trhypochthoniidae: *Archegozetes magnus*) on a frog (Leptodactylidae: *Engystomops pustulosus*). *Int. J. Acarol.* **2013**, *39*, 325–326. [CrossRef]
102. Heethoff, M.; Domes, K.; Laumann, M.; Maraun, M.; Norton, R.A.; Scheu, S. High genetic divergences indicate ancient separation of parthenogenetic lineages of the oribatid mite *Platynothrus peltifer* (Acari, Oribatida). *J. Evol. Biol.* **2007**, *20*, 392–402. [CrossRef] [PubMed]
103. Seniczak, S.; Seniczak, A.; Kaczmarek, S.; Marquardt, T.; Ondoño, E.F.; Coulson, S.J. Morphological ontogeny and ecology of *Platynothrus punctatus* (Acari, Oribatida, Camisiidae), with comments on *Platynothrus* Berlese. *Syst. Appl. Acarol.* **2022**, *27*, 551–580. [CrossRef]
104. Seniczak, S.; Seniczak, A. Morphological ontogeny and ecology of a common peatland mite, *Nanhermannia coronata* (Acari, Oribatida, Nanhermanniidae). *Animals* **2023**, *13*, 3590. [CrossRef] [PubMed]
105. Seniczak, A.; Seniczak, S.; Hassel, K.; Flatberg, K.I. Morphological ontogeny of *Platynothrus troendelagicus* sp. nov. (Acari, Oribatida, Camisiidae) from Norway. *Syst. Appl. Acarol.* **2022**, *27*, 1702–1722. [CrossRef]

animals
MDPI

Article

Evaluation of Genetic Diversity in Quill Mites of the Genus *Syringophiloidus* Kethley, 1970 (Prostigmata: Syringophilidae) with Six New-to-Science Species

Eliza Glowska [1,*], Izabella Laniecka [1], Kamila Ostrowska [1], Christina A. Gebhard [2], Julia Olechnowicz [3] and Miroslawa Dabert [3]

1 Department of Animal Morphology, Faculty of Biology, Adam Mickiewicz University in Poznań, Uniwersytetu Poznańskiego 6, 61-614 Poznan, Poland; izabella.laniecka@gmail.com (I.L.); kamromanowska@gmail.com (K.O.)
2 Division of Birds, Smithsonian Institution, MRC 116, P.O. Box 37012, Washington, DC 20013-7012, USA; gebhardc@si.edu
3 Molecular Biology Techniques Laboratory, Faculty of Biology, Adam Mickiewicz University in Poznań, Uniwersytetu Poznańskiego 6, 61-614 Poznan, Poland; julia.olechnowicz@amu.edu.pl (J.O.); miroslawa.dabert@amu.edu.pl (M.D.)
* Correspondence: glowska@amu.edu.pl

Simple Summary: Morphology and barcode data were used to estimate the diversity and genetic variability of fourteen putative species of the genus *Syringophiloidus* Kethley, 1970. In most cases, both sources of information were consistent. The only exception was *S. amazilia* Skoracki, 2017, which according to our results is most likely a population of *S. stawarczyki* Skoracki, 2004, and probably should be treated as its junior synonym. The further findings of our study are six new-to-science species described herein. We indicate that both the host phylogeny and distribution can drive the evolution of quill mites. Our results increase the knowledge of quill mite diversity and provide some premises to formulate and further test evolutionary, ecological, and epidemiological inquiries.

Abstract: Quill mites (Acariformes: Syringophilidae) are poorly explored bird parasites. *Syringophiloidus* Kethley, 1970, is the most specious and widespread genus in this family. It is believed to contain mono-, steno- and poly-xenous parasites and thus seems to be an exemplary for studies on biodiversity and host associations. In this work, we applied the DNA barcode marker (mitochondrial cytochrome c oxidase subunit I gene fragment, COI) to analyze the species composition and host specificity of representatives of fifteen *Syringophiloidus* populations parasitizing fifteen bird species. The neighbor joining analyses distinguished thirteen monophyletic lineages, almost completely corresponding to seven previously known species recognized based on morphological features, and six new-to-science species. The only exception is *S. amazilia* Skoracki, 2017, which is most likely conspecific with *Syringophiloidus stawarczyki* Skoracki, 2004. The intraspecific distances of all species were not higher than 0.9%, whilst the interspecific diversity ranged from 5.9% to 19.2% and 6.3–22.4%, inferred as the distances *p* and K2P, respectively. Although all putative species (except *S. amazilia*) are highly supported, the relationships between them have not been fully resolved and only faintly indicate that both host phylogeny and distributions influence the phylogenetic structure of quill mite taxa.

Keywords: quill mites; bird parasites; molecular taxonomy; DNA barcoding; COI

Citation: Glowska, E.; Laniecka, I.; Ostrowska, K.; Gebhard, C.A.; Olechnowicz, J.; Dabert, M. Evaluation of Genetic Diversity in Quill Mites of the Genus *Syringophiloidus* Kethley, 1970 (Prostigmata: Syringophilidae) with Six New-to-Science Species. *Animals* **2023**, *13*, 3877. https://doi.org/10.3390/ani13243877

Academic Editor: Alexis Ribas

Received: 28 October 2023
Revised: 25 November 2023
Accepted: 6 December 2023
Published: 16 December 2023

1. Introduction

Quill mites (Acariformes: Syringophilidae) are widespread permanent bird ectoparasites. To date, 417 species have been described [1,2], although their actual number is estimated to be several times higher, probably reaching up to 5000 species [3]. Although the knowledge about syringophilid diversity and host associations has been growing recently [4–6], they remain one of the least understood bird parasites. This is due to their small body size, poorly accessible habitats (bird's feather quill), and low prevalence [7]. Further difficulties are caused by weakly informative morphology and relatively few diagnostic characters [8,9]. Moreover, the vast majority of species were described only on the basis of female features, and the consequence is that males, nymphs, and larvae are virtually unidentifiable. To overcome the limitations of morphology, molecular methods have recently come into use in mite taxonomic studies [9–11]. DNA barcoding is an approach employing a short fragment of the mitochondrial cytochrome c oxidase subunit I (COI) sequence. It is commonly used as an effective marker in the process of species identification in many groups of animals (Hebert 2003, 2004) [12,13], including quill mites [14]. Although only a very small fraction of those parasites have been barcoded so far, this approach has proven reliable in such systematic inquiries as female dimorphism or phenotypic plasticity [8,9]. It has also been successfully used for the estimation of host spectrum [15] and cryptic species detection [16].

Precise and unambiguous species diagnosis is crucial for any other research, including that on quill mites' parasitological and epidemiological importance. This is particularly important in the context of recent reports that mites are the host of unique phylogenetic lineages of bacteria of the genera *Wolbachia* and *Spiroplasma*. In addition, the presence of *Bartonella* and *Brucella* taxa has been detected in syringophilids, which makes them potentially important in the process of circulation of pathogens among birds [17].

The *Syringophiloidus* Kethley, 1970, is the most specious and widely distributed genus of quill mites with 48 known species widespread around the world. This taxon has been recorded from 80 avian host species, belonging to 29 families and five orders [1,2]. Since the species of this genus are known to have various (mono-, steno- and poly-xenous) associations with hosts, they seem to be a representative material for research on diversity and host associations.

In this paper, we supplement the morphology with DNA barcode coverage to evaluate the species composition and host specificity of representatives of fifteen *Syringophiloidus* populations parasitizing fifteen selected bird species.

2. Materials and Methods

2.1. Animal Material and Morphological Analysis

Mite material used in the study (Table 1) was acquired from several sources: (i) the collection of feathers deposited in the Smithsonian Institution, National Museum of Natural History, Department of Vertebrate Zoology, Division of Birds, Washington, DC, USA (USNM) (September 2014), and bird specimens originally collected in Gabon (2009), Namibia (2009), and Peru (2009); (ii) the Biocenter Grindel and Zoological Museum (University of Hamburg), and bird specimens originally collected in Tanzania; (iii) mite samples collected in Mexico (field no. SVM 08-0506-1/4) (2008) and Brazil (2010); (iv) mites obtained from dead birds (due to probable collisions with the window glass) found at the AMU campus, Poznań, Poland (2009).

Table 1. Mites and sequences used in the molecular study.

Mite Species	Host Species	Host Order and Family	Location	Specimen and DNA Code	GenBank Access No.
Syringophiloidus calamonastes sp. n.	Barred Wren-Warbler *Calamonastes fasciolatus* (Smith)	Passeriformes: Cisticolidae	Namibia	KR043	OR721880
				KR045	OR721881
S. paludicolae sp. n.	Plain Martin *Riparia paludicola* (Vieiilot)	Passeriformes: Hirundinidae	Namibia	KR055	OR723490
				KR056	OR723491
S. ripariae sp. n.	Bank Swallow *Riparia riparia* (L.)	Passeriformes: Hirundinidae	Poland	EG079	OR723492
				EG080	OR723493

Table 1. *Cont.*

Mite Species	Host Species	Host Order and Family	Location	Specimen and DNA Code	GenBank Access No.
S. atlapetes sp. n.	White-headed Brushfinch *Atlapetes albiceps* (Taczanowski)	Passeriformes: Passerellidae	Peru	EG974	OR827223
				EG975	OR827224
				EG976	OR827229
				EG977	OR827227
S. campephilus sp. n.	Guayaquil Woodpecker *Campephilus gayaquilensis* (Lesson)	Piciformes: Picidae	Peru	EG964	OR723494
				EG971	OR723495
S. mahali sp. n.	White-browed Sparrow-Weaver *Plocepasser mahali* Smith	Passeriformes: Ploceidae	Namibia	KR047	OR827226
				KR048	OR827228
				KR052	OR827225
S. sporophila Skoracki, 2017	Cinnamon-rumped Seedeater *Sporophila torqueola* (Bonaparte)	Passeriformes: Thraupidae	Mexico	EG362	OR829593
				EG363	OR829592
				EG364	OR829594
				EG365	OR829596
				EG366	OR829595
				EG688	OR829597
				EG689	OR829598
				EG690	OR829599
				EG691	OR829601
				EG692	OR829602
* *S. stawarczyki* Skoracki, 2004	Blue Dacnis *Dacnis cayana* (Linnaeus)	Passeriformes: Thraupidae	Brazil	EG854	OR829600
				EG855	OR829606
* *S. amazilia* Skoracki, 2017	White-bellied Emerald *Chlorestes candida* (Bourcier and Mulsant)	Caprimulgiformes: Trochilidae	Mexico	EG880	OR829607
S. picidus Skoracki, Klimovičová, Muchai and Hromada, 2014	Cardinal Woodpecker *Chloropicus fuscescens* (Vieillot)	Piciformes: Picidae	Namibia	KR031	OR730469
				KR033	OR730471
S. plocei Glowska, Broda, Gebhard and Dabert, 2016	Village Weaver *Ploceus cucullatus* (St. Muller)	Passeriformes: Ploceidae	Gabon	GE041	OR829603
				GE042	KU646845.1
	Vieillot's Black Weaver *Ploceus nigerrimus* (Vieillot)	Passeriformes: Ploceidae	Gabon	GE038	OR829605
				GE039	OR829604
S. pseudonigritae Glowska, Dragun-Damian and Dabert, 2012	Gray-headed Social-Weaver *Pseudonigrita arnaudi* (Bonaparte)	Passeriformes: Ploceidae	Tanzania	EG545	OR829610
				EG546	OR829608
				EG547	OR829609
S. glandarii (Fritsch, 1958)	Hooded Crow *Corvus corone cornix* L.	Passeriformes: Corvidae	Poland	EG519	OR829611
				EG522	OR829612
S. parapresentalis Skoracki, 2011	*Redwing Turdus iliacus* L.	Passeriformes: Turdidae	Poland	EG019	OR829613
				EG061	OR829614
				EG062	OR829615
				EG063	OR829616
Stibarokris phoeniconaias Skoracki & OConnor, 2010 outgroup	American flamingo *Phoenicopterus ruber* L.	Galliformes: Phasianidae	Germany	EG642	OR726320

* According to our results, *S. amazilia* and *S. stawarczyki* are conspecific.

Drawings were made with an Olympus BH2/BX41/BX53 microscopes with differential interference contrast (DIC) optics and a camera lucida. All measurements are in micrometers (μm). Idiosomal setation follows that of [18] with modifications adapted for Prostigmata by [19]. The nomenclature of leg chaetotaxy follows that proposed by [20]. The application of this chaetotaxy to Syringophilidae was recently provided by [21] with a few changes by [22]. Latin and common names of the birds follow [23].

Material depositories and abbreviations: AMU—Adam Mickiewicz University, Poznań, Poland; USNM—Smithsonian Institution, National Museum of Natural History, Washington, DC, USA. The voucher slides and corresponding DNA samples are deposited in the collection of the AMU and USNM under the identification numbers indicated below. The sequences are deposited in GenBank under accession nos. specified in Table 1.

2.2. Molecular Data and Analysis

Total genomic DNA was extracted from single specimens using DNeasy Blood & Tissue Kit (Qiagen GmbH, Hilden, Germany) as described by [24]. The COI gene fragment was amplified via PCR with degenerate primers: Aseq01F (GGAACRATATAYTTTATTTTAGA) and Aseq03R (GGATCTCCWCCTCCWGATGGATT) [9]. PCR amplifications were carried out in 10 μL reaction volumes containing 5 μL of Type-it Microsatellite Kit (Qiagen), 0.5 μM of each primer, and 4 μL of DNA template using a thermocycling profile of one cycle of 5 min at 95 °C followed by 35 steps of 30 s at 95 °C, 1 min at 50 °C, and 1 min at 72 °C, with a final step of 5 min at 72 °C. After amplification, PCR products were diluted two-fold with water, and 5 μL of the sample was analyzed via electrophoresis on 1.0% agarose gel. Samples containing visible bands were purified with thermosensitive Exonuclease I and FastAP Alkaline Phosphatase (Fermentas, Thermo Scientific, Waltham, MA, USA). The amplicons (585 bp) were sequenced in one direction using the Aseq01F primer. Sequencing was performed with BigDye Terminator v3.1 on ABI Prism 3130XL Analyzer (Applied Biosystems, Foster City, CA, USA). Sequence chromatograms were checked for accuracy and edited using Geneious R11 (Biomatters Ltd., Auckland, New Zealand).

Phylogenetic associations between the studied taxa were estimated with the neighbor joining (NJ) method implemented in MEGA7 [25]. Support for the recovered trees was evaluated with 1000 (NJ) non-parametric bootstrap replicates [26]. Pairwise distances between nucleotide COI sequences were calculated using Kimura's two-parameter (K2P) and distance *p* models [27] for all codon positions with MEGA7. *Stibarokris phoeniconaias* Skoracki and OConnor, 2010, was chosen as an outgroup to root the tree. Tree visualizations were prepared using tree editing tools in MEGA7 and Figtree v.1.4.2—[28] (http://tree.bio.ed.ac.uk/) URL (accessed on 15 October 2023).

3. Results

3.1. Systematics

Family Syringophilidae Lavoipierre
Subfamily Syringophilinae Lavoipierre
Genus *Syringophiloidus* Kethley

3.2. Molecular Analysis

We provided DNA barcode coverage for the representation of fifteen populations of *Syringophiloidus* ssp. recorded from fifteen bird species. The COI alignment was 552 bp long and comprised 43 sequences of *Syringophiloidus* mites (ingroup) and one sequence of *Stibarokris phoeniconaias* Skoracki & OConnor, 2010 (outgroup). The number of sequences obtained from each mite population varied from 1 to 10. The alignment contained 242 variable sites, 196 of which were parsimony informative.

The neighbor joining phylogenetic analyses (K2P and distance *p*) distinguished thirteen monophyletic lineages, among which seven lineages exactly correspond to seven previously known and species that are morphologically distinguished here. The only exception in the obtained pattern is presented by *S. amazilia*, which is very close to that of *S. stawarczyki* and most likely represents a population or subspecies of this species (Figures A1 and 1).

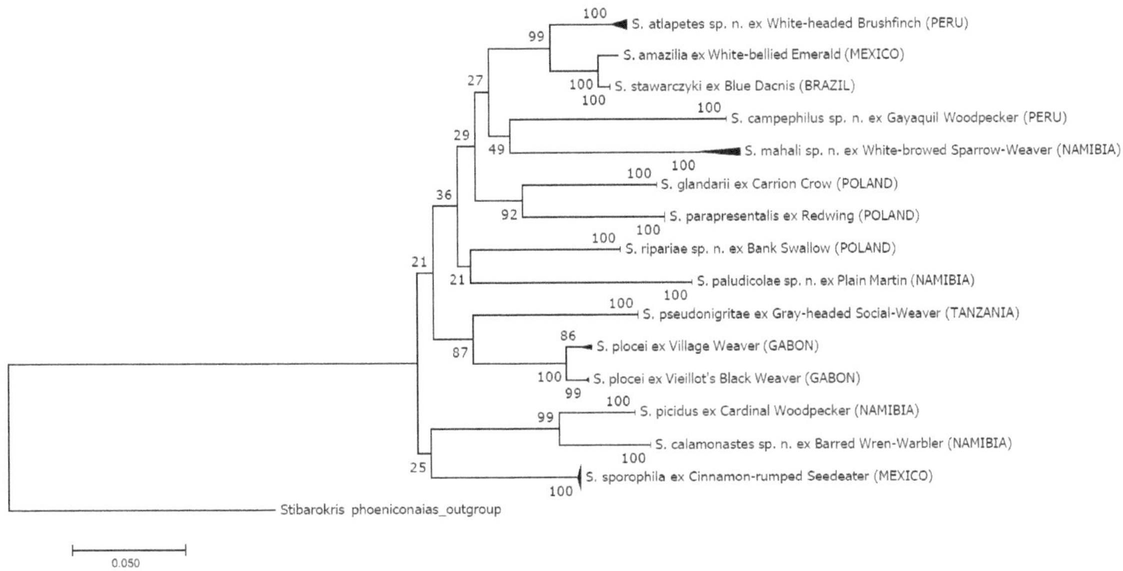

Figure 1. Neighbor joining phylogenetic tree of the *Syringophiloidus* species based on the K2P model. The tree was constructed in Mega v.7. and rooted by *Stibarokris phoeniconaias*.

This assumption is also supported by the genetic distance between the two populations (1.4 and 1.5% of *distance p* and K2P) (Table 2), which is lower than that between bihost *S. plocei* populations (2.1%) and comparable to the previously reported intraspecific values within other quill mites [8]. Although all putative species are highly supported with bootstrap values (100%) and, as they predictably delineate the morphospecies, the relationships between them have not been fully resolved and only weakly suggest various evolutionary scenarios.

The genetic distances were compared at intra- and inter-specific levels according to both the *distance p* and the K2P model. The integrity and separateness of particular taxa were proven for almost all populations resulting in the recognition of seven previously known species and six species new to science. The intraspecific distances of all species were not higher than 0.9%, whilst the interspecific diversity ranged from 5.9% to 19.2% and 6.3–22.4% for genetic distances *p* and K2P, respectively (Table 2).

Table 2. Estimates of evolutionary divergences between COI sequences of *Syringophiloidus* populations based on K2P (and *p*) distances.

Mite Species	Distance *p* (Lower Left) and K2P (Upper Right) (%)																
	Within Groups	Between Groups															
		1.	2.	3.	4.	5.	6.	7.	8.	9.	10.	11.	12.	13.	14.	15.	16.
1. *S. ripariae* sp. n. ex_Bank_Swallow	0.0		16.2	16.0	14.4	16.9	16.3	15.4	15.8	14.2	16.3	17.7	16.0	17.2	17.4	16.2	39.9
2. *S. sporophila* ex Cinnamon-rumped Seedeater	0.2	14.5		17.0	15.6	21.8	15.6	14.2	14.4	16.2	19.6	22.4	15.1	16.6	17.8	17.4	36.8
3. *S. pseudonigritae* ex Gray-headed Social-Weaver	0.0	14.3	15.0		18.7	20.4	15.7	12.6	12.1	18.4	20.2	21.2	19.8	20.2	19.1	18.6	39.4
4. *S. stawarczyki* ex Blue Dacnis	0.0	13.0	14.0	16.3		14.4	6.3	16.2	15.8	1.5	17.8	16.1	15.3	15.5	14.5	13.4	39.5
5. *S. campephilus* sp. n. ex Guayaquil Woodpecker	0.0	15.0	18.7	17.6	12.9		17.3	20.5	21.0	14.6	19.0	19.5	19.0	19.0	19.6	17.5	43.0
6. *S. atlapetes* sp. n. ex White-headed Brushfinch	0.7	14.5	14.0	14.0	5.9	15.1		17.8	17.2	6.3	20.2	18.7	16.4	16.6	15.0	15.2	41.2
7. *S. plocei* ex Village Weaver	0.9	13.9	12.8	11.3	14.5	17.8	15.7		2.1	17.3	18.0	20.1	19.1	19.9	18.3	18.3	37.7
8. *S. plocei* ex Vieillot's Black Weaver	0.2	14.1	13.0	11.0	14.1	18.1	15.2	2.1		16.6	18.0	19.0	19.5	19.6	17.9	18.3	36.2
9. *S. amazilia* ex White-bellied Emerald	n/a	12.9	14.5	16.1	1.4	13.0	5.9	15.3	14.8		18.3	16.0	16.2	15.1	14.2	14.0	39.5
10. *S. paludicolae* sp. n. ex Plain Martin	0.0	14.5	17.2	17.6	15.8	16.7	17.5	15.9	15.9	16.1		21.9	19.0	21.4	18.5	19.8	39.2
11. *S. mahali* sp. n. ex White-browed Sparrow-Weaver	0.0	15.7	19.2	18.4	14.4	17.1	16.4	17.5	16.7	14.4	18.9		18.6	17.6	19.9	20.6	44.6
12. *S. picidus* ex Cardinal Woodpecker	0.0	14.3	13.6	17.2	13.8	16.7	14.6	16.8	17.0	14.5	16.7	16.2		7.3	21.7	19.5	41.4
13. *S. calamonastes* sp. n. ex Barred Wren-Warbler	0.0	15.2	14.8	17.6	13.9	16.7	14.8	17.4	17.1	13.6	18.5	15.5	6.9		19.9	19.2	38.6
14. *S. parapresentalis* ex Redwing	0.0	15.4	15.8	16.7	13.0	17.0	13.5	16.1	15.8	12.9	16.3	17.5	18.7	17.4		12.1	41.2
15. *S. glandarii* ex Hooded Crow	0.0	14.5	15.4	16.3	12.1	15.4	13.6	16.1	16.1	12.7	17.2	17.9	17.0	16.8	11.1		42.3
16. *Stibarokris phoeniconaias* (outgroup)	-	30.4	28.8	30.1	30.3	32.1	31.1	29.3	28.4	30.3	30.1	33.0	31.2	29.7	31.2	31.7	

3.3. *Morphological Systematics*

3.3.1. Descriptions

Syringophiloidus atlapetes sp. n. (Figures 2 and 3)

For females (holotype and three paratypes; range in parentheses) (Figure 2A–E), the total body length is 605 (600–620). For *Gnathosoma,* the infracapitulum is punctate. Each medial and lateral branch of peritremes has 2–3 and 9–11 chambers, respectively (Figure 2C). The stylophore is punctate and has a body length of 165 (150–155). For *Idiosoma,* the propodonotal shield is rounded anteriorly and sparsely punctate on the entire surface. The length ratio of setae *vi:ve:si* is 1:1:3.3–4.6. The hysteronotal shield is clearly visible and punctate in anterior and posterior parts. The pygidial shield is punctate and distinctly sclerotized in the area bearing bases of setae *f1* and *f2*. Setae *f1* and *h1* are subequal in length. The length ratio of setae *ag1:ag2:ag3* is 1.1:1:1.2. For *Legs,* Coxal fields I–IV are sparsely punctate. Setae *3c* is 3.4–3.6 times longer than *3b*. Fan-like setae *p′* and *p″* of legs III–IV have seven tines (Figure 2E). Setae *tc″* is 1.3–1.6 times longer than *tc′*. Lengths of setae are as follows: *vi* 15 (20); *ve* 15 (20); *si* 70 (65–75); *c2* 160 (185); *se* 205 (195–225); *c1* 215; *d2* 175 (165); *d1* 145 (145); *e2* 125 (140–170); *f1* 20 (25); *f2* 180 (205); *h1* 25 (20); *h2* 295 (285); *ag1* 125; *ag2* 115 (125–135); *ag3* 160; *g1*, *g2* 25 (25); *ps1* 12 (12); *ps2* 17 (17), *tc′* (30–40); *tc″* (50); *l′RIII* 35 (40–45); *l′RIV* (25); *3b* 25 (15); *3c* 85 (55–75); *4b* 20 (20); *4c* 85 (55).

For males (paratype) (Figure 3A–E), the total body length is 400. For Gnathosoma, the infracapitulum is apunctate. The stylophore is apunctate and 130 long. Each medial branch of peritremes has four chambers, and each lateral branch has 10 chambers (Figure 3C). For Idiosoma, the propodonotal shield is weakly sclerotized, bearing bases of setae *vi, ve, si, se,* and *c1,* and sparsely punctate near bases of setae *vi, ve* and *si*. Striation is clearly visible on the entire surface. The length ratio of setae *ve:si* is 1:1. The hysteronotal shield is weakly sclerotized, and the striae are visible, not fused to a pygidial shield, and apunctate. Setae *d1, d2,* and *e2* are subequal in length. The pygidial is shield small, restricted to bases of setae *f2* and *h2,* and to the genito-anal region or only to the genito-anal region; it is apunctate. Genital setae *g1* is situated anterior to the level of setae *g2,* and both pairs are subequal in length. Pseudanal setae *ps1* and *ps2* are subequal in length. Length ratios of setae *ag1:ag2* and *f2:h2* are 1.3:1 and 1:11.5, respectively. Coxal fields I–IV are punctate. Setae *3c* is four times longer than *3b*. For legs, fan-like setae *p′* and *p″* of legs III and IV have 6 tines (Figure 3E). The length ratio of setae *tc′III–IV:tc″III–IV* is 1:1.7. The lengths of setae are as follows: *ve* 15, *si* 15, *se* 100, *c1* 100, *c2* 55, *d1* 13, *d2* 13, *e2* 13, *f2* 10, *h2* 115, *ag1* 45, *ag2* 35, *3b* 10, *3c* 40, *l′RIII* 13, *l′RIV 15 tc′III–IV* 15, and *tc″III–IV* 25.

Host and Distribution

Birds of the family Passerellidae: the white-headed brushfinch, *Atlapetes albiceps* (Taczanowski) from Peru.

Type Material

The type material included a female holotype, and seven female and one male paratypes from the quill of the white-headed brushfinch, *Atlapetes albiceps* (Taczanowski) (Passeriformes: Passerellidae), PERU, Tumbes, Parque Nacional Cerros de Amotape, El Platano, 4 07 46 S, 80 37 13 W, 11, 13 July 2009, coll. Milensky, C. M, (USNM 643973). Mites were sampled by Glowska E.; the vouchers and DNA codes are as follows: EG974–977. DNA barcode GenBank accession numbers as specified in Table 1.

Type Material Deposition

The female holotype (USNMENT acc. number: USNMENT01967000) and four paratypes (three females and one male) (USNMENT01967001–USNMENT01967004) are deposited in the USNM, and four female paratypes are deposited in the AMU (EG23-0628-003.01-04).

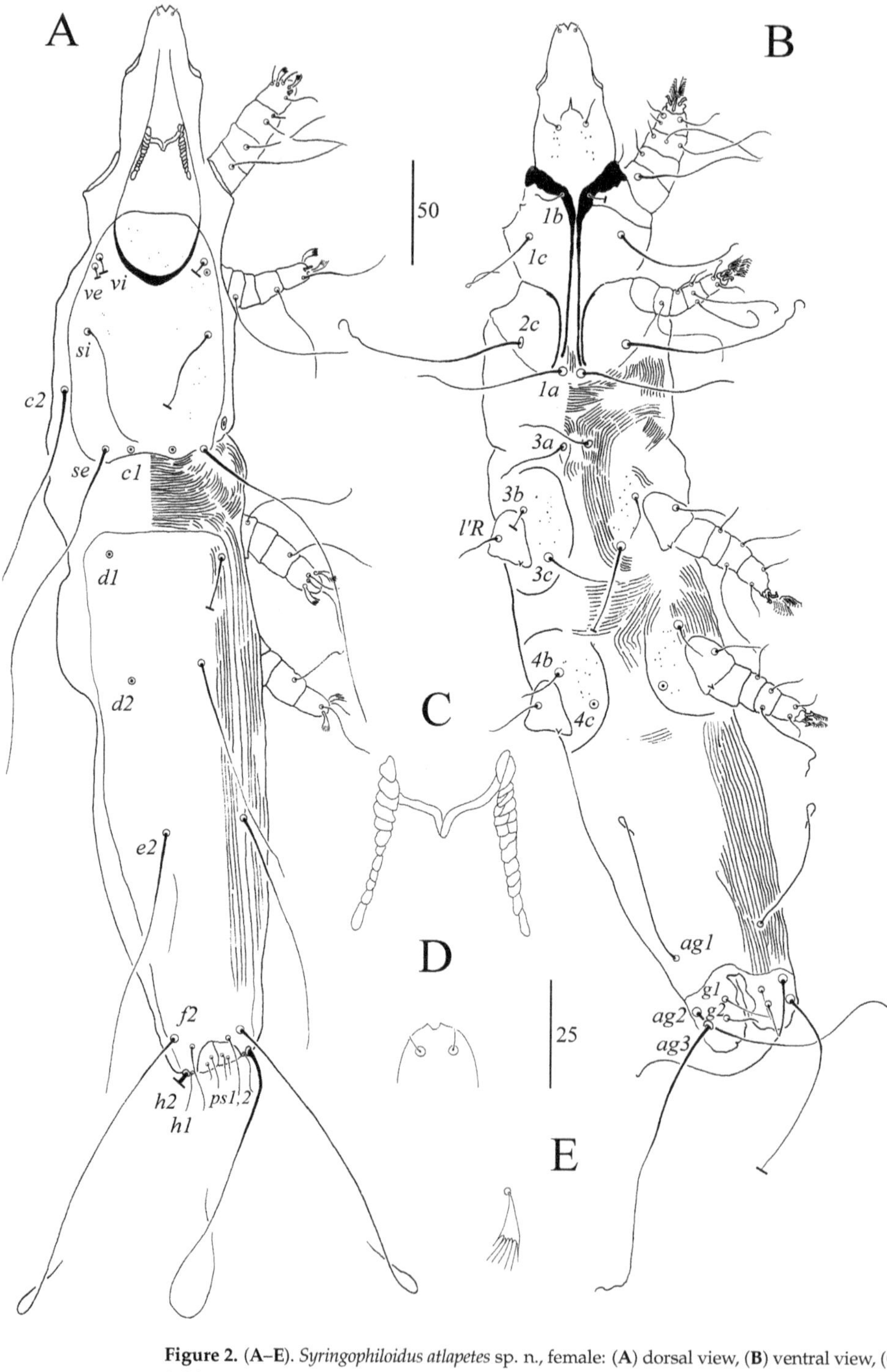

Figure 2. (**A–E**). *Syringophiloidus atlapetes* sp. n., female: (**A**) dorsal view, (**B**) ventral view, (**C**) peritremes, (**D**) hypostomal apex, and (**E**) fan-like setae *p′* of leg III. Scale bars: (**A**,**B**) = 50 μm; (**C–E**) = 25 μm.

Differential Diagnosis

Syringophiloidus atlapetes sp. n. is morphologically most similar to *S. stawarczyki* Skoracki, 2004, described from the golden-rumped euphonia *Euphonia cyanocephala* (Vieillot) (Passeriformes: Fringillidae) and additionally recorded from the white-lined tanager, *Tachyphonus rufus* (Boddaert), and the blue dacnis, *Dacnis cayana* (L.) (Passeriformes: Thraupidae) [29,30]. Females of both species have a similar number of peritremal chambers, punctate dorsal shields, and Coxal fields I–IV, as well as similar or nearly coinciding lengths of most setae. Females of *S. atlapetes* sp. n. differ from those of *S. stawarczyki* in terms of a stylophore length of 150–165 (vs. that of 170–195 in *S. stawarczyki*), lengths of the setae *se* of 200–225 (vs. 165–170) and *d2* 165–175 (vs. 115–125), and the sparse punctation of the dorsal and Coxal shields (vs. dense punctation). The genetic distance between both species is 6.3% of K2P and 5.9% of distance *p*. *S. atlapetes* sp. n. is also very similar to *S. coccothraustes* Skoracki, 2011, described from the hawfinch *Coccothraustes coccothraustes* (L.) (Passeriformes: Fringillidae) [22]. Females of both species have a similar number of peritremal chambers, punctate dorsal shields, and Coxal fields I-IV, as well as coinciding lengths of most setae. Females of *S. atlapetes* sp. n. differ from those of *S. coccothraustes* in terms of the lengths of setae *vi* of 15–20 (vs. 25–35), *ve* of 15–20 (vs. 25–35), *h2* of 285–295 (vs. 305–330), *g1* and *g2* of 25 (vs. 35–40), and *tc″* of 50 (vs. 70).

Etymology

The name is taken from the generic name of the host and is a noun in apposition.

Syringophiloidus calamonastes sp. n. (Figure 4A–E)

Female (holotype and 7 paratypes; range in parentheses). Total body length 715 (645). *Gnathosoma*. Infracapitulum apunctate. Each medial and lateral branch of peritremes with 6–8 and 8–10 chambers, respectively (Figure 4D). Stylophore apunctate, 155 (155) long. *Idiosoma*. Propodonotal shield rounded anteriorly and apunctate. Length ratio of setae *vi:ve:si* 1:1.4–2.3:2.2–3.5. Hysteronotal shield strongly sclerotized and apunctate, fused to pygidial shield. Pygidial shield punctate, distinctly sclerotized in the area bearing bases of setae *f1* and *f2*. Setae *h1* 1.2–1.5 longer than *f1*. Length ratio of setae *ag1:ag2:ag3* 1–1.2:1:1.3–1.6. Genital plate present, bearing bases of setae *ag2* and *ag3*. *Legs*. Coxal fields I–IV apunctate. Setae *3c* 2.6–3.2 times longer than *3b*. Fan-like setae *p′* and *p″* of legs III–IV with 6–7 tines (Figure 4E). Setae *tc″* 2–2.8 times longer than *tc′*. *Lengths of setae*: *vi* 25 (20–25); *ve* 35 (35–45); *si* 55 (55–70); *c2* 155 (130–170); *se* 170 (165–205); *c1* 200 (170–180); *d2* 145 (130–180); *d1* 105 (105–145); *e2* 155 (145–155); *f1* 20 (25); *f2* 145 (160); *h1* 30 (30–35); *h2* 340 (260–295); *ag1* 125 (115–130); *ag2* 110 (95–130); *ag3* 160 (150–165); *g1*, *g2* 30 (30–35); *ps1,2* 15 (15–20); *tc′* 20 (20–25); *tc″* 55 (45–60); *l′RIII* 35 (25); *l′RIV* 35 (35); *3b* 25 (20–25); *3c* 65 (65–75); *4b* 35 (25–30); *4c* 60 (70–80).

Male: not found.

Host and Distribution

Birds of the family Cisticolidae: southern barred warbler *Calamonastes fasciolatus* (Smith) from Namibia.

Type Material

Female holotype and 10 female paratypes from the quill of the Southern Barred Warbler *Calamonastes fasciolatus* (Smith) (Passeriformes: Cisticolidae), NAMIBIA, Erongo, Tubusis, 21 39 46 S, 15 23 44 E, 4 Sep 2009, bird specimen coll. Gebhard, C. A. (USNM 642616), mites sampled by Glowska E. (15 Sep 2013); vouchers and DNA codes: KR043, KR045; DNA barcode GenBank accession numbers as specified in Table 1.

Type Material Deposition

Female holotype (USNMENT acc. number: USNMENT01967005) and 5 paratypes (USNMENT01967006–USNMENT01967010) are deposited in the USNM and 5 female paratypes in the AMU (EG23-0628-001.01–05).

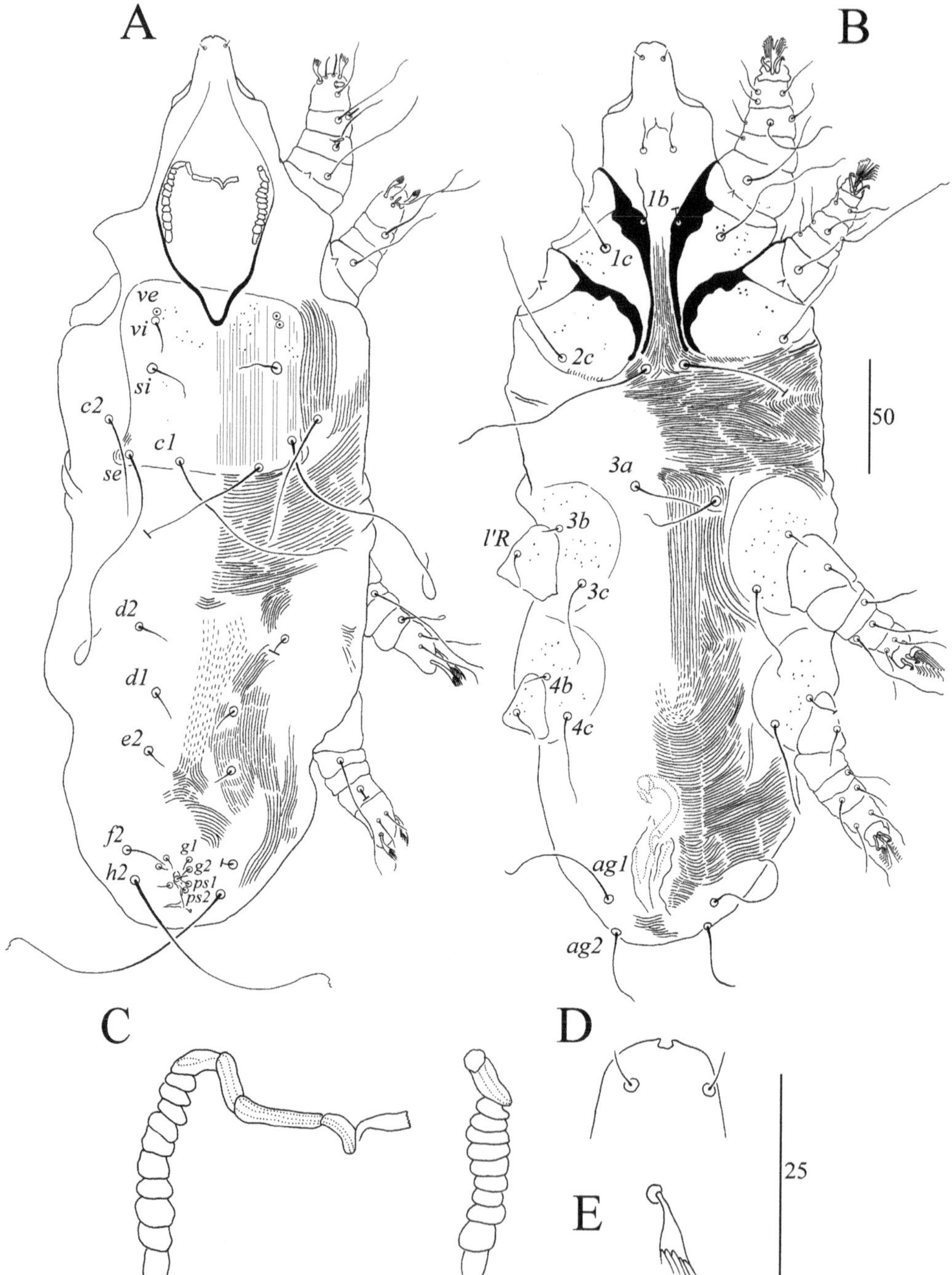

Figure 3. (**A–E**). *Syringophiloidus atlapetes* sp. n., male: (**A**) dorsal view, (**B**) ventral view, (**C**) peritremes, (**D**) hypostomal apex, and (**E**) fan-like setae *p′* of leg III. Scale bars: (**A**,**B**) = 50 μm; (**C–E**) = 25 μm.

Differential Diagnosis

Syringophiloidus calamonastes sp. n. is morphologically most similar to *S. picidus* Skoracki, Klimovičová, Muchai and Hromada, 2014 described from the cardinal woodpecker *Dendropicos fuscescens* (Vieillot) (Piciformes: Picidae) and recorded in Kenya, Tanzania and Uganda [31]. Females of both species have a similar number of peritremal chambers, all propodonotal setae serrated, fused, posteriorly punctate hystero-pygidial shield, and pseudanal setae *ps1* and *ps2* subequal in length. Females of *S. calamonastes* sp. n. differ from *S. picidus* by apunctate infracapitulum, propodonotal shield and coxal fields (vs. punctate in *S. picidus*), length ratio of setae *vi:ve:si* 1:1.4–2.3:2.2–3.5 (vs. 1:1.2–1.3:1.7–2.3) and equal genital setae (vs. *g1* 1.2 longer than *g2*). Our molecular analysis revealed that both species differ by 7.3% of K2P (and 6.9% of distance *p*). *S. calamonastes* sp. n. is also very similar to *S. minor* (Berlese, 1887) described from the house sparrow *Passer domesticus* (L.) (Passeriformes: Passeridae) from Europe and additionally recorded from several species and localities around the world [22]. Females of both species have a similar number of peritremal chambers, fused hysteronotal and pygidial shields, fan-like setae *p′* and *p″* of legs III–IV with 6–7 tines and same lengths of most setae. Females of *S. calamonastes* sp. n. differ from *S. minor* by apunctate infracapitulum and propodonotal shield (vs. punctate in *S. minor*) and the lengths of setae *se* 165–205 (vs. 150–160), *e2* 145–155 (vs. 105–135), *tc′* 20–25 (vs. 40), and *tc″* 45–60 (vs. 75–80).

Etymology

The name is taken from the generic name of the host and is a noun in apposition.

Syringophiloidus campephilus sp. n. (Figure 5A–E)

Female (holotype and 7 paratypes; range in parentheses). Total body length 665 (650–655). *Gnathosoma.* Infracapitulum apunctate or sparsely punctate. Each medial and lateral branch of peritremes with 2–3 and 8–10 chambers, respectively (Figure 5C). Stylophore apunctate, 150 (150) long. *Idiosoma.* Propodonotal shield weakly sclerotized punctate around bases of setae *ve*. Length ratio of setae *vi:ve:si* 1:1–1.6:1–1.6. Hysteronotal shield weakly sclerotized and apunctate, fused to pygidial shield. Pygidial shield sparsely punctate. Setae *f1* and *h1* subequal in length. Length ratio of setae *ag1:ag2:ag3* 1:1.1–1.9:1.5–1.9. *Legs.* Coxal fields I–II sparsely punctate, III–IV punctate. Setae *3c* 3.3–5 times longer than *3b*. Fan-like setae *p′* and *p″* of legs III–IV with 6–7 tines (Figure 5E). Setae *tc″* 1.4–1.6 times longer than *tc′*. *Lengths of setae*: *vi* 15 (15); *ve* 25 (20–25); *si* 25 (15–20); *c2* 145 (145–170); *se* 180 (195); *c1* 180 (170–195); *d2* 15 (10–15); *d1* 70 (65–80); *e2* (70–90); *f1* 15 (10–20); *f2* 205 (215–235); *h1* 15 (15–20); *h2* 285; *ag1* 65 (55–60); *ag2* 70 (75–105); *ag3* 100 (105); *g1*, *g2* 15 (15); *ps1* 15 (15); *tc′* 25 (20–25); *tc″* 35 (30–40); *l′RIII* 25 (20–25); *l′RIV* 25 (15–20); *3b* 10 (10–15); *3c* 50 (35–50); *4b* 10 (10–15); *4c* 50 (35–50).

Host and Distribution

Birds of the family Picidae: guayaquil woodpecker *Campephilus gayaquilensis* (Lesson) from Peru.

Type Material

Female holotype and 7 female paratypes from the quill of the guayaquil woodpecker *Campephilus gayaquilensis* (Lesson) (Piciformes: Picidae) (USNM 643881), PERU, Tumbes, El Caucho Biological Station, 3 49 25 S, 80 15 37 W, 9 Jun 2009, bird coll. Vargas, W.; mites sampled by Glowska E.; vouchers and DNA codes: EG964, EG971; DNA barcode GenBank accession numbers as specified in Table 1.

Type Material Deposition

Female holotype (USNMENT acc. number: USNMENT01967011) and 3 paratypes (USNMENT01967012–USNMENT01967014) are deposited in the USNM and 4 female paratypes in the AMU (EG23-0628-004.01–04).

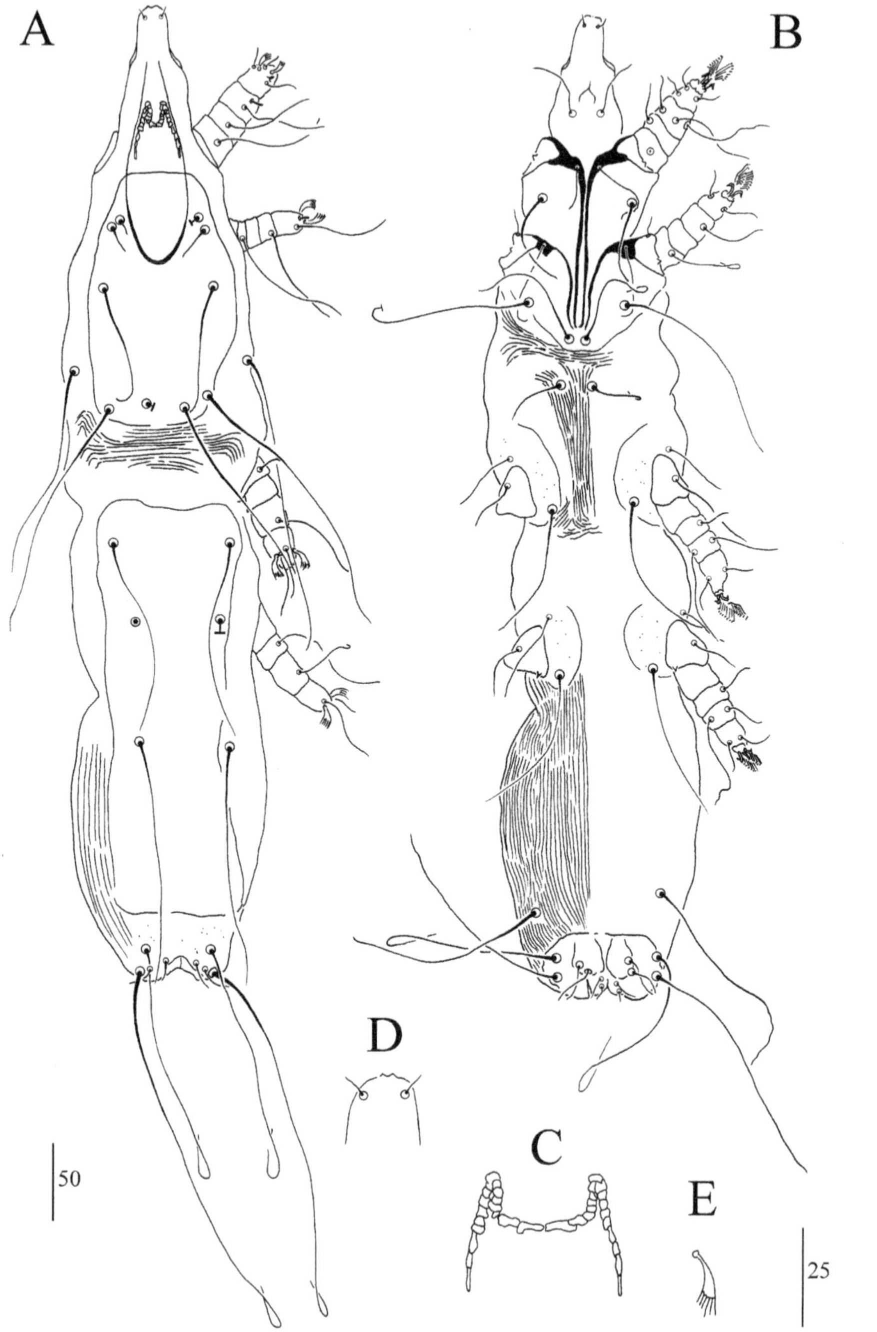

Figure 4. (**A**–**E**). *Syringophiloidus calamonastes* sp. n., female: (**A**)—dorsal view, (**B**)—ventral view, (**C**)—peritremes, (**D**)—hypostomal apex, (**E**)—fan-like setae p' of leg III. Scale bars: (**A**,**B**) = 50 μm; (**C**–**E**) = 25 μm.

Differential Diagnosis

Syringophiloidus campephilus sp. n. is morphologically most similar to *S. atlapetes* sp. n. described from the white-headed brushfinch *Atlapetes albiceps* (Taczanowski) (Passeriformes: Passerellidae) from Peru. Females of both species have a similar number of peritremal chambers, punctate pygidial shields and fan-like setae with 6–7 tines. Females of *S. campephilus* sp. n. differ from *S. atlapetes* sp. n by the apunctate stylophore (vs. punctate in *S. atlapetes*) and lengths of setae *si* 15–25 (vs. 65–75), *d2* 10–15 (vs. 165–175), *d1* 65–80 (vs. 145), *e2* 70–90 (vs. 125–170), *ag1* 55–65 (vs. 125) and *ag3* 100–105 (vs. 160). The genetic distance between these species equals 17.3% of K2P (and 15.1% of distance *p*). *S. campephilus* sp. n. is also very similar to *S. dendrocittae* Fain, Bochkov and Mironov, 2000 described from the rufous treepie *Dendrocitta rufa* Baker (Passeriformes: Corvidae) from East Asia [32]. Females of both species have a similar number of peritremal chambers and fan-like setae *p′* and *p″* of legs III–IV with 6–8 tines. Females of *S. campephilus* sp. n. differ from *S. dendrocittae* by the lengths of the setae *vi* 15 (vs. 24), *ve* 20–25 (vs. 45), *d2* 10–15 (vs. 94), *d1* 65–80 (vs. 157), *e2* 70–90 (vs. 132), *ag1* 55–65 (vs. 128–157), *ag2* 70–105 (vs. 135), *ag3* (vs. 166), *g1,2* 15 (vs. 33), *ps1,2* 15 (vs. 27).

Etymology

The name is taken from the generic name of the host and is a noun in apposition.

Syringophiloidus mahali sp. n. (Figure 6A–E)

In terms of females (a holotype and six paratypes; range in parentheses), the total body length is 785 (715–785). For *Gnathosoma*, the infracapitulum is sparsely punctate. Each medial and lateral branch of peritremes has 5–6 and 10–11 chambers, respectively (Figure 6C). The stylophore is apunctate and 185 (170–180) long. For *Idiosoma*, the propodonotal shield is anteriorly concave and apunctate. The length ratio of setae *vi:ve:si* is 1:1.1–1.3:1.1–1.8. The hysteronotal shield is apunctate and fused to the pygidial shield. The pygidial shield is distinctly sclerotized and punctate in the area bearing bases of setae *f1* and *f2*. Setae *h1* is 1.2–1.3 times longer than *f1*. The length ratio of setae *ag1:ag2:ag3* is 1–1.2:1–1.2:1.1–1.3. Setae *ps2* is 1.3–1.4 longer than *ps1*. Setae *g1* and *g2* are subequal in length. For *Legs*, Coxal fields I–IV are sparsely punctate. Setae *3c* is 3–3.6 times longer than *3b*. Fan-like setae *p′* and *p″* of legs III–IV have six to seven tines (Figure 6E). Setae *tc″* is 1.5–2 times longer than *tc′*. Lengths of setae are as follows: *vi* 30 (25–30); *ve* 40 (30–35); *si* 55 (35–45); *c2* 155 (135–160); *se* 185 (175–195); *c1* 170 (180); *d2* 150 (155–180); *d1* 130 (145–155); *e2* 150 (155–170); *f1* 20 (15–20); *f2* 120 (130–150); *h1* 25 (20–25); *h2* 285 (290–295); *ag1* 150 (110–130); *ag2* 120 (115–130); *ag3* 130 (145–150); *g1*, *g2* 25 (20); *ps1* 15 (10–15); *ps2* 20 (15–20); *tc′* 35 (20–30); *tc″* 55 (40–45); *l′RIII* 30 (30–35); *l′RIV* 30 (25); *3b* 25 (25); *3c* 80 (75–90); *4b* 25 (20–25); *4c* 90 (85–65).

Host and Distribution

Birds of the family Ploceidae: the white-browed sparrow-weaver, *Plocepasser mahali* Smith from Namibia.

Type Material

The type material consisted of a female holotype and six female paratypes from the quill of the white-browed sparrow-weaver, *Plocepasser mahali* Smith (Passeriformes: Ploceidae) (USNM 642639), Namibia, Hardap, Aukens, 25 09 03 S, 16 32 00 E, 29 Aug 2009, coll. Mughongora, V. K. Mites were sampled by Glowska E.; vouchers and DNA codes are as follows: KR047-048 and KR052. DNA barcode GenBank accession numbers are specified in Table 1.

Type Material Deposition

A female holotype (USNMENT acc. number: USNMENT01967015) and three paratypes (USNMENT01967016–USNMENT01967018) are deposited in the USNM; three female paratypes are deposited in the AMU (EG23-0628-005.01–03).

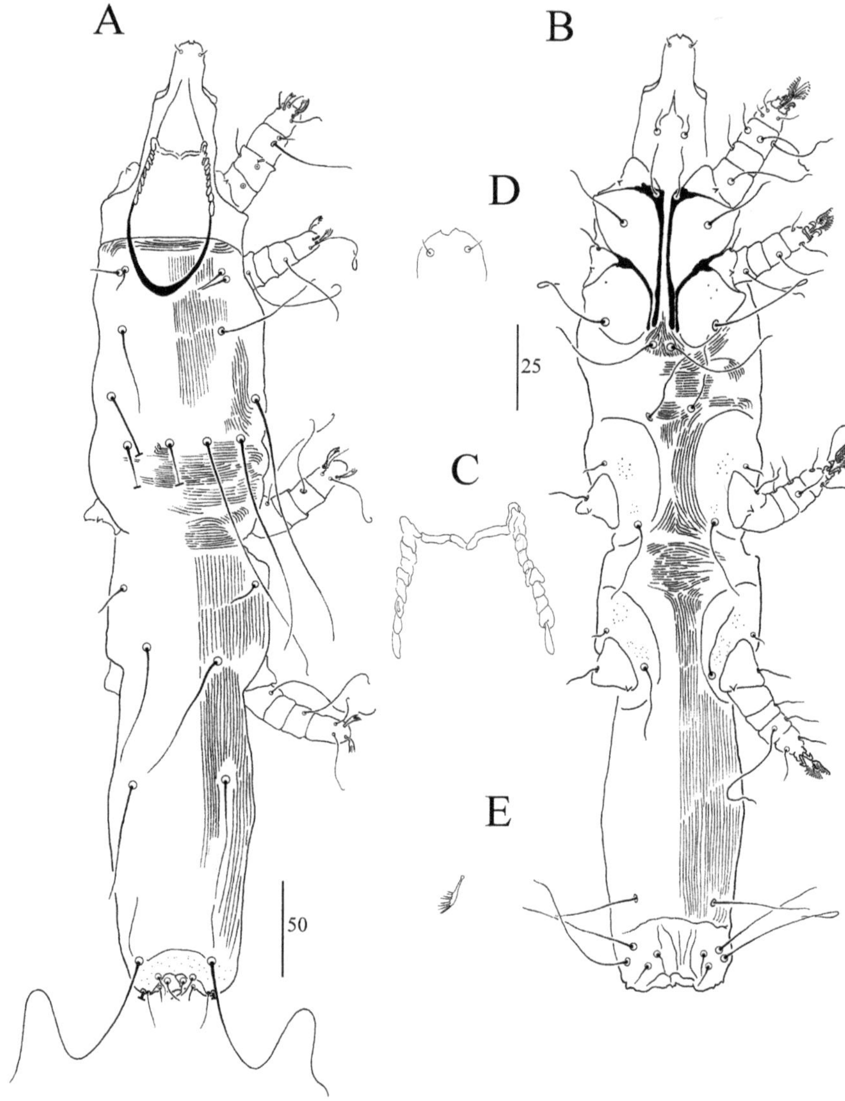

Figure 5. (**A–E**). *Syringophiloidus campephilus* sp. n., female: (**A**)—dorsal view, (**B**)—ventral view, (**C**)—peritremes, (**D**)—hypostomal apex, (**E**)—fan-like setae *p′* of leg III. Scale bars: (**A**,**B**) = 50 µm; (**C–E**) = 25 µm.

Differential Diagnosis

Syringophiloidus mahali sp. n. is morphologically most similar to *S. picidus* Skoracki, Klimovičová, Muchai and Hromada, 2014, described from the cardinal woodpecker *Dendropicos fuscescens* and recorded in Kenya, Tanzania and Uganda [31]. Females of both species have a similar number of peritremal chambers, a hysteronotal shield fused to the pygidial shield, a punctate pygidial shield in the posterior part, and punctate Coxal fields I–IV. Females of *S. mahali* sp. n. differ from those of *S. picidus* in terms of the length of the stylophore, which is 170–185 (vs. 155–170 in *S. picidus*), the setae *ps2*, which is 1.3–1.4 longer than *ps1* (vs. *ps1,2* subequal in length), and the lengths of setae *si*, which are 35–55 (vs. 60–80), of *c1*, which are 170–180 (vs. 210–215), of *f2*, which are120–150 (vs. 150–180), and of *h2*, which are 285–295 (vs. 315–395). *S. mahali* sp. n. is also very similar to *S. philomelosus* Skoracki, 2011, described from the song thrush *Turdus philomelos* Brehm (Passeriformes: Turdidae) from Jordan [22]. Females of both species have a similar number of chambers in the lateral branches, fan-like setae *p′* and *p″* of legs III–IV with six to seven tines and lengths of most setae. Females of *S. mahali* sp. n. differ from those of *S. philomelosus* in terms of the number of chambers of the medial branch of peritremes (5–6 and 8–10 in *S. mahali* sp. n. and *S. philomelosus*, respectively), fused hysteronotal and pygidial shields (vs. not fused) and lengths of setae *c1* of 170–180 (vs. 220–225), *c2* of 135–160 (vs. 175–180), *f1* of 15–20 (vs. 30), *f2* of 120–150 (vs. 190–200), *h2* of 285–295 (vs. 345), *tc′* of 20–35 (vs. 40–45), and *tc″* of 40–55 (vs. 65).

Etymology

The name is taken from the generic name of the host and is a noun in apposition.

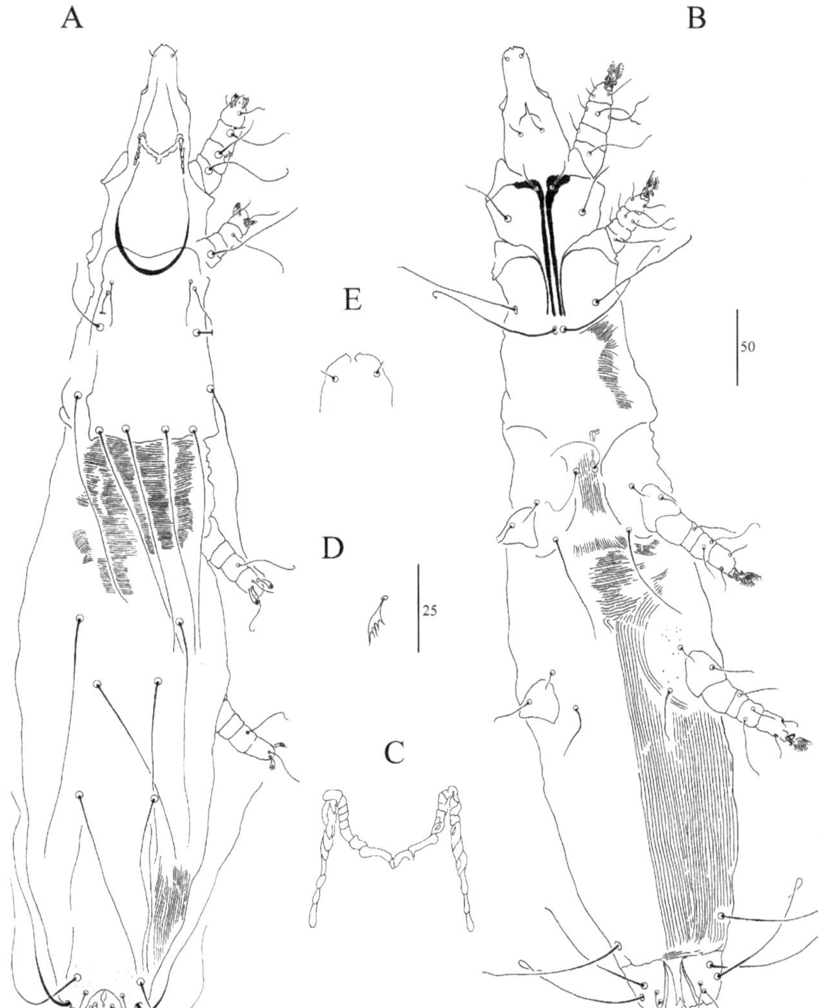

Figure 6. (**A–E**). *Syringophiloidus mahali* sp. n., female: (**A**) dorsal view, (**B**) ventral view, (**C**) peritremes, (**D**) hypostomal apex, and (**E**) fan-like setae *p′* of leg III. Scale bars: (**A,B**) = 50 μm; (**C–E**) = 25 μm.

Syringophiloidus paludicolae sp. n. (Figure 7A–E)

In terms of females (a holotype and seven paratypes; range in parentheses), the total body length is 800 (770–835). For *Gnathosoma,* the infracapitulum is sparsely punctate. Each medial and lateral branch of peritremes has one to two and seven to eight chambers, respectively (borders are poorly marked) (Figure 7C). The stylophore is apunctate and has a length of 160 (150–155). For *Idiosoma,* the propodonotal shield is weakly sclerotized and apunctate. The length ratio of setae *vi:ve:si* is 1:1.5–1.6:5.3–6. The hysteronotal shield is weakly sclerotized (striation is clearly visible on the entire surface) and apunctate. The pygidial shield is distinctly sclerotized and sparsely punctate in the area bearing bases of setae *f1* and *f2,* while the upper part is weakly sclerotized. Setae *h1* is 1.1 times longer than *f1.* The length ratio of setae *ag1:ag2:ag3* ois 1.1–1.3:1:1.1–1.6. For *Legs,* Coxal fields I–IV are sparsely punctate. Setae *3c* is 2.5–2.9 times longer than *3b*. Fan-like setae *p′* and *p″* of legs III–IV have seven to eight tines (Figure 7E). Setae *tc″* is 1.6–2.4 times longer than *tc′*. Lengths of setae are as follows: *vi* 30 (30); *ve* 50 (45); *si* (160–180); *c2* 230 (220–250); *se* 270 (255–265); *c1* 270 (270–285); *d2* 115 (105); *d1* 195 (170–180); *e2* 180 (160–180); *f1* 35 (35–40); *f2* (305–330); *h1* 40 (40–45); *h2* 420 (400–410); *ag1* 175 (170–195); *ag2* 155 (130–175); *ag3* 205 (195–230); *g1, g2* 40 (40–50); *ps1* 20 (25); *ps2* 35 (40); *tc′* 35 (25–35); *tc″* 50 (55–60); *l′RIII* 55 (40–55); *l′RIV* 30 (25–35); *3b* 40 (40–50); *3c* 110 (115–125); *4b* 35 (30–45); *4c* 120 (95–135).

Male: not found.

Host and Distribution

Birds of the family Hirundinidae: the plain martin, *Riparia paludicola* (Vieiilot) from Namibia.

Type Material

The type material consists of a female holotype and seven female paratypes from the quill of the plain martin, *Riparia paludicola* (Vieiilot) (Passeriformes: Hirundinidae) (USNM 642532), NAMIBIA, Karas, Sandfontein near Orange River, 28 51 45 S, 18 33 08 E, 17 Aug 2009, bird specimen coll. Gebhard C. A., mites are sampled by Glowska E.; vouchers and DNA codes are as follows: KR055-056. DNA barcode GenBank accession numbers are specified in Table 1.

Type Material Deposition

A female holotype (USNMENT acc. number: USNMENT01967019) and three paratypes (USNMENT01967020–USNMENT01967022) are deposited in the USNM, and four female paratypes are deposited in the AMU (EG23-0628-002.01–04).

Differential Diagnosis

Syringophiloidus paludicolae sp. n. is morphologically most similar to *S. tarnii* Skoracki and Sikora, 2002, described from the huet huet *Pteroptochos tarni* (King) (Passeriformes: Rhinocryptidae) from Argentina [33]. Females of both species have a punctate infracapitulum, a weakly sclerotized and apunctate hysteronotal shield, fan-like setae *p′* and *p″* of legs III–IV with six to eight tines, and similar lengths of most setae. Females of *S. paludicolae* sp. n. differ from those of *S. tarnii* in terms of the number of peritremal chambers, i.e., one to two and seven to eight in medial and lateral branches (vs. three to four and nine), and lengths of setae *c2* of 220–250 (vs. 155–205), *se* of 255–270 (vs. 165–225), *c1* of 270–285 (vs. 190–240), *d1* of 170–195 (vs. 125–145), *e2* of 160–180 (vs. 115–155), and *f2* of 305–330 (vs. 250–280). *S. paludicolae* sp. n. is also very similar to *S. ripariae* sp. n., described from the sand martin, *Riparia riparia* (L.) (Passeriformes: Hirundinidae) from Poland (p.p.). Females of both species are similar in length and weakly sclerotized, have a similar number of peritremal chambers, and setae *g1* and *g2* that are subequal in length. Females of *S. paludicola* sp. n. differ from those of *S. ripariae* sp. n. in terms of the lengths of setae *ve* of 45–50 (vs. 35 in *S. paludicolae*), *d2* of 105–115 (vs. 180), *h1* of 40–45 (vs. 30), and *ps2* of 35–40 (vs. 25). The genetic distance between these species is 15.3% of K2P (and 13.7 of distance *p*).

Etymology

The name is taken from the specific name of the host and is a noun in the genitive case.

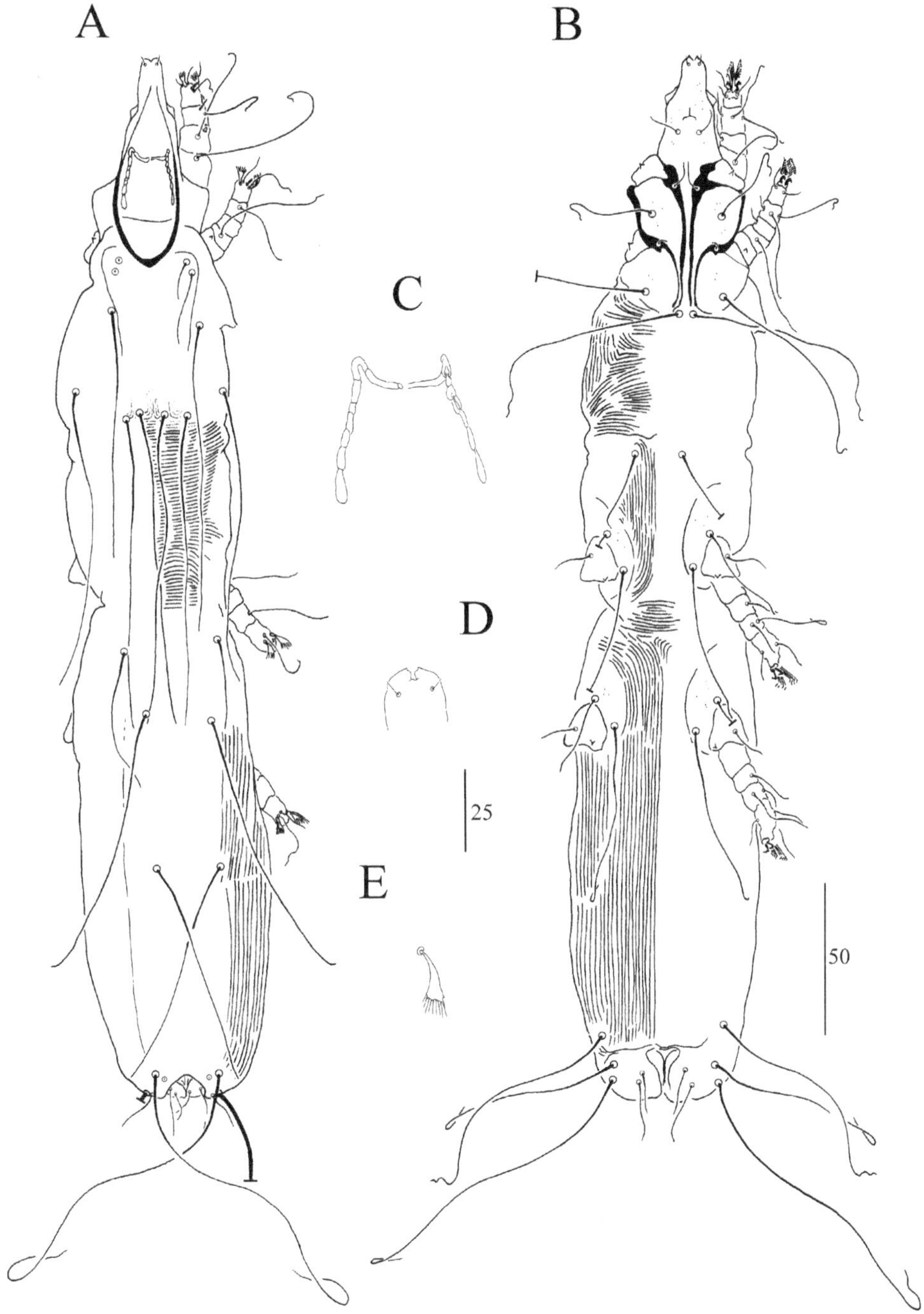

Figure 7. (**A**–**E**). *Syringophiloidus paludicolae* sp. n., female: (**A**) dorsal view, (**B**) ventral view, (**C**) peritremes, (**D**) hypostomal apex, and (**E**) fan-like setae *p′* of leg III. Scale bars: (**A**,**B**) = 50 μm; (**C**–**E**) = 25 μm.

Syringophiloidus ripariae sp. n.

Female (holotype). Total body length 820. *Gnathosoma*. Infracapitulum punctate. Each medial and lateral branch of peritremes with 2 and 7 chambers, respectively. Stylophore apunctate, 180 long. *Idiosoma*. Propodonotal shield weakly sclerotized and apunctate. Length ratio of setae *vi:ve* 1:1.4. Hysteronotal shield weakly sclerotized (striation visible on the entire surface) and apunctate. Pygidial shield sparsely punctate, distinctly sclerotized in the area bearing bases of setae *f1* and *f2*, upper part weakly sclerotized. Setae *f1* 1.1 longer than *h1*. Length ratio of setae *ag1:ag2:ag3* 1.3–1:1.7 *Legs*. Coxal fields I sparsely punctate, III–IV apunctate. Setae *3c* 1.4 times longer than *3b*. Fan-like setae *p′* and *p″* of legs III–IV with 6 tines. *Lengths of setae*: *vi* 25; *ve* 35; *c2* 205; *se* 260; *c1* 275; *d2* 180; *d1* 180; *e2* 180; *f1* 32; *f2* 300; *h1* 28; *h2* 355; *ag1* 170; *ag2* 130; *ag3* 220; 40; *ps1* 20, *ps2* 25; *l′RIII* 45; *l′RIV* 40; *3b* 55; *3c* 80; *4b* 35; *4c* 100.

Male: not found.

Host and Distribution

Birds of the family Hirundinidae: the sand martin *Riparia riparia* (L.) from Poland.

Type Material

The type material was a female holotype from the quill of the sand martin, *Riparia riparia* (L.) (Passeriformes: Hirundinidae), Poznań, POLAND, 52.4672007265976, 16.924954974622207, April 2009, coll. Glowska E.; the voucher and DNA code are as follows: EG079 (holotype). DNA barcode GenBank accession numbers are specified in Table 1.

Type Material Deposition

The holotype was accidentally crushed after species diagnosis was carried out and before the specimen was drawn.

Differential Diagnosis

Syringophiloidus ripariae sp. n. is morphologically most similar to *S. tarnii* Skoracki and Sikora, 2002, described from the huet huet *Pteroptochos tarnii* (Passeriformes: Rhinocryptidae) from Argentina [33]. Females of both species have a punctate infracapitulum, a weakly sclerotized and apunctate hysteronotal shield, fan-like setae *p′* and *p″* of legs III–IV with six tines, and similar lengths of most setae. Females of *S. ripariae* sp. n. differ from those of *S. tarnii* in terms of the number of peritremal chambers, i.e., two and seven in the medial and lateral branches (vs. 3–4 and 9), and lengths of setae *se* of 260 (vs. 165–225), *c1* of 275 (vs. 190–240), *d2* of 180 (vs. 130), *d1* of 180 (vs. 125–145), *e2* of 180 (vs. 115–155), and *ag3* of 220 (vs. 145–185). *Syringophiloidus ripariae* sp. n. is also very similar to *S. paludicolae* sp. n. described from the plain martin, *Riparia paludicola* (Vieillot), from Namibia (p.p.). See the *S. paludicolae* sp. n. differential diagnosis that is given above.

Etymology

The name is taken from the specific name of the host and is a noun in the genitive case.

3.3.2. Other Species

Syringophiloidus amazilia Skoracki, 2017

Syringophiloidus amazilia Skoracki, 2017: 181.

Type host: *Chlorestes candida* (Bourcier and Mulsant) (Apodiformes: Trochilidae). Type locality: Mexico.

Host and Distribution

Birds of the family Trochilidae: the white-bellied emerald, *Amazilia candida* (Bourcier and Mulsant), from Mexico [34].

Material Examined

The material examined included one female from the quill of the white-bellied emerald, *Amazilia candida* (Bourcier and Mulsant) (Apodiformes: Trochilidae), Mexico, Veracruz, Los Tuxtlas, 9 May 2008, coll. S.V. Mironov (SVM 08-0509-8/4). Specimen vouchers and DNA codes are as follows: EG880. DNA barcode GenBank accession nos. are given in Table 1.

Material Deposition

Material deposited in the AMU (EG23-0628-008.01).

Remark

Our results revealed that *S. stawarczyki* and *S. amazilia* are conspecific, and as a consequence, *S. amazilia* could be treated as a junior synonym of *S. stawarczyki*. Although our results are precise, they are based on a relatively small sample. This is due to the limited availability of the mite material. For this reason, we do not formally synonymize these species, but only formulate a premise for further systematic research on the populations covering a more significant number of individuals.

Syringophiloidus glandarii (Fritsch,1958)

Syringophilus minor glandarii Fritsch, 1958: 235.
Syringophilus glandarii as incertae sedis Kethley 1970: 65.
Syringophiloidus glandarii Bochkov and Mironov 1998: 14.
Type host: *Garrulus glandarius* L. (Passeriformes: Corviidae)
Type locality: Germany.

Host and Distribution

Birds of the family Corvidae: the eurasian jay, *Garrulus glandarius* (L)., eurasian magpie, *Pica pica* (L.), eurasian jackdaw, *Corvus monedula* L., rook *Corvus frugilegus* L. [22], American crow, *Corvus brachyrhynchos* Brehm, steller's jay, *Cyanocitta stelleri* (Gmelin) [35], and hooded crow, *Corvus corone* L. (p.p.) from Germany [36], Russia, Kazakhstan, Japan [22], USA [35], and Poland (p.p.).

Material Examined

Four females from the quill of the hooded crow *Corvus corone cornix* L. (Passeriformes: Corvidae) were used, and the material obtained from dead birds (due to probable collisions with a window) was found on the AMU campus, Poznań, Poland (8 May 2009), coll, Glowska E. Specimen vouchers and DNA codes are as follows: EG519 and EG522. DNA barcode GenBank accession nos. are given in Table 1.

Material Deposition

Material is deposited in the AMU (EG23-0628-13.01–04).

Syringophiloidus parapresentalis Skoracki, 2011

Syringophiloidus parapresentalis Skoracki, 2011: 63.
Type host: *Turdus merula* L. (Passeriformes: Turdidae)
Type locality: Poland.

Host Range and Distribution

Birds of the family Turdidae: the Eurasian blackbird, *Turdus merula* L., fieldfare, *T. pilaris* L., black-throated thrush, *T. atrogularis* Jarocki [22], and redwing *T. iliacus* L. ([22], p.p.) from Slovakia, Kazakhstan, Russia and Jordan [22], and Poland ([22], p.p.).

Material Examined

Five females from the quill of the redwing *Turdus iliacus* L. (Passeriformes: Turdidae) made up the material examined, and the material was obtained from dead birds (due probable collision with glass) found on the AMU campus, Poznań, Poland (16 July 2009), coll, Glowska E. Specimen vouchers and DNA codes are as follows: EG019, EG061-063. DNA barcode GenBank accession nos. are given in Table 1.

Material Deposition

Material is deposited in the AMU (EG23-0628-14.01–05).

Syringophiloidus picidus Skoracki, Klimovičová, Muchai and Hromada, 2014

Syringophiloidus picidus Skoracki, Klimovičová, Muchai and Hromada, 2014: 184.
Type host: *Dendropicos fuscescens* (Vieillot) (Piciformes: Picidae)
Type locality: Kenya.

Host and Distribution

Birds of the family Picidae: the cardinal woodpecker, *Dendropicos fuscescens* (Vieillot), from Kenya, Tanzania, Uganda [31], Namibia (p.p).

Material Examined

Two females from the quill of the cardinal woodpecker, *Dendropicos fuscescens* (Vieillot) (Piciformes: Picidae), NAMIBIA, 14 August 2009, Karas, Oas, 27 29 43 S, 19 13 14 E, bird coll. Gebhard C. A., mite coll. Glowska E (USNM 642511), were examined. Specimen vouchers and DNA codes are as follows: KR031; KR033. DNA barcode GenBank accession nos. are given in Table 1.

Material Deposition

One female deposited in the USNM (USNMENT acc. number: USNMENT01967023) and one female in the AMU (EG23-0628-009.01).

Syringophiloidus plocei Glowska, Broda, Gebhard and Dabert, 2016

Syringophiloidus plocei Glowska, Broda, Gebhard and Dabert, 2016: 563.
Type host: *Ploceus cucullatus* (St. Muller) (Passeriformes: Ploceidae).
Type locality: Gabon.

Host and Distribution

Birds of the family Ploceidae: the village weaver, *Ploceus cucullatus* (Müller), and Vieillot's black weaver, *Ploceus nigerrimus* Vieillot [15].

Material Examined

Four females from the quill of the village weaver, *Ploceus cucullatus* (St. Muller) (Passeriformes: Ploceidae) GABON, Ogooue Maritime Province, Gamba Complex of Protected Areas, near the mouth of Nyanga River, 22 October 2009, bird host coll. C.A. Gebhard, were sampled; mites were sampled by E. Glowska (September 2014) (USNM 642906). Four females from the Vieillot's black weaver, *Ploceus nigerrimus* Vieillot (Ploceidae), GABON, Estuaire Province, Cap Esterias, National Forestry School (ENEF), 3 November 2009, bird host coll. C.A. Gebhard, were also sampled; mites were sampled by E. Glowska (USNM 642955). Specimen vouchers and DNA codes are as follows: GE038-039; GE041-042. DNA barcode GenBank accession nos. are given in Table 1.

Material Deposition

Two females from each species (the village weaver and the Vieillot's black weaver) are deposited in the USNM (USNMENT acc. number: USNMENT01967024–USNMENT01967027) and in the AMU (EG23-0628-011.01–04).

Syringophiloidus pseudonigritae Glowska, Dragun-Damian and Dabert, 2012

Syringophiloidus pseudonigritae Glowska, Dragun-Damian and Dabert, 2012.
Type host: *Pseudonigrita arnaudi* (Bonaparte) (Passeriformes: Ploceidae).
Type locality: Tanzania.

Host and Distribution

Birds of the family Ploceidae: the grey-headed social weaver, *Pseudonigrita arnaudi* (Bonaparte), from Tanzania (Glowska et al., 2012) [10].

Material Examined

Four females from the quill of the frozen specimen of the grey-headed social weaver, *Pseudonigrita arnaudi* (Bonaparte) (Passeriformes: Ploceidae), were examined; the bird host was initially collected from the wild in Tanzania and imported to Hamburg in 1990 where it was housed in the Biozentrum Grindel and Hamburg Zoological Museum in the University of Hamburg, Germany, coll. E. Glowska, November 2010.

Specimen vouchers and DNA codes are as follows: EG545-547. DNA barcode GenBank accession nos. are given in Table 1.

Material Deposition

Material is deposited in the AMU (EG23-0628-12.01–04).

Syringophiloidus sporophila Skoracki, 2017

Syringophiloidus sporophila Skoracki, 2017: 184.

Type host: *Sporophila torqueola* (Bonaparte) (Passeriformes: Thraupidae). Type locality: Mexico.

Host and Distribution

Birds of the family Thraupidae: the cinnamon-rumped seedeater, *Sporophila torqueola* (Bonaparte), from Mexico (Skoracki 2017) [34].

Material Examined

Ten females from the quill of the cinnamon-rumped seedeater, *Sporophila torqueola* (Bonaparte) (Passeriformes: Thraupidae), Mexico, Veracruz, Los Tuxtlas, 6 May 2008, coll. S.V. Mironov (SVM 08-0506-1/4), were used. Specimen vouchers and DNA codes are as follows: EG362–366, and EG688–692. DNA barcode GenBank accession nos. are given in Table 1.

Material Deposition

Material deposited in the AMU (EG23-0628-006.01–10).

Syringophiloidus stawarczyki Skoracki, 2004

Syringophiloidus stawarczyki Skoracki, 2004: 291.

Type host: *Euphonia cyanocephala* (Vieillot) (Passeriformes: Emberizidae). Type locality: Brazil.

Host and Distribution

Birds of the families Emberizidae and Thraupidae: the golden-rumped euphonia, *Euphonia cyanocephala* (Vieillot) (type host), white-lined tanager, *Tachyphonus rufus* (Boddaert) [29], and blue dacnis, *Dacnis cayana* (L.) [30].

Material Examined

Two females from the quill of the blue dacnis *Dacnis cayana* (Linnaeus) (Passeriformes: Thraupidae), Brazil, Minas Gerais, Nova Lima, APP do Condomínio Miguelão, 20°07′17.2″ S 43°58′03.1″ W, 8 September 2010, coll. S.V. Mironov, F.A. Hernandes & M.P. Valim (field no. SVM 10-0908-1–2), were examined. Specimen vouchers and DNA codes are as follows: EG854–855. DNA barcode GenBank accession nos. are given in Table 1.

Material Deposition

Materials are deposited in the AMU (EG23-0628-007.01–02).

4. Discussion

Both topologies of the phylogenetic trees and genetic distances revealed thirteen strongly supported monophyletic lineages which are in most cases in accordance with the morphological identifications. The only exception is *S. amazilia*, which very close to the *S. stawarczyki* clade and most likely represents a population of this species. This result is further supported by the genetic distance between the two lineages (1.4% and 1.5% of distances *p* and K2P, respectively), which is lower than that between bihost *S. plocei* populations (2.1%) and comparable to the previously reported intraspecific values within other quill mites [8,16]. Also, a morphological analysis of the type material of both species showed that they are almost indistinguishable and share most diagnostic characteristics (both qualitative and quantitative). The differences between the alleged "species" are very subtle and manifest only in the length of setae *d2* (135–170 in females of *S. amazilia* vs. 115–125 in *S stawarczyki*), *f2* (175 vs. 220), *ag1* (105–120 vs. 130–135), and *ag2* (100–110 vs. 125–135). It is very likely that the differences are caused by the fact that both species were described based on a few specimens only (seven and three females of *S. amazilia* and *S. stawarczyki*, respectively) [29,34]. This is a common practice when researchers work with hard-to-reach and low-prevalence material. It seems, however, that more individuals' availability would fill the metric data gap between *S. amazilia* and *S. stawarczyki* and show the continuity of the divergent characters. The presence of the same mite species on two phylogenetically distant hosts (representatives of different orders, i.e., Apodiformes and Passeriformes) can be explained by horizontal transfer since the ranges of both hosts overlap in Central America. At the moment, we do not have sufficient data to point the direction of the transfer. To carry this out, more individuals representing more populations of both hosts should be analyzed The cases of the host switching of quill mites have already been reported and our result supports the earlier assumption that this phenomenon is not incidental but rather one of the possible scenarios for the dispersion and evolution of this group of parasites [16,37].

In all other cases, the analysis of molecular data (NJ and genetic distances) confirmed the morphological separateness of previously known and newly described species. The intraspecific distances of all tested taxa were not higher than 0.9% and were comparable to the interpopulation values, i.e., 1.5% between *S. plocei* from the vieillot's black weaver and village weaver. All these values are similar to those previously observed in other stenoxenous quill mites (0.0–2.3) [8,14]. Also, interspecific diversity, which ranged from 5.9% to 19.2% and 6.3–22.4% based on distance *p* and K2P, respectively (Table 2), is comparable to that among the species in other previously barcoded syringophilid genera [16].

Although all putative species (except *S. amazilia*) are highly supported with bootstrap values (100%), the relationships between them have not been fully resolved and only faintly indicate that both the host phylogeny and distributions may influence the phylogenetic structure of mites. For example, *S. ripariae* sp. n. from Poland and *S. paludicolae* sp. n. from Namibia were both recorded from hirundinid birds. Their populations show clear intraspecific integrity as well as species separateness measured via genetic distance (16.3% and 14.5% of K2P and *p*, respectively). Even though both species come from geographically distant locations, they form a sister group on the phylogenetic tree. This may suggest a parallel evolution of mites with avian hosts. Another example of a co-phylogenetic relationship is shown by *S. plocei* found on two ploceid species in Namibia. This clade forms a sister group with *S. pseudonigritae*, a parasite of another ploceid bird, the grey-headed social weaver in Tanzania. This result confirmed our earlier observations for these taxa (Glowska et al. 2016) [15]. Another factor that may shape the phylogenetic structure of mites is geographical distribution. Two species, *S. glandarii* and *S. parapresentalis*, form a statistically well-supported sister group. Although they were obtained from birds from different families (Corvidae and Turdidae, respectively), they have a common location (Poland). Analogously, two species parasitize separate bird orders, *S. calamonastes* sp. n. and *S. picidus* form the "Namibian cluster". The same can be observed with the clearly distinct clade represented by mites from Mexico and South America (*S. atlapetes*; *S. Stawarczyki-S. amazilia*).

In this work, we used morphological and barcode data to estimate the diversity and genetic variability of fifteen populations of the genus *Syringophiloidus*. In most cases, both sources of information were consistent. The only exception was *S. amazilia*, which seems to be a population of *S. stawarczyki* and formally should be treated as its junior synonym. The further findings of our study are six now-to-science species, described herein. We indicate that both host phylogeny and distribution can drive the evolution of quill mites. However, we treat our results as a starting point for further in-depth research on these issues. Our results increase the knowledge about mite diversity and demonstrate the usefulness of the parallel use of morphological and molecular methods in solving systematic puzzles in this group of parasites.

5. Conclusions

Even though there has been progress in understanding quill mite systematics, little is known about their global diversity and host associations. This is mainly due to the weakly informative morphology and relatively few diagnostic characters. To address this challenge, a combination of classical morphology and DNA barcodes is used to increase the efficiency of species identification. This approach has been proven to be a reliable tool for this purpose, regardless of sex or developmental stage. It is also helpful for estimating genetic diversity and host specificity issues or revealing phenomena resulting from the incorrect interpretation of morphological characters, such as phenotypic plasticity, polymorphisms, or cryptic species.

Accurate species diagnosis is essential for further research, particularly in understanding quill mites' epidemiological importance. Recent reports suggest that mites host unique phylogenetic lineages of bacteria, such as *Wolbachia* and *Spiroplasma*. Additionally, they are believed to spread diseases by ingesting food (sucking the host's bodily fluids), although their epidemiological significance has not yet been well studied. Our findings contribute to knowledge about mite diversity and provide a basis for further evolutionary, ecological, and epidemiological investigations.

Author Contributions: Conceptualization, E.G. and M.D.; data curation, E.G.; formal analysis, E.G.; funding acquisition, E.G.; investigation, E.G., I.L., K.O., J.O. and M.D.; methodology, E.G. and M.D.; project administration, E.G.; resources, E.G. and C.A.G.; software, E.G.; supervision, E.G.; validation, E.G.; visualization, E.G. and I.L.; writing—original draft, E.G.; writing—review and editing, E.G. and M.D. All authors have read and agreed to the published version of the manuscript.

Funding: This study was supported by National Science Centre of Poland grant 2015/19/D/NZ8/00191 (EG).

Institutional Review Board Statement: Not applicable.

Informed Consent Statement: Informed consent was obtained from all subjects involved in the study.

Data Availability Statement: The data presented in this study are available on request from the corresponding author.

Acknowledgments: We thank the reviewers for their valuable comments on the manuscript. We are very grateful to the late Andre V. Bochkov, Pavel Klimov, and Sergey V. Mironov, who collected the material and made it available for our research. The collection of mites in Mexico was conducted under the license no. FAUT-0209 issued by Dirección General de Vida Silvestre (Mexico); the field expedition was supported by a grant from the U.S. National Science Foundation (DEB-0613769) to Barry M. OConnor (Museum of Zoology, University of Michigan, Ann Arbor, USA). We also thank the staff of the Department of Vertebrate Zoology, Division of Birds, Smithsonian Institution, National Museum of Natural History, Washington, D.C., USA (USNM), for making the collection of feathers available for this study.

Conflicts of Interest: The authors declare no conflict of interest.

Appendix A

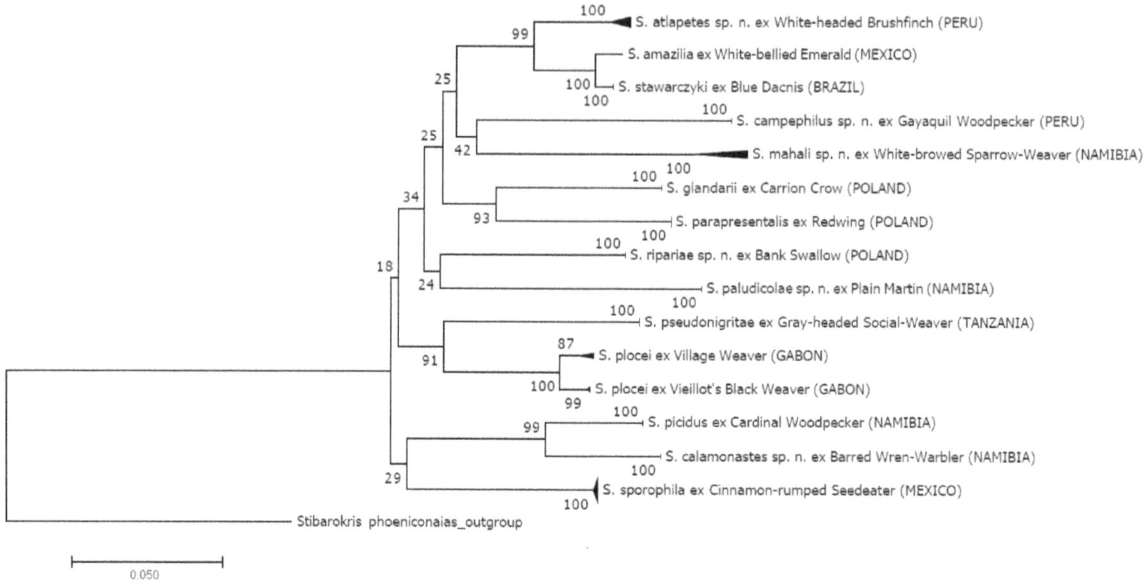

Figure A1. Neighbor-joining phylogenetic tree of the *Syringophiloidus* species based on the *distance p* model. The tree was constructed in Mega v.7. and rooted by *Stibarokris phoeniconaias*.

References

1. Glowska, E.; Chrzanowski, M.; Kaszewska, K. Checklist of the quill mites (Acariformes: Syringophilidae) of the World. *Zootaxa* **2015**, *3968*, 1–81. [CrossRef] [PubMed]
2. Zmudzinski, M.; Skoracki, M.; Sikora, B. An Updated Checklist of Quill Mites of the Family Syringophilidae (Acariformes: Prostigmata). 2021. Available online: https://figshare.com/articles/dataset/An_updated_checklist_of_quill_mites_of_thefamily_Syringophilidae_Acariformes_Prostigmata_/16529574/1 (accessed on 4 May 2023).
3. Johnston, D.E.; Kethley, J.B. A numerical phenetic study of the quill mites of the family Syringophilidae (Acari). *J. Parasitol.* **1973**, *59*, 520–530. [CrossRef]
4. Skoracki, M.; Sikora, B.; Unsoeld, M.; Hromada, M. Mite Fauna of the Family Syringophilidae (Acariformes: Prostigmata) Parasitizing Darwin's Finches in Galápagos Archipelago. *Diversity* **2022**, *14*, 585. [CrossRef]
5. Sikora, B.; Mahamoud-Issa, M.; Unsoeld, M.; Hromada, M.; Skoracki, M. Species Composition of Parasitic Mites of the Subfamily Picobiinae (Acariformes: Syringophilidae) Associated with African Barbets (Piciformes: Lybiidae). *Animals* **2023**, *13*, 2007. [CrossRef] [PubMed]
6. Glowska, E.; Schmidt, B.K. New quill mites (Cheyletoidea: Syringophilidae) parasitizing the black-headed paradise-flycatcher Terpsiphone rufiventer (Passeriformes: Monarchidae) in Gabon. *Zootaxa* **2014**, *3786*, 57–64. [CrossRef]
7. Skoracki, M.; Michalik, J.; Sikora, B. Prevalence and habitat preference of quill mites (Acari, Syringophilidae) parasitizing forest passerine birds in Poland. *Acta Parasitol.* **2010**, *55*, 188–193. [CrossRef]
8. Glowska, E.; Dragun-Damian, A.; Dabert, J. DNA-barcoding contradicts morphology in quill mite species *Torotrogla merulae* and *T. rubeculi* (Prostigmata: Syringophilidae). *Folia Parasitol.* **2013**, *60*, 51–60. [CrossRef]
9. Glowska, E.; Dragun-Damian, A.; Broda, L.; Dabert, J.; Dabert, M. DNA barcodes reveal female dimorphism in syringophilid mites (Actinotrichida: Prostigmata: Cheyletoidea): *Stibarokris phoeniconaias* and *Ciconichenophilus phoeniconaias* are conspecific. *Folia Parasitol.* **2014**, *61*, 272–276. [CrossRef]
10. Glowska, E.; Dragun-Damian, A.; Dabert, J. A new quill mite *Syringophiloidus pseudonigritae* sp. nov. (Prostigmata, Syringophilidae) parasitizing *Pseudonigrita arnaudi* (Passeriformes, Ploceidae)—A combined description using morphology and DNA barcode data. *Zootaxa* **2012**, *3532*, 64–68. [CrossRef]
11. Glowska, E.; Dragun-Damian, A.; Dabert, J. *Picobia dziabaszewskii* sp. nov. (Acari, Syringophilidae)—Combined description (morphology with DNA barcode data) of a new quill mite species parasitizing *Garrulax formosus* (Passeriformes: Leiothrichidae). *Zootaxa* **2012**, *3224*, 57–61. [CrossRef]
12. Hebert, P.D.; Ratnasingham, S.; deWaard, J.R. Barcoding animal life: Cytochrome c oxidase subunit 1 divergences among closely related species. *Proc. Biol. Sci.* **2003**, *270* (Suppl. S1), S96–S99. [CrossRef] [PubMed]

13. Hebert, P.D.; Penton, E.H.; Burns, J.M.; Janzen, D.H.; Hallwachs, W. Ten species in one: DNA barcoding reveals cryptic species in the neotropical skipper butterfly Astraptes fulgerator. *Proc. Natl. Acad. Sci. USA* **2004**, *101*, 14812–14817. [CrossRef] [PubMed]
14. Glowska, E.; Romanowska, K.; Schmidt, B.K.; Dabert, M. Combined description (morphology with DNA barcode data) of a new quill mite *Torotrogla paenae* n. sp. (Acariformes: Syringophilidae) parasitising the Kalahari scrub-robin *Cercotrichas paena* (Smith) (Passeriformes: Muscicapidae) in Namibia. *Syst. Parasitol.* **2018**, *95*, 863–869. [CrossRef] [PubMed]
15. Glowska, E.; Broda, L.; Dabert, M. A new quill mite *Syringophiloidus plocei* sp. nov. (Prostigmata: Syringophilidae) parasitizing ploceid birds (Passeriformes) in Gabon—A combined description using morphology and DNA barcode data. *Acta Parasitol.* **2016**, *61*, 562–566. [CrossRef] [PubMed]
16. Glowska, E.; Broda, L.; Dabert, M. Insight into the species diversity of the quill mite genus Betasyringophiloidus Skoracki, 2011 (Prostigmata: Syringophilidae) on the basis of the DNA barcodes. *Folia Parasitol.* **2019**, *66*, 2019.009. [CrossRef] [PubMed]
17. Glowska, E.; Filutowska, Z.K.; Dabert, M.; Gerth, M. Microbial composition of enigmatic bird parasites: Wolbachia and Spiroplasma are the most important bacterial associates of quill mites (Acariformes: Syringophilidae). *MicrobiologyOpen* **2020**, *9*, e964. [CrossRef]
18. Grandjean, F. Les segments postlarvaires de l'hysterosoma chez les oribates (Acariens). *Bull. Soc. Zool. Fr.* **1939**, *64*, 273–284.
19. Kethley, J.B. Acarina: Prostigmata (Actinedida). In *Soil Biology Guide*; Dindal, D.L., Ed.; John Wiley & Sons: New York, NY, USA, 1990; pp. 667–754.
20. Grandjean, F. Observations sur les acariens de la famille Stigmaeidae. *Arch. Sci. Phys. Nat.* **1944**, *26*, 103–131.
21. Bochkov, A.V.; OConnor, B.M.; Wauthy, G. Phylogenetic position of the family Myobiidae within the Prostigmata (Acari: Acariformes). *Zool. Anz.* **2008**, *247*, 15–45. [CrossRef]
22. Skoracki, M. Quill mites (Acari: Syringophilidae) of the Palaearctic region. *Zootaxa* **2011**, *2840*, 1–414. [CrossRef]
23. Clements, J.F.; Schulenberg, T.S.; Iliff, M.J.; Roberson, D.; Fredericks, T.A.; Sullivan, B.L.; Wood, C.L. The eBird/Clements Checklist of Birds of the World: v2022. Available online: http://www.birds.cornell.edu/clementschecklist/download/ (accessed on 14 October 2023).
24. Dabert, J.; Ehrnsberger, R.; Dabert, M. *Glaucalges tytonis* sp. n. (Analgoidea, Xolalgidae) from the barn owl Tyto alba (Strigiformes, Tytonidae): Compiling morphology with DNA barcode data for taxon descriptions in mites (Acari). *Zootaxa* **2008**, *1719*, 41–52.
25. Kumar, S.; Stecher, G.; Tamura, K. MEGA7: Molecular Evolutionary Genetics Analysis Version 7.0 for Bigger Datasets. *Mol. Biol. Evol.* **2016**, *33*, 1870–1874. [CrossRef] [PubMed]
26. Felsenstein, J. Confidence limits on phylogenies: An approach using the bootstrap. *Evolution* **1985**, *39*, 783–791. [CrossRef] [PubMed]
27. Kimura, M. A simple method for estimating evolutionary rate of base substitutions through comparative studies of nucleotide sequences. *J. Mol. Evol.* **1980**, *16*, 111–120. [CrossRef] [PubMed]
28. Rambaut, A. Tree Figure Drawing Tool Version 1.4.2. Institute of Evolutionary Biology, University of Edinburgh. 2014. Available online: http://tree.bio.ed.ac.uk/ (accessed on 20 October 2023).
29. Skoracki, M. A review of quill mites of the genus Syringophiloidus KETHLEY, 1970 parasitizing quills of passeriform birds, with descriptions of four new species (Acari: Prostigmata: Syringophilidae). *Genus* **2004**, *15*, 281–300.
30. Skoracki, M.; Mironov, S.V.; Hernandes, F.A.; Valim, M.P. Syringophilid quill mites (Acari: Syringophilidae) parasitizing passerines (Aves: Passeriformes) in Brazil. *Int. J. Acarol.* **2016**, *42*, 252–257. [CrossRef]
31. Skoracki, M.; Klimovičová, M.; Muchai, M.; Hromada, M. New taxa of the family Syringophilidae (Acari: Prostigmata) from African barbets and woodpeckers (Piciformes: Lybiidae, Picidae). *Zootaxa* **2014**, *3768*, 178–188. [CrossRef]
32. Fain, A.; Bochkov, A.V.; Mironov, S.V. New genera and species of quill mites of the family Syringophilidae (Acari: Prostigmata). *Bull. L'institut R. Des Sci. Nat. Belg.* **2000**, *70*, 33–70.
33. Skoracki, M.; Sikora, B. New ectoparasitic mites of the family Syringophilidae (Acari: Prostigmata: Cheyletoidea) associated with birds from Argentina. *Zootaxa* **2002**, *27*, 1–8. [CrossRef]
34. Skoracki, M. Quill mites (Acariformes: Syringophilidae) associated with birds of Mexico. *Zootaxa* **2017**, *4282*, 179–191. [CrossRef]
35. Skoracki, M.; Spicer, G.S.; OConnor, B.M. A systematic review of the subfamily Syringophilinae (Acari: Syringophilidae) of the Nearctic region. Part 1: Quill mites associated with passerines (Aves: Passeriformes). *Zootaxa* **2016**, *4084*, 451–494. [CrossRef] [PubMed]
36. Fritsch, W. Die milbengattung Syringophilus Heller, 1880 (subordo Trombidiformes, Fam. Myobiidae Megnin, 1877). *Zoologische Jahrbücher Systematik* **1958**, *86*, 227–234.
37. Broda, L.; Dabert, M.; Glowska, E. *Aulonastus similis* n. sp.—combined description (morphology with DNA barcode data) of a new quill mite species (Syringophilidae) parasitising passeriform birds (Tyrannidae, Cardinalidae) in Mexico. *Syst. Parasitol.* **2016**, *93*, 715–719. [CrossRef] [PubMed]

animals

MDPI

Article

Taxonomy of the Family Teneriffiidae (Acari: Prostigmata: Anystoidea): Generic Synonymies with the Key to World Species of the Family

Jawwad Hassan Mirza, Muhammad Kamran and Fahad Jaber Alatawi *

Department of Plant Protection, College of Food and Agriculture Sciences, King Saud University, Riyadh 11451, Saudi Arabia; jmirza@ksu.edu.sa (J.H.M.); murafique@ksu.edu.sa (M.K.)
* Correspondence: falatawi@ksu.edu.sa

Simple Summary: The generic divisions of the family Teneriffiidae have been dealt with superficially, by which different morphological features were introduced over time to justify the addition of apparently unnecessary genera. The present research provides thorough and detailed insight into the taxonomy of the family Teneriffiidae, and different morphological characters were evaluated. As a result, two genera, *Teneriffia* Thor and *Parateneriffia* Thor, were considered valid based on persistent morphological character/s. The other existing genera were synonymized, and a diagnostic key to genera and species of the family Teneriffiidae was developed while four species were synonymized.

Abstract: The family Teneriffiidae Thor has an equivocal and patchy generic history due to a lack of proper diagnostic character/s, causing the addition of an over-sufficient number of genera (i.e., nine) for the 28 described species. The present study aimed to resolve those taxonomic uncertainties related to generic divisions and species assignments by thoroughly reviewing all the published literature of the family, identifying key diagnostic character/s for generic divisions while debating on previously used morphological features. In the present research, only two genera, *Teneriffia* Thor and *Parateneriffia* Thor, are considered valid genera in the family Teneriffiidae, based on the absence and presence of palpgenu oncophysis, respectively. The previously used other generic diagnostic characters such as coxal setal formula, pectination strength of leg claws, absence or presence of genital papillae, genital discs, and pedal solenidion have been argued for their inconsistencies. A total of four species were synonymized with the closely related species, while additional notes for six poorly described species are given. Moreover, the key to the genera and species of the family Teneriffiidae is provided.

Keywords: palpgenu oncophysis; species synonymy; history; scientific gaps; literature review; character strength; distribution

Citation: Mirza, J.H.; Kamran, M.; Alatawi, F.J. Taxonomy of the Family Teneriffiidae (Acari: Prostigmata: Anystoidea): Generic Synonymies with the Key to World Species of the Family. *Animals* **2023**, *13*, 3736. https://doi.org/10.3390/ani13233736

Academic Editors: Monika Fajfer and Maciej Skoracki

Received: 7 November 2023
Revised: 26 November 2023
Accepted: 28 November 2023
Published: 2 December 2023

1. Introduction

The members of the family Teneriffiidae Thor (Acari: Prostigmata: Anystoidea) are moderate-sized fast-walking mites, usually found in terrestrial (trees, rocks, caves, mountains, etc.) and occasionally in marine habitats [1–3]. They are predatory, feeding on small arthropods [1,4]. After hatching from eggs, individuals undergo four immature stages, including larva, protonymph, deutonymph, and tritonymph, before molting into adults [5]. The biology and ecology of teneriffids are poorly studied, with a single observation of an immobile pre larva enclosed in an eggshell [6].

The diagnostic morphological features of the family Teneriffiidae include the presence of bothridial setae on prodorsum with a rosette-patterned base, disc-like palp tarsus, strong and simple palp tibial claw, subtended by two smaller, straight spurs, oncophysis on palpgenu absent or present, strongly bipectinated claws of at least leg I, and claw-like empodium present on legs III–IV [4]. Currently, there are about 28 globally reported species belonging to nine genera [3]. However, these figures have been contrastingly reported in some recent publications [2,3,7,8].

For the number of described species in the family Teneriffiidae, the number of genera erected has been previously questioned [9,10]. As a result of this, different taxonomic revisions were made where genera were either synonymized or reinstated [5,9–15]. Some useful taxonomic information was presented by McDaniel et al. [10], Judson [13,14], Schmölzer [15], and Ueckermann et al. [16].

Even after these revisions, the comprehensive literature review of the family Teneriffiidae has shown that the taxonomic history of its genera have scientific uncertainties and research gaps due to different reasons, including lost type specimens, generic additions or revisions with missing references, overlooked valid species, immatures being considered as adults, etc. This has led to the dire need for significant taxonomic revision of the family where all species and their assigned genera must be re-evaluated based on distinct and persistent morphological characters. The aims of the present study were to highlight and resolve scientific uncertainties related to generic divisions and species assignments in the family Teneriffiidae by assessing the previously defined genera and species and identifying key diagnostic character/s for generic divisions while debating on previously used morphological features of generic division. A diagnostic key to the genera and reported species of the family Teneriffiidae is also provided.

2. Materials and Methods

The taxonomic literature of all nine genera and 28 teneriffid species were critically studied, and the diagnostic characters of the genera were compared. For the differentiation among different developmental stages, McDaniel et al. [10] and Judson [13] were followed. The tables for comparative morphologies and addition of genera and species over time were constructed based on the available published literature. The strength of each morphological character was evaluated for its suitability at the generic level. The key to species of the family Teneriffiidae is provided based on persistent and fixed characteristics.

3. Historical Background of the Family Tenerifiidae

The family Teneriffiidae was erected by Thor in 1911 with two monotypic genera: *Teneriffia* Thor (type genus; type species *T. quadrapapillata*) and *Parateneriffia* Thor (type species *P. bipectinata*) [17] (Table 1). In 1924, Hirst erected the third genus *Neoteneriffiola* (type species *N. luxoriensis*) [18], while the fourth genus *Heteroteneriffia* (type species *H. marina*) was added in 1925 [19]. All four genera were distinguished based on a number of coxal setae I–IV, state of coxal segments, presence or absence of oncophysis on palp genu, and strength of pectination on claws of legs I–IV (Tables 1 and 2).

In 1935, Womersley [20] added the fifth genus *Austroteneriffia* (type species *A. hirsti*) and considered it closely related to the genus *Heteroteneriffia* based on the presence of genital discs (papillae) (absent in *Teneriffia* and *Neoteneriffiola* genera), the differing claws of leg I–II which strongly pectinated (on only leg I in *Heteroteneriffia*), and not having a definite row of setae on anterior margins of the coxae in *Austroteneriffia* (Table 1).

Later, two more monotypic genera (sixth and seventh in series) were added to the family, namely *Mesoteneriffia* Irk [21] (type species *M. steinbocki*) and *Mesoteneriffiola* Schmölzer [22] (type species *M. alpina*). The genus *Mesoteneriffia* was considered close to the genus *Parateneriffia* due to the presence of palpgenu oncophysis and all leg coxae lying close together. These two were separated due to the absence of a genital clasping organ in *Mesoteneriffia*. The genus *Mesoteneriffiola* was separated from *Mesoteneriffia* mainly based on the number of setae on coxae I–IV (4-4-4-4 vs. 3-3-3-1) (Table 1).

Table 1. Diagnostic characters of different genera of the family Teneriffiidae.

	Teneriffia	*Parateneriffia*	*Neoteneriffiola*	*Heteroteneriffia*	*Austroteneriffia*	*Mesoteneriffia*	*Mesoteneriffiola*	*Sinoteneriffia*	*Himalteneriffia*
Thor [17]	(i) Coxae II not reaching level of coxae III (ii) Oncophysis absent	(i) Coxae II reaching level of coxae IV (ii) Oncophysis present	-	-	-	-	-	-	-
Hirst [18]	(i) Coxae close to each other (ii) Coxae setae numerous (iii) Oncophysis absent	(i) Coxae close to each other (ii) Coxae setae numerous (iii) Oncophysis present	(i) Coxae further apart (ii) Coxae setae less numerous (iii) Oncophysis present	-	-	-	-	-	-
Hirst [19]	-	(i) Coxae close to each other (ii) Oncophysis present (iii) Only leg I–II claw bipectinate (iv) Coxal setae numerous	(i) Coxae further apart (ii) Oncophysis present (iii) Only leg I–II claw bipectinate (iv) Coxal setae less numerous	(i) Coxae further apart (ii) Oncophysis absent (iii) Only leg I claw bipectinate (iv) Coxal setae numerous	-	-	-	-	-
Womersly [20]	(i) Genital disc absent	-	(i) Genital disc absent (based on illustration)	(i) Genital disc absent (based on illustration) (ii) Claws leg I bipectinate (iii) Presence of definite setal row on anterior margin of coxae	(i) Gential disc present (ii) Claws leg I–II bipectinate (iii) Absence of definite setal row on anterior margin of coxae	-	-	-	-
Irk [21]	(i) Oncophysis absent	(i) Oncophysis present (ii) Coxae IV close (iii) Coxae I with backward chitinous process. Before the genital opening a "bracket field". All four pairs of legs with combs claws.	(i) Oncophysis present (ii) Coxae IV apart	(i) Oncophysis absent	(i) Oncophysis present	(i) Oncophysis present (ii) Coxae IV close (iii) Coxae I without chitinous process. Without "clasp field". Claws I and II with large and distinct crests, claws of legs III and IV with much smaller ones and only clear at higher magnification	-	-	-
Schmölzer [22]	-	-	-	-	-	(i) Coxal setae 4-4-4-4	(i) Coxal setae 3-3-3-1	-	-

Table 1. *Cont.*

	Teneriffia	*Parateneriffia*	*Neoteneriffiola*	*Heteroteneriffia*	*Austroteneriffia*	*Mesoteneriffia*	*Mesoteneriffiola*	*Sinoteneriffia*	*Himalteneriffia*
McDaniel et al. [10]	(i) Numerous ventral opisthosomal setae (ii) Palpal genu distal process absent (iii) Lengths subequal to or shorter than body length (idiosoma + gnathosoma); each coxa with 4 or more setae arranged along anterior margin and 1 longer seta on posterior margin	(i) Few ventral opisthosomal setae (ii) Palpal genu with or without a distal process (iii) Lengths subequal or longer than body (including gnathosoma): coxae with 7 or fewer setae, usually 3 or 4; anterior tarsal claws strongly bipectinate; posterior tarsal claws simple or weakly bipectinate, with empodial claws	Synonymized				Not Mentioned	-	-
Yin et al. [23] (did not consider McDaniel et al. [10])	-	-						(i) 1 dorsal plate on the forefoot body and no plate on the back of the hind body (ii) Suckers and no setae on the genital valve (iii) Coxae 4-3-4-2 (iv) 2 pairs of setae on both sides of the reproductive valve	-
Judson [13]	-	-	Reinstated: (i) Called work of McDaniel premature (ii) The strong, linear neotrichy of the pedal solenidia (iii) The relatively large dorsal plates of the opisthosoma (iv) The reduced form of the peritremes	Remained Synonymized				-	-

Table 1. *Cont.*

	Teneriffia	*Parateneriffia*	*Neoteneriffiola*	*Heteroteneriffia*	*Austroteneriffia*	*Mesoteneriffia*	*Mesoteneriffiola*	*Sinoteneriffia*	*Himalteneriffia*
Judson [14]	-	-	-	Remained Synonymized	Reinstated: With detailed diagnosis but no comparison with other genera. He added many characters in the original diagnosis, important as are commented	Remained Synonymized		-	-
Schmölzer [15]	-	-	-		-			-	1. Idiosoma dorsal with a structureless one that reaches almost half the length of the idiosoma Dorsal field. 2. Claw ridges strong on the first two pairs of legs, on legs III and IV still clear, but becoming weaker and weaker. 3. Number of coxal bristles on legs I–IV according to the scheme 4-6-7-5.

At this point, the overwhelming number of genera for the number of species described (seven genera for eight species) was first time criticized, but no work on generic revision was performed [9] (Table 2). Later, new synonymies were proposed, recognizing only two valid genera in the family Teneriffiidae, i.e., *Teneriffia* (genus *Heteroteneriffia* synonymized) and *Parateneriffia* (three genera; *Neoteneriffiola*, *Austroteneriffia*, and *Mesoteneriffiola* synonymized) [10] (Table 1).

Table 2. Chronological information for the genera and species in the family Teneriffiidae.

(a) Species and the genera they previously belonged to		
Genus (as reported in the literature)	**Species**	**Year**
Teneriffia	*quadripapillata*	Thor [17]
Parateneriffia	*bipectinata*	Thor [17]
Neoteneriffiola	*luxoriensis*	Hirst [18]
Heteroteneriffia	*marina*	Hirst [19]
Austroteneriffia	*hirsti*	Womersley [20]
Mesoteneriffia	*steinbocki*	Irk [21]
Mesoteneriffiola	*alpina*	Schmölzer [22]
Neoteneriffiola	*uta*	Tibbets [24]
Austroteneriffia	*japonica*	(Ehara [11])
Austroteneriffia	*tadjikistanica*	(Wainstein [12])
Austroteneriffia	*hojoensis*	(Shiba and Furukawa [5])
Austroteneriffia	*littorina*	(Shiba and Furukawa [5])
Teneriffia	*mexicana*	McDaniel et al. [10]
Teneriffia	*mortoni*	(Luxton [25])
Neoteneriffiola	*coineaui*	Judson [13]
Sinoteneriffia	*nuda*	Yin et al. [23]
Austroteneriffia	*leei*	Judson [14]
Sinoteneriffia	*kunmingensis*	Youzhen et al. [26]
Neoteneriffiola	*yunnanensis*	Youzhen et al. [27]
Austroteneriffia	*kamalii*	Ueckermann and Khanjani [7]
Himalteneriffia	*riccabonai*	Schmölzer [15]
Austroteneriffia	*zamaniani*	Khanjani et al. [28]
Neoteneriffiola	*xerophila*	Bernardi et al. [1]
Austroteneriffia	*shiraziensis*	Khanjani et al. [29]
Austroteneriffia	*khorramabadiensis*	Khanjani et al. [30]
Teneriffia	*sebahatae*	Ueckermann and Durucan [7]
Teneriffia	*aethiopica*	Zmudzinski et al. [2]
Teneriffia	*hajiqanbari*	Paktinat-Saeij and Kazemi [8]

Table 2. *Cont.*

(b) Number of genera and species previously added over time		
Year	**Number of Genera**	**Number of Species**
1911	2	2
1924	3	3
1925	4	4
1935	5	5
1939	6	6
1955	7	7
1958	7	8
1965	7	10
1969	7	11
1975	7	12
1976	2	13
1993	2	14
1994	3	16
1995	4	17
1996	4	18
1997	5	19
2002	6	21
2011	6	22
2012	6	23
2013	6	24
2014	6	25
2020	6	26
2021	6	27
2022	6	28

The genus *Neoteneriffiola* (third after McDaniel et al. [10]) was later reinstated, while the previous synonymy was criticized, stating the reasons as lack of paratype observation and inadequate original description of *Parateneriffia* [13]. Simultaneously, a unique species of the reinstated *Neoteneriffiola* genus was reported, and its significance for the basis of a new genus was highlighted, although none was added.

In the same year, another genus (eighth in series and fourth after McDaniel et al. [10]), *Sinoteneriffia,* was added to the family [23]. This genus was separated from *Neoteneriffiola* based on the number of coxal setae, the number of setae on and around the gential valve, and the number of reproductive suckers.

After almost a year, the types of the *Austroteneriffia* genus were revisited [14] (Table 1) and declared as a valid genus (fifth after McDaniel et al. [10]). Also, some species from the previously reinstated *Neoteneriffiola* genus were transferred to the reinstated genus *Austroteneriffia* [14].

The genus *Himalteneriffia* (type species; *H. riccabonai*) (ninth in series, sixth after McDaniel et al. [10]) was added to the family Teneriffiidae [15]. While defining the genus *Himalteneriffia*, different morphological and geographical aspects of only 8 genera (*Teneriffia, Parateneriffia, Austroteneriffia, Mesoteneriffia, Mesoteneriffiola, Heteroteneriffia* and *Himalteneriffia*) and 14 species of the family Teneriffiidae were studied [15]. After critically evaluating all the published literature on the family Teneriffiidae, it was found in the present study that there are six genera (*Austroteneriffia, Himalteneriffia, Neoteneriffiola, Parateneriffia, Sinoteneriffia* and *Teneriffia*) reported in the family Teneriffiidae which were either originally described or reinstated after McDaniel et al. [10]. These genera are comprised of eleven, two, five, one, two, and seven species, respectively. The status of the genus *Mesoteneriffiola* (and its species *M. alpina*) after McDaniel et al. [10] is still unknown and will be discussed.

4. Results

4.1. Taxonomic Uncertainties and Scientific Gaps in the Literature

Throughout the systematic journey of the family Teneriffiidae, its genera were dealt with superficially, and unstable features were used to erect the teneriffid genera. This has resulted in an overall confused taxonomic perspective towards the strength and reliability of morphological characters to be either used for the generic or species level. This will all be discussed in chronological order, where different taxonomic uncertainties and scientific gaps will be highlighted.

McDaniel et al. [10], for the first time, proposed generic synonymies and an in-depth review in the present research work, highlighting the following five shortcomings:

(i) The important published literature, prior to/close to 1976, was not considered and this concern was also previously raised [14]. The revised diagnoses of *Neoteneriffiola* and *Heteroteneriffia* by Ehara [11] and of *Austroteneriffia* by Shiba and Furukawa [5] were not cited. Due to this, one of the incorrect arguments raised by these authors for the synonymy of *Austroteneriffia* with *Parateneriffia* was stated as "*Also, A. hirsti is terrestrial in habit similar to the Parateneriffia-Neoteneriffiola complex whereas the Teneriffia-Heteroteneriffia complex is littoral*". The authors would not have made this statement if the species, *A. littorina*, reported as littoral, ref. [5] was considered.

(ii) The palpgenu oncophysis was reported missing from the genus *Austroteneriffia* based on the description of species, *A. hirsti*. However, Judson [14] reported the presence of palpgenu oncophysis (the "distal process") in the redescription of *A. hirsti* after observing the type specimens. It further contributes to weakening the proposed synonymy.

(iii) While synonymizing the genus *Neoteneriffiola*, three described species (*N. japonica* Ehara, *N. tadjikistanica* Wainstein, and *N. hojoensis* Shiba and Furukawa) were excluded from the work. This makes the status of these species uncertain.

(iv) The character of coxal setal counts was used in a very general manner while bringing *Austroteneriffia* (i.e., some coxae have 4 or fewer setae) and *Mesoteneriffia* (i.e., only four setae on coxae) close to *Parateneriffia-Neoteneriffiola* complex. This is not true, particularly for *Parateneriffia*, in which coxae III has seven setae as described and illustrated in original work [17] and ironically reported by the authors in the key [10].

(v) Another monotypic genus *Mesoteneriffiola*, which was reported close to *Mesoteneriffia* was not even mentioned during this review. The validity of this genus was uncertain as only two valid genera were recognized, i.e., *Teneriffia* and *Parateneriffia*.

During the reinstatement of the genera *Neoteneriffiola* [13] and *Austroteneriffia* [14], morphology-based comparisons were not provided and it was left for the readers to figure out the diagnostic characters of these reinstated genera. However, based on the emended diagnosis, the characters which could be considered distinguishing for the genus *Austroteneriffia* were a low number of pedal solenidia and holotrichous aggenital chaetotaxy [14]. Interestingly, these characters were already present in the diagnosis of the previously reinstated genus *Neoteneriffiola*, (except the species, *N. coineaui*; neotrichy of pedal solenidia). This raises reservations on the overall generic reinstatement. Also, some species described in *Neoteneriffiola* were moved to the genus *Austroteneriffia* without the provision of compelling morphology-based remarks.

The new genus, *Himalteneriffia*, was added to the family Teneriffiidae [15] without citing the important previously published taxonomic literature. Not only a genus (*Sinoteneriffia*) was missed in the generic analysis, but the genera *Heteroteneriffia* and *Mesoteneriffia* were considered valid without any remarks after previous synonymies of McDaniel et al. [10] and the work of Judson [13]. Also, the previously described eight species were overlooked [15].

Also, it is important to mention that uncertainties can still be found in the recently published work of the family Teneriffiidae. Ueckermann and Durucan [7] mentioned there are eight genera in the family where *Heteroteneriffia* (three species) and *Mesoteneriffia* (two species) were added in the generic count while genus *Mesoteneriffiola* was excluded. Zumudzinski et al. [2] believed in the presence of about 20 species in 9 genera. These authors considered those three genera as valid. Paktinat-Saeij and Kazemi [8] also reported 27 species in 9 genera. It is worth mentioning that even though the genus *Heteroteneriffia* has been considered valid, the provided number of species is incorrect. Shiba and Furukawa [5] synonymized the species *T. tokiokai* (Ehara) with *T. marina* (Hirst). Lastly, Beron [3] provided the catalogue for the family Teneriffiidae. Although the correct number of species in each genus was provided, the author still considered the three genera as valid.

As a result of the thorough literature review in the present study, it became evident that taxonomic ranks were treated sloppily in the family Teneriffiidae. Different morphological characters were used without measuring their taxonomic significance and the possibility of variability in the character states. The missing references in the published works and lack of comparative morphological analysis of genera and species has only further downgraded the situation. It is crucial to validate the significance of each character at different taxonomic ranks.

4.2. Strength of Morphological Characters for Generic Divisions

During 1911–1925, the genera were separated based on intercoxal distances, the presence or absence of palpgenus oncophysis, the number of setae on coxae I–IV, and pectination strength of leg tarsal claws (Table 1). Womersley [20] introduced the absence and presence of a gential disc and definite setal row on the anterior margin of coxae. Eller and Strandtmann [9] debated on the character of genital disks, attributing it as a sexual difference. Irk [21] again used the characters of palpgenu oncophysis, intercoxal distances, and further added chitinous process on coxa I and the presence or absence of bracket field (translated from original German description "*Vor der Genitalöffnung ein „Spangenfeld"*). McDaniel [10], while synonymizing the genera, considered the number of ventral opisthosomal setae, the presence or absence of palpgenu oncophysis, the length of legs comparative to body, and the number of setae on coxae I–IV. Judson [13,14], during the reinstatement of two genera, placed emphasis on the neotrichy of pedal solenidia, the size of dorsal opisthosomal shield and relatively large dorsal plates, and the reduced form of peritremes. Schmölzer [15] also considered dorsal shield size, the ridges on leg claws I–IV, and the number of setae on coxae I–IV as generic character.

Throughout the taxonomic history of adding, synonymizing, and reinstating the genera of the family Teneriffiidae, two morphological characters, i.e., palpgenus oncophysis and the number of setae on coxae I–IV, were found to be repeatedly used. The number of coxal setae appear unreliable as it has been reported to be variable not only among different populations of a species but even in one population of single species (Table 3). However, in two genera out of nine, *Austroteneriffia* (eleven species) and *Neoteneriffiola* (five species) this character is quiet stable among all the described species. On the other hand, the palpgenu oncophysis is a very persistent and stable character among all the described species and genera of the family Teneriffiidae, with only two states, i.e., present or absent.

Interestingly, the character of pectination strength on leg claws appeared once to differentiate the genus *Heteroteneriffia*. However, this genus is still under synonymy with the genus *Teneriffia* [13]. The number of setae on ventral opisthosoma near genital region being numerous belong to two genera, i.e., *Heteroteneriffia* and *Teneriffia*. The character of genital discs, as mentioned earlier, cannot be used for generic differentiations as it differs between female and male [9]. In the present study, based on these two characters, the synonymy of *Heteroteneriffia* with *Teneriffia* is considered valid.

Table 3. Diagnostic characters of two genera proposed in this study, their species and distribution.

Species	Genus	Distribution	Year	Species Characters						
				Prodorsal Shield	Dorsal Setae	Ventral Setae around G	Tarsi III–IV	Coxae I–IV	Genu Oncophysis	Length of c2
mexicana	*Teneriffia* Thor	Mexico	1976	present	on cuticle	multiples	divided	7/10-7/12-7/10-6/9	absent	crossing d
quadripapillata		Spain	1911	present	on sclerite	15 pairs	undivided	7-8-6-6	absent	crossing d
sebahatae		Turkey	2020	present	on cuticle	23 pairs	divided IV	7-8-6-6	absent	reaching d
hajiqanbari		Iran	2022	present	on cuticle	17–20 pairs	divided	6/7/8-6/7-6/7-5	absent	reaching d
kamalii		Iran	2002	present	on cuticle	6 pairs	not described	4-3-4-3	absent	reaching f
zamaniani		Iran	2011	present	on cuticle	6 pairs	divided	4-3-4-3	absent	reaching e
littorina		Japan	1975	inconspicuous	on cuticle	6 pairs	not described	4-3-4-3	absent	reaching d
riccabonai		India	2002	present	on cuticle	5 pairs	divided	4-6-7-5	absent	subequal to all
marina		Japan, Malaysia	1925	absent	on cuticle	more than 30 pairs	divided	6/7-7/10-7/8-5/8	absent	subequal to all
mortoni		Japan	1993	absent	on cuticle	atleast 40	divided	8-7/8-8/9-8	absent	reaching d
aethiopica	*Parateneriffia* Thor	Ethiopia	2021	present	on sclerite	6–7 pairs	divided	7-6-6-5	present	reaching e
coineaui		Namibia	1994	present	on sclerite	5 pairs	divided	4-3-4-3	present	subequal to all
xerophila		Brazil	2012	present	on sclerite	5 pairs	not described	3-4(6)-4-3	present	subequal to all
uta		Mexico, USA	1958	present	on cuticle	5 pairs	divided	4-3-4-3	present	reaching f
hojoensis		Japan	1975	present	on cuticle	6 pairs	divided	4-3-4-3	present	crossing d
hirsti		Australia	1935	present	on cuticle	6 pairs	divided	4-3-4-3	present	crossing d
khorramabadiensis		Iran	2014	present	on cuticle	6 pairs	divided	4-3-4-3	present	crossing h
shiraziensis		Iran	2013	present	on cuticle	6 pairs	divided	4-3-4-3	present	reaching h
leei		Australia	1995	present	on cuticle	6 pairs	divided	4-3-4-3	present	not described
bipectinata		Paraguay	1911	not described	not described	not described	not described	3-3-7-4	present	not described
steinbocki		Austria, Switzerland	1939	present	on cuticle	not described	divided	4-4-4-4	present	reaching d
alpina		France	1955	present	on cuticle	4 pairs	divied	3-3-3-1	present	reaching e
tadjikistanica		Tadjikistan, Yemen	1969	present	on cuticle	6 pairs	divided	4-3-4-3	present	longer than other
luxoriensis		Egypt	1924	inconspicuous	on cuticle	5–6 pairs	divided	4-3-4-3	present	reaching d

After analyzing all the morphological characters ever used for the generic differentiation in the present study, it became suitable and convenient to place the finger on the most persistent morphological character, i.e., palpgenu oncophysis. This character is found in all the described stages and in both females and males, and it could be the most suitable for the generic divisions.

4.3. Generic Division

Among the 28 described species in the family Teneriffiidae, different species were described either from male or female or both (Table 4). The male descriptions and illustrations were provided for only 18 species (63%), while females are described and illustrated from all the species (100%). After the detailed study of the published literature of all teneriffid species, two genera, *Teneriffia* Thor and *Parateneriffia* Thor, are considered as valid in this study, for all the described teneriffid species based on the presence and absence of palpgenu oncophysis in females (Table 3). The genera *Heteroteneriffia*, *Himalteneriffia*, and *Sinoteneriffia* are hereby synonymized with the genus *Teneriffia* (absence of palpgenu oncophysis). The genera *Austroteneriffia*, *Neoteneriffiola*, *Mesoteneriffia* and *Mesoteneriffiola* are synonymized with the genus *Parateneriffia* (presence of palpgenus oncophysis). Out of the 28 species described up to now, 24 species are assigned between these two genera (excluding four proposed species synonymies).

Table 4. List of species in the family Teneriffiidae and their developmental stages (the green color represents the stage/s described).

Genus	Species	Larva	Nymph			Adult	
			Proto-	Deuto-	Trito-	Male	Female
Teneriffia	*quadripapillata*						■
Parateneriffia	*bipectinata*						■
Neoteneriffiola	*luxoriensis*						■
Heteroteneriffia	*marina*						■
Austroteneriffia	*hirsti*					■	■
Mesoteneriffia	*steinbocki*						■
Mesoteneriffiola	*alpina*						■
Neoteneriffiola	*uta*	■	■	■	■	■	■
Austroteneriffia	*japonica*					■	
Austroteneriffia	*tadjikistanica*						■
Austroteneriffia	*hojoensis*		■	■		■	■
Austroteneriffia	*littorina*					■	■
Teneriffia	*mexicana*					■	■
Heteroteneriffia	*mortoni*						■
Neoteneriffiola	*coineaui*				■	■	■
Sinoteneriffia	*nuda*					■	■
Austroteneriffia	*leei*				■	■	■
Sinoteneriffia	*kunmingensis*					■	■
Neoteneriffiola	*yunnanensis*					■	■
Austroteneriffia	*kamalii*					■	■
Himalteneriffia	*riccabonai*						■
Austroteneriffia	*zamaniani*						■
Neoteneriffiola	*xerophila*					■	■
Austroteneriffia	*shiraziensis*		■	■		■	■
Austroteneriffia	*khorramabadiensis*					■	■
Teneriffia	*sebahatae*					■	■
Teneriffia	*aethiopica*					■	■
Teneriffia	*hajiqanbari*			■		■	■

Family Teneriffiidae Thor
Teneriffiidae Thor 1911:179 [17]
Teneriffiolidae Hirst, 1924: 1078 [18]
Teneriffiinae Womersley, 1935: 334 [20]
Type genus: *Teneriffia* Thor, 1911 [17]
Diagnosis:

The diagnosis of the family has been provided by several authors [4,9,20,25]. In the present study, a precisely updated family diagnosis is provided.

Naso present, small and without setae, prodorsal bothridial setae with rosette patterned base, palp tarsus reduced; disc like, palp tibial claw strong with two small spurs at the base, chelicerae with sickle like chelae, not fused proximally, pretarsal empodial claws absent on legs I–II while present on legs III–IV, the true claws on at least leg I highly pectinated, peritremes not emargant, multichambered and present anterolateraly.

Genus *Teneriffia* Thor
Teneriffia Thor 1911:172 [17]
Heteroteneriffia Hirst 1925:1278 [19]. Type species *H. marina* Hirst 1925 [19]
Sinoteneriffia Yin et al. 1994:443 [23]. Type species *S. nuda* Yin et al. 1994 [23]
Himalteneriffia Schmölzer 2002:133 [15]. Type species *H. riccabonai* Schmölzer 2002 [15]

Type species: by original designation, *T. quadripapillata*, Thor 1911:173 [17], Ueckermann et al. 2022: 789 [16].

Diagnosis: Palpgenu oncophysis absent, prodorsal shield either present or absent.

Number of species included: 10 (Table 3)

Distribution: Mexico, Spain, Turkey, Iran, Japan, India, Malaysia

Remarks: This genus is retained based on original designation by Thor [17] as type genus of the family. Ueckermann et al. [16] recollected a number of specimens from type locality. This genus was originally described with the palpgenu oncophysis absent which is endorsed in the present study.

Genus *Parateneriffia* Thor

Parateneriffia Thor 1911:176 [17]

Neoteneriffiola Hirst 1924:1078 [18]. Type species *N. luxoriensis* Hirst 1924 [18]

Austroteneriffia Womersley 1935:334 [20]. Type species *A. hirsti* Womersley 1935 [20]

Mesoteneriffia Irk 1939:220 [21]. Type species *M. steinbocki* Irk 1939 [21]

Mesoteneriffiola Schmölzer 1955:36 [22]. Types species *M. alpina* Schmölzer 1955 [22]

Type species: *Parateneriffia bipectinata* Womersely

Diagnosis: palpgenu oncophysis present, prodorsal shield always present.

Number of species included: 14 (Table 3)

Distribution: Ethiopia, Namibia, Egypt, Mexico, USA, Brazil, China, Japan, Iran, Tadjikistan, Yemen, Australia, Paraguay, Austria, Switzerland, France

Remarks: This genus is retained based on its original designation by Thor [17] as the second genus in the family Teneriffiidae. It was originally diagnosed by the presence of palpgenu oncophysis, which is endorsed in the present study. The original type of the genus was *P. bipectinata* [17]. This species was criticized due to the loss of its type specimens and an inadequate original description and illustration [13].

4.4. On the Suggested Synonymy of Some Species

The species, *P. hojoensis* (Shiba and Furukawa) was originally distinguished from *P. japonica* (Ehara) based on the presence or absence of a solenidion on leg genu I–IV, i.e., leg genu I–IV solenidotaxy as 1-1-1-0 and 0-0-0-0, respectively [5,11]. Later, a short description of *A. japonica* reported the presence of solenidion on leg genu I–II [14]. Here, in this study, a critical review of the descriptions of both the species revealed a few differences as in leg chaetotaxy and solenidotaxy. Other than that, these two species are morphologically resembling. The species *P. japonica* was originally described from two males while *P. hojoensis* was originally described from more than ten individuals of male, female, deutonymph, and protonymph. Additionally, both species were reported from Japan. In the present study, these two species belong to the genus *Parateneriffia* (presence of palpgenu oncophysis). However, based on the argument provided above, *P. hojoensis* is suggested as junior synonym of *P. japonica*.

Youzhen et al. [27] described the species *Neoteneriffiola yunnanesis* based on the male, with few morphological characters which were typical of the genus. The original remarks placed this species close to *P. japonica* and *P. tadjikistanica* and differentiated it based on body length and number of setae on genu IV. Also, the original description did not include the *P. hojoensis* in the key. It became clear upon comparing the original descriptions of these three species that *N. yunnanensis* resembles *P. japonica* and *P. hojoensis* and it is suggested as synonym of *P. japonica*.

There are two species described under the genus *Sinoteneriffia* by Yin et al. [23]. As argued earlier the genus and its type species *S. nuda* were diagnosed based on deutonymphal characters (genital shield without setae, two setae around genital shield) and hence are not valid. Similarly, the second species, *S. kunmingensis*, described by Youzhin et al. [26] was also diagnosed on the supposed male but has similar characters to the deutonymphal stage. Hence, the genus *Sinoteneriffia*, as stated above, and its two species are not valid because both species were described based on deutonymphs.

4.5. Key to Genera and Species of the Family Teneriffiidae Based on Females

1. Palpgenu oncophysis absent .... genus *Teneriffia* Thor 2
- Palpgenus oncophysis present .... genus *Parateneriffia* Thor 11
2. Gnathosoma ventrally with clasp organ .... *T. quadripapillata* Thor
- Gnathosoma without ventral clasp organ 3
3. Only claws of leg I heavily pectinated 4
- Claws of leg I–II heavily pectinated 5
4. Naso punctate, spurs of hypostome elongate, tarsus II with three solenidia ....
.... *T. marina* (Hirst) comb. nov.
- Naso not punctate, spurs of hypostome squat, tarsus II with four solenidia ....
.... *T. mortoni* (Luxton) comb. nov.
5. Venter with five to six pairs of setae surrounding the genital valve 8
- Venter with numerous pairs of setae (15 to more than 30 pairs) surrounding the genital valve 6
6. Venter with >30 pairs of setae; tarsi III and IV with 1–2 and 2–3 solenidia, respectively, ...
.... *T. mexicana* (McDaniel et al.) comb. nov.
- Venter with 17–23 pairs of setae; tarsi III and IV with 0 and 1 solenidion, respectively, ... 7
7. Venter with 17–20 pairs of setae; seven pairs of genital setae present ....
.... *T. hajiqanbari* (Paktinat-Saeij and Kazemi) comb. nov.
- Venter with 23 pairs of setae; six pairs of genital setae present ....
.... *T. sebahatae* (Ueckermann and Durucan) comb. nov.
8. Setae c_2 almost extending to base of seta *d* 9
- Setae c_2 reaching to over the base of setae *e* or *f* 10
9. Prodorsal shield weakly distinct with thin and close longitudinal striations, coxal formula 4-3-4-3 .... *T. littorina* (Shiba and Furukawa) comb. nov.
- Prodorsal shield smooth, distinctly defined and greatly extended reaching upto half of dorsum, coxal formula 4-6-7-5 .... *T. riccabonai* (Schmölzer) comb. nov.
10. Basifemur I with five setae, tibia II nine setae ....
.... *T. zamaniani* (Khanjani, Asali Fayaz and Ueckermann) comb. nov.
- Basifemur I with four setae, tibia II 10 setae ....
.... *T. kamalii* (Ueckermann and Khanjani) comb. nov.
11. Dorsal setae c_1 and c_2 subequal in length 12
- Dorsal setae c_2 distinctly longer than c_1 13
12. Dorsocentral setae c_1 inserted on over extended prodorsal shield ....
.... *P. coineaui* (Judson) comb. nov.
- Dorsocentral setae c_1 present on the integument .... *P. xerophila* (Bernardi et al.) comb. nov.
13. All opisthosomal setae on small sclerites .... *P. aethiopica* (Zmudzinski et al.) comb. nov.
- All opisthosomal setae on integument 14
14. Dorsocentral setae shorter than or equal to the distance between the consecutive setae ....
.... *P. hojoensis* (Shiba and Furukawa) comb. nov.
- Dorsocentral setae long, crossing the bases of the setae next in line 15
15. Genu IV with a solenidion 16
- Genu IV without a solenidion 17
16. Basifemur I with five setae; telofemur III with five setae .... *P. hirsti* (Womersley) comb. nov.
- Basifemur I with four setae; telofemur III with four setae .... *P. leei* (Judson) comb. nov.
17. Trochanter IV with 2 setae .... *P. khorramabadiensis* (Khanjani et al.) comb. nov.
- Trochanter IV with 3 setae .... *P. shiraziensis* (Khanjani et al.) comb. nov.

4.6. Additional Notes on the Status of Some Teneriffid Species Excluded from the Key

Among the 28 described species of the family Teneriffiidae so far, six species have incomplete descriptions, insufficient illustrations, and inappropriate species comparisons based on variable morphological characters. These species were excluded from the key and comments have been provided; meanwhile, four species were considered as suggested synonyms due to variable characters used as species diagnosis. These species are as follows.

Parateneriffia bipectinata **Thor**

Parateneriffia bipectinata Thor, 1911:177 [17], McDaniel et al., 1976:532 [10]

The species, *P. bipectinata*, was designated as the type species of the monotypic genus *Parateneriffia*, reported from Paraguay [17]. The original description and illustrations of the species are insufficient, such that important morphological characters for the species differentiation could not be inferred. The author did not illustrate dorsum, gnathosoma, and legs, nor were these body segments described comprehensively. McDaniel et al. [10] provided a very short complementary description and also illustrated only the venter of this species. The most distinct feature provided could be the presence of two transverse sclerotized cleft anterior to the genital slit [10,17]. This character has not been reported since in any of the recently published teneriffid species. Ironically, it now cannot be confirmed as these types of the species have been reported as "lost" [13]. Hence, it was not possible to place it in the diagnostic key provided in the present study.

***Parateneriffia steinbocki* (Irk)** comb. nov.

Parateneriffia steinbocki (Irk) McDaniel et al. 1976:536 [10]

Mesoteneriffia steinbocki Irk 1939:222 [21]; Strandtmann, 1965:261 [31]

The monotypic genus *Mesoteneriffia* with its type, *M. steinbocki*, was added in the family Teneriffiidae, by Irk [21] from Ötztal Alps, Austria. The authors provided detailed diagnosis of this genus based on inconsistent (setal arrangement on leg coxae, integument with small pores, absence of genital palps, etc.) and overlapping morphological characters (structure and shape of palp including palp tarsus presence of palp oncophysis, etc.). The type species, *M. steinbocki*, was also insufficiently described and illustrated.

In the present study, the species *P. steinbocki* comb. nov., is placed in the genus *Parateneriffia* (presence of palponcophysis) and strikingly resembles the species *P. uta* comb. nov., and *P. japonica* comb. nov. It is difficult to discern from later species as leg chaetotaxy, along with other important morphological characters, were not provided in the original description [10,21]. The apparent differences between *P. steinbocki* comb. nov. and *P. uta* comb. nov. could be the length of setae c_2. Ironically, this character cannot be considered as it was found variable between the two different descriptions of *P. uta* comb. nov. [9,31]. The possible differences between *P. steinbocki* comb. nov. and *P. japonica* comb. nov. could be coxal setal formula as 4-4-4-4 vs. 4-3-4-3, respectively. This character in particular is insufficient based on the discussion provided above. Due to morphological similarities and poor descriptions and illustrations, the species, *P. steinbocki* comb. nov. is excluded from the key.

***Parateneriffia alpina* (Schmölzer)** comb. nov.

Mesoteneriffiola alpina Schmölzer 1955:36 [22]

The monotypic genus *Mesoteneriffiola* was added in the family based on the collection from "*Unterhalb d. Roche d'Alvau*" [22]. Its species *P. alpina* comb. nov. was designated close to the species *P. steinbocki* comb. nov. and was differentiated from the latter based on the number of coxal setae (Table 1) and position of third pair of prodorsal seta on the prodorsal shield. Similar to *P. steinbocki*, the species *P. alpina* morphologically resembles the species *P. japonica* comb. nov. Although the description and illustration of *P. alpina* comb. nov. are poor, the number of coxal setae are by far the lowest reported in any of the Teneriffid species, i.e., coxae I–IV 3-3-3-1. Other than this, it is difficult to morphologically discern it from the closely related species.

As a result of new generic divisions proposed in this study, *P. alpina* comb. nov. is placed in the genus *Parateneriffia* but has been excluded from the key due to insufficient morphological description.

***Parateneriffia luxoriensis* (Hirst)** comb. nov.

Parateneriffia luxoriensis (Hirst) McDaniel et al. 1976:532 [10]

Neoteneriffiola luxoriensis Hirst 1924:1078 [18]

The species *P. luxoriensis* (Hirst) comb. nov. was the type species of the genus *Neoteneriffiola* and is currently placed in the genus *Parateneriffia*. Due to incomplete description, this species is excluded from the key. The closely related species, *P. uta* comb. nov. (later described in 1958) was distinguished based on length of dorsocentral setae and number of setae on palptarsus [31]. Originally, the pedal chaetotaxy and solenidotaxy is neither described nor illustrated [18].

***Parateneriffia uta* (Tibbets)** comb. nov.

Parateneriffia uta (Tibbets) McDaniel et al. 1976:532 [10]

Neoteneriffiola uta Tibbets 1958:44 [24]

This species, *P. uta* (Tibbets) comb. nov., was originally described as closely related to the species *P. luxoriensis* comb. nov. The differential characters used were comparative lengths of dorsocentral setae and number of setae on palp tarsus [24]. The species' redescription and the key to species provided by Eller and Strandtmann [9] used similar morphological characters. However, McDaniel et al. [10] disagreed with this, stating that inter-setal lengths of dorsocentral setae are variable subject to the state of slide-mounted specimen. Instead, they used the length of leg I vs. body length character in the key. Although the number of setae on palp tarsus was repeatedly used as differential feature, it is unclear if this number in both species includes the solenidion or not [9,10,19,24]. Due to an incomplete description and ambiguity in the diagnostic characters, this species is excluded from the key.

Parateneriffia tadjikistanica **(Wainstein)** comb. nov.

Neoteneriffiola tadjikistanica Wainstein 1969:1250 [12]; Wainstein 1978:202 [32]

Austroteneriffia tadjikistanica (Wainstein) Judson 1995:838 [14]

Based on presence of genu palp oncophysis, this species belongs to the genus *Parateneriffia* as proposed in the present study. This species has been reported as morphologically similar to *P. japonica* comb. nov., but this is difficult to discern due to ambiguous leg chaetotaxy [14]. For this reason, the species *P. tadjikistanica* comb. nov. is not included in the presented key.

5. Conclusions

Morphological features, which can be used as the generic diagnostic character, must be carefully evaluated. In the family Teneriffiidae, different morphological characters were used over time for generic differentiation, which has led to the unnecessary addition of different genera in the family. In the present research, two genera viz; *Teneriffia* (palpgenus oncophysis absent) and *Parateneriffia* (palpgenus oncophysis present), are recognized in the family Teneriffiidae. This character was found to have been used constantly as one of the generic diagnostic characters since the family Teneriffiidae was recognized [17]. It represents the strength and stability of the character. Through the extensive research performed in the present paper, it is emphasized that such morphological characters must be carefully avoided as they may result in the addition of different genera for a fewer number of species. In contrast, morphological features which provide clear generic differentiations and are persistent even in newly described species must be used.

Author Contributions: Conceptualization, J.H.M. and F.J.A.; methodology, J.H.M., M.K. and F.J.A.; formal analysis, J.H.M.; data curation, J.H.M.; writing—original draft preparation, J.H.M.; writing—review and editing, M.K. and F.J.A.; supervision, F.J.A.; funding acquisition, F.J.A. All authors have read and agreed to the published version of the manuscript.

Funding: The authors would like to extend their sincere appreciation to the researchers supporting project number (RSPD2023R807), King Saud University, Riyadh, Saudi Arabia.

Institutional Review Board Statement: Not applicable.

Informed Consent Statement: Not applicable.

Data Availability Statement: All necessary data for the manuscript is provided.

Acknowledgments: The authors would like to appreciate the funding provided by the Researchers Supporting Project at King Saud University, Riyadh, Saudi Arabia. The authors are also thankful for three anonymous referees for their constructive feedback.

Conflicts of Interest: The authors declare no conflict of interest.

References

1. Bernardi, L.F.O.; Pellegrini, T.G.; Ferreira, R.L. New species of *Neoteneriffiola* (Acari: Trombidiformes: Teneriffiidae) from Brazilian caves: Geographical distribution and ecological traits. *Int. J. Acarol.* **2012**, *38*, 410–419. [CrossRef]
2. Zmudzinski, M.; Skoracki, M.; Friedrich, S. A new species of Teneriffiidae (Acariformes: Prostigmata) from Ethiopia. *Int. J. Acarol.* **2021**, *47*, 317–326. [CrossRef]
3. Beron, P. *ACARORUM CATALOGUS X. Trombidiformes, Prostigmata, Superfamilia Labidostommatoidea (Labidostommatidae), Superfamilia Eupodoidea,(Eupodidae, Dendrochaetidae, Rhagidiidae, Eriorhynchidae, Pentapalpidae, Penthalodidae, Penthaleidae, Proterorhagiidae, Strandtmanniidae), Superfamilia Tydeoidea, Ereynetidae, Superfamily Paratydeoidea, Paratydeidae, Superfamilia Anystoidea, (Anystidae, Erythracaridae, Teneriffiidae, Pseudocheylidae, Stigmocheylidae), Superfamilia Caeculoidea (Caeculidae), Superfamilia*; Advanced Books; Pensoft: Sofia, Bulgaria; Moscow, Russia, 2022; p. e68612.
4. Krantz, G.W.; Walter, D.E. *A Manual of Acarology*, 3rd ed.; Texas Tech University Press: Lubbock, TX, USA, 2009; p. 807.
5. Shiba, M.; Furukawa, M. Studies on the family Teneriffiidae (Acarina: Prostigmata) in Japan. *Rep. Res. Matsuyama Shinome Jr. Coll.* **1975**, *7*, 111–126.
6. Otto, J.C. Observations on prelarvae in Anystidae and Teneriffiidae. In *Acarology IX, Volume I, Proceedings*; Mitchell, R., Horn, D.J., Needham, G.R., Welbourn, W.C., Eds.; Ohio Biological Survey: Columbus, OH, USA, 1997; pp. 343–354.
7. Ueckermann, E.A.; Durucan, F. *Teneriffia sebahatae* sp. nov. (Acari: Trombidiformes: Teneriffiidae), the first teneriffiid mite from Turkey. *Syst. Appl. Acarol.* **2020**, *25*, 1139–1146.
8. Paktinat-Saeij, S.; Kazemi, S. *Teneriffia hajiqanbari* sp. nov. (Acari: Trombidiformes: Teneriffiidae), first record of the genus from Iran, with a key to world species of Teneriffia. *Acarologia* **2022**, *62*, 262–269. [CrossRef]
9. Eller, R.; Strandtmann, R.W. Notes on Teneriffiidae (Acarina: Prostigmata). *Southwest. Nat.* **1963**, *8*, 23–31. [CrossRef]
10. McDaniel, B.; Morihara, D.; Lewis, J.K. The Family Teneriffiidae Thor, with a New Species from Mexico. *Ann. Entomol. Soc. Am.* **1976**, *69*, 527–537. [CrossRef]
11. Ehara, S. Two new species of Teneriffiidae from Japan, with notes on the genera *Heteroteneriffia* and *Neoteneriffiola* (Acarina: Prostigmata). *Publ. Seto Mar. Biol. Lab.* **1965**, *13*, 221–229. [CrossRef]
12. Wainstein, B.A. A new species from the family Teneriffiidae (Acariformes: Prostigmata). *Zool. Zhurnal* **1969**, *48*, 1250–1252. (In Russian, with English Summary).
13. Judson, M. Studies on the Morphology and Systematics of the Teneriffiidae (Acari, Prostigmata) I. A New Species of *Neoteneriffiola* from Namibia. *Acarologia* **1994**, *35*, 115–134.
14. Judson, M. Studies on the Teneriffiidae (Acari: Anystoidea). II. A Review of the Genus Austroteneriffia. *Invertebr. Taxon.* **1995**, *9*, 827–839. [CrossRef]
15. Schmölzer, K. Über Teneriffiidae, sowie Beschreibung einer neuen Gattung und Art aus dem Himalaya (Acarina, Trombidifomes). *Ber. Nat.-Med. Verein Innsbruck* **2002**, *89*, 123–136.
16. Ueckermann, E.A.; De La Paz, J.C.; Hernández-Teixidor, D.; Durucan, F. Rediscovery and redescription of *Teneriffia quadripapillata* Sig Thor (Acari: Trombidiformes: Teneriffiidae). *Acarologia* **2022**, *62*, 786–797. [CrossRef]
17. Thor, S. Eine neue Acarinenfamilie (Teneriffiidae) und zwei neue Gattungen, die eine von *Teneriffa*, die andere aus Paraguay. *Zool. Anz.* **1911**, *38*, 171–179.
18. Hirst, S. On Three new Acari belonging to the Superfamily Trombidioidea (Erythraeidae and Teneriffiolidae). In *Proceedings of the Zoological Society of London*; Blackwell Publishing Ltd.: Oxford, UK, 1924; Part 4; pp. 1075–1080.
19. Hirst, S. On some new genera and species of Arachnida. *J. Zool.* **1925**, *95*, 1271–1280. [CrossRef]
20. Womersley, H. On the occurrence in Australia of Acarina of the Family Teneriffiidae (Trombidoidea). *Rec. South Aust. Mus.* **1935**, *5*, 333–338.
21. Irk, V. Drei neue Milbenarten aus dem Tiroler Hochgebirge. *Zool. Anz.* **1939**, *128*, 217–223.
22. Schmölzer, K. Landmilben aus dem Dauphiné (Acarina terrestria). *Öst. Zool. Zeitschr.* **1955**, *6*, 542–565.
23. Yin, S.; Bei, N.; Li, X. A new genus and new species of Teneriffiidae from China (Acari: Prostigmata). *Acta Zootaxonomica Sin.* **1994**, *19*, 433–437.
24. Tibbetts, T. A new species in the genus *Neoteneriffiola* from Utah (Acarina: Anystoidea: Teneriffiidae). *Great Basin Nat.* **1958**, *18*, 2.
25. Luxton, M. The genus *Heteroteneriffia* Hirst, 1925 (Acarina: Prostigmata: Teneriffiidae). *Zool. Anz.* **1993**, *230*, 103–109.
26. Youzhen, L.; Suigong, Y.; Yidong, Z.; Bin, C. A new species of *Sinoteneriffia*. *Yunnan Nong Ye Da Xue Xue Bao J. Yunnan Agric. Univ.* **1996**, *11*, 243–247.
27. Youzhen, L.; Suigong, Y.; Bin, C.; Yidong, Z. A new species of *Neoteneriffiola* from China (Acarina: Teneriffidae). *Yunnan Nong Ye Da Xue Xue Bao J. Yunnan Agric. Univ.* **1997**, *12*, 87–91.
28. Khanjani, M.; Yazdanpanah, S.; Fayaz, B.A. *Austroteneriffia shiraziensis* sp. nov. (Acari: Teneriffiidae) from southwestern Iran, with description of male and immature stages. *Zootaxa* **2013**, *3683*, 35–50. [CrossRef]
29. Khanjani, M.; Fayaz, B.F.; Ueckermann, B.F. A new species of the genus *Austroteneriffia* (Acari: Anystina: Teneriffiidae) from western Iran. *Int. J. Acarol.* **2011**, *37*, 550–555. [CrossRef]
30. Khanjani, M.; Hoseini, M.A.; Fayaz, B.A. *Austroteneriffia khorramabadiensis* n. sp. (Acari: Teneriffiidae) as new species from southwestern Iran, with description of male. *Acarologia* **2014**, *54*, 69–79. [CrossRef]

31. Strandtmann, R.W. Additional notes on Teneriffiidae (Acarina: Prostigmata) with two previously unpublished plates by A.C. Oudemans. *J. Kans. Entom. Soc.* **1965**, *38*, 258–261.
32. Wainstein, B.A. Fam. Teneriffiidae Thor. In *Keys for Determination of Trombidiformes*; Gilyarov, M.S., Ed.; Nauka: Moskow, Russian, 1978; pp. 202–203, 271p. (In Russian)

animals

MDPI

Article

Morphological Ontogeny and Ecology of a Common Peatland Mite, *Nanhermannia coronata* (Acari, Oribatida, Nanhermanniidae)

Stanisław Seniczak [1] and Anna Seniczak [2,*]

[1] Department of Evolutionary Biology, Faculty of Biological Sciences, Kazimierz Wielki University, 85-093 Bydgoszcz, Poland
[2] Faculty of Applied Ecology, Agricultural Sciences and Biotechnology, Inland Norway University of Applied Sciences, 2318 Elverum, Norway
* Correspondence: anna.seniczak@inn.no

Simple Summary: *Nanhermannia coronata* is a common and abundant oribatid species in peatlands, but it can be easily mistaken for *N. sellnicki* as an adult. The identity of adults of *N. coronata* investigated herein from several sites in Norway and Ireland was supported by the COI sequence data. Based on this material, the morphological ontogeny of *N. coronata* was investigated, and some characters were found that clearly differentiate *N. coronata* from *N. sellnicki*, like the number of setae on femora of adults and tritonymphs, the shape of insertions of prodorsal seta *in* and all gastronotal and adanal setae of juveniles. Our ecological observations confirm a common occurrence of *N. coronata* in raised bogs, a high percentage of juvenile stages in populations and a preference of this species for humid microhabitats, whereas *N. sellnicki* is less common than *N. coronata* and occurs in drier habitats.

Abstract: *Nanhermannia coronata* Berlese, 1913, is a common and abundant oribatid species in peatlands but can be easily mistaken for *N. sellnicki* Forsslund, 1958, as an adult. Therefore, the identity of adults of *N. coronata* from several sites in Norway and Ireland was supported by the COI sequence data, and based on this material, the morphological ontogeny of this species is described and illustrated to highlight the differences between *N. coronata* and *N. sellnicki*. In all juvenile stages of *N. coronata*, the bothridial seta is absent, but two pairs of exobothridial setae are present, including short *exp* and *exa* reduced to its alveolus. In the larva, seta f_1 is setiform, but in the nymphs, it is reduced to its alveolus. Most prodorsal and gastronotal setae of larva are short, and of nymphs they are long. In all instars, the leg segments are oval in cross section and relatively thick, and many setae on tarsi are relatively short, thick and conical, except for longer apical setae. Seta *d* accompanies solenidion σ on all genua, φ_1 on tibia I and φ on other tibiae. We found some morphological characters that clearly differentiate *N. coronata* from *N. sellnicki*, like the number of setae on femora of adults and tritonymphs, the shape of insertions of prodorsal seta *in* and all gastronotal and adanal setae of juveniles; in *N. sellnicki*, these setae are inserted in small individual depressions, whereas in *N. coronata*, these depressions are absent. Our ecological observations confirm a common occurrence of *N. coronata* in raised bogs, a high percentage of juvenile stages in its populations and a preference of this species for humid microhabitats, whereas *N. sellnicki* is less common than *N. coronata* and occurs in drier habitats.

Keywords: oribatid mites; juveniles; leg setation; stage structure; ecology; integrated taxonomy approach

Citation: Seniczak, S.; Seniczak, A. Morphological Ontogeny and Ecology of a Common Peatland Mite, *Nanhermannia coronata* (Acari, Oribatida, Nanhermanniidae). *Animals* **2023**, *13*, 3590. https://doi.org/10.3390/ani13223590

Academic Editor: Alexis Ribas

Received: 27 September 2023
Revised: 8 November 2023
Accepted: 10 November 2023
Published: 20 November 2023

1. Introduction

Nanhermannia Berlese, 1913, *sensu stricto* (*N. nanus* Nicolet, 1855, as the type species) comprises 31 nominative species [1]; all are of medium size as adults. The diagnosis of adults of this genus was given by Seniczak et al. [2] as follows: Adults are medium-sized (450–660 μm), elongated, brown to dark brown and narrow, with bothridia and bothridial setae situated on top of the prodorsum; there are protuberances on the posterior part of the prodorsum with sclerotized tubercles, often extending above the anterior part of the notogaster. The notogaster is cylindrical, with a coarse structure of pits and 15 pairs of long setae, curved and appressed to body, extended ventrally and connected ventro-medially; an arched suture delimits the genital and aggenital area laterally (diagastry). The aggenital plate is fused with the epimere, and the adanal plate is fused with the notogaster, but still recognizable; the formula of the epimeral setae is usually 3-1-3-3 or 3-1-3-4, but in some species, hypertrichy on some epimeres occurs; 7–10 pairs of genital setae, two pairs of aggenital setae, two pairs of anal setae and three pairs of adanal setae are present. Legs have one claw.

Identification of *Nanhermannia* species is not easy because they are relatively similar to each other by having a similar body shape, structure of pits and shape of notogastral setae. The species differ from one another by the shape of protuberances on the posterior margin of the prodorsum and the number of sclerotized tubercles, which are considered diagnostic in *Nanhermannia*, but these characters vary in some species [3–10], including *N. coronata* Berlese, 1913, being a source of confusion. For example, during the revision of the twenty-year-old oribatid mite collection of the Institute of Biology, University of Latvia, a high discrepancy in the identification of *N. coronata* was detected, and ca. 50% of specimens (out of 40 studied) were wrongly identified [11]. Therefore, more investigations on the morphology of *Nanhermannia* species are required, including the juvenile stages and molecular investigations to improve the diagnosis of species of this genus, and in some cases to support the identity of individuals within species.

Systematic problems in *Nanhermannia* also occur. For example, in the past, *N. coronata* was confused with *N. nana* (Nicolet, 1855) *sensu* Willmann [12], which was clarified by Forsslund [13], and it was confirmed by Solhøy [14,15] that in Norwegian oligotrophic bogs, *N. coronata* was present. Another example is considering *N. coronata* by Subías [1] as a junior synonym of *N. dorsalis* (Banks, 1896), whereas Weigmann [8] and Norton and Ermilov [16] treated it as a separate species, and we agree with the latter opinion.

Identification of *Nanhermannia* species can also be problematic in ecological investigations. For example, *N. coronata* can be mistaken for *N. sellnicki* Forsslund, 1958. An identification of *N. coronata* is commonly based on the shape of protuberances on the posterior margin of the prodorsum and the number of sclerotized tubercles, which vary within this species, so the ecology of *N. coronata* given in some papers can be imprecise and needs improving. These species have undoubtedly different ecological preferences; *N. sellnicki* is less common than *N. coronata* and occurs in drier habitats [13], like birch forests, especially with understory formed by *Vaccinium* and *Empetrum*, while *N. coronata* is found in moist habitats, especially in raised bogs [15,17–22]. The latter species can be very abundant and dominant among the Oribatida [15,19,20,23] and among the mites [20]. For example, in western Norway, among nearly 60,000 mites collected from different peatland microhabitats, and represented by 154 species from all mite orders (Mesostigmata, Trombidiformes and Sarcoptiformes), *N. coronata* was the most abundant species (it made up 18% of all mite specimens) and occurred in about 90% of samples [20,22]. Such an abundant and common occurrence of *N. coronata* in peatlands requires better knowledge of the morphology of adults and juveniles, which justifies the need for the current study. The more so that in populations of this species, the juveniles are often very abundant, e.g., in peatlands in Norway, they constituted about 40% of individuals [20,22], so it is very important to include them in ecological analyses.

The morphology of juveniles of *Nanhermannia* is insufficiently known. According to Norton and Ermilov [15] and Seniczak et al. [2], the full morphological ontogeny of *N. comitalis* Berlese, 1916, *N.* cf. *coronata*, *N. nana* and *N. sellnicki* is known, which constitutes nearly 13% of all species. The morphological ontogeny of *N.* cf. *coronata* was investigated by Ermilov and Łochyńska [9], but these authors treated the leg setation generally, without labelling of leg setae, which we consider species-specific. Moreover, we found some differences in the morphology of juveniles of this species investigated by these authors and those studied herein, which probably illustrates the morphological variability of the species.

The aim of this paper is to describe the morphological ontogeny of *N. coronata* and compare it with that of congeners. The identity of specimens from several sites in Norway and Ireland is supported by the COI sequence data. We also add some data on the ecology of this species.

2. Material and Methods

2.1. Morphological and Biological Studies

The adults and juveniles of *N. coronata* used in morphological and biological studies were collected on 29 and 30 June 2020 (leg. A. Seniczak, K.I. Flatberg, K. Hassel and S. Roth) from an Atlantic raised bog located in Hitra (Hitra municipality, Trøndelag, Norway, 63°29′21.7″ N, 8°52′25.1″ E, 82 m a. s. l.). In total, 26 samples were collected. These samples were transported in plastic bags in cool boxes for four days to the laboratory of Bydgoszcz University of Science and Technology, Bydgoszcz, Poland, and extracted with Tullgren funnels for ten days into 90% ethanol. In these samples, *N. coronata* was the only member of *Nanhermannia*, and therefore, we considered the juveniles to belong to this species. The morphological ontogeny of *N. coronata* investigated herein is based on the abundant individuals from the transition zone between hummock and hollow in Hitra, but the morphological characters of instars were checked with those from other sites studied herein. We investigated the stage structure of mites, and based on 30 randomly selected adults, we determined the sex ratio, number of gravid females and carried eggs. We also measured the total body length (tip of the rostrum to the posterior edge of the notogaster) in the lateral aspect and the body width (widest part of the notogaster) in the dorsal aspect. In a similar way, we measured the morphological characters of juvenile and adult instars of *N. coronata* given in Table 1, as well as the size of anal and genital openings and setae perpendicularly to their length in μm. We used the microscopy Nikon Eclipse Ni.

Table 1. Information about sequenced specimens of *Nanhermannia coronata* and other oribatid species used in this study; ad—adult, juv—juveniles.

Species	Sequence Code at BOLD	Stage	GeneBank Access No.	Locality	Coordinates	Elevation (m a.s.l.)	Collection Data
N. coronata	UMNFO663-18	ad	OR732229	NO: Vestland, Lydehorn	60.370, 5.244	216.8	6 October 2018, Seniczak, A.
	UMNFO664-18	ad	OR732235	NO: Vestland, Lydehorn	60.370, 5.244	216.8	6 October 2018, Seniczak, A.
	UMNFO665-18	ad	OR732231	NO: Vestland, Lydehorn	60.370, 5.244	216.8	6 October 2018, Seniczak, A.
	MARBN105-21	ad	OR732225	NO: Trondelag, Hitra	63.489, 8.874	48.2	29 July 2020, leg. A. Seniczak, K.I. Flatberg, K. Hassel, S. Roth
	MARBN106-21	ad	OR732221	NO: Trondelag, Hitra	63.489, 8.874	48.2	29 July 2020, leg. A. Seniczak, K.I. Flatberg, K. Hassel, S. Roth
	MARBN158-21	ad	OR732223	NO: Trondelag, Høstadmyra	63.405, 10.12	110.0	30 July 2020, leg. A. Seniczak, K.I. Flatberg, K. Hassel, S. Roth
	MARBN159-21	ad	OR732230	NO: Trondelag, Høstadmyra	63.405, 10.12	110.0	30 July 2020, leg. A. Seniczak, K.I. Flatberg, K. Hassel, S. Roth
	MARBN160-21	ad	OR732234	NO: Trondelag, Høstadmyra	63.405, 10.12	110.0	30 July 2020, leg. A. Seniczak, K.I. Flatberg, K. Hassel, S. Roth

Table 1. *Cont.*

Species	Sequence Code at BOLD	Stage	GeneBank Access No.	Locality	Coordinates	Elevation (m a.s.l.)	Collection Data
	MARBN516-22	ad	OR732224	NO: Nordland, Kummeren	67.043, 14.216	28.1	26 July 2021, Seniczak, A., Flatberg, K.I.
	MARBN517-22	ad	OR732218	NO: Nordland, Kummeren	67.043, 14.216	28.1	26 July 2021, Seniczak, A., Flatberg, K.I.
	MARBN519-22	ad	OR732227	NO: Nordland, Kummeren	67.043, 14.216	28.1	26 July 2021, Seniczak, A., Flatberg, K.I.
	MARBN343-21	ad	OR732226	IR: Leinster, Lullymore	53.270, −6.949	77.4	9 December 2014, leg. A. Seniczak, T. Bolger
	MARBN344-21	ad	OR732233	IR: Leinster, Lullymore	53.270, −6.949	77.4	9 December 2014, leg. A. Seniczak, T. Bolger
	MARBN345-21	ad	OR732232	IR: Leinster, Lullymore	53.270, −6.949	77.4	9 December 2014, leg. A. Seniczak, T. Bolger
	MARBN346-21	ad	OR732228	IR: Leinster, Lullymore	53.270, −6.949	77.4	9 December 2014, leg. A. Seniczak, T. Bolger
	FINOR987-16	ad	MZ608481	FIN: Varsinais-Suomi: Paimio, Jaervessuo	60.451, 22.775	49.9	9 October 2014, leg. R. Penttinen
	FINOR988-16	ad	MZ609187	FIN: Varsinais-Suomi: Paimio, Jaervessuo	60.451, 22.775	49.9	9 October 2014, leg. R. Penttinen
	FINOR989-16	ad	MZ611116	FIN: Varsinais-Suomi: Paimio, Jaervessuo	60.451, 22.775	49.9	9 October 2014, leg. R. Penttinen
N. nana	GBMYR861-15	ad	In BOLD	GE:			
	UZINS170-23	ad	In BOLD	SL: Bratislava, Sitina	48.171, 17.0656	213.3	1 June 2022, leg. Mangova, B.
	UZINS171-23	ad	In BOLD	SL: Bratislava, Sitina	48.171, 17.0656	213.3	1 June 2022, leg. Mangova, B.
Camisia foveolata	MARBN334-21	ad	OR732222	NO: Svalbard, Longyearbyen, Endalen	78.209, 15.711	22.6	5 June 2018, Roth, S.
	MARBN335-21	juv	OR732220	NO: Svalbard, Longyearbyen, Endalen	78.209, 15.711	22.6	5 June 2018, Roth, S.
	MARBN341-21	ad	OR773185	NO: Vestland, Finse	60.593, 7.432	1352.0	7 September 2018, Seniczak, A.
Platynothrus punctatus	MARBN097-21	ad	OL671034	NO: Vestland, Finse	60.593, 7.432	1352.0	22 September 2019, Seniczak, A.
	MARBN098-21	ad	OL671021	NO: Vestland, Finse	60.593, 7.432	1352.0	22 September 2019, Seniczak, A.
	MARBN365-21	juv	OL671024	SP: Andalusia, Borreguil de la Virgen	37.087, −3.374	2500.7	18 August 2017, Seniczak, A., F. Ondoño, E.

The illustrations are limited to the body regions that show substantial differences between instars and were prepared from individuals mounted temporarily on slides in lactic acid. In the text and figures, we used the following abbreviations: rostral (*ro*), lamellar (*le*), interlamellar (*in*) and exobothridial (*exa, exp*) setae, bothridium (*bo*), bothridial seta (*bs*), notogastral or gastronotal setae (*c-, d-, e-, f-, h-, p*-series), cupules or lyrifissures (*ia, im, ip, ih, ips, iad, ian*), cheliceral seta (*cha, chb*), Trägårdh organ (*Tg*), palp setae (*sup, l, cm, acm, vt, ul, su*) and solenidion ω, epimeral setae (*1a–c, 2a, 3a–c, 4a–d*), genital setae (*g*), adanal and anal setae (*ad-, an*-series), leg solenidia (σ, φ, ω), famulus (ε) and setae (*bv, ev, d, l, v, ft, tc, pv, a, s, p, u*). The leg setae *l* on femora, and *l* and *v* on tarsi were labelled according to their appearance in the ontogeny. The terminology used follows that of Grandjean [24–28] and Norton and Behan-Pelletier [29]. The species nomenclature follows partly Subías [1], Weigmann [8] and Norton and Ermilov [14].

For scanning electron microscopy (SEM), the mites were air-dried and coated with Au/Pd in a Polaron SC502 sputter coater and placed on Al-stubs with double-sided sticky carbon tape. Observations and micrographs were made with a QUANTA FEG 450 scanning electron microscope.

2.2. DNA Barcoding

For molecular studies, we used the specimens of *N. coronata* collected in raised bogs in southern, central and northern Norway, and Ireland (Table 1). In all locations, samples of *Sphagnum* mosses of 500 cm^3 each were collected and extracted into 90% ethanol in the same way as described above. Additionally, we used publicly available DNA sequences of *N. coronata* from Finland and some other species of *Nanhermannia* identified (*N. nana* from Germany and Slovakia) and *N. comitalis* from Germany. We used species of putatively close genera as outgroups, *P. punctatus* (L. Koch, 1879) and *Camisia foveolata* Hammer, 1955.

Specimens of *N. coronata* from different locations were sent for DNA barcoding to the Canadian Centre for DNA Barcoding (CCDB) in Guelph, Canada. Each specimen was photographed, and the photos are the vouchers that are available in the Barcode of Life Data System (BOLD, http://boldsystems.org, accessed on 1 October 2023). The specimens were subsequently placed in a well containing 50 mL of 90% ethanol in a 96-well microplate and sent to the CCDB. Mites were sequenced for the barcode region of the COI gene according to standard protocols at the CCDB (www.ccdb.ca, accessed on 1 October 2023, using either LepF1/LepR1 [30] or LCO1490/m HCO2198 [31] primer pairs. The DNA extracts were placed in archival storages at −80 °C, most at the CCDB, and some (with sequencing code starting with UMNFO) at the University Museum of Bergen (ZMBN). Fifteen COI sequences that met the criteria of animal barcodes (sequence length ≥ 500 bp) were obtained. These sequences were blasted against GenBank in order to detect and exclude possible contaminations and were further used in the analyses. The sequences are available in GenBank (accessions numbers in Table 1).

Sequence variation within *N. coronata* specimens and between species was calculated in BOLD using the Kimura 2 Parameter distance model, pairwise deletion and BOLD Aligner (Amino Acid based HMM). The sequences were aligned by eye and neighbor-joining trees were constructed using MEGA6 [32]. Joint neighborhood topologies were visualized in FigTree 1.4.2 (available at http://tree.bio.ed.ac.uk/software/figtree).

2.3. Ecological Studies

Ecological studies on *N. coronata* were carried out in Atlantic raised bogs located in Hitra (coordinates were given above) and Høstadmyra (Trondheim municipality, Trøndelag, Norway, 63°24′19.4″ N, 10°07′13.5″ E, 110 m a. s. l.). Hitra is an island and forms a separate municipality, which is characterized by the mild oceanic climate, with a mean annual temperature of 8 °C and an annual precipitation of 917.8 mm. In the coldest month (January), the average temperature is 1 °C, and in the warmest month (July), it is 16 °C. Høstadmyra is characterized by a slightly colder and drier climate than in Hitra. The mean annual temperature is 6 °C, and annual precipitation is 575.6 mm. The average temperature in the coldest month (January) is −1 °C, and in the warmest month (July), it is 15 °C (https://www.timeanddate.no, accessed on 1 October 2023). In total, 63 samples of *Sphagnum* mosses of 500 cm^3 each were collected on 29 and 30 June 2020 (26 from Hitra and 37 from Høstadmyra) from the following microhabitats: hummocks (31 samples; 12 from Hitra and 19 from Høstadmyra), lawns (14 samples; 4 from Hitra and 10 from Høstadmyra), low part of hummocks (4 samples; 2 from Hitra and 2 from Høstadmyra) and hollows (14 samples; 8 from Hitra and 6 from Høstadmyra). The method of extraction of samples was described above.

Populations of *N. coronata* from Hitra and Høstadmyra, and from studied microhabitats (hummocks, lawns, low part of hummocks and hollows) were characterized by abundance

(*A* in 500 cm^3). The basic statistical descriptors included the mean values and standard deviation. Equality of variance was tested with the Levene test, and normality of the distribution was tested with the Kolmogorov–Smirnov test. As the assumptions of variance analysis were not met, non-parametric tests were employed. Kruskal–Wallis ANOVA by ranks was utilized to test for significant differences between means [33]. The significance level for all analysis was accepted $\alpha = 0.05$. These calculations were carried out with STATISTICA 12.5 software.

3. Results

3.1. Morphological Ontogeny of Nanhermannia coronata Berlese, 1913 (Figures 1–23)

3.1.1. Diagnosis

Adults are of medium size (length 450–660), with characters of *Nanhermannia* given by Seniczak et al. [2]. The prodorsal seta *in* is thin, longer than the bothridial seta; protuberances on the posterior part of the prodorsum are highly sclerotized, with 5–7 small posterior tubercles and five pairs of light spots between setal pair *in*. Seta *exp* is reduced to its alveolus. The ratio of body length/width is 2.3:1, and the notogastral setae are long, c_1, c_3, d_2 and e_1 reaching insertions of setae d_1, d_2, e_1 and h_1, respectively. The formula of the epimeral setae is 3-1-3-4. Seta *d* accompanying solenidion σ on all genua, φ_1 on tibia I and φ on other tibiae are present.

Juveniles are elongated, the body unpigmented and with pits, the hysterosoma cylindrical and the central part of the prodorsum, epimeres and legs light brown. The bothridium is small, the bothridial seta absent, seta *exp* short and *exa* reduced to its alveolus. The larva has 12 pairs of short gastronotal setae, including f_1 and h_2; nymphs have 15 pair of long setae, excluding f_1 reduced to its alveolus, all setae smooth. Leg segments are relatively thick and oval in cross section, and many setae on tarsi are relatively short, thick and conical, except for the longer apical setae. Seta *d* accompanying solenidion σ on all genua, φ_1 on tibia I and φ on other tibiae are present.

The formula of the genital setae is 1-4-6-9 (protonymph to adult), and femora of deutonymph are 4-4-2-2 (leg I–IV), tritonymph 5-(5-6)-(2-3)-3 and adult 5-7-3-3.

Figure 1. *Nanhermannia coronata*, adult, legs partially drawn, scale bars 50 µm. Dorsal aspect.

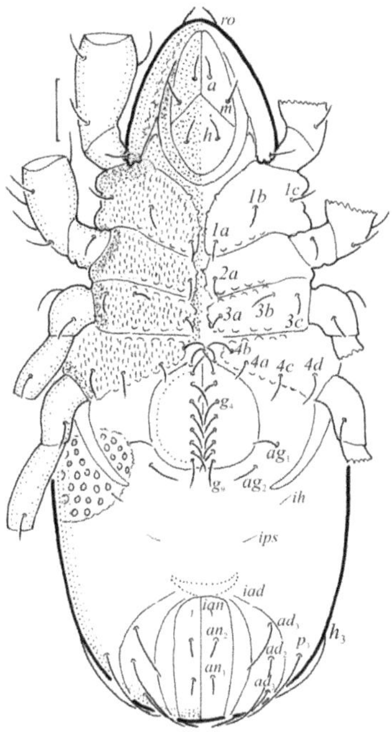

Figure 2. *Nanhermannia coronata*, adult, legs partially drawn, scale bars 50 μm. Ventral aspect.

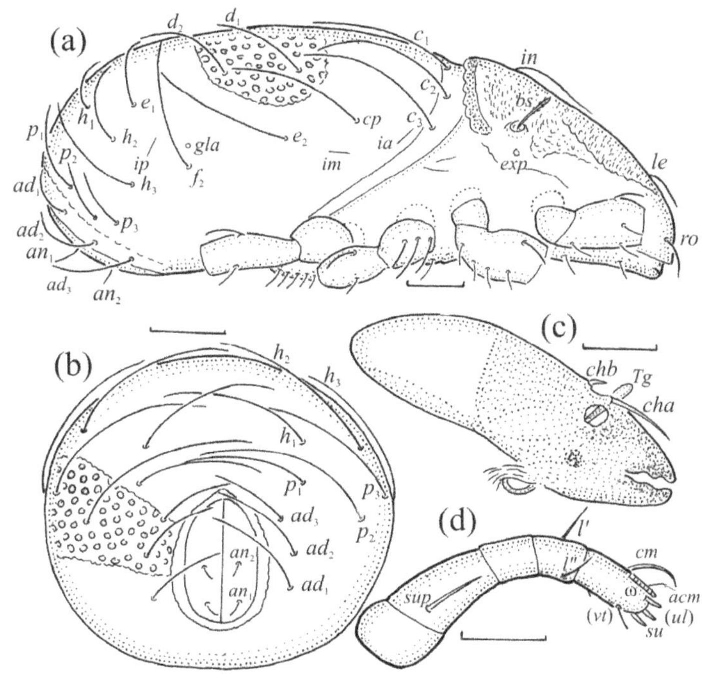

Figure 3. *Nanhermannia coronata*, adult. (**a**) Lateral aspect, legs partially drawn. (**b**) Posterior part of notogaster, posterior aspect. Mouthparts, right side, antiaxial aspect. (**c**) Chelicera. (**d**) Palp. Scale bars (**a–c**) 50 μm, (**d**) 20 μm.

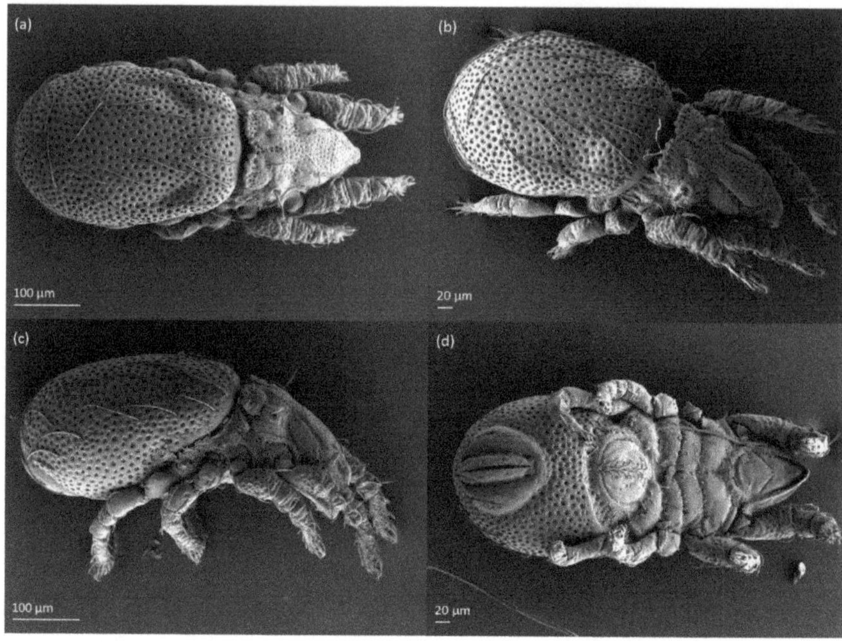

Figure 4. *Nanhermannia coronata,* adult, SEM micrographs. (**a**) Dorsal view, (**b**) dorsolateral view, (**c**) lateral view, (**d**) ventral view.

Figure 5. *Nanhermannia coronata,* adult, SEM micrographs. (**a**–**c**) Anterior part of body, dorsal view, (**d**) bothridial seta, lateral view.

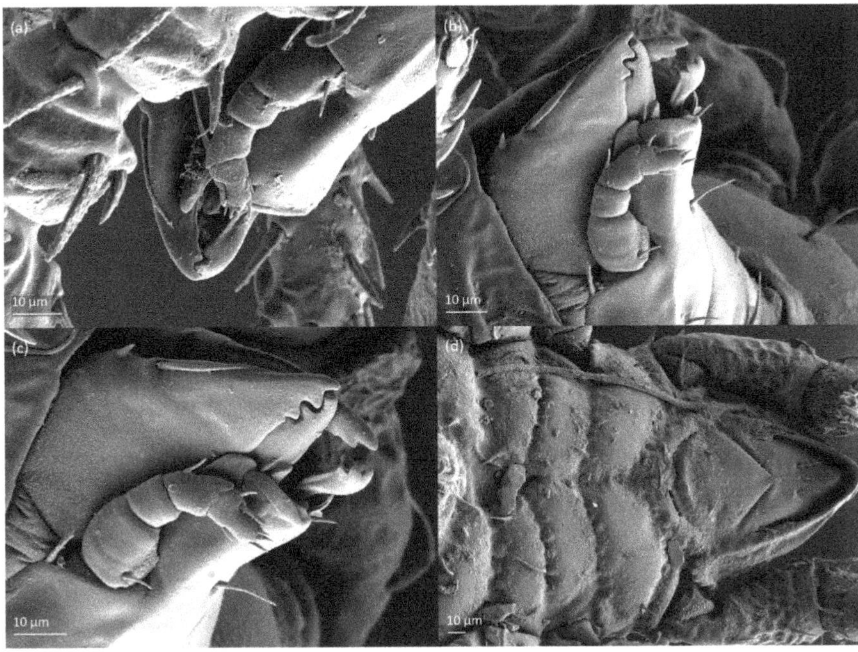

Figure 6. *Nanhermannia coronata,* adult, SEM micrographs. (**a–c**) Mouthparts, lateral view, (**d**) anterior part of body, ventral view.

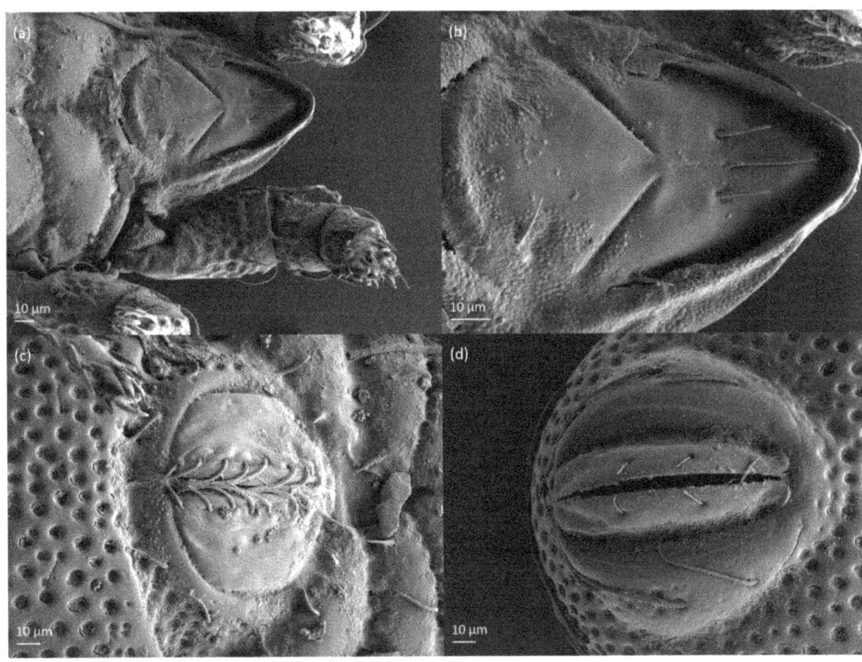

Figure 7. *Nanhermannia coronata,* adult, SEM micrographs. Ventral view, (**a**,**b**) anterior part of body, (**c**) genital plates, (**d**) anal plates.

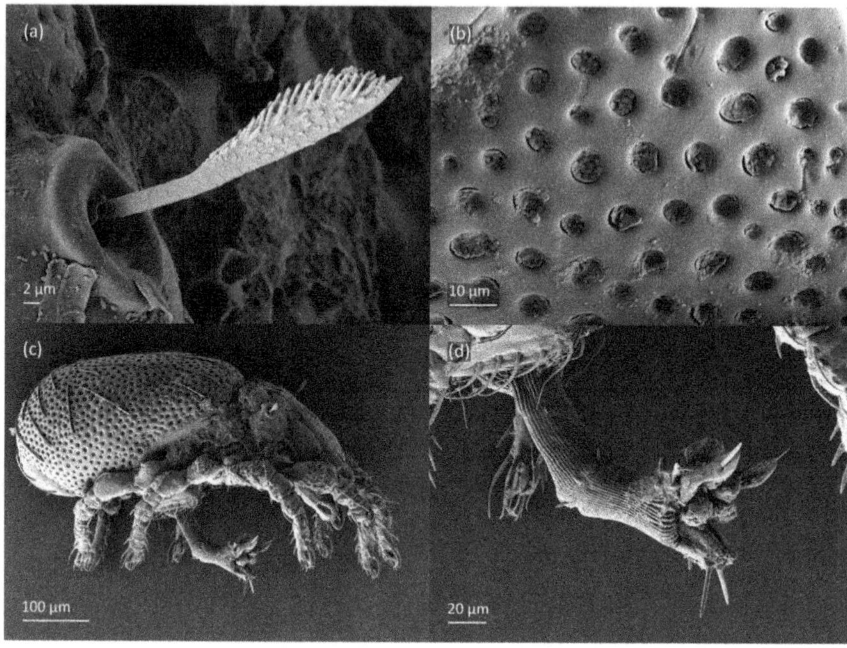

Figure 8. *Nanhermannia coronata*, adult, SEM micrographs. (**a**) Bothridial seta, lateral view; (**b**) pattern of notogaster; lateral view, (**c**) adult with rejected ovipositor, (**d**) ovipositor.

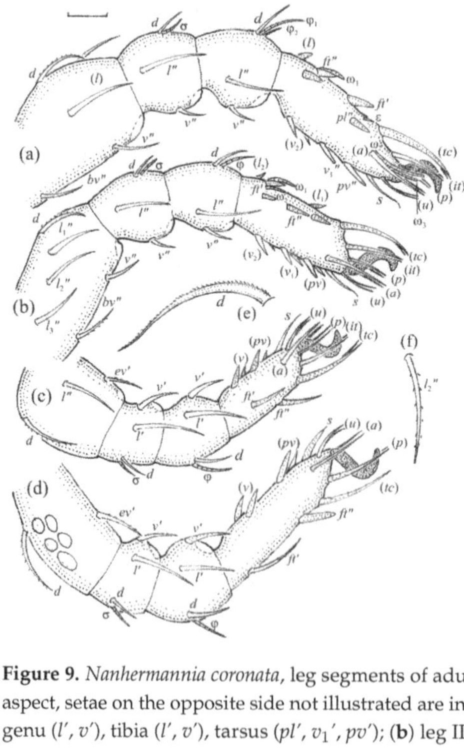

Figure 9. *Nanhermannia coronata*, leg segments of adult (part of femur to tarsus), right side, antiaxial aspect, setae on the opposite side not illustrated are indicated in the legend, scale bar 20 µm. (**a**) Leg I, genu (l', v'), tibia (l', v'), tarsus (pl', v_1', pv'); (**b**) leg II, femur (l_1'), genu (l', v'), tibia (l', v'); (**c**) leg III, tibia (v''); (**d**) leg IV, tibia (v''); (**e**) seta *d* on leg IV; (**f**) seta l_2'' on femur II (e, f enlarged).

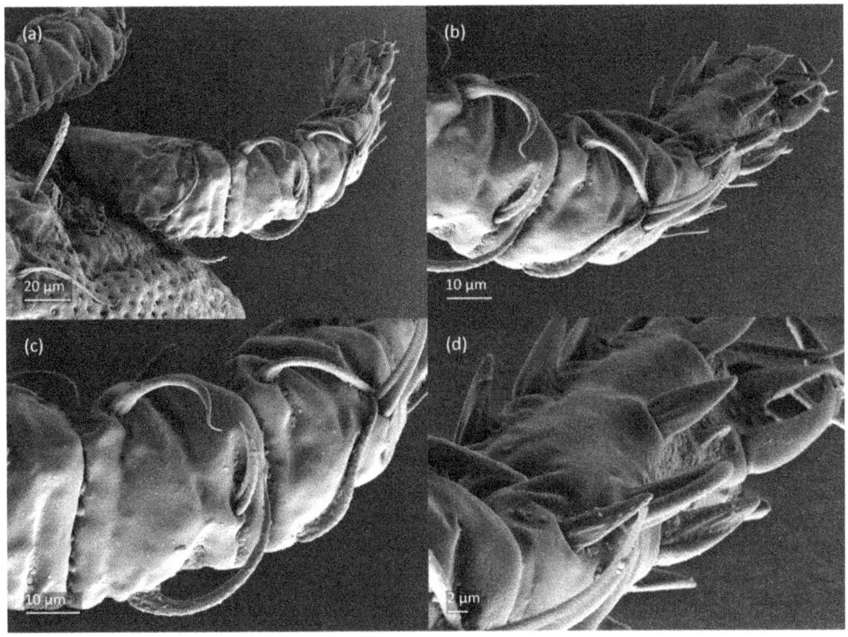

Figure 10. *Nanhermannia coronata*, adult, SEM micrographs. Dorsal view, (**a**) leg I; (**b–d**) parts of leg I.

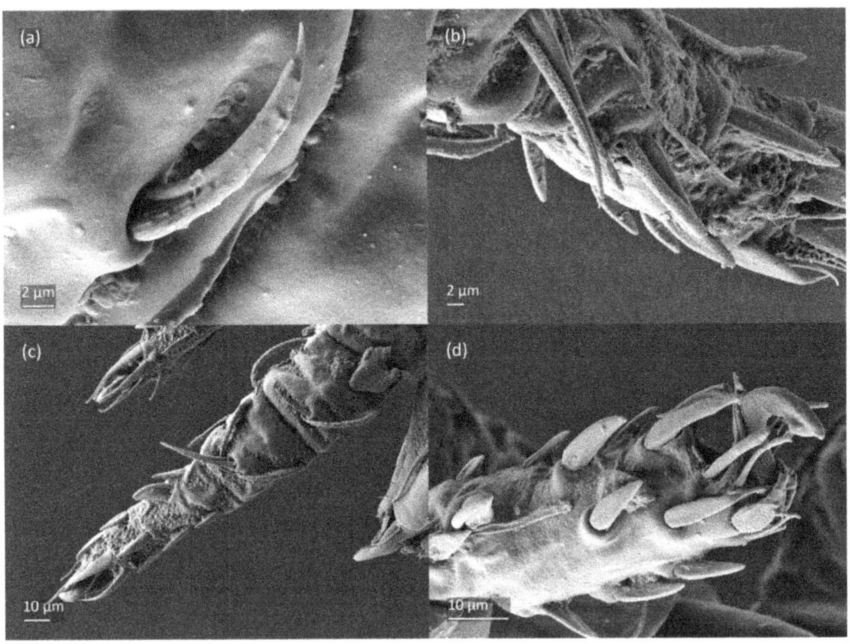

Figure 11. *Nanhermannia coronata*, leg I of adult, SEM micrographs. Dorsal view, (**a**) solenidion σ and seta *d* on genu I, (**b**) part of genu, tibia and tarsus I, (**c**) part of leg I; (**d**) part of tarsus I, lateral view.

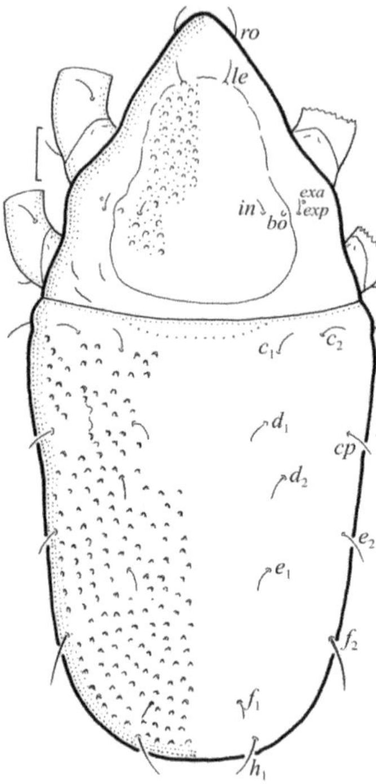

Figure 12. *Nanhermannia coronata*. Larva, dorsal aspect, legs I and II partially drawn, scale bar 20 µm.

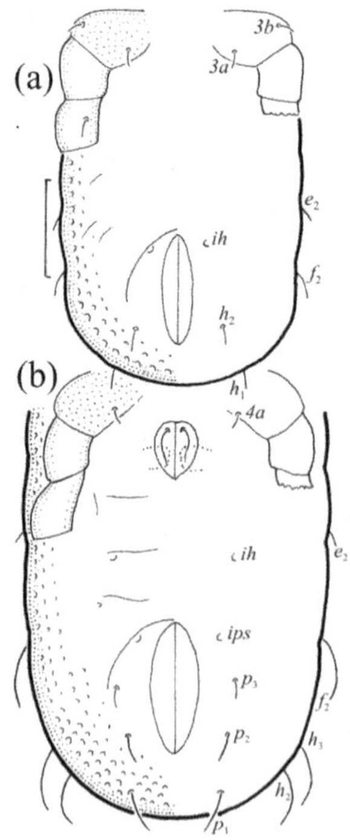

Figure 13. *Nanhermannia coronata*. Ventral part of hysterosoma, legs III and VI partially drawn, scale bar 50 µm, (**a**) larva, (**b**) protonymph.

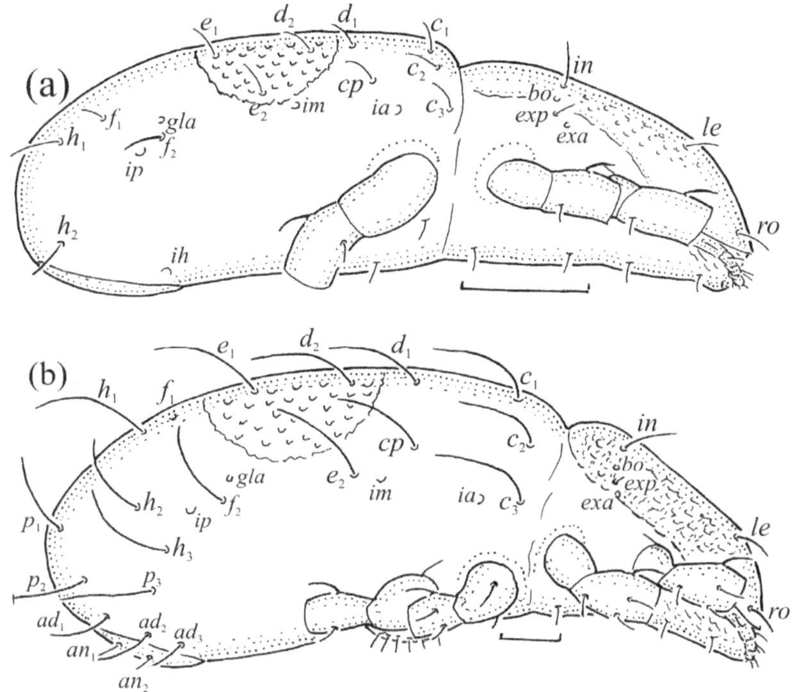

Figure 14. *Nanhermannia coronata*, lateral aspect, legs partially drawn, scale bars 50 μm. (**a**) Larva, (**b**) tritonymph.

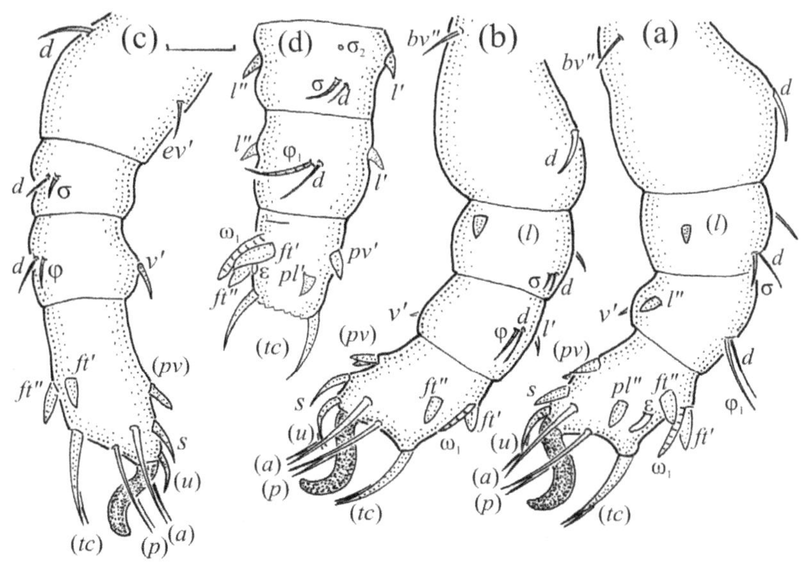

Figure 15. *Nanhermannia coronata*, leg segments of larva (part of femur to tarsus), right side, scale bar 20 μm. Antiaxial aspect, (**a**) leg I, tarsus (*pl'* not illustrated); (**b**) leg II; (**c**) leg III; dorsal aspect, (**d**) leg I, genu, tibia and part of tarsus.

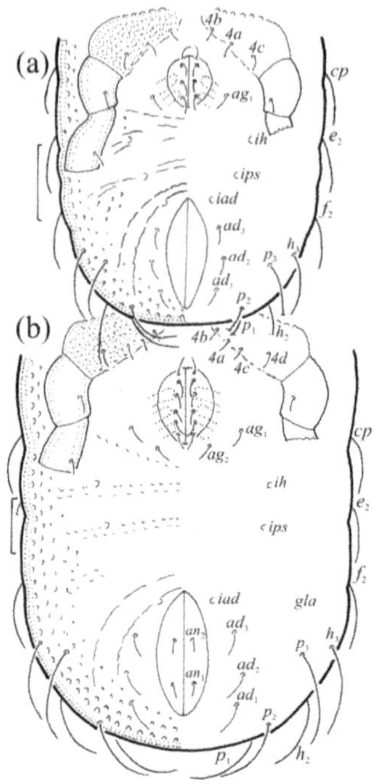

Figure 16. *Nanhermannia coronata*, scale bars 50 µm. Ventral part of hysterosoma, legs IV partially drawn, (**a**) deutonymph, (**b**) tritonymph.

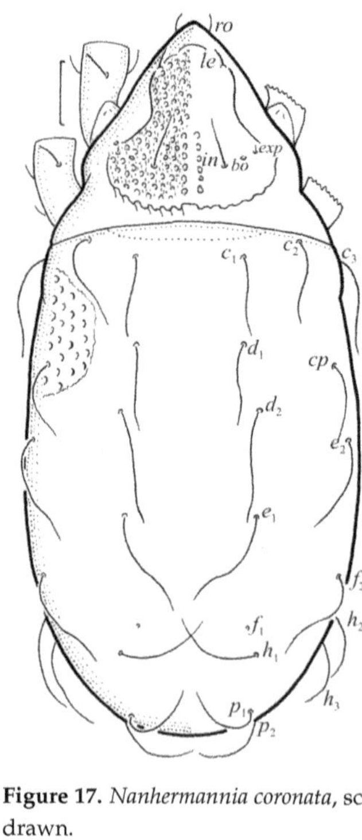

Figure 17. *Nanhermannia coronata*, scale bars 50 µm. Tritonymph, dorsal aspect, legs I and II partially drawn.

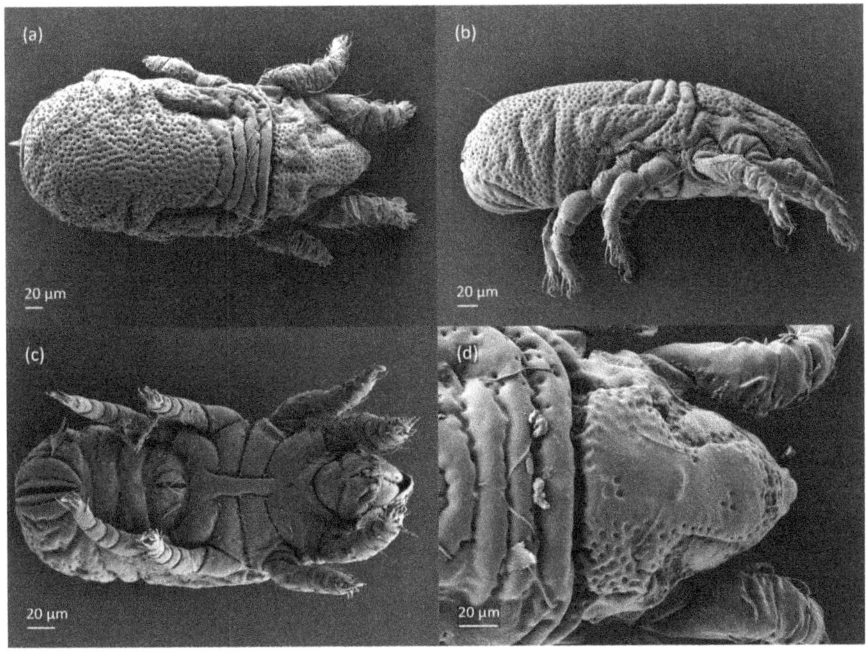

Figure 18. *Nanhermannia coronata*, tritonymph, SEM micrographs. (**a**) Dorsal view, (**b**) lateral view, (**c**), ventral view, (**d**) anterior and medial part of body.

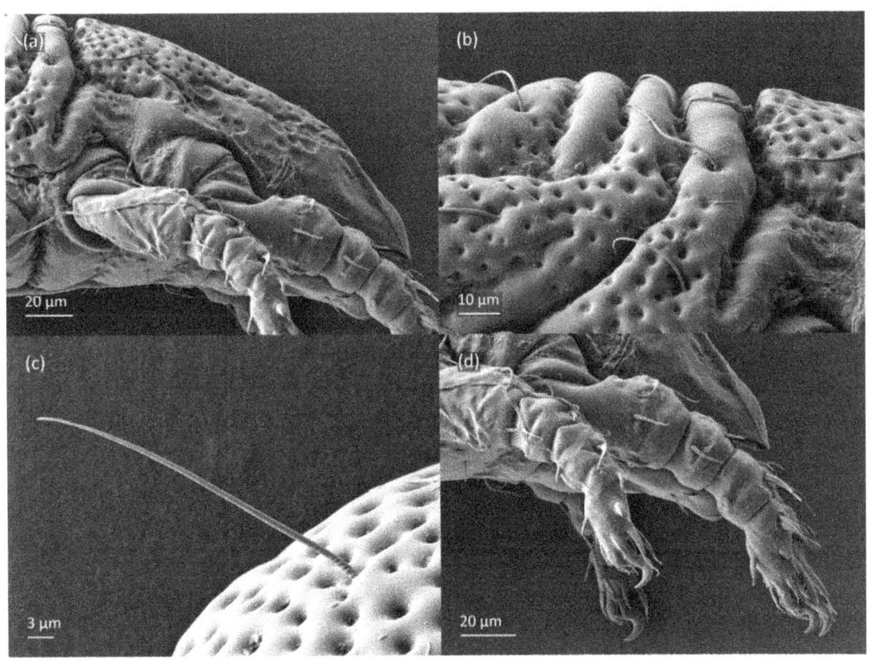

Figure 19. *Nanhermannia coronata*, tritonymph, SEM micrographs. Lateral view, (**a**) anterior part of body, (**b**) medial part of body, (**c**) seta h_3, (**d**) leg I and II.

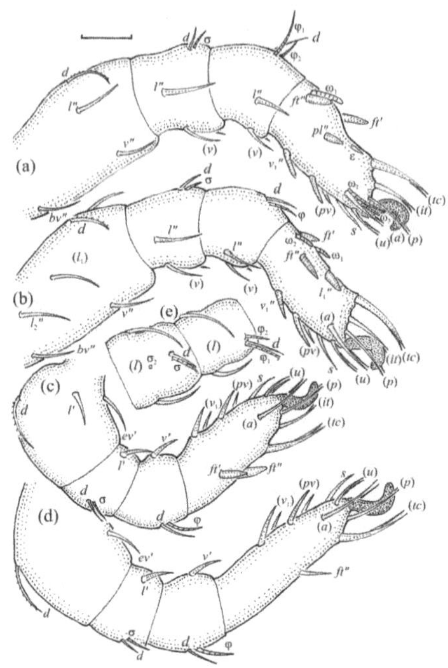

Figure 20. *Nanhermannia coronata*, leg segments of tritonymph (part of femur to tarsus), right side, seta on the opposite side not illustrated are indicated in the legend, scale bar 10 µm. Antiaxial aspect, (**a**) leg I, femur (*l'*), genu (*l'*), tibia (*l''*), tarsus (*pl'*, *v'*); (**b**) leg II, genu (*l'*), tibia (*l'*), tarsus ($l_1{}'$, $v_1{}'$); (**c**) leg III, tibia (*v''*); (**d**) leg IV, tibia (*v''*); dorsal aspect, (**e**) genu and tibia I, dorsal view.

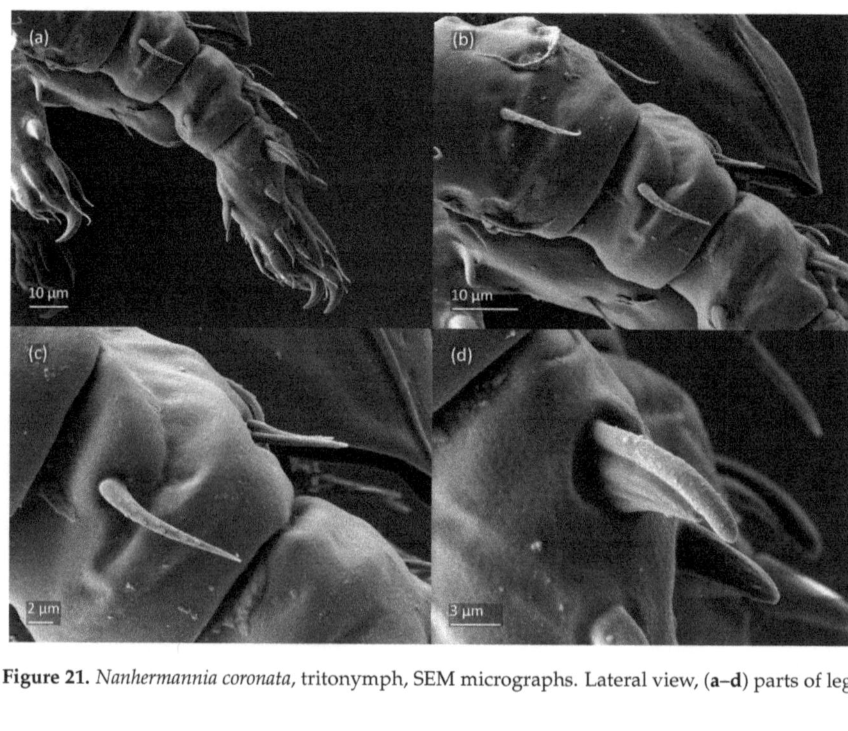

Figure 21. *Nanhermannia coronata*, tritonymph, SEM micrographs. Lateral view, (**a**–**d**) parts of leg I.

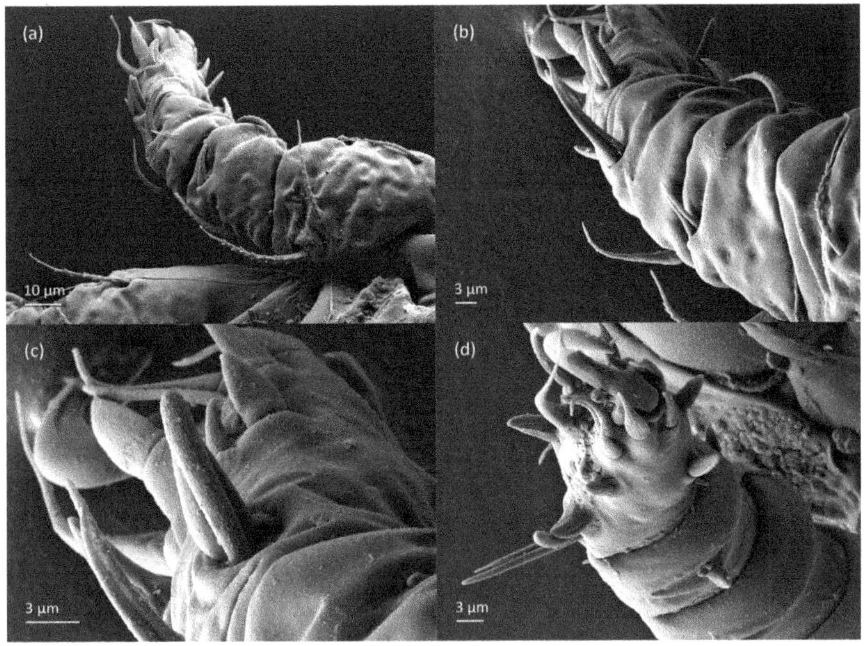

Figure 22. *Nanhermannia coronata*, tritonymph, SEM micrographs. Parts of leg I, (**a**–**c**) dorsal view; (**d**) ventral view.

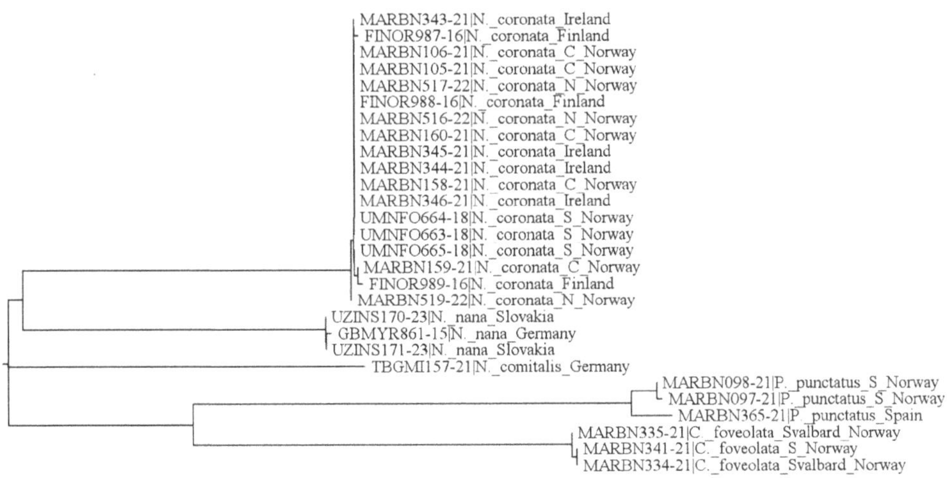

Figure 23. Neighbor-joining tree based on COI sequences of *Nanhermannia coronata* from southern, central and northern Norway (S Norway, C Norway and N Norway, respectively), Ireland, and Finland, and some other *Nanhermannia* species; *Platynothrus punctatus* and *Camisia foveolata* were used as outgroups. Details on sampling locations are given in Table 1.

3.1.2. Morphology of Adult

The adult is elongated (Figures 1, 2, 3a and 4), similar to that described by Berlese [34], but see Remarks. The mean length (and range) of females is 541.7 ± 14.0 (520–556, n = 30); the mean width (and range) is 252.0 ± 6.3 (228–260); males absent. The prodorsal seta *in* is long and thin, longer than the bothridial seta; protuberances on the posterior part of the prodorsum are highly sclerotized, with 5–7 small posterior tubercles (Figures 1, 3a, 4a–c and 5 and Table 2), and six pairs of light spots present between setal pair *in*. The ratio of body length/width is 2.3:1. Seta *exp* short, *exa* is reduced to its alveolus. The prodorsum and hysterosoma are characterized by pits, observed in SEM figures as small holes (Figures 4, 5, 7c,d, 8b,c and 10a). The notogastral setae are long and smooth, c_1, c_3, d_2 and e_1 reaching insertions of setae d_1, d_2, e_1 and h_1, respectively (Figures 1, 3a,b, 4a–c, 5a–c and 8c and Table 2). Lyrifissure *ia* is posterior to seta c_3, *im* anterior to seta e_2, *ip* anterior to seta h_2, *ian* anterior to seta an_2, *iad* anteromedial to seta ad_3, and *ips* and *ih* pushed anterior and anterolateral to seta ad_3, respectively (Figures 2 and 3a). Opisthonotal gland opening *gla* is not observed among pits. The chelicera is chelate-dentate, and seta *cha* is located anterior to *chb* and clearly longer than *chb*, both smooth (Figures 3c and 6a–c). Most palpal setae are relatively short and smooth (Figures 3d and 6a–c), solenidium ω and eupathidia are short, and the formula of setae (trochanter to tarsus + solenidion ω) is 1-0-2-7(1). Hypostomal and epimeral setae are short and smooth (Figures 2, 4d, 6c,d and 7a,b). Aggenital setae (two pairs) are short, and genital setae (nine pairs) are slightly longer, all smooth (Figures 2, 4d, 7c,d and 8c,d). Adanal setae (three pairs) are long, and anal setae are short, all smooth (Figures 2, 3a,b, 4d, 7d and 8c). The ovipositor is relatively thick, with relatively thick setae (Figure 8c,d). The legs are relatively thick, cuticle with ornamentation (Figures 4, 5a–c, 6c, 7a, 8c, 9, 10 and 11) and all femora oval in cross section. Seta *l* on trochanter II and *d* on all femora are barbed, while other leg setae are smooth or finely barbed. Solenidion ω_1 on tarsus I is located medial to seta *ft*″, whereas solenidia ω_2 and ω_3 are located anterior to seta *a*″. Solenidia ω_1 and ω_2 on tarsus II are located medial and lateral to seta *ft*′, respectively. Seta *d* accompanying solenidion σ on all genua, φ_1 on tibia I and φ on other tibiae are present (Figures 9–11 and Table 3). In all tarsi, hypertrichy occurs; setae on the basal and medial part of tarsi are conical, those on the distal part normal. The formulae of leg setae (and solenidia), trochanter to tarsus, are I–1-5-5(1)-5(2)-24(3), II–1-7-5(1)-5(1)-23(2), III–4-3-3(1)-4(1)-17 and IV–1-3-3(1)-4(1)-(14-15). Leg tarsi are monodactylous.

Table 2. Measurements of some morphological characters of juvenile stages and adult of *Nanhermannia coronata* (mean measurements of 10 specimens in µm); nd—not developed.

Morphological Characters	Larva	Protonymph	Deutonymph	Tritonymph	Adult
Body length	284	351	416	572	559
Body width	125	155	172	305	251
Length of prodorsum	104	120	152	171	215
Length of:					
seta *ro*	16	17	21	32	39
seta *le*	15	17	19	30	35
seta *in*	19	23	30	45	75
seta *bs*	nd	nd	nd	nd	67
seta c_1	12	19	30	57	129
seta c_3	14	22	31	82	136
seta *cp*	10	20	32	95	142
seta d_1	12	21	33	70	125
seta d_2	13	22	32	93	135
seta e_1	10	20	35	88	130
seta e_2	11	21	38	83	134
seta f_1	8	lost	lost	lost	lost
seta f_2	24	32	43	96	133

Table 2. *Cont.*

Morphological Characters	Larva	Protonymph	Deutonymph	Tritonymph	Adult
seta h_1	20	22	30	80	120
seta h_3	nd	23	31	95	133
seta p_1	nd	8	19	72	122
seta p_3	nd	16	30	83	130
genital opening	nd	26	33	46	83
anal opening	52	65	77	123	114

Table 3. Ontogeny of leg setae (Roman letters) and solenidia (Greek letters) in *Nanhermannia coronata*.

Leg	Trochanter	Femur	Genu	Tibia	Tarsus
Leg I					
Larva	–	d, bv''	$d, (l), \sigma$	$(l), v', d, \varphi_1$	$(ft), (tc), (pl), (p), (u), (a), s, (pv), \varepsilon, \omega_1$
Protonymph	–	–	–	–	ω_2
Deutonymph	v'	(l)	–	φ_2	–
Tritonymph	–	v''	v', v''	v''	$(v_1), (it), \omega_3$
Adult	–	–	–	–	$(l), (v_2)$
Leg II					
Larva	–	d, bv''	$d, (l), \sigma$	l', v', d, φ	$(ft), (tc), (p), (u), (a), s, (pv), \omega_1$
Protonymph	–		–	l''	–
Deutonymph	v'	(l_1)	–	–	ω_2
Tritonymph	–	l_2'', v'' [1]	v', v''	v''	$l_{1'}, (v_1), (it)$
Adult	–	$l_{3'}$	–	–	$l_1'', (l_2), (v_2)$
Leg III					
Larva	–	d, ev'	d, σ	v', d, φ	$(ft), (tc), (p), (u), (a), s, (pv)$
Protonymph	v'	–	–	–	–
Deutonymph	l'_1	–	–	–	–
Tritonymph	l'_2	l' [1]	v'	v''	$(it), (v_1)$
Adult	l'_3	–	–	l'	–
Leg IV					
Protonymph		–	–	–	$ft'', (pv), (p), (u)$
Deutonymph	–	d, ev'	l', d, σ	v', d, φ	$(tc), (a), s$
Tritonymph	v'	l'	v'	v''	(v_1)
Adult	–	–	–	l'	ft' [2]

Note: structures are indicated where they are first added and are present through the rest of ontogeny; pairs of setae are in parentheses; dash indicates no additions; [1] is added in some individuals; if not, it is added in the next stage; [2] in some individuals is absent.

Remarks. The mean body length and width of individuals studied herein are larger than those described by Berlese [34]—length 490, width 220—but smaller than those investigated by Sitnikova [5]—length 575, width 250—and Weigmann [8]—length 480–570. Some authors [5,6,9,10] observed on protuberance of *N. coronata*, 4–5 posterior tubercles of different shape, whereas our adults have 5–7 small tubercles.

3.1.3. Description of Juvenile Stages

The larva is elongated (Figures 12, 13a and 14a), the body unpigmented and with pits and the central part of the prodorsum, epimeres and legs light brown. The prodorsum is subtriangular, the central part punctate and with small pits. Prodorsal setae *ro*, *le*, *in* and *exp* are short and smooth (Figures 12 and 14a and Table 2), and seta *exa* is reduced to its alveolus. The mutual distance between setal pair *le* is slightly shorter than that between setal pair *ro*, and the mutual distance between setal pair *in* is about two times longer than that between setal pair *ro*. The opening of the bothridium is small and rounded, and the bothridial seta is absent. The prodorsum and hysterosoma of the larva have small pits.

The hysterosoma of the larva is cylindrical, and the gastronotum has 12 pairs of setae, including dorsal f_1 and ventral h_2, inserted lateral to the posterior part of the anal valves (Figures 13a and 14a); most are short and smooth, except for slightly longer f_2 and

h_1 (Table 2). Cupule *ia* is posterior to seta c_3, cupule *im* anterolateral to seta e_2, cupule *ip* posterolateral to seta f_2 and cupule *ih* lateral to the anterior part of the anal valves (Figures 13a and 14a). The opisthosomal gland opening is medial to seta f_2. The anal valves of the larva (segment P) are glabrous. The leg segments are relatively thick, and most leg setae are short, thick and conical, except for the longer apical setae on tarsi (Figure 15). Seta *d* accompanying solenidion σ on all genua, φ_1 on tibia I and φ on other tibiae are present.

The shape and color of nymphs and prodorsal setae are as in the larva, but seta *in* is clearly longer, and pits in central part of prodorsum denser than in the larva. The bothridium is weakly developed and bothridial seta absent. The gastronotum of the protonymph has small pits and 15 pairs of setae because seta f_1 has been lost and only the alveolus of this seta remains, and setae h_3 and *p*-series appear and remain in the deutonymph and tritonymph (Figures 13b, 14b, 16, 17 and 18a–c). The prodorsum and hysterosoma of nymphs have pits, observed in SEM figures as small holes (Figures 18 and 19). All gastronotal setae are long (Table 2) and smooth. In the protonymph, one pair of seta appears on the genital valves, and three pairs are added in the deutonymph and two pairs in the tritonymph (Figures 13b, 14b and 16), all short and smooth. In the deutonymph, one pair of aggenital setae appears, and one pair is added in the tritonymph, all short and smooth (Figures 14b and 16a,b). Anal valves of the protonymph and deutonymph (segments AD and AN, respectively) are glabrous, and those of the tritonymph have two pairs of short and smooth setae. In the tritonymph, the opisthonotal gland opening and cupules *ia*, *im* and *ip* are as in the larva, cupule *iad* lateral to the anal valves, cupules *ips* and *ih* pushed anterolateral to cupule *iad* (Figures 14b and 16b). The leg segments of the tritonymph relatively thick, and most leg setae are short, thick or conical, except longer apical setae on tarsi (Figures 18, 19a,d and 20–22). Seta *d* accompanying solenidion σ on all genua, φ_1 on tibia I and φ on other tibiae are present (Table 3).

3.1.4. Summary of Ontogenetic Transformations

In the larva of *N. coronata*, the prodorsal setae *ro*, *le*, *in* and *exp* are short, and in the nymph, seta *in* is relatively longer, whereas in the adult, setae *ro* and *le* are short, *in* is long and *exp* is reduced to its alveolus. In all juveniles, the bothridium is weakly developed, and the bothridial seta is absent, whereas in the adult, the bothridium is well developed, with a small, rounded opening, and the bothridial seta is setiform, with a slightly thicker, barbed head. The larva has 12 pairs of gastronotal setae, including f_1 and h_2, whereas the nymphs and adult have 15 pairs (in the protonymph, f_1 is reduced to its alveolus, and h_3 and *p*-series are added). The formula of the gastronotal setae of *N. coronata* is 12-15-15-15-15 (larva to adult, excluding alveolar f_1). The formulae of the epimeral setae are 3-1-2 (larva, including scaliform *1c*), 3-1-3-2 (protonymph), 3-1-3-3 (deutonymph) and 3-1-3-4 (tritonymph and adult). The formula of the genital setae is 1-4-6-9 (protonymph to adult), the aggenital setae is 1-2-2 (deutonymph to adult) and the formula of setae of segments PS–AN is 03333-0333-022. The ontogeny of leg setae and solenidia is given in Table 3.

3.2. Results of DNA Barcoding

A neighbor-joining tree based on cytochrome oxidase I (COI) nucleotide sequences confirmed morphological observations that the adults from all included localities (southern, central and northern Norway; Ireland; and Finland) represented the same species (Figure 23). The maximum mitochondrial DNA variation within *N. coronata* was 0.31%, while the minimum distance to compared representatives of putatively close genera was 26.53%.

3.3. Ecology and Biology

Our data on the ecology of *N. coronata* indicate that this species is common in raised bogs; it was present in 70% of collected samples. This study also shows wider ecological tolerance of *N. coronata* towards moisture; this species was found in different bog microhabitats: hummocks, lawns, transition zone between hummocks and hollows, and in hollows (Figure 24). It seems to prefer intermediate moisture conditions; in hummocks, it occurred in 87% of samples and was particularly abundant in the lower zone between hummocks and hollows (Figure 24). Results of Kruskal–Wallis ANOVA by ranks show significant differences between microhabitats ($H = 9.27$, $p = 0.03$).

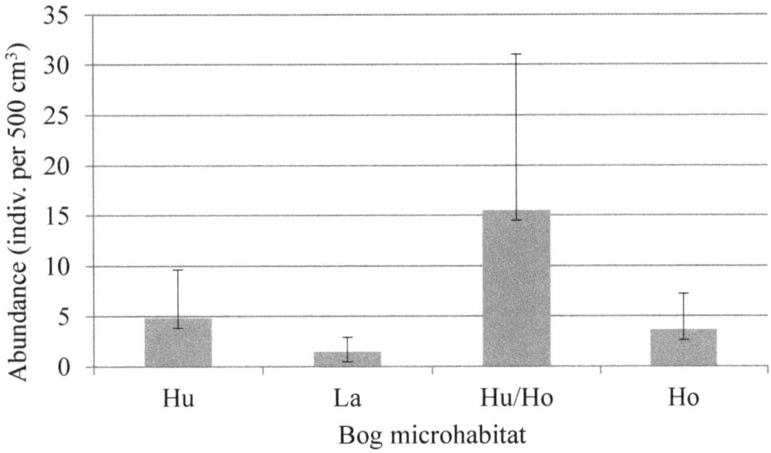

Figure 24. Average abundance of *Nanhermannia coronata* (individuals per 500 cm^3) (bars) with standard deviation (whiskers) in selected microhabitats of bogs in Norway: Hu—hummock, La—lawn, Hu/Ho—transition zone between hummock and hollow, Ho—hollow; numbers above bars present number of samples collected from a certain microhabitat and percentage of samples where the species was present (constancy index).

In Hitra, the average abundance of *N. coronata* was higher (on average 4.42 specimens per 500 cm^3) than in Høstadmyra (on average 2.73 specimens per 500 cm^3) (Figure 25), but significant differences were observed only in deutonymphs (Kruskal–Wallis ANOVA by ranks: $H = 3.96$, $p = 0.04$). In Hitra, the average abundance of deutonymphs was 0.73 specimens per 500 cm^3, while in Høstadmyra, it was 0.19 specimens per 500 cm^3. In Hitra, the percentage of juveniles was higher than in Høstadmyra (39% and 32% of all individuals of species, respectively).

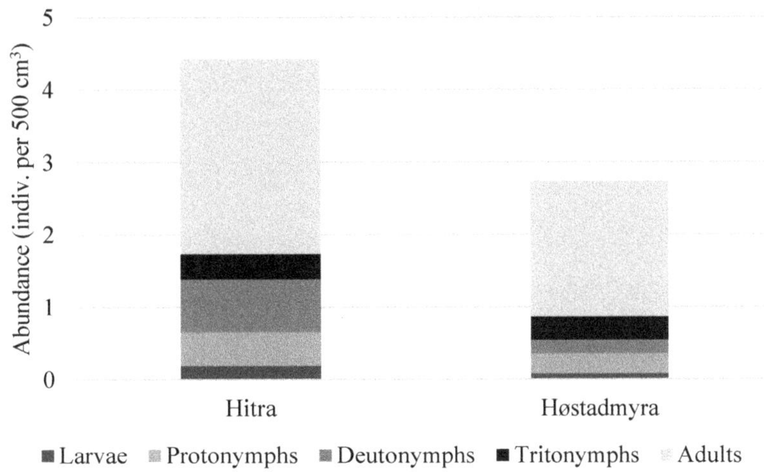

Figure 25. Average abundance of developmental stages of *Nanhermannia coronata* in two bogs in Norway.

The stage structure of *N. coronata* we investigated in both populations of this species from Høstadmyra and Hitra. In total, 213 individuals of *N. coronata* were found, including 7 larvae, 24 protonymphs, 24 deutonymphs, 21 tritonymphs and 137 adults. Among 30 individuals investigated, all were females, and 25% were gravid, carrying one or rarely two large eggs (215 × 119), constituting about 40% of the total body length of females.

3.4. Comparison of Morphology of *Nanhermannia coronata* with Congeners

Seniczak et al. [2] compared the morphology of adults of *Nanhermannia* species, which differ from one another mainly in the shape of posterior prodorsal protuberances and tubercles (from prominent tubercles to small ones), and the length of some setae on the prodorsum and notogaster. Some species differ also from one another by the number of genital setae and formula of epimeral setae, but in most species, these characters are unknown. The shape of posterior prodorsal protuberance and the number of tubercles are considered diagnostic in *Nanhermannia*, but in some species, the number of tubercles varies [3–10], lowering their diagnostic values. The adult of *N. coronata* differs from that of *N. sellnicki* by its darker body color and larger number of posterior tubercles (6–7 tubercles) than that of *N. sellnicki* (3–4 tubercles), but some authors [5,8–10] observed on this protuberance 4–5 posterior tubercles, so the number of these tubercles in these species overlaps and has a small diagnostic value.

In Table 4, we compared selected morphological characters of adults, tritonymphs and larvae of *N. coronata*, *N. comitalis*, *N. nana* and *N. sellnicki*. The adult of *N. coronata* differs from that of *N. sellnicki* by the length of setae in and d_1 and the formula of setae on femora ([2], Table 4). The juveniles of these species differ from one another by the length of some setae, the number of posterior tubercles and the formula of femora of the tritonymph. The juveniles of *N. coronata* differ from those of *N. sellnicki* by the formula of femora of tritonymph and the shape of insertion of prodorsal seta in and all gastronotal and adanal setae. In the latter species, these setae are in small individual depressions [2], whereas in *N. coronata*, these depressions are absent. The juveniles of *N. nana* have clearly longer gastronotal setae than other species, both in the larva and nymphs.

Table 4. Comparison of selected morphological characters of some instars of *Nanhermannia coronata*, *N. comitalis*, *N. nana* and *N. sellnicki*.

Characters	*N. coronata*	*N. comitalis* [1]	*N. nana* [1]	*N. sellnicki* [2]
Adult				
Length of seta *in*	Longer than *bs*	Longer than *bs*	As long as *bs*	As long as *bs*
Posterior tubercles [3]	5–7	4	1	3–4
Length of seta d_1 [4]	No	Yes	Yes	No
Formula of femora	5-7-3-3	unknown	unknown	5-6-3-2
Tritonymph				
Length of seta *in*	As long as c_1	As long as c_1	Shorter than c_1	As long as c_1
Posterior tubercles [3]	6–7	5	6	4–5
Length of seta c_1 [5]	No	Yes	Yes	No
Length of seta d_1 [4]	No	Yes	Yes	No
Length of seta e_1 [6]	No	No	Yes	No
Formula of femora	5-(5-6)-(2-3)-3	unknown	unknown	5-6-3-2
Larva				
Length of seta *in*	Shorter than c_1	Longer than c_1	Shorter than c_1	As long as c_1
Posterior tubercles [3]	Absent	2–3	2	Absent
Length of seta c_1 [5]	No	Yes	Yes	No
Length of seta d_1 [4]	No	Yes	Yes	No

[1] According to Seniczak [35], [2] Seniczak et al. [2], [3] on each posterior prodorsal protuberance, [4] reaches insertion of seta e_1, [5] reaches insertion of seta d_1, [6] reaches insertion of seta h_1.

4. Discussion

Nanhermannia coronata is a common and abundant oribatid species in peatlands, but the adult of this species is often mistaken for *N. sellnicki*, which deteriorates our knowledge on the ecology of both species. According to our observations, the adult of *N. coronata* differs from that of *N. sellnicki* by having a darker body color [2], which is diagnostic only for darker adults of *N. coronata*. Freshly emergent adults of both species are light brown. In our studies, the adult of *N. coronata* differs from that of *N. sellnicki* by the length of seta *in* and the number of posterior tubercles on prodorsal protuberance (5–7 and 3–4 tubercles, respectively), but this character also has a small diagnostic value. Some authors [5,6,9,10] observed on prodorsal protuberance of *N. coronata* 4–5 posterior tubercles, which overlap with the diagnostic character of *N. sellnicki*. Diagnostic for these species is the formula of setae on femora, which clearly differentiates the adult and tritonymph of *N. coronata* studied herein from those instars of *N. sellnicki* ([2], Table 4). All juvenile stages of *N. coronata* differ from those of *N. sellnicki* by the shape of insertion of prodorsal seta *in* and all gastronotal and adanal setae. In the latter species, these setae are inserted in small individual depressions, whereas in *N. coronata*, these depressions are absent.

In the juveniles of *N. coronata* studied herein, some morphological differences were observed, comparing to those investigated by Ermilov and Łochyńska [9]. In the larva studied herein, seta f_2 is slightly longer and seta h_2 slightly shorter than in that studied by Ermilov and Łochyńska [9], whereas the nymphs have most gastronotal setae slightly longer than those investigated by these authors. The nymphs studied herein have clear posterior prodorsal protuberance, which was not observed by the mentioned authors [9]. These differences may reflect either the geographic variability of juveniles of *N. coronata* or different methods of preparation of figures.

In *N. coronata*, seta f_1 is lost in the protonymph, as in *N. sellnicki* [2] and most genera of Crotonioidea, except for *Hermannia* Nicolet, 1855, *Phyllhermannia* Berlese, 1916, and *Nothrus* C.L. Koch, 1836, in which this seta is retained in all instars [36–41]. The larva of *N. coronata* has a pit on dorsal part of genu I, as that of *N. sellnicki* [2], which Grandjean [27] considered an alveolus of second solenidion σ. Two solenidia on genu I occur in some groups of lower Oribatida, for example in Lohmanniidae [42,43].

The leg segments, setae and solenidia of *N. coronata* are generally similar as in *N. sellnicki* except for the pattern of sculpture on femora, the shape of some setae and the number of setae on femur II and tarsi III and IV [2]. In both species, most leg setae are short, thick or conical, and most solenidia are blunt. In these species, an additional seta l occurs on femora II and III, and l and v on most tarsi, as in *Platynothrus coulsoni* A. and S. Seniczak, 2022, *P. punctatus* (L. Koch, 1879) and *P. troendelagicus* Seniczak et al., 2022 [21,44,45]. However, in *N. coronata* and *N. sellnicki*, setal pairs l_1 and l_2 on tarsus II are separated by solenidia ω_1 and ω_2 and setae *ft*, whereas in *Platynothrus* species, solenidia ω_1 and ω_2 and setae *ft* are located in anterior position, and all setae l are added posterior to them. In all species of *Nanhermannia* and *Platynothrus*, the number of setae on femora of the deutonymph, tritonymph and adult, and the number of setae on tarsi of the adult is species-specific.

The shape of the chelicera and palp of adults of *N. coronata* is similar to *N. sellnicki* and other *Nanhermannia* species discussed by Seniczak et al. [2]. The chelicera is chelate-dentate and has short and thick seta *chb* located posterior to longer seta *cha*, and the palpal setae are short and thick, and solenidion ω is separated from seta *acm*.

Nanhermannia coronata is known from the Holarctic region [46]. It is hygrophilous [47], with some range of tolerance towards moisture [48], but according to Rajski [23], with narrow tolerance towards pH and preferences of its low values. The acidity is considered the main factor for abundant occurrence of *N. coronata*, and this species has been mainly reported from raised bogs and swamps [15,17–20,22,49–57], less from other acidic forest soils and heaths [8], and it was absent from eutrophic mires [49,58]. Some studies show however that it has been abundant in beech forests [59], deciduous and birch forests [60], and in Scots pine forests [61]. In contrast, in broadleaf forests in Norway, it was few and rare [62–64]. In forests and open heath, it was strongly negatively affected by reduced soil moisture [65] and clearly associated with soil (not litter) [66]. It was also found in fruiting bodies of bracket fungi, and its abundance and frequency increased with the degree of decay of fungi [67]. Single individuals were also found in bird plumage and bird nests [68]. It was dominant in 14 Danish spring areas, which may reflect its tolerance to low levels of calcium [69]. In studied gradients in mires, its abundance was neither affected by the moisture nor by the pH, when it ranged between 3.68 and 5.75 [70].

Nanhermannia coronata is a secondary decomposer and fungivorous feeder (feeds partly on fungi and partly on litter) [71,72]. It reproduces parthenogenetically [46]. Ermilov and Łochyńska [9] cultured *N.* cf. *coronata* in laboratory conditions (22–23 °C, 100% air humidity) and fed it with algae (*Protococcus* sp.) and raw potatoes. The adults fed on both types of food, whereas the juveniles preferred algae. The total development of this species lasted 105–124 days, and each stage developed during the following days (+ immovable stage between instars): egg (6–8), larva [(13–20) + (4–8)], protonymph [(11–31) + (5–11)], deutonymph [(12–26) + (6–12)] and tritonymph [(9–26) + (10–13)] to obtain the adult. The maturation time of *N.* cf. *coronata* lasts 21 weeks at 20 °C and 16 weeks at 22.5 °C. Its tolerance to temperature was tested in another experiment, and it was similar in adults and juvenile specimens: 38 °C for 4 h and 36 °C for 12 h [73].

Our ecological observations confirmed a common occurrence of *N. coronata* in raised bogs, with a high percentage of juvenile stages in populations. In two studied locations, Hitra and Høstadmyra, we observed significant differences in the abundance of deutonymphs and proportion of the juvenile stages was higher in Hitra than in Høstadmyra, which can be explained by the warmer (average annual temperature higher by 2 °C) and milder, oceanic climate in Hitra compared to Høstadmyra. More advanced development of species and higher proportion of juveniles has also been observed in another oribatid species, *Ceratozetes parvulus* Sellnick, 1922, collected from the same bogs [74]. In wetter microhabitats, *N. coronata* was not abundant and occurred with a low constancy. This is consistent with the observations from inundated bog habitats—edges of water bodies—where *N. coronata* was absent or very few in number [75]. In degraded, drier bogs, it was more abundant [76,77], but in completely dry bogs, it was absent [78]. When the bog was dried, the abundance of *N. coronata* decreased in the hummocks but, at the same time, increased in the hollows [79].

Nanhermannia coronata prefers acid and humid microhabitats [23], whereas *N. sellnicki* is less common than *N. coronata* and occurs in drier habitats [13], like birch forests, especially with understory formed by *Vaccinium* and *Empetrum*. In investigated habitats, like hummocks, lawns, transition zone between hummocks and hollows, and in hollows, *N. coronata* was particularly abundant in between hummocks and hollows, with abundant juvenile stages.

According to the literature, this species can also be abundant in beech forests [59], deciduous and birch forests [60], Scots pine forests [61], bird plumage and bird nests [68], but these data need confirmation using the diagnostic characters of species given herein.

5. Conclusions

1. *Nanhermannia coronata* is an abundant and common oribatid mite in raised bogs, with a high percentage of juveniles, and has a preference for humid microhabitats, whereas *N. sellnicki* is less common than *N. coronata* and occurs in drier habitats.

2. *Nanhermannia coronata* differs clearly from *N. sellnicki* by the following morphological characters: the number of setae on femora I–IV of the adult (*N. coronata* 5-7-3-3, *N. sellnicki* 5-6-3-2) and tritonymph [*N. coronata* 5-(5-6)-(2-3)-3, *N. sellnicki* 5-6-3-2], and the shape of insertions of prodorsal seta *in* and all gastronotal and adanal setae of juveniles; in *N. sellnicki*, these setae are inserted in small individual depressions, whereas in *N. coronata*, these depressions are absent.

Author Contributions: Conceptualization, S.S. and A.S.; methodology, S.S. and A.S.; line drawings, S.S.; DNA barcoding and SEM imaging, A.S.; writing—original draft, S.S. and A.S.; writing—review and editing, S.S. and A.S. All authors have read and agreed to the published version of the manuscript.

Funding: Norwegian Taxonomy Initiative (Grant No. 6-20, 70184243). The sequencing was financed by Norwegian Barcode of Life (NorBOL).

Institutional Review Board Statement: Not applicable.

Informed Consent Statement: Not applicable.

Data Availability Statement: Ecological data on *N. coronata* are stored at DataverseNO open archive (https://doi.org/10.18710/F7LO2Y).

Conflicts of Interest: The authors declare no conflict of interest.

References

1. Subías, L.S. Listado sistemático, sinonímico y biogeográfico de los Ácaros Oribátidos (Acariformes, Oribatida) del mundo (1758–2002). *Graellsia* **2004**, *60*, 540. [CrossRef]
2. Seniczak, A.; Seniczak, S.; Hagen, S.B.; Klütsch, F.C. Morphological ontogeny of *Nanhermannia sellnicki* (Acari, Oribatida, Nanhermanniidae), and comments on *Nanhermannia* Berlese. *Zootaxa* **2023**, *5324*, 110–132. [CrossRef]
3. Kunst, M. *Nanhermannia komareki* n. sp., eine neue Oribatiden-Art aus Böhmen (Acarina, Oribatei). *Acta Soc. Zool. Bohem.* **1956**, *20*, 267–272.
4. Karppinen, E. Mitteilungen über einige für Finnland neue Oribatiden (Acari). *Ann. Entom. Fenn.* **1958**, *24*, 192–196.
5. Sitnikova, L.G. Superfamily Nanhermannoidea. In *Key to Soil-Inhabiting Mites—Sarcoptiformes*; Ghilarov, M.S., Ed.; Nauka Publisher: Moscow, Russia, 1975; pp. 277–303. (In Russian)
6. Fujikawa, T. Oribatid mites from *Picea glehni* forest at Mo-Ashoro, Hokkaido (3). Two new species of the family Nanhermanniidae. *Edaphologia* **1990**, *43*, 5–15.
7. Fujikawa, T. Five species of Nanhermanniidae (Acari: Oribatida) from Nippon. *Edaphologia* **2003**, *71*, 1–8.
8. Weigmann, G. Hornmilben (Oribatida). In *Die Tierwelt Deutschlands. Part 76*; Dahl, F., Ed.; Series Founder; Goecke & Evers: Keltern, Germany, 2006; pp. 1–520.
9. Ermilov, S.G.; Łochyńska, M. The morphology of juvenile stages and duration of the development of *Nanhermannia* cf. *coronata* Berlese, 1913 (Acari, Oribatida, Nanhermannidae). *Acarologia* **2008**, *48*, 61–68.
10. Mahunka, S.; Mahunka-Papp, L. A new survey of the oribatid-fauna of Maramureş (Romania, Transylvania) (Acari, Oribatida). *Stud. Univ. Vasile Goldis* **2008**, *18*, 365–378.
11. Kagainis, U. Revision and renovation of oribatid mites (Acari: Oribatida) specimen collection of Institute of Biology, Latvia. *Environ. Exp. Biol.* **2012**, *10*, 67–75.
12. Willmann, C. Moosmilben oder Oribatiden (Cryptostigmata). In *Die Tierwelt Deutschlands*; Dahl, F., Ed.; Gustav Fischer: Jena, Poland, 1931; Volume 2, pp. 79–200.
13. Forsslund, K.H. Notizen über Oribatei (Acari). II. *Entomol. Tidskr.* **1958**, *79*, 75–86.
14. Solhøy, T. Species composition of Oribatei (Acari) on oceanic mountain ground in western Norway. *Norw. J. Entomol.* **1976**, *23*, 17–22.
15. Solhøy, T. Oribatids (Acari) from an oligotrophic bog in western Norway. *Fauna Nor.* **1979**, *26*, 91–94.
16. Norton, R.A.; Ermilov, S.G. Catalogue and historical overview of juvenile instars of oribatid mites (Acari: Oribatida). *Zootaxa* **2014**, *3833*, 1–132. [CrossRef] [PubMed]
17. Borcard, D.; Geiger, W.; Matthey, W. Oribatid mite assemblages in a contact zone between a peat-bog and a meadow in the Swiss Jura (Acari, Oribatei): Influence of landscape structure and historical processes. *Pedobiologia* **1995**, *39*, 318–330.
18. Borcard, D.; Vaucher-von Ballmoos, C. Oribatid mites (Acari, Oribatida) of a primary peat bog-pasture transition in the Swiss Jura Mountain. *Ecoscience* **1997**, *4*, 470–479. [CrossRef]
19. Lehmitz, R.; Haase, H.; Otte, V.; Russell, D. Bioindication in peatlands by means of multi-taxa indicators (Oribatida, Araneae, Carabidae, Vegetation). *Ecol. Indic.* **2020**, *109*, 105837. [CrossRef]
20. Seniczak, A.; Seniczak, S.; Iturrondobeitia, J.C.; Solhøy, T.; Flatberg, K.I. Diverse *Sphagnum* mosses support rich moss mite communities (Acari, Oribatida) in mires of western Norway. *Wetlands* **2020**, *40*, 1339–1351. [CrossRef]
21. Seniczak, A.; Seniczak, S.; Hassel, K.; Flatberg, K.I. Morphological ontogeny of *Platynothrus troendelagicus* sp. nov. (Acari, Oribatida, Camisiidae) from Norway. *Syst. Appl. Acarol.* **2022**, *27*, 1702–1722. [CrossRef]
22. Seniczak, A.; Seniczak, S.; Iturrondobeitia, J.C.; Marciniak, M.; Kaczmarek, S.; Makol, J.; Kazmierski, A.; Zawal, A.; Schwarzfeld, M.D.; Flatberg, K.I. Inclusion of juvenile stages improves diversity assessment and adds to our understanding of mite ecology—A case study from mires in Norway. *Ecol. Evol.* **2022**, *12*, e9530. [CrossRef] [PubMed]
23. Rajski, A. Autecological-zoogeographical analysis of moss mites (Acari, Oribatei) on the basis of fauna in the Poznan Environs. Part, I. *Pol. J. Entomol.* **1967**, *37*, 69–166.
24. Grandjean, F. Etude sur le développement des Oribates. *Société Zool. Fr.* **1933**, *58*, 30–61.
25. Grandjean, F. La notation des poils gastronotiques et des poils dorsaux du propodosoma chez les Oribates (Acariens). *Bull. Soc. Zool. Fr.* **1934**, *59*, 12–44.
26. Grandjean, F. Les segments post-larvaires de l'hystérosoma chez les Oribates (Acariens). *Bull. Soc. Zool. Fr.* **1939**, *64*, 273–284.
27. Grandjean, F. Observations sur les Oribates (13e série). *Bull. Mus. Natl. Hist. Nat.* **1940**, *12*, 62–69.

28. Grandjean, F. Essai de classification des Oribates (Acariens). *Soc. Zool. Fr.* **1953**, *78*, 421–446.
29. Norton, R.A.; Behan-Pelletier, V.M. Suborder Oribatida. In *A Manual of Acarology*; Krantz, G.W., Walter, D.E., Eds.; Texas Tech University Press: Lubbock, TX, USA, 2009; pp. 430–564.
30. Hebert, P.D.N.; Cywinska, A.; Ball, S.L.; de Waard, J.R. Biological identifications through DNA barcodes. *Proc. R. Soc. Lond. B.* **2003**, *270*, 313–321. [CrossRef]
31. Folmer, O.; Black, M.; Hoeh, W.; Lutz, R.; Vrijenhoek, R. DNA primers for amplification of mitochondrial cytochrome c oxidase subunit I from diverse metazoan invertebrates. *Mol. Mar. Biol. Biotech.* **1994**, *3*, 294–299.
32. Tamura, K.; Stecher, G.; Peterson, D.; Filipski, A.; Kumar, S. MEGA6: Molecular evolutionary genetics analysis version 6. *Mol. Biol. Evol.* **2013**, *30*, 2725–2729. [CrossRef]
33. Stanisz, A. *Easy Course of Statistic Using Statistica PL and Medicine Examples, Basic Statistic*; StatSoft Polska: Kraków, Poland, 2006.
34. Berlese, A. Acari nuovi, Manipoli VII–VIII. *Redia* **1913**, *9*, 77–111.
35. Seniczak, S. The morphology of juvenile stages of moss mites of the family Nanhermanniidae (Acari: Oribatida). I. *Zool. Anz.* **1991**, *227*, 319–330.
36. Seniczak, S. The morphology of juvenile stages of moss mites of the family Nothridae (Acari: Oribatida). I. *Zool. Anz.* **1992**, *229*, 134–148.
37. Seniczak, S.; Żelazna, E. The morphology of juvenile stages of moss mites of the family Nothridae (Acari: Oribatida). II. *Zool. Anz.* **1992**, *229*, 149–162.
38. Seniczak, S.; Norton, A. The morphology of juvenile stages of moss mites of the family Nothridae (Acari: Oribatida). III. *Zool. Anz.* **1993**, *230*, 19–33.
39. Colloff, M.J. New species of the oribatid mite genus *Phyllhermannia* Berlese, 1916 (Acari, Oribatida, Hermanniidae) from wet forests in south-eastern Australia show a high diversity of morphologically-similar, short-range endemics. *Zootaxa* **2011**, *2770*, 1–60. [CrossRef]
40. Seniczak, S.; Seniczak, A.; Coulson, S.J. Morphological ontogeny, distribution, and descriptive population parameters of *Hermannia reticulata* (Acari: Oribatida: Hermanniidae), with comments on Crotonioidea. *Intern. J. Acarol.* **2017**, *43*, 52–72. [CrossRef]
41. Seniczak, S.; Seniczak, A.; Coulson, S.J. Morphological ontogeny and distribution of *Hermannia scabra* (Acari: Oribatida: Hermanniidae) in Svalbard and descriptive population parameters. *Acarologia* **2017**, *57*, 877–892. [CrossRef]
42. Seniczak, S.; Seniczak, A.; Kaczmarek, S.; Haq, M.A.; Marquardt, T. Morphological ontogeny of *Heptacarus hirsutus* (Acari, Oribatida, Lohmanniidae), with comments on *Heptacarus* Piffl. *Syst. Appl. Acarol.* **2018**, *23*, 911–924. [CrossRef]
43. Seniczak, S.; Ivan, O.; Marquardt, T.; Seniczak, A. Morphological ontogeny of *Perlohmannia nasuta* (Acari, Oribatida, Perlohmanniidae), with comments on *Perlohmannia* Berlese. *Zootaxa* **2021**, *5086*, 29–048. [CrossRef]
44. Seniczak, A.; Seniczak, S. Morphological ontogeny of *Platynothrus coulsoni* sp. nov. (Acari, Oribatida, Camisiidae) from Spitsbergen (Norway). *Syst. Appl. Acarol.* **2022**, *27*, 1436–1453. [CrossRef]
45. Seniczak, S.; Seniczak, A.; Kaczmarek, S.; Marquardt, T.; Fernández Ondoño, E.; Coulson, S.J. Morphological ontogeny and ecology of *Platynothrus punctatus* (Acari, Oribatida, Camisiidae), with comments on *Platynothrus* Berlese. *Syst. Appl. Acarol.* **2022**, *27*, 551–580. [CrossRef]
46. Domes, K.; Norton, R.A.; Maraun, M.; Scheu, S. Reevolution of sexuality breaks Dollo's law. *Proc. Natl. Acad. Sci. USA* **2007**, *104*, 7139–7144. [CrossRef] [PubMed]
47. Schatz, H.U. Ordn.: Oribatei, Hornmilben. In *Catalogus Faunae Austriae*; Verlag der Österreichischen Akademie der Wissenschaften: Vienna, Austria, 1983; p. 118.
48. Strenzke, K. Untersuchungen über die Tiergemeinschaften des Bodens: Die Oribatiden und ihre Synusien in den Böden Norddeutschlands. *Zoologica* **1952**, *104*, 1–173.
49. Druk, A.Y. Armored mites in some moor-types of the Moscow region. In *Soil Living Evertebrates of the Moscow Region*; Nauka: Moscow, Russia, 1982; pp. 72–77.
50. Starý, J. Moss mites (Acari, Oribatida) from some peat bogs in Sumava Mts., South Bohemia. *Acta Mus. Bohem. Merid. Sci. Nat.* **1988**, *28*, 99–107. (In Czech)
51. Starý, J. Contribution to the knowledge of the oribatid mite fauna (Acari, Oribatida) of peat bogs in Bohemian Forest. *Silva Gabreta* **2006**, *12*, 35–47.
52. Borcard, D. Les Oribates des tourbières du Jura suisse (Acari, Oribatei). Ecologie, I. Quelques aspects de la communauté d'Oribates des sphaignes de la tourbière du Cachot. *Rev. Suisse Zool.* **1991**, *98*, 303–317. [CrossRef]
53. Borcard, D. Les Oribates des tourbières du Jura suisse (Acari, Oribatei): Ecologie II. Les relations Oribates-environnement à la lumière du test de Mantel. *Rev. Ecol. Biol. Sol* **1991**, *28*, 323–339.
54. Borcard, D. Les Oribates des tourbières du Jura suisse (Acari, Oribatei): Ecologie III. Comparaison a posteriori de nouvelles récoltes avec un ensemble de données de référence. *Rev. Suisse Zool.* **1991**, *98*, 521–533. [CrossRef]
55. Weigmann, G.; Horak, F.; Franke, K.; Christian, A. Verbreitung und Ökologie der 707 Hornmilben (Oribatida) in Deutschland. Senckenberg, Museum für Naturkunde, Görlitz. *Peckiana* **2015**, *10*, 171.
56. Minor, M.A.; Ermilov, S.G.; Philippov, D.A.; Prokin, A.A. Relative importance of local habitat complexity and regional factors for assemblages of oribatid mites (Acari: Oribatida) in Sphagnum peat bogs. *Exp. Appl. Acarol.* **2016**, *70*, 275–286. [CrossRef]
57. Leonov, V.D. The first report on the oribatid mites (Acari: Oribatida) in tundra of the Chunatundra Mountains on the Kola Peninsula, Russia. *Acarologia* **2020**, *60*, 722–734. [CrossRef]

58. Sidorchuk, E.A. Oribatid mites (Acari, Oribatei) of three fens in the northern part of European Russia. *Zool. Zhurnal* **2008**, *87*, 626–631. [CrossRef]
59. Migge, S.; Maraun, M.; Scheu, S.; Schaefer, M. The oribatid mite community (Acarina) on pure and mixed stands of beech (*Fagus sylvatica*) and spruce (*Picea abies*) at different age. *Appl. Soil Ecol.* **1998**, *9*, 115–121. [CrossRef]
60. Huhta, V.; Räty, M.; Ahlroth, P.; Hänninen, S.M.; Mattila, J.; Penttinen, R.; Rintala, T. Soil fauna of deciduous forests as compared with spruce forests in central Finland. *Memo. Soc. Fauna Flora Fenn.* **2005**, *81*, 52–70.
61. Caruso, T.; Melecis, V.; Kagainis, U.; Bolger, T. Population asynchrony alone does not explain stability in species-rich soil animal assemblages: The stabilizing role of forest age on oribatid mite communities. *J. Anim. Ecol.* **2020**, *89*, 1520–1531. [CrossRef] [PubMed]
62. Seniczak, A.; Bolger, T.; Roth, S.; Seniczak, S.; Djursvoll, P.; Jordal, B.H. Diverse mite communities (Acari: Oribatida, Mesostigmata) from a broadleaf forest in western Norway. *Ann. Zool. Fenn.* **2019**, *56*, 121–136. [CrossRef]
63. Seniczak, A.; Seniczak, S.; Graczyk, R.; Kaczmarek, S.; Jordal, B.H.; Kowalski, J.; Djursvoll, P.; Roth, S.; Bolger, T. A forest pool as a habitat island for mites in a limestone forest in Southern Norway. *Diversity* **2021**, *13*, 578. [CrossRef]
64. Seniczak, A.; Seniczak, S.; Starý, J.; Kaczmarek, S.; Jordal, B.H.; Kowalski, J.; Roth, S.; Djursvoll, P.; Bolger, T. High Diversity of Mites (Acari: Oribatida, Mesostigmata) Supports the High Conservation Value of a Broadleaf Forest in Eastern Norway. *Forests* **2021**, *12*, 1098. [CrossRef]
65. Brooker, R.W.; Osler, G.H.R.; Gollisch, J. Association of vegetation and soil mite assemblages with isolated Scots pine trees on a Scottish wet heath. *Landscape Ecol.* **2008**, *23*, 861–871. [CrossRef]
66. Osler, G.H.R.; Korycinska, A.; Cole, L. Differences in litter mass change mite assemblage structure on a deciduous forest floor. *Ecogeography* **2006**, *29*, 811–818. [CrossRef]
67. Gdula, A.K.; Konwerski, S.; Olejniczak, I.; Rutkowski, T.; Skubała, P.; Zawieja, B.; Gwiazdowicz, D.J. Pathogens as creators of biodiversity. A study on influence of decayed bracket fungi on alpha diversity of microarthropods in the Karkonosze National Park, Poland. *Sylwan* **2022**, *166*, 7–40.
68. Lebedeva, N.V.; Lebedev, V.D. Transport of Oribatid Mites to the Polar Areas by Birds. In *Integrative Acarology, Proceedings of the 6th European Congress, Montpellier, France, 21–25 July 2008*; Bertrand, M., Kreiter, S., McCoy, K.D., Migeon, A., Navajas, M., Tixier, M.S., Vial, L., Eds.; European Association of Acarologists: Montpellier, France, 2008; pp. 359–367.
69. Gjelstrup, P.; Hansen, P.; Warncke, E. Moss mites (Oribatida, Acari) in mosses from some Danish spring areas. *Nat. Jutl.* **1991**, *23*, 33–44.
70. Minor, M.A.; Ermilov, S.G.; Philippov, D.A. Hydrology-driven environmental variability determines abiotic characteristics and Oribatida diversity patterns in a *Sphagnum* peatland system. *Exp. Appl. Acarol.* **2019**, *77*, 43–58. [CrossRef] [PubMed]
71. Schneider, K.; Migge, S.; Norton, R.A.; Scheu, S.; Langel, R.; Reineking, A.; Maraun, M. Trophic niche differentiation in oribatid mites (Oribatida, Acari): Evidence from stable isotope ratios (15N/14N). *Soil Biol. Bioch.* **2004**, *36*, 1769–1774. [CrossRef]
72. Magilton, M.; Maraun, M.; Emmerson, M.; Caruso, T. Oribatid mites reveal that competition for resources and trophic structure combine to regulate the assembly of diverse soil animal communities. *Ecol. Evol.* **2019**, *9*, 8320–8330. [CrossRef] [PubMed]
73. Malmström, A. Temperature tolerance in soil microarthropods: Simulation of forest-fire heating in the laboratory. *Pedobiologia* **2008**, *51*, 419–426. [CrossRef]
74. Seniczak, A.; Seniczak, S. Morphological ontogeny and ecology of *Ceratozetes parvulus* (Acari: Oribatida: Ceratozetidae). *Zootaxa* **2021**, *5086*, 111–134. [CrossRef]
75. Seniczak, A.; Seniczak, S.; Graczyk, R.; Waldon-Rudzionek, B.; Nowicka, A.; Pacek, S. Seasonal Dynamics of Oribatid Mites (Acari, Oribatida) in a Bog in Poland. *Wetlands* **2019**, *39*, 853–864. [CrossRef]
76. Seniczak, A.; Seniczak, S.; Maraun, M.; Graczyk, R.; Mistrzak, M. Oribatid mite species numbers increase, densities decline and parthenogenetic species suffer during bog degradation. *Exp. Appl. Acarol.* **2016**, *68*, 409–428. [CrossRef] [PubMed]
77. Seniczak, A.; Seniczak, S.; Iturrondobeitia, C.; Gwiazdowicz, D.J.; Waldon-Rudzionek, B.; Flatberg, K.I.; Bolger, T. Mites (Oribatida and Mesostigmata) and vegetation as complementary bioindicators in peatlands. *Sci. Total Envir.* **2022**, *851*, 158335. [CrossRef]
78. Kehl, C. Die Hornmilbenzönosen verschieden stark degradierter Moorstandorte (Acari, Oribatida). *Zool. Beiträge* **1997**, *38*, 3–10.
79. Markkula, I. Comparison of the communities of the oribatids (Acari: Cryptostigmata) of virgin and forest-ameliorated pine bogs. *Ann. Entomol. Fenn.* **1986**, *23*, 33–38.

animals

MDPI

Article

Communities of Uropodina (Acari: Mesostigmata) in Nest Boxes Inhabited by Dormice (*Glis glis* and *Muscardinus avellanarius*) and Differences in Percentages of Nidicoles in Nests of Various Hosts

Jerzy Błoszyk [1,2], Grzegorz Hebda [3], Marta Kulczak [1], Michał Zacharyasiewicz [1], Tomasz Rutkowski [1] and Agnieszka Napierała [1,*]

[1] Department of General Zoology, Faculty of Biology, Adam Mickiewicz University, Uniwersytetu Poznańskiego 6, 61-614 Poznań, Poland; bloszyk@amu.edu.pl (J.B.); markul12@amu.edu.pl (M.K.); zacharyasiewicz@gmail.com (M.Z.); tomasz.rutkowski@amu.edu.pl (T.R.)

[2] Natural History Collections, Faculty of Biology, Adam Mickiewicz University, Uniwersytetu Poznańskiego 6, 61-614 Poznań, Poland

[3] Institute of Biology, Opole University, Oleska 22, 45-040 Opole, Poland; grzesio@uni.opole.pl

[*] Correspondence: agan@amu.edu.pl

Simple Summary: Hanging nest boxes, which are used by various groups of animals such as birds or mammals (e.g., dormice and bats), increase the number of shelters and breeding places for these often rare animals. Nest boxes not only become habitats for the host, but are also inhabited by various groups of invertebrates, including insects, spiders, millipedes, snails, and also small arachnids, which are mites. In this article, we present an analysis of the community of one of the groups of mites—Uropodina—which also inhabit nest boxes. In the examined boxes, five species belonging to the discussed group were found, out of which only one (*Leiodinychus orbicularis*) is a nidicole, i.e., a species that inhabits the nests of various animals. This article also analyses the habitat preferences of the mentioned species and another Uropodina species associated with nests—*Apionoseius infirmus*. It was proven that *L. orbicularis* clearly dominated both in the examined dormouse, bat, and bird boxes, whereas *A. infirmus*, which was less numerous in the communities, preferred natural nests, including the nests of birds of prey. The clear dominance of *L. orbicularis* in the examined boxes can be explained by the specific microclimate, such as very low humidity, which prevails in the boxes.

Abstract: Bird and mammal nests and nest boxes constitute microenvironments in which various groups of invertebrates can live, including mites from the suborder Uropodina (Acari: Mesostigmata). The main aim of the current study was to ascertain the characteristics of mite communities from the suborder Uropodina, which inhabit the nests of dormice (Gliridae) built in nest boxes. The second aim of the study was to compare the habitat preferences of *Leiodinychus orbicularis* (C. L. Koch) and *Apionoseius infirmus* (Berlese), i.e., two typically nest-dwelling species of Uropodina. The material for the study was collected from nest boxes in six forest complexes in southwestern Poland. The conducted research revealed the presence of five species of Uropodina, with a total number of 559 specimens, in the examined boxes. *Leiodinychus orbicularis* was found in almost half of all of the examined boxes and was a superdominant species in the communities. The analysis of the habitat preferences of the two nest species of Uropodina showed that *A. infirmus* preferred old natural nests, in which the communities were formed from a larger number of species, without a significant statistical prevalence of one species. On the other hand, *L. orbicularis* occurred sporadically in open bird nests, but was very numerous and frequent in nest boxes. The significant dominance of *L. orbicularis* in nest boxes can probably be explained by the specific conditions prevailing in this type of microhabitat, including the very low humidity and food resources that this mite species prefers compared to other species of Uropodina.

Citation: Błoszyk, J.; Hebda, G.; Kulczak, M.; Zacharyasiewicz, M.; Rutkowski, T.; Napierała, A. Communities of Uropodina (Acari: Mesostigmata) in Nest Boxes Inhabited by Dormice (*Glis glis* and *Muscardinus avellanarius*) and Differences in Percentages of Nidicoles in Nests of Various Hosts. *Animals* **2023**, *13*, 3567. https://doi.org/10.3390/ani13223567

Academic Editor: Theo de Waal

Received: 27 September 2023
Revised: 8 November 2023
Accepted: 14 November 2023
Published: 18 November 2023

Keywords: *Apionoseius infirmus*; bird nests; dormice; *Leiodinychus orbicularis*; mammal nests; merocenose; nidicoles

1. Introduction

"Ecological niche" is a term for the position of a species within an ecosystem, describing both the range of conditions necessary for the persistence of the species and its ecological role in the ecosystem [1]. Every species strives to maximise the use of the available niches and populate them with individuals. That is why the number of inhabited environments or microenvironments and the number of local populations can be considered as indicators of a species' evolutionary success.

The nesting abilities of secondary cavity nesters, bats, and Gliridae mammals depend on the presence of natural cavities that are necessary for establishing nests [2–5]. The availability of nesting sites for species inhabiting natural cavities is limited, especially in younger commercial forests. The number of nesting sites is regularly increased by creating "artificial cavities", that is, by hanging nest boxes for particular groups of animals (e.g., for birds [2–5]; for dormice [6,7]; and for bats [8]). By creating artificial shelters and breeding places for birds and some endangered mammals, humans contribute to the creation of new niches for many invertebrate species. Bird nest boxes, bat boxes, and less commonly encountered boxes intended for mammals from the Gliridae family are specific types of microenvironments (merocenose) inhabited by diverse groups of invertebrates including Arachnida and Insecta (esp. species from orders such as Coleoptera, Diptera, Siphonaptera, Hemiptera, Hymenoptera, and Lepidoptera), and even invertebrates that are not typically associated with nest boxes on trees such as Isopoda, Gastropoda, and Myriapoda [9–16]. Much scholarly attention was paid to the presence of ectoparasites in such places related to their hosts (fleas, ticks, Diptera: Protocalliphoridae, and some mites) [17–24]. Among the groups that are frequently observed in nest boxes, the most interesting phenomena is the presence of typical nidicoles, for whom nests are the proper type of habitat [25,26].

Nest boxes may contain different materials of organic origin. Typically, their most considerable portion consists of nest material, which may be composed of both plant (stems, leaves, the roots of plants, and mosses) and animal components (feathers, fur, and hairs), and other remains left by the host, including faeces, pellets, food storages, and remnants after broods (egg shells and dead juveniles) [27]. Such diverse nest box contents can attract organisms presenting different foraging strategies. These organisms are saprophagous species that feed on decomposing nest materials or bird and mammal prey and dropping remains [28,29]; scavengers and carnivores, which feed on all developmental stages of other invertebrates commonly living in the nest [16,26,30]; and vertebrate ectoparasites, which spend at least part of their lives buried in nest material [18,31]. Apart from the direct trophic benefits for organisms inhabiting nest boxes, other species can also be associated with more favourable conditions occurring in nest boxes than in natural conditions, for example, some groups of hymenopterans like ants, bumblebees, and social wasps [14,32].

Previous studies have shown that the microhabitats of bird and mammal nests are often also inhabited by mites from the suborder Uropodina (Acari: Mesostigmata). Uropodina mites were found both in the nests and nest boxes of various bird species [25,30,33–43], mole nests and badger burrows [44–46], and bat boxes [47]. The results obtained so far have shown that nests constitute various environments for Uropodina. The community structure of Uropodina in these microhabitats depends on different factors, such as the nesting host ecology, the duration of the nest existence, and the location of the nest. As far as the time of the nest's existence is concerned, communities of Uropodina have been examined so far in nest boxes [34,47], one-year natural nests [39,43], and perennial nests of birds of prey [36,38,41,48–53]. Other important microhabitats for Uropodina communities are perennial nests of mammals, such as burrows of small and medium mammals (including mole (*Talpa europaea* L.), marmot (*Marmota marmota latirostris* Kratochvíl), and badger

(*Meles meles* L.)) [45,46]. These nests, especially badger burrows, can exist for a very long time [54], which enables the formation of diverse Uropodina communities [45]. The factor of nest existence is also very important because of the slow rate of colonisation observed in Uropodina species in this type of unstable microhabitat. The most frequent method of colonising nests as well as other types of microhabitats used by Uropodina mites is phoresy [55].

It is also worth mentioning that most of the research conducted so far has focused mainly on Uropodina communities inhabiting arboreal or aboveground bird nests [36,39,48–53,56,57]. However, recent studies on Uropodina communities found in nests of the wood warbler (*Phylloscopus sibilatrix* (Bechstein)), the passerine species, which builds its nests on the ground, have revealed that there is a lack of typical nidicoles in such nests and that the community structure is very similar to those found in the soil and nests of the common mole [35,46] compared to those recorded in other nests of birds.

The Uropodina species, which is most frequently and abundantly found in the nests of various bird and mammal species, is *Leiodinychus orbicularis* [55], described by Koch in 1839. The typical nidicoles associated with mammal nests are *Phaulodiaspis rackei* (Oudemans), *Ph. advena* (Trägårdh), and *Ph. borealis* (Sellnick), which inhabit underground nests of the mole, the marmot, and the European water vole (*Arvicola amphibius* L.) [35]. In nests of various bird species, *Apionoseius infirmus* (Berlese) is often and numerously found, while *Nenteria pandioni* Wiśniewski *et* Hirschmann occurs exclusively in the nests of the white-tailed eagle (*Haliaeetus albicilla* L.). Nest boxes inhabited by dormouse mammals have not yet been studied for the presence of Uropodina mites.

Dormice (Gliridae) are a family of mammals from the suborder Sciuromorpha in the order Rodentia. In Poland, they are represented by four species, the European edible dormouse (*Glis glis* L.), the garden dormouse (*Eliomys quercinus* L.), the forest dormouse (*Dryomys nitedula* (Pallas)), and the hazel dormouse (*Muscardinus avellanarius* L.), all of which are legally protected, with some still requiring active protection. The presence of dormice is associated with the presence of deciduous and mixed forests of high natural value, with an availability of trees with hollows, in which they shelter, reproduce, and store food [58]. The lack of old deciduous tree stands with numerous trees with hollows, which are also the natural habitats of dormice, creates, like in the case of birds, the need to hang special boxes that serve as their substitute shelters. Indeed, dormice readily occupy nest boxes, and providing these artificial shelters has become the basic method in studies of many aspects of dormice biology [6,7,59–61] and the impact of Gliridae on hole-nesting birds [62,63]. For this reason, we decided to study the communities of Uropodina mites inhabiting nest boxes occupied by dormice.

The collection of material from several nest boxes inhabited by dormice in southwestern Poland allowed us, for the first time, to characterise the communities of Uropodina inhabiting these nest boxes. That is why the aim of this study was to ascertain the characteristics of mite communities from the suborder Uropodina, which inhabit the nests of dormice (Gliridae) built in nest boxes. In addition, an analysis of the occurrence and ratio of two nest species of Uropodina, i.e., *L. orbicularis* and *A. infirmus*, in the examined nests of different hosts was also carried out.

2. Materials and Methods

2.1. Study Area

This study was conducted in six forest complexes in the central and southern parts of Opolskie voivodeship (south and southwestern Poland). Two sites are located in the Opawskie Mountains, and the four other sites are located in the lowland part of the Opole region, that is, in the Stobrawski Landscape Park and Niemodlin Forest (see characteristics: Table 1 and Figure 1A). Forest complexes were predominantly deciduous and mixed old forests, with multiple horizontal layers and beech and oak as the dominant tree species, which makes them attractive for dormice. In each of the four lowland forest complexes, groups of 24 designed dormice nest boxes were provided in 2020. In the two mountain forests, only bird nest boxes were present.

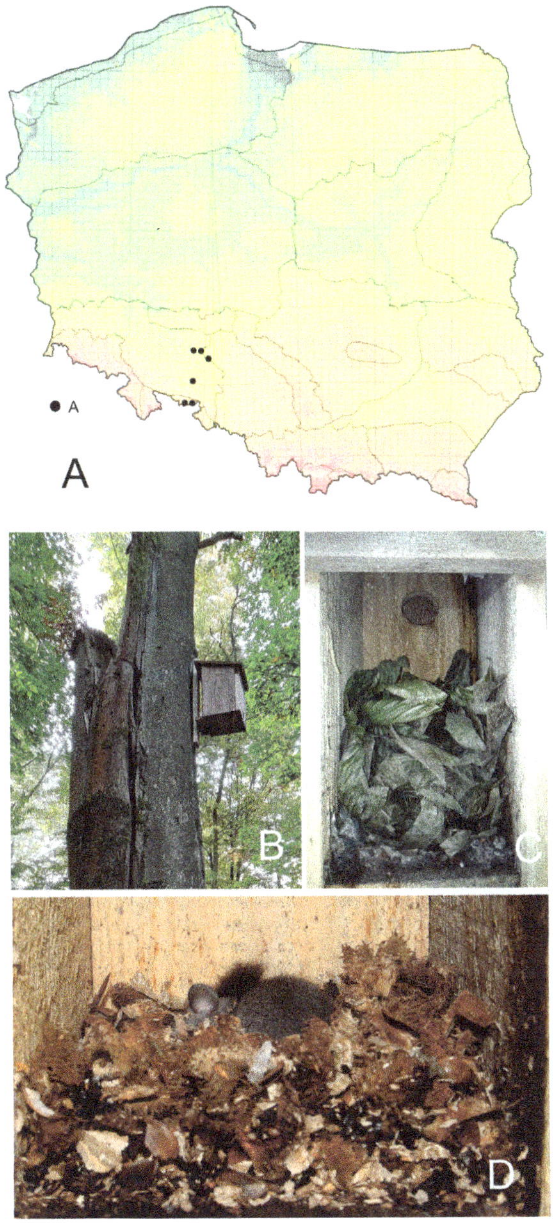

Figure 1. (**A**) Location of the study sites in Poland (black dots) (UTM 10 × 10 km). (**B**) Typical designed dormice nest box placed on beech; the opening is facing the tree trunk. (**C**) Designed dormice nest box with a bulk of leaves and droppings of European edible dormouse. (**D**) Designed dormice nest box with sleeping European edible dormouse and gnawed beech seeds.

Table 1. Characteristics of six study sites with nest boxes; number of examined nest boxes in brackets.

Study Site; GPS of Central Point of the Study Site	Habitat, Dominated Tree Species; Type of Nest Boxes	Dormice Species: Nest Box Content during Sampling
Opawskie Mts: Pokrzywna GPS: 50.2780, 17.4514	Deciduous forest, beech; bird nest boxes (19)	Edible dormice: leaves of trees, gnawed beech and oak seeds, droppings
Opawskie Mts: Dębowiec GPS: 50.2787, 17.5398	Deciduous forest, beech; bird nest boxes (4)	Edible dormice: leaves of trees, gnawed oak seeds, droppings
Stobrawski Landscape Park: Kup GPS: 50.8613, 17.9292	Mixed forest, pine, and beech; dormice nest boxes (6)	Edible dormice: leaves of trees, gnawed beech seeds, droppings
Stobrawski Landscape Park: Lubsza GPS: 50.9328, 17.5658	Deciduous forest, beech, oak; dormice nest boxes (4)	Hazel dormice: nests
Stobrawski Landscape Park: Kozuby GPS: 50.9381, 17.8124	Deciduous forest, oak; dormice nest boxes (3)	Edible and hazel dormice: nests, gnawed oak and beech seeds, leaves of trees, droppings
"Niemodlin Forest": Goszczowice GPS: 50.5784, 17.6084	Deciduous forest, beech, dormice nest boxes (2)	Edible dormice: leaves of trees, gnawed beech seeds, droppings

2.2. Data Collection

The samples were collected from wooden nest boxes, both those designed for dormice and those designed for birds (for nest boxes, see Table 1). All dormice and bird nest boxes were checked yearly since 2020 and cleaned before the next season. The selected nest boxes were examined once between the 12th of October 2022 and the 19th of January 2023.

In this study, we analysed material from only 38 nest boxes where typical remnants of dormice presence were left: 15 nest boxes designed for dormice (Figure 1B–D) and 23 typical bird nest boxes (Table 1). The dimensions of the examined dormice boxes were as follows: diameter of opening: 4.5 cm, bottom: 14 × 16 cm, distance from the opening to the bottom: 25 cm, distance from the top to the bottom: 33 cm. The openings of the dormice boxes were facing the tree trunk, and the box was fixed to the tree at a distance of 4.5 cm from the wooden pole. The dimensions of the bird boxes were more variable, and they were as follows: diameter of opening: 3.5–4.5 cm, bottom: 14 × 14–16 cm, distance from the opening to the bottom: 17–25 cm, distance from the top to the bottom: c. 25–35 cm. The dormice and bird nest boxes were placed on trees, c. 4–5 m above the ground, and only such nest boxes were examined, where typical remnants of edible or hazel dormouse were left, including nests, aggregations of leaves, droppings, gnawed nuts of hazels, or beech and oak seeds.

During the box examination, its rough qualitative content characteristics were registered, including the presence of bird or mammal nest remains, faeces, or food storage. The entire contents of the examined nest boxes were placed into sealed plastic bags with labels describing the box's location and date of collection.

The samples were then immediately transferred to Berlese -Tullgren funnels for mite extraction. This process lasted 72–96 h for each sample, depending on its volume. The extracted specimens were collected in Eppendorf tubes filled with c. 70–80% ethanol alcohol and labelled. The mite specimens were sorted and identified with a stereoscopic microscope (Olympus SZX 16), and the identification of the extracted species was conducted by the first author on the basis of the publications by Karg [64], Błoszyk [65], and Mašán [66]. The extracted specimens were stored in the Natural History Collections (Faculty of Biology) at Adam Mickiewicz University in Poznań.

2.3. Data Analysis

The structures of the analysed mite communities are characterised with the index of dominance (D) and the frequency of occurrence (F). The scale has the following classes: dominance D5 eudominants (>30.0%), D4 dominants (15.1–30.0%), D3 subdominants (7.1–15.0%), D2 recedents (3.0–7.0%), and D1 subrecedents (<3.0%); frequency F5 euconstants (>50.0%), F4 constants (30.1–50.0%), F3 subconstants (15.1–30.0%), F2 accessory species (5.0–15.0%), and F1 accidents (<5.0%) [43]. The average number of specimens in positive samples, presented in Table 2, includes only the nest boxes occupied by dormice. The data used to compare the occurrences of *L. orbicularis* and *A. infirmus* in nests of different hosts (Table 3) were stored in the computer database in the Natural History Collections (Faculty of Biology).

Table 2. Species composition, number of specimens (N), dominance (%), frequency (F%), and average number of specimens in positive samples from nests of dormice. SD—standard deviation; F—females; M—males; D—deutonymphs; P—protonymphs; L—larvae.

Species	N	D%	F%	Average ± SD	Max.	F	M	D	P	L
Leiodinychus orbicularis (C.L. Koch)	559	99.1	46.0	32.9 ± 69.1	285	184	153	165	57	-
Trachytes aegrota (C.L. Koch)	2	0.5	2.7	2.0	2	2	-	-	-	-
T. irenae Pecina	1	0.2	2.7	1.0	1	1	-	-	-	-
Neodiscopoma splendida (Kramer)	1	0.2	2.7	1.0	1	1	-	-	-	-
Nenteria sp.	1	0.2	2.7	1.0	1	-	1	-	-	-
Total	564	100.0	50.0	29.7 ± 65.8	285					

Table 3. Occurrence of *L. orbicularis* and *A. infirmus* in nests of different hosts: A—bat boxes; B—nest boxes occupied by dormice (*Glis glis* (L.) and *Muscardinus avellanarius* (L.)); C—nests of tits (Paridae sp.) and flycatchers (*Muscicapa* sp.) in boxes; D—nests of starlings (Sturnidae sp.) in boxes; E—nests of white storks (*Ciconia ciconia* (L.)); F—nests of thrushes (Turdinae sp.); G—nests of black storks (*Ciconia nigra* (L.)); H—nests of kites (*Milvus* sp.); I—burrows of various mammals; J—mole nests (*Talpa europaea* L.); K—tree nests of various birds of prey; L—wood warbler nests (*Phylloscopus sibilatrix* (Bechstein)); M—badger (*Meles meles* (L.)) burrows. Bold—highest dominance and frequency in examined communities.

	A	B	C	D	E	F	G	H	I	J	K	L	M
Number of boxes or nests	58	38	170	103	38	47	39	52	23	132	34	66	32
Number of Uropodina species	2	5	3	2	11	15	11	11	24	15	11	14	16
Number of specimens	119	564	453	1525	2827	275	373	942	782	4718	925	595	413
					L. orbicularis								
Number of specimens	118	559	443	1281	904	42	49	7	2	5	0	0	0
Dominance (%)	**99**	**99**	**98**	**84**	32	15	13	>1	>1	>1			
Frequency (%)	19	46	11	21	**74**	6	5	4	4	>1			
Average number of specimens in a nest ± SD	10 ± 14	32.9 ± 69.1	23 ± 52	58 ± 151	32 ± 91	14 ± 14	24 ± 16	0.1 ± 0.7	2	5			
					A. infirmus								
Number of specimens	0	0	0	244	26	0	270	225	1	0	289	0	4
Dominance (%)				16	>1		21	24	>1		10		1
Frequency %				10	**26**		**31**	18	4		**32**		9
Average number of specimens in a nest ± SD				24 ± 28	2 ± 2		22 ± 141	22 ± 140	1		26 ± 53		1

2.4. Statistical Analysis

We used non-parametric tests (Fisher's exact test and Mann–Whitney U test). The established significance level in the statistical analysis was $p < 0.05$. All probability values shown here are two-tailed. All statistical analyses followed the formulae in STATISTICA 12.0 [67].

3. Results

3.1. Characteristics of Uropodina Communities in Nests of Species from Gliridae Family

Mites from the Uropodina group were present in the nest boxes in all six forest complexes, and 50% of the examined boxes for the purpose of this study contained at least one specimen of Uropodina (Table 2). In the 38 nest boxes, the presence of five species of Uropodina was recorded. *Leiodinychus orbicularis* turned out to be the most numerous species. A total of 559 specimens of this species were found, including 184 females, 153 males, 165 deutonymphs, and 57 protonymphs. Moreover, *L. orbicularis* was the superdominant species in the examined community and was present in nearly half of the examined nest boxes. It was present both in bird nest boxes (11 boxes) and dormice nest boxes (6); there were no differences in the type of box selection (Fisher's exact test, $p = 0.56$). The mean number of *L. orbicularis* also did not differ between the types of nest boxes (Mann–Whitney U test; $U = 23.0$, $p = 0.48$). In individual boxes, the presence of 1 to 285 specimens was recorded (on average, in one nest, there were 29.7 specimens $\pm$ 65.8) (Table 2).

3.2. Frequency and Abundance of Nidicoles Leiodinychus Orbicularis and Apionoseius Infirmus in Material from Nests of Different Hosts

The analysis of the species composition of the examined Uropodina communities found in nests and nest boxes inhabited by various species of mammals and birds showed that *L. orbicularis* occurs in most of the merocenoses of this type that have been analysed so far (Table 3). However, the percentage of this species in Uropodina communities in such microhabitats varies (Table 3). A very high percentage of this species (even >90%) was recorded in boxes for birds, bats, and dormouse mammals. These are the communities with a small number of species, which means that *L. orbicularis* is a superdominant species in such cases. On the other hand, *A. infirmus* has not been found so far in most nests found in boxes; it only occurred in boxes inhabited by starlings, but the percentage of this species and the frequency were small there. This species also did not occur in the nests of thrushes, the nests of wood warblers, and in underground mole nests.

4. Discussion

In the examined boxes inhabited by dormice, the occurrence of five species of Uropodina was recorded, of which only one, i.e., *L. orbicularis*, can be typically considered as nest species [25,34,55]. The other species that were found there, such as two species from genus *Trachytes*, and species from genus *Nenteria*, such as *N. splendida*, were soil species, associated with the litter and soils of different forests or open environments [65]. Occasional adult specimens (lack of juvenile forms) probably found themselves there accidentally with the nesting material or food collected from the ground. Previous studies have shown that *L. orbicularis* is a nidicole associated with various types of bird nests, mammal nests, and nest boxes [25,34]. The boxes inhabited by dormice are another microhabitat, in which the presence of juvenile forms of this species, especially protonymphs, shows that *L. orbicularis* can live and reproduce in such places. The research also did not reveal any differences in the preferences of *L. orbicularis* in relation to the type of nest box host (bird vs. dormice). This means that the specific microclimate in the boxes is the factor that attracts this mite species.

The comparison of Uropodina communities from the nests of other birds and mammals showed that in artificial, human-made microhabitats, such as bird nest boxes, dormouse boxes, and bat boxes, this species has the largest percentage in the whole community (even above 90%). However, these are usually communities with a low number of species, in which *L. orbicularis* monopolises all available resources (Table 3). In typical open bird nests, it occurs rather sporadically. The only exception in this respect are nests of the white stork, in which it was found to be relatively numerous and frequent [39]. It seems that this species avoids the nests of sparrows located on the ground (as seen in the lack of the species in nests of wood warblers [43], mammal burrows (mole and badger), and tree nests of birds of prey [39]) (see Table 3).

The absence of this species in the nests of birds of prey (e.g., eagles and the white-tailed eagle (*Haliaeetus albicilla* (L.))), ground nests of the wood warbler, and underground burrows of the badger, and the low percentage of the specimens in other underground mammal nests is probably due to the method of dispersion of the species, namely the deutonymphs of *L. orbicularis* (with pedicels), which are found in nests and are carried by insects [41]. However, no carrier species have been identified yet, though it is assumed that they are probably carried by beetles (unpublished data). The peculiar habitat preferences of *L. orbicularis* determine the preferred type of the nests inhabited by this species, excluding those located underground. It is worth mentioning that the discussed species is characterised by a wide ecological valence, which allows it to occupy various niches, including those created by humans, and for this reason, it was also found in stored products [68]. It is possible that the numerous occurrences of nests built in nest boxes are related to the possibility of colonising environments of anthropogenic origin.

The second of the analysed nidicoles, i.e., *A. infirmus*, was not found in the analysed material from the boxes inhabited by dormice. As for the boxes, the species was only found in the nests of the starling (Table 3). Besides this, it occurred mainly in the nests of birds of prey, in the nests of kites, and in the nests of both species of the stork. In the nests of the black stork, black kite (*Milvus migrans* (Boddaert)), and red kite (*Milvus milvus* (L.)), the abundance of this species was higher than that of *L. orbicularis*. Generally, it can be stated that unlike *L. orbicularis*, *A. infirmus* avoids nests built in nest boxes, but it is more often present in old natural nests, where there are usually more species in the community, without a clear statistical prevalence of one species.

5. Conclusions

Apparently, little is known about the method of dispersion of nidicoles and the routes by which they reach isolated microhabitats, such as nest boxes. The presence of phoretic deutonymphs suggests that the discussed species of Uropodina are carried by insects, probably beetles (like most phoretic Uropodina). However, specific vector species have not been found yet. The clear dominance of *L. orbicularis* in the examined boxes (regardless of the host that inhabited them) can be explained by the specific microclimate that prevails in the boxes [69]. This species tolerates very low humidity, which is seen in most boxes, better than other Uropodina species. Finally, it cannot be ruled out that under such conditions in the boxes, or more precisely, in the nesting material, specific fungi can grow, which are probably the food of this mite species.

Author Contributions: Conceptualisation, G.H. and J.B.; Methodology, G.H. and J.B.; Software, J.B.; Validation, J.B.; Formal Analysis, J.B. and A.N.; Investigation, G.H. and T.R.; Resources, J.B.; Data Curation, J.B. and M.Z.; Writing—Original Draft Preparation, A.N., G.H., J.B., M.K., and T.R.; Writing—Review and Editing, A.N., M.K., M.Z., and T.R.; Visualisation, J.B.; Supervision, J.B. and A.N.; Project Administration, A.N. and J.B. All authors have read and agreed to the published version of the manuscript.

Funding: This research received no external funding.

Institutional Review Board Statement: Not applicable.

Informed Consent Statement: Not applicable.

Data Availability Statement: The data presented in this study are stored in an Invertebrate Fauna Bank (Natural History Collections, Faculty of Biology, Adam Mickiewicz University, Poznań, Poland).

Acknowledgments: We are very grateful to Tomasz Biwo and Dominik Rithaler (Opawskie Góry and Góra Świętej Anny Landscape Park, Opole Region, Poland) for their participation in the fieldwork, including finding the nest boxes and collecting the samples. The map of the study area was prepared using the free program, UTM ver. 5.2 (author: G. Gierlasiński, www.heteroptera.us.edu.pl/mapautm.html, accessed on 15 March 2023)).

Conflicts of Interest: The authors declare no conflict of interest.

References

1. Polechová, J.; Storch, D. Ecological Niche. *Encycl. Ecol.* **2019**, *3*, 72–80.
2. Aitken, K.E.H.; Martin, K. The importance of excavators in hole-nesting communities: Availability and use of natural tree holes in old mixed forests of western Canada. *J. Ornithol.* **2007**, *148* (Suppl. S2), 425–434. [CrossRef]
3. Cockle, K.L.; Martin, K.; Robledo, G. Linking fungi, trees, and hole-using birds in a Neotropical tree-cavity network: Pathways of cavity production and implications for conservation. *For. Ecol. Manag.* **2012**, *264*, 210–219. [CrossRef]
4. Walankiewicz, W.; Czeszczewik, D.; Stański, T.; Sahel, M.; Ruczyński, I. Tree cavity resources in spruce-pine managed and protected stands of the Białowieża Forest, Poland. *Nat. Areas J.* **2014**, *34*, 423–428. [CrossRef]
5. Zawadzka, D. Dziuple w ekosystemach leśnych: Formowanie, rozmieszczenie, znaczenie ekologiczne i wskazania ochronne. *Sylwan* **2018**, *162*, 509–520. [CrossRef]
6. Morris, P.A.; Bright, P.W.; Woods, D. Use of nestboxes by the dormouse *Muscardinus avellanarius*. *Biol. Conserv.* **1990**, *51*, 1–13. [CrossRef]
7. Koppmann-Rumpf, B.; Heberer, C.; Schmidt, K.H. Long term study of the reaction of the edible dormouse *Glis glis* (Rodentia: Gliridae) to climatic changes and its interactions with hole-breeding passerines. *Acta Zool. Acad. Sci. Hung.* **2003**, *49* (Suppl. S1), 69–76.
8. Rueegger, N. Bat boxes—A review of their use and application, past, present and future. *Acta Chiropt.* **2016**, *18*, 279–299. [CrossRef]
9. Nordberg, S. Biologisch-ökologische Untersuchungen über die Vogelnidicolen. *Acta Zool. Fenn.* **1936**, *21*, 1–168.
10. Woodroffe, G.E. An ecological study of the insects and mites in the nests of certain birds in Britain. *Bull. Entomol. Res.* **1953**, *44*, 739–772. [CrossRef]
11. McComb, W.C.; Noble, R.E. Invertebrate use of natural tree cavities and vertebrate nest boxes. *Am. Midl. Nat.* **1982**, *107*, 163–172. [CrossRef]
12. Tajovský, K.; Mock, A.; Krumpál, M. Millipedes (Diplopoda) in birds' nests. *Eur. J. Soil Biol.* **2001**, *37*, 321–323. [CrossRef]
13. Turienzo, P.; Di Iorio, O.; Mahnert, V. Global checklist of pseudoscorpions (Arachnida) found in birds' nests. *Rev. Suisse Zool.* **2010**, *117*, 557–598.
14. Broughton, R.K.; Hebda, G.; Maziarz, M.; Smith, K.W.; Smith, L.; Hinsley, S.A. Nest-site competition between bumblebees (Bombidae), social wasps (Vespidae) and cavity-nesting birds in Britain and the Western Palearctic. *Bird Study* **2015**, *62*, 427–437. [CrossRef]
15. Boyes, D.H.; Lewis, O.T. Ecology of Lepidoptera associated with bird nests in mid-Wales, UK. *Ecol. Entomol.* **2018**, *44*, 1–10. [CrossRef]
16. Jaworski, T.; Gryz, J.; Krauze-Gryz, D.; Plewa, R.; Bystrowski, C.; Dobosz, R.; Horák, J. My home is your home: Nest boxes for birds and mammals provide habitats for diverse insect communities. *Insect Conserv. Divers.* **2022**, *15*, 461–469. [CrossRef]
17. Heeb, P.; Kolliker, M.; Richner, H. Bird-Ectoparasite Interactions, Nest Humidity, and Ectoparasite Community Structure. *Ecology* **2000**, *81*, 958–968. [CrossRef]
18. Eeva, T.; Andersson, T.; Berglund, Å.M.; Brommer, J.E.; Hyvönen, R.; Klemola, T.; Laaksonen, T.; Loukola, O.; Morosinotto, C.; Rainio, K.; et al. Species and abundance of ectoparasitic flies (Diptera) in pied flycatcher nests in Fennoscandia. *Parasites Vectors* **2015**, *8*, 648. [CrossRef] [PubMed]
19. Sosnina, E.F. Parasites of *Glis glis caspicus* Satun. in the Caucasus National Park. *Proc. Leningrad. Univ. (Biol.)* **1949**, *101*, 128–144.
20. Skuratowicz, W.A.; Bartkowska, K. Pchły (Siphonaptera) zebrane w Jugosławii. *Fragm. Faunist.* **1977**, *23*, 51–65. [CrossRef]
21. Merino, S.; Potti, J. Weather dependent effects of nest ectoparasites on their bird hosts. *Ecography* **1996**, *19*, 107–113. [CrossRef]
22. Mašán, P.; Fenda, P. *A Review of the Laelapid Mites Associated with Terrestrial Mammals in Slovakia, with a Key to the European Species*; Institute of Zoology, Slovak Academy of Sciences: Bratislava, Slovakia, 2010.
23. Medvedev, S.G.; Tretyakov, K.A. Fleas of small mammals in St. Petersburg. *Parazitologiia* **2014**, *48*, 302–314. [CrossRef] [PubMed]
24. Kirillov, A.A.; Kirillova, N.Y.; Ruchin, A.B. Parasites, Bacteria and Viruses of the Edible Dormouse *Glis glis* (Rodentia: Gliridae) in the Western Palaearctic. *Diversity* **2022**, *14*, 562. [CrossRef]
25. Błoszyk, J.; Gwiazdowicz, D.J.; Kupczyk, M.; Książkiewicz-Parulska, Z. Parasitic mesostigmatid mites (Acari)—Common inhabitants of the nest boxes of starlings (*Sturnus vulgaris*) in a Polish urban habitat. *Biologia* **2016**, *71*, 1034–1037. [CrossRef]
26. Cosandey, V.; Séchaud, R.; Béziers, P.; Chittaro, Y.; Sanchez, A.; Roulin, A. Nidicolous beetle species richness is driven by Barn Owl's nests occupancy and landscape structure. *J. Ornithol.* **2021**, *162*, 857–864. [CrossRef]
27. Hansell, M. *Bird Nests and Construction Behaviour*; Cambridge University Press: Cambridge, UK, 2000.
28. Roy, L.; Bouvier, J.-C.; Lavigne, C.; Galès, M.; Buronfosse, T. Impact of pest control strategies on the arthropodofauna living in bird nests built in nestboxes in pear and apple orchards. *Bull. Entomol. Res.* **2013**, *103*, 458–465. [CrossRef]
29. Boyes, D.H. Natural history of Lepidoptera associated with bird nests in mid-Wales. *Entomol. Rec. J. Var.* **2018**, *130*, 249–259.
30. Krištofík, J.; Mašán, P.; Šustek, Z.; Nuhličková, S. Arthropods (Acarina, Coleoptera, Siphonaptera) in nests of hoopoe (*Upupa epops*) in Central Europe. *Biologia* **2013**, *68*, 155–161. [CrossRef]
31. Rendell, W.B.; Verbeek, N.A.M. Are avian ectoparasites more numerous in nest boxes with old nest material? *Can. J. Zool.* **1996**, *74*, 1819–1825. [CrossRef]
32. Lambrechts, M.M.; Schatz, B.; Bourgault, P. Interactions between ants and breeding Paridae in two distinct Corsican oak habitats. *Folia Zool.* **2008**, *57*, 264–268.

33. Mašán, P.; Krištofík, J. Mesostigmatid mites (Acarina, Mesostigmata) in the nests of penduline tit (*Remiz pendulinus*). *Biologia* **1995**, *50*, 481–485.
34. Błoszyk, J.; Olszanowski, J. Materiały do znajomości fauny roztoczy gniazd i budek lęgowych ptaków. II. Różnice w liczebności i składzie gatunkowym populacji Uropodina (Acari, Anactinotrichida) budek lęgowych na Mierzei Wiślanej na podstawie dwuletnich obserwacji. *Przegl. Zool.* **1986**, *30*, 63–66.
35. Błoszyk, J.; Bajaczyk, R. The first record of *Phaulodinychus borealis* (Sellnick, 1949) phoresy (Acari, Uropodina) associated with fleas (Siphonaptera) in mole nests. In *Soil Zoology in Central Europe*; AS CR Institute of Soil Biology: České Budějovice, Czech Republic, 1999; pp. 13–17.
36. Bajerlein, D.; Błoszyk, J.; Gwiazdowicz, D.J.; Ptaszyk, J.; Halliday, B. Community structure and dispersal of mites (Acari, Mesostigmata) in nests of the white stork (*Ciconia ciconia*). *Biologia* **2006**, *61*, 525–530. [CrossRef]
37. Mašán, P.; Krištofík, J. Mites and ticks (Acarina: Mesostigmata et Ixodida) from the nests of *Riparia riparia* L. in South Slovakia. *Biologia* **1993**, *48*, 155–162.
38. Gwiazdowicz, D.J.; Błoszyk, J.; Bajerlein, D.; Halliday, R.B.; Mizera, T. Mites (Acari: Mesostigmata) inhabiting nests of the white-tailed sea eagle *Haliaeetus albicilla* (L.) in Poland. *Entomol. Fennica* **2006**, *17*, 366–372. [CrossRef]
39. Błoszyk, J.; Bajerlein, D.; Gwiazdowicz, D.J.; Halliday, R.B.; Dylewska, M. Uropodine mite communities (Acari: Mesostigmata) in birds' nests in Poland. *Belg. J. Zool.* **2006**, *136*, 145–153.
40. Błoszyk, J.; Gwiazdowicz, D.J.; Halliday, B.; Dolata, P.T.; Gołdyn, B. Nests of the black stork *Ciconia nigra* as a habitat for mesostigmatid mites (Acari: Mesostigmata). *Biologia* **2009**, *64*, 962–968. [CrossRef]
41. Błoszyk, J.; Dražina, T.; Gwiazdowicz, D.; Halliday, R.B.; Gołdyn, B.; Napierała, A.; Rybska, E. Mesostigmatic mites (Acari: Mesostigmata) in nests of the Eurasian griffon vulture (*Gyps fulvus*) in Croatia. *Biologia* **2011**, *66*, 335–339. [CrossRef]
42. Błoszyk, J.; Hebda, G.; Adamski, Z.; Zacharyasiewicz, M. Redescription of *Chiropturopoda nidiphila* Wiśniewski & Hirschmann (Acari: Uropodina) from a woodpecker's tree holes, including all development stages and first notes on its ecology. *Syst. Appl. Acarol.* **2021**, *26*, 1867–1899. [CrossRef]
43. Napierała, A.; Maziarz, M.; Hebda, G.; Broughton, R.K.; Rutkowski, T.; Zacharyasiewicz, M.; Błoszyk, J. Lack of specialist nidicoles as a characteristic of mite assemblages inhabiting nests of the ground-nesting wood warbler, *Phylloscopus sibilatrix* (Aves: Passeriformes). *Exp. Appl. Acarol.* **2021**, *84*, 149–170. [CrossRef]
44. Mašán, P.; Kalúz, S.; Babjaková, A. Mites (Acarina) from the winter nests of the common mole (*Talpa europaea* L.) in South Slovakia. *Biologia* **1994**, *49*, 667–673.
45. Kurek, P.; Nowakowski, K.; Rutkowski, T.; Ważna, A.; Cichocki, J.; Zacharyasiewicz, M.; Błoszyk, J. Underground diversity: Uropodina mites (Acari: Mesostigmata) from European badger (*Meles meles*) nests. *Exp. Appl. Acarol.* **2020**, *82*, 503–513. [CrossRef]
46. Napierała, A.; Mądra, A.; Leszczyńska-Deja, K.; Gwiazdowicz, D.J.; Gołdyn, B.; Błoszyk, J. Community structure variability of Uropodina mites (Acari: Mesostigmata) in nests of the common mole, *Talpa europaea*, in Central Europe. *Exp. Appl. Acarol.* **2016**, *68*, 429–440. [CrossRef] [PubMed]
47. Błoszyk, J.; Rutkowski, T.; Wojtaszyn, G.; Książkiewicz-Parulska, Z.; Zacharyasiewicz, M.; Napierała, A. *Leiodinychus orbicularis* (CL Koch, 1839) in bat boxes in Poland. *Eur. J. Biol. Res.* **2020**, *10*, 150–155. [CrossRef]
48. Gwiazdowicz, D.J.; Mizera, T.; Skorupski, M. Mites in greater spotted eagle nests. *J. Raptor Res.* **1999**, *33*, 257–260.
49. Gwiazdowicz, D.J.; Mizera, T.; Skorupski, M. Mites (Acari, Gamasida) from the nests of birds of prey in Poland. *Buteo* **2000**, *11*, 97–100.
50. Gwiazdowicz, D.J. Mites (Acari: Mesostigmata) occurring in the nests of birds of prey (Falconiformes) and owls (*Strigi formes*). *Acarina* **2003**, *11*, 235–239.
51. Gwiazdowicz, D.J. Mites (Acari, Mesostigmata) appearing in Poland, in the birds's nests of *Passeriformes*, *Falconiformes* and *Strigiformes* orders. In *Kształtowanie i Ochrona Środowiska Leśnego*; Miler, A.T., Ed.; Wydawnictwo Akademii Rolniczej: Poznań, Poland, 2003; pp. 562–572.
52. Gwiazdowicz, D.J.; Błoszyk, J.; Mizera, T.; Tryjanowski, P. Mesostigmatic mites eagle nests (*Haliaeetus albicilla*). *J. Raptor Res.* **2005**, *39*, 60–65.
53. Bloszyk, J.; Gwiazdowicz, D.J.; Bajerlein, D.; Halliday, R.B. Nests of the white stork *Ciconia ciconia* (L.) as a habitat for mesostigmatic mites (Acari, Mesostigmata). *Acta Parasitol.* **2005**, *50*, 171–175.
54. Kurek, P.; Piechnik, Ł.; Wiatrowska, B.; Ważna, A.; Nowakowski, K.; Pardavila, X.; Cichocki, J.; Seget, B. Badger *Meles meles* as Ecosystem Engineer and Its Legal Status in Europe. *Animals* **2022**, *12*, 898. [CrossRef]
55. Napierała, A.; Błoszyk, J. Unstable microhabitats (merocenoses) as specific habitats of Uropodina mites (Acari: Mesostigmata). *Exp. Appl. Acarol.* **2013**, *60*, 163–180. [CrossRef]
56. Fend'a, P.; Lengyel, J. Roztoče (Acarina, Mesostigmata) v hniezdach orliaka morského (*Haliaeetus albicilla*) na Slovensku [Mites (Acarina, Mesostigmata) in nests of the white-tailed sea eagle *Haliaeetus albicilla* (L.) in Slovakia]. *Entomofauna Carpath.* **2007**, *19*, 48–50.
57. Gwiazdowicz, D.J.; Mizera, M. Preliminary research on mites (Acari, Gamasida) occurring in the pellets of birds of prey and owls. *Sci. Pap. Agric. Univ. Pozn. Anim. Sci.* **2002**, *4*, 117–125.
58. Fedyń, I.; Figarski, T.; Kajtoch, Ł. Overview of the impact of forest habitats quality and landscape disturbances on the ecology and conservation of dormice species. *Eur. J. For. Res.* **2021**, *140*, 511–526. [CrossRef]

59. Schlund, W.; Scharfe, F.; Ganzhorn, J.U. Long-term comparison of food availability and reproduction in the edible dormouse (*Glis glis*). *Mamm. Biol.* **2002**, *67*, 219–232. [CrossRef]
60. Bakó, B.; Hecker, K. Factors determining the distribution of coexisting dormouse species (Gliridae, Rodentia). *Pol. J. Ecol.* **2006**, *54*, 379–386.
61. Trout, R.C.; Brooks, S.; Morris, P. Nest box usage by old edible dormice (*Glis glis*) in breeding and non-breeding years. *Folia Zool.* **2015**, *64*, 320–324. [CrossRef]
62. Juškaitis, R. Interactions between dormice (Gliridae) and hole-nesting birds in nestboxes. *Folia Zool.* **2006**, *55*, 225–236.
63. Adamík, P.; Král, M. Nest losses of cavity nesting birds caused by dormice (Gliridae, Rodentia). *Acta Theriol.* **2008**, *53*, 185–192. [CrossRef]
64. Karg, W. Acari (Acarina) Milben, Unterordnung Parasitiformes (Anactinochaeta). Uropodina Kramer, Schildkrötenmilben. In *Die Tierwelt Deutschlands und der Angrenzenden Meeresteile*; Senglaub, K., Hannemann, H.-J., Schumann, H., Eds.; Gustav Fischer Verlag: Jena, Germany, 1989; Volume 67, p. 203.
65. Błoszyk, J. *Geograficzne i Ekologiczne Zróżnicowanie Zgrupowań Roztoczy z Kohoryty Uropodina (Acari: Mesostigmata) w Polsce: Uropodina Lasów Grądowych (Carpinion betuli)*, 1st ed.; Kontekst: Poznań, Poland, 1999; p. 245.
66. Mašan, P. *Mites of the Cohort Uropodina (Acarina, Mesostigmata) in Slovakia*, 1st ed.; Annotationes Zoologicae et Botanicae: Bratislava, Slovakia, 2001; p. 320.
67. StatSoft Inc. *STATISTICA (Data Analysis Software System)*, version 12; StatSoft Inc.: Tulsa, OK, USA, 2014.
68. Radinovsky, S. The biology and ecology of Granary Mites the Pacific Northwest. III. Life history and development of *Leiodinychus krameri* (Acarina: Uropodidae). *Ann. Entomol. Soc. Am.* **1965**, *58*, 259–267. [CrossRef] [PubMed]
69. Maziarz, M.; Broughton, R.K.; Wesołowski, T. Microclimate in tree cavities and nest-boxes: Implications for hole-nesting birds. *For. Ecol. Manag.* **2017**, *389*, 306–313. [CrossRef]

animals

Article

Effect of Different Host Plants on Life Type Characteristics of Three Spider Mite Pests (Acari: Prostigmata: Tetranychidae)

Hafiz Muhammad Saqib Mushtaq, Hafiz Muhammad Sajid Ali, Muhammad Kamran and Fahad Jaber Alatawi *

Department of Plant Protection, College of Food and Agriculture Sciences, King Saud University, Riyadh 11451, Saudi Arabia; hmushtaq@ksu.edu.sa (H.M.S.M.); hsajid@ksu.edu.sa (H.M.S.A.); murafique@ksu.edu.sa (M.K.)
* Correspondence: falatawi@ksu.edu.sa

Simple Summary: Some spider mite species are economically important agricultural pests, attacking both annual and perennial host plants. They usually produce silken threads of varying densities on the surface of the leaves of inhabiting plants to perform various biological/behavioral activities. In the present study, field-collected leaf samples and laboratory-infested leaves were used to evaluate the effect of different plants on the web-associated behavioral characteristics (life type) of three spider mite species, namely, *Tetranychus urticae*, *Eutetranychus orientalis*, and *Eutetranychus palmatus*. Both annual and perennial plants for *T. urticae* and only perennial plants for *E. orientalis* and *E. palmatus* were used. Two spider mites, *E. orientalis* and *E. palmatus*, showed persistence in life type characteristics on different plant species. In contrast, some behavioral characteristics of *T. urticae* varied by changing the host plants. Although *T. urticae* showed variations in some behavioral characteristics, it did not change its life type, which shows its high adaptability to utilizing the host plant resources. The variations observed in the life type characteristics of *T. urticae* could be helpful in applied pest management.

Abstract: The present study evaluated the host plant effect on life type characteristics of three important spider mite pest species, *Tetranychus urticae* Koch, *Eutetranychus orientalis* (Klein), and *E. palmatus* Attiah (Acari: Prostigmata: Tetranychidae), based on both field and laboratory observations. The polyphagous species, *T. urticae* with complicated web (CW-u) life type, occupying unstable habitats, showed variations in the sites for quiescence (SQ), sites for oviposition (SO), sites for defecation (SD), and webbing density (WD) on different annual/perennial host plants. The SQ, SO, and SD of *T. urticae* were observed either on the leaf, web threads, or trichomes. *Tetranychus urticae* constructed the lowest WD on tomato plants and the highest WD on maize/mulberry plants. Two spider mite species of the genus *Eutetranychus* Banks, the polyphagous *E. orientalis* and the oligophagous *E. palmatus*, inhabit stable host plants, depicted in the little web (LW-j) life types with persistency in all characteristics on different plants. It is concluded that polyphagous spider mites have restricted their life types, showing their high adaptability to utilize the resources of different host plants for survival with slight variation in some important life type characteristics.

Keywords: citrus brown mite; life types; silken threads; two-spotted spider mite; webbing behavior

Citation: Mushtaq, H.M.S.; Ali, H.M.S.; Kamran, M.; Alatawi, F.J. Effect of Different Host Plants on Life Type Characteristics of Three Spider Mite Pests (Acari: Prostigmata: Tetranychidae). *Animals* **2023**, *13*, 3433. https://doi.org/10.3390/ani13223433

Academic Editors: Maciej Skoracki and Monika Fajfer

Received: 4 October 2023
Revised: 1 November 2023
Accepted: 3 November 2023
Published: 7 November 2023

1. Introduction

Spider mites belonging to the family Tetranychidae Donnadieu (Acari: Prostigmata: Tetranychidae) inhabit various annual and perennial host plants [1]. Some members of the subfamily Tetranychinae Berlese are notorious agricultural pests [2,3] and usually produce silken web threads of varying densities and complexities [4–6]. Such web structures serve various biological purposes and represent an adaptation for the mite survival on the inhabited host plants [6].

The webbing structures and associated behavioral characteristics of spider mite species on the surface of host plants are technically termed "life types", and are mainly categorized as little web (LW), complicated web (CW), and woven nest (WN) [4,5]. Further, these three main life types are subdivided into various subtypes based on persistent differences in associated characteristics [4,5]. Among different life type characteristics, the type of the host plant inhabited by the spider mite species is also considered the defining characteristic of a subtype [4].

The idea of classifying spider mites based on their life types was initially proposed over 40 years ago, and numerous studies on the peculiarities of these life types have since been documented [7–9]. The life types of different spider mite species belonging to different genera of Tetranychidae have been studied so far [4,5,10–14]. Among them, members of the genera *Tetranychus* Dufour and *Eutetranychus* Banks are severe threats to many different economic host plants [3]. The two spotted spider mite, *Tetranychus urticae* Koch and the citrus brown mite *Eutetranychus orientalis* (Klein), are polyphagous in feeding habits with CW-u and LW-j life types, respectively [4,5]. The oligophagous species, *Eutetranychus palmatus* Attiah, has been reported from two plant families of Arecacea and Malvaceae [14] and is considered a pest of date palm [15]; however, its life type still remains unexplored and needs to be studied.

The "type" of the life type of a spider mite species could depend on its feeding habit (oligophagous or polyphagous) and the type of host plant (annual or perennial) inhabited [4–6]. It has been argued that the spider mite species infesting annual host plants (unstable habitat) tend to have a fixed life type and mites infesting perennial host plants (stable habitat) exhibit a different life type [4]. Therefore, this study was designed to assess the life type characteristics of two polyphagous spider mite pests, *T. urticae* and *E. orientalis*, feeding on different host plant (annual and perennial) leaves. Also, the life type of the oligophagous pest *E. palmatus* was characterized for the first time.

2. Materials and Methods

2.1. Collection Sites and Mite Rearing

The populations of *Tetranychus urticae*, *Eutetranychus orientalis*, and *E. palmatus* were collected from infested leaves of *Solanum melongena* L., (Solanaceae), *Citrus* sp. (Rutaceae), and *Washingtonia filifera* (Lindl.) (Arecaceae) plants grown within the vicinity of the King Saud University (KSU), Riyadh, Saudi Arabia (SA), respectively, between 2020 and 2021. In addition, a small colony of each spider mite species was reared separately by making leaf arenas of their respective host plants, based on the rearing methods of Mirza et al. [9], with slight modifications. Only the size (3.5–4.5 $\times$ 4.5–5.5 cm^2) and shape (either rectangular or circular) of mites-rearing arenas were modified due to the differences in leaf morphology of respective host plants. All spider mite cultures were kept in a climate-controlled growth chamber (Binder, Tuttlingen, Germany) and maintained at 28 $\pm$ 2 °C, 35 $\pm$ 10% RH, and L14: D10 photoperiod throughout the experimental duration.

2.2. Spider Mites' Identification

The specimens of each tested spider mite species were mounted on glass slides in Hoyer's medium under the SZX10 stereomicroscope (Olympus, Tokyo, Japan). In addition, these specimens were taxonomically identified as species using a BX51 fluorescence microscope (Olympus, Tokyo, Japan) with the help of the published taxonomic literature [16–18]. Finally, the voucher specimens of each identified/tested species were preserved in the Acarology section of the King Saud University Museum of Arthropods, Riyadh, SA.

2.3. Experimental Procedure to Study the Annual and Perennial Host Plants' Effect on Life Type Characteristics of Polyphagous/Oligophagous Spider Mite Pests in the Laboratory

The experiment was conducted in the Biological Control Laboratory, Department of Plant Protection, KSU, during 2020–2021. The life type characteristics of (a) *T. urticae* were examined on six (four annual and two perennial) host plants, *Capsicum annum* L., *S. melongena*, *S. lycopersicum* L., (Solanaceae), *Morus alba* L. (Moraceae), *Zea mays* L. (Poaceae), and *Ziziphus jujuba* L. (Rhamnaceae); (b) *E. orientalis* was examined on three perennial plants, *Citrus* sp., *Ricinus communis* L. (Euphorbiaceae), and *Z. jujube*; and (c) *E. palmatus* was examined on two perennial plants, *Phoenix dactylifera* L. and *W. filifera* (Arecaceae). Each treatment (=host plant) was replicated 10 times. The life type characteristics of tested spider mite species were studied on the leaf arenas of size (3.5–4.5 × 4.5–5.5 cm^2) prepared with the leaves of host plant species mentioned above following Mirza et al. [12], in a climate-controlled chamber at (28 ± 2 °C, 35 ± 5% RH, and L14: D10 photoperiod).

In the laboratory experiments, the newly matured females along with conspecific males were released, separately for each of three tested spider mite species, viz., *T. urticae*, *E. orientalis*, and *E. palmatus*, into their respective experimental arenas. After 3 to 4 days, the mated/gravid females were then used in the experiment. Mites were released near leaf-midrib with the help of a fine camel hairbrush. The life type characteristics of (a) *T. urticae* were examined on six (four annual and two perennial) host plants, *Capsicum annum* L., *S. melongena*, *S. lycopersicum* L., (Solanaceae), *Morus alba* L. (Moraceae), *Zea mays* L. (Poaceae), and *Ziziphus jujuba* L. (Rhamnaceae); (b) *E. orientalis* was examined on three perennial plants, *Citrus* sp., *Ricinus communis* L. (Euphorbiaceae), and *Z. jujube*; and (c) *E. palmatus* was examined on two perennial plants, *Phoenix dactylifera* L. and *W. filifera* (Arecaceae). The experimental arenas were set according to the size (3.5–4.5 × 4.5–5.5 cm^2) and shape (either rectangular or circular) of each respective plant leaf. Each treatment (=host plant) was replicated 10 times. The experimental arenas were kept in a climate-controlled growth chamber for 10 days and maintained at 28 ± 2 °C, 35 ± 5% RH, and L14: D10 photoperiod. To ensure the establishment of spider mite colonies, all experimental arenas were observed after the 3rd day of mite release under an M165 C Stereomicroscope (LEICA, Wetzlar, Germany). However, the final observational data were recorded on the 10th day of mite release. Due to the biological activities (e.g., feeding and defecation) of spider mites, when time passed, leaf color was slightly changed (green to pale yellow) in some experimental arenas. The leaf side (adaxial or abaxial) of each host plant leaf for each tested spider mite species was selected based on the leaf side of natural infestation. The life type characteristics of *T. urticae* and *E. orientalis* were evaluated on the adaxial leaf sides of all tested host plants, whereas *E. palmatus* was assessed on adaxial and abaxial sides of *P. dactylifera* and *W. filifera*, respectively.

2.4. Observations of Life Type Characteristics on Field-Infested Leaf Samples

The leaf samples of the following mentioned plant species infested naturally with *T. urticae*, *E. orientalis*, and *E. palmatus* were collected from the field and brought to the laboratory. The life type characteristics of the spider mite pest (a) *T. urticae* on five (four annual and one perennial) host plants, *C. annum*, *S. melongena*, *S. lycopersicum*, *Z. mays*, and *Z. jujuba*; (b) *E. orientalis* on three perennial plants, *Citrus* sp., *R. communis*, and *Z. jujuba*; and (c) *E. palmatus* on a perennial host plant, *W. filifera*, were observed from these naturally infested leaf samples collected from the field under stereomicroscope in the laboratory, grown within the vicinity of KSU. As compared to the laboratory tests, the life type characteristics of *T. urticae* and *E. palmatus* could not observed on the field-infested leaves of *M. alba* and *P. dactylifera* due to the unavailability of natural infestation, respectively. Randomly, five leaves (=replicates) were collected from each host plant, preserved separately in polyethylene bags, and brought to the Acarology and Biological Control R&D Labs., College of Food and Agriculture Sciences, KSU. Each spider mite species on each host plant leaf was examined under a stereomicroscope to investigate its life type characteristics.

2.5. Data Reading and Statistical Analysis

A total of 10 characteristics of spider mite life type were observed, i.e., host plant type (HP; annual, perennial, etc.), leaf side (LS; upper, lower, etc.) inhabited, webbing structure (WS) and density (WD), sites for oviposition (SO), defecation (SD), and quiescence (SQ), spinning during walking (SW), site for feeding and walking (SFW), and egg cover (EC; guy ropes, dense web, etc.) produced by females.

Following Mushtaq et al. [10], the observed life type characteristics for each spider mite species were comparatively investigated on different host plant leaves, and the obtained results were expressed in percentages. Each examined life type characteristic was considered and recorded as 10% and 20% per replicate in the laboratory and field observations, respectively. Moreover, the observational data were separated into supposed ranks (1–6, as in Table S1) for statistical analysis. Additionally, photographs related to observations on some life type characteristics were captured using an Olympus Microscope Camera (DP72) attached to a stereomicroscope. The ranked data (Table S1) were statistically analyzed either through the Kruskal–Wallis test and Wilcoxon two-sample test (Mann–Whitney U-test), and mean scores were ranked by Wilcoxon rank-sums test using the SAS computer program v.9.4 [19].

The webbing density on field-collected and laboratory-infested leaves was quantified by using the methodology adopted by Sabelis [20] and Lemos et al. [21], with slight modifications, i.e., five different webbing density levels: no webbing (0%), low webbing (1–25%), medium webbing (26–50%), high webbing (51–75%), and extremely high webbing (76–100%) were proposed based on differences in obtained WD percentages (Table S1). To quantify the webbing density (%), a white sheet with a 1 cm^2 hole was placed over the infested leaf. An accurately measured quantity of 3 mg sand was sprinkled through a 1 cm^2 hole on the web surface. Some of the sand particles were passed through the silken strands of the web and landed on the leaf, while some adhered to the web threads. The webbing density (WD) was calculated by using the following equation.

$$\text{Webbing density (WD) \%} = \text{SW}/\text{TSP} \times 100$$

whereas

SW = Sand particles adhere on/within web threads.

TSP = Total number of sand particles (SW + SL).

SL = Sand particles on the leaf surface, do not adhere to web threads.

3. Results

3.1. Life Type Characteristics of Tetranychus urticae on Some Annual and Perennial Host Plant Leaves

The results confirmed that the polyphagous spider mite species *T. urticae* did not change its life type (CW-u) either on four annual (unstable habitat) or two perennial (stable habitat) host plant leaves. Some behavioral characteristics, i.e., WS, EC, SFW, and SW, of *T. urticae* remained persistent on tested annual/perennial plant leaves in laboratory experiments and on field-collected leaves samples (Tables 1 and 2). However, the SQ, SO, SD, and WD were found to be variable (Tables 1 and 2; Figures 1–3). In laboratory experiments, four life type characteristics, i.e., SQ (H = 38.268), SO (H = 49.820), SD (H = 56.420), and WD (H = 39.826) (all df = 5, $p < 0.05$) of *T. urticae* showed significant differences among six tested host plant leaves (Table 3). The SQ was significantly different in *S. melongena* (vs. *Z. mays*, *M. alba*, *S. lycopersicum*), *C. annum* (vs. *Z. mays*, *M. alba*, *S. lycopersicum*), *Z. mays* (vs. *Z. jujuba*), and *M. alba* (vs. *Z. jujuba*) (Table 3). The SO was significantly different in *S. melongena* (vs. *C. annum*, *Z. mays*, *M. alba*, *S. lycopersicum*, *Z. jujuba*), *C. annum* (vs. Z. mays, *S. lycopersicum*, *Z. jujuba*; Table), *Z. mays* (vs. M. alba), and *M. alba* (vs. *S. lycopersicum*, *Z. jujuba*) (Table 3). The SD was significantly different in *S. melongena* (vs. *C. annum*, *Z. mays*, *M. alba*, *S. lycopersicum*, *Z. jujuba*) and *C. annum* (vs. *Z. mays*, *M. alba*, *S. lycopersicum*, *Z. jujuba* (Table 3). The WD was significantly different in *S. melongena* (vs. *C. annum*, *S. lycopersicum*), *C. annum* (vs. *Z. mays*, *M. alba*, *S. lycopersicum*, *Z. jujuba*; Table 3), *Z. mays* (vs. *S. lycopersicum*), *M. alba* (vs. *S. lycopersicum*), and *S. lycopersicum* (vs. *Z. jujuba*) (Table 3).

Similarly, the field observations also showed significant differences in SQ (H = 23.073), SO (H = 23.073), SD (H = 23.073), and WD (H = 20.947) (all df = 4, $p < 0.05$) of *T. urticae* among five (four annual and one perennial) tested host plants (Table 4). The SQ, SO, and SD were significantly different in *S. melongena* (vs. *C. annum*, *Z. mays*, *S. lycopersicum*, *Z. jujuba*),

C. annum (vs. *Z. mays*, *S. lycopersicum*, *Z. jujuba*), and *Z. mays* (vs. *S. lycopersicum*, *Z. jujuba*) (Table 4), whereas the WD was significantly different in *S. melongena* (vs. *C. annum*, *Z. mays*, *S. lycopersicum*), *C. annum* (vs. *S. lycopersicum*, *Z. jujuba*), *Z. mays* (vs. *S. lycopersicum*, *Z. jujuba*), and *S. lycopersicum* (vs. *Z. jujuba*) (Table 4).

Figure 1. Variations were observed in site for quiescence (SQ) shown by *Tetranychus urticae* on the adaxial leaf sides of different (annual and perennial) host plant leaves (*Solanum melongena*, *Capsicum annum*, *Zea mays*, *Morus alba*, *S. lycopersicum*, and *Ziziphus jujuba*) in the laboratory and/or field observations. *S. melongena*: SQ on (**a**) leaf, (**b**) web, and (**c**) trichome; *C. annum*: on (**d**) leaf and (**e**) web; *Z. mays*: on (**f**) leaf and (**g**) web; *M. alba:* on (**h**) leaf and (**i**) web; *S. lycopersicum*: on (**j**) leaf and (**k**) web; and *Z. jujuba*: on (**l**) leaf and (**m**) web.

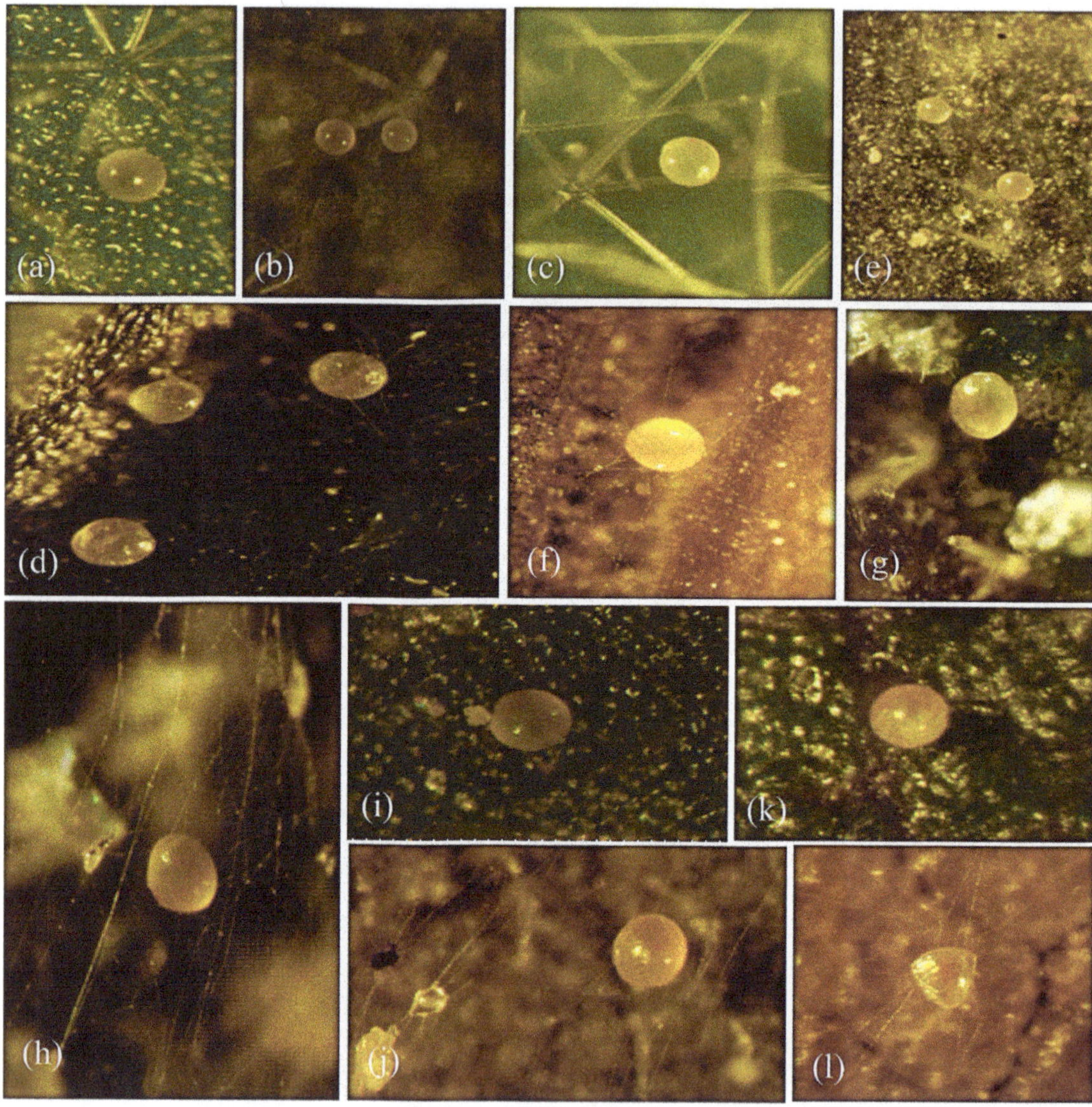

Figure 2. Variations were observed in the site for oviposition (SO) exhibited by *Tetranychus urticae* on the adaxial leaf sides of different (annual and perennial) host plant leaves (*Solanum melongena*, *Capsicum annum*, *Zea mays*, *Morus alba*, *S. lycopersicum*, and *Ziziphus jujuba*) in the laboratory and field observations. *S. melongena*: SO on (**a**) leaf, (**b**) web, and (**c**) trichome; *C. annum*: on (**d**) leaf and web; *Z. mays*: on (**e**) leaf and (**f**) web; *M. alba:* on (**g**) leaf and (**h**) web; *S. lycopersicum*: on (**i**) leaf and (**j**) web; and *Z. jujuba*: on (**k**) leaf and (**l**) web.

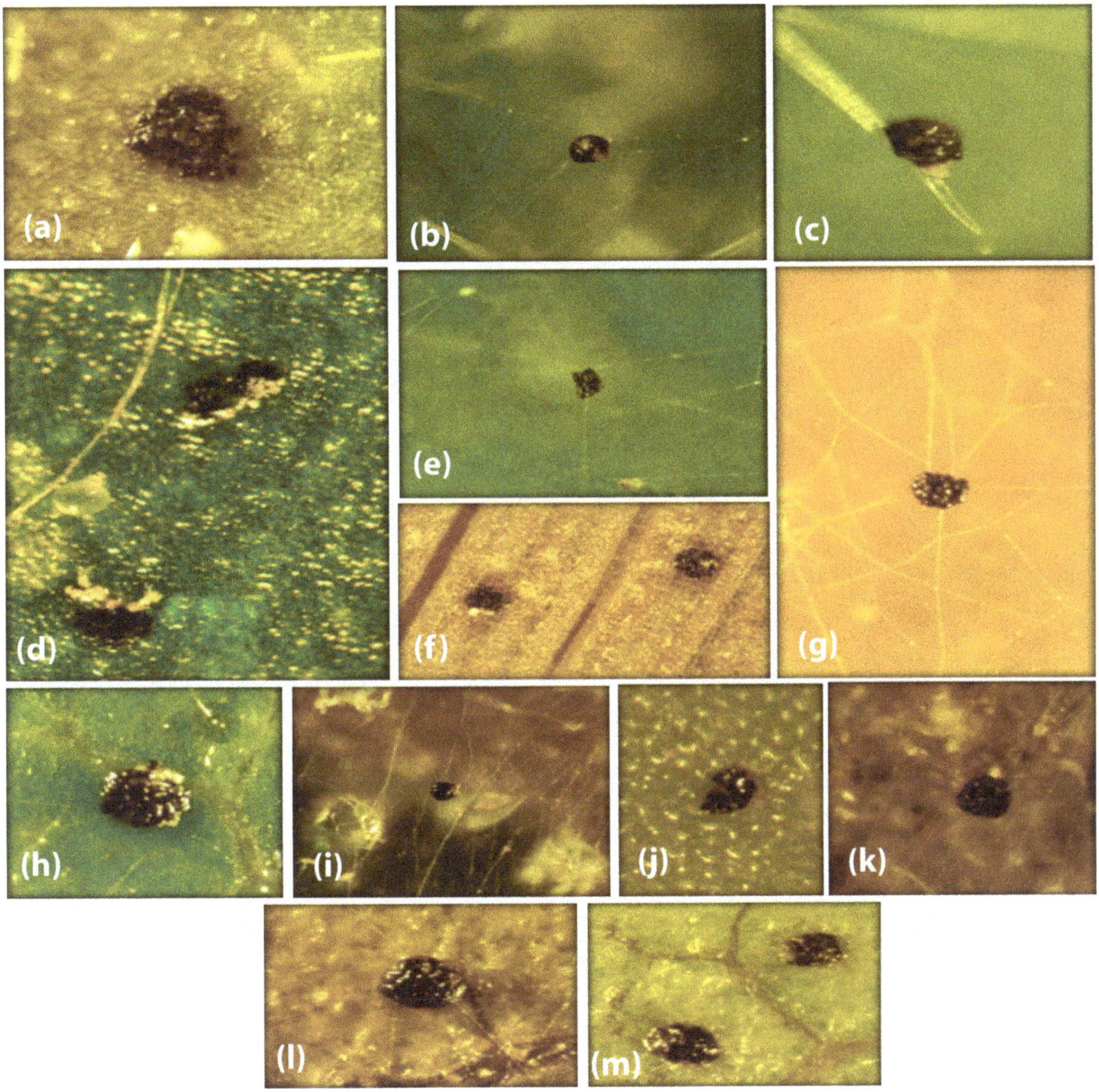

Figure 3. Variations were observed in the site for defecation (SD) shown by *Tetranychus urticae* on the adaxial leaf sides of different (annual and perennial) host plant leaves (*Solanum melongena*, *Capsicum annum*, *Zea mays*, *Morus alba*, *S. lycopersicum*, and *Ziziphus jujuba*) in the laboratory and field observations. *S. melongena*: SD on (**a**) leaf, (**b**) web, and (**c**) trichome; *C. annum*: on (**d**) leaf and (**e**) web; *Z. mays*: on (**f**) leaf and (**g**) web; *M. alba:* on (**h**) leaf and (**i**) web; *S. lycopersicum*: on (**j**) leaf and (**k**) web; and *Z. jujuba*: on (**l**) leaf and (**m**) web.

Table 1. Differences and variations (estimated in %) were observed in some important life type characteristics of *Tetranychus urticae*, *Eutetranychus orientalis*, and *E. palmatus* on adaxial or abaxial leaf sides of different host plants under laboratory conditions.

Spider Mite Species	Host Plants	Host Plant Type (HP)	No. of Obs. (N)	Life Type Characteristics																						
				Site for Quiescence (SQ) %						Site for Oviposition (SO) %						Site for Defecation (SD) %						Webbing Density (WD) %				
				W	L	WL	WT	T	WTL	W	L	WL	WT	T	WTL	W	L	WL	WT	T	WTL	NW	LW	MW	HW	EHW
T. urticae	*Solanum melongena*	Annual	10	0	100	0	0	0	0	0	0	0	0	0	100	0	0	0	0	0	100	0	0	0	40	60
	Capsicum annum	Annual	10	0	100	0	0	0	0	0	100	0	0	0	0	0	100	0	0	0	0	0	10	80	10	0
	Zea mays	Annual	10	0	0	100	0	0	0	0	0	100	0	0	0	0	0	100	0	0	0	0	0	10	20	70
	Morus alba	Perennial	10	0	0	100	0	0	0	0	70	30	0	0	0	0	0	100	0	0	0	0	0	10	20	70
	S. lycopersicum	Annual	10	0	20	80	0	0	0	0	20	80	0	0	0	0	10	90	0	0	0	0	100	0	0	0
	Ziziphus jujuba	Perennial	10	30	20	50	0	0	0	0	0	100	0	0	0	0	0	100	0	0	0	0	10	10	40	40
E. orientalis	*Ricinus communis*	Perennial	10	0	100	0	0	0	0	0	100	0	0	0	0	0	100	0	0	0	0	100	0	0	0	0
	Citrus sp.	Perennial	10	0	100	0	0	0	0	0	100	0	0	0	0	0	100	0	0	0	0	100	0	0	0	0
	Z. jujuba	Perennial	10	0	100	0	0	0	0	0	100	0	0	0	0	0	100	0	0	0	0	100	0	0	0	0
E. palmatus	*Phoenix dactylifera*	Perennial	10	0	100	0	0	0	0	0	100	0	0	0	0	0	100	0	0	0	0	100	0	0	0	0
	Washingtonia filifera	Perennial	10	0	100	0	0	0	0	0	100	0	0	0	0	0	100	0	0	0	0	100	0	0	0	0

W = on/within web threads, L = on leaf surface, WL = on/within web threads and on leaf surface, WT = on/within web threads and on trichrome, T = on trichrome, WTL = on/within web threads, on trichrome and on leaf surface, NW = no webbing, LW = low webbing, MW = moderate webbing, HW = high webbing, EHW = extremely high webbing.

Table 2. Differences and variations (estimated in %) were observed in some important life type characteristics of *Tetranychus urticae*, *Eutetranychus orientalis*, and *E. palmatus* on adaxial or abaxial sides of field-infested leaves of different host plants.

Spider Mite Species	Host Plants	Host Plant Type (HP)	No. of Obs. (N)	Life Type Characteristics																						
				Site for Quiescence (SQ) %						Site for Oviposition (SO) %						Site for Defecation (SD) %						Webbing Density (WD) %				
				W	L	WL	WT	T	WTL	W	L	WL	WT	T	WTL	W	L	WL	WT	T	WTL	NW	LW	MW	HW	EHW
T. urticae	*Solanum melongena*	Annual	5	0	0	0	0	0	100	0	0	0	0	0	100	0	0	0	100	0	0	0	0	0	0	100
	Capsicum annum	Annual	5	100	0	0	0	0	0	100	0	0	0	0	0	100	0	0	0	0	0	0	0	40	60	0
	Zea mays	Annual	5	0	80	20	0	0	0	0	80	20	0	0	0	0	80	20	0	0	0	0	0	60	40	0
	S. lycopersicum	Annual	5	0	0	100	0	0	0	0	0	100	0	0	0	0	0	100	0	0	0	0	100	0	0	0
	Ziziphus jujuba	Perennial	5	0	0	100	0	0	0	0	0	100	0	0	0	0	0	100	0	0	0	0	0	0	40	60
E. orientalis	*Ricinus communis*	Perennial	5	0	100	0	0	0	0	0	100	0	0	0	0	0	100	0	0	0	0	100	0	0	0	0
	Citrus sp.	Perennial	5	0	100	0	0	0	0	0	100	0	0	0	0	0	100	0	0	0	0	100	0	0	0	0
	Z. jujuba	Perennial	5	0	100	0	0	0	0	0	100	0	0	0	0	0	100	0	0	0	0	100	0	0	0	0
E. palmatus	*Washingtonia filifera*	Perennial	5	0	100	0	0	0	0	0	100	0	0	0	0	0	100	0	0	0	0	100	0	0	0	0

W = on/within web threads, L = on leaf surface, WL = on/within web threads and on leaf surface, WT = on/within web threads and on trichrome, T = on trichrome, WTL = on/within web threads, on trichrome and Pl on leaf surface, NW = no webbing, LW = low webbing, MW = moderate webbing, HW = high webbing, EHW = extremely high webbing.

Table 3. Wilcoxon two-sample test results of host plants' effect on some important life type characteristics (SQ, SO, SD, and WD) of *Tetranychus urticae* on adaxial leaf sides of different host plants in the laboratory.

Host Plants Comparison	Site for Quiescence (SQ)			Site for Defecation (SO)			Site for Oviposition (SD)			Webbing Density (WD)		
	U	DF	*p*	*U*	DF	*p*	*U*	DF	*p*	*U*	DF	*p*
Solanum melongena vs. *Capsicum annum*	0.000	1	1.0000	19.000	1	<0.0001	19.000	1	<0.0001	14.702	1	0.0001
S. melongena vs. *Zea mays*	19.000	1	<0.0001	19.000	1	<0.0001	19.000	1	<0.0001	0.074	1	0.7864
S. melongena vs. *Morus alba*	19.000	1	<0.0001	17.195	1	<0.0001	19.000	1	<0.0001	0.074	1	0.7864
S. melongena vs. *S. lycopersicum*	12.667	1	0.0004	17.593	1	<0.0001	18.182	1	<0.0001	16.964	1	<0.0001
S. melongena vs. *Zizipus jujuba*	0.745	1	0.3880	19.000	1	<0.0001	19.000	1	<0.0001	1.378	1	0.2404
C. annum vs. *Z. mays*	19.000	1	<0.0001	19.000	1	<0.0001	19.000	1	<0.0001	12.794	1	0.0003
C. annum vs. *M. alba*	19.000	1	<0.0001	3.353	1	0.0671	19.000	1	<0.0001	12.794	1	0.0003
C. annum vs. *S. lycopersicum*	12.667	1	0.0004	12.667	1	0.0004	15.546	1	<0.0001	15.000	1	0.0001
C. annum vs. *Z. jujuba*	0.745	1	0.3880	19.000	1	<0.0001	19.000	1	<0.0001	7.234	1	0.0072
Z. may vs. *M. alba*	0.000	1	1.0000	10.231	1	0.0014	0.000	1	1.0000	0.000	1	1.0000
Z. may vs. *S. lycopersicum*	2.111	1	0.1462	2.111	1	0.1462	1.000	1	0.3173	17.148	1	<0.0001
Z. may vs. *Z. jujuba*	6.209	1	0.0127	0.000	1	1.0000	0.000	1	1.0000	1.700	1	0.1923
M. alba vs. *S. lycopersicum*	0.146	1	2.1111	4.798	1	0.0285	1.000	1	0.3173	17.148	1	<0.0001
M. alba vs. *Z. jujuba*	6.209	1	0.0127	10.231	1	0.0014	0.000	1	1.0000	1.700	1	0.1923
S. lycopersicum vs. *Z. jujuba*	2.587	1	0.1078	2.111	1	0.1462	1.000	1	0.3173	14.119	1	0.0002

Table 4. Wilcoxon two-sample test results of host plants' effect on some important life type characteristics (SQ, SO, SD, and WD) of *Tetranychus urticae* on adaxial leaf sides of field-infested leaves of different host plants.

Host Plants Comparison	Site for Quiescence (SQ)			Site for Oviposition (SO)			Site for Defecation (SD)			Webbing Density (WD)		
	U	DF	*p*	*U*	DF	*p*	*U*	DF	*p*	*U*	DF	*p*
Solanum melongena vs. *Capsicum annum*	9.000	1	0.003	9.000	1	0.003	9.000	1	0.003	8.036	1	0.005
S. melongena vs. *Zea mays*	8.333	1	0.004	8.333	1	0.004	8.333	1	0.004	8.036	1	0.005
S. melongena vs. *S. lycopersicum*	9.000	1	0.003	9.000	1	0.003	9.000	1	0.003	9.000	1	0.003
S. melongena vs. *Zizipus jujuba*	9.000	1	0.003	9.000	1	0.003	9.000	1	0.003	2.250	1	0.134
C. annum vs. *Z. mays*	8.333	1	0.004	8.333	1	0.004	8.333	1	0.004	0.360	1	0.549
C. annum vs. *S. lycopersicum*	9.000	1	0.003	9.000	1	0.003	9.000	1	0.003	8.036	1	0.005
C. annum vs. *Z. jujuba*	9.000	1	0.003	9.000	1	0.003	9.000	1	0.003	4.641	1	0.031
Z. may vs. *S. lycopersicum*	6.000	1	0.014	6.000	1	0.014	6.000	1	0.014	8.036	1	0.005
Z. may vs. *Z. jujuba*	6.000	1	0.014	6.000	1	0.014	6.000	1	0.014	5.400	1	0.020
S. lycopersicum vs. *Z. jujuba*	0.000	1	1.000	0.000	1	1.000	0.000	1	1.000	8.036	1	0.005

3.2. Life Type Characteristics of Eutetranychus orientalis

The results revealed that life type behavioral characteristics of polyphagous *E. orientalis* remained persistent on tested perennial plants, both in the laboratory and in field observations (Tables 1 and 2). The life type of *E. orientalis* also did not change on different perennial plants.

3.3. Life Type Characteristics of Eutetranychus palmatus

The life type characteristics of *E. palmatus* were investigated for the first time in the present study, and its life type also remained the same on two perennial host plants. It was observed that all the behavioral characteristics of *E. palmatus* remained persistent both in the laboratory and in field observations (Tables 1 and 2).

In the laboratory and field observations, mobile stages of *E. palmatus* neither spun a web while walking on the leaf surface nor showed dragging behavior. The quiescent stages/exuviae, eggs, and feces were consistently observed on the leaf surface (Figure 4), near to or away from the leaf midrib. The female of *E. palmatus* constructed egg covers as dense webs (Figure 4b) and showed weaving behavior. Females, males, nymphs, and larvae were randomly observed during feeding, walking, and resting on the leaf surface near to or away from the midrib. *Eutetranychus palmatus* showed an LW life type and LW-j subtype based on the observed behavioral characteristics on the leaves of *P. dactylifera* and *W. filifera*.

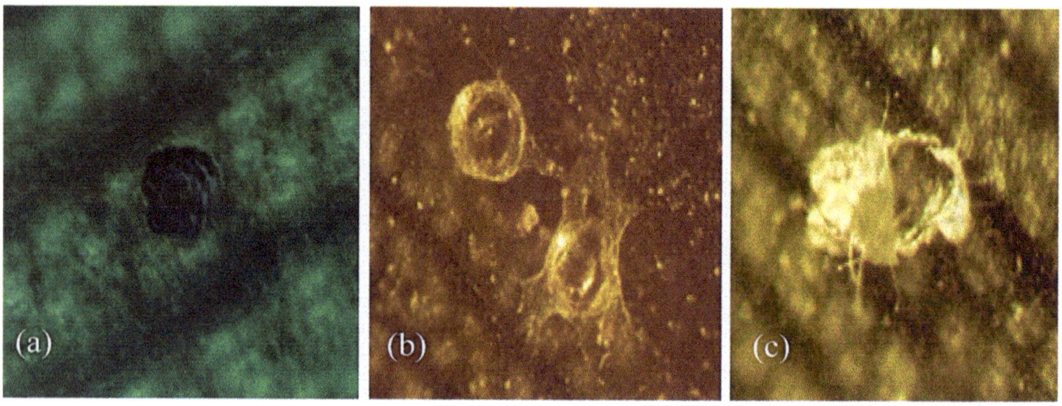

Figure 4. Some important life type (LW-j) characteristics of *Eutetranychus palmatus* on the adaxial and abaxial leaf sides of *P. dactylifera* and *W. filifera* (perennial) were observed in field and laboratory observations. (**a**) Feces on the leaf surface; (**b**) egg covers as dense web; (**c**) quiescence stage on the leaf surface.

4. Discussion

In the present study, the host plant effect was assessed on the life type characteristics of three spider mite pests, *T. urticae*, *E. orientalis*, and *E. palmatus*. The four life type characteristics, i.e., site for quiescence (SQ), site for oviposition (SO), site for defecation (SD), and webbing density (WD) of *T. urticae* were observed either on the leaf surface, web threads, or trichomes and varied within and between different host plants (annual and perennial) (Tables 1 and 2, Figures 1–3). The life type characteristics of *T. urticae* were not previously studied regarding the annual type of host plants. However, the CW-u life type was determined for *T. urticae* on *Sambucus sieboldiana* (perennial), where the preferred SQ, SO, and SD were on threads of an irregularly complicated web [4]. Such variations in the life type characteristics could be due to the change in the microhabitat, leaf structure (glabrous or pubescence), and mite population density [4,10,22].

Tetranychus urticae is a polyphagous pest with >1100 annual and perennial hosts [14,15]. Although Saito [4] indicated that polyphagous mite species inhabiting annual and perennial host plants exhibit fixed and diverse life type, respectively. In contrast, *T. urticae* showed significant variations in the SQ, SO, SD, and WD on different annual and perennial plant leaves in the present study (Tables 1 and 2, Figures 1–3). Similarly, an oligophagous species, *Oligonychus afrasiaticus* (McGregor) having CW-d life type, showed variations in SO and SQ when tested on annual (*Sorghum bicolor* (Poaceae)) and perennial (*Saccharum officinarum* L. (Poaceae)) host plants without changing their life type [13]. It has been reported that some polyphagous spider mite pests change their life type on perennial plants [4,11,22]. For example, *Eotetranychus tilliarium* (Hermann) showed the CW-r life type on the hairy leaves of *Alnus hirsuta* Turcaz. (Betulaceae), and the WN-t life type on glabrous leaves of *A. japonica* Steud [4]. Likewise, *E. asiaticus* Ehara changed its life type characteristics on two different perennial plants by depicting the WN-t life type on strawberry leaves [23], and the CW-g life type when inhabiting the galls created by an insect species, *Trioza cinnamomi* (Boselli) (Hemiptera: Triozidae) on leaves of *Cinnamomum japonicum* Siebold (Laureacea) [11].

The amount of webbing (WD, an essential characteristic of CW-life type) produced by *T. urticae* was reported to be affected due to changes in environmental conditions, host plants inhabited, and mite population density [23–25]. In the present study, the WD of *T. urticae* varied on annual host plants from low (on tomato) to extremely high (e.g., on brinjal) (Tables 1 and 2). On the other hand, the WD range was high to extremely high on the tested perennial plants. Similarly, *T. urticae* showed variations in the amount of web deposition when tested on seven perennial host plants, i.e., Algerian ivy, bean, cotton, castor bean, hibiscus, rose, and sweet potato [26,27]. These variations in the WD of *T. urticae* are probably due to the differences in the physical structure of the leaves of tested host plants. Because leaf depressions (e.g., along leaf midribs) play a crucial role as the basis for the construction of complicated web structures, females of *T. urticae* showed aggregation behavior near such depressions [6].

In the present study, two congeneric spider mite pests, the polyphagous *E. orientalis* (>200 hosts) and the oligophagous *E. palmatus* (six hosts) [14], did not show variations/differences in any life type characteristics on different perennial host plants (Tables 1 and 2). The life type of *E. palmatus* (LW-j) was investigated for the first time in this study. Previously, the LW-j life type was also detected for *E. orientalis* on the leaves of a perennial plant, *Manihot glaziovii* Müll. Arg. (Euphorbiaceae) [10]. In the present study, these two *Eutetranychus* species showed persistency in life type characteristics on different perennial host plant leaves. It could be due to the fact that LW is the primary and simplest life type [4,6], and is observed in less advanced Tetranychinae genera, e.g., *Aponychus* Rimando, *Eurytetranychus* Oudemans, *Eutetranychus* Banks, *Panonychus* Yokoyama, *Stylophoronychus* Prasad, and *Yezonychus* Ehara [4]. The members (e.g., *E. orientalis* and *E. palmatus*) of the tribe Eurytetranychini Reck are considered more primitive than the members (e.g., *T. urticae* and *Oligonychus* spp.) of the tribe Tetranychini Reck [6]. It could be one of the reasons why, in the LW-j life type of *E. orientalis* and *E. palmatus*, pest species never spin threads as dense webs while walking on the leaf surface (dragging behavior), but females produce dense web covers on eggs (weaving behavior) to protect their progeny [28–32].

5. Conclusions

It is concluded that the tested spider mites have restricted life types with variations in some life type characteristics, which shows their high adaptability to utilize the host plant resources. The variations observed in the life type characteristics of *T. urticae* could be helpful in applied pest management (e.g., in the proper selection of potential biological control agents) when infesting various economic plants.

Supplementary Materials: The following supporting information can be downloaded at: https://www.mdpi.com/article/10.3390/ani13223433/s1, Table S1: Coding for statistical analysis of laboratory and field observations of life type characteristics.

Author Contributions: Conceptualization, H.M.S.M.; methodology, H.M.S.M. and H.M.S.A.; validation, M.K. and F.J.A.; formal analysis, H.M.S.M., H.M.S.A., F.J.A. and M.K.; investigation, H.M.S.M., H.M.S.A. and F.J.A.; data curation, H.M.S.M. and H.M.S.A.; writing—original draft preparation,

H.M.S.M.; writing—review and editing, M.K. and F.J.A.; supervision, F.J.A.; funding acquisition, F.J.A. All authors have read and agreed to the published version of the manuscript.

Funding: The authors would like to extend their sincere appreciation to the researchers supporting project number (RSPD2023R807), King Saud University, Riyadh, Saudi Arabia.

Institutional Review Board Statement: Not applicable.

Informed Consent Statement: Not applicable.

Data Availability Statement: All necessary data for the manuscript is provided.

Acknowledgments: We are thankful to Jawwad Hassan Mirza (Department of Plant Protection, King Saud University) for editing and reviewing the manuscript with his constructive and useful comments that greatly improved it. Special thanks are given to Muhammad Waleed Shakoor and Nasreldeen Ahmed Elgoni (Department of Plant Protection, King Saud University) for their help in the experimentation.

Conflicts of Interest: The authors declare no conflict of interest.

References

1. Bolland, H.R.; Gutierrez, J.; Flechtmann, C.H.W. *World Catalogue of the Spider Mite Family (Acari: Tetranychidae)*; Brill Academic Publishers: Leiden, The Netherlands, 1998.
2. Jeppson, L.R.; Keifer, H.H.; Baker, E.W. *Mites Injurious to Economic Plants*; University of California Press: Berkeley, CA, USA, 1975.
3. Hoy, M.A. *Agricultural Acarology. Introduction to Integrated Mite Management*; CRC Press: Boca Raton, FL, USA, 2011.
4. Saito, Y. The concept of 'life types' in Tetranychinae. An attempt to classify the spinning behavior of Tetranychinae. *Acarologia* **1983**, *24*, 377–391.
5. Saito, Y. Life types of spider mites. In *Spider Mites: Their Biology, Natural Enemies and Control*, 1st ed.; Helle, W., Sabelis, W., Eds.; Elsevier: Amsterdam, The Netherlands, 1985; Volume 1A, pp. 253–264.
6. Saito, Y. *Plant Mites and Sociality. Diversity and Evolution*; Springer: Tokyo, Japan, 2010.
7. Aponte, O.; McMurtry, J.A. Damage on 'Hass' avocado leaves, webbing and nest behaviour of *Oligonychus perseae* (Acari: Tetranychidae). *Exp. Appl. Acarol.* **1997**, *21*, 265–272. [CrossRef]
8. Mori, K.; Saito, Y.; Sakagami, T. Effects of the nest web and female attendance on survival of young in a subsocial spider mite, *Schizotetranychus longus* (Acari: Tetranychidae). *Exp. Appl. Acarol.* **1999**, *23*, 411–418. [CrossRef]
9. Oku, K.; Yano, S.; Takafuji, A. Spider mite's use of a refuge during the quiescent stage in the presence of a predator. *Entomol. Exp. Appl.* **2003**, *108*, 71–74. [CrossRef]
10. Saito, Y. Sociobiological aspects of spider mite life types. *J. Acarol. Soc. Jpn.* **1995**, *4*, 55–67. [CrossRef]
11. Saito, Y.; Lin, J.Z.; Zhang, Y.X.; Ito, K.; Liu, Q.Y.; Chittenden, A.R. Two new species and four new life types in Tetranychidae. *Ann. Entomol. Soc. Am.* **2016**, *109*, 463–472. [CrossRef]
12. Mirza, J.H.; Kamran, M.; Alatawi, F.J. Webbing life type and behavioral response of the date palm mite, *Oligonychus afrasiaticus*, to webbing residues on leaves and fruits of date palm. *Exp. Appl. Acarol.* **2018**, *76*, 197–207. [CrossRef]
13. Mushtaq, H.M.S.; Kamran, M.; Alatawi, F.J. Two new life types and assessment of web-associated behavioral characteristics of some *Oligonychus* species on various host plants. *Exp. Appl. Acarol.* **2021**, *83*, 211–227. [CrossRef]
14. Spider Mites Web: A Comprehensive Database for the Tetranychidae. Available online: http://www1.montpellier.inra.fr/CBGP/spmweb (accessed on 17 August 2023).
15. Zélé, F.; Altıntaş, M.; Santos, I.; Cakmak, I.; Magalhães, S. Inter-and intraspecific variation of spider mite susceptibility to fungal infections: Implications for the long-term success of biological control. *Ecol. Evol.* **2020**, *10*, 3209–3221. [CrossRef]
16. Palevsky, E.; Lotan, A.; Gerson, U. Evaluation of *Eutetranychus palmatus* (acari: Tetranychidae) as a pest of date palms in Israel. *Isr. J. Plant Sci.* **2010**, *58*, 43–51. [CrossRef]
17. Alatawi, F.J.; Kamran, M. Spider mites (Acari: Tetranychidae) of Saudi Arabia: Two new species, new records and a key to all known species. *J. Nat. Hist.* **2018**, *52*, 429–455. [CrossRef]
18. Kamran, M.; Khan, E.M.; Alatawi, F.J. The spider mites of the genus *Eutetranychus* BBanks (Acari, Trombidiformes, Tetranychidae) from saudi arabia: Two new species, a re-description, and a key to the world species. *ZooKeys* **2018**, *799*, 47–88. [CrossRef] [PubMed]
19. Krantz, G.W. *A Manual of Acarology*, 2nd ed.; Oregon State University Book Stores, Inc.: Corvallis, OR, USA, 1978.
20. Der, G.; Everitt, B.S. *A Handbook of Statistical Analyses Using Sas*, 3rd ed.; Chapman and Hall/CRC Press: New York, NY, USA, 2008.
21. Sabelis, M.W. Biological Control of Two-Spotted Spider Mites Using Phytoseiid Predators. Part I. Pudoc. Ph.D. Thesis, Wageningen University and Research, Wageningen, The Netherlands, 1981.
22. Lemos, F.; Sarmento, R.A.; Pallini, A.; Dias, C.R.; Sabelis, M.W.; Janssen, A. Spider mite web mediates anti-predator behaviour. *Exp. Appl. Acarol.* **2010**, *52*, 1–10. [CrossRef] [PubMed]
23. Oloo, G.W.; Ogol, C.K.; Kambona, K.O. Biotaxonomy of cassava green spider mites, *Mononychellus* spp. (Tetranychidae): 'Life type' as a possible biological criterion for their identification. *Int. J. Trop. Insect. Sci.* **1987**, *8*, 995–999. [CrossRef]

24. Osakabe, M.H. Which predatory mite can control both a dominant mite pest, *Tetranychus urticae,* and a latent mite pest, *Eotetranychus asiaticus*, on strawberry? *Exp. Appl. Acarol.* **2002**, *26*, 219–230. [CrossRef]
25. Hazan, A.; Gerson, U.; Tahori, A.S. Spider mite webbing. I. The production of webbing under various environmental conditions. *Acarologia* **1974**, *16*, 68–84.
26. Gerson, U.; Aronowitz, A. Spider mite webbing. V. The effect of various host plants. *Acarologia* **1981**, *22*, 277–281.
27. Osakabe, M.H.; Hongo, K.; Funayama, K.; Osumi, S. *Amensalism* via webs causes unidirectional shifts of dominance in spider mite communities. *Oecologia* **2006**, *150*, 496–505. [CrossRef]
28. Sato, Y.; Saito, Y. Nest sanitation in social spider mites: Interspecific differences in defecation behavior. *Ethology* **2006**, *112*, 664–669. [CrossRef]
29. Shimoda, T.; Kishimoto, H.; Takabayashi, J.; Amano, H.; Dicke, M. Comparison of thread-cutting behavior in three specialist predatory mites to cope with complex webs of *Tetranychus spider mites. Exp. Appl. Acarol.* **2009**, *47*, 111–120. [CrossRef]
30. Sabelis, M.W.; Bakker, F.M. How predatory mites cope with the web of their Tetranychid prey—A functional view on dorsal chaetotaxy in the Phytoseiidae. *Exp. Appl. Acarol.* **1992**, *16*, 203–225. [CrossRef]
31. Yano, J.; Saito, Y.; Chittenden, A.R.; Sato, Y. Variation in counterattack success against a phytoseiid predator between two forms of the social spider mite *Stigmaeopsis miscanthi. J. Ethol.* **2011**, *29*, 337–342. [CrossRef]
32. Yano, S. Cooperative web sharing against predators promotes group living in spider mites. *Behav. Ecol. Sociobiol.* **2012**, *66*, 845–853. [CrossRef]

animals

MDPI

Article

Tenuipalpus Sensu Lato Donnadieu (Acari: Prostigmata: Tenuipalpidae); New Species Groups, a New Species, and Keys to the World Species †

Nasreldeen Ahmed Elgoni, Muhammad Kamran and Fahad Jaber Alatawi *

Department of Plant Protection, College of Food and Agriculture Sciences, King Saud University, Riyadh 11451, Saudi Arabia; nabdelkarim@ksu.edu.sa (N.A.E.); murafique@ksu.edu.sa (M.K.)
* Correspondence: falatawi@ksu.edu.sa

† urn:lsid:zoobank.org:pub:6BF0C116-51FC-479A-BA72-AD6D496A2162.

Simple Summary: Castro et al. divided the genus Tenuipalpus into two groups, i.e., Tenuipalpus sensu stricto and Tenuipalpus sensu lato. Four new species groups of the Tenuipalpus sensu lato group are proposed in this study, considering the total number of dorsal opisthosomal setae. Additionally, diagnostic keys to new species groups and the world species of Tenuipalpus sensu lato are developed.

Abstract: Four new species groups of the *Tenuipalpus* sensu lato group are proposed in the present study based on the total number of dorsal opisthosomal setae, namely, *carolinensis* with ten pairs of setae (214 species), *dubinini* with nine pairs of setae (33 species), *granati* with eight pairs of setae (29 species), and *barticanus* with seven pairs of setae (7 species) Additionally, diagnostic keys to species groups and 273 species of the *Tenuipalpus* sensu lato are provided. Three species, *T. lustrabilis* Chaudhri, *T. guptai* Sadana and Gupta, and *T. solanensis* Sadana and Gupta, are synonymized with *T. punicae* Pritchard and Baker. One species, *T. rodionovi* Chalilova, is suggested as a junior synonym of *T. granati* Sayed, and eight species, *T. chiococcae* De Leon, *T. costarricensis* Salas and Ochoa, *T. ephedrae* Livschitz and Mitrofanov, *T. molinai* Evans, *T. santae* Manson, *T. simplychus* Cromroy, *T. tetrazygiae* De Leon, and *T. oxalis* (Flechtmann), belonging to the *carolinensis* species group, are not included in the key. Furthermore, a new species of *Tenuipalpus* sensu lato, *T. jazanensis* sp. nov., is described and illustrated based on females collected from the *Chamaerops* spp. (Arecaceae).

Keywords: opisthosomal setae; tropical regions; divisions; *Chamaerops* spp.; related species; geographical distribution

Citation: Elgoni, N.A.; Kamran, M.; Alatawi, F.J. *Tenuipalpus* Sensu Lato Donnadieu (Acari: Prostigmata: Tenuipalpidae); New Species Groups, a New Species, and Keys to the World Species. *Animals* **2023**, *13*, 3278. https://doi.org/10.3390/ani13203278

Academic Editors: Monika Fajfer and Maciej Skoracki

Received: 17 August 2023
Revised: 14 October 2023
Accepted: 18 October 2023
Published: 20 October 2023

1. Introduction

Tenuipalpus Donnadieu is the largest genus in the family Tenuipalpidae Berlese (Acari: Prostigmata: Tetranychoidea) and consists of more than 320 species, distributed worldwide, especially in the tropical and subtropical regions [1–3]. The members of this genus are mostly oligophagous in their feeding habits [2]. The three species, i.e., *T. granati* Sayed, *T. punicae* Pritchard and Baker, and *T. eriophyoides* Baker, are serious pests of fruit trees worldwide [4].

The genus *Tenuipalpus* was erected by Donnadieu in 1875 [5]. Based on the number of dorsal setae and palp segmentation, Reck [6] and Mitrofanov [7] erected six genera by transferring some of *Tenuipalpus* species, *Extenuipalpus* Reck [6], *Aegyptopalpus* Mitrofanov [7], *Deleonipalpus* Mitrofanov [7], *Gnathopalpus* Mitrofanov [6], *Tuttlepalpus* Mitrofanov [7], and *Ultratenuipalpus* Mitrofanov [7]. Meyer [8] synonymized five of those genera with the genus *Tenuipalpus* by using the same morphological characters and proposed six species groups, namely, *albae*, *caudatus*, *elegans*, *granati*, *quadrisetosus*, and *trisetosus*. Later, Baker and Tuttle [9] recognized only two species groups, *caudatus* and *proteae*, based on the presence and absence of setae *f2* (seven and six pairs of dorsolateral setae), respectively. Further, they divided the *caudatus* group into three species subgroups based on the number of intercoxal setae (*3a* and *4a*). Meyer [10] followed the concept of Baker and Tuttle [9] and divided the *proteae* into three species subgroups and added two more species subgroups to the *caudatus* group.

Collyer [11] made the world key of 102 *Tenuipalpus* species. Additionally, some local keys were developed over time by Meyer [10], which included around 90 species from Africa; Baker and Tuttle [9] included 20 species from Mexico; Al-Gboory [12] included seven species from Iraq; Khanjani et al. [13] included nine species from Iran; Castro and Feres [14] included 12 species from Brazil; and Xu et al. [15] included 25 species from China.

For more than three decades, *Tenuipalpus* species groups, as proposed by Baker and Tuttle [9] and Meyer [10], were consistent. However, recently, Castro et al. [1] divided the genus into two groups by using a combination of characters: *Tenuipalpus* sensu stricto (36 species), with a pair of lateral projections associated with setae *c3* and only one pair of setae *4a*, and *Tenuipalpus* sensu lato (287 species), without a lateral projection associated with setae *c3* and one to four pairs of setae *4a*. Also, a world key to the species of *Tenuipalpus* sensu stricto was provided [1,16]. However, no diagnostic key to the world species of *Tenuipalpus* sensu lato has been developed yet.

The aims of the present study were to (i) classify all species of *Tenuipalpus* sensu lato into new species groups based on distinct morphological characters; (ii) develop diagnostic keys to species groups and 273 species of the *Tenuipalpus* sensu lato; and (iii) examine specimens of *Tenuipalpus* species collected from different regions of Saudi Arabia.

2. Materials and Methods

The published taxonomic literature of all known 287 species belonging to the *Tenuipalpus* sensu lato group was collected using different resources: the Acarology Laboratory King Saud University, Google Scholar, ResearchGate, and different acarological journals, and through personal communication (Dr. Qing Hai Fan, Fujian Agriculture and Forestry University, China; and Dr. Elizeu B. Castro, São Paulo State University, São José do Rio Preto campus, São Paulo, Brazil). Diagnostic keys to species groups and 273 species are provided. All specimens of the genus that were collected by the Acarology lab have been examined. The mounted specimens of *Tenuipalpus* species were examined and identified under a phase contrast microscope (DM2500, Leica, Wetzlar, Germany). Different mite body parts were pictured using an Auto-Montage software system v4.0.1.1 (Syncroscopy, Cambridge, UK) and drawn with Adobe Illustrator v27.7 (Adobe System Inc., San Jose, CA, USA). All measurements are in micrometers. The terminology used in the research follows that of Mesa et al. [2]. The specimens were deposited at the King Saud University Museum of Arthropods (KSMA, Acarology Section), Department of Plant Protection, College of Food and Agriculture Sciences, King Saud University, Riyadh, Saudi Arabia.

3. Results

The *Tenuipalpus* sensu lato group was divided into four species groups based on the total number of dorsal opisthosomal setae, namely, the *carolinensis* group—with ten pairs of setae (214 species), the *dubinini* group—with nine pairs of setae (33 species), the *granati* group—with eight pairs of setae (29 species), and the *barticanus* group—with seven pairs of setae (seven species). This proposed division did not consider whether a specific setae was absent or not (i.e., any setae among the opisthosomal setae can be absent). The diagnostic keys to those four new species groups and 273 species of *Tenuipalpus* sensu lato were also developed.

Among the *carolinensis* group, eight species were not included in the key: *T. chiococcae* De Leon, *T. costarricensis* Salas and Ochoa, *T. ephedrae* Livschitz and Mitrofanov, *T. molinai* Evans, *T. santae* Manson, *T. simplychus* Cromroy, *T. tetrazygiae* De Leon, and *T. oxalis* (Flechtmann) (Table 1). Three species; *T. guptai* Sadana and Gupta, *T. solanensis* Sadana and Gupta, and *T. lustrabilis* Chaudhri, were synonymized with *T. punicae* Pritchard and Baker.

Two species, *T. simplex* Vitzthum and *T. jasmini* Khan were poorly described; they were tentatively placed in the new groups of *carolinensis* and *granati*, respectively. The species *T. rodionovi* Chalilova was suggested as a junior synonymy of *T. granati* Sayed.

Table 1. List of species not included in the diagnostic key of the new species group *carolinensis*.

Species (Geographic Distribtuion)	Host Plant	Related Species (Geographical Distribution)	Host Plant
T. chiococcae [17] **(USA)**	*Chiococca pinetorum*	*T. pigrus* Pritchard and Baker (USA)	*Umbellularia californica*
T. costarricensis [18] **(Costa Rica)**	*Cedrela* sp.	*T. granati* Sayed (Worldwide)	Polyphagous
T. ephedrae [19] **(Ukraine)**	*Ephedra distachya*	-	-
T. molinai [20] **(Honduras)**	Asteraceae: Unidentified plant	-	-
T. oxalis [21] **(Brazil)**	*Oxalis* sp.	-	-
T. santae [22] **(Costa Rica)**	Unidentified (fence tree)	*T. celtidis* Pritchard and Baker (USA)	*Celtis* sp.
T. simplychus [23] **(Puerto Rico)**	*Cordia sulcata*	*T. knorri* Baker and Pritchard (Argentina)	-
T. tetrazygiae [17,24] **(India and USA)**	*Anacardium occidentale* and *Etrazygia bicolor*	-	-

Among the examined Tenuipalpus specimens collected from different regions of Saudi Arabia, a new species, *Tenuipalpus jazanensis* sp. nov., belonging to the Tenuipalpus sensu lato group, resulted. The new species is hereby fully described and illustrated based on females collected from the European fan palm, *Chamaerops* spp. (Arecaceae) from the Jazan region (Figures 1–4).

3.1. Family Tenuipalpidae Berlese, 1913 [25]

Genus *Tenuipalpus* Donnadieu, 1875 [5].

Type species:*Tenuipalpus palmatus* Donnadieu, 1875 [5] (=*Tenuipalpus caudatus* Dugès, 1834) [26].

Diagnosis: (modified after Castro et al. [16]): Prodorsum have three pairs of setae (*v2*, *sc1*, *sc2*) except the species *T. elegans* (Collyer) with two pairs of prodorsum setae (*sc1*, *sc2*), setae *v2* absent, opisthosoma with 7–10 pairs of setae; (*c3*, *d3*, *e3*, *f3*, *h1*, *h2* present; *c2*, *d2*, e2 absent; *c1*, *d1*, *e1*, *f2* present or absent (*d1*, *e1* rarely absent), setae *h2* elongate, flagellate; palp one- to three-segmented; venter with one to two pairs of setae *3a* (*3a1* always present; *3a2* present or absent) and one to five pairs of setae *4a* (*4a1* always present; *4a2*, *4a3*, *4a4*, *4a5* present or absent); two pairs of pseudanal setae *ps1-2* (three pairs, *ps1-3*, rarely present).

3.2. Divisions of the Tenuipalpus Sensu Lato Group

Tenuipalpus sensu lato Castro et al. [1]

Diagnosis: (modified after Castro et al. [1]). Opisthosoma without one pair of lateral body projections associated with setae *c3*, venter with one to five pairs of intercoxal setae *4a*.

3.2.1. *T. carolinensis* Species Group

Diagnosis: (based on female). Opisthosoma with ten pairs of dorsal setae. This group consists of 214 species.

3.2.2. *T. dubinini* Species Group

Diagnosis: (based on female). Opisthosoma with nine pairs of dorsal setae. This group consists of 33 species.

3.2.3. *T. granati* Species Group

Diagnosis: (based on female). Opisthosoma with eight pairs of dorsal setae. This group consists of 29 species.

3.2.4. *T. barticanus* Species Group

Diagnosis: (based on female). Opisthosoma with seven pairs of dorsal setae. This group consists of seven species.

3.3. New Species

Tenuipalpus (sensu lato Castro et al. [1])

T. carolinensis group

Tenuipalpus jazanensis sp. nov.

urn:lsid:zoobank.org:act:ACDFDCC0-A7D3-4B3B-8618-27E9F14A6A44

Diagnosis: (based on Female). Propodosoma without lateral lobes anterior marginally; propodosoma with transverse striate medially and laterally, reticulated sublaterally. Hysterosoma medially with few reticulations in the area between setae *c1-c1* and sublaterally; one pair of seta *3a* and five pairs of setae *4a* (*4a1*−5) present; rostrum reaching to the middle of femur I; palp three-segmented; legs setal counts on coxae, 2–2–1–1 trochanters, 1–1–2–1; femora 4–4–2–0; genua 3–2–1–1; tibiae 5–5–3–3 and tarsus 8 (1)–8 (1)–5–5.

Description Female (n = 5)

Dorsum (Figure 1): Anterior margin of prodorsum deeply incised, depth of notch 25 (24–27), propodosoma without lateral lobes anterior marginally; propodosoma with transverse striate medially and laterally, reticulated sublaterally. Hysterosoma medially with few reticulations in the area between setae *c1* and sublaterally; dorsal body setae lanceolate serrate, almost equal in length, except setae *h2*; opisthosomal pores present; distance between setae *v2–h1* 339 (335–355), *sc2–sc2* 180 (160–240). Prodorsum slightly wider than proximal section of opisthosoma; anterior central of prodorsum with transverse striations; lateral regions with reticulations. All dorsal setae short (except *h2*), not more than 23 μm long; setae *v2*, *sc1*, and *sc2* serrate, *sc2* longer than *v2* and subequal in length with *sc1*; opisthonotum with reticulations; opisthosomal setae serrate; setae *e3* shorter than *f2* and *f3*, *h2* flagelliform and elongate serrate. Setal lengths: *v2* 16 (13–16), *sc1* 23 (20–23), *sc2* 23 (19–23), *c1* 13 (13–16), *c3* 13 (12–15), *d1* 12 (12–13), *d3* 14 (12–16), *e1* 13 (10–15), *e3* 12 (12–15), f_2 14 (14–16), f_3 14 (14–16), *h1* 14 (14–20), *h2* 233 (230–239). Distance between dorsal setae: *v2–v2* 43 (41–43), *sc1–sc1* 107 (95–107), *c1–c1* 57 (52–59), *c3–c3* 195 (192–197), *d1–d1* 32 (30–39), *d3–d3* 156 (160–169), *e1–e1* 25 (12–25), *e3–e3* 82 (82–90), *f2–f2* 71 (66–72), *f3–f3* 56 (52–57), *h2–h2* 41 (36–43), *h1–h1* 20 (20–23), *c1–d1* 45 (40–45), *d1–e1* 60 (60–62), *c1–c3* 66 (66–69), *d1–d3* 64 (59–65) *e1–e3* 41 (40–44), *c3–d3* 38 (30–38), *d3–e3* 121 (120–127), *e3–f3* 30 (28–36), *f3–f2* 10 (12–15), *h1–h2* 7 (7–10).

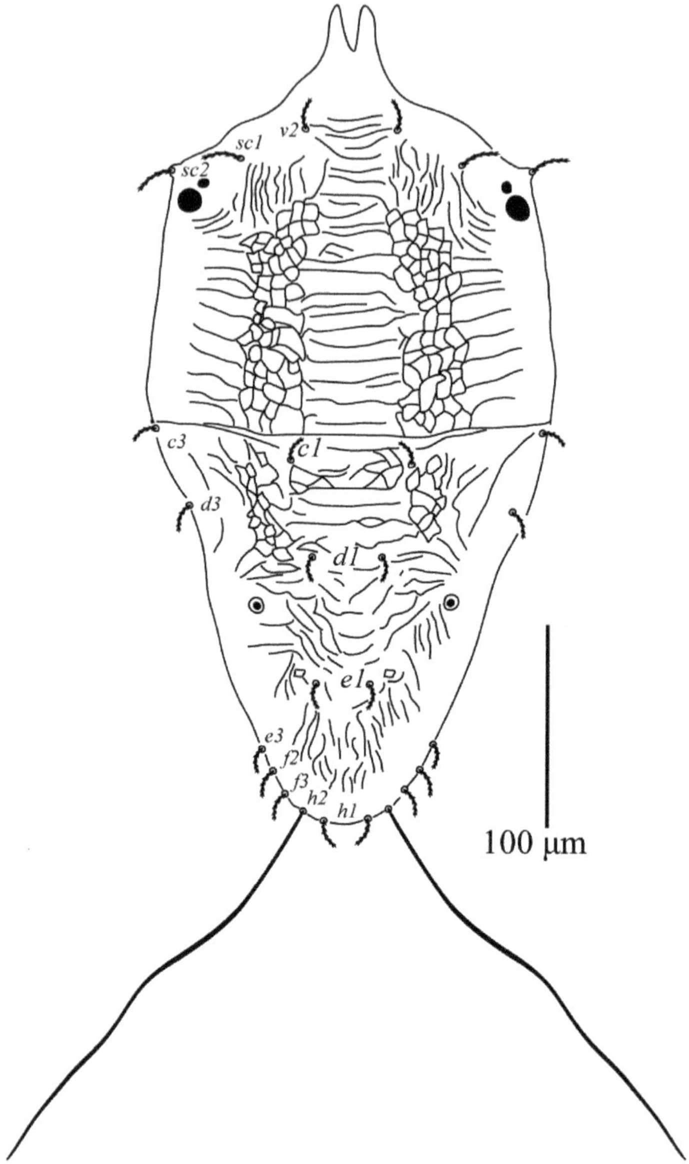

Figure 1. *Tenuipalpus jazanensis* sp.nov. Female. Dorsum. Scale bar: 100 µm.

Venter (Figure 2): Ventral cuticle with broken longitudinal striations. Area posterior setae *4a* with broken transverse striate. All ventral setae smooth, setae *1a*, *4a* flagelliform, and elongate. Setal lengths: *1a* 82 (82–92), *1b* 30 (22–31), *1c* 30 (26–30), *2b* 33 (23–33), *2c* 33 (23–33), *3a* 25 (25–31), 3*b* 30 (21–31), *4a1* 90 (90–100), *4a*2 83 (82–91), *4a*3 83 (80–86), *4a4* 76 (76–84), *4a5* 73 (73–102), *4b* 33 (30–33), *ag* 36 (26–36), *g1* 33 (26–34), *g2* 33 (25–33), *ps1* 23 (18–25), *ps2* 23 (18–23). Distance between ventral setae: *1a–1a* 23 (19–24), *3a–3a* 30 (30–32), *4a1–4a1* 7 (5–7), *4a2–4a2* 21 (19–22), *4a3–4a3* 35 (31–36), *4a4–4a4* 49 (42–49), *4a5–4a5* 60 (57–62), *1a–3a* 108 (102–110), *3a–4a1* 57 (57–60), *1b–1c* 14 (13–19), *2b–2c* 20 (19–25).

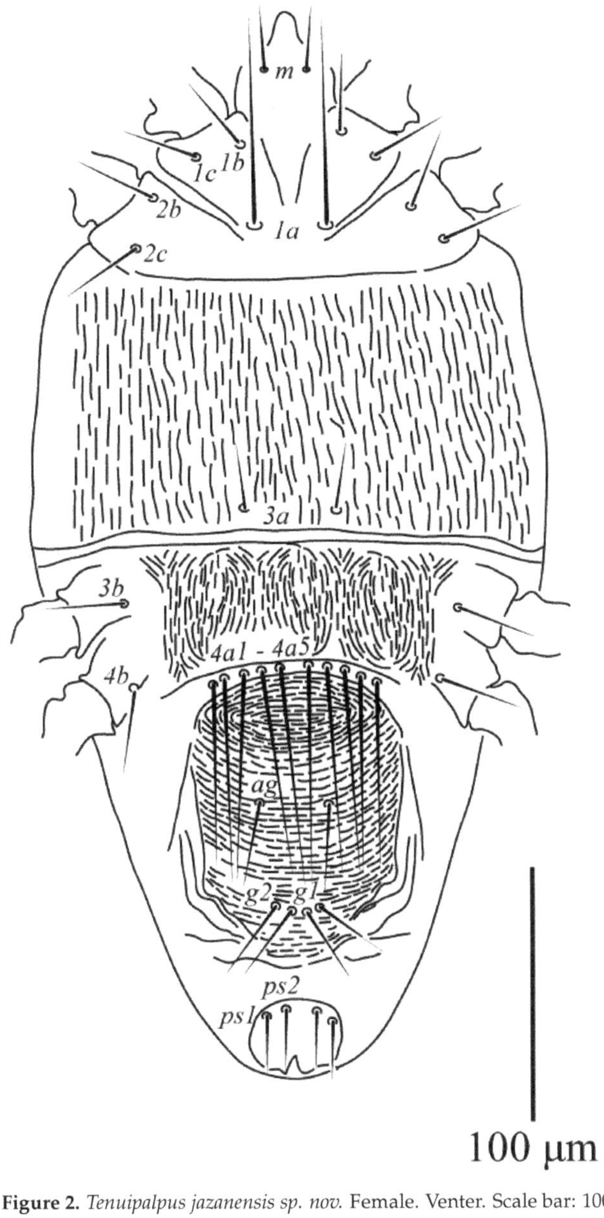

Figure 2. *Tenuipalpus jazanensis sp. nov.* Female. Venter. Scale bar: 100 μm.

Gnathosoma (Figure 3): Ventral setae *m* 25 (14–25); distance between setae *m–m* 15 (10–15) Palps 3-segmented (Figure 3), median segment elongates, bearing one barbed seta *d* 15 (12–20); distal segment short, with eupathidium *ul′* 4 (4–7) *ul″* 15 (12–15).

Legs (Figure 4): Setae on legs as follows: coxa 2–2–1–1; trochanters 1–1–2–1; femora 4–4–2–0; genu 3–2–1–1; tibiae 5–5–3–3; tarsus 8 (1ω)–8 (1ω)–5–5; femur IV without seta. Leg I–IV setal count as follows (solenidia in parenthesis).

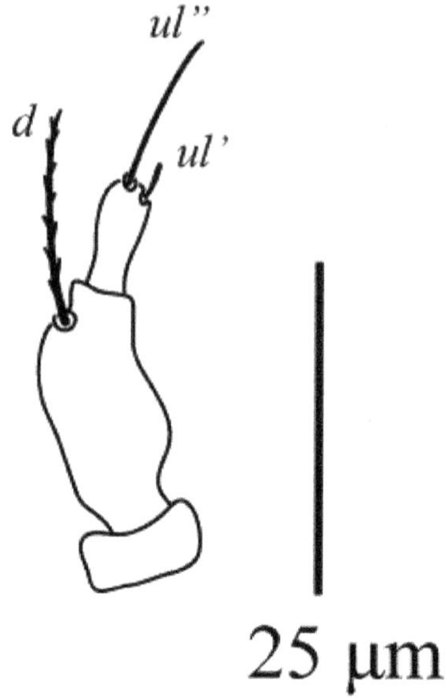

Figure 3. *Tenuipalpus jazanensis* sp. nov. Female. Palp. Scale bar: 25 µm.

Males and immature. Unknown

Type Materials. Holotype female, four paratype females, from *Chamaerops* sp. (Arecaceae) Wadi Baydh, Jazan, 17°37.559′ N, 42°22.196′ E, 10 October 2020, coll. J. H. Mirza, H.M.S. Mushtaq and E.M. Khan.

Etymology: The specific epithet (*jazanensis*) is derived from the type region, Jazan.

Remarks: The new species, *Tenuipalpus jazanensis* sp. nov., belongs to the *carolinensis* species group. This species group is distinguished from other species groups of *Tenuipalpus* sensu lato by having ten pairs of opisthosomal setae. *Tenuipalpus jazanensis* sp. nov resembles *T. pareriophyoides* Meyer and Gerson and *T. eriophyoides* Baker by having more than three pairs of intercoxal setae *4a*, one pair of setae *3a*, dorsal setae lanceolate serrate and dorsum with irregular striate laterally. The new species differs from *T. pareriophyoides* and *T. eriophyoides* by the following characters; prodorsum medially with transverse striations; area in between setae *c1*, *d1* and *d3* with few reticulations (prodorsum medial area mostly smooth or punctate; area in between setae *c1*, *d1* and *d3* with rugose patten or striate); genu I with three setae (genu I with two setae); setae *g* and *ag* smooth (setae *g* and *ag* serrate in *T. pareriophyoides*; setae *g* and *ag* serrate in *T. eriophyoides*).

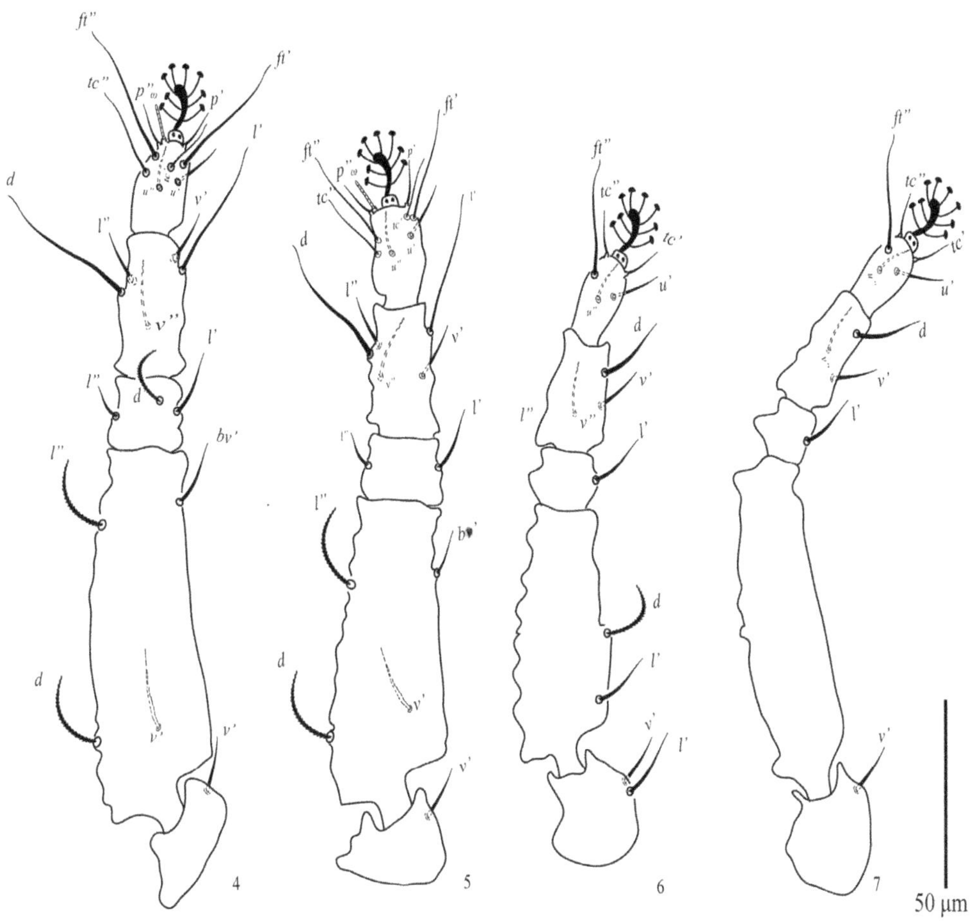

Figure 4. *Tenuipalpus jazanensis* sp. nov. Female. 4, Leg I; 5, leg II; 6, leg III; 7, leg IV. Scale bar: 50 μm.

3.4. Key to the Groups and Species Groups of Tenuipalpus (Based on Females)

1. Dorsum with one pair of lateral projections associated with setae *c3* and another pair of lateral projections anterior to setae *sc2*; lateral setae *sc2*, *c3*, *e3*, *f2*, *f3*, and *h1* variable in shape from lanceolate, obovate to ovate; femora I and II with setae d inserted in lateral position on tubercles . . . . *Tenuipalpus* sensu stricto (sensu Castro et al. [1])

1′ Dorsum always without a pair of lateral projections associated with setae *c3* and usually without the lateral projection anterior to setae *sc2*; lateral setae not as mentioned above and usually setiform or minute; femora I and II with setae *d* usually inserted in dorsal position .... *Tenuipalpus* sensu lato (sensu Castro et al. [1]) 2

2. Dorsum with ten pairs of opisthosomal setae . . . . *carolinensis* species group

2′ Dorsum with less than ten pairs of opisthosomal setae 3

3. Dorsum with nine pairs of opisthosomal setae . . . . *dubinini* species group

3′ Dorsum with less than nine pairs of opisthosomal setae 4

4. Dorsum with eight pairs of opisthosomal setae . . . . *granati* species group

4′ Dorsum with seven pairs of opisthosomal setae . . . . *barticanus* species group

3.5. Key to the World Species of the T. carolinensis Species Group (Based on Females)

1 Venter with one pair of *4a* 2

1′ Venter with more than one pair of *4a* … 3
2 Venter with one pair of *3a* … 6
2′ Venter with two pairs of *3a* … 141
3 Venter with two pairs of *4a* … 4
3′ Venter with four or five pairs of *4a* … 5
4 Venter with one pair of *3a*. … 169
4′ Venter with two pairs of *3a* … 197
5 Venter with four pairs of 4a and one pair of *3a* … 201
5′ Venter with five pairs of 4a and one pair of *3a* … *T. jazanensis* sp. nov.
6 Opisthosoma with lateral bulges … 7
6′ Opisthosoma without lateral bulges … 17
7 Venter with three pairs of anal setae *ps1*, *ps2*, and *ps3* … *T. venustus* Collyer
7′ Venter with two pairs of anal setae *ps1* and *ps2* … 8
8 Propodosoma with transverse striae medially … 9
8′ Propodosoma without transverse striae medially … 10
9 Palp with one seta on terminal segment; propodosoma striated sublaterally… *T. orilloi* Rimando
9′ Palp with two seta on terminal segment; lateral sublateral area of propodosoma smooth … *T. frondosus* Cromroy
10 Palp two segmented … *T. dominguensis* De Leon
10′ Palp three segmented … 11
11 Genua I and II each with one seta … *T. tabebuiae* De Leon
11′ Genua I and II each with more than one seta … 12
12 Median area of propodosoma with longitudinal striae; femur II with three setae … *T. crescentiae* De Leon
12′ Median area of propodosoma not as mentioned above … 13
13 Venter of propodosoma with transverse striae medially … 14
13′ Venter of propodosoma without transverse striae medially … 15
14 Setae *g1-2* crossing the bases of setae *ps1-2* … *T. placitus* Chaudhri
14′ Setae *g1-2* not crossing the bases of setae *ps1-2*… *T. couroapiupita* De Leon
15 Palp with one seta on terminal segment; rostral with two lateral lobes well developed … *T. apichai* Castro and Feres
15′ Palp with two setae on terminal segment … 16
16 Trochanter I with one seta; femora I and II each with four setae … *T. haripuriensis* Akbar and Chaudhri
16′ Trochanter I without setae; femora I and II each with two setae … *T. zizyphae* Mohanasundaram
17 Second pair of propodosomal setae *sc1* longer than first and third pairs *v2* and *sc2* … 18
17′ Second pair of propodosomal setae *sc1* equal or shorter than first and third pairs *v2* and *sc2* … 32
18 Dorsocentral setae *d1* elongate crossing the bases of setae *e1* … 19
18′ Dorsocentral setae *d1* not crossing the bases of setae *e1* … 22
19 Dorsocentral setae *c1* elongate crossing the bases of setae *d1* … 20
19′ Dorsocentral setae *c1* not crossing the bases of setae *d1* … 21
20 Dorsolateral setae *d3* longer than the distance between setae *c3–d3*, setae *e3* reaching or crossing the bases of setae *f3* … *T. bagdadensis* Al-Gboory
20′ Dorsolateral setae *d3* shorter than the distance between setae *c3–d3*, setae *e3* not reaching the bases of setae *f3* … *T. gatoomensis* Meyer
21 Dorsocentral setae *e1* crossing the bases of setae *h2* … *T. galpiniae* Meyer
21′ Dorsocentral setae *e1* not reaching to the bases of setae *h2*… *T. ovalis* Meyer and Ryke
22 Dorsocentral setae *c1*, *d1* and *e1* spatulate … *T. magalismontani* Meyer
22′ Dorsocentral setae *c1*, *d1* and *e1* setiform or lanceolate … 23

23 Palpus two segmented .... *T. acritus* Meyer
23′ Palpus three segmented 24
24 Dorsum covered with longitudinal striae 25
24′ Dorsum not covered with longitudinal striae 26
25 Dorsolateral setae *e3*, *f2*, *f3*, and *h1* subspatulate .... *T. africanus* Meyer
25′ Dorsolateral setae *e3*, *f2*, *f3*, and *h1* lanceolate .... *T. striolatus* Meyer
26 Genua I and II each with two setae 27
26′ Genua I and II each with three setae 28
27 Setae *4a* crossing the bases of setae *g1-2*; rostral shield with well-developed lateral lobes; tarsi I and II each with eight tactile setae .... *T. protectus* Meyer
27′ Setae *4a* not crossing the bases of setae *g1*-2; rostral shield with small lateral lobes; tarsi I and II each with seven tactile setae .... *T. pieteri* Meyer
28 Femur II with three setae; tarsi I and II each with six setae .... *T. trifoliatae* Mohanasundaram
28′ Femur II with four setae 29
29 Genua III and IV each with one seta .... *T. combreti* Meyer
29′ Genua III and IV each without setae 30
30 Dorsolateral setae *e3*, *f2*, and *f3* short not reaching the bases of next setae; femur IV with one setae .... *T. ariauae* Feres and Hernandes
30′ Dorsolateral setae *e3*, *f2*, and *f3* reaching or crossing the bases of next setae 31
31 Dorsum pitted medially, with few transverse striae posterior setae *e1*; setae *ag* not reaching the bases of setae *g1-2* .... *T. ueckermanni* Meyer
31′ Dorsum with transverse to irregular striae medially; setae *ag* reaching the bases of setae *g1-2* .... *T. prunioides* Meyer
32 First pair of propodosomal setae *v2* reaching or crossing to the bases of second pair *sc1* 33
32′ First pair of propodosomal setae *v2* not reaching the bases of second pair *sc1*........ 39
33 Palp one or two segmented 34
33′ Palp three segmented 35
34 Palp one segmented .... *T. senecionis* Collyer
34′ Palp two segmented .... *T. rangiorae* Collyer
35 Dorsocentral setae *c1* crossing the bases of setae *d1*; femora I and II each with three setae .... *T. laminasetae* Mohanasundaram
35′ Dorsocentral setae *c1* not reaching the bases of setae *d1* 36
36 Genua I and II each with two setae .... *T. portulacae* Parsi, Khosrowshahi and Farid
36′ Genua I and II each with three setae 37
37 Tibiae I and II each with five setae .... *T. morianus* Meyer
37′ Tibiae I and II each with four setae 38
38 Setae in trochanters I–IV 2-2-1-1; femora I and II each with four setae .... *T. acacii* Maninder and Ghai
38′ Setae in trochanters I–IV 1-1-2-2; femora I and II each with five setae .... *T. fici* Maninder and Ghai
39 Dorsolateral setae *e3* reaching or crossing to the bases of setae *f2* 40
39′ Dorsolateral setae *e3* not reaching to the bases of setae *f2* 77
40 Dorsocentral setae *c1* reaching or crossing bases of setae *d1* 41
40′ Dorsocentral setae *c1* not reaching bases of setae *d1* 45
41 Femora I and II each with six setae .... *T. faresianus* Maninder and Ghai
41′ Femora I and II each with less than six setae 42
42 Femora I and II each with five setae .... *T. ixorae* Maninder and Ghai
42′ Femora I and II each with less than five setae 43
43 Dorsal setae oblanceolate smooth, setae *e1* short; femora II and IV with four and one setae respectively .... *T. szarvasensis* Bozai
43′ Dorsal setae laminate or lanceolate 44

44 Dorsal setae laminate serrate; femur II with three setae .... *T. erythrinae* Mohanasundaram
44′ Dorsal setae lanceolate; femur II with four setae .... *T. baeri* Reck
45 Dorsum almost smooth or with few striations 46
45′ Dorsum not as mentioned above 53
46 Setae *sc2* associated with projection or sit on tubercles 47
46′ Setae *sc2* not associated with projection or sit on tubercles; dorsolateral setae *c3* and *d3* setiform .... *T. trichiliae* De Leon
47 Palp one segmented 48
47′ Palp more than one segmented 49
48 Dorsolateral setae *e3*, *f2*, *f3*, and *h1* spatulate or ovate; venter pitted laterally .... *T. terminaliae* De Leon
48′ Dorsolateral setae *e3*, *f2*, *f3*, and *h1* lanceolate; venter completely smooth .... *T. bucidae* De Leon
49 Dorsolateral setae *c3* broadly spatulate or leaf-like.......... 50
49′ Dorsolateral setae *c3* not as mentioned above 51
50 Propodosoma with longitudinal striae medially; dorsolateral setae *c3* broadly spatulate .... *T. austrocedri* Gonzalez
50′ Propodosoma smooth; dorsolateral setae *c3* leaf-like .... *T. kapoki* De Leon
51 Dorsolateral setae *d3* spatulate, serrate .... *T. lawrencei* Baker and Pritchard
51′ Dorsolateral setae *d3* setiform minute 52
52 Setae *sc2* crossing bases of setae *sc1*; area posterior setae *4a* with broken striae .... *T. tapirirae* De Leon
52′ Setae *sc2* not reaching bases of setae *sc1*; area posterior setae *4a* smooth, few striae posterior setae *g1-2* .... *T. zanthus* De Leon
53 Palp with two eupathidium setae on terminal segment 54
53′ Palp with one eupathidium setae on terminal segment 58
54 Genua I and II each with one seta; tarsi III and IV each with six setae .... *T. malligai* Mohanasundaram
54′ Genua I and II each with more than one seta 55
55 Genua I and II each with two setae 56
55′ Genua I and II each with three setae 57
56 Tibiae I and II each with four setae .... *T. metis* Hasan, Akbar, and Bashir
56′ Tibiae I and II each with five setae .... *T. velitor* Hasan, Akbar, and Bashir
57 Setae *4a* crossing bases of setae *ps1-2*; setae ag not crossing bases of setae *g1-2* .... *T. insularis* Meyer
57′ Setae *4a* not crossing bases of setae *ps1-2*; setae *ag* crossing bases of setae *g1-2* and reaching to the bases of *ps2* .... *T. zuluensis* Meyer
58 Opisthosoma reticulated medially 59
58′ Opisthosoma without reticulation 61
59 Genua I and II each with one setae .... *T. ghaii* Mohanasundaram
59′ Genua I and II each with two setae 60
60 Opisthosoma with two pairs of pores; genua III and IV each with one seta .... *T. boninens* Ehara
60′ Opisthosoma with one pair of pores; Genua III and IV each without setae .... *T. decus* Chaudhri
61 Palp one segmented 62
61′ Palp more than one segmented 63
62 Dorsum with irregular ridges, caudolateral setae *e3*, *f2*, *f3*, and *h1* narrowly lanceolate .... *T. metopii* De Leon
62′ Dorsum pitted, caudolateral setae *e3*, *f2*, *f3*, and *h1* subspatulate ... *T. unimerus* De Leon
63 Genu I with one seta 64
63′ Genu I with more than one seta 67
64 Genu II with one seta 65

64′ Genu II without setae; dorsum striated with smooth patches .... *T. celtidis* Pritchard and Baker
65 Palp three segmented 66
65′ Palp two segmented .... *T. pagesae* Rimando
66 Dorsum covered with wavy lines; tibiae I and II each with three setae .... *T. vriddagiriensis* Mohanasundaram
66′ Dorsum ornamented; tibiae I and II each with five setae .... *T. heveae* Baker
67 Genu II with two setae 68
67′ Genu II with three setae 72
68 Genu III with one seta 6
68′ Genu III without seta 71
69 Genu IV with one seta; dorsum smooth laterally .... *T. antipodus* Collyer
69′ Genu IV without setae; dorsum with broken striae laterally .... *T. knorri* Baker and Pritchard
70 Dorsum with short wavy lines, dorsocentral setae not minute .... *T. leguminae* Mohanasundaram
70′ Dorsum with transverse striations, dorsocentral setae minute .... *T. knorri* Baker and Pritchard
71 Tibia I with four setae; setae *ag* not reaching bases of setae *g1* .... *T. bacuri* Flechtmann and Noronha
71′ Tibia I with five setae; setae *ag* reaching bases of setae *g1* .... *T. platycaryae* Wang
72 Opisthosoma with irregular transverse elevation between setae *c3–c3* .... *T. attiahi* Baker and Pritchard
72′ Opisthosoma without irregular transverse elevation between setae *c3–c3* 73
73 Dorsum with transverse ridges; setae *v2* and *sc1* leaf-like, spatulate, serrate .... *T. tepicanus* De Leon
73′ Dorsum without transverse ridges 74
74 Setae *sc2* about three times or more longer than the length of setae *v2* and *sc1* 75
74′ Setae *sc2* twice or smaller than the length of setae *v2* and *sc1* .... *T. jamaicensis* De Leon
75 Dorsal setae *c1*, *d1*, *e1*, *c3*, and *d3* very small, setiform, minute serrate 76
75′ Dorsal setae *c1*, *d1*, *e1*, *c3*, and *d3* lanceolate serrate .... *T. namaensis* Meyer
76 Coxae III and IV each with one setae; tibia IV with three setae .... *T. oliveirai* Flechtmann
76′ Coxae III and IV each with two setae; tibia IV with two setae .... *T. odoratus* Souza, Castro, and Oliveira
77 Palp one segmented .... *T. rhizophorae* De Leon
77′ Palp more than one segmented 78
78 First of dorsocentral setae *c1* reaching have or more than have distance between setae *c1-d1* 79
78′ First of dorsocentral setae *c1* less than have distance between setae *c1-d1* 81
79 Trochanters I–IV, 2-2-2-2; femur I with five setae .... *T. indicus* Maninder and Ghai
79′ Trochanters I–IV, 1-1-2-1; femur I with four setae 80
80 Tarsi I and II each with four setae; setae *4a* short not reaching to the bases of setae *ag* .... *T. garciniae* Meyer and Bolland
80′ Tarsi I and II each with four setae; setae *4a* crossing the bases of setae *ps1-2* .... *T. eucleae* Meyer
81 Dorsum almost smooth except few striations 82
81′ Dorsum not as mentioned above 86
82 Genua I and II each with one seta .... *T. vieirae* Castro, Ramos, and Feres
82′ Genua I and II each with more than one seta 83
83 Setae *4a* extend behind to the bases of setae *ag* 84
83′ Setae *4a* not extend behind to the bases of setae *ag* 85
84 Dorsolateral setae *c3* lanceolate to spatulate; area between setae *3a* and *4a* smooth .... *T. simarubae* De Leon

84′ Dorsolateral setae *c3* setiform; area between setae *3a* and *4a* striated .... *T. mourerae* De Leon
85 Venter striated; dorsolateral setae *f2, f3,* and *h1* narrowly lanceolate .... *T. hurae* De Leon
85′ Venter almost smooth; dorsolateral setae *f2, f3,* and *h1* broadly lanceolate .... *T. guamensis* Baker
86 Genua I and II each with one seta 87
86′ Genua I and II each with more than one seta 96
87 Genua III and IV each with one seta .... *T. amygdalusae* Maninder and Ghai
87′ Genua III and IV each without setae 88
88 Propodosoma with reticulation pattern 89
88′ Propodosoma without reticulation pattern 91
89 Trochanters I and II each without setae; tibiae I and II each with four setae .... *T. zhengzhouensis* Xu and Yin
89′ Trochanters I and II each with one seta; tibiae I and II each with five setae 90
90 Area between *c1* and *d1* and area posterior setae *e1* with reticulation pattern .... *T. taonicus* Ma and Yuan
90′ Area between *c1* and *d1* and area posterior setae *e1* without reticulation pattern .... *T. muguanicus* Ma and Yuan
91 Opisthosoma with reticulation pattern 92
91′ Opisthosoma without reticulation pattern 93
92 Propodosoma bordered by longitudinal striae; area between setae *c1* and *d1* with complete or incomplete reticulation .... *T. japonicas* Nishio
92′ Propodosoma not bordered by longitudinal striae; area between setae *c1* and *d1* with transverse irregular .... *T. zhizhilashviliae* Reck
93 Propodosoma strongly ridged medially; area between setae *c1, d1* and *d1* almost smooth .... *T. guettardae* De Leon
93′ Propodosoma without ridge 94
94 Tibiae III and IV each with two setae .... *T. anoplus* Baker and Pritchard
94′ Tibiae III and IV each with three setae 95
95 Tarsi I and II each with one solenidion setae .... *T. cedrelae* De Leon
95′ Tarsi I and II each without solenidion setae .... *T. anoplomexus* Baker and Tuttle
96 Genua III with one seta 97
96′ Genua III without setae 109
97 Genu IV with one seta 98
97′ Genu IV without setae 101
98 Femur IV with one seta 99
98′ Femur IV without setae .... *T. toowongi* Smiley and Gerson
99 Femur I with three setae .... *T. danxianensis* Yin, Cui, and Lin
99′ Femur I with four setae 100
100 Hysterosoma having strong transversal wrinkle posterior to setae *d1* and a "U-shaped" pattern posterior to setae *e1* .... *T. gneti* Xu, Fan, Huang, and Zhang
100′ Dorsum covered with irregular striae .... *T. qingchengensis* Wang
101 Palp with one seta on terminal segment 102
101′ Palp with two setae on terminal segment 106
102 First and second pairs of propodosomal setae *v2* and *sc1* short, subequal in length; third pair *sc2* elongate 103
102′ First pair of propodosomal seta *v2* short, second and third pairs *sc1* and *sc2* subequal in length and longer than first pair *v2* .... *T. uvae* De Leon
103 Dorsocentral setae *c1, d1,* and *e1* spatulate; propodosoma provided with a definite horseshoe ridge medially .... *T. comptus* Meyer
103′ Dorsocentral setae *c1, d1,* and *e1* lanceolate; propodosoma withoth a definite horseshoe ridge 104
104 Trochanter IV with two setae .... *T. disparilis* Wang and Cui

104′ Trochanter IV with one seta 105

105 Tarsi I and II each with eight tactile setae .... *T. toropi* Castro, Ramos, and Feres

105′ Tarsi I and II with five and four tactile setae .... *T. sharmai* Sadana and Gupta

106 Dorsocentral setae *c1* spatulate; area between setae *c1*, *d1* and *e1* with longitudinal striae .... *T. falcatus* Meyer

106′ Dorsocentral setae *c1* lanceolate or setiform; area between setae *c1*, *d1* and *e1* without longitudinal striae 107

107 Genua I and II each with two setae .... *T. ferosus* Akbar and Chaudhri

107′ Genua I and II each with three setae 108

108 Tarsi I and II each with six tactile setae .... *T. lunatus* Meyer

108′ Tarsi I and II each with seven tactile setae .... *T. elongatus* Meyer

109 Genua I and II each with two setae 110

109′ Genua I and II each with three setae 127

110 Trochanter III with one seta; tibia IV with two setae 111

110′ Trochanter III with two setae; tibia IV with two or three setae 112

111 Tarsi III and IV each with five setae .... *T. flacourtiae* Meyer

111′ Tarsi III and IV each with four setae .... *T. emeticae* Meyer

112 Tibia IV with two setae; all dorsal setae short .... *T. moraesi* Feres and Hernandes

112′ Tibia IV with three setae 113

113 Femur II with three setae .... *T. mopaneae* Meyer

113′ Femur II with four setae 114

114 Propodosoma with reticulation pattern sublaterally and striated medially 115

114′ Propodosoma not as mentioned above 116

115 Opisthosoma reticulated between setae *d1* and *e1* .... *T. jianfengensis* Ma and Yuan

115′ Opisthosoma without reticulation .... *T. sanyaensis* Yin, Cui, and Lin

116 Propodosoma framed by longitudinal striae or ridge 117

116′ Propodosoma as mentioned above 120

117 Dorsal setae *v2*, *sc1*, *sc2*, *c1*, *d1*, and *e1* subspatulate to spatulate .... *T. dumus* Meyer

117′ Dorsal setae *v2*, *sc1*, *sc2*, *c1*, *d1*, and *e1* lanceolate or setiform 118

118 Setae *4a* crossing the basis of setae *g1-2* 119

118′ Setae *4a* not crossing the basis of setae *g1-2*; venter of propodosoma with longitudinal, broken striae laterally .... *T. smithi* Meyer

119 Setae *ag* crossing the bases of setae *g1-2*; area between setae *3a* and *4a* with transverse striae .... *T. zeyheri* Meyer

119′ Setae *ag* not crossing the bases of setae *g1-2*; area between setae *3a* and *4a* with longitudinal striae .... *T. dombeyae* Meyer

120 Palp with one seta on terminal segment 121

120′ Palp with two setae on terminal segment 125

121 Opisthosoma with transverse folds between dorsocentral setae *d1* and *e1* .... *T. burserae* De Leon

121′ Opisthosoma without transverse folds between dorsocentral setae *d1* and *e1* 122

122 Genital plate with complete to incomplete reticulated; genital setae *g1-2* strongly serrate .... *T. melhaniae* Meyer

122′ Genital plate without reticulation pattern 123

123 Venter smooth .... *T. aboharensis* Sadana and Chhabra

123′ Venter not smooth 124

124 Dorsum irregularly striate-rugose with smooth patches, complete and incomplete reticulations sublaterally; venter strongly rogues .... *T. budensis* Ueckermann and Ripka

124′ Dorsum irregularly, broken striate without smooth patches; venter with smooth areas .... *T. hornotinus* Chaudhri

125 Tarsi III and IV each with four setae; dorsum with irregular broken longitudinal striae .... *T. umarii* Hasan, Wakil, and Bashir

125′ Tarsi III and IV each with five setae; dorsum with reticulation pattern 126
126 Third pair of propodosomal setae *sc2* as long as the distance between setae *sc1* and *sc2*; ventral cuticle with smooth regions anterior to setae *3a* and *4a* .... *T. punicae* Pritchard and Baker
126′ Third pair of propodosomal setae *sc2* less than the distance between setae *sc1* and *sc2*; ventral cuticle without smooth regions anterior to setae *3a* and *4a* .... *T. shishehbouri* Khanjani, Khanjani, and Seeman
127 Median area of propodosoma smooth except few traverse striae posterior and anterior medially .... *T. jussiaeae* De Leon
127′ Median area of propodosoma not as mentioned above 128
128 Tibiae I and II each with four setae; dorsum covered with polygonal reticulate pattern .... *T. vitexi* Meyer
128′ Tibiae I and II each with five setae 129
129 Dorsum with reticulation sublaterally, medially with irregular striae .... *T. menglunensis* Yin and Cui
129′ Dorsum without reticulation 130
130 Propodosoma with longitudinal furrow medially, rest of dorsum covered with innumerable ridge .... *T. sanblasensis* De Leon
130′ Propodosoma without longitudinal furrow medially 131
131 Setae *4a* extend behind to the bases of setae *g1-2* 132
131′ Setae *4a* not extend behind to the bases of setae *g1-2* 136
132 Coxa IV and femur IV each with two setae .... *T. panici* De Leon
132′ Coxa IV and femur IV each with one seta 133
133 Median area of propodosoma framed by longitudinal striae 134
133′ Median area of propodosoma not framed by longitudinal striae 135
134 Area between setae *3a* and *4a* with longitudinal striae .... *T. crocopontensis* Meyer
134′ Area between setae *3a* and *4a* smooth .... *T. sclerocaryae* Meyer
135 Palp with one seta on terminal segment; rostral shield with two lateral lobes strongly angulate .... *T. bauchani* Castro, Feres, Mesa, and Moraes
135′ Palp with two setae on terminal segment; rostral shield with two lateral lobes not strongly angulate .... *T. bellulus* Meyer
136 Tarsi I and II each with seven tactile setae 137
136′ Tarsi I and II each with eight tactile setae 138
137 Dorsum with few transverse, irregular striae medially .... *T. sophiae* Meyer
137′ Dorsum with incomplete reticulation or areolae medially .... *T. leonorae* Meyer
138 Tarsi III and IV each with four setae 139
138′ Tarsi III and IV each with five setae 140
139 Setae *g1-2* reaching or crossing to the bases of setae *ps1-2*; median area of propodosoma not framed by longitudinal striae .... *T. lycioides* Meyer
139′ Setae *g1-2* not reaching to the bases of setae *ps1-2*; median area of propodosoma framed by longitudinal striae .... *T. auriculatae* Meyer
140 Median area of propodosoma framed by longitudinal striae .... *T. rusapensis* Meyer
140′ Median area of propodosoma not framed by longitudinal striae .... *T. lanceae* Meyer
141 Dorsocentral setae *c1* reaching or crossing have distance between setae *c1* and *d1* 142
141′ Dorsocentral setae *c1* short not reaching have distance between setae *c1* and *d1* ... 145
142 Propodosoma setae *v2* and *sc1* short not reaching have distance between their basis, setae *sc2* longer than *v2* and *sc1*; propodosoma smooth medially and with transverse striae laterally .... *T. hondurensis* Evans
142′ Propodosoma setae *v2* and *sc1* long reaching or crossing have distance between their bases 143
143 Genua I and II with two setae, genu IV with two setae; opisthosoma with a few broken longitudinal mediolaterally.......................... *T. jawadii* Hasan, Wakil, and Bashir
143′ Genua I and II with three setae 144

144 Venter with transverse striations pattern broadly spaced medially; dorsocentral setae slender serrate . . . *T. etemadii* Mahdavi and Asadi

144′ Venter with transverse striations medially except area around setae *3a* smooth; dorsocentral setae setiform serrate . . . *T. ortus* Chaudhri

145 Dorsum with few reticulations or incomplete reticulations medially . . . 146

145′ Dorsum striated . . . 150

146 Palp with two setae on terminal segment . . . 147

146′ Palp with one seta on terminal segment . . . 148

147 Area between coxae III and IV with longitudinal broken striation; ventral of propodosoma smooth medially . . . *T. aurantiacus* Wang

147′ Area between coxae III and IV with transverse striations medially and few longitudinal striations laterally . . . 149

148 Prodorsal setae *sc2* nearly twice as long as prodorsal setae *v2* and *sc1*; dorsal setae subspatulate . . . *T. angolensis* Meyer

148′ Prodorsal setae *sc2* about as long as prodorsal setae *v2* and *sc1*; dorsal body setae spatulate . . . *T. nigerianus* Meyer

149 Venter with longitudinal striation laterally and smooth medially; propodosoma with incomplete reticulations sublaterally . . . *T. spatulatus* Wang

149′ Venter with incomplete reticulations laterally and smooth medially; propodosoma striated sublaterally . . . *T. obvelatus* Wang

150 Propodosoma smooth medially . . . 151

150′ Propodosoma with different pattern of striations medially. . . 152

151 Opisthosoma covered with broken striations; area posterior setae *e1* with transverse broken striations . . . *T. leptadeniae* Mohanasundaram

151′ Opisthosoma smooth medially, sublateral area with few longitudinal broken striations; area posterior setae *e1* smooth . . . *T. kobachidzei* Reck

152 Propodosoma with a few wavy lines with smooth area medially; setae on femora I–IV 3-3-2-1 . . . *T. coimbatorensis* Mohanasundaram

152′ Propodosoma with irregular or broken striations . . . 153

153 Dorsolateral setae spatulate, broadly spatulate or broadly lanceolate . . . 154

153′ Dorsolateral setae setiform, lanceolate or slender or ovate . . . 157

154 Ventral of propodosoma smooth medially . . . 155

154′ Ventral of propodosoma with transverse striations medially . . . 156

155 Propodosomal setae *v2* and *sc1* spatulate; opisthosoma with longitudinal striations forming a thick ridge-like structure medially . . . *T. waqasii* Hasan, Wakil, and Bashir

155′ Propodosomal setae *v2* and *sc1* ovate; opisthosoma striae forming star-shaped pattern around setae *d1* . . . *T. bakerdeleonorum* Evans

156 Genua I and II with two setae; dorsolateral setae *e3* setiform; ventral propodosoma pitted laterally . . . *T. orchidofilo* Moraes and Freire

156′ Genua I and II with three setae; dorsolateral setae *e3* broadly lanceolate serrate; ventral propodosoma with longitudinal striations laterally . . . *T. erasus* Chaudhri

157 Propodosoma with transverse striations medially . . . 158

157′ Propodosoma without transverse striations medially . . . 161

158 Area between setae *1c* and *d1* smooth; propodosoma with few striations posteriorly . . . *T. lygodii* De Leon

158′ Area between setae *1c* and *d1* with transverse or irregular striations . . . 159

159 Genua I and II with two setae; propodosoma anterior medially with few transverse striations . . . *T. chinariensis* Akbar and Chaudhri

159′ Genua I and II with three setae . . . 160

160 Opisthosoma with transverse broken striations and few longitudinal striations posteriorly . . . *T. mustus* Chaudhri

160′ Opisthosoma with longitudinal broken striations laterally, sublaterally with incomplete reticulations pattern . . . *T. carolinensis* Baker

161 Tibiae I and II with four setae; propodosoma with irregular broken longitudinal and transverse striae .... *T. cissampelosa* Maninder and Ghai
161′ Tibiae I and II with five setae 162
162 Propodosoma with distinct, longitudinal rugose pattern or with large central raised region of weakly colliculate cuticle flanked by series of fine longitudinal folds, becoming oblique laterally..163
162′ Propodosoma with irregular striation 164
163 Opisthosomal setae *f2* and *f3* ovate, other opisthosomal setae lanceolate; setae *sc2* oblanceolate; area between setae *c1-c1* mostly smooth .... *T. crassulus* Baker and Tuttle
163′ All opisthosomal setae lanceolate; setae *sc2* oblanceolate; area between setae *c1-c1* weakly reticulated .... *T. sarcophilus* Welbourn and Beard
164 Venter completely smooth; opisthosoma with lateral bulge anterior coxa III .... *T. shanxiensis* Qian, Yuan, and Ma
164′ Venter not as mentioned above 165
165 Ventral propodosoma smooth laterally; area between setae *c1* and *d1* with few longitudinal striations; dorsolateral setae slightly lanceolate; dorsocentral setae setiform .... *T. pernicis* Chaudhri, Akbar, and Rasool
165′ Ventral propodosoma with longitudinal or broken striations or reticulated laterally 166
166 Anterolateral ventral cuticle with broken longitudinal striations laterally 167
166′ Anterolateral ventral cuticle with longitudinal striations or reticulations 168
167 Propodosomal setae *v2*, *sc1*, and *sc2* lanceolate, propodosoma outlined by longitudinal striae and provided with few irregular striae inside this area .... *T. geigeriae* Meyer
167′ Propodosomal setae *v2*, *sc1*, and *sc2* setiform; propodosoma with irregular broken striations .... *T. eremitus* Chaudhri
168 Dorsolateral setae narrowly lanceolate; anterolateral ventral cuticle entirely striate, ventral cuticle between setae *1a–4a* with transverse striae .... *T. daneshvari* Khosrowshahi and Arbabi
168′ Dorsolateral setae setiform; anterolateral ventral cuticle reticulate laterally; ventral cuticle with smooth regions anterior to setae *3a* ...... *T. parsii* Khosrowshahi and Arbabi
169 Propodosoma smooth medially; area posterior setae *3a* with transverse, broken striae, genua I and II with each with one setae .... *T. ludhianaensis* Sadana and Chhabra
169′ Propodosoma not smooth medially 170
170 Dorsum covered with reticulation, incomplete reticulations, or polygonal cells medially 171
170′ Dorsum without reticulation, incomplete reticulations, or polygonal cells medially 181
171 Genua I and II each with one seta 172
171′ Genua I and II each with more than one seta 180
172 Propodosoma ventrally with reticulations at the bases of coxa II, ventral shield reticulated .... *T. reticulus* Siddiqui and Chaudhri
172′ Propodosoma ventrally without reticulations 173
173 Venter with transverse striae medially 174
173′ Venter smooth or smooth medially 176
174 Area anterior setae *3a* smooth, few transverse striae at the bases of coxa II; dorsum with reticulations medially .... *T. kesari* Sadana, Gupta, and Goyal
174′ Area anterior setae *3a* covered with transverse striae 175
175 Dorsum entirely covered with polygonal reticulation; opisthosoma with a distinct band of transverse striae level with setae *d1* .... *T. kamalii* Khosrowshahi and Arbabi
175′ Dorsosublateral region of propodosoma with irregular, incomplete reticulation; dorsosublateral region of opisthosoma with polygonal reticulation; opisthosoma without distinct band of transverse striae ....*T. euonymi* Khosrowshahi

176 Ventral propodosoma with longitudinal, broken striae laterally; opisthosoma with transverse lines near second pair of dorsocentral setae *d1* .... *T. raptor* Akbar and Chaudhri
176′ Ventral propodosoma smooth 177
177 Opisthosoma with a transverse non-reticulated band posterior central setae *d1* 178
177′ Opisthosoma without a transverse non-reticulated band 179
178 Femur IV with one seta; tibiaa I and II each with five setae .... *T. persicae* Sadana, Chhabra and Gupta
178′ Femur IV with two setae; tibiae I and II each with four setae .... *T. pyrusae* Maninder and Ghai
179 Trochanter IV with one seta; tibiae I and II each with five setae .... *T. dimensus* Chaudhri
179′ Trochanter IV with two setae; tibiae I and II each with four setae .... *T. pruni* Maninder and Ghai
180 Setae in genua I-IV 2-2-0-0 181
180′ Setae in genua I–IV 3-3-1-1; tibia III with two setae; dorsum with wavy diagonal laterally .... *T. tectonae* Mohanasundaram
181 Palp three-segmented; setae in tibiae I–IV with 5-5-3-3; dorsum with longitudinal latterally .... *T. trisegmentus* Siddiqui and Chaudhri
181′ Palp two-segmented; setae in tibiae I–IV with 4-4-3-3; dorsum with few reticulations, and broken striae medially .... *T. jandialensis* Kauser, Akbar, and Naz
182 Dorsum covered with sub-areolate rogues 183
182′ Dorsum without sub-areolate rogues 186
183 Genua I and II each with two setae 184
183′ Genua I and II each with three setae 185
184 Opisthosoma with sublateral grooves; dorsocentral setae *c1*, *d1* and *e1* setiform to lanceolate .... *T. karrooi* Meyer
184′ Opisthosoma without sublateral grooves; dorsocentral setae *c1*, *d1* and *e1* subspatulate .... *T. kraussianae* Meyer
185 Opisthosoma setae *d1*, *e1*, and *d3* subspatulate; setae *ag* crossing have distance between setae *ag-g* .... *T. abutiloni* Meyer
185′ Opisthosoma setae *d1*, *e1*, and *d3* lanceolate; setae *ag* reaching have distance between setae *ag-g* .... *T. ombrensis* Meyer
186 Dorsum covered with longitudinal thick striae; setae in genua I-IV 3-3-1-1; venter with longitudinal striae medially .... *T. pagina* Chaudhri, Akbar, and Rasool
186′ Dorsum without longitudinal thick striae 187
187 Genua I and II each with one seta 188
187′ Genua I and II each with more than one seta 191
188 Propodosoma or opisthosoma with reticulations pattern medially 189
188′ Propodosoma or opisthosoma without reticulations pattern medially............. 190
189 Palp with two segments; propodosoma with reticulation elements anteriorly; dorsum with longitudinal striae laterally .... *T. kenos* Hasan, Wakil, and Bashir
189′ Palp with three segments; opisthosoma with incomplete reticulation up to second pair of dorsocentral setae *d1*. .... *T. mandraensis* Hasan, Wakil, and Bashir
190 Third pair of propodosomal setae *sc2* subspatulate longer thane second pair *sc1*; dorsum roguse .... *T. heteropyxis* Meyer
190′ Second and third pairs of propodosomal setae *sc1* and *sc2* lanceolate and subequal in length; dorsum wrinkled .... *T. tapiae* Castro and Feres
191 Genua I and II each with two setae 192
191′ Genua I and II each with three setae 195
192 Femur IV with one seta 193
192′ Femur IV with two setae 194
193 Propodosoma medially with incomplete to complete areolate; second and third pairs of propodosomal setae *sc1* and *sc2* subequal in length .... *T. mkuziensis* Meyer

193′ Propodosoma medially irregular to transverse striae; third pair of propodosomal setae *sc2* nearly twice as long as first and second pairs *v2* and *sc1* .... *T. microphylli* Meyer

194 Third pair of propodosmal setae *sc2* about twice as long as first and second pairs *v2* and *sc1* .... *T. acaciae* Ryke and Meyer

194′ Third pair of propodosmal setae *sc2* about three times as long as first and second pairs *v2* and *sc1* .... *T. pyroides* Meyer

195 Femur IV with one seta; setae in genua I-IV 3-3-1-0; dorsum with transverse, irregular striae medially .... *T. carlosflechtmanni* Feres and Hernandes

195′ Femur IV with two setae 196

196 Setae in genua I–IV 3-3-0-0; dorsal setae subspatulate; area posterior setae *4a* with longitudinal striae .... *T. aethiopicus* Meyer

196′ Setae in genua I–IV 3-3-1-1; dorsal setae broadly lanceolate; area posterior setae *4a* with transverse striae .... *T. legatus* Chaudhri

197 Palp with two segments; propodosoma with a wide U-shape medially .... *T. Stefani* Meyer

197′ Palp with three segments 198

198 Propodosoma with irregular transverse striae medially and with longitudinal striae mediolaterally; opisthosoma with dorsolateral reticulations .... *T. lulinicus* Ma and Yuan

198′ Propodosoma with various patterns of striations or rugose 199

199 Trochanter IV without setae; ventral propodosoma smooth laterally .... *T. stativus* Chaudhri, Akbar, and Rasool

199′ Trochanter IV with one seta 200

200 Dorsum with longitudinal, broken striae, and few tranvers striae posterior setae *d1* .... *T. pisinnus* Chaudhri, Akbar, and Rasool

200′ Dorsum without striation posterior setae *d1* .... *T. pacificus* Baker

201 Tibiae I and II each with 4 setae; genito-ventral plate smooth .... *T. yarensis* Hasan, Bashir, and Wakil

201′ Tibiae I and II each with five setae; genito-ventral plate with transverse striation or with small rounded structures 202

202 Seta *v2* about as long as seta *sc1* and about half as long as seta *sc2* 203

202′ Setae *v2*, *sc1*, and *sc2* subequal; aggenital and genital setae barbed 204

203 All dorsal setae lanceolate, serrate; femur IV with one seta; genu IV with one seta .... *T. eriophyoides* Baker

203′ All dorsal setae subspatulate to spatulate serrate; femur IV with two setae; genu IV without setae .... *T. scitulus* Meyer

204 Ventral cuticle of idiosoma mostly with small rounded structures, except for the central region between setae *1a-3a*, where the rounded structures become elongate or fuse to each other to form wavy, broken longitudinal ridges, and for the regions laterad of ventrigenital plate and between genital and anal openings, smooth; genito-ventral plate also with small rounded structures; genua I–IV with 3-3-2-2 setae .... *T. omani* Moraes, Al-Shanfari, and Silva

204′ Ventral cuticle of idiosoma mostly with wavy, broken longitudinal lines .... *T. pareriophyiodes* Meyer and Gerson

3.6. Key to the World Species of the T. dubinini Species Group (Based on Females)

1 Venter with one pair of *4a* 7

1′ Venter with more than one pair of *4a* 2

2 Venter with two pairs of *4a* 3

2′ Venter with more than two pairs of *4a* 4

3 Venter with one pair of *3a* 19

3′ Venter with two pairs of *3a* 32

4 Venter with three or four pairs of 4a 5

4′ Venter with five or six pairs of *4a* .... *T. dubinini* Reck
5 Setae *d1* absent .... *trisetosus* Baker and Tuttle
5′ Setae *d1* present 6
6 Venter with one pair of *3a* .... *T. rosae* Kadzhaja
6′ Venter with two pairs of *3a* .... *T. crassus* Andre
7 Setae *d3* present 9
7′ Setae *d3* absent 8
8 Dorsum with transverse striae medially .... *T. yousefi* Nassar and Ghai
8′ Dorsum without transverse striae medially .... *T. gumbolimbonis* De Leon
9 First pair of dorsocentral setae *c1* reaching or crossing bases of next setae *d1* 10
9′ First pair of dorsocentral setae *c1* short not reaching bases of next setae *d1* 13
10 Palp with one or two segments 11
10′ Palp with three segments 12
11 Palp with one segment; setae in genua I–IV 3-3-1-1; dorsocentral setae *d1* and *e1* short not reaching bases of setae *h1* and *h2* .... *T. grevilleae* Gutierrez and Schicha
11′ Palp with two segments; setae in genua I–IV 2-2-0-0; dorsocentral setae *d1* and *e1* very long crossing bases of setae *h1* and *h2* .... *T. banksiae* Gutierrez and Schicha
12 Tibiae I and II each with four setae; femur IV with one seta; dorsocentral setae *e1* crossing bases of setae *h1* and *h2* .... *T. comatus* Meyer
12′ Tibiae I and II each with five setae; femur IV with two setae; dorsocentral setae *e1* not reaching bases of setae .... *T. leucospermi* Meyer
13 Trochanters I and II without setae; tibiae I and II each with four setae; dorsom almost smooth, with few lines .... *T. mallotae* Mohanasundaram
13′ Trochanters I and II with one seta 14
14 Tibia IV with two setae; dorsum coarsely reticulate to reticulate or rugose to areolate 15
14′ Tibia IV with three setae 17
15 Venter with broken striae medially 16
15′ Venter with transverse striae medially; dorsum strongly rugose to areolate; setae in tibiae I–IV 4-4-2-2 .... *T. leipoldti* Meyer
16 Setae in tibiae I–IV 4-4-2-2; tarsi III-IV each with four setae .... *T. rhusi* Meyer
16′ Setae in tibiae I–IV 5-5-3-2; tarsi III-IV each with five setae .... *T. oribiensis* Meyer
17 Palp with three segments 18
17′ Palp with one segments .... *T. tortulus* Meyer
18 Setae in trochanters I–IV 1-1-1-1; genu I with two setae; dorsal setae *sc1, sc2, c3, f2, f3,* and *h1* leaf-like and spiculate .... *T. flechtmanni* Mesa, Moraes, and Ochoa
18′ Setae in trochanters I–IV 1-1-2-1; genu I with three setae; dorsal setae minute lanceolate .... *T. proteae* Meyer
19 Opisthosoma with two pairs of dorsocentral setae *c1* and *e1*; dorsum rugose; setae in genua I–IV 2-2-0-0 .... *T. annonae* De Leon
19′ Opisthosoma with three pairs of dorsocentral setae *c1, d1* and *e1* 20
20 Propodosoma reticulated medially, and with longitudinal striae laterally; venter smooth; setae in genua I–IV 3-3-1-0 .... *T. tauricus* (Mitrofanov and Strunkova)
20′ Propodosoma without reticulation medially 21
21 Genua I and II each with one or two setae 22
21′ Genua I and II each with three setae 25
22 Genua I and II each with one seta; dorsum with longitudinal to irregular striae .... *T. feliciae* Meyer
22′ Genua I and II each with two setae 23
23 Palp with one segment; setae in trochanters 1-1-1-1; dorsum with transverse ornamentation medially.......... *T. punjabensis* (Maninder and Ghai)
23′ Palp with three segments 24
24 Tibiae I and II each with four setae; dorsum covered with broken wavy striae; area posterior setae *4a* with transverse, broken striae .... *T. ilocanus* Corpuz-Raros

24′ Tibiae I and II each with five setae; dorsum covered with irregular, wrinkles medially .... *T. maai* Xu, Fan, Huang, and Zhang
25 Genu III without setae, palp two segmented .... *T. berkheyae* Meyer
25′ Genu III with one seta 26
26 Propodosoma almost smooth medially 27
26′ Propodosoma not as mentioned above 28
27 Opisthosoma almost smooth between dorsocentral setae *c1-e1*; femur IV with two setae; dorsal setae narrowly lanceolate .... *T. neokeiensis* Khan, Kamran, and Alatawi
27′ Opisthosoma covered with elongate cells; femur IV with one seta; dorsal setae lanceolate to oblanceolate .... *T. kermanicus* Khadem, Asadi, and Seeman
28 Palp with one segment 29
28′ Palp with more than one segments 30
29 Coxa I with two setae; tibiae I and II each with four setae .... *T. jagatkhanaens* Sadana and Gupta
29′ Coxa I with one seta; tibiae I and II each with three setae .... *T. bassiae* Mohanasundaram
30 Femur IV with one seta 31
30′ Femur IV with two setae; dorsum with longitudinal striae and reticulate elements medially; dorsal setae broadly spatulate .... *T. alhagus* Khan, Kamran, and Alatawi
31 Dorsum with transverse striae medially; ventral propodosoma with broken longitudinal striae laterally .... *T. keiensis* Meyer
31′ Dorsum covered with longitudinal, irregular striae; venter of propodosoma with longitudinal striae laterally .... *T. clematidos* Wang
32 Palp with two segments; genu III with one seta .... *T. nenaxi* Meyer
32′ Palp with three segments; genu III without setae .... *T. xerocolus* Meyer

3.7. Key to the World Species of the T. granati Species Group (Based on Females)

1 Venter with one pair of *4a* 3
1′ Venter with two pairs of *4a* 2
2 Venter with one pair of *3a* 22
2′ Venter with two pairs of *3a* .... *T. iranicus* Khadem, Asadi, and Seeman
3 Propodosoma with two pairs of dorsal setae *sc1* and *sc2* present, setae *v2* absent .... *T. elegans* (Collyer)
3′ Propodosoma with three pairs of dorsal setae *sc1*, *v2* and *sc2* present 4
4 Genua I–IV without setae .... *T. philippinensis* (Corpuz-Raros)
4′ Genua I–IV not as mentioned above 5
5 Dorsocentral setae *c1* present 6
5′ Dorsocentral setae *c1* absent 10
6 Dorsocentral setae *d1* and *e1* absent 7
6′ Dorsocentral setae *e1* absent 8
7 Dorsum with polygonal reticulation; palp with three segments; genua I and II each with three setae .... *T. cyatheae* Gerson and Collyer
7′ Dorsum with irregular striae; palp with two segments; genua I and II each with two setae .... *T. pariae* Hasanvand, Jafari, Khanjani, and Khanjani
8 Second pair of dorsocentral seta *d1* lanceolate to elliptic lanceolate .... *T. transvaalensis* Meyer
8′ Second pair of dorsocentral seta *d1* min, setiform 9
9 Opisthosoma with lateral projection anterior coxa III .... *T. robustae* Meyer
9′ Opisthosoma without lateral projection anterior coxa III .... *T. capparis* Meyer
10 Opisthosoma with lateral lobes; palp with one segment 11
10′ Opisthosoma without lateral lobes 14
11 Setae in trochanters I–IV 1-1-0-0 12
11′ Setae in trochanters I–IV 1-1-1-1 13

12 Genua III and IV each with one seta; pregenital area finely areolate, setae *ag* crossing bases of setae *g1-2* .... *T. calcarius* Meyer
12′ Genua III and IV each without setae; pregenital and genital areas partly punctate and striate, setae *ag* not reaching bases of setae *g1-2* .... *T. protumidus* Meyer
13 Genua I and II each with two setae; tibiae I and II each with four setae; dorsal setae lanceolate serrate .... *T. jonkeri* Meyer
13′ Genua I and II each with three setae; tibiae I and II each with five setae; dorsal setae linear lanceolate to setiform .... *T. athrixiae* Meyer
14 Genua I and II each with two setae 15
14′ Genua I and II each with three setae 18
15 Palp with three segments; setae in trochanters I–IV 1-0-1-0; tibiae I and II each with four setae opisthosoma with cross lines .... *T. acuminatae* Mohanasundaram
15′ Palp with one segment 16
16 Femur IV with one seta; tibia II with four setae; propodosoma with three or more conspicuous sclerotic rings .... *T. palosapis* Corpuz-Raros
16′ Femur IV with two setae; tibia II with five setae 17
17 Dorsal integument coarsely areolate; dorsal setae coarsely barbed; area anterior setae *4a* with longitudinal striae .... *T. engelbrechti* Meyer
17′ Dorsal integument finely areolate; dorsal setae finely barbed; area anterior setae *4a* with broken striae .... *T. faveolus* Meyer
18 Dorsum with three oblong propodosomal plates and three rounded opisthosomal plates like structures medially with punctate inside pattern these plates; dorsum with longitudinal striae laterally .... *T. jordaani* Meyer
18′ Dorsum without three oblong propodosomal plates 19
19 Tibia IV with two setae 20
19′ Tibia IV with three setae 21
20 Second pair of propodosomal *sc1* setae setiform minute; tibiae I and II each with five setae .... *T. vernoniae* Meyer
20′ Second pair of propodosomal *sc1* setae sub spatulate; tibiae I and II with four and three setae respectively .... *T. amatikulensis* Meyer
21 Dorsum areolate; opisthosoma with sublateral grooves .... *T. albae* Meyer
21′ Propodosoma with transvers, broken striae medially; opisthosoma without sublateral grooves .... *T. caledonicus* Meyer
22 Dorsum with one pair of dorsocentral setae *c1* present, setae *d1* and *e1* absent 23
22′ Dorsum with two pairs of dorsocentral setae *c1* and *e1* present, setae *d1* absent .... *T. tamarixi* Mahdavi and Asadi
23 Dorsum smooth or almost smooth medially, propodosoma with few striae laterally 24
23′ Dorsum with irregular striae 26
24 Trochanter III with one setae; dorsum almost smooth medially; opisthosoma with few faint, transverse striae; propodosoma with few striae laterally .... *T. citus* Chaudhri, Akbar, and Rasool
24′ Trochanter III with two setae 25
25 Genu III with one seta; femur IV with two setae; area between setae *3a* and *4a* with transverse striae .... *T. myrtus* Al-Gboory
25′ Genu III without setae; femur IV with one seta; area between setae *3a* and *4a* smooth .... *T. viticola* Al-Gboory
26 Venter with transverse striae medially; setae *ag* reaching or crossing bases of setae *g1-2* 27
26′ Venter without transverse striae medially; setae *ag* not reaching bases of setae *g1-2* .... *T. lineosetosus* Wang
27 Setae *4a* reaching to the bases of setae *g1-2* .... *T. granati* Sayed

27′ Setae *4a* reaching to the bases of setae *g1-2* .... *T. populi* Al-Gboory

3.8. Key to the World Species of the T. barticanus Species Group (Based on Females)

1 Venter with one pair of *4a* 3
1′ Venter with two pairs of *4a* 2
2 Venter with one pair of *3a* 5
2′ Venter with two pairs of *3a* .... *T. sparsus* Chaudhri, Akbar, and Rasool
3 Palp with three segments .... *T. capassae* Meyer
3′ Palp with two segments 4
4 First and second pairs of propodosmal setae *v2* and *sc1* subequal in length; opisthosoma with lateral lobes .... *T. chelinus* Meyer
4′ Second pair of propodosmal setae *sc1* longer than first pair *v2*; opisthosoma without lateral lobes .... *T. papiothalensis* Meyer
5 Dorsum smooth; venter with transverse striae medially, propodosoma completely smooth .... *T. salicis* Al-Gboory
5′ Dorsum striated 6
6 Genu I with one seta; dorsum strongly ridged; third pair of propodosomal setae *sc2* short not reaching bases of second pair *sc1* .... *T. barticanus* De Leon
6′ Genu I with two setae; propodosomal with transverse striae medially; third pair of propodosomal setae *sc2* elongate crossing bases of second pair *sc1* .... *T. isabelae* Mesa, Moraes, and Ochoa

4. Discussion

The number of setae (dorsocentral and dorsolateral setae) has been used by different authors to erect tenuipalpid genera by transferring species from the genus *Tenuipalpus*, e.g., *Aegyptopalpus* Mitrofanov [7], *Deleonipalpus* Mirofanov [7], *Gnathopalpus* Mitrofanov [7], and *Tuttlepalpus* Mitrofanov [7]. Later, some of those genera were synonymized with the genus *Tenuipalpus* [8].

Different species groups were proposed based on the number of dorsocentral setae or dorsolateral setae [8–10]. According to the number of dorsolateral setae, two species groups, *caudatus* and *proteae*, based on the presence and absence of setae *f2* (with seven and six pairs of dorsolateral setae), respectively. Later, these two species groups were divided into species subgroups based on the number intercoxal setae *3a* and *4a* were recognized [9,10]. Also, the character of dorsocentral and dorsolateral setae have been used as a first couplet in the different keys to classify the *Tenuipalpus* species (Al-Gboory [12]; Khanjani et al. [13]; Castro et al. [16]; Xu et al. [15]). The literature of 287 species of *Tenuipalpus* sensu lato signified the importance of dorsal opisthosomal setae as a prominent diagnostic morphological character. However, using the presence and absence of dorsolateral setae *f2* (seven and six pairs of setae, respectively) could not be helpful, because some species (i.e., *T. clematidos* Wang, *T. flechtmanni* Mesa, Moraes, and Ochoa, *T. isabelae* Mesa, Moraes, and Ochoa, and *T. salicis* Al-Gboory) have six of dorsolateral setae (*f2* present), but the dorsolateral setae (*d3*) is absent. Hence, those species can not be placed in any species groups that proposed Baker and Tuttle [9] and Meyer [10] because of these grouping *caudatus* and *proteae* is based on presence and absense of setae *f2*. However, some previous works are considering the number of dorsolateral setae to distinguish these two species groups, regardless of which dorsolateral seta is absent. For example, Mahdavi and Asadi [27] included two species (*T. clematidos* Wang and *T. salicis* Al-Gboory) in the *proteae* species group, which have setae *f2* present. Therefore, we found that using the total number of opisthosomal setae is proper for dividing species groups, while the absence or presence of certain opisthosomal setae may be used for species differentiation in diagnostic keys. Hence, in this study, the four new species groups were proposed based on the number of opisthosomal setae.

4.1. Further Notes on the Poorly Described and Illustrated Species of the Species Group Carolinensis

The following remarks and additional notes are about those eight species belonging to the species group *carolinensis*, which were not included in the key due to poor/incomplete descriptions and illustrations. Their related species and a possible place in the diagnostic key are provided as follows:

Tenuipalpus chiococcae De Leon

The species *T. chiococcae* was originally described as close to *T. pigrus* Pritchard and Baker [17]. The recent classification by Castro et al. [1] placed *T. pigrus* in *Tenuipalpus* sensu stricto since this species bears a pair of lateral body projections associated with setae *c3*. The result of our study placed *T. chiococcae* among five species in the diagnostic key, i.e., *T. aurantiacus* Wang, *T. angolensis* Meyer, *T. nigerianus* Meyer, *T. spatulatus* Wang, and *T. obvelatus* Wan.

Tenuipalpus costarricensis Salas and Ochoa

Previously, *T. costarricencis* was considered close to *T. granati* Sayed [18]. However, in our study *T. granati* is placed in the group *granati*, while *T. costarricensis* came closer to three species; *T. eucleae* Meyer, *T. garciniae* Meyer and Bolland, and *T. indicus* Maninder and Ghai in the species group *carolinensis*.

Tenuipalpus ephedrae Livschitz and Mitrofanov

For the species *T. ephedrae*, the morphological characters were obtained from a poor redescription [19], and the original description was not found. The taxonomic information available in the redescription only helped to place this species in the group *carolinensis*.

Tenuipalpus molinai Evans

The species *T. molinai* was originally described as close to *T. pedrus* Manson [20]. The closely related *T. pedrus* Manson had been transferred to the genus *Colopalpus* Pritchard and Baker, previously [28]. However, the available morphological characters indicate that the species *T. molinai* belongs only to the species group *carolinensis*, but could not be assigned in the key.

Tenuipalpus oxalis Flechtmann

The species *T. oxalis* is poorly described and illustrated, and the related species was not provided [21]. The available morphological characters helped to designate the species only to the level of the species group *carolinensis*.

Tenuipalpus santae Manson

The species *T. santae* is morphologically close to *T. celtidis* Pritchard and Baker in the original description [22]. *Tenuipalpus santae* could not be placed due to missing leg cheatotaxy and other diagnostic characters.

Tenuipalpus simplychus Cromroy

The species *T. simplychus* was described as closely related to *T. knorri* Baker and Pritchard in the original description [23]. Due to missing leg chaetotaxy information, *T. simplychus* could not be assigned a certain place in the diagnostic key.

Tenuipalpus tetrazygiae De Leon

The species *T. tetrazygiae*, it was distinguished by the author from other described species of that time by the shape of dorsocentral setae and dorsum covered with irregular ridges [17]. Although this species has been placed in the *caroliensis* group, no certain place could be identified in the diagnostic key.

4.2. Synonymy of Some Species of the Carolinensis Species Group

The species *T. lustrabilis* was previously reported as a suspected junior synonym of *T. punicae* Pritchard and Baker [2,13]. We reviewed the original description and illustration [29] as well as the characters of *T. lustrabilis* in the published key by Meyer [8]. *T. lustrabilis* is hereby synonymized with *T. punicae* based on the number of shared characteristics, i.e., leg chaetotaxy, palp segmentation, shape and number of dorsal setae, pattern of dorsal reticulations, as well as its geographic distribution (Pakistan) and host plant (*Punica granatum*).

Similarly, the species *T. guptai* Sadana and Gupta was also suggested as a junior synonym of *T. solanensis* Sadana and Gupta [2,13]. We found that both species share most of their morphological features except for the number of setae on tarsi I-II (5-5 in *T. guptai* vs 7-7 in *T. solanensis*). This character has been commented to be a miscalculation [13], especially that *T. guptai* was described based on a single holotype female while *T. solanensis* only from three females. Moreover, both species were described based on specimens collected from the same host plant (*P. granatum*), the same type locality (India), on the same collection date (22-VI-1981), and have been mounted on the same slide (slide#91) [30].

Interestingly, the related species of *T. solanesis* is *T. lustrabilis*, which is declared as junior synonym of *T. punicae*. A detailed comparison of available description for both of these species show they are very similar; sharing leg chaetotaxy, palp segmentation, shape and number dorsal setae, pattern of dorsal reticulations, and host plant. Hence, the two species (*T. guptai* and *T. solanensis*) are also synonymized with *T. punicae*.

Tenuipalpus rodionovi Chalilova was described poorly without illustrations [31]. Therefore, it is neither assigned to any of the four species groups and not placed in the diagnostic keys. However, it was mentioned in the original description that it resembles three species: *T. granati* Sayed, *T. zhizhilashviliac* Reck, and *T. kobachidzei* Reck. The latter two species belong to the species group *carolinensis*, while the former one belongs to the *granati* new species group. Pritchard and Baker [32] and Wainstein [33] suspected this species as a junior synonym of *T. granati*. There is a need to check the type specimens of this species to validate its status. Hence, *T. rodionovi* is considered as a suggested synonym of *T. granati* until further studies are made.

5. Conclusions

The history of the genus *Tenuipalpus* is complicated due to taxonomic and classification-based modifications. This research indicates that the genus *Tenuipalpus* needs more taxonomical studies to raise the level of groups and species groups to higher taxonomical ranking, by using persistent and strong morphological characters. This may even direct future research in the family Tenuipalpidae to study the other closely related genera, in order to validate their status.

Author Contributions: Conceptualization, N.A.E. and M.K.; methodology, N.A.E. and M.K.; validation, M.K. and F.J.A.; formal analysis, N.A.E., F.J.A. and M.K.; investigation, N.A.E. and M.K.; data curation, N.A.E. and M.K.; writing—original draft preparation, N.A.E.; writing—review and editing, M.K. and F.J.A.; supervision, F.J.A.; funding acquisition, F.J.A. All authors have read and agreed to the published version of the manuscript.

Funding: The authors extend their sincere appreciation to the researchers supporting project number (RSPD2023R807), King Saud University, Riyadh, Saudi Arabia.

Institutional Review Board Statement: Not applicable.

Informed Consent Statement: Not applicable.

Data Availability Statement: All the relevant data has been provided here in the manuscript.

Acknowledgments: The authors wish to thank the researchers supporting project, at King Saud University, Riyadh, Saudi Arabia.

Conflicts of Interest: The authors declare no conflict of interest.

References

1. Castro, E.B.; Kane, E.C.; Feres, R.J.; Ochoa, R.; Bauchan, G.R. Definition of *Tenuipalpus* sensu stricto (Acari, Tenuipalpidae), with redescription of *Tenuipalpus caudatus* (Dugès) and description of a new species from Costa Rica. *Int. J Acarol.* **2016**, *42*, 106–126. [CrossRef]
2. Mesa, N.C.; Ochoa, R.; Welbourn, W.C.; Evans, G.A.; De Moraes, G.J. A catalog of the Tenuipalpidae (Acari) of the World with a key to genera. *Zootaxa* **2009**, *2098*, 1–85. [CrossRef]
3. Castro, E.B.; Feres, R.J.; Mesa, N.C.; de Moraes, G.J. A new flat mite of the genus *Tenuipalpus* Donnadieu (Trombidiformes: Tenuipalpidae) from Brazil. *Syst. Appl. Acarol.* **2022**, *27*, 368–380. [CrossRef]

4. Hoy, M.A. *Agricultural Acarology: Introduction to Integrated Mite Management*; CRC Press: Boca Raton, FL, USA, 2011; Volume 7.
5. Donnadieu, A.L. *Recherches pour Servir a l'Histoire des Tétranyques—Thèses*; Faculté des Sciences de Lyon: Lyon, France, 1875; pp. 1–131.
6. Reck, G. On the chaetological basis of the systematics of tetranychid mites. *Soobshch. Akad. Nauk. Gruz. SSR* **1959**, *23*, 465–471.
7. Mitrofanov, V.I. Revision of the system of phytophagous mites of the subfamily Tenuipalpinae s. str. (Trombidiformes, Tenuipalpidae). *Zool. Zhurnal* **1973**, *52*, 1315–1320.
8. Meyer, M.K.P. *The Tenuipalpidae (Acari) of Africa. With Keys to the World Fauna*; In Entomology Memoir; Department of Agricultural Technical Services: Johannesburg, South Africa, 1979; Volume 50, 135p.
9. Baker, E.W.; Tuttle, D.M. *The False Spider Mites of Mexico (Tenuipalpidae: Acari)*; Technical Bulletin Number 1706; United States Department of Agriculture—Agricultural Research Service: Washington, DC, USA, 1987; Volume 1706, 236p.
10. Meyer, M.K.P. A revision of the genus *Tenuipalpus* Donnadieu (Acari: Tenuipalpidae) in the Afrotropical region. In *Entomology Memoirs*; Department of Agricultural Technical Services: Johannesburg, South Africa, 1993; Volume 88, 84p.
11. Collyer, E. New species of the genus *Tenuipalpus* (Acari: Tenuipalpidae) from New Zealand, with a key to the world fauna. *N. Z. J. Sci.* **1973**, *16*, 915–955.
12. Al-Gboory, I. Taxonomic Studies of False Spider Mites (Acari: Tenuipalpidae) in Central Iraq. Ph.D. Thesis, University of Bonn, Bonn, Germany, 1987.
13. Khanjani, M.; Khanjani, M.; Seeman, O.D. The flat mites of the genus *Tenuipalpus* Donnadieu (Acari: Tenuipalpidae) from Iran. *Int. J. Acarol.* **2013**, *39*, 97–129. [CrossRef]
14. Castro, E.B.; Feres, R.J. New species of *Tenuipalpus* (Acari: Tenuipalpidae) from Semidecidual Forest remnants in the State of São Paulo, Brazil. *Zootaxa* **2013**, *3716*, 475–493. [CrossRef]
15. Xu, Y.; Fan, Q.H.; Huang, J.; Zhang, F.P. Two new species of *Tenuipalpus* and re-description of *Tenuipalpus lineosetus* Wang, 1983 (Acari: Tenuipalpidae) from China. *Syst. App. Acarol.* **2018**, *23*, 539–580. [CrossRef]
16. Castro, E.B.; Feres, R.J.; Ochoa, R.; Bauchan, G.R. A new species of *Tenuipalpus* sensu stricto (Acari: Tenuipalpidae) from Brazil, with ontogeny and a key to the known species. *Zootaxa* **2016**, *4088*, 355–378. [CrossRef]
17. De Leon, D. Six new false spider mites from southern Florida (Acarina: Tenuipalpidae). *Fla. Entomol.* **1956**, *39*, 55–60. [CrossRef]
18. Salas, L.A.; Ochoa, R. Una especie nueva de ácaro plano *Tenuipalpus costarricensis* (Acari: Tenuipalpidae), en Costa Rica. *Agron. Costarricense* **1986**, *10*, 203–205.
19. Livshitz, I.Z.; Mitrofanov, V.I. New species of mites (Tenuipalpidae, Acariformes). *Zool. Zhurnal* **1970**, *49*, 787–789.
20. Evans, G.; Cromroy, H.L.; Ochoa, R. The Tenuipalpidae of Honduras (Tenuipalpidae: Acari). *Fla. Entomol.* **1993**, *76*, 126–155. [CrossRef]
21. Flechtmann, C.H.W. Preliminary report on the false spider mites (Acari: Tenuipalpidae) from Brazil and Paraguay. *Entomol. Soc. Wash.* **1976**, *78*, 58–64.
22. Manson, D.C. Seven new species of false spider mites (Tenuipalpidae: Acarina). *Acarologia* **1963**, *5*, 213–224.
23. Cromroy, H.L. A preliminary survey of the plant mites of Puerto Rico. *J. Agric. Univ. Puerto Rico* **1958**, *42*, 74–87.
24. Gupta, S.K.; Ghosh, S.K. Some prostigmatid mites (Acarina) from Andaman and Nicobar Islands. *Rec. Zool. Surv. India* **1980**, *77*, 189–213. [CrossRef]
25. Berlese, A. *Acarotheca Italica*; Tipografia M. Ricci: Firenze, Italy, 1913; 221p.
26. Duges, A. Recherches pour servir a l'ordre des Acariens en general et la famille des Trombidies en particulier. *Ann. Sci. Nat.* **1834**, *1*, 5–46.
27. Mahdavi, S.M.; Asadi, M. New flat mite species of the genus *Tenuipalpus* (Acari: Tenuipalpidae) from Iran. *Sysl. Appl. Acarol.* **2018**, *23*, 2184–2191. [CrossRef]
28. Castro, E.B.; Ochoa, R.; Feres, R.J.; Beard, J.J.; Bauchan, G.R. Reinstatement of the genus *Colopalpus* Pritchard and Baker (1958) and re-description of *Colopalpus matthyssei* Pritchard and Baker (1958), the type species of the genus (Acari, Tenuipalpidae). *Int. J. Acarol.* **2015**, *41*, 310–328. [CrossRef]
29. Chaudhri, W.M.; Akbar, S.; Rasool, A. *Taxonomic Studies of the Mites Belonging to the Families Tenuipalpidae, Tetranychidae, Tuckerellidae, Caligonellidae, Stigmaeidae and Phytoseiidae*; Project No. A17-Ent-26; University of Agriculture: Lyallpur, Pakistan, 1974; 250p.
30. Sadana, G.L.; Gupta, B.K. New species of the genus *Tenuipalpus* Donnadieu (Tenuipalpidae, Acarina) from India. *Entomon* **1984**, *9*, 141–147.
31. Chalilova, S.G. Opisanie devuk novick tetranickovick kleschei ez semeist Bryobiidae I Trichadenidae. *Soobshch. Akad. Nauk. Gruz. SSR* **1953**, *14*, 549–551.
32. Pritchard, A.E.; Baker, E.W. The false spider mites (Acarina: Tenuipalpidae). *Uni. Calif. Publ. Entomol.* **1958**, *14*, 175–274.
33. Wainstein, B.A. Tetranychoid mites of Khazakhstan (with revision of the family). *Trudy Naucno-Issledovatelskogo Instituta Zashiti Rastenij* **1960**, *5*, 231–248.

animals

Article

Mites Living in the Nests of the White Stork and Black Stork in Microhabitats of the Forest Environment and Agrocenoses

Radomir Graczyk [1,*], Piotr Indykiewicz [1], Adam Olszewski [2] and Marcin Tobółka [3]

1 Department of Biology and Animal Environment, Faculty of Animal Breeding and Biology, Bydgoszcz University of Science and Technology, Mazowiecka 28 Str., 85-084 Bydgoszcz, Poland; piotr.indykiewicz@pbs.edu.pl
2 Kampinos National Park, Tetmajera 38 Str., 05-080 Izabelin, Poland; aolszewski@kampinoski-pn.gov.pl
3 Department of Zoology, Poznań University of Life Sciences, Wojska Polskiego 71c Str., 60-625 Poznan, Poland; tobolkamarcin@gmail.com
* Correspondence: graczyk@pbs.edu.pl; Tel.: +48-52-374-9383

Simple Summary: Mites are one of the most diverse groups of invertebrates that inhabit a wide range of environments. The acarofauna, and in particular Oribatida, inhabiting the nests of the White Stork and the Black Stork has not been thoroughly explored so far. The material collected from White and Black Stork nests in Poland was analyzed. This study presents original data on species diversity, abundance, density, and the age structure of Oribatida mites inhabiting the nests of two stork species that breed in Poland. Of the mites, the most numerous group was Mesostigmata. The average number of Oribatida (80.2 individuals in 500 cm^3) was several times higher in the Black Stork nests than in the White Stork nests. Also, the species diversity of oribatid mites was greater in the Black Stork nests (47 species). The species diversity of oribatid mites was also greater in the Black Stork nests. In addition, we noted the potential importance of White and Black Stork nests for mite dispersion and the evolution of interspecies interactions.

Abstract: The White Stork (*Ciconia ciconia*) and the Black Stork (*Ciconia nigra*) are well-known model organisms for the study of bird migration, as well as the selectivity of nesting sites and the choice of living environment. The former breeds mainly in open areas, while the latter inhabits forest areas. The acarofauna, and in particular Oribatida, inhabiting the nests of these species, has not been thoroughly explored so far. Therefore, we analyzed the material collected from 70 White Stork nests and 34 Black Stork nests in Poland, between Poznań and Rawicz, and in Kampinos National Park. Our research has increased the faunal and ecological knowledge of the mite fauna inhabiting the nests of large migratory bird species. Oribatida constituted 5–12% of the total mites identified in the nests of White and Black Storks. Their average number was several times higher in the Black Stork nests (80.2 individuals in 500 cm^3). Also, the species diversity of moss mites was greater in the Black Stork nests (47 species). In total, the nests of the two stork species were inhabited by 62 moss mite species, with only 22 recorded in both the White and the Black Storks' nests. The most numerous species included *Ramusella clavipectinata*, *R. fasciata*, *Oppiella subpectinata*, *Acrogalumna longipluma*, and *Scheloribates laevigatus*. In addition, we found that juvenile oribatid mites accounted for 0.6% of all the mites in the White Stork nests, with tritonymphs having the largest share, while juveniles in the Black Stork nests comprised 1.4%, of which larvae and protonymphs had the largest share. Our research shows that the nests of large migratory birds provide living space for many mite species. In addition, we noted the potential importance of White and Black Stork nests for mite dispersion and the evolution of interspecies interactions.

Keywords: Oribatida; bird nests; microhabitats; storks

Citation: Graczyk, R.; Indykiewicz, P.; Olszewski, A.; Tobółka, M. Mites Living in the Nests of the White Stork and Black Stork in Microhabitats of the Forest Environment and Agrocenoses. *Animals* **2023**, *13*, 3189. https://doi.org/10.3390/ani13203189

Academic Editor: Jukka Jokimäki

Received: 4 September 2023
Revised: 7 October 2023
Accepted: 8 October 2023
Published: 12 October 2023

1. Introduction

Mites are among the most diverse groups of invertebrates, inhabiting a wide range of environments. Some of them form periodic associations with vertebrates, especially mammals [1–4] and birds [5–8]. According to Proctor and Owens [9], at least 2500 species of mites from 40 families periodically reside on the bodies of birds or their nests. Mites present in the burrows and nests of birds function as free-living predators [10–12], ectoparasites [13–16], or coprophilous or edaphic organisms, thus becoming an accompanying fauna that is associated with the micro-environment of the nest or burrow rather than with the birds themselves [5,17,18]. However, bird nests are unstable microhabitats (merocenoses) characterized by specific food, physicochemical, and microclimatic conditions [5,19]. Depending on the type of nest and the bird species it is used by, nests are inhabited by different groups and species of mites, as evidenced, inter alia, by the results of species composition analyses of mites identified in cup nests of the Barn Swallow (*Hirundo rustica*) [20], Red-backed Shrike (*Lanius collurio*) and Great Grey Shrike (*Lanius excubitor*) [15,21] and ground cup nests of the Wood Warbler (*Phylloscopus sibilatrix*) [22], in natural cavities used by the Red-cockaded Woodpecker (*Leuconotopicus borealis*) [23], in nest boxes occupied by the Saddleback (*Philesturnus carunculatus rufusater*) [24] and Starling (*Sturnus vulgaris*) [16] or in platform nests of the Greater Spotted Eagle (*Clanga clanga*) and White-tailed Sea Eagle (*Haliaeetus albicilla*) [25,26]. It is known, however, that the variability of microhabitat conditions in bird nests is determined, among others, by the shape and size of the nest, the type of building material [27,28], the duration of nest occupancy [25], and the setting/location of the nest [13]. These, in turn, significantly affect the composition and abundance of the mite fauna. For example, studies of platform nests built by the White-tailed Eagle have demonstrated that the number of invertebrates present in nests used by birds for many breeding seasons was significantly greater than in nests utilized during one season only [26,29].

The White Stork and the Black Stork build platform nests equal in size to those of eagles. These species breed in Europe in different environments and are characterized by a slightly different biology depending on the breeding season. The areas preferred by the White Stork during the breeding season consist of a mosaic of agrocenoses with a significant proportion of meadows and pastures in river valleys or lake districts and with rural buildings [30,31]. These birds build relatively large platform nests; in Poland, their average diameter is 141 cm (range: 80–230 cm) [32]. The structure of the nest is composed of sticks and branches, usually 3–4 cm thick, arranged in the form of a ring. It is lined with hay, straw, fragments of sod, couch grass, rags, pieces of foil and paper, and sometimes manure [33–35]. Storks use the nests for many years (even more than 100 years) [30], building them up and supplementing them with new material almost throughout the breeding season, which means that a single nest can weigh several hundred kilograms or even more than 1 ton [32]. White Stork nests are usually built on power line poles, roofs of houses, chimneys, and trees [31] (Figure 1). Quite frequently, the empty niches located in the base of White Storks' nests are used as breeding sites by other bird species, e.g., House Sparrows (*Passer domesticus*), Eurasian Tree Sparrows (*Passer montanus*), or Common Starlings [36–38]. The food brought to the nestlings for about 8–9 weeks, whose remains are left in the nest, is usually obtained from grassy meadows, fields, and shallow swamps located a short distance from the nest, and sometimes also from landfills or slaughterhouse waste [39–42]. The White Stork is an opportunistic feeder, having a diet composed of earthworms (*Lumbricidae*), insects (mainly beetles *Coleoptera* and locusts *Orthoptera*), as well as fish, amphibians, and small mammals (mainly voles *Microtus* sp.) [34,43–45].

Figure 1. Black Stork nests located in the branches of old trees (top row, fot. Adam Olszewski) and White Stork nests located on a power pole, building roof, and chimney (fot. Marcin Tobółka).

Unlike the White Stork, the Black Stork is a woodland species that, during the breeding season, prefers large patches of moist deciduous or mixed forests, alder trees, and moist coniferous forests. However, it is also found in fresh coniferous forests and coniferous swamps [46,47]. It prefers nesting areas a short distance from rivers, oxbow lakes, streams, and peat bogs [48,49]. It chooses 100+-year-old trees for nest sites, usually oaks (*Quercus* sp.), pines (*Pinus* sp.), and black alders (*Alnus glutinosa*); occasionally, it places its nest on the tops of wooden towers or on the roofs of hunting pulpits [50–53]. Black Storks can have more than one nest in their breeding area, in which case they change them every few years. They place their nests at a height between 3 and 25 m, but in almost half of the cases, no higher than 15 m above the ground [51,54,55] (Figure 1). Most often, the Black Stork nests in trees with crowns large enough to keep the nest away from the main trunk and, at the same time, in the lower part of the tree crown to ensure good access. The nests are built of branches and sticks, as a rule, no thicker than 3 cm, and the lining consists of dry grass, moss, sod, animal hair, leaves, soil, and clay [54–56]. In common with the White Stork, the Black Stork uses its nest for several decades and, in each breeding season, expands it by adding another layer of branches and lining, as a result of which the nest ranges 49–115 cm in diameter, has a height of up to 1.55 m, and may weigh more than 1 ton [57].

Unlike the White Stork, the food brought to the Black Stork's nestlings is not very varied. For the first 7–9 weeks of their lives, nestlings are fed almost exclusively fish and amphibians, with only a marginal proportion of invertebrates in their diet. Fish account for up to 65% of prey items and more than 85% of the total weight of prey [58,59].

It might be worth mentioning that the micro-environmental conditions in the nests of the two stork species are subject to significant periodic changes. This is because storks are migratory birds, and each year, they use their nests only during the breeding season, i.e., usually from the end of March or April until July or August, and sometimes even until September. During that period, the microclimate and nutritional conditions created by the adult birds incubating their eggs and later by the nestlings (food remains, fragments of feathers, feces, soil, and plants) are far more favorable for the mite fauna than in the autumn and winter periods, when the nest remains empty and the weather conditions are much more severe.

Bird nests as microarthropod habitats have long been of interest to many researchers [2–4,27,60–63]. Until now, most of these studies, including those concerning stork nests, have focused on Mesostigmata mites [35,64,65]. More recently, however, more and more attention has been given to Oribatida mites, both those inhabiting migratory bird nests [7,28,66–69] and those found in the feathers of these birds [70–72]. That latter aspect is important because it concerns the hitherto insufficiently explored role of birds in carrying microarthropods over long distances, e.g., from wintering to breeding grounds, and thus the role of birds in increasing the diversity of mites in northern latitudes and expanding their ranges [27,70].

The present study was conducted to compare the species diversity, abundance, and density, as well as the age structure of Oribatida mites inhabiting the nests of White and Black Storks that breed in different environments, i.e., in agrocenoses and forest communities. In addition, we want to verify the hypothesis that the species composition of mites in the nests of the two species of storks is significantly different due to the fact that the Black Stork and the White Stork enter reproduction in different environments, i.e., in forest communities and agrocenoses (different building materials and food are available). Our research was designed to verify the hypothesis that stork nests provide optimal microenvironmental conditions for the development of Oribatida juveniles. As our research is limited (spatially and numerically), we want to indicate, based on the factual data collected, the direction of future research on Oribatida, including the revision of species found in the national populations of the White and Black Stork.

2. Materials and Methods

The material for the study was collected from 70 White Stork nests and 34 Black Stork nests between 6 May and 2 July 2015 as part of an annual nest in central Poland along a north-south transect between Poznań and Rawicz (51°59′59″ N, 16°52′20″ E) (hereinafter referred to as "Poznań") and within the boundaries of Kampinos National Park (52°19′1″ N, 20°34′1″ E) (hereinafter referred to as "KPN") (Figure 2).

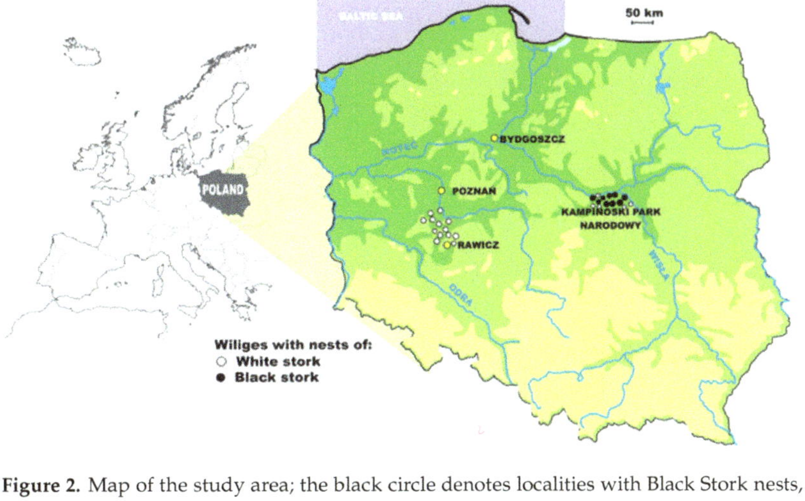

Figure 2. Map of the study area; the black circle denotes localities with Black Stork nests, and the white circle denotes localities with White Stork nests; shades of green indicate terrain.

The samples with mites, each with a volume of 500 cm^3, were obtained from the central part of the nest, from the upper layer of the lining (from a depth of no more than 7 cm), and contained raw organic matter (plant fragments, branches, leaves, feces, etc.) [35,64,73–75]. The samples were taken by hand, without mechanical instruments, and then subjected to the extraction process in the Tullgren funnel for 14 days. The Tullgren funnels have glass funnels, each with a diameter of 12 cm. The heating source is 250-watt, 1.0-m-long heaters, two heaters for eighteen stations, and has adjustable height relative to the funnels. Alcohol vials, as a preservative, into which the mites fall, are cooled in the housing and closed; there is no exchange with the temperature of the room. Baskets are composed of plastic and have a height of 7cm.

The extracted mites were preserved in 90% ethanol. The adult and juvenile stages of Oribatida were identified with accuracy to species or genus [76–89], while the remaining mites were identified to order [90]. The mites were characterized using the parameters of abundance (*A*, in individuals in 500 cm^3), the Shannon index (*H'*), and the Jaccard index [91–94]. In Section 3, Results, the name White Stork is replaced by the abbreviation WS and Black Stork by the abbreviation BS. Functional groups of Oribatida are given after Weigmann [95], Schatz [96], Bernini et al. [97], Domes-Wehner [98], Fischer et al. [99,100], Weigmann and Schatz [101], and Schatz and Fischer [102].

The basic statistical descriptors included the mean values and standard deviation. Normality of the distribution was tested with the W Shapiro–Wilka test, while the equality of variance in different samples, with the Levene test. To find significant differences between the means, the analysis of variance was conducted [103,104]. The level of significance for all statistical tests was accepted at $\alpha = 0.05$. The above calculations were carried out with MS Excel 2019 software (Microsoft, Redmond, WA, USA, 2019) and STATISTICA 13.1 (Dell, Round Rock, TX, USA, 2022) software.

3. Results

Based on the research and analysis conducted, it was established that of the 71.72 thou. individuals of mites identified in the nests of White (WS) and Black (BS) Storks, a significantly greater number was found in the nests of the former (respectively: WS—49.55 thou. mites, BS—22.18 thou. mites) (Table 1).

Table 1. Density [individuals in 500 cm^3 ± SD (standard deviation)] of mites in the nests of the White Stork and the nests of the Black Stork.

Group	White Stork				Black Stork					ANOVA	
	Mean	SD	Total	%	Mean	SD	Total	%	Total Number of Individuals	*F*	*p*
Oribatida A [1]	17.1	27.3	2387	4.8	71.4	167.0	2429	11.0	4816	13.59	<0.001
Oribatida L	0.1	0.4	8	0.02	3.4	9.0	114	0.5	122	19.06	<0.001
Oribatida PN	0.1	0.5	12	0.02	3.8	14.1	128	0.6	140	9.63	0.002
Oribatida DN	0.3	1.0	46	0.1	0.8	3.6	27	0.1	73	1.72	0.191
Oribatida TN	1.7	4.6	239	0.5	0.9	2.3	31	0.1	270	0.97	0.327
Oribatida Juv	2.2	5.5	305	0.6	8.8	24.4	299	1.3	604	8.68	0.004
Oribatida Tot	19.2	29.9	2692	5.4	80.2	182.6	2728	12.3	5420	14.29	<0.001
Mesostigmata	184.6	286.7	25,850	52.2	299.7	447.4	10,191	46.0	36,041	3.46	0.065
Other	150.0	361.6	21,003	42.4	272.2	783.6	9256	41.7	30,259	1.83	0.178
Acari	353.9	483.2	49,545	100	652.2	933.5	22,175	100	71,720	6.84	0.010

[1] A—adults, L—larvae, PN—protonymphs, DN—deutonymphs, TN—tritonymphs, Tot—totally.

The most numerous group of mites inhabiting the stork nests were Mesostigmata, with a similar share in the total population of Acari in both cases (WS—52% and BS—46%). Although there were 2.5 times more Mesostigmata individuals found in the nests of the

White Stork compared with the nests of the Black Stork, their proportion to Oribatida was different; specifically, in the Black Stork nests, the proportion of Oribatida relative to Mesostigmata was 12% to 46%, and in the White Stork nests it was 5% to 52% (Table 1).

Apart from soil mites, also present in significant numbers were groups of ectoparasitic mites (in both stork species—42% each), pest mites (WS—16%, BS—10%), and *Dermanyssus* (storage) mites.

In addition, it was found that, among the identified mites, the proportion of Oribatida in the entire population ranged from about 5% (in WS nests) to more than 12% (in BS nests). It should be mentioned, however, that the nests of both bird species were inhabited by a similar number of Oribatida (Table 1).

It was also established that the proportion of juvenile Oribatida forms was 11% in the nests of both stork species. In the nests of White Storks, the predominant juvenile Oribatida forms were tritonymphs and deutonymphs (78% and 15%, respectively). In contrast, in the case of Black Stork nests, larvae and protonymphs were the most numerous (38% and 43%, respectively) (Table 1).

In the nests of both stork species, 62 Oribatida species were found, including 22 common species and a relatively large number of exclusive species. In the case of White Stork nests, there were 15 (40%), and in Black Stork nests, there were 25 such species (53%) (Table 2). In addition, in 16 species (26%) of all the identified Oribatida, both adult individuals and juvenile forms were found to be present. Jaccard's similarity for Oribatida adults identified in White and Black Stork nests equals 47.4%, and for Oribatida juveniles equals 11%.

Table 2. Number of species (*S*) of Oribatida and Shannon index (*H'*) in the nests of the White Stork and the nests of the Black Stork.

	White Stork		Black Stork
Total number of species		62	
S	37 (59.7%)		47 (75.8%)
Common species		22 (35.5%)	
Exclusive species	15 (40.5%)		25 (53.2%)
Number of species with juveniles	10 (1 [1])		11 (3 [1])
H'	2.465		1.952

[1] the numbers of exclusive species.

Most of the Oribatida identified in the nests of both stork species were eurytopic species that prefer grassland habitats, although there were also species typical of woodland and arboreal communities (Table S1). Nearly half of the Oribatida species found belonged to the panphytophage group (29 species, 47.5%). Other groups represented were microphytophages (15 species, 24.6%), macrophytophages (8 species, 13.1%), necrophages (2 species, 3.3%), and coprophages 1 (1.6%) (Table S1).

Furthermore, analyses revealed that the following species were among the most abundant in the White Stork nests: *Scheloribates laevigatus, Ramusella fasciata, Punctoribates punctum, Tectocepheus velatus, Oribatula exilis,* and *Liebstadia similis* (Table 3). It might be worth mentioning that all the above species were also found in the nests of other stork species. However, the most numerous species in the Black Stork nests were *Ramussela clavipectinata, Oppiella subpectinata,* and *Acrogalumna longipluma,* which were also species found exclusively in the Black Stork nests (Table 3).

It is noteworthy that, of the mite species found in the nests of the two stork species, three were represented only by juvenile forms. These were *P. peltifer* that were found in the nests of both stork species, *N. silvestris* (exclusively in the White Stork nests), and *A. longipluma* (exclusively in the Black Stork nests) (Table 4).

Table 3. Average density [individuals in 500 cm^3 ± SD (standard deviation)] and total number of individuals species of Oribatida in the nests of the White Stork and the nests of the Black Stork.

Taxon	White Stork			Black Stork		
	Mean	SD	Total	Mean	SD	Total
Scheloribates laevigatus (C. L. Koch, 1835)	3.8	11.6	531	1.7	3.0	59
Ramusella fasciata (Paoli, 1908)	3.2	14.8	447	0.4	2.6	15
Punctoribates punctum (C.L. Koch, 1839)	2.9	7.9	412	1.4	4.0	49
Tectocepheus velatus (Michael, 1880)	1.3	2.3	180	0.9	2.2	31
Oribatula exilis (Nicolet, 1855)	1.1	3.0	148	2.8	10.0	94
Liebstadia similis (Michael 1888)	0.9	2.5	127	1.0	3.6	33
Oppia denticulata (Canestrini, 1882)	0.9	3.1	120	1.6	8.8	55
Oribatula pannonica (Willmann, 1949)	0.9	6.6	120	nf [1]	nf	nf
Trichoribates trimaculatus (C. L. Koch, 1835)	0.8	1.5	107	0.4	1.8	14
Eupelops occultus (C. L. Koch, 1835)	0.7	1.7	95	0.1	0.5	5
Galumna obvia (Berlese, 1915)	0.5	2.3	71	0.1	0.3	2
Achipteria nitens (Nicolet, 1855)	0.5	3.5	70	0.6	2.2	21
Achipteria coleoptrata (Linné, 1758)	0.3	0.7	40	2.2	6.3	74
Platynothrus peltifer (C.L. Koch, 1839)	0.3	1.2	36	0.8	2.3	27
Scheloribates palidulus (C.L. Koch, 1841)	0.2	0.7	33	nf	nf	nf
Tectoribates ornatus (Schuster, 1958)	0.2	0.9	24	nf	nf	nf
Trichoribates incisellus (Kramer, 1897)	0.1	0.6	19	nf	nf	nf
Pergalumna nervosa (Berlese, 1914)	0.1	0.7	13	1.1	4.0	39
Chamobates cuspidatus (Michael, 1884)	0.1	0.5	13	0.3	1.1	10
Neoribates aurantiacus (Oudemans, 1914)	0.1	0.5	12	nf	nf	nf
Diapterobates humeralis (Hermann, 1804)	0.1	0.3	10	nf	nf	nf
Eupelops subuliger (Berlese, 1916)	0.1	0.4	10	nf	nf	nf
Oppiella nova (Oudemans, 1902)	<0.1	0.4	5	2.2	9.8	75
Eniochtchonius minutissimus (Berlese, 1903)	<0.1	0.3	5	nf	nf	nf
Ceratozetes gracillis (Michael, 1884)	<0.1	0.3	5	nf	nf	nf
Carabodes labyrinthicus (Michael, 1879)	<0.1	0.2	4	0.6	1.6	21
Liacarus coracinus (C.L. Koch, 1841)	<0.1	0.2	4	0.1	0.3	2
Spatiodamaeus verticilipes (Nicolet, 1855)	<0.1	0.2	4	0.1	0.3	2
Eupelops plicatus (C.L. Koch, 1836)	<0.1	0.2	4	nf	nf	nf
Nothrus silvestris (Nicolet, 1855)	<0.1	0.3	4	nf	nf	nf
Minutozetes pseudofusiger (Schweizer, 1922)	<0.1	0.2	3	0.2	0.9	8
Phthiracarus sp. (Perty, 1841)	<0.1	0.2	3	<0.1	0.2	1
Punctoribates hexagonus (Berlese, 1908)	<0.1	0.1	3	nf	nf	nf
Ramusella furcata (Willmann, 1928)	<0.1	0.1	3	nf	nf	nf
Peloptulus phenotu (C. L. Koch, 1844)	<0.1	0.2	3	nf	nf	nf
Nanhermannia nana (Nicolet, 1855)	<0.1	0.2	2	0.2	0.6	6
Adoristes ovatus (C.L. Koch, 1839)	<0.1	0.2	2	nf	nf	nf
Ramusella calvipectinata (Michael, 1885)	nf	nf	nf	38.5	110.5	1308
Oppiella subpectinata (Oudemans, 1900)	nf	nf	nf	9.6	56.3	328
Acrogalumna longipluma (Berlese, 1904)	nf	nf	nf	8.4	29.1	287

Table 3. *Cont.*

Taxon	White Stork			Black Stork		
	Mean	SD	Total	Mean	SD	Total
Oribella pectinata (Michael, 1885)	nf	nf	nf	1.8	8.6	61
Suctobelbella subtrigona (Oudemans, 1916)	nf	nf	nf	0.4	2.2	13
Suctobelbella sarekensis (Forsslund, 1941)	nf	nf	nf	0.4	1.6	13
Autogneta longilamellata (Michael, 1885)	nf	nf	nf	0.2	1.0	8
Phthiracarus italicus (Oudemans, 1906)	nf	nf	nf	0.2	1.4	8
Scheloribates initialis (Berlese, 1908)	nf	nf	nf	0.2	1.0	8
Hypochthonius rufulus (C.L. Koch, 1835)	nf	nf	nf	0.2	0.9	7
Liebstadia humerata (Sellnick, 1928)	nf	nf	nf	0.2	0.8	7
Steganacarus carinatus (C.L. Koch, 1841)	nf	nf	nf	0.2	0.5	6
Subiasella quadrimaculata (Evans, 1952)	nf	nf	nf	0.2	0.6	6
Phauloppia rauschenensis (Sellnick, 1908)	nf	nf	nf	0.1	0.7	4
Microppia minus (Paoli, 1908)	nf	nf	nf	0.1	0.4	4
Carabodes willmani (Bernini, 1975)	nf	nf	nf	0.1	0.5	3
Licneremaeus licnophorus (Michael, 1882)	nf	nf	nf	0.1	0.5	3
Licnodamaeus pulcherimus (Paoli, 1908)	nf	nf	nf	0.1	0.3	2
Eueremaeus oblongus (C.L. Koch, 1835)	nf	nf	nf	0.1	0.2	2
Oribatella reticulata (Berlese, 1916)	nf	nf	nf	0.1	0.2	2
Carabodes ornatus (Štorkán, 1925)	nf	nf	nf	0.1	0.3	2
Furcoribula furcillata (Nordenskiöld, 1901)	nf	nf	nf	<0.1	0.2	1
Metabelba pulverosa (Strenzke, 1953)	nf	nf	nf	<0.1	0.2	1
Zetorchestes falzonii (Coggi, 1898)	nf	nf	nf	<0.1	0.2	1
Fuscozetes fuscipes (C. L. Koch, 1844)	nf	nf	nf	<0.1	0.2	1

[1] nf—not found.

Table 4. Age structure [average density of individuals in 500 cm^3 ± SD (standard deviation) and total number of individuals] of Oribatida species with identified juveniles in the nests of the White Stork and the nests of the Black Stork.

Taxon	Symbol [1]	White Stork			Black Stork		
		Mean	SD	Total	Mean	SD	Total
Scheloribates laevigatus	Juv	0.7	2.5	97	nf [2]	nf	nf
	Tot	3.8	11.6	531	1.7	3.0	59
Punctoribates punctum	Juv	0.3	1.2	44	nf	nf	nf
	Tot	2.9	7.9	412	1.4	4.0	49
Platynothrus peltifer	Juv	0.3	1.2	36	0.5	1.9	17
	Tot	0.3	1.2	36	0.8	2.3	27
Liebstadia similis	Juv	0.2	1.1	30	0.2	1.0	6
	Tot	0.9	2.5	127	1.0	3.6	33
Trichoribates trimaculatus	Juv	0.2	0.7	29	0.2	1.0	6
	Tot	0.8	1.5	107	0.4	1.8	14

Table 4. *Cont.*

Taxon	Symbol [1]	White Stork			Black Stork		
		Mean	SD	Total	Mean	SD	Total
Galumna obvia	Juv	0.2	1.0	21	nf	nf	nf
	Tot	0.5	2.3	71	0.1	0.3	2
Eupelops occultus	Juv	0.2	0.6	21	nf	nf	nf
	Tot	0.7	1.7	95	0.1	0.5	5
Oribatula exilis	Juv	0.1	0.8	16	1.0	5.0	35
	Tot	1.1	3.0	148	2.8	10.0	94
Tectocepheus velatus	Juv	0.1	0.5	7	0.2	0.6	6
	Tot	1.3	2.3	180	0.9	2.2	31
Nothrus silvestris	Juv	<0.1	0.3	4	nf	nf	nf
	Tot	<0.1	0.3	4	nf	nf	nf
Acrogalumna longipluma	Juv	nf	nf	nf	5.9	22.1	200
	Tot	nf	nf	nf	8.4	29.1	287
Achipteria coleoptrata	Juv	nf	nf	nf	0.4	1.6	12
	Tot	0.3	0.7	40	2.2	6.3	74
Pergalumna nervosa	Juv	nf	nf	nf	0.4	1.5	12
	Tot	0.1	0.7	13	1.1	4.0	39
Hypochthonius rufulus	Juv	nf	nf	nf	0.1	0.4	3
	Tot	nf	nf	nf	0.2	0.9	7
Chamobates cuspidatus	Juv	nf	nf	nf	<0.1	0.2	1
	Tot	0.1	0.5	13	0.3	1.1	10
Eueremaeus oblongus	Juv	nf	nf	nf	<0.1	0.2	1
	Tot	nf	nf	nf	0.1	0.2	2

[1] Juv—juveniles, Tot—totally, [2] nf—not found.

4. Discussion

In this study, we present for the first time some original data on the mites of the Oribatida group inhabiting the nests of two stork species during the breeding season. We show here not only the species diversity and abundance of these mites but also the age structure with the respective proportions of the individual juvenile stages. Of the 62 species we found, as many as 16 (26%) species were represented by juvenile forms. One of the reasons why this is important is that the presence of juvenile forms of oribatid mites can determine the development and survival of predatory species of Mesostigmata. Another reason is that, because of their more abundant intestinal microflora, juveniles show higher metabolic activity in the decomposition of organic matter than adults [105–107].

We identified 47 species of Oribatida in the nests of the Black Stork, and a similar or greater number of these mite species have been found so far in relatively poor European forest communities and in fertile deciduous forests [75,108–113]. The species diversity of Oribatida in the nests of the White Stork we analyzed was similar in open, moist, and extensively used grasslands [110,111,114,115].

Verifying the hypothesis of environmental influence on species diversity, we found that the greater species diversity discovered in the nests of Black Storks compared with the nests of White Storks may be because Black Stork nests are an integral part of the forest environment since they are set in trees just below the wide crown, and the building and lining material is obtained from the immediate vicinity of the nest. Meanwhile, in the case of the White Stork, nests are set on anthropogenic elements of agrocenoses (buildings, chimneys, poles), which have a natural or direct contact with grassland microhabitats or cultivated fields. As a result, mites have an impediment to vertical migration into the nest.

The majority of the oribatid mites identified in stork nests are eurytopic species, and nearly half of them are representatives of the groups of panphytophages, microphytophages, macrophyphages, necrophages, and coprophages. As is well known, their presence is directly related to the fact that decomposing organic matter of plant and animal origin, together with soil microorganisms and saprotrophic mycelia brought by storks to the nest as building and lining material or food for the nestlings, constitutes a basic diet for the majority of Oribatida [116–120].

Furthermore, the results of our research, particularly the age structure of selected species of Oribatida we have identified, prove that the presence of adult storks and their chicks in nests may alter the living conditions and development of the individual species of mites. Specifically, the presence of juvenile forms in the nests in June may prove the birds' role in the change in seasonal dynamics of the mite population. However, it cannot be ruled out that the age structure of Oribatida observed in stork nests may be a consequence of dramatic climate changes. Nevertheless, verifying each of the above hypotheses would require in-depth research over multiple seasons.

An intriguing problem that needs further research is the response of Oribatida to an increasing carbon, nitrogen, and phosphorus content in their living environment. This change has been reported to cause an increase in the number of Oribatida in forest soil [121,122], and mixed-species leaf litter [123,124]. Therefore, the nestlings' excrement with the remaining undigested food residues present in the nest may be expected to periodically increase the nitrogen and phosphorus content and thus affect the abundance of Oribatida. However, the results of studies carried out in the breeding colonies of Great Cormorants (*Phalacrocorax carbo*) proved that the birds' excrement, which increases the concentration of nitrogen, phosphorus, and organic matter in the soil under the nests, does not cause an increase in the abundance of Oribatida [125]. It may be worth adding here that nests used by White and Black Storks for many breeding seasons, and thus regularly supplemented with organic matter, contained significantly higher numbers of Oribatida than nests used by these birds during a single season only [64,65]. Finally, it might be worth pointing out that although White and Black Storks are migratory species, we found in their nests no live or dead representatives of African mite species that inhabit the wintering grounds of these birds.

When planning future research, it seems appropriate to focus on determining the seasonal dynamics of mites in stork nests. To achieve this goal, it is necessary, among other things, to collect material at least four times during the season, i.e., before the birds return from the wintering grounds to their nests (in the second half of March), during overbuilding, replenishment of nesting material, and laying of eggs (May), during the rearing of chicks (June), and after the birds leave the nests (August). In addition, it would be necessary to take into account the size, mass, and structure of the nest, determining the microclimate and thus affecting the diversity and abundance of mites.

5. Conclusions

This study presents original data on species diversity, abundance and density, as well as on the age structure of Oribatida mites inhabiting the nests of two stork species that breed in Poland.

The species diversity of Oribatida identified in the nests of both stork species was considered to be average compared to that found in forest communities and agrocenoses. Most of these are eurytopic species typical of the above environments, representing the groups of panphytophages, microphytophages, macrophytophages, necrophages and coprophages.

Scheloribates laevigatus, *Ramusella fasciata*, *Punctoribates punctum*, *Tectocepheus velatus*, *Oribatula exilis* and *Liebstadia similis*, were found to be most numerous in the white stork nests, while the most abundant species in the black stork nests included *Ramusella clavipectinata*, Oppiella subpectinata and *Acrogalumna longipluma*.

Of all the Oribatida species, only three were represented exclusively by juvenile forms: *Nothrus silvestris* and *Platynothrus peltifer* in white stork nests, and (also) *P. peltifer* and *Acrogalumna longipluma* in black stork nests.

Supplementary Materials: The following supporting information can be downloaded at: https://www.mdpi.com/article/10.3390/ani13203189/s1. Table S1: List of *Oribatida taxons* and their preferences, found in the nests of the White Stork and the nests of the Black Stork.

Author Contributions: Conceptualization R.G. and P.I.; methodology R.G. and P.I.; formal analysis R.G. and P.I.; sampling A.O. and M.T.; writing—original draft preparation R.G. and P.I.; writing—review and editing R.G. and P.I. All authors have read and agreed to the published version of the manuscript.

Funding: This research received no external funding.

Institutional Review Board Statement: The animal study protocol was approved by the Institutional Review Board of Regional Director for Environmental Conservation in Poznań (Permit No: WPN-II.6401.167.2015.AS.2) and the Local Ethical Commission in Poznań (43/2010, 44/2015).

Informed Consent Statement: Not applicable.

Data Availability Statement: Data sharing not applicable.

Acknowledgments: We thank the editor and anonymous reviewers for constructive and helpful suggestions that considerably improved the scientific value of this paper.

Conflicts of Interest: The authors declare no conflict of interest.

References

1. Mašan, P.; Stanko, M. Mesostigmatic mites (Acari) and fleas (Siphonaptera) associated with nests of mound-building mouse, Mus spicilegus Petenyi, 1882 (Mammalia, Rodentia). *Acta Parasitol.* **2005**, *50*, 228–234.
2. Napierała, A.; Mądra, A.; Leszczyńska-Deja, K.; Gwazdowicz, D.J.; Gołdyn, B.; Błoszyk, J. Community structure variability of Uropodina mites (Acari: Mesostigmata) in nests of the common mole, Talpa europaea, in Central Europe. *Exp. Appl. Acarol.* **2016**, *68*, 429–440. [CrossRef] [PubMed]
3. Kaminskienė, E.; Radzijevskaja, J.; Balčiauskas, L.; Gedminas, V.; Paulauskas, A. Laelapidae mites (Acari: Mesostigmata) infesting small rodents in the Curonian Spit Lithuania. *Biologija* **2017**, *63*, 169–176. [CrossRef]
4. Celebias, P.; Melke, A.; Gwiazdowicz, D.J.; Przewoźny, M.; Komosiński, K.; Baraniak, E.; Winnicka, K.; Melosik, I.; Ziomek, J. Species composition, diversity, and the abundance of arthropods inhabiting burrows of the common hamster (*Cricetus cricetus* L.). *Bull. Entomol. Res.* **2019**, *109*, 781–793. [CrossRef] [PubMed]
5. Błoszyk, J.; Bajerlein, D.; Gwiazdowicz, D.; Halliday, R.; Dylewska, M. Uropodine mite communities (Acari: Mesostigmata) in birds' nests in Poland. *Belgian J. Zool.* **2006**, *136*, 145–153.
6. Davidova, R.D.; Vasilev, V.M. Mite Fauna (Acari: Parasitiformes) in Nests of Eurasian Blue Tit Cyanistes caeruleus (Linnaeus, 1758) (Passeriformes: Paridae) and a Comparison with Two Other Passerine Bird Species. *Acta Zool. Bul.* **2020**, *72*, 217–224.
7. Napierała, A.; Maziarz, M.; Hebda, G.; Broughton, R.K.; Rutkowski, T.; Zacharyasiewicz, M.; Błoszyk, J. Lack of specialist nidicoles as a characteristic of mite assemblages inhabiting nests of the ground-nesting wood warbler, Phylloscopus sibilatrix (Aves: Passeriformes). *Exp. Appl. Acarol.* **2021**, *84*, 149–170. [CrossRef]
8. Malekhina, E.N.; Korolev, A.N.; Selivanova, N.P. Oribatid Mites (Oribatida) Associated with Nests of Hollow-Nesting Birds, on the Example of a Model Species, the European Pied Flycatcher (*Ficedula hypoleuca*), in the Taiga Forests of the European North-East of Russia. *Diversity* **2023**, *15*, 765. [CrossRef]
9. Proctor, H.; Owens, I. Mites and birds: Diversity, parasitism and coevolution. *Trends Ecol. Evol.* **2000**, *15*, 358–364. [CrossRef]
10. Fain, A.; Vangeluwe, D.; Defreef, M.; Wauthy, G. Observations on mites inhabiting nests of *Bubo bubo* (L.) (Strigiformes, Strigidae) in Belgium. *Belg. J. Zool.* **1993**, *123*, 3–26.
11. Solarz, K.; Szilman, P.; Szilman, E.; Krzak, M.; Jagla, A. Some allergenic species of astigmatid mite (Acari, Acaridida) from different synanthropic environments in southern Poland. *Acta Zool. Cracov.* **2004**, *47*, 125–145. [CrossRef]
12. Ardeshir, F. A preliminary study on mite fauna of bird nests in Iran. *Podoces* **2010**, *5*, 112–115.
13. Ambros, M.; Krištofik, J.; Šustek, Z. The mites (Acari, Mesostigmata) in the birds' nests in Slovakia. *Biologia* **1992**, *47*, 369–381.
14. Philips, J.R. A review and checklist of the parasitic mites (Acarina) of the Falconiformes and Strigiformes. *J. Raptor Res.* **2000**, *34*, 210–231.
15. Mašan, P.; Fend'a, P.; Krištofik, J.; Halliday, B. A review of the ectoparasitic mites (Acari: Dermanyssoidea) associated with birds and their nests in Slovakia, with notes on identification of some species. *Zootaxa* **2014**, *3893*, 77–100. [CrossRef]
16. Błoszyk, J.; Gwiazdowicz, D.J.; Kupczyk, M.; Książkiewicz-Parulska, Z. Parasitic mesostigmatid mites (Acari)—Common inhabitants of the nest boxes of starlings (Sturnus vulgaris) in a Polish urban habitat. *Biologia* **2016**, *71*, 1034–1037. [CrossRef]

17. Fendá, P.; Schniererová, E. Mites (Acarina: Mesostigmata) in the nests of *Acrocephalus* spp. and inneighbouring reeds. *Biologia* **2004**, *59* (Suppl. S15), 41–47.
18. Napierała, A.; Konwerski, S.; Gutowski, J.M.; Błoszyk, J. Species diversity of Uropodina communities (Acari: Parasitiformes) in soil and selected microhabitats in the Białowieża Primeval Forest. In *Mites (Acari) of the Białowieża Primeval Forest*; Błoszyk, J., Napierała, A., Eds.; Kontekst: Poznań, Poland, 2020; pp. 11–60.
19. Castaño-Vázquez, F.; Merino, S. Differential effects of environmental climatic variables on parasite abundances in blue tit nests during a decade. *Integr. Zool.* **2022**, *17*, 511–529. [CrossRef] [PubMed]
20. Møller, A.P. Effects of parasitism by a haematophagous mite on reproduction in the Barn Swallow. *Ecology* **1990**, *71*, 2345–2357. [CrossRef]
21. Komczyc, J.; Teodorowicz, E.; Gwiazdowicz, D.J. Mites (Acari, Mesostigmata) in the red-backed shrike *Lanius collurio* and great grey shrike *Lanius excubitor* nests. *Acta Sci. Pol. Silv.* **2011**, *10*, 37–42.
22. Laska, A.; Puchalska, E.; Mikołajczyk, M.; Gwiazdowicz, D.J.; Kaźmierski, A. Mites inhabiting nests of wood warbler, Phylloscopus sibilatrix (Aves: Passeriformes), in the Wielkopolska National Park in western Poland. *Exp. Appl. Acar.* **2023**, *89*, 393–416. [CrossRef] [PubMed]
23. Pung, O.J.; Carlile, L.D.; Whitlock, J.; Vives, S.P.; Durden, L.A.; Spadgenske, E. Survey and host fitness effects of red-cockaded woodpecker blood parasites and nest cavity arthropods. *J. Parasitol.* **2000**, *86*, 506–510. [CrossRef]
24. Stamp, R.K.; Brunton, D.H.; Walter, B. Artificial nest box use by the North Island Saddleback: Effects of nest box design and mite infestations on nest site selection and reproductive success. *N. Z. J. Zool.* **2002**, *29*, 285–292. [CrossRef]
25. Gwiazdowicz, D.J.; Mizera, T.; Skorupski, M. Mites in greater spotted eagle nests. *J. Raptor. Res.* **1999**, *33*, 257–260.
26. Gwiazdowicz, D.J.; Błoszyk, J.; Mizera, T.; Tryjanowski, P. Mesostigmatic mites (Acari: Mesostigmata) in white-tailed sea eagle nests (*Haliaeetus albicilla*). *J. Raptor. Res.* **2005**, *39*, 60–65.
27. Pilskog, H.E.; Solhoy, T.; Gwiazdowicz, D.J.; Grytnes, J.A.; Coulson, S.J. Invertebrate communities inhabiting nests of migrating passerine, wild fowl and sea birds breeding in the High Arctic, Svalbard. *Polar Biol.* **2014**, *37*, 981–998. [CrossRef]
28. Gwiazdowicz, D.J.; Niedbała, W.; Skarżyński, D.; Zawieja, B. Occurrence of mites (Acari) and springtails (Collembola) in bird nests on King George Island (South Shetland Islands, Antarctica). *Polar Biol.* **2022**, *45*, 1035–1044. [CrossRef]
29. Gwiazdowicz, D.J.; Błoszyk, J.; Bajerlein, D.; Halliday, R.B.; Mizera, T. Mites (Acari: Mesostigmata) inhabiting nests of the while-tailed sea eagle *Haliaeetus albicilla* (L.) in Poland. *Entomol. Fennica* **2006**, *8*, 366–372. [CrossRef]
30. Indykiewicz, P. *White Stork Ciconia ciconia (L.)*; LOGO: Bydgoszcz, Poland, 2004; p. 96.
31. Guziak, R.; Jakubiec, Z. White stork *Ciconia ciconia* (L.) in Poland in 2004. In *Results of the VIth International White Stork Census*; PTPP Pro Natura: Wrocław, Poland, 2006; p. 432.
32. Zbyryt, A.; Dylewski, Ł.; Neubauer, G. Mass of white stork nests predicted from their size: Online calculator and implications for conservation. *J. Nat. Conservation* **2021**, *60*, 125967. [CrossRef]
33. Creutz, G. *Der Weissstorch Ciconia ciconia*; NBB 375; A. Ziemsen Verlag: Lutherstadt, Germany, 1985; p. 216.
34. Schulz, H. *Ciconia ciconia* White Stork. *BWP Update* **1998**, *2*, 69–105.
35. Bajerlein, D.; Błoszyk, J.; Gwiazdowicz, D.J.; Ptaszyk, J.; Halliday, B. Community structure and dispersal of mites (Acari, Mesostigmata) in nests of the white stork (*Ciconia ciconia*). *Biologia* **2006**, *61*, 525–530. [CrossRef]
36. Indykiewicz, P. House Sparrow Passer domesticus, Starling Sturnus vulgaris, Tree Sparrow Passer montanus and other residents of the White Stork Ciconia ciconia. In *The White Stork in Poland: Studies in Biology, Ecology and Conservation*; Tryjanowski, P., Sparks, T.H., Jerzak, L., Eds.; Bogucki Wydawnictwo Naukowe: Poznań, Poland, 2006; pp. 225–235.
37. Tobółka, M. Roosting of tree sparrow (*Passer montanus*) and house sparrow (*Passer domesticus*) in white stork (*Ciconia ciconia*) nests during winter. *Turk. J. Zool.* **2011**, *35*, 6. [CrossRef]
38. Zbyryt, A.; Jakubas, D.; Tobółka, M. Factors determining presence of passerines breeding within White Stork Ciconia ciconia nests. *Sci. Nat.* **2017**, *104*, 71. [CrossRef] [PubMed]
39. Tortosa, F.S.; Manez, M.; Barcell, M. Wintering white storks (*Ciconia ciconia*) in South West Spain in the years 1991 and 1992. *Die Vogelwarte* **1995**, *38*, 41–45.
40. Tortosa, F.S.; Pérez, L.; Hillström, L. Effect of food abundance on laying date and clutch size in the White Stork *Ciconia ciconia*. *Bird Study* **2003**, *50*, 112–115. [CrossRef]
41. Johst, K.; Brandl, R.; Pfeifer, R. Foraging in a Patchy and Dynamic Landscape: Human Land Use and the White Stork. *Ecol. Appl.* **2001**, *11*, 60–69. [CrossRef]
42. Kruszyk, R.; Ciach, M. White Storks, *Ciconia ciconia*, Forage on Rubbish Dumps in Poland—A Novel Behaviour in Population. *Eur. J. Wild Res.* **2010**, *56*, 83–87. [CrossRef]
43. Cramp, S.; Simmons, K.E.L. *The Birds of the Western Palearctic*; Oxford University Press: Oxford, UK, 1977; Volume 1.
44. Antczak, M.; Konwerski, S.Z.; Grobelny, S.; Tryjanowski, P. The food composition of immature and non-breeding White stork in Poland. *Waterbirds* **2002**, *25*, 424–428. [CrossRef]
45. Kosicki, J.Z.; Profus, P.; Dolata, P.T.; Tobółka, M. Food composition and energy demand of the White Stork *Ciconia ciconia* breeding population. Literature survey and preliminary results from Poland. In *The White Stork in Poland: Studies in Biology, Ecology and Conservation*; Tryjanowski, P., Sparks, T.H., Jerzak, L., Eds.; Bogucki Wydawnictwo Naukowe: Poznań, Poland, 2006.
46. Del Hoyo, J.; Elliot, A.; Sargatal, J. *Handbook of the Birds of the World. Volume 1: Ostrich to Ducks*; Lynx Edicions: Barcelona, Spain, 1992; Volume 1.

47. Snow, D.W.; Perrins, C.M. *The Birds of the Western Palearctic, Volume 1: Non-Passerines*; Oxford University Press: Oxford, UK, 1998.
48. Keller, M.; Profus, P. Present situation, reproduction and food of the Black Stork in Poland. In *Les Cigognes d'Europe, Actes du Colloque International*; Meriaux, J.L., Schierer, A., Tombal, C., Tombal, J.C., Eds.; Institut Europeen d'Ecologie: Metz, France, 1992; pp. 227–236.
49. Profus, P.; Wojciak, J. Bocian czarny Ciconia nigra [Black Stork *Ciconia nigra*]. In *The Atlas of Breeding Birds in Poland 1985–2004*; Sikora, A., Rohde, Z., Gromadzki, M., Neubauer, G., Chylarecki, P., Eds.; Bogucki Wyd. Nauk: Poznań, Poland, 2007; pp. 126–127.
50. Kołudzki, Z.; Wężyk, M.; Kociniak, M.; Zieliński, P. Occurrence of the Black Stork *Ciconia nigra* in the former Piotrków Voivodship in the years 1994–2001. *Chrońmy Przyr. Ojczystą* **2003**, *59*, 5–19.
51. Lõhmus, A.; Sellis, U. Nest trees—A limiting factor for the Black Stork (*Ciconia nigra*) population in Estonia. *Aves* **2003**, *40*, 84–91.
52. Cano Alonso, L.S.; Franco, C.; Pacheco, C.; Reis, S.; Rosa, G.; Fernandez-Garcia, M. The breeding population of black stork Ciconia nigra in the Iberian Peninsula. *Biota* **2006**, *7*, 15–23.
53. Vlachos, C.G.; Bakaloudis, D.E.; Alexandrou, O.G.; Bontzorlos, A.; Papakosta, M.A. Factors affecting the nest site selection of the black stork, Ciconia nigra in the Dadia-Lefkimi-Soufli National Park, north-eastern Greece. *Folia Zool.* **2008**, *57*, 251–257.
54. Hancock, J.; Kushlan, J.A.; Kahl, M.P. *Storks, Ibises and Spoonbills of the World*; Bloomsbury Publishing: London, UK, 2010; pp. 70–74.
55. Pottetz, L.; Barowski, A. *Black Stork—White Shadow*; Pekrul and Sohn GbR, Godern: Pinnow, Germany, 2016; Volume 3.
56. Janssen, G.; Hormann, M.; Rohde, C. *Der Schwarzstorch. Die Neue Brehm-Bücherei Bd. 468*; Westarp Wissenschaften: Hohenwarsleben, Germany, 2004; p. 414.
57. Strazds, M. Longevity of Black Stork (*Ciconia nigra*) nests and nest site protection in Latvia. *Aves* **2003**, *40*, 69–71.
58. Hampl, R.; Bures, S.; Balaz, P.; Bobek, M.; Pojer, F. Food provisioning and nestling diet of the black stork in the Czech Republic. *Waterbirds* **2005**, *28*, 35–40. [CrossRef]
59. Kamiński, M.; Bańbura, J.; Janic, B.; Marszał, L.; Minias, P.; Zieliński, P. Intra-Seasonal and Brood-Size Dependent Variation in the Diet of Black Stork (*Ciconia nigra*) Nestlings. *Waterbirds* **2018**, *41*, 268–275. [CrossRef]
60. Kristofik, J.; Sustek, Z.; Masan, P. Arthropods (Pseudoscorpionida, Acari, Coleoptera, Siphonaptera) in the nests of red-backed shrike (*Lanius collurio*) and lesser grey shrike (*Lanius minor*). *Biologia* **2002**, *57*, 603–613.
61. Kristofik, J.; Masan, P.; Sustek, Z. Arthropods in the nests of marsh warblers (*Acrocephalus palustris*). *Biologia* **2005**, *60*, 171–177.
62. Kaźmierski, A.; Marciniak, M.; Sikora, B. Tydeinae mites (Acariformes: Prostigmata: Tydeidae) from bird nests with description of three new species. *Syst. Appl. Acarol.* **2018**, *23*, 803–823. [CrossRef]
63. Kaźmierski, A.; Laniecka, I.; Laniecki, R. A review of the genus Primotydeus (Acariformes: Tydeoidea: Iolinidae). *Syst. Appl. Acarol.* **2021**, *26*, 2320–2337. [CrossRef]
64. Błoszyk, J.; Gwiazdowicz, D.J.; Bajerlein, D.; Halliday, R.B. Nests of the white stork *Ciconia ciconia* (L.) as a habitat for mesostigmatic mites (Acari, Mesostigmata). *Acta Parasitol.* **2005**, *50*, 171–175.
65. Błoszyk, J.; Gwiazdowicz, D.J.; Halliday, R.d.B.; Dolata, P.T.; Gołdyn, B. Nests of the black stork Ciconia nigra as a habitat for mesostigmatid mites (Acari: Mesostigmata). *Biologia* **2009**, *64*, 962–968. [CrossRef]
66. Shakhab, S.V. Oribatid mites (Oribatei, Acariformes) in nests of passerine birds. *Entomol. Rev.* **2006**, *86*, 173–176. [CrossRef]
67. Coulson, S.J.; Moe, B.; Monson, F.; Gabrielsen, G.W. The invertebrate fauna of High Arctic seabird nests: The microarthropod community inhabiting nests on Spitsbergen, Svalbard. *Polar Biol.* **2009**, *32*, 1041–1046. [CrossRef]
68. Makarova, O.L.; Osadtchy, A.V.; Melnikov, M.V. Gamasid Mites (Parasitiformes, Mesostigmata) in Nests of Passerine Birds on the Arctic Seven Islands Archipelago, the Barents Sea. *Entomol. Rev.* **2010**, *90*, 643–649. [CrossRef]
69. Melekhina, E.N.; Matyukhin, A.V.; Glazov, P.M. Oribatid mites in nests of the Lapland Bunting (Calcarius lapponicus) on the arctic island of Vaygach (with analysis of the islands fauna). *Proc. Karelian Sci. Cent. Russ. Acad. Sci.* **2019**, *8*, 108–122.
70. Lebedeva, N.V.; Krivolutsky, D.A. Birds Spread Soil Microarthropods to Arctic Islands. *Doklady Biol. Sci.* **2003**, *391*, 329–332. [CrossRef]
71. Krivolutsky, D.A.; Lebedeva, N.V. Oribatid mites (Oribatei) in bird feathers. Part 2. Passeriformes. *Acta Zool. Litu.* **2004**, *14*, 19–38. [CrossRef]
72. Krivolutsky, D.A.; Lebedeva, N.V.; Gavrilo, M.V. Soil microartrhropods in the feathers of Antarctic birds. *Doklady Biol. Sci.* **2004**, *397*, 342–345. [CrossRef]
73. Napierała, A.; Błoszyk, J. Unstable microhabitats (merocenoses) as specific habitats of Uropodina mites (Acari: Mesostigmata). *Exp. Appl. Acarol.* **2013**, *60*, 163–180. [CrossRef]
74. Lebedeva, N.; Poltavskaya, M. Oribatid mites (Acari, Oribatida) of plain area of the Southern European Russia. *Zootaxa* **2013**, *3709*, 101–133. [CrossRef] [PubMed]
75. Skubała, S. Microhabitats and oribatid fauna: Comparison of 2 sampling approaches. *Biol. Lett.* **2016**, *53*, 31–47. [CrossRef]
76. Seniczak, S. Revision of the family Oppiidae Grandjean 1953 (Acarina, Oribatei). *Acarologia* **1975**, *17*, 331–345.
77. Seniczak, S. The morphology of juvenile stages of soil mites of the family Achipteriidae (Acari: Oribatei), I. *Ann. Zool.* **1978**, *34*, 89–99.
78. Seniczak, S. The morphology of the juvenile stages of moss mites of the family Scheloribatidae Grandjean, 1953 (Acari, Oribatei), I. *Acta Zool. Cracov.* **1980**, *24*, 487–500.
79. Seniczak, S. The morphology of juvenile stages of moss mites of the subfamily Trichoribatinae (Acari: Oribatei), II. *Ann. Zool.* **1980**, *35*, 221–231.

80. Seniczak, S. The morphology of juvenile stages of moss mites of the family Pelopidae Ewing (Acarida, Oribatida), II. *Ann. Zool.* **1988**, *41*, 383–393.
81. Seniczak, S. The morphology of juvenile stages of moss mites of the family Scheloribatidae (Acarida, Oribatida), II. *Ann. Zool.* **1990**, *43*, 299–308.
82. Seniczak, S. The morphology of juvenile stages of moss mites of the family Nanhermanniidae (Acari: Oribatida). I. *Zool. Anz.* **1991**, *227*, 300–319.
83. Seniczak, S. The morphology of juvenile stages of moss mites of the family Trhypochthoniidae (Acari: Oribatida), I. *Zool. Jahr.* **1992**, *229*, 413–423.
84. Seniczak, S. The morphology of juvenile stages of moss mites of the family Nothridae (Acari: Oribatida). I. *Zool. Anz.* **1992**, *229*, 134–148.
85. Seniczak, S. The morphology of juvenile stages of moss mites of the subfamily Trichoribatinae (Acari, Oribatida). IV. *Zool. Anz.* **1993**, *230*, 137–151.
86. Seniczak, S.; Klimek, A. The morphology of juvenile stages of moss mites of the family Camisiidae (Acari, Oribatida). I. *Zool. Anz.* **1990**, *225*, 71–86.
87. Seniczak, S.; Seniczak, A. Morphology of juvenile stages of three species of the genus Punctoribates Berlese, 1908 (Acari: Oibatida: Mycobatidae). *Ann. Zool.* **2008**, *58*, 473–485. [CrossRef]
88. Seniczak, S.; Iturrondobeitia, J.C.; Seniczak, S. The ontogeny of morphological traits in three species of Galumnidae (Acari: Oribatida). *Int. J. Acar.* **2012**, *38*, 612–638. [CrossRef]
89. Ermilov, S.G.; Weigmann, G.; Tolstikov, A.V. Morphology of adult and juvenile instars of Galumna obvia (Acari, Oribatida, Galumnidae), with discussion of its taxonomic status. *ZooKeys* **2013**, *357*, 11–28. [CrossRef] [PubMed]
90. Lindquist, E.E.; Krantz, G.W.; Walter, D.E. Classification. In *A Manual of Acarology*; Krantz, G.W., Walter, D.E., Eds.; TTU Press: Lubbock, TX, USA, 2009; pp. 97–103.
91. Niedbała, W. *Mosses-Mites of Terrestrial Ecosystems*; PWN: Warszawa, Poland, 1980. (In Polish)
92. Odum, E.P. *Basics of Ecology*; PWRiL: Warszawa, Poland, 1982. (In Polish)
93. Boczek, J. *Outline of Agricultural Acarology. A Textbook for Students of Plant Protection*; PWN: Warszawa, Poland, 1999; p. 358. (In Polish)
94. Czachorowski, S. Describing Biocenosis—Zoocenology. Master's Thesis, University of Warmia and Mazury, Olsztyn, Poland, 2006.
95. Weigmann, G. *Hornmilben (Oribatida). Die Tierwelt Deutschlands 76 Teil*; Goecke & Evers: Keltern, Germany, 2006.
96. Schatz, H. *Catalogus Fauna Austriae Teil IXi, U. Ordn.: Oribatei, Hornmilben*; Akademie der Wissenschaften: Vienna, Austria, 1983.
97. Bernini, F.; Avanzati, A.M.; Bernini, S. Notulae Oribatologicae XXXVII. Gli Acari Oribatei del Massiccio del Pollino (Italia Meridionale): Aspetti faunistici e biogeografici. *Biogeogr. J. Integr. Biogeogr.* **1987**, *10*, 379–488. [CrossRef]
98. Domes-Wehner, K. Parthenogenesis and Sexuality in Oribatid Mites, Phylogeny Mitochondrial Genome Structure and Resource Dependence. Ph.D. Thesis, Fachbereich Biologie der Technischen Universität, Darmstadt, Germany, 2009.
99. Fischer, B.M.; Schatz, H.; Maraun, M. Community structure, trophic position and reproductive mode of soil and bark-living oribatid mites in an alpine grassland ecosystem. *Exp. Appl. Acarol.* **2010**, *52*, 221–237. [CrossRef]
100. Fischer, B.M.; Meyer, E.; Maraun, M. Positive correlation of trophic level and proportion of sexual taxa of oribatid mites (Acari: Oribatida) in alpine soil systems. *Exp. Appl. Acarol.* **2014**, *63*, 465–479. [CrossRef] [PubMed]
101. Weigmann, G.; Schatz, H. Redescription of Coronoquadroppia monstruosa (Hammer, 1979) (Acari, Oribatida, Quadroppiidae) from Java and variability of the species in Europe. *Zootaxa* **2015**, *3926*, 329–350. [CrossRef] [PubMed]
102. Schatz, H.; Fischer, B.M. Hornmilben (Acari, Oribatida). In *Tag der Artenvielfalt 2015 in Weissenbach (Gemeinde Sarntal, Südtirol, Italien)*; Schatz, H., Wilhalm, T., Eds.; Gredleriana 2016; Volume 16, pp. 113–132. Available online: https://www.zobodat.at/pdf/Gredleriana_016_0181-0234.pdf (accessed on 3 September 2023).
103. McDonald, J.H. *Handbook of Biological Statistics*, 2nd ed.; Sparky House Publishing: Baltimore, MD, USA, 2009; p. 319.
104. Stanisz, A. Easy Course of Statistic Using Statistica PL and Medicine Examples, 1. In *Basic Statistic*; StatSoft Polska: Kraków, Poland, 2006; p. 532. (In Polish)
105. Seniczak, S. *Juvenile Stages (Acai, Oribatei) as an Important Component of the Groupings of These Mites that Process Soil Organic Matter*; Scripts and Supporting Texts; UMK Toruń: Toruń, Poland, 1978. (In Polish)
106. Seniczak, S.; Stefaniak, O. The microflora of the effect of the alimentary canal of Oppia nitens (Acarina, Oribatei). *Pedobiologia* **1978**, *18*, 110–119. [CrossRef]
107. Siepel, H.; De Ruiter-Dijkman, E.M. Feeding guilds of oribatid mites based on their carbohydrase activities. *Soil Biol. Biochem.* **1993**, *25*, 1491–1497. [CrossRef]
108. Seniczak, S.; Graczyk, R.; Seniczak, A.; Faleńczyk-Koziróg, K.; Kaczmarek, S.; Marquardt, T. Microhabitat preferences of Oribatida and Mesostigmata (Acari) inhabiting lowland beech forest in Poland and the trophic interactions between these mites. *Eur. J. Soil Biol.* **2018**, *87*, 25–32. [CrossRef]
109. Graczyk, R. *Moss Mites (Acari, Oribatida) of Selected Microhabitats of One-Storey Dry Coniferous Forests Cladonio-Pinetum on Forest and Post-Agricultural Soil*; University of Technology and Life Sciences Publishing House: Bydgoszcz, Poland, 2020; p. 204. (In Polish)
110. Olszanowski, Z.; Rajski, A.; Niedbała, W. *Catalog of the Fauna of Poland. Mites (Acari), Mosses (Oribatida)*; Sorus: Poznań, Poland, 1996; Volume 34, pp. 1–243. (In Polish)

111. Norton, R.A.; Behan-Pelletier, V.M. Suborder Oribatida. In *A manual of Acarology*, 3rd ed.; Krantz, G.W., Walter, D.E., Eds.; Texas Tech University Press: Lubbock, TX, USA, 2009; pp. 430–564.
112. Sokołowska, M.; Duras, M.; Skubała, P. *Oribatid Mites Communities (Acari: Oribatida) in Dead Wood of Protected Areas under Strong Anthropogenic Pressure Contributions to Soil Zoology in Central Europe III*; Tajovský, K., Schlaghamerský, J., Pižl, V., Eds.; ISB BC AS CR, v.v.i.: České Budějovice, Czech Republic, 2009; pp. 151–155.
113. Skubała, S.; Sokołowska, M. Oribatid fauna (Acari, Oribatida) in fallen spruce trees in the Babia Góra National Park. *Biol. Lett.* **2006**, *43*, 243–248.
114. Rajski, A. Ecological and faunal study of mosses (Acari, Oribatei) in several plant communities. I. *Ekologia. Prace Kom. Biol. PTPN.* **1961**, *25*, 123–283.
115. Rajski, A. Autecological—Zoogeographical analysis of moss mites (Acari, Oribatei) on the basis of fauna in the Poznań environs. Part II. *Fragm. Faun.* **1968**, *12*, 277–405.
116. Hag, M.A. Role of Oribatid Mites in Soil Ecosystem. In *Ecology and Biology of Soil Organisms*; Agrotech Publishing Academy: Udaipur, India, 1994; pp. 143–177.
117. Labandeira, C.C.; Phillips, T.L.; Norton, R.A. Oribatid mites and the decomposition of plant tissues in Palaeozoic coal swamp forests. *Palaios* **1997**, *12*, 319–353. [CrossRef]
118. Smrž, J. Interactions between oribatids and micro-organisms: A complex method of study. *Appl. Soil Ecol.* **1998**, *9*, 109–110. [CrossRef]
119. Walter, D.E.; Proctor, H.C. *Mites: Ecology, Evolution and Behavior: Life at a Microscale*, 2nd ed.; Springer: Dordrecht, The Netherlands; Heidelberg, Germany; New York, NY, USA; London, UK, 2013.
120. Vu, Q.M. *The Oribatid Mite Fauna (Acari: Oribatida) of Vietnam—Systematics, Zoogeography and Formation*; Pensoft: Sofia, Bulgaria; Moscow, Russia, 2015.
121. Salamon, J.A.; Alphei, J.; Ruf, A.; Schaefer, M.; Scheu, S.; Schneider, K.; Sührig, A.; Maraun, M. Transitory dynamic effects in the soil invertebrate community in a temperate deciduous forest: Effects of resource quality. *Soil Biol. Biochem.* **2006**, *38*, 209–221. [CrossRef]
122. Huhta, V. The role of soil fauna in ecosystems: A historical review. *Pedobiologia* **2007**, *50*, 489–495. [CrossRef]
123. Gartner, T.B.; Cardon, Z.G. Decomposition dynamics in mixed-species leaf litter. *Oikos* **2004**, *104*, 230–246. [CrossRef]
124. Wardle, D.A. The influence of biotic interactions on soil biodiversity. *Ecol. Lett.* **2006**, *9*, 870–886. [CrossRef] [PubMed]
125. Oszust, M.; Klimaszyk, P. Soil conditions under cormorant colonies favor for mites excepting Oribatida. *Acarologia* **2022**, *62*, 974–988. [CrossRef]

animals

Article

Diversity and Distribution of Mites (ACARI) Revealed by Contamination Survey in Public Genomic Databases

Jiazheng Xie * and Yi Zhang

Chongqing Key Laboratory of Big Data for Bio Intelligence, Chongqing University of Posts and Telecommunications, Chongqing 400065, China
* Correspondence: xiejz@cqupt.edu.cn

Simple Summary: Mites are a group of minute animals ubiquitously distributed on the planet. They have close ecological ties with other species, such as plants, insects and vertebrates. With the development of sequencing technology, the genomic data have increased dramatically. Although the contaminations of microbial symbionts in public genomic databases have been explored to reveal the interactions between microbes and hosts, no similar study has been carried out to the microscopic mites. Here, we present a survey and analysis of the contamination of mites in Genbank genomic resources for the first time. The results showed that contamination of mites in public databases is not rare. Based on these contaminated contigs, the host associations and evolution of mites are discussed.

Abstract: Acari (mites and ticks) are a biodiverse group of microarthropods within the Arachnida. Because of their diminutive size, mites are often overlooked. We hypothesized that mites, like other closely related microorganisms, could also contaminate public genomic database. Here, using a strategy based on DNA barcodes previously reported, we scanned contaminations related to mites (Acari, exclusive of Ixodida) in Genbank WGS/TSA database. In 22,114 assemblies (17,845 animal and 4269 plant projects), 1717 contigs in 681 assemblies (3.1%) were detected as mite contaminations. Additional taxonomic analysis showed the following: (1) most of the contaminants (1445/1717) were from the specimens of Magnoliopsida, Insecta and Pinopsida; (2) the contamination rates were higher in plant or TSA projects; (3) mite distribution among different classes of hosts varied considerably. Additional phylogenetic analysis of these contaminated contigs further revealed complicated mite-host associations. Overall, we conducted a first systemic survey and analysis of mite contaminations in public genomic database, and these DNA barcode related mite contigs will provide a valuable resource of information for understanding the diversity and phylogeny of mites.

Keywords: Acari; genomic contamination; diversity; distribution; evolution; DNA barcode

Citation: Xie, J.; Zhang, Y. Diversity and Distribution of Mites (ACARI) Revealed by Contamination Survey in Public Genomic Databases. *Animals* **2023**, *13*, 3172. https://doi.org/10.3390/ani13203172

Academic Editors: Theo de Waal, Maciej Skoracki and Monika Fajfer

Received: 14 August 2023
Revised: 24 September 2023
Accepted: 9 October 2023
Published: 11 October 2023

1. Introduction

Acari (mites and ticks) are a highly speciose group of animals within the Arthropoda [1]. With nearly 55,000 described species and up to one million species awaiting discovery or description [2,3], mites can be found widely across various microhabitats around the world, from terrestrial to aquatic or oceanic environments, and even underground niches. Not surprisingly, their lifestyles are also highly diverse, from detritivorous, phytophagous, pollinivorous, fungivorous and predaceous in nonparasitic members to obligate ectoparasitism [1]. They have also multifaceted roles in ecosystems, such as pests of crops (e.g., spider mites and gall mites), parasites on birds and mammals (e.g., quill mites, scabies mites and follicle mites), vectors capable of transmitting notorious viruses and sources of allergens (e.g., house dust mites) [4]. Meanwhile, some of them can be beneficial to humans as biocontrol agents of pests and weeds. Although of great economic and ecological importance, our knowledge of mites is usually fragmentary which is focused on a particular mite taxon at a local scale [1,5], and many gaps still exist in our understanding of the distribution, diversification and evolution of mites.

The phylogenetic relationship among the main lineages of Acari was still a contentious issue [6–8]. In the current NCBI taxonomic system [9], Acari are comprised of two major lineages that have either monophyletic [10,11] or diphyletic origins [12]: the superorder Parasitiformes (Holothyrida + Ixodida (ticks) + Mesostigmata) and Acariformes (Trombidiformes + Sarcoptiformes). The Trombidiformes order contains a small suborder Sphaerolichida and a larger suborder Prostigmata which constists of three large clades (Eleutherengona, Anystina and Eupodina), and the Sarcoptiformes order includes three suborders (Endeostigmata, Oribatida and Astigmata).

Microbiologists have long been aware of contaminations in genomic databases caused by symbiotic bacteria, fungi or protists, and have utilized them as treasures to study the host-microbe interactions [13–17]. However, contaminations of the microscopic mites in genomic databases have not been studied. Our assumption is as follows: the ubiquitous mites, with very small size (mostly 0.4–0.8 mm) [2] and close associations to plants/animals, may go unnoticed in the field samples and have contaminated the public databases. Thus, we modified our previously published pipeline for protistan contaminations to survey mite contaminations in Genbank whole genome shotgun (WGS) genomes and transcriptome shotgun assemblies (TSA) based on DNA barcodes. DNA barcodes (e.g., the mitochondrial cytochrome c oxidase I, COI) are usually used in DNA barcoding experiments because such short sequences can produce accurate species identifications [18]. Our pipeline took advantage of this attribute, and was reliable to detect contaminations related to DNA barcodes in large genomic databases [13].

The aims of current study were as follows: (1) survey possible contaminations of mites in animal and plant genomic data; (2) compare the contamination rates between different sequencing methods (WGS against TSA), or among specimens of different host classes; (3) assess the various host associations of different mites, by calculating the distribution of mite contaminations among different host classes; (4) explore the phylogenetic origins of these contaminated contigs. Given the wide geographic scope and the breadth of organisms covered by Genbank WGS/TSA genomic database, we expect our findings will provide a broad illustration of the distribution and biodiversity of mites.

2. Materials and Methods

2.1. Database Retrieval

A total of 14,523 WGS and 7591 TSA assemblies within the taxonomic groups of "Metazoa (Animals) or Embryophyta (Land Plants), but not Acari" were downloaded from Genbank [19] (https://www.ncbi.nlm.nih.gov/Traces/wgs, accessed on 30 June 2023) (Information listed in Spreadsheet S1). Among them, there are 17,845 animal and 4269 plants assemblies, with 2.39 billion contigs (16 trillion bp).

The nonredundant BLAST nucleotide (or Genbank nt) database was downloaded from (https://ftp.ncbi.nlm.nih.gov/blast/db/, accessed on 27 December 2022).

The Barcode of Life Data System (BOLD) database [20], was downloaded from (http://www.boldsystems.org/index.php/datapackages, accessed on 7 July 2023). It includes 9,401,906 DNA barcodes from 9,090,674 specimens.

2.2. Pipeline of Mite Contamination Survey

We modified our pipeline designed for scanning protistan contamination [13] by using mite barcodes as inclusion set and nonmite barcodes as exclusion set to scan mite contaminations (Figure 1). As Genbank WGS/TSA database is too large to be analyzed routinely, we sequentially eliminated candidate sequences by four steps that (1) were too long (>100,000 bp); (2) have no similarity to mite barcodes; (3) have more similarity to nonmite barcodes; (4) aligned with the best hit outside of Acari (exclusive of Ixodida) in the Genbank nt database, or with less than 80% identity.

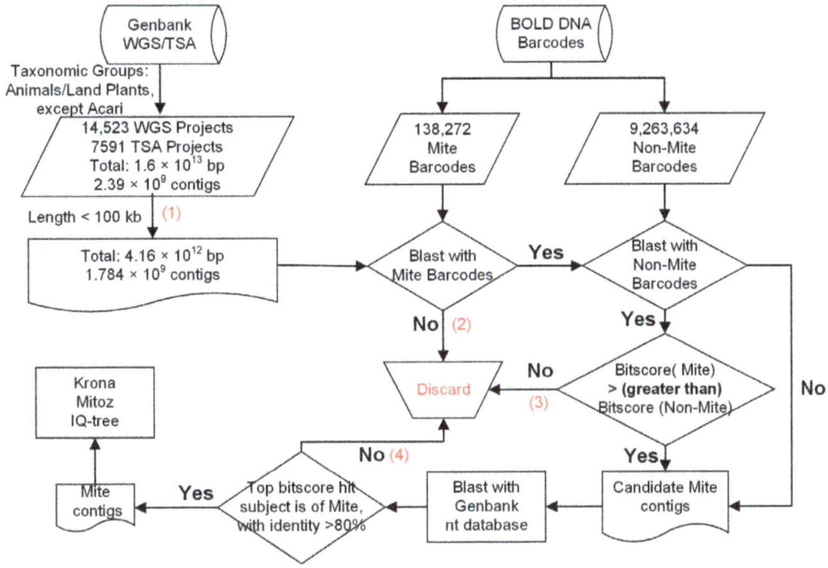

Figure 1. Pipeline to scan mite contamination. The four steps (1–4) to eliminate candidate sequences are marked in red font.

Considering the huge size of Genbank WGS/TSA and the limitation of computational resources, we filtered contigs more than 100 kb based on the reason that all RefSeq mitochondrial genomes of the Acari are less than 25 kb (Figure S1a), and 98.5% of mite barcodes in BOLD library are COI related (Section 3.1 presents the detail); therefore, most of detected mite contaminations were mitochondrial-derived and shorter than 100 kb (Figure S1b).

2.3. Taxonomic Analysis of Mite Contaminated Contigs

To correctly assign the mite contaminated contigs to family, genus or even species level, the thresholds need be more restrictive. It has been reported that the DNA barcodes enable family taxonomic assignments in the Acari with strict similarity thresholds (Sarcoptiformes 89.9%, and Trombidiformes 91.4%) [21]. Thus, we further assigned the output contaminated contigs with mite origin to family level with a similarity threshold of 91.4%, according to the top best-score hit against nt database. The abundance of contaminated contigs was further plotted by Krona [22]. Additionally, the relative abundances were calculated as the percentages of contaminations with different mite family origins across different host classes, and plotted by means of the matplotlib library.

2.4. Phylogenetic Analysis of Contaminated Contigs

The phylogenetic markers COI were predicted with MitoZ [23]. The predicted COI with length more than 80 amino acids, plus reference sequences retrieved from GenBank (Table S1), were aligned with MAFFT v7.310 with the following option: mafft -maxiterate 10,000. The maximum likelihood (ML) tree was generated by IQ-tree V2.0.3 [24] with ultrafast bootstrap (UFBoot) [25] setting, and the following options: iqtree -m MFP -B 1000 -alrt 1000. The best-fit model according to Bayesian information criterion (BIC) score was mtInv+R7 [26]. The velvet spider *Stegodyphus mimosarum* [27] and Manchurian scorpion *Mesobuthus martensii* [28] were used as outgroups [11,29]. Phylogenetic tree was edited with FigTree V1.44 (https://github.com/rambaut/figtree/, accessed on 27 December 2022). All analyses were run on a high-performance computer server with dual Intel Xeon Platinum 8375C CPUs and 512 GB RAM.

3. Results

3.1. Mite DNA Barcodes in BOLD Database

Using a Python script with a regular expression ('.*\ | Animalia,Arthropoda,Arachnida,(Trombidiformes | Sarcoptiformes | Mesostigmata | Holothyrida)') to match the sequence id, 138,272 DNA barcodes belonging to mites were extracted from the BOLD database to form the inclusion set, and the rest nonmite barcodes were used to build the exclusion set.

To get a better understanding of these mite barcodes, we plot the percentage of these barcodes by mite taxa (Figure 2a), and by genes (Figure 2b). The distribution of barcodes among mite taxa is as follows: Trombidiformes (65,183, 47%), Sarcoptiformes (45,819, 33%), Mesostigmata (27,268, 20%) and Holothyrida (2, 0%). The ratios are congruent with the numbers of described species in the taxa constituting the subclass Acari: i.e., Trombidiformes (25,797); Sarcoptiformes (16,299); Mesostigmata (11,424) and Holothyrida (27) [2].

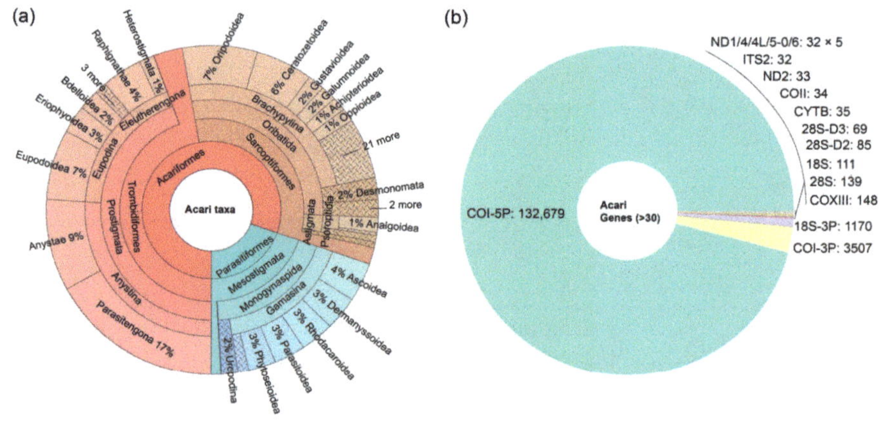

Figure 2. (**a**) Krona plot displaying the distribution of mite DNA barcodes at various Acari taxonomic levels in BOLD database. (**b**) Pie chart of mite DNA barcodes to different gene markers in BOLD database.

As for the distribution among genes, COI-5P (132,679, 96%) plus COI-3P (3507, 2.5%) account for 98.5% of all the barcodes. The COI has long been used to discriminate the small mites, and to resolve the diversity of mite fauna in large-scale surveys [30,31]. It can overcome the shortage of external diagnostic characters of mites in traditional identification through morphology [32,33].

3.2. Mite Contaminations in Genbank nt Database

A substantial fraction of sequences in Genbank database appear to be contaminated [34]. Undetected mite contaminations in the Genbank nt database would lead to false negatives in the fourth step (Figure 1) of eliminating candidate sequences. However, our pipeline [13] could discriminate mite contaminations in the nt database, by checking those records that have 100% identity in the best match against misidentified sequences from the source species, but with the second-best match to mite sequences.

After running the pipeline, it output four misidentified sequences (mite contaminants) (Table 1) in the Genbank nt database. XM_022085578.1–XM_022085580.1 are annotated to be mitochondrial genes of *Zootermopsis nevadensis* (Dictyoptera, Termopsidae), but actually they are contaminations derived from the Acaroidea mite; and XR_002707260.1 is predicted to *Onthophagus taurus* small subunit rRNA, but the real source of this sequence is the Macrochelidae mite. Thus, we must be careful when using *COI*-like genes with the '-like' suffix to identify species, because these genes are likely to be contaminants propagated from contaminations in Genbank WGS database.

Table 1. Misidentified sequences in Genbank nt database, which are actually sourced from mites.

Accession No [1] (WGS Prefix)	Matched Subject (Identity)	Len [2]	Description of Subject Sequence [3]
XM_022085578.1 (AUST01)	XM_022085578.1 (100)	1731	PREDICTED: *Zootermopsis nevadensis* *COX1*-like (LOC110840501), mRNA
	MN857505.1 (80.977)	1719	*Tyrophagus putrescentiae* voucher UMMZ BMOC 17-0108-002 mitochondrion, complete genome
XM_022085579.1 (AUST01)	XM_022085579.1 (100)	1321	PREDICTED: *Zootermopsis nevadensis* *COX3*-like (LOC110840502), mRNA
	MW784238.1 (77.51)	1245	*Lardoglyphus konoi* mitochondrion, complete genome
XM_022085580.1 (AUST01)	XM_022085580.1 (100)	760	PREDICTED: *Zootermopsis nevadensis* *COX2*-like (LOC110840503), mRNA
	NC_038058.1 (81.659)	687	*Rhizoglyphus robini* mitochondrion, complete genome
XR_002707260.1 (JHOM02)	XR_002707260.1 (100)	1790	PREDICTED: *Onthophagus taurus* Eukaryotic small subunit rRNA (LOC111421936)
	AY620939.1 (97.452)	1766	*Macrocheles* sp. AL5995 18S rRNA gene, partial sequence

[1] The misidentified sequences in the Genbank nt database were blasted against nt database; the top two best matches were listed, with the first record to itself and the second to mite sequence. [2] Alignment length. [3] Abbreviation: 'cytochrome c oxidase subunit', *COX*; 'ribosomal RNA', rRNA.

3.3. Distribution of Mite Contaminations in Genbank WGS/TSA

In 22,114 assemblies (14,523 WGS and 7591 TSA projects), our modified pipeline resulted 1717 mite contaminated contigs (Figure S2, Fastafile S1) in 681 assemblies (220 WGS and 461 TSA projects). Thus, the contamination rate of TSA (6.1%) is higher than that of WGS (1.5%).

Next, we calculated the mite contig numbers, and contamination rates in specimens from different hosts (Figure 3a). The results showed that the richness of contaminations varied greatly among different host classes. The top three host classes with the largest number of contaminated contigs were as follows: Magnoliopsida (730 contigs), Insecta (562 contigs) and Pinopsida (148 contigs). Although the contamination rates of Pinopsida (30/138) and Magnoliopsida (290/4047) were higher than average (681/22,114), contamination rate of Insecta was not (223/6224).

To further reveal the distribution of mites, we assigned these contigs to mite families and plotted the relative abundance among different host classes (Figure 3b). Using a similarity threshold of 91.4%, 1041 contigs were successfully assigned to mite families. The distribution can be concluded as follows:

Contaminations in the order Mesostigmata are mostly from plant or insect specimens. For example, in the family Phytoseiidae which harbors most common plant inhabiting predatory mites [35], 38/48 of contaminated contigs are from projects of Magnoliopsida.

In the hyporder Parasitengona (Trombidiformes, Anystina), insect specimens are the predominant sources of contamination, although just a few contigs were detected in the following four families: Trombidiidae (7 contigs), Arrenuridae (7 contigs), Hydrachnidae (7 contigs) and Erythraeidae (12 contigs). Interestingly, in the family Erythraeidae, half of the contigs are from Arachnida assemblies. This is consistent with the reports that Parasitengona larvae can parasite on arthropods, such as larvae of Erythraeidae parasitic on spiders (Arachnida, Araneae) [36] and Harvestmen (Arachnida, Opiliones) [37].

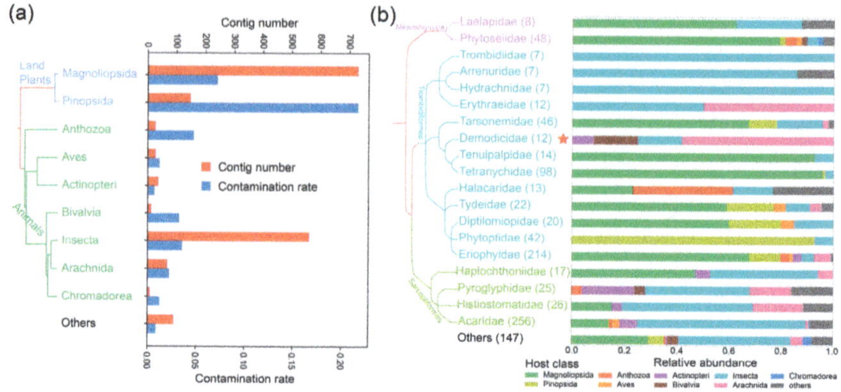

Figure 3. Distribution of mite contaminations among different host classes or mite families (**a**). Numbers of mite contigs or contamination rates among projects of different host classes. (**b**) Relative richness according to the percentages of contigs of different host classes to different mite families. The contig numbers are list in parentheses, and the host classes were indicated at the bottom of the plot. The host/mite cladogram trees were generated by taxtree (https://github.com/nongxinshengxin/taxtree, accessed on 6 August 2023) based on NCBI taxonomy. The artificial contamination with human *Demodex* (Demodicidae) is marked with a star symbol.

For families in Eleutherengona, the detected contigs are modest: Tarsonemidae (46 contigs), Demodicidae (12 contigs), Tenuipalpidae (14 contigs) and Tetranychidae (98 contigs). Apart from Demodicidae, contaminations of these families are mostly associated with the class Magnoliopsida. Tetranychidae (spider mites) and Tenuipalpidae (false spider mites) are phytophagous and include major agricultural pests, thus are mainly found on plants. In the family Tarsonemidae (white mites), *Steneotarsonemus spinki* Smiley (rice mite) is a serious pest of rice crops, whereas some other genus/species are found associated with bark beetles [38,39]. We here found a modest percentage of contigs from Pinopsida and Insecta in Tarsonemidae. Demodicidae mites are ubiquitous skin parasites in mammals [40]. However, all 12 Demodicidae contigs here were related to nonmammal. After carefully checking these contigs, we found that all of them had high identities (96–100%) to the human mites (*Demodex folliculorum* or *Demodex brevis*) (Table S2); thus, we regard these Demodicidae contigs as fortuitous contaminations by human *Demodex* mites, and they should not be considered for further mite–host association analysis.

As for the supercohort Eupodina, (Diptilomiopidae (20 contigs) + Eriophyidae (214 contigs) + Phytoptidae (42 contigs) + Tydeidae (22 contigs) + Halacaridae (13 contigs)), most of them are phytophagous; thus, vagrant on host plants. Hence, most of the contaminants of Eupodina are found in assemblies of plants, except in the Halacaridae family. Notably, there were about 40% Halacaridae (marine mites) contigs from Anthozoa; and over 90% of Pinopsida in Phytoptidae.

Finally, in the order Sarcoptiformes, the most numerous of these contaminations were related to Insecta, followed by plants. Interestingly, of these, there are several contigs from the Actinopteri (bony fishes) assemblies (Table S3). This is consistent with the report that Histiostomatidae mites can attack fishes [41].

3.4. Phylogenetic Analysis of the Mite Contaminants

To further understand the phylogenetic origins of these contaminants, the contigs were annotated with MitoZ, and the predicted COI with a length more than 80 amino acids were used to infer a phylogenetic tree (Figure 4). The clades are colored according to the taxa of mite references retrieved from Genbank, and the host taxa of the contigs are derived from the project/assembly information (Spreadsheet S1) and indicated with symbols. As the preceding subsection revealed, similar host–mite associations can also be deduced from this smaller COI dataset.

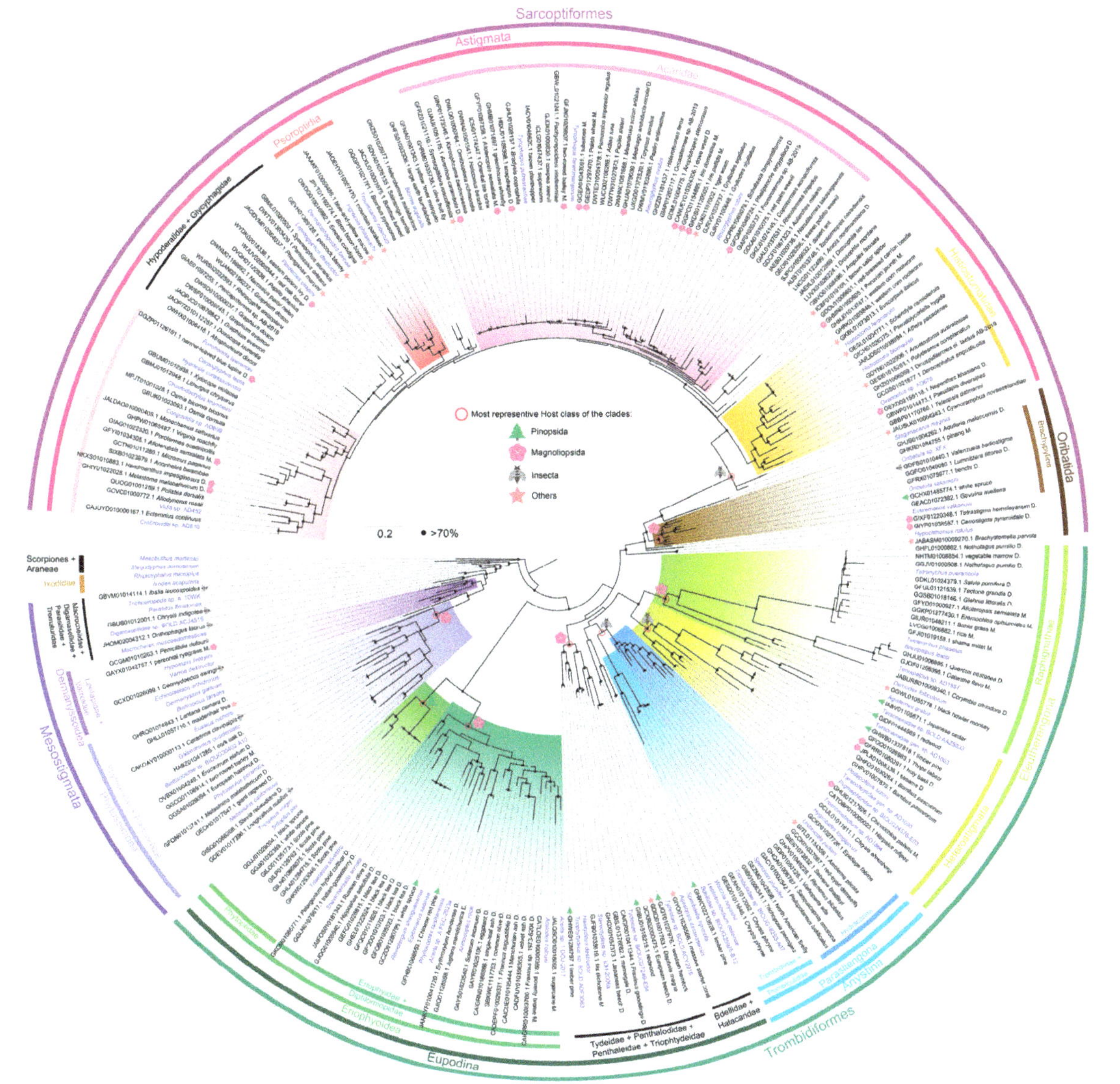

Figure 4. Phylogenetic tree of COI predicted from mite contaminated contigs. The species names of the mite references retrieved from Genbank were colored in blue font. The contaminated WGS/TSA contigs were named with accession numbers following the host names, with host classes represented by a symbol in the nodes (most representative class of that clade), or symbols after exceptional branches individually. The D. following names indicates the host taxon is dicots, and M. indicates monocots. Nodes with bootstrap values (BSP) $\geq$ 70% are marked with a black dot.

According to the phylogenetic tree, conclusions can be drawn as follows: (1) the super-cohort Anystina is monophyletic with low support, whereas the Eupodina is paraphyletic; (2) two superfamilies, Phytoseioidea (Blattisociidae + Phytoseiidae) and Eriophyoidea

(Eriophyidae + Diptilomiopidae + Phytoptidae), were both recovered as monophyletic; (3) the monophylies of two clades, Parasitengona (Anystina) and Eleutherengona were also observed, but with low support in the clade of Parasitengona; (4) a monophyletic Hydracarina (Parasitengona) is strongly supported; (5) a close phylogenetic relationship of Parasitengona to a clade uniting Halacaridae and terrestrial predacious superfamily Bdelloidea was observed; (6) we also observed astigmatid mites nested in oribatid mites. These are consistent with the phylogenetic relationships of major mite groups reported before [8,12,42–47].

Next, we investigated the contamination by clades as follows:

Manure-inhabiting (Coprophilous) Mesostigmata mites are important biological control agents of pests that feed on the eggs or larvae of pests [48]. In Dung Beetles (*Onthophagus taurus*), a contig (JHOM02004312.1) was found related to the Macrochelidae (Mesostigmata) mite. And in this assembly, there was another contig related to rRNA (JHOM02004223.1) which was misidentified (XR_002707260.1) in the nt database (Table 1).

In Eriophyoidea clade [49], the hosts of contaminations can be divided into two groups: the dominant Magnoliopsida (angiosperms) (21/23) in the clade of (Eriophyidae + Diptilomiopidae), and Pinopsida (gymnosperms) (7/7) in Phytoptidae [49]. Interestingly, consistent with multiple host shift reported previously [8], in the clade of (Eriophyidae + Diptilomiopidae), there were two contigs from Pinopsida (gymnosperms) (GCZO01 and GFHB01) and a clade of monocots (JALQSO01 and CATLOE01), which are phylogenetically closest to mites that found in monocots before [50].

As for the aquatic mites, we found a contig (GIYO01) in massive starlet coral (Anthozoa) to Halacaridae clade. It has been reported that cold water coral reefs harbour a diverse Halacaridae fauna [51]. In the Hydracarina (Water mites) clade, there are four contigs from stoneflies (Plecoptera): *Setvena bradleyi* (GIEI01), *Remenus bilobatus* (GHPV01), *Viehoperla ada* (GIDP01) and *Sasquacapnia missiona* (GHQA01); two contigs from damselflies (Odonata, Zygoptera): *Epallage fatime* (GCKP01) and red-eyed damselfly (GCCK01); and one contig from caddisflies (Trichoptera): *Philopotamus ludificatus* (GACV01). The three orders Plecoptera, Odonata and Trichoptera are three major aquatic insect taxa [52]. This is consistent with the lifestyle of Hydracarina that harvest larvae and parasitize adults of aquatic insects [53]. Interestingly, one contig from *Amblema plicata* (Mollusca, GITL01) is closest to *Unionicola parkeri* mite (Hydracarina, Hygrobatoidea, Unionicolidae), which is a common symbiont of molluscs, by living on the gills or mantle and foot of their hosts [54,55].

In the Tetranychoidea (Eleutherengona, Raphignathae) clade, all the contaminated contigs are from Magnoliopsida; among them, the ratio of dicots to monocots is 8:6. There were two clades of Demodicidae (Raphignathae) and Stigmaeidae (Raphignathae) close to the Tetranychoidea. In the Demodicidae clade, the contig is from the black howler monkey (GGWL01), with 83.4% nucleotide identity to *Demodex folliculorum* (Table S2), a known mite parasite that inhabits the skin of humans [40]. In Stigmaeidae, it was a contig from Japanese cedar (Pinopsida; IABV01).

Oribatida are primarily soil dwelling, but also occur on trees [56]. For example, Eremaeidae *Eueremaeus trionus* was found on bases of branches of Siberian pine trees (*Pinus sibirica*) [57]. Thus, in the clade of Oribatida, we found most of the contigs are from Magnoliopsida (7/10). Interestingly, there was a contig from *Brachystomella parvula* (Collembola, JABASM01) which is closest to *Hypochthonius rufulus* (Oribatida, Hypochthoniidae). Springtails (Collembola) are also microarthropods that live below ground as Oribatida mites, and they are usually used together to reveal effects of the environmental change on soil microarthropod populations [58].

In Astigmata clade, the contigs are mostly sourced from Insecta or Magnoliopsida, except in Analgoidea (Psoroptidia). In Psoroptidia clade, there are two contigs from Ave: the mountain parakeet (JAOEHY01) and the blue-and-yellow macaw (JAAAKF01), and one contig from Mammalia: *Bison bison* (JPYT01). They are closest to the feather mite (Analgoidea) *Ingrassia philomachi* or *Dermatophagoides farinae* [59] in the phylogenetic tree.

Histiostomatoidea are typically associated with wet environments, and believed to be the earliest derivative Astigmata [45]. In Histiostomatoidea clade, the contigs are most from Insecta (8/12). The exceptions are as follows: contigs from *Euscorpius italicus* (Scorpiones; GKBL01), *Schendyla carniolensis* (Chilopoda; GESL01), *Polydesmus complanatus* (Diplopoda; GESI01) and *Nepenthes khasiana* (Magnoliopsida; GEXD02).

4. Discussion

Distribution and host associations of mites are complex because of their remarkable diversity of trophic preferences and habitats. Moreover, crossovers often occur (e.g., predators may feed on plants; free-living mites switch to parasitic or phoretic on other animals; and litter-inhabiting mites move onto plants) [1]. Thus, it is very challenging to summarise the distribution and host-interactions of mites.

Fundamental advances in sequencing technology and bioinformatics made en masse biodiversity assessments of microscopic organisms possible [60]. In this study, we applied a bioinformatics method to excavate mite contaminations in Genbank WGS/TSA database with acceptable computational costs and draw some conclusions that are in line with our expectations and mite–host associations concluded in traditional studies. However, we would like to emphasize some limitations of our study:

First, this study was not intended to survey all contaminated contigs related to all mite genes, but just those related to DNA barcodes. The reason was that the huge size and rapid growing of the Genbank database surpasses the limit of our computational resources, as we mentioned before [13].

Second, the mite contaminations detected by this study still have biases. The greatest number of mite species is found in soils [61]. However, we detected relatively few contigs of Oribatida and Endeostigmata (many of which live in deep soil). The reason is that Genbank WGS/TSA does not contain soil environmental data. Besides, the environmental specimens are not suitable for host association study because of the obscure host information.

Third, although BOLD barcode library is largely complete for vertebrate species, it remains poorly developed for invertebrates, especially mites [62]. Since our pipeline relied heavily on the BOLD and Genbank nt databases, we suppose there are still undetected mite contaminations related to unrecognized species. As the BOLD database is growing, it will provide sufficiently available barcodes to allow more precise resolution of the contaminated mites.

Lastly, as mites are so speciose, the contaminated contigs detected in this study still cannot cover all mite or host taxa. Hence, there are some mite families or host classes missed in our deduced distribution pattern. However, as the Genbank database growing, the mite contaminations will increase, and would provide more comprehensive information for mite distribution study.

5. Conclusions

In this study, we systematically studied the mite distribution based on contaminations in the Genbank WGS/TSA database, which covered a large cohort of species (animals: 10,240; plants: 1970; Spreadsheet S1). The results suggest that mite-derived contaminations are common in genomic databases, with three in a hundred of assemblies contaminated by mites. Thus, apart from commonly known microbial contaminations, we should also be aware of the contaminations derived from minuscule mites to avoid erroneous interpretation of the genomic data. Based on these valuable contaminated contigs, host associations of mites were concluded, such as Parasitengona mites on arthropods and Phytoseiidae, Tetranychidae, Tenuipalpidae and Eriophyoidea on plants. Further phylogenetic analysis of the predicted COI derived from these contigs corroborated the mite origination and heterogeneous distribution of the contaminated contigs. Overall, our study provides valuable insights into the global biodiversity and distribution of mites.

Supplementary Materials: The following supporting information can be downloaded at: https://www.mdpi.com/article/10.3390/ani13203172/s1, Figure S1: Length distribution of (a) 59 mitochondrion genomes of mites in Refseq database, and (b) 1717 mite contaminated contigs in Genbank WGS/TSA database detected in this study; Figure S2: Relative abundance of mite contaminated contigs at various Acari taxonomic levels detected in Genbank WGS/TSA database; Table S1: List of protein sequences retrieved from Genbank for the phylogenetic analysis; Table S2: Contaminations of family Demodicidae; Table S3: Contaminations related to the order Sarcoptiformes in fish (Actinopteri) assemblies; Spreadsheet S1: WGS & TSA assembly info.xls; Fastafile S1: Mite contaminated contigs.fasta.

Author Contributions: Conceptualization, J.X.; formal analysis, J.X. and Y.Z.; investigation, Y.Z.; writing—original draft preparation, J.X. All authors have read and agreed to the published version of the manuscript.

Funding: This research was funded by the National Natural Science Foundation of China (grant No. 31900152) and the Science and Technology Research Program of Chongqing Municipal Education Commission (grant No. KJQN202100632).

Institutional Review Board Statement: Not applicable.

Informed Consent Statement: Not applicable.

Data Availability Statement: The Genbank WGS/TSA datasets for this study can be downloaded in [GenBank] (https://www.ncbi.nlm.nih.gov/GenBank, accessed on 30 June 2023). The Barcode of Life Data System library can be found in [Bold] (http://www.boldsystems.org/index.php/datapackages, accessed on 7 July 2023). The bioinformatic code is available at (https://github.com/xiebio/DBCscan, accessed on 13 August 2023).

Conflicts of Interest: The authors declare no conflict of interest.

References

1. Krantz, G.W.; Walter, D.E. *A Manual of Acarology*, 3rd ed.; Texas Tech University Press: Lubbock, TX, USA, 2009.
2. Stork, N.E. How Many Species of Insects and Other Terrestrial Arthropods Are There on Earth? *Annu. Rev. Entomol.* **2018**, *63*, 31–45. [CrossRef] [PubMed]
3. Zhang, Z.-Q. *Animal Biodiversity: An Outline of Higher-Level Classification and Survey of Taxonomic Richness*; Magnolia Press: Auckland, New Zealand, 2011.
4. Hammad, H.; Chieppa, M.; Perros, F.; Willart, M.A.; Germain, R.N.; Lambrecht, B.N. House dust mite allergen induces asthma via Toll-like receptor 4 triggering of airway structural cells. *Nat. Med.* **2009**, *15*, 410–416. [CrossRef] [PubMed]
5. Gan, H.; Zak, D.R.; Hunter, M.D. Scale dependency of dispersal limitation, environmental filtering and biotic interactions determine the diversity and composition of oribatid mite communities. *Pedobiologia* **2019**, *74*, 43–53. [CrossRef]
6. Xue, X.-F.; Dong, Y.; Deng, W.; Hong, X.-Y.; Shao, R. The phylogenetic position of eriophyoid mites (superfamily Eriophyoidea) in Acariformes inferred from the sequences of mitochondrial genomes and nuclear small subunit (18S) rRNA gene. *Mol. Phylogenetics Evol.* **2017**, *109*, 271–282. [CrossRef]
7. Klimov, P.B.; Oconnor, B.M.; Chetverikov, P.E.; Bolton, S.J.; Pepato, A.R.; Mortazavi, A.L.; Tolstikov, A.V.; Bauchan, G.R.; Ochoa, R. Comprehensive phylogeny of acariform mites (Acariformes) provides insights on the origin of the four-legged mites (Eriophyoidea), a long branch. *Mol. Phylogenetics Evol.* **2018**, *119*, 105–117. [CrossRef]
8. Xue, X.F.; Yao, L.F.; Yin, Y.; Liu, Q.; Li, N.; Hoffmann, A.A.; Sun, J.T.; Hong, X.Y. Macroevolutionary analyses point to a key role of hosts in diversification of the highly speciose eriophyoid mite superfamily. *Mol. Phylogenetics Evol.* **2023**, *179*, 107676. [CrossRef]
9. Federhen, S. The NCBI Taxonomy database. *Nucleic Acids Res.* **2012**, *40*, D136–D143. [CrossRef]
10. Lozano-Fernandez, J.; Tanner, A.R.; Giacomelli, M.; Carton, R.; Vinther, J.; Edgecombe, G.D.; Pisani, D. Increasing species sampling in chelicerate genomic-scale datasets provides support for monophyly of Acari and Arachnida. *Nat. Commun.* **2019**, *10*, 2295. [CrossRef]
11. Zhang, Y.-X.; Chen, X.; Wang, J.-P.; Zhang, Z.-Q.; Wei, H.; Yu, H.-Y.; Zheng, H.-K.; Chen, Y.; Zhang, L.-S.; Lin, J.-Z.; et al. Genomic insights into mite phylogeny, fitness, development, and reproduction. *BMC Genom.* **2019**, *20*, 954. [CrossRef]
12. Pepato, A.R.; Costa, S.G.d.S.; Harvey, M.S.; Klimov, P.B. One-way ticket to the blue: A large-scale, dated phylogeny revealed asymmetric land-to-water transitions in acariform mites (Acari: Acariformes). *Mol. Phylogenetics Evol.* **2022**, *177*, 107626. [CrossRef]
13. Xie, J.; Tan, B.; Zhang, Y. A Large-Scale Study into Protist-Animal Interactions Based on Public Genomic Data Using DNA Barcodes. *Animals* **2023**, *13*, 2243. [CrossRef]
14. Orosz, F. Presence of p25alpha-Domain in Seed Plants (Spermatophyta): Microbial/Animal Contaminations and/or Orthologs. *Life* **2023**, *13*, 1664. [CrossRef]
15. Twort, V.G.; Blande, D.; Duplouy, A. One's trash is someone else's treasure: Sequence read archives from Lepidoptera genomes provide material for genome reconstruction of their endosymbionts. *BMC Microbiol.* **2022**, *22*, 209. [CrossRef]

16. Borner, J.; Burmester, T. Parasite infection of public databases: A data mining approach to identify apicomplexan contaminations in animal genome and transcriptome assemblies. *BMC Genom.* **2017**, *18*, 100. [CrossRef] [PubMed]
17. Lopes, R.J.; Merida, A.M.; Carneiro, M. Unleashing the Potential of Public Genomic Resources to Find Parasite Genetic Data. *Trends Parasitol.* **2017**, *33*, 750–753. [CrossRef] [PubMed]
18. Min, X.J.; Hickey, D.A. DNA Barcodes Provide a Quick Preview of Mitochondrial Genome Composition. *PLoS ONE* **2007**, *2*, e325. [CrossRef]
19. Benson, D.A.; Karsch-Mizrachi, I.; Lipman, D.J.; Ostell, J.; Sayers, E.W. GenBank. *Nucleic Acids Res.* **2009**, *37*, D26–D31. [CrossRef]
20. Ratnasingham, S.; Hebert, P.D.N. BOLD: The Barcode of Life Data System (http://www.barcodinglife.org). *Mol. Ecol. Notes* **2007**, *7*, 355–364. [CrossRef] [PubMed]
21. Young, M.R.; deWaard, J.R.; Hebert, P.D.N. DNA barcodes enable higher taxonomic assignments in the Acari. *Sci. Rep.* **2021**, *11*, 15922. [CrossRef]
22. Ondov, B.D.; Bergman, N.H.; Phillippy, A.M. Interactive metagenomic visualization in a Web browser. *BMC Bioinform.* **2011**, *12*, 385. [CrossRef]
23. Meng, G.; Li, Y.; Yang, C.; Liu, S. MitoZ: A toolkit for animal mitochondrial genome assembly, annotation and visualization. *Nucleic Acids Res.* **2019**, *47*, e63. [CrossRef] [PubMed]
24. Minh, B.Q.; Schmidt, H.A.; Chernomor, O.; Schrempf, D.; Woodhams, M.D.; von Haeseler, A.; Lanfear, R. IQ-TREE 2: New Models and Efficient Methods for Phylogenetic Inference in the Genomic Era. *Mol. Biol. Evol.* **2020**, *37*, 1530–1534. [CrossRef] [PubMed]
25. Hoang, D.T.; Chernomor, O.; von Haeseler, A.; Minh, B.Q.; Vinh, L.S. UFBoot2: Improving the Ultrafast Bootstrap Approximation. *Mol. Biol. Evol.* **2018**, *35*, 518–522. [CrossRef]
26. Kalyaanamoorthy, S.; Minh, B.Q.; Wong, T.K.F.; von Haeseler, A.; Jermiin, L.S. ModelFinder: Fast model selection for accurate phylogenetic estimates. *Nat. Methods* **2017**, *14*, 587–589. [CrossRef] [PubMed]
27. Johannesen, J.; Lubin, Y.; Smith, D.R.; Bilde, T.; Schneider, J.M. The age and evolution of sociality in Stegodyphus spiders: A molecular phylogenetic perspective. *Proc. R. Soc. B Biol. Sci.* **2007**, *274*, 231–237. [CrossRef]
28. Choi, E.H.; Park, S.J.; Jang, K.H.; Hwang, W. Complete mitochondrial genome of a chinese scorpion *Mesobuthus martensii* (Chelicerata, scorpiones, buthidae). *DNA Seq.* **2007**, *18*, 459–471. [CrossRef]
29. Liu, Q.; Deng, Y.; Song, A.; Xiang, Y.; Chen, D.; Wei, L. Comparative analysis of mite genomes reveals positive selection for diet adaptation. *Commun. Biol.* **2021**, *4*, 668. [CrossRef] [PubMed]
30. Young, M.R.; Proctor, H.C.; deWaard, J.R.; Hebert, P.D.N. DNA barcodes expose unexpected diversity in Canadian mites. *Mol. Ecol.* **2019**, *28*, 5347–5359. [CrossRef]
31. deWaard, J.R.; Ratnasingham, S.; Zakharov, E.V.; Borisenko, A.V.; Steinke, D.; Telfer, A.C.; Perez, K.H.J.; Sones, J.E.; Young, M.R.; Levesque-Beaudin, V.; et al. A reference library for Canadian invertebrates with 1.5 million barcodes, voucher specimens, and DNA samples. *Sci. Data* **2019**, *6*, 308. [CrossRef]
32. Yin, Y.; Yao, L.-F.; Hu, Y.; Shao, Z.-K.; Hong, X.-Y.; Hebert, P.D.N.; Xue, X.-F. DNA barcoding uncovers cryptic diversity in minute herbivorous mites (Acari, Eriophyoidea). *Mol. Ecol. Resour.* **2022**, *22*, 1986–1998. [CrossRef]
33. Pérez-Sayas, C.; Pina, T.; Sabater-Muñoz, B.; Gómez-Martínez, M.A.; Jaques, J.A.; Hurtado-Ruiz, M.A. DNA Barcoding and Phylogeny of Acari Species Based on ITS and COI Markers. *J. Zool. Syst. Evol. Res.* **2022**, *2022*, 5317995. [CrossRef]
34. Steinegger, M.; Salzberg, S.L. Terminating contamination: Large-scale search identifies more than 2,000,000 contaminated entries in GenBank. *Genome Biol.* **2020**, *21*, 115. [CrossRef] [PubMed]
35. Demite, P.R.; McMurtry, J.A.; De Moraes, G.J. Phytoseiidae Database: A website for taxonomic and distributional information on phytoseiid mites (Acari). *Zootaxa* **2014**, *3795*, 571–577. [CrossRef]
36. Makol, J.; Felska, M. New records of spiders (Araneae) as hosts of terrestrial Parasitengona mites (Acari: Actinotrichida: Prostigmata). *J. Arachnol.* **2011**, *39*, 352–354. [CrossRef]
37. Gabrys, G.; Felska, M.; Klosinska, A.; Starega, W.; Makol, J. Harvestmen (Opiliones) as hosts of Parasitengona (Acari: Actinotrichida, Prostigmata) larvae. *J. Arachnol.* **2011**, *39*, 349–351. [CrossRef]
38. Karmakar, K. Steneotarsonemus spinki Smiley (Acari: Tarsonemidae)—A yield reducing mite of rice crops in West Bengal, India. *Int. J. Acarol.* **2008**, *34*, 95–99. [CrossRef]
39. Khaustov, A.A.; Petrov, A.V.; Kolesnikov, V.B. A new genus and two new species of Tarsonemidae (Acari: Heterostigmata) associated with bark beetles (Coleoptera: Curculionidae: Scolytinae) from Peru. *Zootaxa* **2021**, *4966*, 41–53. [CrossRef] [PubMed]
40. Palopoli, M.F.; Minot, S.; Pei, D.; Satterly, A.; Endrizzi, J. Complete mitochondrial genomes of the human follicle mites Demodex brevis and D. folliculorum: Novel gene arrangement, truncated tRNA genes, and ancient divergence between species. *BMC Genom.* **2014**, *15*, 1124. [CrossRef]
41. Halliday, R.B.; Collins, R.O. *Histiostoma papillata* sp. n. (Acari: Histiostomatidae), a mite attacking fish in Australia. *Aust. J. Entomol.* **2002**, *41*, 155–158. [CrossRef]
42. Dabert, M.; Proctor, H.; Dabert, J. Higher-level molecular phylogeny of the water mites (Acariformes: Prostigmata: Parasitengonina: Hydrachnidiae). *Mol. Phylogenetics Evol.* **2016**, *101*, 75–90. [CrossRef]
43. Pepato, A.R.; Klimov, P.B. Origin and higher-level diversification of acariform mites–evidence from nuclear ribosomal genes, extensive taxon sampling, and secondary structure alignment. *BMC Evol. Biol.* **2015**, *15*, 178. [CrossRef]
44. Li, W.-N.; Shao, R.; Zhang, Q.; Deng, W.; Xue, X.-F. Mitochondrial genome reorganization characterizes various lineages of mesostigmatid mites (Acari: Parasitiformes). *Zool. Scr.* **2019**, *48*, 679–689. [CrossRef]

45. Norton, R.A. Morphological evidence for the evolutionary origin of Astigmata (Acari: Acariformes). *Exp. Appl. Acarol.* **1998**, *22*, 559–594. [CrossRef]
46. Li, W.-N.; Xue, X.-F. Mitochondrial genome reorganization provides insights into the relationship between oribatid mites and astigmatid mites (Acari: Sarcoptiformes: Oribatida). *Zool. J. Linn. Soc.* **2019**, *187*, 585–598. [CrossRef]
47. Dabert, M.; Witalinski, W.; Kazmierski, A.; Olszanowski, Z.; Dabert, J. Molecular phylogeny of acariform mites (Acari, Arachnida): Strong conflict between phylogenetic signal and long-branch attraction artifacts. *Mol. Phylogenetics Evol.* **2010**, *56*, 222–241. [CrossRef] [PubMed]
48. Farahi, S.; Shishehbor, P.; Nemati, A.; Perotti, M.A. Mesostigmata diversity by manure type: A reference study and new datasets from southwestern Iran. *Exp. Appl. Acarol.* **2022**, *86*, 517–534. [CrossRef]
49. Li, H.-S.; Hoffmann, A.A.; Guo, J.-F.; Zuo, Y.; Xue, X.-F.; Pang, H.; Hong, X.-Y. Identification of two lineages of host-associated eriophyoid mites predisposed to different levels of host diversification. *Mol. Phylogenetics Evol.* **2016**, *105*, 235–240. [CrossRef]
50. Chetverikov, P.E.; Fedorov, D.S.; Letukhova, V.Y.; Romanovich, A.E. Description of Cecidophyes fibigiae n. sp., new combinations, records, and DNA barcodes of eriophyid mites (Eriophyoidea, Eriophyidae) from Karadag Nature Reserve (Crimea). *Syst. Appl. Acarol.* **2021**, *26*, 818–828. [CrossRef]
51. Bartsch, I. Lohmannella (Acari, Halacaridae) from a cold-water coral reef off Norway, description of two new and a list of North Atlantic species. *Zootaxa* **2020**, *4722*, 277–286. [CrossRef]
52. Sanchez-Bayo, F.; Wyckhuys, K.A.G. Worldwide decline of the entomofauna: A review of its drivers. *Biol. Conserv.* **2019**, *232*, 8–27. [CrossRef]
53. Vasquez, A.A.; Kabalan, B.A.; Ram, J.L.; Miller, C.J. The Biodiversity of Water Mites That Prey on and Parasitize Mosquitoes. *Diversity* **2020**, *12*, 226. [CrossRef]
54. Edwards, D.D.; Vidrine, M.F.; Ernsting, B.R. Phylogenetic relationships among Unionicola (Acari: Unionicolidae) mussel-mites of North America based on mitochondrial cytochrome oxidase I sequences. *Zootaxa* **2010**, *2537*, 47–57. [CrossRef]
55. Edwards, D.D.; Jackson, L.E.; Johnson, A.J.; Ernsting, B.R. Mitochondrial genome sequence of Unionicola parkeri (Acari: Trombidiformes: Unionicolidae): Molecular synapomorphies between closely-related Unionicola gill mites. *Exp. Appl. Acarol.* **2011**, *54*, 105–117. [CrossRef]
56. Schaffer, S.; Koblmuller, S.; Krisper, G. Revisiting the Evolution of Arboreal Life in Oribatid Mites. *Diversity* **2020**, *12*, 255. [CrossRef]
57. Salavatulin, V. Microhabitat distribution of arboreal oribatid mites (Oribatida), associated with the Siberian pine (*Pinus sibirica*) of Western Siberia. *Exp. Appl. Acarol.* **2019**, *78*, 469–483. [CrossRef]
58. Zhu, D.; Bi, Q.-F.; Xiang, Q.; Chen, Q.-L.; Christie, P.; Ke, X.; Wu, L.-H.; Zhu, Y.-G. Trophic predator-prey relationships promote transport of microplastics compared with the single *Hypoaspis aculeifer* and *Folsomia candida*. *Environ. Pollut.* **2018**, *235*, 150–154. [CrossRef]
59. Klimov, P.B.; Oconnor, B.M. Improved tRNA prediction in the American house dust mite reveals widespread occurrence of extremely short minimal tRNAs in acariform mites. *BMC Genom.* **2009**, *10*, 598. [CrossRef]
60. Bik, H.M.; Porazinska, D.L.; Creer, S.; Caporaso, J.G.; Knight, R.; Thomas, W.K. Sequencing our way towards understanding global eukaryotic biodiversity. *Trends Ecol. Evol.* **2012**, *27*, 233–243. [CrossRef]
61. Arribas, P.; Andujar, C.; Moraza, M.L.; Linard, B.; Emerson, B.C.; Vogler, A.P. Mitochondrial metagenomics reveals the ancient origin and phylodiversity of soil mites and provides a phylogeny of the Acari. *Mol. Biol. Evol.* **2019**, *37*, 683–694. [CrossRef]
62. Trebitz, A.S.; Hoffman, J.C.; Grant, G.W.; Billehus, T.M.; Pilgrim, E.M. Potential for DNA-based identification of Great Lakes fauna: Match and mismatch between taxa inventories and DNA barcode libraries. *Sci. Rep.* **2015**, *5*, 12162. [CrossRef]

animals

MDPI

Article

Life Stages and Phylogenetic Position of the New Scale-Mite of the Genus *Neopterygosoma* (Acariformes: Pterygosomatidae) from Robert's Tree Iguana †

Monika Fajfer [1,*] and Maciej Skoracki [2]

[1] Department of Molecular Biology and Genetics, Institute of Biological Sciences, Cardinal Stefan Wyszynski University, Wóycickiego 1/3, 01-938 Warsaw, Poland
[2] Department of Animal Morphology, Faculty of Biology, Adam Mickiewicz University, Uniwersytetu Poznańskiego 6, 61-614 Poznań, Poland; skoracki@amu.edu.pl
* Correspondence: m.fajfer@uksw.edu.pl
† Zoo Bank: http://www.zoobank.org/0A6EC28E-D6C5-484A-A9F2-4D81CBA7E78A.

Simple Summary: This research presents a description of a new ectoparasitic scale-mite species, *Neopterygosoma robertmertensi* sp. n., collected from a Robert's tree iguana (*Liolaemus robertmertensi*) from Argentina. For the first time, the description of females was accompanied by the description of the male and juvenile stages. The morphology of all post-embryonic stages of this species was analyzed in detail using scanning electron microscopy. Additionally, we conducted a phylogenetic analysis to determine its position within the genus and created an updated identification key for all *Neopterygosoma* species. The findings show that *N. robertmertensi* sp. n. is a part of the *chilensis* group and is a sister taxon to all *Neopterygosoma* spp. collected from *Liolaemus pictus* and *L. chiliensis*.

Abstract: A new pterygosomatid mite species, *Neopterygosoma robertmertensi* sp. n. (Acariformes: Pterygosomatidae) was collected from two specimens of *Liolaemus robertmertensi* (Liolaemidae) from Argentina. This new species is described based on active stages: adults (female and male) and juveniles (deutonymphs, protonymphs, and larvae) and quiescent stages (nymphchrysalis, deutochrysalis and imagochrysalis). The changes in morphological characters that occur during the ontogeny of *N. robertmertensi* have been analyzed in detail. A difference in larval sex morphology was observed for the first time in the family Pterygosomatidae (female larvae differ from male larvae in terms of the shape and size of the idiosoma and the position of the genital area). This new mite species is most similar to *N. cyanogasteri* but can be distinguished by the presence of different leg chaetotaxy patterns of genua IV and femora IV, four to six genital setae, three to five dorsomedial setae, and two to three ventromedial setae. Phylogenetic analysis was conducted based on 120 morphological characters of all *Neopterygosoma* spp. and four outgroup species using the maximum parsimony approach. The results indicated that this species is nested within mites of the *chilensis* group of *Neopterygosoma* associated with host species of the section *chiliensis* of *Liolaemus* s. str. An updated diagnosis of the *chilensis* group of *Neopterygosoma* and an identification key for all species of this genus has been provided.

Keywords: scale-mites; Acari; phylogeny; ontogeny; *Liolaemus*

Citation: Fajfer, M.; Skoracki, M. Life Stages and Phylogenetic Position of the New Scale-Mite of the Genus *Neopterygosoma* (Acariformes: Pterygosomatidae) from Robert's Tree Iguana. *Animals* **2023**, *13*, 2809. https://doi.org/10.3390/ani13172809

Academic Editor: Theo de Waal

Received: 9 July 2023
Revised: 5 August 2023
Accepted: 28 August 2023
Published: 4 September 2023

1. Introduction

Mites of the genus *Neopterygosoma* are permanent ectoparasites, with all life stages living on the hosts. They are associated with endemic South American iguanian lizards of the genus *Liolaemus* (Sauria: Liolaemidae), and until recently, they were placed in the genus *Pterygosoma* [1,2]. The first species of this genus was described by Dittmar de la Cruz et al. [2] from tree lizards (Liolaemidae) in Argentina, exceeding the geographical range of the genus *Pterygosoma*. Later on, Fajfer and González–Acuña [1] described six new species from Chilean tree lizards and established a group *ligare* for mites associated with liolaemids. Nevertheless, the phylogenetic trees constructed by Fajfer [3] clearly showed that the genus *Pterygosoma* was paraphyletic; therefore, a new genus *Neopterygosoma* was erected for mites associated with liolaemids lizards [3]. Since then, only one new species, *N. schroederi* Fajfer, 2020, has been described [4].

Currently, mites of the genus *Neopterygosoma* are represented by eight species associated with the lizards of the genus *Liolaemus*. They are divided into two groups: *chilensis*, represented by monoxenous species associated with lizards from Chile, and *patagonica*, represented by a single oligoxenous species, *N. patagonica* (Dittmar de la Cruz, Morando and Avila, 2004), recorded on several *Liolaemus* spp. from Argentina [2,5].

Although eight species have been described in the genus *Neopterygosoma* so far, most of these descriptions are based only on a few adult females. This was necessitated by the fact that most of the described mite material was accidentally collected by herpetologists during the investigation of lizards or was taken from museum-preserved specimens, which were often washed before being fixed in formalin or alcohol. It should be emphasized that to gain a complete understanding of the mite taxonomy, phylogeny, ecology, and biology, it is essential to study both immature instars and males. In Pterygosomatidae, as in other mites, the description of juvenile stages enabled the detection of homologous features and establishment of the nomenclature used during species description [6,7]. So far, only immatures of one species, *N. schroederi*, and a male of *N. patagonica* have been described [2,4]. Nonetheless, the original description of the male was insufficient, as it only presented the idiosoma's width and length, chaetotaxy of trochanter-tibiae I–IV, and a vague figure of the idiosoma dorsum without any details. Moreover, the type series (syntypes) consists of five males, all of which have been designated as holotypes (!), and five females. However, exact locality data were not provided; only the provinces and host species were listed separately.

In this paper, we describe a new species, *Neopterygosoma robertmertensi* sp. n., from *Liolaemus robertmertensi* from Argentina, including a first comprehensive description of the male within the genus. We extensively examine the post-embryonic stages using scanning electron microscopy, and we note differences between larval males and females for the first time within this family. Additionally, we infer the phylogenetic position of *N. robertmertensi* based on morphological data. Considering both morphology and phylogeny, this new species belongs to the *chilensis* group (the first record of Argentinian host species within the group) and is a sister taxon to Chilenian mite species associated with *Liolaemus pictus* and *L. chiliensis*. Additionally, we have revised the diagnosis of the *chilensis* group and provided an updated identification key for the genus (based on females).

2. Materials and Methods

2.1. Mite Sampling

The mite specimens were collected from the geckos housed in the herpetological collection of HUJ (abbreviations of the institutions are presented below). All lizards were kept in separate jars with 75% ethyl alcohol and were examined for mites, which were then removed from the lizards under a stereomicroscope (Nikon SMZ745 (Nikon Corporation, Tokyo, Japan). Then, the mites were placed in small vials (2 mL) containing 75% ethyl alcohol.

2.2. Morphological Analysis

Before mounting in Hoyer's medium, mite specimens were cleared and softened in Nesbitt's solution at +45 °C for 8–48 h. All specimens were mounted as vouchers using Hoyer's medium on a glass slide following the standard method [8].

Specimens destined for scanning electron microscopy (SEM) were dehydrated in ethanol, covered with gold, and examined using a Carl Zeiss AG–EVO®40 electron microscope (Carl Zeiss Microscopy, Oberkochen, Germany) at the Institute of Plant Protection of the National Research Institute in Poznan (IPP NRI), Poland. Additionally, the mites were studied and measured using a Leica DMD108 microscope (Leica Microsystems, Mannheim, Germany). All measurements, including scale bars, are given in micrometers (μm). In species descriptions, measurements (ranges) of paratypes are given in parentheses, following the data of the holotype.

2.3. Terminology

In the species descriptions, names of the leg and idiosomal setae followed Grandjean [9,10], as described by Norton [6], whereas those of the palpal setae followed Grandjean [11]. Grandjean's nomenclature [9,10] has been applied to the family Pterygosomatidae by Bochkov and O'Connor [7]. The scientific names of the lizards follow the Reptile Database [12]. All of the specimens were deposited in the arachnid collections of HUJ and CSWU. The type material of the *Neopterygosoma* spp. was loaned from the AMU.

2.4. Abbreviations for Museums and Collections

AMU—Department of Animal Morphology, Adam Mickiewicz University, Poznan, Poland;
CSWU—Department of Molecular Biology and Genetics, Institute of Biological Sciences, Cardinal Stefan Wyszynski University in Warsaw, Poland;
HUJ—National Natural History Collections of the Hebrew University of Jerusalem, Israel;
NHM—Natural History Museum, London, the United Kingdom;
ZSM—Bavarian State Collection of Zoology, Munich, Germany.

2.5. Phylogeny Reconstructions Methods

For the analysis of phylogenetic relationships between *Neopterygosoma* species, all species from the genus were used. The outgroup taxa were selected based on the analyses of Reference [3]. As a distant outgroup, *Pimeliaphilus podapolipophagus* Trägårdh, 1905 was designated, and as a close outgroup, the representatives of the genus *Geckobia* (3 spp.) of the family Pterygosomatidae were selected. We chose *G. nitidus* because it was a sister taxon to *Neopterygosoma* in the analyses of Fajfer [3], and *G. gerrhopygus* and *G. hirsti* because they were grouped separately in the analysis (see Figures 6 and 7 in Reference [3]).

2.6. Cladistic Analysis

All of the characters were unordered and unweighted. In total, 13 species and 120 morphological characters of adult females were included in the analysis (List S1, Table S1). Preparing and editing of the data matix were completed using NEXUS Data Editor 0.5.0 [13]. The missing states were designated as "?" and inapplicable characters as "-". The reconstruction of phylogenetic relationships was performed in PAUP 4.0.a 147 for Microsoft Windows [14]. The branch-and-bound option was used for maximum parsimony analysis. Nodal support was evaluated using the Bremer indices calculated using PRAP2 (http://bioinfweb.info/Software/PRAP2) [15]. Analysis of the charactersdistributions and the drawing and editing of the trees were performed using FigTree v1.4.3 [16], and the final illustrations were made in Adobe Illustrator CS6.

3. Results

3.1. Systematics

The new species described here was assigned to the *chilensis* group of the genus *Neopterygosoma* Fajfer, 2019 of the family Pterygosomatidae Oudemans, 1910, based on morphological and phylogenetic evidence. It possesses the diagnostic morphological features of the *chilensis* group (see below) and is phylogenetically nested within the *chilensis* group of *Neopterygosoma*, but with weak support (Bremer = 1).

3.1.1. Description

Species group *chilensis*
Diagnosis

Body much wider (1.5–1.8 times) than long. Posteromedial part of idiosomal dorsum and venter with 3–22 pairs of dorsomedial setae or 2–21 pairs of ventromedial setae, respectively. Peripheral setae numerous and much longer than dorsal and ventral setae situated anteriorly, medially, and laterally. Setae *tc′* and *tc″* of legs II–IV serrate.

Microhabitat

Under the scales of the whole body.

Distribution and host range

This group is associated with tree lizards of the genus *Liolaemus* (Sauria: Liolaemidae) from Chile and Argentina.

Species included

Neopterygosoma chilensis (Fajfer and González–Acuña, 2013), *N. cyanogasteri* (Fajfer and González–Acuña, 2013), *N. formosus* (Fajfer and González–Acuña, 2013), *N. levissima* (Fajfer and González–Acuña, 2013), *N. ligare* (Fajfer and Gonzlez–Acuña, 2013), *N. ovata* (Fajfer and González–Acuña, 2013), *N. schroederi* Fajfer, 2020, *Neopterygosoma robertmertensi* sp. n.

Neopterygosoma robertmertensi sp. n. (Figures 1–14).

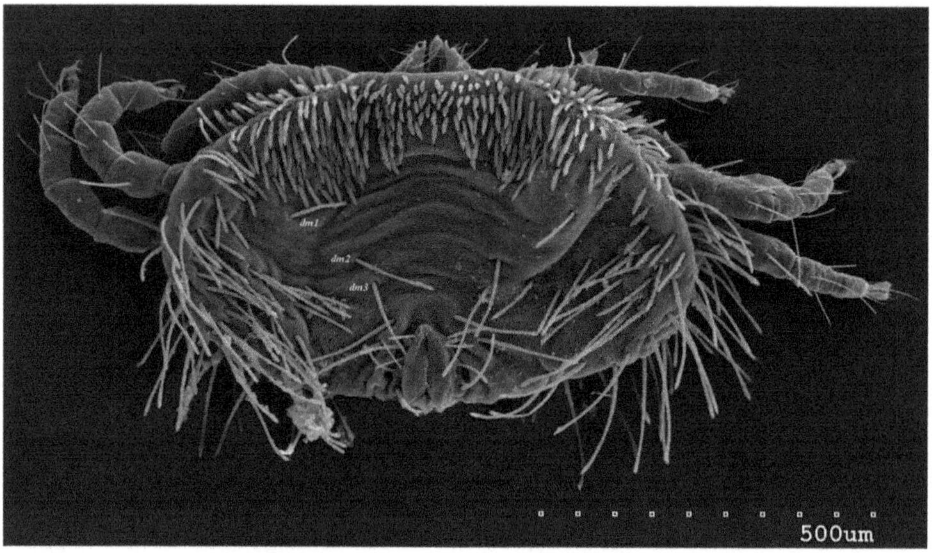

Figure 1. *Neopterygosoma robertmertensi* sp. n., female in dorsal view.

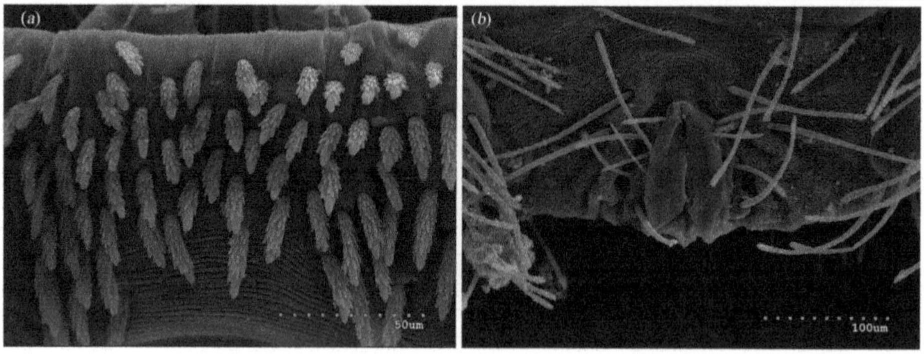

Figure 2. *Neopterygosoma robertmertensi* sp. n., female details: (**a**) propodonotal shield (**b**) genital region.

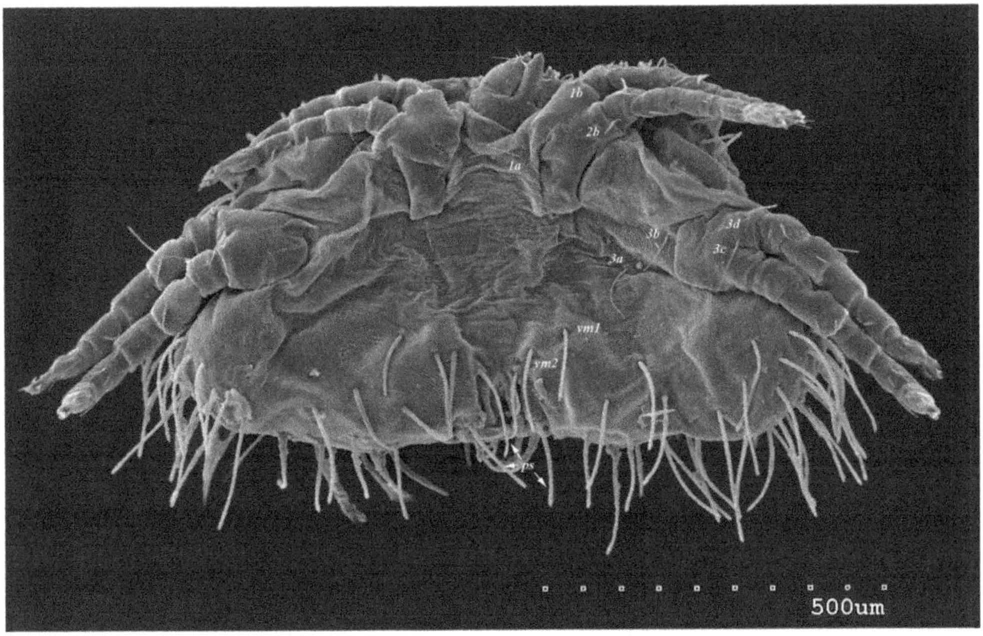

Figure 3. *Neopterygosoma robertmertensi* sp. n., female in ventral view.

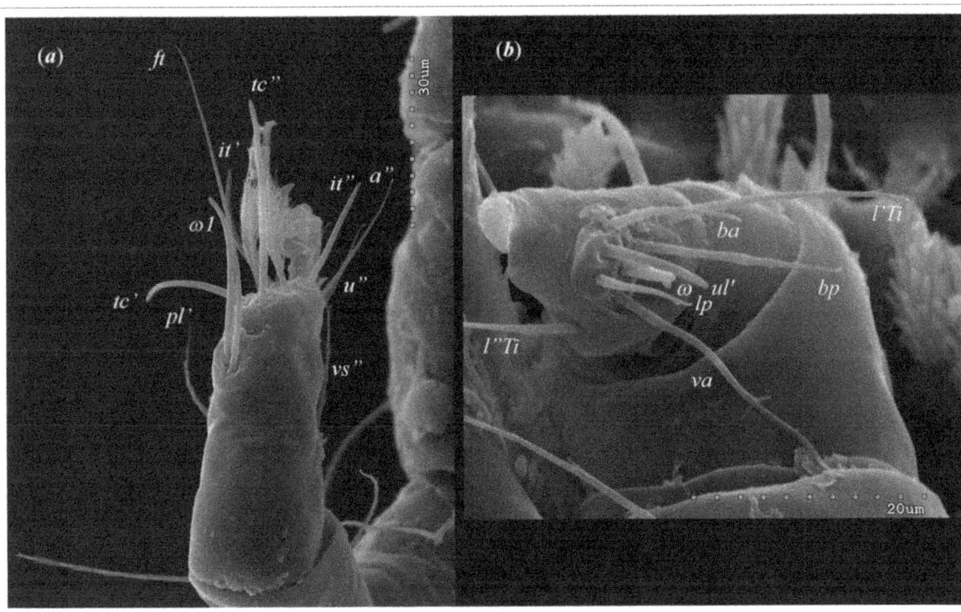

Figure 4. *Neopterygosoma robertmertensi* sp. n., female details: (**a**) tarsi I in dorsal view; (**b**) palps in ventral view.

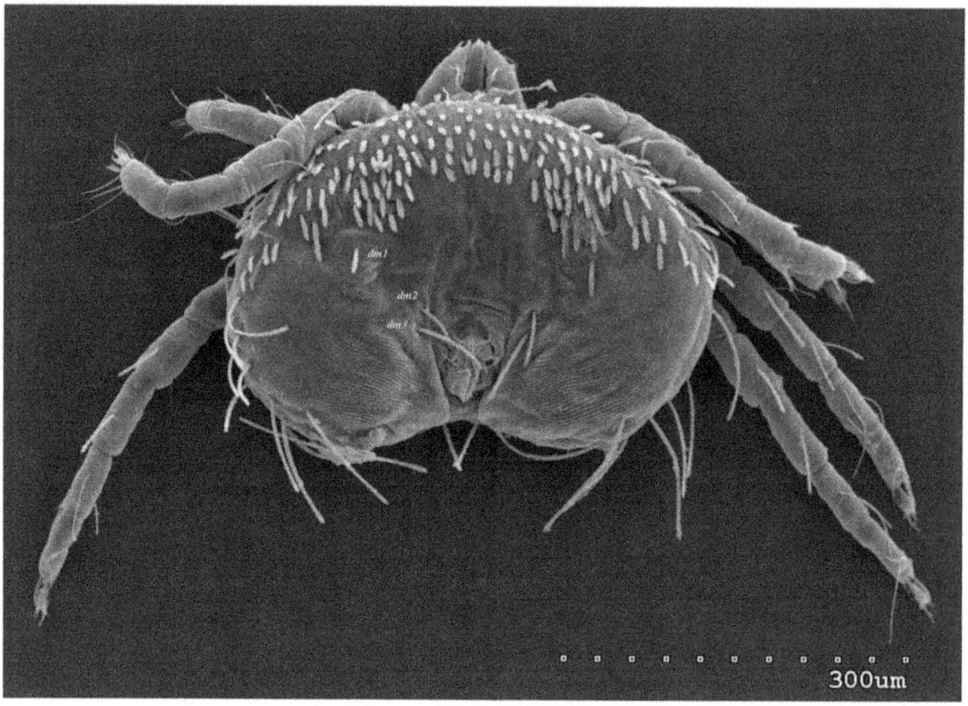

Figure 5. *Neopterygosoma robertmertensi* sp. n., male in dorsal view.

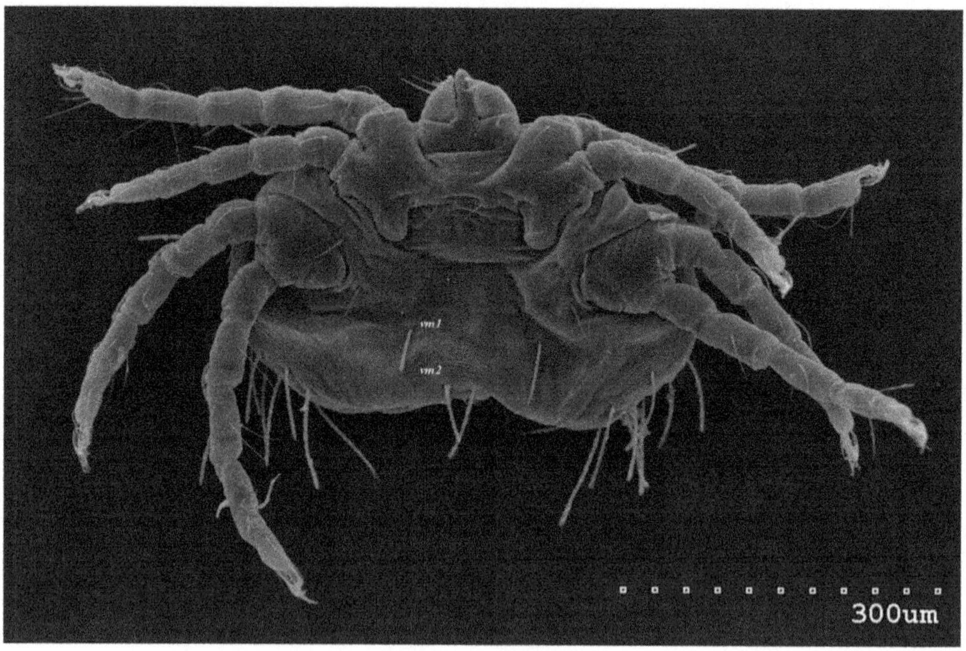

Figure 6. *Neopterygosoma robertmertensi* sp. n., male in ventral view.

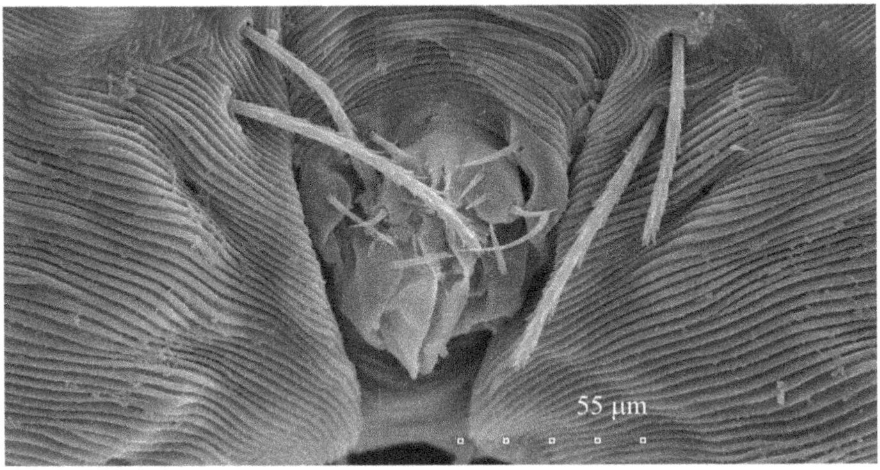

Figure 7. *Neopterygosoma robertmertensi* sp. n., male, genital area, enlarged.

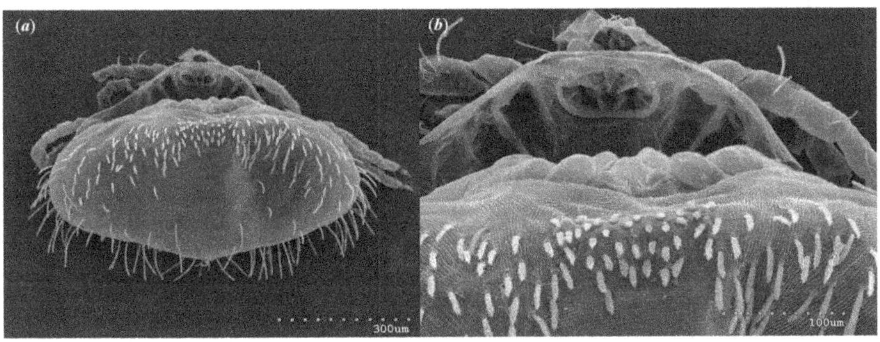

Figure 8. *Neopterygosoma robertmertensi* sp. n. (**a**) imagochrysalis in the exoskeleton of deutonymph, dorsal view; (**b**) reduced gnathosoma, peritremes and coxae I–II of imagochrysalis, enlarged.

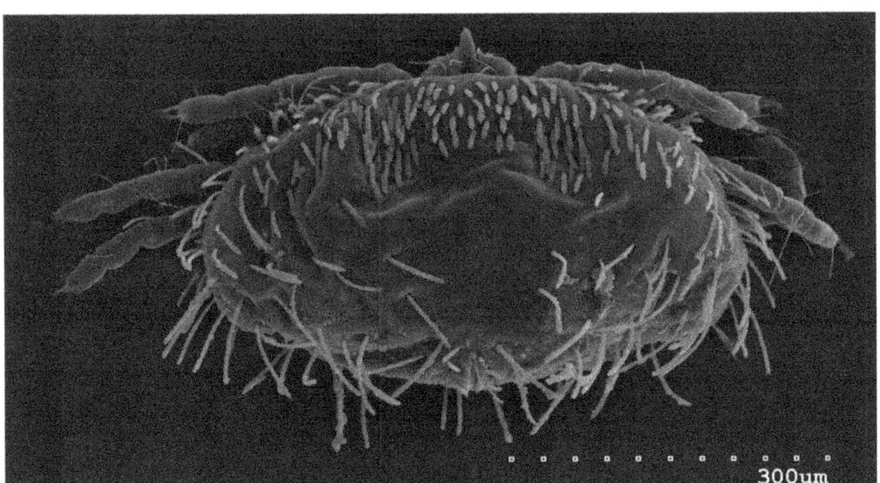

Figure 9. *Neopterygosoma robertmertensi* sp. n., deutonymph in dorsal view.

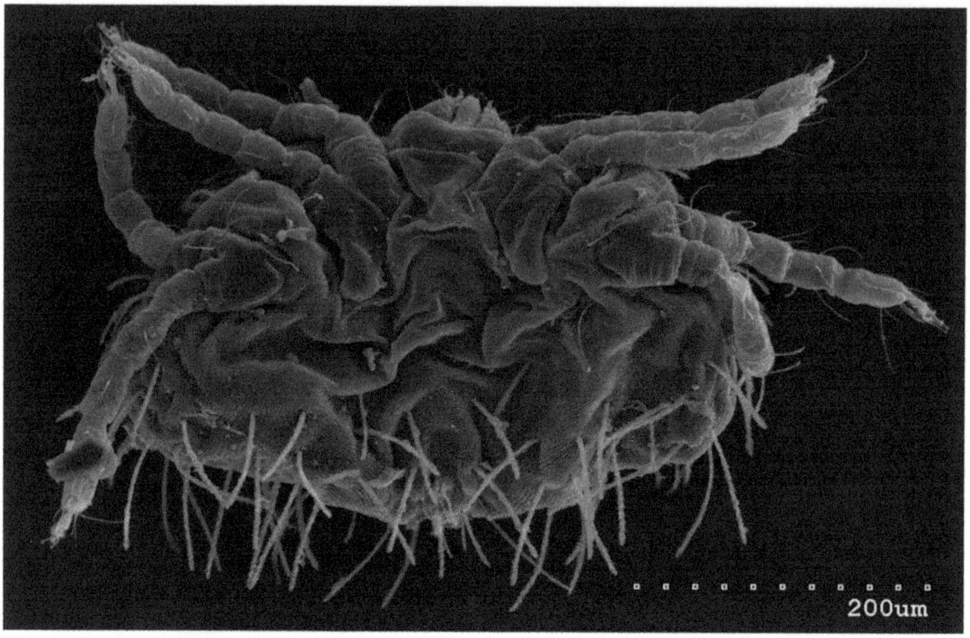

Figure 10. *Neopterygosoma robertmertensi* sp. n., deutonymph in ventral view.

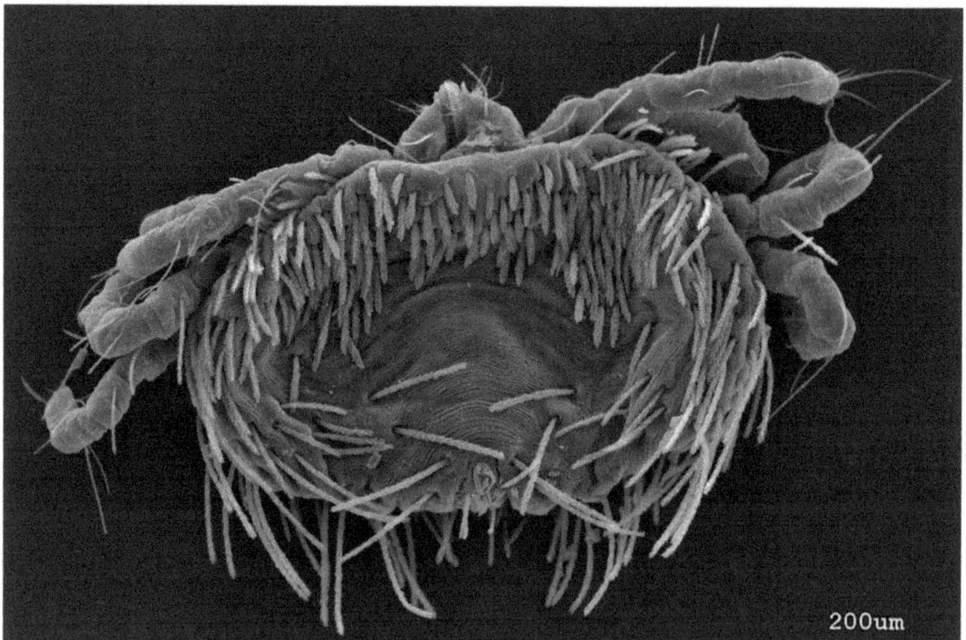

Figure 11. *Neopterygosoma robertmertensi* sp. n., protonymph in dorsal view.

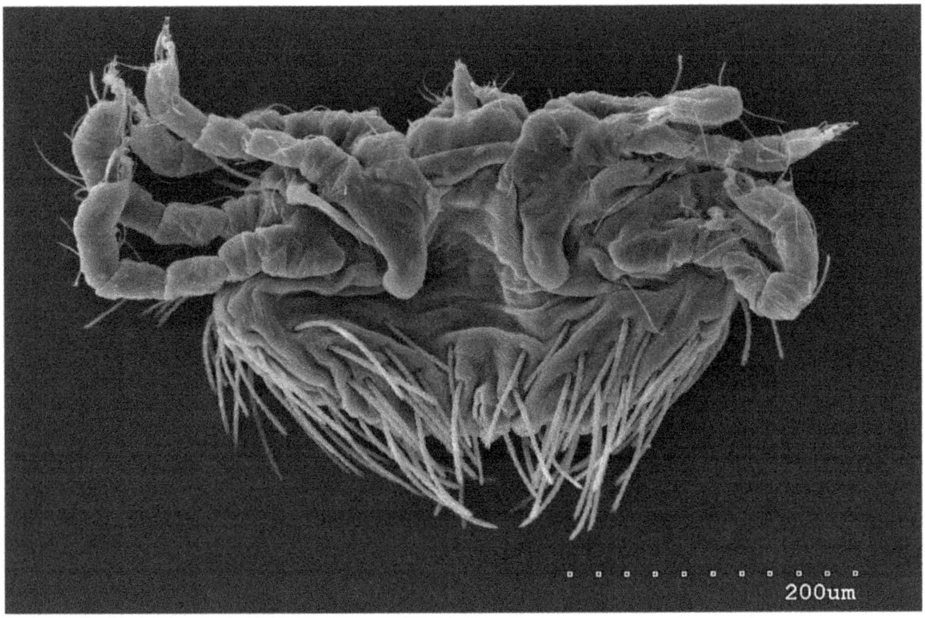

Figure 12. *Neopterygosoma robertmertensi* sp. n., protonymph in ventral view.

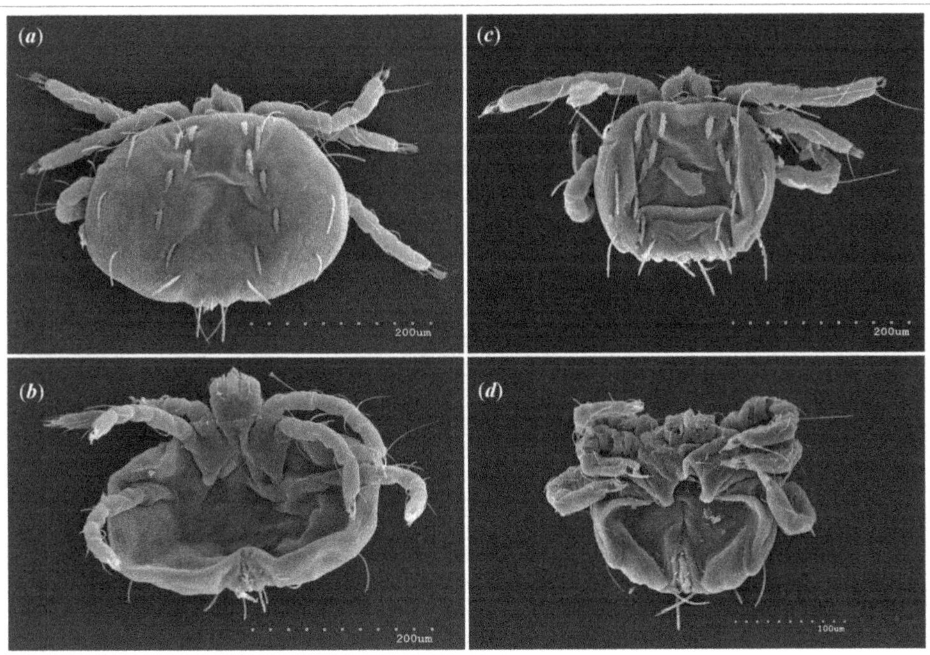

Figure 13. *Neopterygosoma robertmertensi* sp. n. (**a**) female larva in dorsal view; (**b**) female larva in ventral view; (**c**) male larva in dorsal view; (**d**) male larva in ventral view.

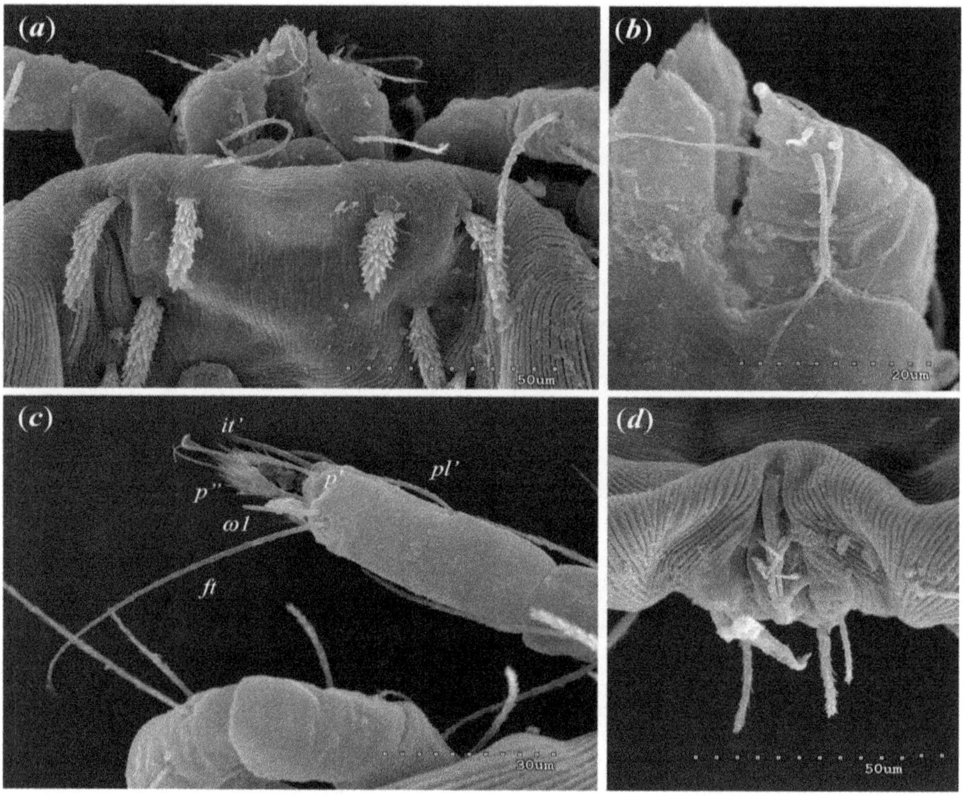

Figure 14. *Neopterygosoma robertmertensi* sp. n. larva, details (**a**) dorsal setae; (**b**) part of gnathosoma in ventral view; (**c**) tarsi I in dorso-lateral view; (**d**) genital region.

Female (holotype, range for 14 paratypes). *Gnathosoma.* Chelicerae 145 (145–150) long. Swollen, proximal part of cheliceral base and slender distal half subequal in length, about 75 (70–75) long. Fixed cheliceral digit spinous, about 10 (10) long. Palpal femur and genu with serrate dorsal seta *dF* and *dG*, 75 (75–80) and 55 (45–60) long, respectively. Palpal tibia with slightly serrate lateral setae *l′Ti* and *l″Ti*, and with barely serrate ventral seta *vTi*. Palpal tarsi with 5 setae and solenidion (Figure 4b). Hypostome with rounded apex. Peritremes with clearly visible chambers, about 85 (85–90) long. Subcapitular seta *n* smooth or with barely discernible serration, 70–75 (75–85) long. *Idiosoma* 535 (405–550) long and 973 (715–975) wide. Dorsum (Figure 1) with an antero-mid cluster of 56 (53–60) plumose setae (20–30 long) that slightly increase in length from anterior to posterior part of this cluster. These setae are situated on smooth, weakly sclerotized propodonotal shield (Figure 2a). Laterally to this cluster about 100 (98–110) setae, 30–40 long, on each side present. About 25 (20–25) of these setae are inserted ventrally, and among them, small eyes present. Mediolateral and postero-lateral parts of idiosoma with 48 (45–50) pairs of setae that increase in length from anterior to posterior part, 40–135 long. Dorsomedial part with 3 (3–5) pairs of serrate dorsomedial setae (*dm*). Setae *dm1* 75 (75–80) long and situated mostly anteriorly, setae *dm2–dm5* 90–125 (110–125) long and situated antero-laterally to the genital area. Peripheral part of body with about 30 pairs of serrate setae, 10–155 long, inserted dorsally (10–12 pairs) or ventrally (18–23 pairs). Venter (Figure 3) with 2 or 3 (2–3) pairs of serrate setae *vm*, about 80–95 long, situated laterally to genital area (Figure 2b). Genital series represented by 5 pairs of serrate setae *g1–g5*, 55–60 (55–60) long, 50–55 (60–65) long, 35–45 (55), 85–90 (75–95), and 70–75 long, respectively. Setae *g1–g4* densely serrate and situated dorsally, setae *g5* slightly serrate and situated terminally. In 3 paratypes unpaired setae *g3* present and in 5 paratypes 6 genital setae present (seta *g3* is doubled). Pseudanal setal series represented by 5 pairs of setae *ps1–ps5*, 75–120 long. Setae *ps1–ps3* situated terminally and *ps4–ps5* ventrally. *Legs.* Coxal setation *1a, 1b, 2b, 3a, 3b, 3c* and *3d* arranged in formula 2–1–4–0. Setae *1a, 3a, 3b* situated outside coxal plates. All coxal setae smooth except for slightly serrate setae *3d*. Setae of trochanters I–IV: 1–1–1–1, femora I–IV: 5–4–3–2, genua I–IV: 5–4–3–2 and tibiae I–IV: 5–5–5–5. Setae *vTrI–IV, vFI–FIII, v″FI–II* filiform and smooth, *v′GI, v″GI–II, v′TiI–IV, v″TiI–IV, v′FIV, vGIV* with barely discernible serration, *d′FI–IV, d″FI–III, l′FI, d′GI–IV, d″GI–III, l′GI, dTiI, l′Ti–IV, l″TiI–IV* serrate. Setation of tarsi: I 14 setae (*ft, tc′, tc″, p′, p″, it′, it″, a′, a″, u′, u″, vs′, vs″, pl′*) and solenidion *ω1* (Figure 4a); II 10 setae (*tc′, tc″, p′, p″, a′, a″, u′, u″, vs′, vs″*) and *ω1*; III and IV with 10 setae each (*tc′, tc″, p′, p″, a′, a″, u′, u″, vs′, vs″*). Setae *tc′, tc″, it′* and *it″* of leg I represented by eupathidia; all setae *p′* and *p″* fan-like; setae *a′, a″, u′, u″* of legs I–IV and *tc′, tc″* of leg II with barely discernible serration; setae *tc′* and *tc″* of legs III–IV serrate.

Male (range for 13 paratypes). *Gnathosoma.* Chelicerae 95 long, swollen cheliceral part 40–50 long; slender distal part 45 long. Setae *dF* filiform and smooth, 50–65 long; setae *dG* filiform with barely discernible serration, 30–45 long. Supcapitular seta *n* filiform and smooth, 35–50 long. Each branch of the peritremes is about 50–70 long. *Idiosoma* 255–320 long and 435–480 wide. Dorsum (Figure 5) with barely visible propodonotal shield bearing plumose setae grouped in anterior mid-dorsal cluster (34–38 setae); these setae, 10–20 long, progressively elongate from the anterior to posterior parts of this cluster. Numerous, slightly longer plumose setae, 25–40 long, situated laterally to this cluster. Between them, small inconspicuous eyes present laterally near one long seta, about 80 long. In the medial part of the dorsum, 3 pairs of serrate setae present: *dm1–dm3* about 30–40 long, 45–65 and 60–90 long, respectively. In the lateral and posterior parts of the idiosoma, about 12 pairs of longer slightly serrate setae, 50–100 long, present, most of them situated dorsally; 2–4 pairs situated ventrally in the posterior part of the idiosoma. Aedeagus 130–140 long. Genital area with 3 pairs of setae, 5–10 long, situated on the anal valve and 3 pairs of genital papillae, 10–25 long, situated laterally to the anal valve (as in Figure 7). Venter with two pairs of ventromedial setae *vm1* and *vm2*. Setae *vm1* 40–65 long and setae *vm2* 70–75 long. *Legs.* Coxae in formula: 2–1-4–0 and all setae filiform and smooth. Setae *1a, 3a, 3b* outside coxal plates. Chaetotaxy of legs I–IV as in female except for lack of setae on tarsi IV. Setae *dTiI–IV, l′TiI–IV, l″TiI–IV, v′TiI–IV, v″TiI–IV, dGI, l′GI, l″GI, v′GI, v″GI, dGII, vGII, l′FII–IV, vFIII–IV, lTrI–IV* smooth; setae *l′GII, l″GII, l″FII* and *l′FIII* with barely discernible serration; setae *l′FII, l″FI, l′FIII–IV, dFI–III* and *vFI–II* serrate.

Imagochrysalis (tritonymph, based on 1 female and 1 male paratype). *Gnathosoma* reduced, peritremes barely visible (Figure 8b). Legs absent, only coxae I–IV visible. Idiosoma of female chrysalis (inside deutonymphal exoskeleton) 625 long and 690 wide (inside imagochrysalis fully developed coiled female with idiosoma 615 long and 685 wide present). Idiosoma of male imagochrysalis (inside larval integument) 320 long and 425 wide (inside imagochrysalis coiled fully developed male with idiosoma 295 long and 395 wide present).

Deutonymph (range for 9 paratypes). Gnathosoma as in female. Chelicerae about 90–95 long. Slender cheliceral part and swollen distal part subequal in length, about 45–50 long. Setae *dF* and *dG* slightly serrate, 55 and 40, respectively. Subcapitular setae *n* slightly serrate and 50 long. Peritremes 55 long. Idiosoma 305–330 long and 530–560 wide. Dorsum (Figure 9) with smooth propodonotal shield covered with antero-mid cluster of 26–34 setae, about 25 long. Laterally to this cluster about 26 shorter antero-lateral setae, 25–30 long, situated more anteriorly; about 30 longer antero-lateral setae, 45–60 long, situated more posteriorly; and about 10 antero-lateral short setae inserted ventrally (among them one pair of small eyes present). Dorsomedial setae represented by 3 pairs: *dm1* about 35 long, *dm2* about 50 long, and *dm3* 65 long. Peripheral setae situated dorsally (7–8 pairs) and ventrally (11–12 pairs) and about 105 long. Venter (Figure 10) with 2 ventromedial setae *vm1* and *vm2*. Genital region with 3 setae *g1–g3*. Setae *g1* and *g2* 20–25 long, setae *g3*

35–45 long. Pseudanal setal series represented by 3 pairs of setae *ps1–ps3*, 70–75 long. Legs as in female, except for lack of setae *vTrIV*.

Deutochrysalis (based on 2 paratypes in exoskeleton of protonymph). Gnathosoma reduced, with barely discernible peritremes. Idiosoma 415–360 long and 620–650 wide. Legs absent, only coxae I–IV present. Inside deutochrysalis fully developed deutonymph present.

Protonymph (range for 5 paratypes). Gnathosoma. Chelicerae 95 long; slender cheliceral part and swollen distal part subequal in length, 45–50 long. Hypostome with rounded apex. Setae *dF* and *dG* slightly serrate, 40–60 and 40–45 long, respectively. Subcapitular seta *n* filiform and smooth, about 50 long. Each branch of peritremes about 60 long. Idiosoma 315–345 long and 535–550 wide. Dorsum (Figure 11) with weakly sclerotized propodonotal shield with densely plumose setae grouped in anterior middorsal cluster (27–42 setae). These setae subequal in length, 20–25 long. Numerous (about 63–67 pairs) of slightly longer plumose setae, 25–40 long, situated laterally to this cluster. Between them small inconspicuous eyes present. In the medial part, 3 pairs of setae *dm1* (30 long), *dm2* (55–65) and *dm3* (60–70) present, and about 20–28 pairs of postero-lateral setae, 40–95 long. Venter (Figure 12) with setae *vm1*, 55 long, and about 29 pairs of serrate peripheral setae in postero-lateral part of the idiosoma, 60–70 long. These setae situated: ventrally (12 pairs), terminally (7–8 pairs), and dorsally (10–11 pairs). Genital area with 3 pairs of genital setae *g1–g3* 30, 15, and 25 long, respectively; and with 3 pairs of densely serrate pseudanal setae *ps1–3*, 70–80 long. Legs. Coxal setae *1a, 1b, 2b, 3a, 3b, 3c* filiform and smooth, setae *3d* slightly serrate. Setae *1a* and *3a* situated outside coxal plates. Chaetotaxy pattern of legs I–IV as in female, except for lack of setae *vTrIV*.

Nymphchrysalis (based on two specimens in larval exoskeleton). Gnathosoma reduced, with barely discernible peritremes. Idiosoma 225–240 long and 350–360 wide with completely developed protonymph inside, about 205 long and 330 wide. Legs absent, only coxae I–IV visible.

Larva (range for 8 larval male paratypes and 11 larval female paratypes). Gnathosoma. Chelicerae about 50 long; swollen cheliceral part 20–25 long and slender distal part about 30 long. Fixed cheliceral digit absent. Tarsi with 5 setae and solenidion (Figure 14b). Each part of peritremal branch 35–40 long. Setae *dG* 20–40 long, setae *dF* 40–50 long. Subcapitular setae *n* absent. Idiosoma wider (290–360 wide) than long (170– 250) in female larvae and almost as long as wide in male larvae (155–200 long and 170–215 wide). Dorsum without propodonotal shield (Figure 14a) and with 11 plumose setae situated as in Figure 13a,c. Five setae situated in anterior part thicker and shorter (15–30 long) than narrower and longer (35–50 long) setae situated in posterior half of idiosoma. Eyes present on lateral margins of idiosoma. Venter (Figure 13b,d) devoid of any setation. Genital area (Figure 14d) with three genital setae *g1–g3*, 10–15 long and two pseudanal setae *ps1* and *ps2*. Setae *ps1* 40–50 long and *ps2* 30–50 long. Legs. Coxae in formula: 2–0–1; setae *1a, 1b, 3a* filiform and smooth. Chaetotaxy of legs I–IV as follows: (5–5–5) (2–2–1) (4–4–3) (0–0–0). Setae *dTiI–III, l′TiI–III, l″TiI–III, vTiI–III, dl′GI–III, dl″GI–II, dl′FI–III, dl″FI–III* filiform and slightly serrate. Setae *vFI–II* with barely discernible serration and setae *dFI–III* serrate. Setation of tarsi: I 11 setae (*ft, p′, p″, it′, a′, a″, u′, u″, vs′, vs″, pl′*) and solenidion $\omega 1$; II 10 setae (*tc′, tc″, p′, p″, a′, a″, u′, u″, vs′, vs″*) and $\omega 1$; III and IV with 10 setae each (*tc′, tc″, p′, p″, a′, a″, u′, u″, vs′, vs″*). Setae *vs′, vs″, a′, a″, pl′* smooth or with barely discernible serration, setae *p′* and *p″* fan-like, setae *tc′, tc″* of legs II–III slightly serrate (*tc′* longer than *tc″*), setae *ft* smooth, setae *it′* in form of eupathidion (Figure 14c).

Eggs 170–180 long 150–160 wide.

Type material

Holotype and 8 female, 12 male, 9 deutonymph, 4 protonymph, 2 imagochrysalis, 1 deutochrysalis, 1 nymphchrysalis, 8 male larvae, and 10 female larvae paratypes from *Liolaemus robertmertensi* Hellmich, 1964 (HUJ no. 17923) (Iguania: Liolaemidae), Argentina, Catamarca, 30 km south of Andalgalá, September 1987, coll. O. Pagaburo and Yehudah L. Werner; 7 female, 1 male, 1 deutonymph, 1 nymph chrysalis, 1 protonymph chrysalis,

1 dutonymph chrysalis, and 1 female larva paratypes from same host (HUJ no. 18091) and location, September 1987, coll. O. Pagaburo and Yehudah L. Werner.

Type Material Deposition

Female holotype and most paratypes were deposited in the HUJ (reg. HUJINV-Acari_Pte00003.1–38 and HUJINV-Acari_Pte00004.1–11), except for six female, three male, three deutonymph, two protonymph, and four larvae paratypes in the CSWU (reg. no. CSWU-Pte20.1.1–16 and Pte20.2.1–2).

Etymology

The species name is derived from the species name of the host.

Differential diagnosis

This species is most similar to *Neopterygosoma cyanogasteri* from *Liolaemus cyanogaster* (Duméril and Bibron) from Chile [1]. In females of both species, the setation of tarsi I–IV, tibiae I–IV, genua I–III, femora I and III, and trochanters I–IV is the same, fixed cheliceral digit is spinous, palp seta *dF* is longer than *dG*, subcapitular seta *n* is smooth or with barely discernible serration, the antero-mid cluster of dorsal setae is represented by about 60 setae, and five pseudanal setae *ps* are present. In *Neopterygosoma robertmertensi* setae *lv'GIV*, *lv'GII* and *ld'FIV* are absent, coxal setae *3a* are smooth, 4–6 pairs of serrate genital setae are present, 3–5 pairs of dorsomedial setae, and 2 or 3 pairs of ventromedial setae are present. In *N. cyanogasteri* setae *lv'GIV*, *lv'GII* and *ld'FIV* are present, coxal setae *3a* are serrate, one smooth genital seta, 17–21 dorsomedial setae, and 14–18 ventromedial setae are present.

Remarks

Our research used scanning electron microscopy to enhance taxonomic descriptions of the new *Neopterygosoma* species. As a result, we noticed that in the original description of *Neopterygosoma* spp. [1], some inaccuracies are mentioned. The detailed photographs revealed that a smooth and weakly sclerotized propodonotal shield is present in all *Neopterygosoma* mites (Figure 4b) (it appears in protonymph).

3.1.2. Key to species of Neopterygosoma (Females) (Based on the Key of Fajfer [4])

1. Body much wider than long (1.5–1.8 times). Setae *tc'* and *tc"* of legs II–IV serrate. Peripheral setae much longer than dorsal and ventral setae situated anteriorly, medially and laterally. . .*chilensis* group 2
   - Body circular, only slightly wider than long (1.1–1.3 times). Setae tc' and tc" of legs II–IV smooth. Peripheral setae subequal with anterior, medial and lateral setae on idiosomal dorsum and venter. . .patagonica group. . .*N. patagonica* (Dittmar de la Cruz, Morando and Avila, 2004)
2. Five setae on genu I and 5 pseudanal setae *ps*. . .3
   - Four setae on genu I and 3 pseudanal setae *ps*. . .*N. formosus* (Fajfer and González–Acuña, 2013)
3. Four setae on femur II. . .4
   - Five setae on femur II. . .5
4. Five pseudanal setae present. Setae *vTrI–IV* densely serrate. Swollen cheliceral part of chelicerae shorter than slender distal part. Subcapitular setae *n* short (45–65 long). . . *N. chilensis* (Fajfer and González–Acuña, 2013)
   - Four pseudanal setae present. Setae *vTrI–IV* smooth. Swollen cheliceral part of chelicerae longer than slender distal part. Subcapitular setae *n* long (about 125 long). . . *N. schroederi* Fajfer, 2019
5. Three setae on femur IV. One pair of genital setae *g1*. Dorsomedial setae *dm* represented by 15–21 pairs of setae. Ventro–medial setae *vm* represented by 10–18 pairs. . .6
   - Two setae on femur IV. Four or five pairs of genital setae. Dorsomedial setae *dm* represented by 3–5 pairs of setae. Ventromedial setae *vm* represented by 1–3 pairs. . .*N. robertmertensi* sp. n.

6. Genital setae smooth. Fixed cheliceral digit spinous, palp setae *dF* serrate only distally, subcapitular setae *n* serrate. . .*N. cyanogasteri* (Fajfer and González-Acuña, 2013)
   - Genital setae serrate. Fixed cheliceral digit reduced to rounded structure, palp setae *dF* serrate on all length, subcapitular setae *n* smooth. . .7
7. Coxal fields I with 2 setae. Gnathosoma situated apically. Free peritremal branch present. Setae *dG* serrate on all length. . .8
   - Coxal fields I with 3 setae. Gnathosoma displaced on dorsal side. Free peritremal branch absent. Setae *dG* serrate only at distal tip. . .*N. ovata* (Fajfer and González-Acuña, 2013)
8. Antero-medial setae increase in length from anterior to posterior part of setal cluster. Setae *a′* and *a″* of tarsi I slightly serrate. Setae *v′TrI–IV* serrate. Setae *3a* smooth and situated outside coxal plates. . .*N. levissima* (Fajfer and González–Acuña, 2013)
   - Antero-medial setae subequal in length. Setae *a′* and a″ of tarsi I smooth. Setae *v′TrI–IV* with barely discernible serration. Setae *3a* slightly serrate and situated on coxal plates. . .*N. ligare* (Fajfer and González-Acuña, 2013).

3.2. Phylogeny

Unweighted Parsimony Analysis

The analysis of the data matrix (Table S1) showed that out of 120 characters (List S1), 85 were informative. The analysis with all characters treated as unordered and unweighted was performed with Paup and produced one parsimonious tree (Figure 15). The tree is 219 steps long and has a consistency index (CI) of 0.64; retention index (RI) of 0.56, and rescaled consistency index (RC) of 0.36.

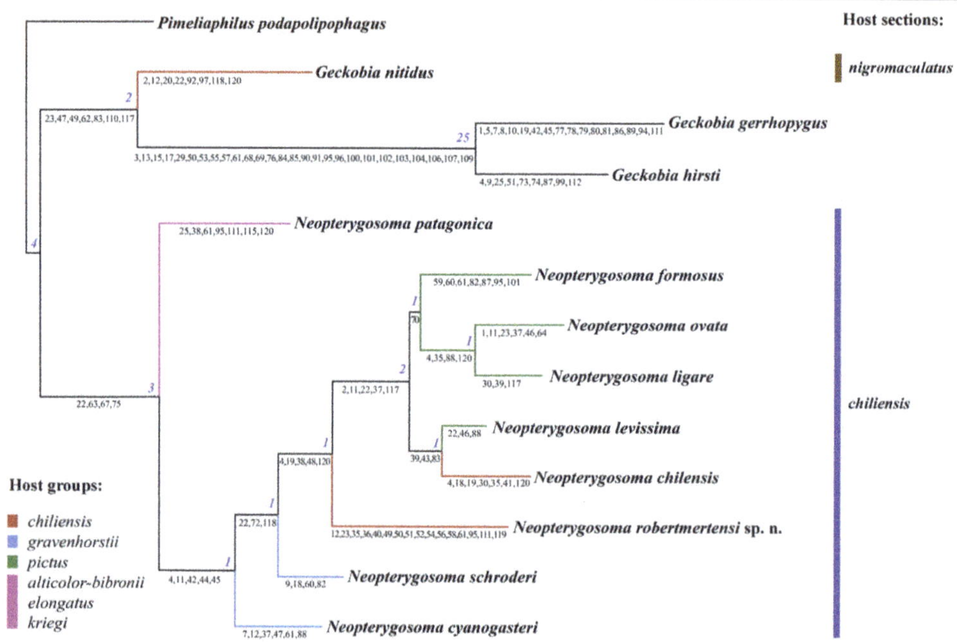

Figure 15. The most parsimonious tree (tree length 219, CI of 0.64, RI of 0.56, RC of 0.36) found using the branch-and-bound search option for the unordered and unweighted dataset. Numbers at nodes are Bremer indices. Numbers below branches are common synapomorphies (character numbers refer to List S1). Distribution of the mite species within host groups and section are marked in different colours.

The monophyly of the genus *Neopterygosoma* is supported by four synapomorphies (Bremer index 3), of which two are unique (length of coxae I, absence of coxal setae *2a* and *4a*). As expected, the resulting topology in this analysis is very similar to that in Fajfer [3]; in that hypothesis, the *P. patagonica* was the sister taxon to three species of Chilean species (*P. chilensis*, *P. ligare*, and *P. formosus*) included in the analysis. Our analysis confirms that *N. patagonica* from Argentina, considered less specialized (it has a circular body shape that is unable to hide under the host's scales), is the sister group to all the other species of the genus from Chile, considered more specialized (their idiosoma is wider than long, therefore, they live completely hidden beneath the scales). Its position is also supported by five common synapomorphies (Bremer index 2), of which three are unique (e.g., the presence of much longer setae in the postero-lateral part and peripheral part of idiosoma than in the anterior half of the dorsum).

The new species, *L. robertmertensi*, is a sister taxon to all species collected from *Liolaemus pictus* (*N. formosus*, *N. ovata*, *N. ligare*, and *N. levissima*) and *L. chiliensis* (*N. chilensis*) and is supported by the presence of five non-unique synapomorphies (Bremer index 1). The node uniting all of the above-mentioned mite species collected from the two host species is supported by five non-unique synapomorphies (Bremer index 2). Within the clade, the relationship within the species is weakly supported: *N. formosus* is a sister taxon to *N. ovata* + *N. ligare* (Bremer index 1), and the three species are a sister group to *N. levissima* + *N. chilensis* (Bremer index 1). Notably, the positions of both *N. schroederi* and *N. cyanogasteri*, are weakly supported by several non-unique synapomorphies (Bremer index of 1).

The only differences between the tree presented in Reference [3] and this tree lay in the position of the outgroup species. In the analysis [3], the genus *Geckobia* was paraphyletic with *G. nitidus* as a sister taxon to representatives of species of the genus *Neopterygosoma*, while *Geckobia gerrhopygus* + *G. hirsti* were as a sister taxon to the genera: *Gerrhosaurobia* + *Zanurobia* + *Ixodiderma* + *Scaphotrix* + *Pterygosoma*. In our analysis, all the outgroup *Geckobia* spp. are grouped in a common clade.

4. Discussion

The genus *Liolaemus* is the most ecologically diverse and species-rich genus distributed in South America from the high Andes of central Perú to the shores of Tierra del Fuego, and it spans an altitudinal range from sea level to over 5000 m [17]. The liolaemid lizards cover various climatic regimes and inhabit a great diversity of habitats (e.g., arid Atacama desert or humid rainforests). Moreover, the lizards exhibit a wide range of reproductive modes, types of diets, coloration patterns, and body sizes [18]. They also have a long evolutionary history dating back to 18–22 million years ago [19,20].

Currently, the genus includes over 280 species [12], but new species are being discovered at a rapid rate every year, e.g., [21,22]; therefore, it is estimated that the actual number of the species may be much higher. The genus is subdivided into two subgenera—*Liolaemus* (sensu stricto) and *Eulaemus* [23]—which appear to have separated at least 12.6 million years ago and are currently each divided into several groups. The presence of *Neopterygosoma* mites has been detected in 12 different species of hosts belonging to *Liolaemus* s. str. living on both sides of the Andes at different elevations, having different types of scales, coloration patterns, etc. [18,24].

As a rule, mites from different pterygosomatid genera are strictly specific with respect to lizard hosts, and cospeciation has a strong influence on the architecture of host–parasite relationships within the family Pterygosomatidae [3]. All representatives of the genus *Neopterygosoma* are monoxenous parasites (the *chilensis* group) except for *N. patagonica* collected from several *Liolaemus* spp. (oligoxenous parasite). Since host species from the same communities (these host taxa distributions partially overlap [17]) do not carry the same set of parasite species, we can expect to observe at least partially parallel evolution of *Neopterygosoma* mites of the *chilensis* group and *Liolaemus* hosts.

Nonetheless, the co-phylogenetic studies require phylogenetic hypotheses or data matrices for both lineages involved in the coevolutionary process. So far, the relationships between *Liolaemus* lizards at the species level are still questionable, e.g., [19,25]. Recently, Troncoso–Palacios et al. [26] conducted a phylogenetic study of the relationship of species of *Liolaemus* s. str. (based on three fragments of the mitochondrial genome); as a result, the species were divided into two main clades named: *chiliensis* and *nigromaculatus* sections. These findings were congruent with the phylogenetical tree (Figure 3 in Reference [17]) based on previous works [19,25,27]. Until now, all *Neopterygosoma* spp. are associated with closely related hosts belonging to the *chiliensis* section, whereas representatives of another pterygosomatid genus, i.e., *Geckobia nitidus* and *G. zapallarensis*, were collected from lizards of the *nigromaculatus* section [28] (marked on Figure 15).

However, not all of the host species groups were recovered monophyletic in the work of Troncoso–Palacios et al. [26]; therefore, Parenza et al. [29] infer a robust phylogeny (based on 541 ultra-conserved elements and 44 protein-coding genes) for a Chilenian clade of *Liolaemus* s. str. using representatives of all thirteen groups. As a result, only the relationship among the major Chilean clade of *Liolaemus* was resolved, as in previous studies [26] (Figure 15). All mites of the *chilensis* group (i.e., monoxenous 'more specialized' mite species) have been associated with closely related hosts belonging to three host groups of [26], i.e., *robertmertensi*, *gravenhorsti*, and *pictus*. The pterygosomatids have been found on all representatives within the two former groups except for *L. sanjuanensis* (*robertmertensi* group) and *L. gravenhorsti* (*gravenhorsti* group), which suggests that checking numerous host specimens of the two species for mites might lead to new species descriptions.

The highest number of *Neopterygosoma* spp. was described from a single host species—*L. pictus* (4 spp.)—belonging to the *pictus* group, including 11 host species. However, the number of species in this group is debatable because a few species have been treated as subspecies of *L. pictus* [30,31] or synonymized with *L. pictus* [32]. This host species has a wide distribution and forms a local population at low elevations (0–1600 a.s.l.) on both sides of the Andes, whereas the remaining *Liolaemus* spp. live either in the eastern or western slopes of the mountains [19]. It is unknown if the mite species occupy the full geographical range of their main host because so far, they have been found only in Isla Mocha (Arauco Province, Chile), although attempts to collect the mites from different localities were made (by M. Fajfer in ZSM and NHM). This could be interpreted as a consequence of the recent evolution of new mite species which are competing on the same host; therefore, further studies may prove that this group of parasites undergoes rapid adaptive radiation.

Our phylogenetic analysis shows that *N. patagonica* is a sister taxon to all monoxenous mites of the *chilensis* group. It agrees with the findings of Fajfer [3]. *P. patagonica* inhabits various host species of three different groups (see Figure 15) [17,26], which might suggest that this mite species' association is not fully recovered, and we can expect even more multi-host associations. *P. patagonica,* due to its circular shape of idiosoma, is morphologically unable to take shelter under the scales; therefore, most of its idiosoma protrudes beyond the scales. This probably allows the mite, by virtue of its effective dispersal abilities, to switch off quickly from a host when the opportunity arises, and then locate and colonize another host. This is especially probable if the host species, as in this case, share the same diet and occur at least partially in the same habitat [17].

The phylogenetical analysis indicates that the newly described species, *N. robertmertensi*, is nested within the mites of the *chilensis* group of *Neopterygosoma* associated with species of the section *chiliensis* of *Liolaemus* s. str. Its placement is also confirmed by a set of morphological features (see Figure 15), although the Bremmer support is only 1. The reason for this may be that *N. robertmertensi* has many unique features (e.g., the number of dorsomedial, ventromedial, or genital setae, i.e., characters 36, 40, and 49–52 in Figure 15, respectively), which do not match the description of the *chilensis* group provided in Reference [4]. Therefore, a revised description of the species group is presented here.

For the first time, we collected all mites from the host specimens that were preserved directly after collection. As a result, we collected hundreds of mites which were completely hidden beneath the lizard's scales. We found 1–12 specimens under a single scale, and the mites inhabited each body part of the host specimens. This lack of topical (habitat) specificity is quite surprising because in pterygosomatids living under the scales (such as *Pterygosoma* or *Geckobia*), a high preference towards a microhabitat on the host body is observed [33,34].

This large number of mites allowed us to observe morphological diversity among juveniles and adults and to illustrate for the first time the complete morphological ontogeny of these mites. For the first time in the family Pterygosomatidae, we were able to determine differences between the sexes of larvae. In male larvae of *N. robertmertensi*, the idiosoma is smaller and almost as long as wide (155–200 long and 170–215 wide), the genital region is situated ventrally, and the male develops directly in chrysalis inside the larval integument. In female larvae, the idiosoma is bigger and wider than long (170–250 long and 290–360 wide), the genital region is situated terminally, and the life cycle of the larva consists of both: active stages that feed on blood (protonymph, deutonymph, and adult) and legless inactive stages (nymphchrysalis, deutochrysalis, and imagochrysalis).

Our study shows that a female larva forms a chrysalis that resembles those found in other pterygosomatids (e.g., see Figure 8C in Reference [35]). Inside the chrysalis, a coiled protonymph develops. After molting, the newly emerged protonymph is larger than larva, and we observe the appearance of four pair of legs with the full set of setae on femora–tarsi IV, numerous idiosomal setae arranged similarly to subsequent stages, subcapitular setae *n*, weakly sclerotized small propodonotal shield, additional setae *ps3* in the genital region, leg setae on coxae II–III (*2b*, *3b*, *3c*, *3d*), genua-trochanter I–III (*v′GI–III*, *v″G–III*, *l′GI*, *vFI*, *vTrI–III)*, and tarsi I, i.e., *it″* (ζ), *tc′* (ζ) and *tc″* (ζ).

In the protonymph integument, we observed a deutochrysalis with a completely formed coiled deutonymph. This stage differs from a protonymph by the presence of much smaller gnathosoma and longer palpal setae (*dF*, *dG*), fewer setae on the mid-dorsal cluster, and the arrangement of setae (fewer in number) that resembles that in females. An adult female develops in the imagochrysalis (tritonymph). It differs from a deutonymph by the size of the idiosoma, the presence of additional two or three genital setae (*g4–g6*) and pseudanal setae (*ps4–ps5*), and ventral setae on trochanter IV. The males develop directly in the chrysalis inside the exoskeleton of larvae.

At this point, it is unclear whether the presence of both male and female larvae is unique for the genus *Neopterygosoma*. In Pterygosomatidae, as a rule, the description of juvenile morphology is often neglected. This could be due to several factors, such as (i) the difficulty of associating juveniles with an adult if the adults are missing in the sample, (ii) a small number of specimens found on hosts in museum collections (the mites might fall off the host during its preservation), (iii) the presence of only female mites on hosts, which may be explained by the short duration of their juvenile stages or (iv) the small size and transparency of the juvenile stages which make them difficult to notice on the hosts.

It is interesting to note that the larvae of *Neopterygosoma* differ from those of other genera, such as *Pterygosoma* or *Geckobia*, due to the absence of setae on tarsi I, specifically *it″* (ζ), *tc′* (ζ) and *tc″* (ζ). In other pterygosomatid larvae, only one fan-like proral setae *p′*, one simple tectal seta *tc′*, and paired iterals *it′* and *it″* in the form of eupathidia are present. Additionally, Norton's description of leg chaetotaxy [6], based on Grandjean's work [10,11], referred to the iterals as "post-larval setae" that are added in the protonymph stage. Yet, in *Neopterygosoma* spp. larvae, there is only one euphatidial setae *it′* while in contrast, the larvae of *Pterygosoma* have a pair of iterals (*it′* and *it″*).

5. Conclusions

In this research, we meticulously described and illustrated the morphology of the new species of pterygosomatid mite, *Neopterygosoma robertmertensi*, using scanning electron microphotography. As a result, we found new morphological features which were not recognized in previous studies of *Neopterygosoma* spp., such as the presence of a weakly sclerotized propodonotal shield. We observed the species morphological ontogeny and analyzed the main morphological differences between juvenile stages. For the first time in Pterygosomatidae, we observed both male and female larvae that differ mainly by the size and shape of idiosoma and from other pterygosomatid larvae by chaetotaxy of tarsi I. Additionally, the phylogenetic analysis showed that this species is nested within the *chilensis* group of *Neopterygosoma*, which was consistent with the morphological analysis. *Neopterygosoma* mites occur only on hosts belonging to three groups of the *chiliensis* section of the subgenus *Liolaemus* s. str., whose distributions partially overlap. Nonetheless, the hosts do not carry the same sets of parasite species. This suggests that mites of the *chilensis* group might be a good fit for cophylogenetic studies, especially if we take into account the fact that some studies conducted on pterygosomatid mites revealed a cophylogenetic pattern [3].

Supplementary Materials: The following supporting information can be downloaded at: https://www.mdpi.com/article/10.3390/ani13172809/s1, List S1: List of morphological characters and character states used in the analyses. Table S1: Matrix of morphological characters used in the phylogenetical analysis.

Author Contributions: Conceptualization, M.F. and M.S.; methodology, M.F. and M.S.; investigation, M.F.; resources and material collection, M.F.; writing—original draft preparation, M.F.; writing—review and editing, M.F. and M.S. All authors have read and agreed to the published version of the manuscript.

Funding: This research was funded by a grant from the European Commission's (FP6) Integrated Infrastructure Initiative programme SYNTHESYS + IL-TAF-TA3-2(2021).

Institutional Review Board Statement: Not applicable.

Informed Consent Statement: Not applicable.

Data Availability Statement: The material is stored in Cardinal Stefan Wyszynski University (Warsaw, Poland) and will be shared upon reasonable request to Monika Fajfer.

Acknowledgments: I want to express my gratitude to Dror Hawlena and Boaz Shacham from HUJ for their invaluable assistance during the mite collection process, as well as to Efrat Gavish-Regev from HUJ for lending us the mites. Additionally, I am extremely appreciative of Magdalena Gawlak from IPP NRI for capturing SEM photos of the mites.

Conflicts of Interest: The authors declare no conflict of interest.

References

1. Fajfer, M.; González-Acuña, D. Pterygosomatid mites of a new species group *ligare* (Acariformes: Pterygosomatidae: *Pterygosoma*) parasitizing tree iguanas (Squamata: Liolaemidae: *Liolaemus*). *Zootaxa* **2013**, *3693*, 301–319. [CrossRef] [PubMed]
2. Dittmar de la Cruz, K.; Morando, M.; Avila, L. Description of a new pterygosomatid mite (Acari: Actinedida: Pterygosomatidae) parasitic on *Liolaemus* spp. (Iguania: Liolaemini) from Argentina. *Zootaxa* **2004**, *521*, 1–6. [CrossRef]
3. Fajfer, M. Systematics of reptile-associated scale mites of the genus *Pterygosoma* (Acariformes: Pterygosomatidae) derived from external morphology. *Zootaxa* **2019**, *4603*, zootaxa.4603.3.1. [CrossRef]
4. Fajfer, M. A systematic revision of the genus *Neopterygosoma* Fajfer, 2019 (Acariformes: Pterygosomatidae) with the description of a new species. *Syst. Parasitol.* **2020**, *97*, 535–551. [CrossRef] [PubMed]
5. Fajfer, M. Redescription of *Pterygosoma patagonica* (Acariformes: Pterygosomatidae) with new host and distribution data. *Int. J. Acarol.* **2014**, *40*, 160–164. [CrossRef]
6. Norton, R.A. A review of F. Grandjean's system of leg chaetotaxy in the Oribatei and its application to the Damaeidae. In *Biology of Oribatid Mites*, 1st ed.; Dindal, D.L., Ed.; SUNY College of Environmental Science and Forestry: Syracuse, NY, USA, 1977; pp. 33–61.
7. Bochkov, A.V.; OConnor, B.M. A review of the external morphology of the family Pterygosomatidae and its systematic position within the Prostigmata (Acari: Acariformes). *Parasitologiya* **2006**, *40*, 201–214.
8. Krantz, G.W.; Walter, D.E. *A Manual of Acarology*; Texas Tech University Press: Lubbock, TX, USA, 2009.
9. Grandjean, F. Les Segments Post-Larvaires de L'hystérosoma Chez Les Oribates (Acariens). *Bull. Soc. Zool. Fr.* **1939**, *64*, 273–284.
10. Grandjean, F. Observations sur les Acariens de la famille des Stigmaeidae. *Arch. Sci. Phys. Nat.* **1944**, *26*, 103–1131.
11. Grandjean, F. Au sujet de l'organe de Claparede, des eupathides multiples et des taenidies mandiubulaires chez les Acariens actinochitineux. *Arch. Sci. Phys. Nat.* **1946**, *28*, 63–87.
12. The Reptile Database. Available online: http://www.reptile-database.org (accessed on 25 June 2023).
13. Page, R.D.M. *NDE, NEXUS Data Editor 0.5.0*; University of Glasgow: Glasgow, UK, 2001.

14. Swofford, D.L. PAUP*. In *Phylogenetic Analysis Using Parsimony (*and Other Methods), Version 4*; Sinauer Associates: Sunderland, MA, USA, 2002; p. 144.
15. Müller, K. PRAP—Computation of Bremer support for large data sets. *Mol. Phylogenet. Evol.* **2004**, *31*, 780–782. [CrossRef]
16. Rambaut, A.; Institute of Evolutionary Biology, University of Edinburgh, Edinburgh. FigTree v1.3.1. 2010. Available online: http://tree.bio.ed.ac.uk/software/figtree/ (accessed on 10 June 2023).
17. Pincheira-Donoso, D.; Scolaro, J.; Sura, P. A monographic catalogue on the systematics and phylogeny of the South American iguanian lizard family Liolaemidae (Squamata, Iguania). *Zootaxa* **2008**, *1800*, 1–85. [CrossRef]
18. Pincheira-Donoso, D.; Tregenza, T.; Hodgson, D.J. Body size evolution in South American *Liolaemus* lizards of the *boulengeri* clade: A contrasting reassessment. *J. Evol. Biol.* **2007**, *20*, 2067–2071. [CrossRef] [PubMed]
19. Schulte, J.A.; Macey, J.R.; Espinoza, R.E.; Larson, A. Phylogenetic relationships in the iguanid lizard genus *Liolaemus*: Multiple origins of viviparous reproduction and evidence for recurring Andean vicariance and dispersal. *Zool. J. Linn. Soc.* **2000**, *69*, 75–102. [CrossRef]
20. Fontanella, F.M.; Olave, M.; Avila, L.J.; Morando, M. Molecular dating and diversification of the South American lizard genus *Liolaemus* (subgenus *Eulaemus*) based on nuclear and mitochondrial DNA sequences. *Zool. J. Linn. Soc.* **2012**, *164*, 825–835. [CrossRef]
21. Avila, L.J.; Perez, C.H.F.; Minoli, I.; Medina, C.D.; Sites, J.W., Jr.; Morando, M. New species of *Liolaemus* (Reptilia, Squamata, Liolaemini) of the *Liolaemus donosobarrosi* clade from northwestern Patagonia, Neuquén province, Argentina. *Zootaxa* **2017**, *4362*, 535–563. [CrossRef]
22. Sánchez, K.I.; Morando, M.; Avila, L.J. A new lizard species of the *Liolaemus kingii* group (Squamata: Liolaemidae) from northwestern Chubut province (Argentina). *Zootaxa* **2023**, *5264*, 235–255. [CrossRef]
23. Laurent, R.F. Contribución al conocimiento de la estructura taxonómica del género *Liolaemus* Wiegmann (Iguanidae). *Bol. Asoc. Herp. Arg.* **1983**, *1*, 16–18.
24. Fernández, M.G.; Abdala, C.S.; Ruiz-Monachesi, M.R.; Semham, R.V.; Quinteros, A.S. Redescription of *Liolaemus robertmertensi*, Hellmich 1964 (Iguania: Liolaemidae) with description of a new species. *Cuad. Herpetol.* **2021**, *35*, 65–78. [CrossRef]
25. Espinoza, R.E.; Wiens, J.J.; Tracy, C.R. Recurrent evolution of herbivory in small, cold-climate lizards: Breaking the ecophysiological rules or reptilian herbivory. *Proc. Natl. Acad. Sci. USA* **2004**, *101*, 16819–16824. [CrossRef]
26. Troncoso-Palacios, J.; Schulte, J.A.; Marambio-Alfaro, Y.; Hiriart, D. Phenotypic variation, phylogenetic position and new distributional records for the poorly known *Liolaemus silvai* Ortiz, 1989 (Iguania: Iguanidae: Liolaemini). *S. Am. J. Herpetol.* **2015**, *10*, 71–81. [CrossRef]
27. Schulte, J.A.; Losos, J.B.; Cruz, F.B.; Núñez, H. The relationship between morphology, escape behaviour and microhabitat occupation in the lizard clade *Liolaemus* (Iguanidae: Tropidurinae: Liolaemini). *J. Evol. Biol.* **2004**, *17*, 408–420. [CrossRef] [PubMed]
28. Fajfer, M. Mites of the new species group *nitidus* (Acariformes: Pterygosomatidae: *Geckobia*), parasites of lizards in South America. *Syst. Parasitol.* **2015**, *90*, 213–222. [CrossRef] [PubMed]
29. Panzera, A.; Leaché, A.D.; D'Elia, G.; Victoriano, P.F. Phylogenomic analysis of the Chilean clade of *Liolaemus* lizards (Squamata: Liolaemidae) based on sequence capture data. *PeerJ* **2017**, *5*, e3941. [CrossRef] [PubMed]
30. Pincheira-Donoso, D.; Núñez, H. Las especies *chilenas* del género *Liolaemus* Wiegmann, 1834 (Iguania Tropiduridae: *Liolaeminae*). Taxonomía, sistemática y evolución. *Mus. Nac. Hist. Nat. Chile Publ. Occ.* **2005**, *59*, 7–486.
31. Ruiz de Gamboa, M. Lista actualizada de los reptiles de Chile. *Bol. Chil. Herp.* **2016**, *3*, 7–12.
32. Lobo, F.; Espinoza, R.E.; Quinteros, S. A critical review and systematic discussion of recent classification proposals for liolaemid lizards. *Zootaxa* **2010**, *2549*, 1–30. [CrossRef]
33. Fajfer, M.; Karanth, P. New morphological and molecular data reveal an underestimation of species diversity of mites of the genus *Geckobia* (Acariformes: Pterygosomatidae) in India. *Diversity* **2022**, *14*, 1064. [CrossRef]
34. Fajfer, M. Three new species of scale mites (Acari: Pterygosomatidae) parasitizing *Agama sankaranica* (Sauria: Agamidae). *Zootaxa* **2013**, *3700*, 271–272. [CrossRef]
35. Fajfer, M. Two new species of the genus *Pterygosoma* (Acariformes: Pterygosomatidae) parasitizing agamid lizards (Sauria: Agamidae) from the Indian subcontinent. *Acta Parasitol.* **2016**, *61*, 343–354. [CrossRef]

animals

MDPI

Article

Tinamiphilopsis temmincki sp. n., a New Quill Mite Species from Tataupa Tinamou, and the Early History of Syringophilid Mites †

Maciej Skoracki [1,*], Monika Fajfer [2], Martin Hromada [3], Jan Hušek [4] and Bozena Sikora [1,*]

1 Department of Animal Morphology, Faculty of Biology, Adam Mickiewicz University, Uniwersytetu Poznańskiego 6, 61-614 Poznań, Poland
2 Department of Molecular Biology and Genetics, Institute of Biological Sciences, Cardinal Stefan Wyszynski University, Wóycickiego 1/3, 01-938 Warsaw, Poland; m.fajfer@uksw.edu.pl
3 Laboratory and Museum of Evolutionary Ecology, Department of Ecology, Faculty of Humanities and Natural Sciences, University of Prešov, 08001 Prešov, Slovakia; hromada.martin@gmail.com
4 National Museum of the Czech Republic, Václavské námestí 68, 11579 Prague, Czech Republic; jan.husek@nm.cz
* Correspondence: skoracki@amu.edu.pl (M.S.); bozena.sikora@amu.edu.pl (B.S.)

† urn:lsid:zoobank.org:act:096661C0-2D05-4998-88E9-CE4032B64109.

Simple Summary: This research presents a description of a new species of quill mite, *Tinamiphilopsis temmincki* sp. n. (Acariformes: Syringophilidae), which was found on a representative of palaeognathous bird species, the Tataupa Tinamou (*Crypturellus tataupa*), in South America. Alongside describing this new species, a phylogenetic analysis was conducted on the primitive syringophilid genera. The results indicate that the genus *Tinamiphilopsis* is nested among the syringophilid genera associated with neognathous birds. This placement has significant implications for understanding the evolutionary relationship between quill mites and their avian hosts.

Abstract: The quill mite fauna of the Syringophilidae family (Acariformes: Prostigmata), which is associated with palaeognathous birds of the Tinamiformes order, remains poorly studied. Thus far, only two species of syringophilid mites have been documented on four species of tinamous. In this study, we present a description of a new species, *Tinamiphilopsis temmincki* sp. n., which was found on the Tataupa Tinamou (*Crypturellus tataupa*) in South America. This newly identified species differs from others in the genus due to the short hysteronotal setae *d2* in females, unlike the long setae *d2* found in females of other *Tinamiphilopsis* species. In addition to describing the new species, we conducted a phylogenetic analysis of the primitive syringophilid genera. The results reveal that the *Tinamiphilopsis* genus does not emerge as a sister group to all other syringophilids. Instead, it is deeply embedded within the radiation of quill mites associated with neognathous birds. This study provided evidence that mites belonging to the genus *Tinamiphilopsis* initially parasitised Neoavian birds before host switching to tinamous birds. This placement carries significant implications for our understanding of the evolution of quill mites and their relationship with their avian hosts.

Keywords: Acari; birds; ectoparasites; phylogeny; Syringophilidae; tinamous

Citation: Skoracki, M.; Fajfer, M.; Hromada, M.; Hušek, J.; Sikora, B. *Tinamiphilopsis temmincki* sp. n., a New Quill Mite Species from Tataupa Tinamou, and the Early History of Syringophilid Mites. *Animals* **2023**, *13*, 2728. https://doi.org/10.3390/ani13172728

Academic Editor: Alexis Ribas

Received: 3 July 2023
Revised: 24 July 2023
Accepted: 23 August 2023
Published: 28 August 2023

1. Introduction

Tinamidae (Tinamous), the only family in the order Tinamiformes, consists of small- to medium-sized birds found in Central and South America. This family comprises 47 species in nine genera and two subfamilies, Tinaminae and Nothurinae [1,2]. Birds of this family are widespread geographically and are associated with woodland and open grassland habitats from Southern Mexico to Patagonia [3,4]. Many studies have established the monophyly of Tinamidae and their connection to flightless ratites (including ostriches, emus, and their relatives). Both groups belong to palaeognaths (Palaeognathae), an early diverging group of modern birds [5–13]. However, there is limited research on the relationships among tinamous themselves [14,15]. The most comprehensive study was recently presented by Almeida et al. [16] and comprised the analysis of tinamous phylogenetic relationships and divergence dates, including both living and extinct species.

Prostigmatan fauna (Acariformes: Trombidiformes: Prostigmata) that is permanently associated with tinamous is represented only by members of the family Syringophilidae, whereas representatives of the other prostigmatan families, including Harpirhynchidae, Cheyletidae (Ornithocheyletini, Metacheyletiini, and Cheletosomatini), and Ereynetidae (Speleognathinae), have never been collected from any of the palaeognathous birds [17–21]. Currently, the family Syringophilidae associated with Tinamidae is represented by two species of the genus *Tinamiphilopsis*, which are recorded on four species of the subfamily Nothurinae, i.e., *Tinamiphilopsis elegans* Skoracki and Sikora, 2004, collected from the elegant crested tinamou *Eudromia elegans* Geoffroy Saint-Hilaire, and *Tinamiphilopsis ariconte* Skoracki et al., 2012, recorded from the red-winged tinamou *Rhynchotus rufescens* (Temminck), the white-bellied nothura *Nothura boraquira* (Spix), and the lesser nothura *Nothura minor* (Spix) [22,23]. Our knowledge encompasses only the four host species mentioned above, representing merely 9% of tinamous diversity, which vividly demonstrates the paucity of our understanding regarding syringophilid mites from this host group.

In this paper, we present the description of a new species of syringophilid mite, *Tinamiphilopsis temmincki* sp. n., collected from a representative of the subfamily Tinaminae, the tataupa tinamou, *Crypturellus tataupa* (Temminck), from South America. We also conducted a phylogenetic analysis to examine the placement of the *Tinamiphilopsis* in relation to the most primitive genera of Syringophilidae. Our findings shed new light on the evolutionary relationships of these mites and provide important insights into their biodiversity.

2. Materials and Methods

2.1. Mites Collection and Description

Mite material was collected from the dry bird skin of *Crypturellus tataupa* deposited in the ornithological collection, which is housed in the National Museum of the Czech Republic, Prague, Czechia (NMP) (Figure 1). Under laboratory conditions, the infected quill (the wing-covert quill) was dissected. Individual mites were removed and placed in Nesbitt's liquid for 36 h at room temperature, and then, they were mounted on slides in Faure's medium [24]. Identifications and drawings of the mite specimens were carried out using a ZEISS Axioscope light microscope (Carl-Zeiss AG, Oberkochen, Germany) equipped with differential interference contrast optics. Drawings of the new quill mite species were made with the drawing attachment (a camera lucida).

Figure 1. Host specimen of the tataupa tinamou *Crypturellus tataupa*, infested by *Tinamiphilopsis temmincki* sp. n.

All measurements in the description are presented in micrometers. The paratypes' measurements are indicated in brackets, appearing after the data for the holotype. The idiosomal setation adheres to Grandjean's classification [25] as adapted for Prostigmata by Kethley [26]. The leg chaetotaxy follows the nomenclature proposed by Grandjean [27], while the morphological terminology is in accordance with Skoracki [24]. The scientific and common names of the birds are based on Clements et al. [2].

Specimen depositories and reference numbers are abbreviated as follows: AMU—Adam Mickiewicz University, Department of Animal Morphology, Poznan, Poland; ZSM—Bavarian State Collection of Zoology, Munich, Germany.

2.2. *Phylogenetic Analysis*

2.2.1. Taxa Selection

Because this study aimed to recognise the phylogenetic relationship of the genus *Tinamiphilopsis*, we included in the ingroup all mite genera that possess a full complement of setae of the idiosoma and legs (plesiomorphic feature). Considering the arguments of Yeates [28] and Prendini [29] that it is preferable to include real species in a cladistic analysis rather than supra-species taxa, the genera or each species group recognised within them is represented by 1–3 species in our analysis.

Because the monophyly of the family Syringophilidae was tested with numerous outgroups and always received high support [30–32], only two outgroups were used in the analyses, a free-living predator *Cheyletus eruditus* (Schrank) and a quill-inhabiting predator *Cheletopsis norneri* (Poppe), both belonging to the sister family Cheyletidae.

2.2.2. Cladistic Analysis

The qualitative characters from the external morphology, such as the presence/absence of a structure or the form of specific morphological features, were used in this analysis. Only adult females were analysed because males and immatures are unknown in many included taxa. In total, 29 OTUs and 49 informative characters were included in the maximum parsimony analysis (Supplementary Table S1). The data matrix was prepared using NEXUS Data Editor 0.5.0 [33] (Supplementary Table S2).

All characters were treated as unordered, and their states were polarised using an outgroup comparison. The plesiomorphic state of each character was designated as '0', the apomorphic states were designated as '1, 2, 3', the missing states were designated as '?', and inapplicable was designated as '-'. Characters with multiple states were treated as polymorphic and not modified into binary characters. The characters, such as the number of tines in the proral setae (p' and p''), the number of chambers in the peritreme branches, and the total body lengths, were divided into multiple states.

The reconstruction of phylogenetic relationships was performed using PAUP 4.0 [34]. The heuristic search option was used for the maximum parsimony analysis. The delayed transformation option favours parallelism over reversal and was applied for a posteriori optimisation of character states and tracing of character changes in lineages. Initially, all characters were unweighted, and then successive weighting was performed according to the rescaled consistency index (RC) to find a maximally consistent tree [35,36].

3. Results

3.1. *Systematic*

Family: Syringophilidae Lavoipierre, 1953.
Subfamily: Syringophilinae Lavoipierre, 1953.
Genus: *Tinamiphilopsis* Skoracki and Sikora, 2004.

3.1.1. Description

Tinamiphilopsis temmincki sp. n.

Female, holotype (Figures 2 and 3): The total body length is 700 (660–750 in 11 paratypes). In the gnathosoma, the stylophore is 250 (230–250) long, and the exposed portion of the stylophore is apunctate and 190 (175–190) long. The infracapitulum is punctate in the anterior part. Each medial branch of the peritremes has one longitudinal chamber and each lateral branch has five chambers. The movable cheliceral digit is 190 (185–190) long. In the idiosoma, the propodonotal shield is well sclerotised and punctate, with a concave posterior margin, and bearing bases of all propodonotal setae except *c2*. The propodonotal setae *vi*, *ve*, and *si* are smooth. The length ratio of setae *vi*:*ve*:*si* is 1:1.4–1.8:1.7–2.4. The bases of setae *c1* and *se* are situated at the same transverse level. The hysteronotal shield is well sclerotised, fused to the pygidial shield, and apunctate, and the bases of setae *d1* are situated on the lateral margin, with the anterior margin reaching the level of setae *d2*. The bases of setae *d1* are situated closer to *d2* than to *e2*. The length ratio of setae *d2*:*d1*:*e2* is 1:2:2.3–2.4. The genital plate is well sclerotised, bearing bases of setae *ag2* and *ag3* on the lateral margins. Setae *ag1* and *ag2* are subequal in length, both slightly shorter than *ag3*. The coxal fields I–IV are well sclerotised and punctate. In the legs, the solenidia are shown in Figure 3B, and there are fan-like setae of legs III and IV with nine or ten tines.

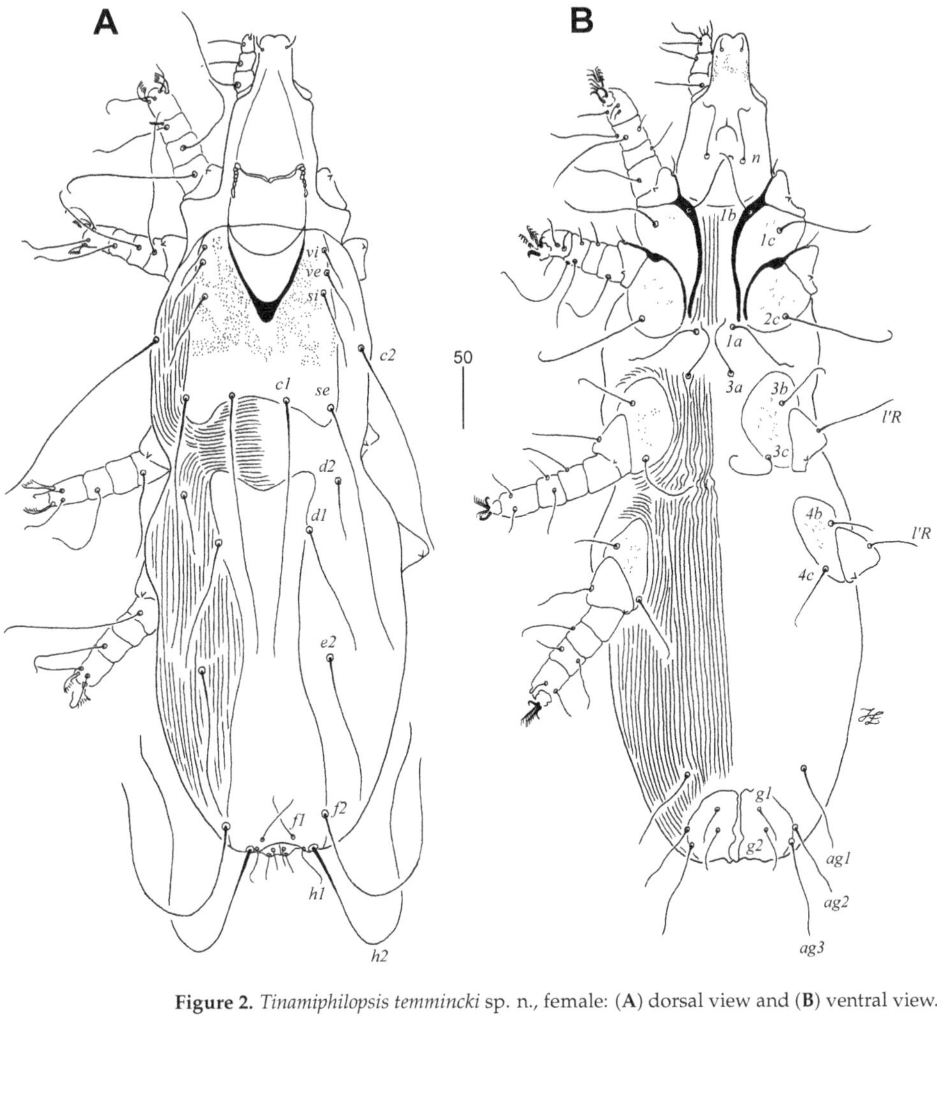

Figure 2. *Tinamiphilopsis temmincki* sp. n., female: (**A**) dorsal view and (**B**) ventral view.

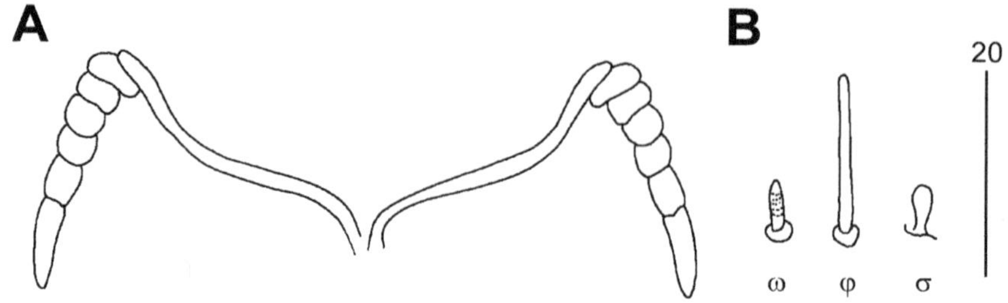

Figure 3. *Tinamiphilopsis temmincki* sp. n., female: (**A**) peritremes and (**B**) solenidia of leg I.

Lengths of setae: *vi* 35 (25–35), *ve* 50 (45–60), *si* 60 (60–65), *se* 200 (175–200), *c1* 205 (185–210), *c2* 185 (195–215), *d1* 105 (100–115), *d2* 50 (50–55), *e2* 120 (120–140), *f1* 40 (35–40), *f2* 215 (180–205), *h1* 30 (25–35), *h2* 285 (300–320), *ag1* 65 (55–60), *ag2* 65 (40–60), *ag3* 80 (65–75), *ps1* and *ps2* 25 (25–30), *g1* and *g2* 35 (30–35), *l'RIII* 55 (45–55), *l'RIV* 30 (35–40), *3b* 50 (40–50), *3c* 65 (50–65), *4b* 40 (35–50), *4c* 55 (45–50), *tc'III–IV* 35 (30–35), and *tc"III–IV* 55 (45–55).

Male (Figure 4): The total body length is 570 in one paratype. In the gnathosoma, the stylophore is 200 long, and an exposed portion of the stylophore is apunctate and 160 long. The infracapitulum is covered with minute punctations in the posterior part. Each medial branch of the peritreme has one chamber and each lateral branch has six chambers. In the idiosoma, the propodonotal shield is entire and punctate, rectangular in shape, and bearing bases of all propodonotal setae except *c2*. The length ratio of setae *vi:ve:si* is 1:2:5.3. The hysteronotal shield is well sclerotised, fused to the pygidial shield, and punctate laterally. Setae *d2* is 3.7 times longer than *d1* and *e2*. Setae *h2* is about 13 times longer than *f2*. The aggenital series are represented by two pairs of setae, with setae *ag1* being slightly (1.2 times) longer than *ag2*. The coxal fields I–IV are well sclerotised and punctate; the anterior margins of coxal fields III reach above the level of setae *3a*. The cuticular striations are shown in Figure 2A,B. In the legs, there are fan-like setae of legs III and IV with nine or ten tines.

Lengths of setae: *vi* 30, *ve* 60, *si* 160, *se* 210, *c1* 190, *c2* 200, *d1* 15, *d2* 55, *e2* 15, *f2* 20, *h2* 255, *ag1* 65, *ag2* 55, *l'RIII* 50, *l'RIV* 35, *3b* 50, and *3c* 70.

Type Material

Female holotype and paratypes: Eleven females and one male were collected from the wing-covert quill of the tataupa tinamou, *Crypturellus tataupa* (Temminck), from South America (host reg. no. NMP P6V-100166), and there are no other data.

Type Material Deposition

The female holotype and most paratypes were deposited in the AMU (reg. no. AMU MS 22-1112-002), except two female paratypes that were deposited in the SNSB-ZSM.

Differential Diagnosis

This new species, collected from a host representative of the subfamily Tinaminae, differs from the other two described species, which were collected from host members of the subfamily Nothurinae, by the presence of short propodonotal setae *si* and hysteronotal setae *d2*. In females of *Tinamiphilopsis temmincki*, the setae *si* and *d2* lengths are 60–65 μm and 50–55 μm, respectively. In females of *Tinamiphilopsis elegans* Skoracki and Sikora, 2004, the lengths of setae *si* and *d2* are 160–205 μm and 150–185 μm, respectively, whereas in females of *Tinamiphilopsis ariconte* Skoracki et al., 2012, the setae *si* and *d2* are 155–165 μm and 110–125 μm long, respectively.

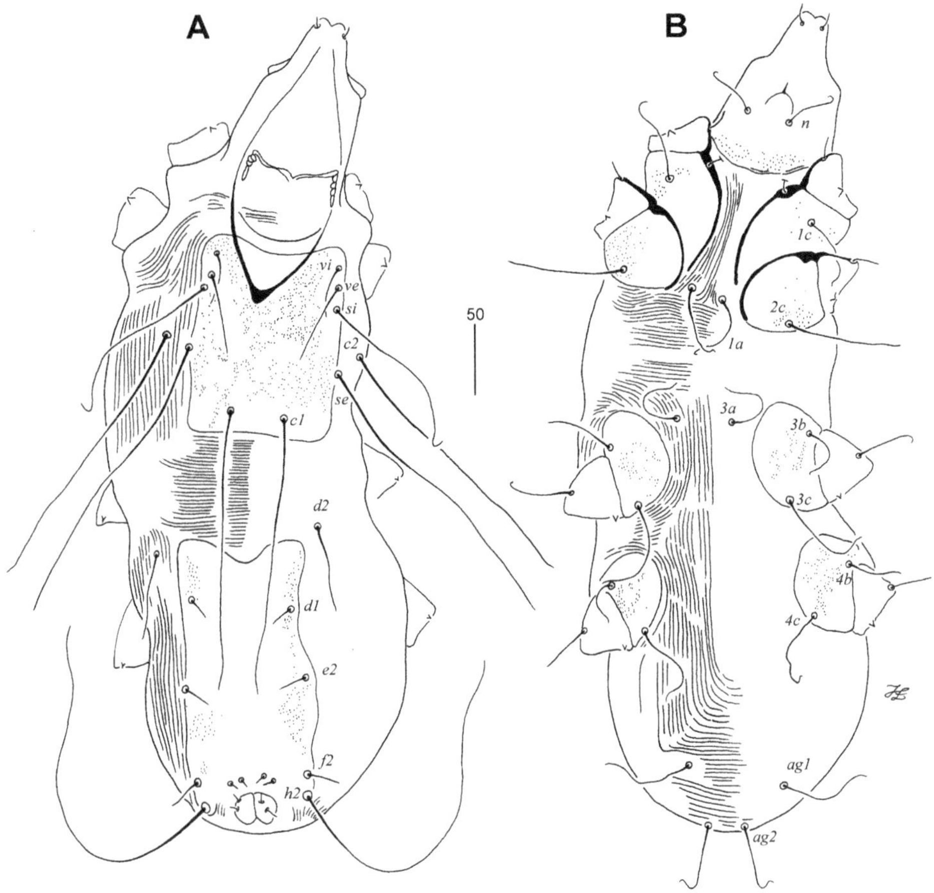

Figure 4. *Tinamiphilopsis temmincki* sp. n., male: (**A**) dorsal view and (**B**) ventral view.

Etymology

The new species is named in honour of the Dutch ornithologist and naturalist Coenraad Jacob Temminck (1778–1858), who made significant contributions to the field of ornithology, particularly in the study and classification of bird species.

3.2. Parsimony Analysis

Three equally parsimonious trees were produced based on the initial analysis, with all characters being treated as unordered and unweighted (tree length 114, consistency index (CI) for phylogenetically informative characters—0.50, retention index (RI)—0.71, and rescaled consistency index (RC)—0.35); the character data and data matrix are presented in Supplementary Table S1. The strict consensus of these trees is shown in Figure 5. The differences between these trees lay only in the position of the genus *Trypetoptila* in relation to the genera *Crotophagisyringophilus*, *Syringophilopsis*, and *Torotrogla* (Figure 6). The successive weighting according to the rescaled consistency index did not change the topology of the strict consensus tree.

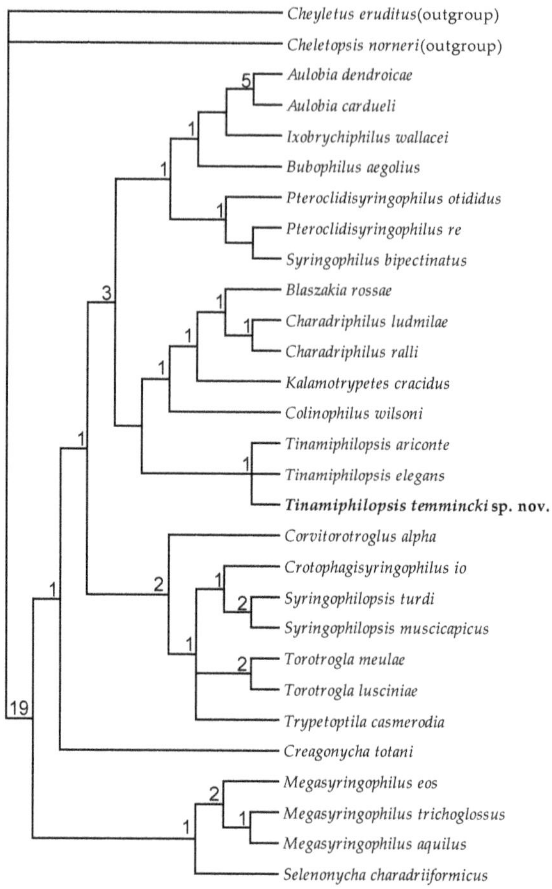

Figure 5. Strict consensus of the three most parsimonious trees (tree length 114, consistency index (CI) for phylogenetically informative characters—0.50, retention index (RI)—0.71, rescaled consistency index (RC)—0.35) found using the heuristic search option for the unordered and unweighted dataset. Numbers at nodes—Bremer indices.

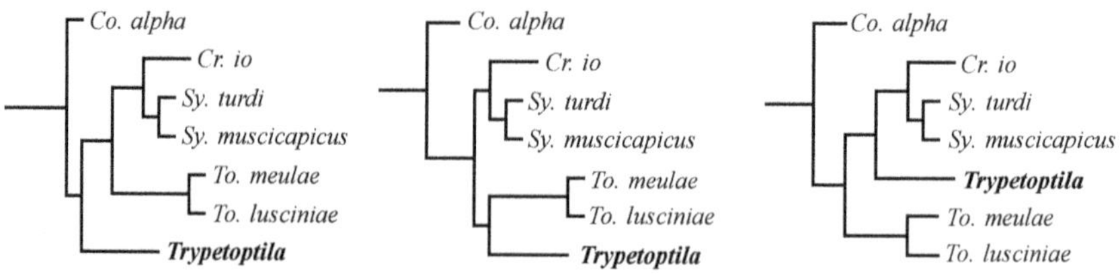

Figure 6. The differences between the topology of the three most parsimonious trees.

4. Discussion

To date, approximately 11,000 existing species are categorised as crown birds (Neornithes) [2]. These birds can be classified into two distinct and monophyletic groups: Palaeognathae (consisting of tinamous and ratites) and Neognathae (encompassing all other bird groups). Among the Neognathae, the Galloanserae (including Galliformes and Anseriformes) is considered the sister group to all other birds, referred to as the Neoaves [37–39]. Currently, syringophilid mites have been documented to inhabit 27 out of 44 orders of extant neognathous and paleognathous birds ([40], current study) (Figure 7).

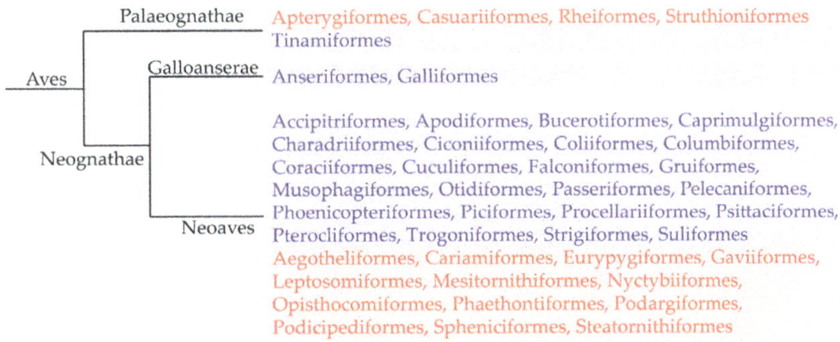

Figure 7. Phylogeny of birds with all extant orders (according to Sangster et al. [41]). Orders of birds on which syringophilid mites have been found are marked in blue; orders on which no syringophilids have been found thus far are marked in red.

4.1. Hypotheses on the Early History of Syringophilid Mites

The origin and the early evolution of birds and syringophilids associations is one of the most interesting aspects of the study of quill mites. It was hypothesised that syringophilid mites, which are similar to members of the family Cheyletidae, evolved from micro-predators that resided in bird nests or even the nests of theropod dinosaurs. Then, they migrated from such nests to feather quills [42,43]. Initially, the ancestors of syringophilids likely preyed upon other mites that inhabited wing vanes, like the modern cheyletid representatives of the tribe Cheletosomatini. It is worth noting that the majority of Cheletosomatini species are obligate predators residing in wing quills; however, mites from the genus *Picocheyletus* or *Metacheyletia* (the sole genus in the Metacheyletiini tribe) are likely parasites rather than predators in quills [44,45].

The "molecular clock" hypothesis suggests that the cheyletids and syringophilids diverged from each other approximately 180–185 million years ago, during the Early Jurassic period [46]. On the other hand, the earliest fossil widely accepted to belong to Neornithes, which includes all extant bird species, is *Vegavis* from the end-Cretaceous (~67 million years ago (Mya) [47]. However, numerous molecular dating studies have indicated that the diversification of Neornithes, which includes all extant bird species, likely started 100–110 million years ago [12,48,49]. In contrast, Prum et al. [8] presented findings, based on molecular clock analysis, that are congruent with the palaeontological record, supporting the major radiation of crown birds in the wake of the Cretaceous–Palaeogene (K–Pg) mass extinction (approximately 66 Mya). The facts mentioned above suggest that syringophilids likely had already formed relationships with the ancestors of birds, theropod dinosaurs, many of which had feathers, e.g., *Archaeopteryx* from the Late Jura [50–52] or *Aurornis* from the Middle-Late Jura [53].

4.2. Distribution of the Primitive Quill Mite Genera on the Host Lineages

The concept of coevolution was formally established as Fahrenholz's rule by Eichler [54,55]. The simplest version of this rule is that "Parasite phylogeny mirrors host phylogeny" [56]. Coevolution is an appealing concept due to its simplicity and elegant explanatory power for the evolution of numerous parasites. Furthermore, in cases where coevolution takes place, the phylogeny of hosts can be inferred from the phylogeny of their parasites, and vice versa. This reciprocal relationship may offer valuable insights into the evolutionary dynamics of both hosts and parasites [57]. The expected similarities between host and parasite phylogenies, however, often do not exceed the similarity expected by chance between two random trees. This is because historical events (host switches, extinctions, etc.) often erode the expected patterns of co-speciation [58].

In 2004, Skoracki and Sikora [22] described the first species of syringophilid mites, *Tinamiphilus elegans*, collected from palaeognathous birds, the elegant crested tinamou.

Eight years later, in 2012, Skoracki et al. [23] described the second species of this genus, *T. ariconte*, which was found on three tinamou hosts: the red-winged tinamou, the white-bellied nothura, and the lesser nothura. Taking into consideration that (i) syringophilids are obligate and permanent parasites; (ii) transmission occurs typically only when hosts come into direct physical contact, and most physical contact between individual hosts is between conspecifics, in particular between mates and between parents and offspring; (iii) many species of quill mite infect only a single or phylogenetically closely related species of host, and moreover, genera of syringophilids often are restricted to a single order of birds; (iv) representatives of the genus *Tinamiphilopsis* exhibit several primitive character conditions (e.g., smooth hypostomal apex, a large gnathosoma, edentate chelicerae, well-developed and sclerotised dorsal idiosomal shields, and full complement of idiosomal and leg setae); and (v) syringophilid mites exhibit high host specificity, the authors suggested that these discoveries support the hypothesis that the ancestor of the Syringophilidae transitioned to parasitism prior to the divergence of birds into the two major clades, Palaeognathae and Neognathae.

In 2013, Skoracki et al. [21] presented the first, but not fully resolved, phylogeny of syringophilid mites, where the genus *Tinamiphilopsis* was placed not as a sister lineage to the other syringophilid genera but in the core of the tree. These results contradicted the previous hypothesis [22,23] that the initial association of the genus *Tinamiphilopsis* was with Tinamiformes. The results obtained in the current study support the latter hypothesis. In the syringophilid tree, mites on the earliest derivate branches, i.e., *Selenonycha* Kethley and *Megasyringophilus* Fain et al., are associated with birds of the advanced clade Neoaves (Charadriiformes and Psittaciformes, respectively). In contrast, the mite genera associated with the earliest derivate clades of extant birds, Tinamiformes (Palaeognathae) and Galloanserae (Anseriformes and Galliformes), are mosaically distributed in the core of the tree (Figure 8). This contradiction between the presumable syringophilid parasitism of the common bird ancestor and the phylogenetic pattern obtained could be explained by the multiple switches (secondary infestation) from hosts of the Neoaves clade to palaeognathous and galloanserae birds and subsequent co-speciation.

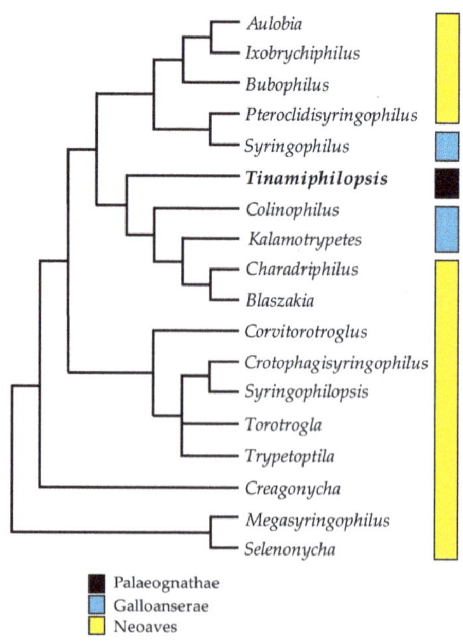

Figure 8. Distribution of the alpha mite genera on the main lineages of birds: Palaeognathae and Neognathae (Galloanserae + Neoaves).

5. Conclusions

In this research, we described a new species of syringophilid mite found in the feather quill of tinamou from the subfamily Tinaminae, the tataupa tinamou. This new species is easily distinguished from the other two species of the genus *Tinamiphilopsis* recorded from the representatives of the subfamily Nothurinae by the presence of the short setae *si* and *d2*. We also reconstructed the phylogeny of the most primitive genera of syringophilid mites, which showed incongruence with modern avian phylogenies. This suggests that host switching could play an important role in the early evolution of this group of mites. Furthermore, this study demonstrated that the mites of the genus *Tinamiphilopsis* originally parasitised Neoavian birds before moving to tinamous birds.

Supplementary Materials: The following supporting information can be downloaded at https://www.mdpi.com/article/10.3390/ani13172728/s1, Table S1: List of characters used in the phylogenetic analysis of the primitive syringophilid genera, i.e., the genera with a full complement of body and leg setae; Table S2: Data matrix.

Author Contributions: Conceptualisation, M.S., M.F. and B.S.; methodology, M.S. and M.F.; validation, M.S., M.F. and B.S.; formal analysis, M.S., M.F. and B.S.; investigation, M.S., M.F. and B.S.; material collection, M.H. and J.H.; data curation, M.S. and B.S.; writing—original draft preparation, M.S., M.F. and B.S.; writing—review and editing, M.S., M.F., M.H., J.H. and B.S.; visualisation, M.S. and J.H.; supervision, M.S.; project administration, B.S.; funding acquisition, B.S. All authors have read and agreed to the published version of the manuscript.

Funding: This research was funded by the Ministry of Education, Science, Research and Sport of the Slovak Republic under grant No. VEGA-1/0876/21, by the Slovak Research and Development Agency under contract No. APVV-22-0440 (to M.S., M.H. and B.S.), and by the National Scholarship Programme, SAIA, n. o. (2022/2023) (to B.S.).

Institutional Review Board Statement: Not applicable.

Informed Consent Statement: Not applicable.

Data Availability Statement: Data are available upon request from the corresponding author.

Acknowledgments: We thank Dr. Izabella Łaniecka (AMU) for the ink drawings of quill mites. We also thank the reviewers for their critical review of the manuscript.

Conflicts of Interest: The authors declare no conflict of interest. The funders had no role in the design of the study; in the collection of data and analyses; or in the decision to publish the results.

References

1. Winkler, D.W.; Billerman, S.M.; Lovette, I.J. Tinamous *(Tinamidae)*, ver. 1.0. In *Birds of the World*; Billerman, S.M., Keeney, B.K., Rodewald, P.G., Schulenberg, T.S., Eds.; Cornell Lab of Ornithology: Ithaca, NY, USA, 2020. [CrossRef]
2. Clements, J.F.; Schulenberg, T.S.; Iliff, M.J.; Roberson, D.; Fredericks, T.A.; Sullivan, B.L.; Wood, C.L. The eBird/Clements Checklist of Birds of the World. Ver. 2022. Available online: http://www.birds.cornell.edu/clementschecklist/download/ (accessed on 20 May 2023).
3. Cabot, J. Order Tinamiformes. In *Handbook of the Birds of the World*; Ostrich to Ducks; del Hoyo, J., Elliot, A., Sargatal, J., Eds.; Lynx Edicions: Barcelona, Spain, 1992; Volume 1, pp. 112–138.
4. Davies, S.J.J.F. *Ratites and Tinamous. Tinamidae, Rheidae, Dromaiidae, Casuariidae, Apterygidae, Struthionidae*; Oxford University Press: New York, NY, USA, 2002.
5. Cracraft, J. Phylogeny and evolution of the ratites birds. *Ibis* **1974**, *116*, 494–521. [CrossRef]
6. Lee, K.; Feinstein, J.; Cracraft, J. The phylogeny of ratite birds: Resolving conflicts between molecular and morphological data sets. In *Avian Molecular Evolution and Systematics*; Mindell, D.P., Ed.; Academic Press: San Diego, CA, USA, 1997; pp. 173–211.
7. Livezey, B.C.; Zusi, R.L. Higher-order phylogeny of modern birds (Theropoda, Aves: Neornithes) based on comparative anatomy. II. Analysis and discussion. *J. Linn. Soc.* **2007**, *149*, 1–95. [CrossRef]
8. Prum, R.O.; Jacob, S.B.; Dornburg, A.; Field, D.J.; Townsend, J.P.; Lemmon, E.M.; Lemmon, A.R. A comprehensive phylogeny of birds (Aves) using targeted next-generation DNA sequencing. *Nat. Lett.* **2015**, *15697*, 569–573. [CrossRef]
9. Hackett, S.J.; Kimball, R.T.; Reddy, S.; Bowie, R.C.K.; Braun, E.L.; Braun, M.J.; Chojnowski, J.L.; Cox, W.A.; Han, K.L.; Harshman, J.; et al. A phylogenomic study of birds reveals their evolutionary history. *Science* **2008**, *320*, 1763–1768. [CrossRef]
10. Haddrath, O.; Baker, B. Multiple nuclear genes and retroposons support vicariance and dispersal of the palaeognaths, and an Early Cretaceous origin of modern birds. *Proc. R. Soc.* **2012**, *279*, 4617–4625. [CrossRef]

11. Smith, J.V.; Braun, E.L.; Kimball, R.T. Ratite nonmonophyly: Independent evidence from 40 novel loci. *Syst. Biol.* **2013**, *62*, 35–49. [CrossRef]
12. Jarvis, E.D.; Mirarab, S.; Aberer, A.J.; Li, B.; Houde, P.; Li, C.; Ho, S.Y.W.; Faircloth, B.C.; Nabholz, B.; Howard, J.T.; et al. Whole-genome analyses resolve early branches in the tree of life of modern birds. *Science* **2014**, *346*, 1320–1331. [CrossRef]
13. Kimball, R.T.; Oliveros, C.H.; Wang, N.; White, N.D.; Barker, F.K.; Field, D.J.; Ksepka, D.T.; Chesser, R.T.; Moyle, R.G.; Braun, M.J.; et al. A Phylogenomic Supertree of Birds. *Diversity* **2019**, *11*, 109. [CrossRef]
14. Bertelli, S.; Giannini, N.P.; Goloboff, P.A. A phylogeny of the tinamous (Aves: Palaeognathiformes) based on integumentary characters. *Syst. Biol.* **2002**, *51*, 959–979. [CrossRef]
15. Bertelli, S.; Chiappe, L.M.; Mayr, G. Phylogenetic interrelationships of living and extinct Tinamidae, volant palaeognathous birds from the New World. *Zool. J. Linn. Soc.* **2014**, *172*, 145–184. [CrossRef]
16. Almeida, F.C.; Porzecanski, A.L.; Cracraft, J.L.; Bertelli, S. The evolution of tinamous (Palaeognathae: Tinamidae) in light of molecular and combined analyses. *Zool. J. Linn. Soc.* **2021**, *195*, 106–124. [CrossRef]
17. Bochkov, A.V.; Mironov, S.V.; Fain, A. Phylogeny and host-parasite relationships of the family Harpirhynchidae (Acari, Prostigmata). *Acarina* **1999**, *7*, 69–87.
18. Bochkov, A.V.; Fain, A. Phylogeny and system of the Cheyletidae (Acari: Prostigmata) special reference to their host-parasite associations. *Bull. Inst. R. Sci. Nat. Belgique* **2001**, *71*, 5–36.
19. Bochkov, A.V.; Klimov, P.B.; Skoracki, M. Morphological phylogenetic conflict in the parasitic mite family Harpirhynchidae (Acariformes: Cheyletoidea) correlates with host associations. *Zool. Anz.* **2017**, *271*, 33–48. [CrossRef]
20. Skoracki, M.; Zabludovskaya, S.A.; Bochkov, A.V. A review of Prostigmata (Acariformes: Trombidiformes) permanently associated with birds. *Acarina* **2012**, *20*, 67–107.
21. Skoracki, M.; Glowska, E.; Bochkov, A.V. Phylogeny of quill mites of the family Syringophilidae (Acari: Prostigmata) based on their external morphology. *Eur. J. Entomol.* **2013**, *110*, 663–675. [CrossRef]
22. Skoracki, M.; Sikora, B. *Tinamiphilopsis elegans* gen. nov. et sp. nov., a first record of the quill mites (Acari: Syringophilidae) from tinamou birds (Tinamiformes: Tinamidae). *Acta Parasitol.* **2004**, *49*, 348–352.
23. Skoracki, M.; Sikora, B.; Ozminski, M. A new quill mite species (Acari: Syringophilidae) parasitising tinamous (Aves: Tinamiformes). *Syst. Parasitol.* **2012**, *81*, 109–113. [CrossRef] [PubMed]
24. Skoracki, M. Quill mites (Acari: Syringophilidae) of the Palaearctic region. *Zootaxa* **2011**, *2840*, 1–416. [CrossRef]
25. Grandjean, F. Les segments post-larvaires de l'hystérosoma chez les Oribates (Acariens). *Bull. Soc. Zool. Fr.* **1939**, *64*, 273–284.
26. Kethley, J.B. Acarina: Prostigmata (Actinedida). In *Soil Biology Guide*; Dindal, D.L., Ed.; John Wiley & Sons: New York, NY, USA, 1990; pp. 667–756.
27. Grandjean, F. Observations sur les Acariens de la famille des Stigmaeidae. *Arch. Sci. Phys. Nat.* **1944**, *26*, 103–131.
28. Yeates, D.K. Groundplans and exemplars: Paths to the tree of life. *Cladistics* **1995**, *11*, 343–357. [CrossRef] [PubMed]
29. Prendini, L. Species or supraspecific taxa as terminals in cladistic analysis? Ground plans versus exemplars revisited. *Syst. Biol.* **2001**, *50*, 290–300. [CrossRef] [PubMed]
30. Bochkov, A.V. Classification and phylogeny of mites of the superfamily Cheyletoidea (Acari: Prostigmata). *Entomol. Obozr.* **2002**, *81*, 488–513. (In Russian)
31. Bochkov, A.V. Mites of the family Cheyletidae (Acari: Prostigmata): Phylogeny, distribution, evolution and analysis of parasite-host relationship. *Parazitologia* **2004**, *38*, 122–138. (In Russian)
32. Bochkov, A.V.; O'Connor, B.M.; Wauthy, G. Phylogenetic position of the family Myobiidae within the Prostigmata (Acari: Acariformes). *Zool. Anz.* **2008**, *247*, 15–45. [CrossRef]
33. Page, R.D.M. *NDE, NEXUS Data Editor 0.5.0*; University of Glasgow: Glasgow, Scotland, 2001.
34. Swofford, D.L. *PAUP. Phylogenetic Analysis Using Parsimony (* and Other Methods), Version 4*; Sinauer Associates: Sunderland, MA, USA, 2001.
35. Platnick, N.I.; Humphries, C.J.; Nelson, G.J.; Williams, D.M. Is Farris optimisation perfect? Three-taxon statements and multiple branching. *Cladistics* **1996**, *12*, 243–252. [CrossRef]
36. Farris, J.S. A successive approximations approach to character weighting. *Syst. Zool.* **1969**, *18*, 374–385. [CrossRef]
37. Cracraft, J. Avian evolution, Gondwana biogeography, and the Cretaceous-Tertiary mass extinction event. *Proc. R. Soc.* **2001**, *268B*, 459–469. [CrossRef]
38. Livezey, B.C.; Zusi, R.L. Higher-order phylogenetics of modern Aves based on comparative anatomy. *Neth. J. Zool.* **2001**, *51*, 179–205. [CrossRef]
39. van Tuinen, M.; Sibley, C.G.; Hedges, S.B. The early history of modern birds inferred from DNA sequences of nuclear and mitochondrial ribosomal genes. *Mol. Biol. Evol.* **2000**, *17*, 451–457. [CrossRef]
40. Zmudzinski, M.; Skoracki, M.; Sikora, B. An Updated Checklist of Quill Mites of the Family Syringophilidae (Acariformes: Prostigmata). 2021. Available online: https://figshare.com/articles/dataset/An_updated_checklist_of_quill_mites_of_the_family_Syringophilidae_Acariformes_Prostigmata_/16529574/1 (accessed on 15 June 2023).
41. Sangster, G.; Braun, E.L.; Johansson, U.S.; Kimball, R.T.; Mayr, G.; Suh, A. Phylogenetic definitions for 25 higher-level clade names of birds. *Avian. Res.* **2022**, *13*, 100027. [CrossRef]
42. Bochkov, A.V. New observations on phylogeny of cheyletoid mites (Acari: Prostigmata: Cheyletoidea). *Proc. Zool. Inst. (St. Petersburg)* **2008**, *312*, 54–73. [CrossRef]

43. Bochkov, A.V. A review of mites of the parvorder Eleutherengona (Acariformes: Prostigmata)—Permanent parasites of mammals. *Acarina* **2009**, *1*, 1–149.
44. Bochkov, A.V.; Skoracki, M. A new cheyletid mite *Metacheyletia ngaii* n. sp. (Acariformes: Cheyletidae) from quills of *Corythaixoides leucogaster* (Musophagidae) from Tanzania. *Acarologia* **2011**, *51*, 93–97. [CrossRef]
45. Skoracki, M. New data on the species *Metacheyletia degenerata* Fain and Bochkov (Acariformes: Cheyletidae). *Ann. Parasitol.* **2016**, *62*, 349–350. [CrossRef] [PubMed]
46. Dabert, M.; Witalinski, W.; Kazmierski, A.; Olszanowski, Z.; Dabert, J. Molecular phylogeny of acariform mites (Acari, Arachnida): Strong conflict between phylogenetic signal and long-branch attraction artifacts. *Mol. Phyl. Evol.* **2010**, *56*, 222–241. [CrossRef] [PubMed]
47. Clarke, J.A.; Tambussi, C.P.; Noriega, J.I.; Erickson, G.M.; Ketcham, R.A. Definitive fossil evidence for the extant avian radiation in the Cretaceous. *Nature* **2005**, *433*, 305–308. [CrossRef]
48. Brown, J.W.; Rest, J.S.; García-Moreno, J.; Sorenson, M.D.; Mindell, D.P. Strong mitochondrial DNA support for a Cretaceous origin of modern avian lineages. *BMC Biol.* **2008**, *6*, 6. [CrossRef] [PubMed]
49. Jetz, W.; Thomas, G.H.; Joy, J.B.; Hartmann, K.; Mooers, A.O. The global diversity of birds in space and time. *Nature* **2012**, *491*, 444–448. [CrossRef]
50. Mayr, G.; Pohl, B.; Peters, D.S. A well-preserved *Archaeopteryx* specimen with theropod features. *Science* **2005**, *310*, 1483–1486. [CrossRef] [PubMed]
51. Xu, X.; Zhou, Z.; Wang, X. Exceptional dinosaur fossils show ontogenetic development of early feathers. *Nature* **2010**, *464*, 1338–1341. [CrossRef] [PubMed]
52. Xu, X.; You, H.; Du, K.; Han, F. An *Archaeopteryx*-like theropod from China and the origin of Avialae. *Nature* **2011**, *475*, 465–470. [CrossRef] [PubMed]
53. Godefroit, P.; Cau, A.; Dong-Yu, H.; Escuillié, F.; Wenhao, W.; Dyke, G. A Jurassic avialan dinosaur from China resolves the early phylogenetic history of birds. *Nature* **2013**, *498*, 359–362. [CrossRef]
54. Fahrenholz, H. Ectoparasiten und abstammungslehre. *Zool. Anz.* **1913**, *41*, 371–374.
55. Eichler, W. Wirtsspezifitat und stammesgeschichliche Gleichläufigkeit (Fahrenholzsche Regel) bei Parasiten im allgemeinen und bei Mallophagen im besonderen. *Zool. Anz.* **1940**, *132*, 254–262.
56. Brooks, D.R. Testing the content and extent of host-parasite coevolution. *Syst. Zool.* **1979**, *28*, 299–307. [CrossRef]
57. Barker, S.C. Phylogeny and classification, origins, and evolution of host associations of lice. *Int. J. Parasitol.* **1994**, *24*, 1285–1291. [CrossRef]
58. Page, R.D.M. Tangled trees. In *Phylogeny, Cospeciation and Coevolution*; The University of Chicago Press: Chicago, IL, USA, 2003.

animals

MDPI

Article

A New Species of *Demodex* (Acariformes: Prostigmata) Observed in the Mouflon, *Ovis aries musimon* (Artiodactyla: Bovidae) with Data on the Parasitism and Occurrence of Other Ectoparasites †

Paulina Kozina, Joanna N. Izdebska * and Leszek Rolbiecki

Department of Invertebrate Zoology and Parasitology, Faculty of Biology, University of Gdańsk, Wita Stwosza 59, 80-308 Gdańsk, Poland; paulina.kozina@ug.edu.pl (P.K.); leszek.rolbiecki@ug.edu.pl (L.R.)

* Correspondence: joanna.izdebska@ug.edu.pl

† urn:lsid: zoobank.org:act:F90A91F9-D87F-4153-9440-040F54CB92B3.

Simple Summary: The aim of the study was to analyse the community of parasitic mites in mouflons *Ovis aries musimon* from the Polish population introduced from the Mediterranean area. It was important to determine whether these parasites were specific to or typical of the genus *Ovis* and whether foreign species had appeared with the mouflons that might pose a parasitological threat to the native fauna or, according to the theory of *parasites lost*, the introduction contributed to the loss of the original fauna of parasitic mites and the subsequent repopulation of the vacant microhabitats by species typical of local bovids, cervids, and other ungulates. Forty-one mouflons, obtained by hunting, were examined. Two species of mites were found, including *Demodex musimonis* sp. nov., previously unknown demodecid mites, probably specific to the mouflon. These discoveries were accompanied by the tick *Ixodes ricinus*, a parasite common in many parts of Europe with a wide host range, and *Lipoptena cervi*, a parasitic fly typical of ungulates. The new species represents the first finding of a skin mite of the family Demodecidae in wild representatives of the caprine Caprinae, and the present study is the first to examine the parasitic arthropods of the mouflon, including ectoparasites and skin mites. *Demodex musimonis* sp. nov. occurred asymptomatically in the mouflon, and as a specific species does not pose a threat to the native fauna. In contrast, the presence of *I. ricinus* clearly indicates a broadening of the reservoir for pathogens transmitted by epidemiologically significant ectoparasite species in Europe.

Abstract: A greater understanding of mite biodiversity and ecology can explain their preference for microhabitats within host bodies, i.e., as ecto-, meso-, and endoparasites. Similarly, learning about the patterns formed by mite communities in wild animals can shed light on the mechanisms of parasitosis development in their domesticated counterparts. Hence, the present study examined the acarofauna of the mouflon, introduced to Poland in the early 20th century from a region of endemic occurrence. Forty-one individuals were examined for the presence of ectoparasites between 2010 and 2013; later, skin fragments were analysed for the presence of skin mites. A new species of Demodecidae, *Demodex musimonis* sp. nov., was discovered in the skin of the pasterns of 14.6% mouflons, as well as the ectoparasitic tick *Ixodes ricinus* (prevalence 29.3%) and ungulate-typical fly *Lipoptena cervi* (34.1%). However, no mites typical for ovine *Ovis* (Psorergatidae, Psoroptidae) were noted, nor any colonisation of microhabitats by species from local ungulates, resulting from the loss of original parasites during the introduction. A comparison of the acarofauna of the mouflon and domestic sheep *Ovis aries aries* suggests that a mite community with a complex structure and the co-occurrence of different families may be formed. However, it is not known whether the acarofauna of domestic sheep is derived from wild ancestors or the process of domestication. Even so, the peculiar Demodecidae must have been part of the natural parasitofauna of wild sheep, of which only *D. musimonis* sp. nov. has been recognised so far.

Citation: Kozina, P.; Izdebska, J.N.; Rolbiecki, L. A New Species of *Demodex* (Acariformes: Prostigmata) Observed in the Mouflon, *Ovis aries musimon* (Artiodactyla: Bovidae) with Data on the Parasitism and Occurrence of Other Ectoparasites. *Animals* **2023**, *13*, 2619. https://doi.org/10.3390/ani13162619

Academic Editors: Monika Fajfer and Maciej Skoracki

Received: 30 July 2023
Revised: 10 August 2023
Accepted: 12 August 2023
Published: 14 August 2023

Keywords: mouflon; *Ovis aries musimon*; *Demodex musimonis*; *Ixodes ricinus*; *Lipoptena cervi*; ectoparasites; skin mites

1. Introduction

An interesting direction for the study of mite biodiversity and ecology is the analysis of their occurrence in specific environments, which for parasites is the host's body. Parasitic mites are able to make optimal use of the host by colonising all available microhabitats; most live as ectoparasites, residing in the fur and on the skin, but others colonise the skin as mesoparasites and others the respiratory tract or digestive system as endoparasites. Some stay on the host stationarily, throughout their lives, using it not only as a source of food and a place of shelter but also for the development of juvenile stages. Others use host resources periodically. Within a host, different mites can co-occur with each other, regardless of whether they are from the same group, with similar habitat requirements, or from different systematic and ecological groups differing in lifestyle, with varied food preferences, developmental cycles, adaptations, and variant parasitism. However, the co-occurrence of different mite species from the same families (e.g., Chirodiscidae, Demodecidae, and Psorergatidae) and even genera has mainly been described in rodents [1–4]. Similarly, most analyses of the co-occurrence of representatives from different systematic groups and parasitism types have concerned small mammals, e.g., [5–10]. Comparatively little data exists on co-occurrence and habitat sharing in the context of the ungulate mammalian Artiodactyla acarofauna.

Among wild animals, more information has been obtained regarding the European bison *Bison bonasus* (Linnaeus, 1758) (Bovidae: Bovinae), in which 11 species of mites from the Ixodidae (Parasitiformes: Ixodida), Psoroptidae, Sarcoptidae (Acariformes: Astigmata), and Demodecidae (Acariformes: Prostigmata) have been found. However, no such studies are available for the wild representatives of the subfamily Caprinae, including the widely distributed genus *Ovis* in Eurasia and North America. So far, such data have only covered the domestic sheep *Ovis aries aries* Linnaeus, 1758, where both temporary parasitic mites (Ixodidae) and stationary skin mites of the Astigmata (Psoroptidae and Sarcoptidae) and Prostigmata (Demodecidae and Psorergatidae) have been recorded, including specific parasites of this host: *Demodex aries* Desch, 1986, *D. ovis* Railliet, 1895, and *Psorobia ovis* (Womersley, 1941). Most information, however, relates to observations of the disease manifestations of the parasitoses they cause (psoroptosis, chorioptosis, sarcoptosis, and demodecosis), whose development is often related to domesticated lifestyle and husbandry conditions [11–16]. In contrast, in the natural environment, such parasitoses are rare, and parasitic mites usually form stable and balanced host–parasite systems with their hosts, where load parasite levels do not exceed host tolerance levels and infestations do not cause disease symptoms [17]. In addition, breeding conditions favour infestations with atypical parasites that are passed from other mammals (neighbouring hosts) under favourable circumstances [18]. Therefore, there is undoubtedly more value in observing the biodiversity and co-occurrence of mites in mammals from wild populations. In addition, learning about the pattern of occurrence of the mite community in wild animals can further our understanding of the mechanisms of parasitoses development in their domesticated counterparts.

In this sense, the mouflon *Ovis aries musimon* (Pallas, 1811) is a particularly interesting species and one whose acarofauna remains relatively unstudied. Its relict populations have inhabited certain Mediterranean islands, including Corsica and Sardinia, for several years. Mouflons have occasionally been introduced into various areas of Europe throughout both historical and modern times. At the beginning of the 20th century, they also found their way to the southern regions of Poland, where they are an alien species. Currently, their numbers in Poland are estimated at 3000 individuals, with the largest population, about 70% of all individuals, living in the Lower Silesia region [19,20].

Existing data on the parasitic mites of mouflon in its various regions of distribution has been limited to a few mentions of the occurrence of ticks [21,22]. Therefore, the purpose of the current study was to determine whether the mite community in mouflons from the Polish population includes specific species or at least those typical of the genus *Ovis*, especially members of the Demodecidae, Psorergatidae, and Psoroptidae. In addition, it aims to confirm whether the introduction of mouflons has brought with it alien species to the fauna of Poland that may pose a potential parasitological threat to the native fauna; it also examines whether, according to the theory of *parasites lost* [23], the introduction contributed to the loss of the natural/primary acarofauna in the mouflon, leaving vacant microhabitats that could be populated by species typical of local ungulates, such as those of the roe deer *Capreolus capreolus* (Linnaeus, 1758). The current study included all ectoparasites, including insects, to account for the possibility of correlation or interaction with parasitic mites sharing the same or neighbouring microhabitats.

2. Materials and Methods

A total of 41 European mouflons, *Ovis aries musimon*, from Poland, Lower Silesia, were examined, including 19 from the Jemna Forestry (Bardo Forest Inspectorate; 50°35′35.97″ N, 16°39′4.28″ E) and 22 from the Jugów Forest Inspectorate (near Sokolec; 50°39′11.69″ N, 16°28′24.25″ E). The animals were obtained between October 2010 and November 2013 from shoots carried out by employees of the forestry districts or from foreign exchange shoots. Ectoparasites living on the body surface and in the fur were collected immediately after shooting [24]. In addition, skin fragments were taken from various areas of the body, including the head/ear pinnae, neck, back, abdomen, groin, and pasterns. These fragments were preserved in a 70% ethanol solution for later studies of skin mite occurrence. For the recovery of demodecid mites, skin fragments were individually digested in a 10% potassium hydroxide solution as described previously [25]. The digested material was decanted and examined under phase-contrast microscopy (Nikon Eclipse 50i) with one 1 cm^2 of skin sample yielding approximately 100 wet preparations. The mites were placed in a polyvinyl-lactophenol solution and the following measurements were taken: total body length = length of gnathosoma, podosoma, and opisthosoma; gnathosomal width (at base); and podosomal and opisthosomal width (maximum width). All measurements are given in micrometres.

The specimen depositories are cited using the following abbreviation: UGDIZP, University of Gdańsk, Department of Invertebrate Zoology and Parasitology, Gdańsk, Poland [26].

The description of the species adopted the nomenclature commonly used for the family Demodecidae [27] and was completed with the nomenclature proposed by Bochkov [28] for the superfamily Cheyletoidea (Acariformes: Prostigmata) and by Izdebska and Rolbiecki [4]. The scientific and common names of the hosts follow Wilson and Reeder [29] and the Integrated Taxonomic Information System [30]. Basic parasitological parameters, i.e., the prevalence (percentage of hosts infected), mean intensity (mean number of parasites in infected hosts), intensity range (minimum and maximum number of parasite individuals per host), mean density (mean number of parasites per unit area; counted only for demodecid mite), and density range (minimum and maximum number of parasite per unit area; counted only for demodecid mite) were measured to determine the host infection level [31].

3. Results

3.1. Overall Mouflon Infestation

Two species of parasitic mites were found—*Ixodes ricinus* Linnaeus, 1758 (41 ♀♀, 31 ♂♂, 6 NN, and 1 L) and *Demodex musimonis* sp. nov. (23 ♀♀, 16 ♂♂, and 2 DN). Among the other parasitic arthropods, the fly *Lipoptena cervi* (Linnaeus, 1758) was also found (37 ♀♀, 25 ♂♂, and 1 egg). The parasites showed a strict topographical preference, with the majority of ticks (53.2%) recorded on the neck, and the remainder on the abdomen (13.9%), groin (11.4%), back (8.9%), anal area (2.5%), eye (2.5%), pastern (2.5%), and ear (1.3%); for three individuals (3.8%), the location was not determined. Also, *L. cervi* preferred the neck area

(48.3%), while the others preferred the back (22.6%), abdomen (19.4%), and groin (9.7%). In turn, all specimens of *D. musimonis* sp. nov. occurred in the skin of the pastern (Figure 1).

Figure 1. Topography preferences of ectoparasites on the mouflon: • *Ixodes ricinus*, • *Lipoptena cervi*, and • *Demodex musimonis* sp. nov.

The overall prevalence (including all parasite species) was 56.1%. In turn, the infestation of *I. ricinus* was 29.3%, 6.6, 1–43, for *L. cervi* 34.1%, 4.4, 1–23, and for *D. musimonis* sp. nov. 14.6%, density of 6.8/1 cm^2, density range of 3–16/1 cm^2.

3.2. Descriptions (Figures 2 and 3, Table 1)

Demodex musimonis Izdebska, Kozina and Rolbiecki, 2023.

Male (n = 15 and holotype). Body elongated, stocky, conical, distinctly separated gnathosoma; 196 (183–207) long and 43 (38–47) wide (holotype, 196 × 43). Gnathosoma trapezoidal, slightly shorter than width at base; on dorsal side in the central part of basal segment, pair of wedge-shaped (rounded at end, posterior edge with bulge) supracoxal spines (setae *elc.p*) present, ca. 4.0 long (holotype, 4.0), directed medially, oblique. Palps three-segmented, terminating in three different shaped spines (one smallest—single, unbifurcated, conical; two bifurcated—larger and smaller) on tibio-tarsus; also, small setae $v''F$ present on middle segment (trochanter-femur-tarsus). On ventral surface of gnathosoma, horseshoe-shaped pharyngeal bulb with pair of conical subgnathosomal setae (setae n), situated anterior on both sides. Podosoma rectangular; four pairs of short legs with coxa integrated into ventral idiosomal wall and five free, overlapping segments (trochanter-tarsus); two bifurcated claws, ca. 5.0 long (holotype, 5.0), with pointed subterminal spur on each tarsus; knob on each femur. Epimeral plates (coxal fields) distinctly sclerotized; all trapezoidal. On the dorsal side of podosoma, podosomal shield with distinctly irregular striation, reaches level of legs III; posterior edge of this shield is convex. Opisthosoma constitutes 63% (59–66%) of body length (holotype, 62%); conical, wide at base, slightly rounded at end. Whole opisthosoma distinctly annulated; annulations reach dorsal podosoma side (pair of legs III); annuli relatively wide at ca. 1.0–1.5 µm. Opisthosomal organ not visible. Aedeagus 22 (18–26) long, on dorsal surface, located at level of epimeral plates I–III. Genital opening located at level of posterior edge of epimeral plate I (on border with epimeral plate II).

Female (n = 23). Body elongated, more slender than male, spindle-shaped, distinctly separated gnathosoma; 236 (208–257) long and 44 (42–47) wide. Gnathosoma shape similar to male. Pharyngeal bulb and morphological details of gnathosoma similar to those in male. Podosoma conical, slightly widening posterior end; legs similar to those in male. Epimeral plates distinctly sclerotized; I pair trapezoidal, II–III rectangular; posterior edges of pair IV form distinctly incision. On the dorsal side of podosoma, podosomal shield, with distinctly irregular striation, reaches level of legs III; posterior edge of this shield is convex. Opisthosoma constitutes 64% (61–69%) of body length; conical, rounded at end. Whole opisthosoma distinctly annulated; annulations reach dorsal podosoma side (pair of legs III); annuli relatively wide at ca. 1.0–1.5 μm. Opisthosomal organ not visible. Vulva 12 (10–14) long, located slightly behind incision of IV epimeral plates.

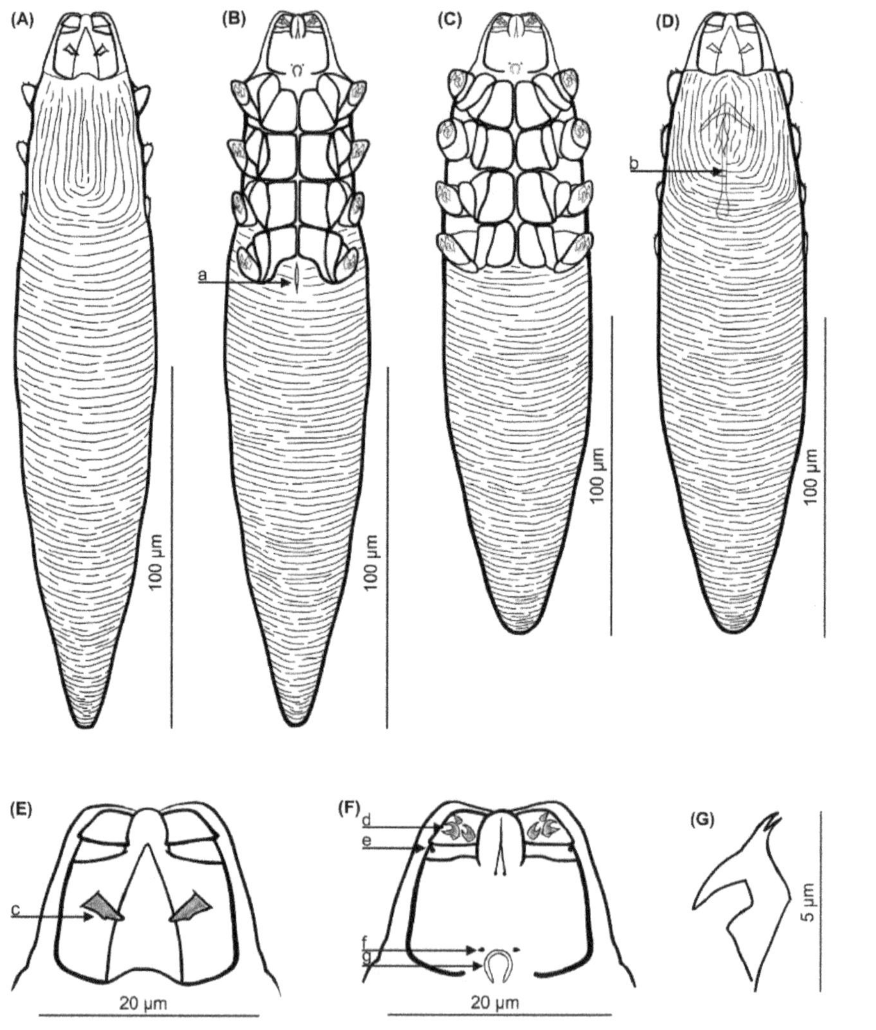

Figure 2. *Demodex musimonis* sp. nov.: female, dorsal view (**A**), female, ventral view (**B**), male, ventral view (**C**), male, dorsal view (**D**), gnathosoma, male, dorsal view (**E**), gnathosoma, male, ventral view (**F**), claw on the leg (**G**); a: vulva, b: aedeagus, c: supracoxal spine (seta *elc.p*), d: spines on palps, e: seta $v''F$, f: subgnathosomal seta (seta n), and g: pharyngeal bulb.

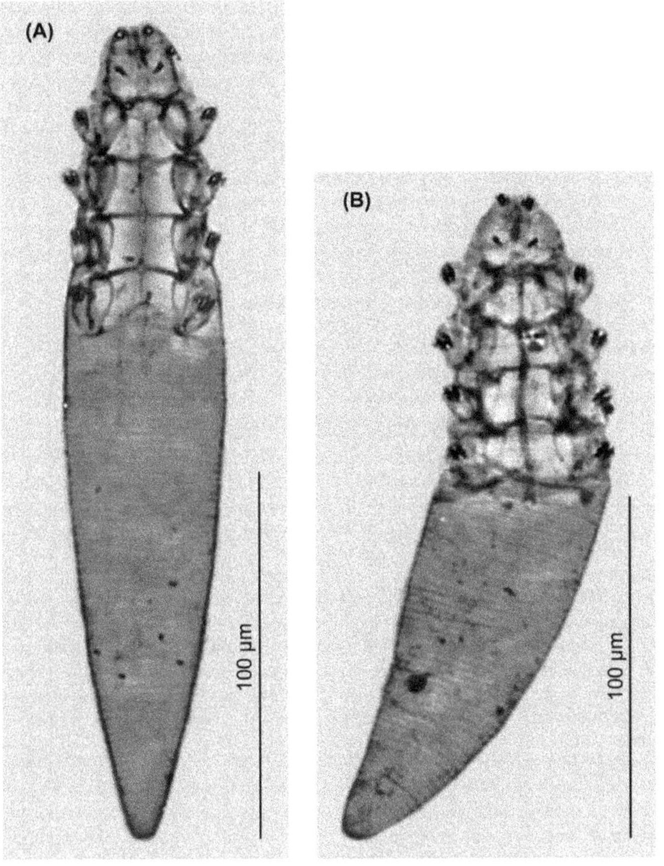

Figure 3. *Demodex musimonis* sp. nov.: female (**A**), male (**B**).

Table 1. Body size (micrometres) for adults of *Demodex musimonis* sp. nov.

Morphological Features	Males (n = 16) Mean (Range) ± SD	Females (n = 23) Mean (Range) ± SD
Length of gnathosoma	18 (17–20), ±1	19 (18–20), ±1
Width of gnathosoma (at base)	21 (20–23), ±1	22 (20–24), ±1
Length of podosoma	55 (51–59), ±2	66 (60–73), ±4
Width of podosoma	43 (38–47), ±2	44 (38–47), ±2
Length of opisthosoma	123 (108–131), ±6	150 (130–172), ±13
Width of opisthosoma	41 (38–45), ±2	44 (42–47), ±1
Aedeagus	22 (18–26), ±2	–
Vulva	–	12 (10–14), ±1
Total length of body	196 (183–207), ±6	236 (208–257), ±14

SD, standard deviation.

Egg (n = 1): 70 long and 22 wide.

Deutonymph (n = 2): 85–235 long and 20–40 wide.

Material deposition: Male holotype (reg. no. UGDIZPBOamDDm04m), 15 male paratypes (reg. no. UGDIZPBOamDDm01m–03m, UGDIZPBOamDDm05m–16m) and 23 female paratypes (reg. no. UGDIZPBOamDDm01f–23); pastern region; host *Ovis aries musimon* (reg. no. MABOam10/2012, MABOam18/2011, MABOam33/2012, MABOam34/2013, MABOam37/2013, and MABOam41/2013); Sokolec and Jemna, Poland; January 2011, November 2012, January 2012 and November 2013; coll. J.N. Izdebska and P. Kozina; the whole-type material (mounted microscope slides with the demodecid mites) deposited within the framework of the Collection of Extant Invertebrates in Department of Invertebrate Zoology and Parasitology, University of Gdańsk, Poland.

Location in the host: *Demodex musimonis* sp. nov. was found exclusively in the pastern region of the examined mouflon.

Etymology: The specific epithet *musimonis* refers to the subspecific name of the host.

Differential diagnosis: the shape of *Demodex musimonis* sp. nov., and some of its features, most closely resemble those of *D. ovis* Hirst, 1919 (redesc. Desch 1986) from the domestic sheep *Ovis aries aries* (Table 2). However, *D. musimonis* sp. nov. is a little larger and has different body proportions. In addition, *D. musimonis* sp. nov. demonstrates a more distinct sexual dimorphism: the body length-to-width ratios being 4.6 in males and 5.3 in females; in contrast, in *D. ovis*, both sexes demonstrate similar proportions (5.2 for males and 5.5 for females). Supracoxal spines on the gnathosoma are large (4 μm) and wedge-shaped in *D. musimonis* sp. nov., while they are smaller (2 μm) and peg-like in *D. ovis*. The spines on the terminal segment of the palpi are similarly shaped in both species but are different in size and more massive in *D. musimonis* sp. nov.; moreover, *v″F* seta are present on the palpi of *D. musimonis* sp. nov. Subgnathosomal setae in *D. musimonis* sp. nov. are located relatively higher (at the level of anterior margin of the pharyngeal bulb) than in *D. ovis* (below the anterior margin of the pharyngeal bulb). The epimeral plates are distinctly sclerotized in *D. musimonis* sp. nov., with clearly outlined edges, but are weakly sclerotized (edges indistinct) in *D. ovis*. In addition, the posterior edges of the epimeral plate IV in *D. musimonis* sp. nov. females possess a large, trapezoidal incision encompassing a vulva, while they are weakly outlined in female *D. ovis*, with a small triangular incision, below which the vulva is located. Furthermore, *D. musimonis* sp. nov. males demonstrate a longer aedeagus (mean 22.0 μm in length) located at level of epimeral plates I–III, with the genital opening at the level of the posterior part of epimeral plate I; in male *D. ovis*, the aedeagus is shorter (mean, 20.0 μm in length) and located at epimeral plates II–III with the genital opening at the level of epimeral plate II. The typical microhabitat is also different: *D. musimonis* sp. nov. was exclusively found in the pastern region, while *D. ovis* was noted throughout the entire body, though preferring the head, flanks, and shoulders.

Table 2. Morphometric comparison between *Demodex musimonis* sp. nov. and *Demodex ovis*.

Feature/Species	***Demodex musimonis* sp. nov.**		***Demodex ovis***	
Source	**Present Study**		**Desch [32]**	
Sex	**Males**	**Females**	**Males**	**Females**
Sample Size	**(*n* = 16)**	**(*n* = 23)**	**(*n* = 20)**	**(*n* = 20)**
Body total length	196 (183–207), SD 6	236 (208–257), SD 14	170 (140–201), SD 17 *	214 (187–274), SD 26 *
Body total width	43 (38–47), SD 2	44 (42–47), SD 1	33 (30–38), SD 3 *	39 (36–52), SD 4 *
Body length-to-width ratio	4.6:1 (3.9–5.1:1), SD 0.3:1	5.3:1 (4.5–5.8:1), SD 0.3:1	5.2:1 **	5.5:1 **
Opisthosoma length to body length ratio (%)	63 (59–66), SD 2	64 (61–69), SD 2	59 **	59 **
Aedeagus length	22 (18–26), SD 2	–	20 (18–22), SD 1 *	–
Vulva length	–	12 (10–14), SD 1	–	6 (4–7), SD 1

SD: standard deviation. * Measurements were rounded to the nearest micrometre with respect to the original results [32]. ** Calculated from measurements by Desch [32].

4. Discussion

Undoubtedly, the most important result of the current research is the discovery and description of *Demodex musimonis* sp. nov., a new species of Demodecidae associated with the mouflon, and one which may be a specific parasite. The Demodecidae exhibit high host specificity, with several monoxenic species often found in individual mammals; they are known to inhabit different microhabitats within the skin or other host structures and tissues [17]. The largest number of such synhospital species has been recognised in rodents, such as seven in the house mouse *Mus musculus* Linnaeus, 1758. However, such coexistence has also been noted among ungulates: three species have been described in domestic cattle *Bos taurus* Linnaeus, 1758, and two each in domestic sheep, European bison, red deer *Cervus elaphus* Linnaeus, 1758, and the domestic horse *Equus caballus* Linnaeus, 1758 [17]. Host-specific demodecid mites, especially those found in wild mammals, often show a high intensity or density of infection while lacking symptoms of parasitosis; as such, they usually accompany their hosts to different regions of occurrence and can be introduced with them to new regions [17,33,34]. However, due to their high peculiarity, they do not present significant threats to native fauna as alien species.

The demodecid mites noted in the present study showed topographical specificity, colonising only the pastern region. Such preference is a phenomenon typical of mites of this group, where only some species (associated with follicles of normal hair) colonise the entire skin surface, albeit usually unevenly [35]. Examples include *D. kutzeri* Bukva, 1987, found in red deer, which has been recorded in skin fragments from more than half of the studied deer, or *D. ovis* from domestic sheep [36–39]. Most species are found in narrow microhabitats, often concentrated in the head region, such as the Meibom's glands of the eyelids (e.g., *D. bisonianus* Kadulski and Izdebska, 1996 from European bison, *D. ghanensis* Oppong, Lee, and Yasin, 1975 from domestic cattle), the eye region (e.g., *D. tauri* Bukva, 1986 from domestic cattle), the hairless skin region of the nose (*D. bialoviensis* Izdebska, Rolbiecki and Bielecki, 2022 from European bison), the ear canals (e.g., *D. conicus* Izdebska and Rolbiecki, 2015 from a house mouse), the vibrissae region (e.g., *Miridex putorii* Izdebska, Rolbiecki, and Rehbein from the polecat *Mustela putorius* Linnaeus, 1758), and the tongue and gums (*Glossicodex musculi* Izdebska and Rolbiecki, 2016 from a house mouse) [4,39–44]. A few species of Demodecidae are known to be restricted to the limb region. Examples include *D. obliquus* Izdebska and Rolbiecki, 2022 from the domestic cat *Felis catus* Linnaeus, 1758, or *D. ponderosus* Izdebska, and Rolbiecki, 2014 from the Norway rat *Rattus norvegicus* (Berkenhout, 1769) [45,46].

In the present study, *D. musimonis* sp. nov. demonstrated a relatively low level of infestation (prevalence 14.6%), with only 41 specimens being found in the studied skin fragments. This may be due to the locality of occurrence (pastern), the survey technique, i.e., the labour-intensive method of digesting mites from selected skin fragments, and also the dispersal (low density of mites in the skin). It is also possible that the study period was not optimal for population development, as indicated by the discovery of only a few juvenile stages.

Among those infecting the genus *Ovis*, demodecid mites have only been described in domestic sheep. *Demodex ovis*, discovered first in sheep, can occur throughout the whole body, preferring the region of the head and trunk, while *D. aries*, described in the late 20th century, is associated with areas of high sebaceous gland density (e.g., the skin around the vulva, foreskin, and nostrils) [32]. In addition, Bukva [36] mentions yet another unidentified species of *Demodex* found in the head/eyelid region.

It is difficult to compare the Demodecidae from the domestic sheep with the demodecid mite described from the mouflon. Although *D. musimonis* sp. nov. morphologically resembles *D. ovis*, it shows different topographical preferences. Perhaps the species share a common pedigree. For such an inference, however, the pedigree of the mouflon itself is problematic, being a wild sheep that is believed to have inhabited the Mediterranean island region for 6000 to 8000 years (according to various sources), and probably originally originated from Asia. However, the origin of mouflons in Europe is controversial, and the pedigree of domestic sheep has not yet been conclusively clarified; it is possible that the mouflon may be a protoplast or one of several possible ancestors [47–51]. In general, questions of the evolution of host–parasite systems in the context of *Ovis–Demodex* require more extensive analyses of material from different hosts and regions of occurrence.

Despite this valuable discovery, the acarofauna of mouflons from Poland should be considered poor; in a representative sample of 41 hosts examined in different seasons, only two species of mites were found. In addition, to the representative of the Demodecidae, a new discovery, only the polyxenic tick *Ixodes ricinus*, common in various regions of Europe, was recorded. Other parasitic arthropods were also found to be poorly represented: only *Lipoptena cervi*, an oligoxenic ectoparasite of ungulates, especially cervids, usually found in large numbers in autumn, was observed [21,52]. The hematophagous temporal ectoparasites, castor bean ticks, and deer flies showed similar topographical preferences, i.e., they often occurred in the neck region, which is probably related to the availability of blood vessels [33].

The significance of the parasites found among the mouflons is difficult to determine. *Demodex musimonis* sp. nov. occurred sparsely and asymptomatically, which is typical of such mite infestations among wild mammals, e.g., [53,54]. *Ixodes ricinus*, on the other hand, probably had no pathogenic significance as a parasite, but its role in mouflon with regard to pathogen transmission is unknown. Undoubtedly, the mouflon represents an increase in reservoir animals for pathogens transmitted by *I. ricinus*, regarded as an ectoparasite with considerable epidemiological significance in Europe.

However, no representatives of the Psorergatidae, Psoroptidae, Sarcoptidae, or even other Ixodidae, which frequently attack ungulates, were found in the mouflon, nor any typical ungulate parasitic insects from the Phthiraptera. This is not a matter of the limited sample size or study period. Analogous observations, using the same research methods, carried out on other ungulates reported a much richer community of mites inhabiting body surfaces and skin. A good object for such comparisons seems to be the European bison, which, like the mouflons found in Poland, was introduced into the environment through human activities. The European bison became extinct under natural conditions at the beginning of the 20th century, and the current restitution of the species resulted from reintroductions or introductions of individuals that survived in breeding conditions and zoos. To date, 14 species of parasitic arthropods have been found in the European bison (considering only parasites *sensu stricto*), including 11 species of mites, including 2 specific Demodecidae, as well as oligoxenic, ungulate-typical Psoroptidae, a polyxenic species of Sarcoptidae and various Ixodidae. In European bison, mites or other parasitic arthropods are recorded in all seasons (most of the studies were conducted in winter), and the prevalence is sometimes high; hence, studies of even a small sample yield findings of at least several species. For example, surveys of 12 bison from the Bieszczady Mountains conducted in the winter from 1998 to 2000 yielded findings of four species of parasitic arthropods (including three mites), while winter surveys of 12 European bison from a reserve in the Bialowieza Forest yielded nine species (including seven mites). Surveys of just six bison conducted in July 2011 yielded the identification of seven species of parasitic arthropods (including six species of mites) [32–34]. However, European bison, apart from the natural specific (monoxenic) parasitofauna preserved in this mammal (including two species of the Demodecidae), could take over the parasitofauna from other ungulates, probably cervids and cattle, during the restoration of the wild populations [33]. Hence, they harbour species of Psoroptidae typical of the Bovidae—*Chorioptes bovis* (Hering, 1845), *Psoroptes equi* (Hering, 1838), the polyxenic *Sarcoptes scabiei* (DeGeer, 1778), or ticks associated with local environments, particularly *I. ricinus* and *Dermacentor reticulatus* Fabricius, 1794. However, *Ch. bovis* or *P. equi* (syn. *P. ovis*—according to Zahler et al. [55]) are also typical parasitofauna of domestic sheep [56], so they could potentially be parasites of natural mouflon populations.

In addition, if any parasites are lost in the introduction process, they could be taken over by local ungulates, such as the roe deer or red deer found in the region, in which these mites are common [52]. Here, however, it should be remembered that the takeover of skin mites has many limitations. Not only does it depend on the ability of the parasite to colonise and adapt to another host, but also on the direct contact between hosts needed for transfer. Interspecies contact is relatively rare and involves, for example, mammalian carnivorans (predator–prey contact).

The presence of other skin mites in the mouflon would undoubtedly shed light on the findings of such mites in mouflons from other populations. Unfortunately, there is a lack of data on parasitic arthropods of mouflons from other areas of distribution, including endemic areas, apart from a few reports of the presence of local tick species [21,22,57,58].

5. Conclusions

A comparison of the acarofauna of the mouflon with that of the domestic sheep suggests the potential formation of a mite community with a complex topical and topographical structure, with the possibility of co-occurrence of representatives of several skin mite families. It is not known, however, whether this structure was adopted by domestic sheep from wild ancestors or formed through domestication, where oligoxenic Psoroptidae may have been adopted from other domesticated ungulates. However, the typically monoxenic/specific Demodecidae, or Psorergatidae, should already constitute the natural parasitofauna of wild sheep.

The current study has so far yielded the discovery of one species from these mite groups associated with wild sheep (mouflons): *D. musimonis* sp. nov. In addition, this is the first finding of mites from the Demodecidae in wild Caprinae; so far, only two species have been described from the domestic sheep and one from the domestic goat, *Capra hircus* Linnaeus, 1758. It is also the first comprehensive study of parasitic arthropods inhabiting the mouflon, including both typical ectoparasites and skin mites.

Author Contributions: Conceptualisation, J.N.I., P.K. and L.R.; sampling, P.K. and J.N.I.; data analysis, J.N.I., P.K. and L.R.; writing—original draft preparation, J.N.I., L.R. and P.K.; writing, review and editing, J.N.I., L.R. and P.K.; supervision, J.N.I. All authors have read and agreed to the published version of the manuscript.

Funding: This research received no external funding.

Institutional Review Board Statement: Not applicable.

Informed Consent Statement: Not applicable.

Data Availability Statement: The data presented in this study are available on request from the corresponding author.

Conflicts of Interest: The authors declare no conflict of interest.

References

1. Bochkov, A.V.; Dubinina, H.V. Mites of the genus *Schizocarpus* (Acariformes: Chirodiscidae) parasitizing the Eurasian beaver *Castor fiber* (Rodentia: Castoridae) in the Voronezh National Reserve. *Acarina* **2011**, *19*, 53–66.
2. Bochkov, A.V.; Saveljev, A.P. Fur mites of the genus *Schizocarpus* Trouessart (Acari: Chirodiscidae) from the Eurasian beaver *Castor fiber tuvinicus* Lavrov (Rodentia: Castoridae) in the Azas River (Tuva Republic, Russia). *Zootaxa* **2012**, *3410*, 1–18. [CrossRef]
3. Cierocka, K.; Izdebska, J.N.; Rolbiecki, L.; Ciechanowski, M. The occurrence of skin mites from the Demodecidae and Psorergatidae (Acariformes: Prostigmata) families in bats, with a description of a new species and new records. *Animals* **2022**, *12*, 875. [CrossRef]
4. Izdebska, J.N.; Rolbiecki, L. A new genus and species of demodecid mites from the tongue of a house mouse *Mus musculus*: Description of adult and immature stages with data on parasitism. *Med. Vet. Entomol.* **2016**, *30*, 135–143. [CrossRef]
5. Haitlinger, R.; Turek, M. Arthropods occurring on *Mus musculus* Linnaeus, 1758 (Mammalia: Rodentia: Muridae) in Poland. *Zesz. Nauk. Uniw. Przyr. Wroc. Biol. Hod. Zwierz.* **2006**, *548*, 43–57.
6. Haitlinger, R.; Jankowska, A. Arthropods occurring on *Rattus norvegicus* (Berkenhout, 1769) (Rodentia: Muridae) in Poland. *Zesz. Nauk. AR Wroc. Zootech.* **2005**, *529*, 35–44.
7. Çicek, H.; Stanyukovich, M.; Yağci, S.; Aktaş, M.; Karaer, Z. Gamasine mite (Parasitiformes: Mesostigmata) infestations of small mammals (Mammalia: Rodentia, Insectivora) in Turkey. *Turk. Parazitolojii Derg.* **2008**, *32*, 65–70.
8. Várfalvyová, D.; Stanko, M.; Miklisová, D. Composition and seasonal changes of mesostigmatic mites (Acari) and fleas fauna (Siphonaptera) in the nests of *Mus spicilegus* (Mammalia: Rodentia). *Biologia* **2011**, *66*, 528–534. [CrossRef]
9. Krasnov, B.R.; Stanko, M.; Lareschi, M.; Khokhlovac, I.S. Species co-occurrences in ectoparasite infracommunities: Accounting for confounding factors associated with space, time, and host community composition. *Ecol. Entomol.* **2020**, *45*, 1158–1171. [CrossRef]

10. Izdebska, J.N.; Rolbiecki, L. Diversity of the parasite fauna of *Mus musculus* L. (Rodentia, Muridae) from different habitats. *Russ. J. Ecol.* **2013**, *44*, 428–432. [CrossRef]
11. Murshed, M.; Al-Quraishy, S.; Mares, M.M. Survey of mange mite infesting sheep in Riyadh region, Saudi Arabia. *Saudi J. Biol. Sci.* **2022**, *29*, 595–600. [CrossRef]
12. Murthy, G.S.S.; Nagesha, A.M.; Gowda, K.H. Therapeutic management of sarcoptic mange in a sheep flock. *J. Parasit. Dis.* **2013**, *37*, 281–282. [CrossRef]
13. Parmar, D.; Chandra, D. Sarcoptic mange infestation in sheep with its therapeutic management. *Int. J. Curr. Microbiol. Appl. Sci.* **2018**, *7*, 845–849. [CrossRef]
14. Jardim Filho, J.O.; Almeida, G.L.; Piazer, J.V.M.; Rodegheri, L.J.; Ruivo, N.B.; Pires, B.S.; Leal, M.L.R. Psoroptes ovis infestation of sheep in São Vicente do Sul, Rio Grande do Sul, Brazil. *Ciência Rural* **2020**, *50*, e20191026. [CrossRef]
15. Yeruham, I.; Rosen, S.; Hadani, A. Sheep demodecosis (*Demodex ovis* Railliet, 1895) in Israel. *Rev. Elev. Med. Vet. Pays. Trop.* **1986**, *39*, 363–365.
16. Johnson, P.W.; Plant, J.W.; Boray, J.C.; Blunt, S.C.; Nicholls, P.J. The prevalence of itchmite, *Psorergates ovis*, among sheep flocks with a history of fleece derangement. *Aust. Vet. J.* **1990**, *67*, 117–120. [CrossRef]
17. Izdebska, J.N.; Rolbiecki, L. The biodiversity of demodecid mites (Acariformes: Prostigmata), specific parasites of mammals with a global checklist and a new finding for *Demodex sciurinus*. *Diversity* **2020**, *12*, 261. [CrossRef]
18. Izdebska, J.N. European bison arthropod parasites from closed Polish breeding facilities. *Acta Parasitol.* **2001**, *46*, 135–137.
19. Szczęśniak, E. Obecność muflonów *Ovis aries musimon* w Polsce–czy to naprawdę konieczne? *Chrońmy Przyr. Ojcz.* **2011**, *67*, 99–117.
20. Nasiadka, P.; Wajdzik, M.; Skubis, J. Aktualny stan badań nad muflonem (*Ovis musimon*) jako podstawa do zarządzania, ochrony lub eliminacji tego gatunku z Polski. *Sylwan* **2015**, *159*, 381–391.
21. Kadulski, S. Występowanie Stawonogów Pasożytniczych na łownych Lagomorpha i Artiodactyla—Próba Syntezy. *Zesz. Nauk. Uniwesytetu Gdańskiego. Rozpr. I Monogr.* **1989**, *132*, 1–140.
22. Alonso, F.D.; Gomis, J.; Ruiz de Ybáñez, M.R.; Gutiérrez, A.; Sierra, J.; Martínez-Carrasco, C. Tick infestation on wild ruminants in the Sierra de Cazorla, Segura y Las Villas Natural Park, Spain. In Proceedings of the XII Congress of the Mediterranean Federation for Health and Production of Ruminants, Istanbul, Turkey, 16–19 September 2004; pp. 1–6.
23. Clay, K. Parasites lost. *Nature* **2003**, *42*, 585–586. [CrossRef] [PubMed]
24. Kadulski, S.; Izdebska, J.N. Game ectoparasites as a potential threat for hunters. In *Arthropods. A Variety of Forms and Interactions*; Buczek, A., Błaszak, C., Eds.; Koliber: Lublin, Poland, 2005; pp. 145–147.
25. Izdebska, J.N. *Demodex* spp. (Acari: Demodecidae) in brown rat (Rodentia: Muridae) in Poland. *Wiad. Parazytol.* **2004**, *50*, 333–335. [PubMed]
26. Zhang, Z.-Q. Repositories for mite and tick specimens: Acronyms and their nomenclature. *Syst. Appl. Acarol.* **2018**, *23*, 2432–2446. [CrossRef]
27. Nutting, W.B. Hair follicle mites (*Demodex* spp.) of medical and veterinary concern. *Cornell Vet.* **1976**, *66*, 214–231.
28. Bochkov, A.V. New observations on phylogeny of cheyletoid mites (Acari: Prostigmata: Cheyletoidea). *Proc. Zool. Inst. RAS* **2008**, *312*, 54–73. [CrossRef]
29. Wilson, D.E.; Reeder, D.M. (Eds.) *Mammals Species of the World: A Taxonomic and Geographic Reference*, 3rd ed.; The Johns Hopkins University Press: Baltimore, MD, USA, 2005. Available online: http://www.departments.bucknell.edu/biology/resources/msw3/ (accessed on 28 July 2023).
30. Taxonomic Information System (ITIS). Available online: http://www.itis.gov (accessed on 28 July 2023).
31. Bush, A.O.; Lafferty, K.D.; Lotz, J.M.; Shostak, A.W. Parasitology meets ecology on its own terms: Margolis et al. revisited. *J. Parasitol.* **1997**, *83*, 575–583. [CrossRef]
32. Desch, C.E. *Demodex aries* sp. nov., a sebaceous gland inhabitant of the sheep, *Ovis aries*, and a redescription of *Demodex ovis* Hirst, 1919. *New Zeal. J. Zool.* **1986**, *13*, 367–375. [CrossRef]
33. Izdebska, J.N. Stawonogi pasożytnicze żubrów z Bieszczad. *Sci. Messenger Lviv State Acad. Vet. Med. S.Z. Gzhytskyj* **2001**, *3*, 208–211.
34. Izdebska, J.N.; Rolbiecki, L.; Bielecki, W. The first data on parasitic arthropods of the European bison in the summer season with a world checklist. *Diversity* **2022**, *14*, 75. [CrossRef]
35. Izdebska, J.N.; Rolbiecki, L. A new species of *Demodex* (Acari, Demodecidae) with data on topical specificity and topography of demodectic mites in the striped field mouse *Apodemus agrarius* (Rodentia, Muridae). *J. Med. Entomol.* **2013**, *50*, 1202–1207. [CrossRef] [PubMed]
36. Bukva, V. Three species of the hair follicle mites (Acai: Demodecidae) parasitizing the sheep, *Ovis aries* L. *Folia Parasitol.* **1990**, *37*, 81–91.
37. Izdebska, J.N.; Fryderyk, S. *Demodex acutipes* Bukva & Preisler, 1988 (Acari, Demodecidae)–a rare parasite of red deer (*Cervus elaphus* L.). *Ann. Parasitol.* **2012**, *58*, 161–166.
38. Izdebska, J.N.; Kozina, P.; Fryderyk, S. The occurrence of *Demodex kutzeri* Bukva, 1987 (Acari, Demodecidae) in red deer (*Cervus elaphus* L.) in Poland. *Ann. Parasitol.* **2013**, *59*, 85–88.
39. Kadulski, S.; Izdebska, J.N. *Demodex bisonianus* sp. nov. (Acari, Demodicidae) a new parasite of the bison (*Bison bonasus* L.). *Wiad. Parazytol.* **1996**, *42*, 103–110.

40. Oppong, E.N.W.; Lee, R.P.; Yasin, S.A. *Demodex ghanensis* sp. nov. (Acari, Demodicidae) parasitic on West African cattle. *Ghana J. Sci.* **1975**, *15*, 39–43.
41. Bukva, V. *Demodex tauri* sp. n. (Acari: Demodicidae), a new parasite of cattle. *Folia Parasitol.* **1986**, *33*, 363–369.
42. Izdebska, J.N.; Rolbiecki, L.; Bielecki, W. *Demodex bialoviensis* sp. nov. (Acariformes, Demodecidae) a new, specific parasite of the European bison *Bison bonasus* (Artiodactyla, Bovidae). *Int. J. Parasitol. Parasites Wildl.* **2022**, *17*, 138–143. [CrossRef]
43. Izdebska, J.N.; Rolbiecki, L. A new species of the genus *Demodex* Owen, 1843 (Acari: Demodecidae) from the ear canals of the house mouse *Mus musculus* L. (Rodentia: Muridae). *Syst. Parasitol.* **2015**, *91*, 167–173. [CrossRef] [PubMed]
44. Izdebska, J.N.; Rolbiecki, L.; Rehbein, S. Morphological and ontogenetic characteristics of *Miridex putorii* (Acariformes: Demodecidae), a new genus and species of skin mite specific to the European polecat Mustela putorius. *Int. J. Parasitol. Parasites Wildl.* **2022**, *18*, 225–231. [CrossRef]
45. Izdebska, J.N.; Rolbiecki, L.; Fryderyk, S. *Demodex murilegi* and *Demodex obliquus*, two new specific skin mites from domestic cat *Felis catus*, with notes on parasitism. *Med. Vet. Entomol.* **2023**, *37*, 263–274. [CrossRef]
46. Izdebska, J.N.; Rolbiecki, L. New species of Demodex (Acari: Demodecidae) with data on parasitism and occurrence of other demodecids of *Rattus norvegicus* (Rodentia, Muridae). *Ann. Entomol. Soc. Am.* **2014**, *107*, 740–747. [CrossRef]
47. Rezaei, H.R.; Naderi, S.; Chintauan-Marquier, I.C.; Taberlet, P.; Virk, A.T.; Naghash, H.R.; Rioux, D.; Kaboli, M.; Pompanon, F. Evolution and taxonomy of the wild species of the genus *Ovis* (Mammalia, Artiodactyla, Bovidae). *Mol. Phylogenetics Evol.* **2010**, *54*, 315–326. [CrossRef] [PubMed]
48. Balicka-Ramisz, A.; Laurans, Ł.; Jurczyk, P.; Kwita, E.; Ramisz, A. Gastrointestinal nematodes and the deworming of mouflon (*Ovis aries musimon*) from Goleniowska Forest in West Pomerania province, Poland. *Ann. Parasitol.* **2017**, *63*, 27–32. [PubMed]
49. Kozakiewicz, B.; Maszewska, I. Występowanie i leczenie parazytoz muflonów (*Ovis musimon* L.) w warunkach hodowli w ośrodku łowieckim w Wielkopolsce. *Med. Weter.* **1984**, *40*, 536–538.
50. Su, R.; Qiao, X.; Gao, Y.; Li, X.; Jiang, W.; Chen, W.; Fan, Y.; Zheng, B.; Zhang, Y.; Liu, Z.; et al. Draft genome of the European mouflon (*Ovis orientalis musimon*). *Front. Genet.* **2020**, *11*, 533611. [CrossRef]
51. Portanier, E.; Chevret, P.; Gélin, P.; Benedetti, P.; Sanchis, F.; Barbanera, F.; Kaerle, C.; Queney, G.; Bourgoin, G.; Devillard, S.; et al. New insights into the past and recent evolutionary history of the Corsican mouflon (*Ovis gmelini musimon*) to inform its conservation. *Conserv. Genet.* **2022**, *23*, 91–107. [CrossRef]
52. Kadulski, S. Ectoparasites of Cervidae in north-east Poland. *Acta Parasitol.* **1996**, *41*, 204–210.
53. Cierocka, K.; Izdebska, J.N.; Rolbiecki, L. *Demodex crocidurae*, a new demodecid mite (Acariformes: Prostigmata) parasitizing the lesser white-toothed shrew and a redescription of *Demodex talpae* from European mole with data on parasitism in Soricomorpha. *Animals* **2021**, *11*, 2712. [CrossRef]
54. Izdebska, J.N.; Rolbiecki, L.; Cierocka, K.; Pawliczka, I. *Demodex phocidi* (Acariformes: Demodecidae) from *Phoca vitulina* (Carnivora: Phocidae)–the second observation in the world and a supplement to the species description. *Oceanol. Hydrobiol. St.* **2020**, *49*, 49–55. [CrossRef]
55. Zahler, M.; Hendrikx, W.M.; Essig, A.; Rinder, H.; Gothe, R. Species of the genus *Psoroptes* (Acari: Psoroptidae): A taxonomic consideration. *Exp. Appl. Acarol.* **2000**, *24*, 213–225. [CrossRef] [PubMed]
56. Bochkov, A.V.; Klimov, P.B.; Hestvik, G.; Saveljev, A.P. Integrated Bayesian species delimitation and morphological diagnostics of chorioptic mange mites (Acariformes: Psoroptidae: Chorioptes). *Parasitol. Res.* **2014**, *113*, 2603–2627. [CrossRef] [PubMed]
57. George, A. *The, Ticks, or Ixodides, of the U.S.S.R.: A Review of the Literature*; Public Health Service Publication, No. 548.; United States Department of Health, Education and Welfare: Rockville, MD, USA, 1957; p. 397.
58. Konjević, D.; Janicki, Z.; Severin, K.; Stanko, M.; Zivicnjak, T.; Slavica, A.; Staresina, V. An outbreak of tick paralysis in free-ranging mouflon (*Ovis ammon musimon*). *J. Zoo. Wildl. Med.* **2007**, *38*, 585–587. [CrossRef] [PubMed]

animals

MDPI

Article

Ptyctimous Mites (Acari, Oribatida) of Peru with the Description of an Extraordinary New Phthiracaroid Mite from the Peruvian Andes †

Wojciech Niedbała [1,*], Zbigniew Adamski [2], Ronald Laniecki [1] and Wojciech L. Magowski [1]

1 Department of Animal Taxonomy and Ecology, Faculty of Biology, Adam Mickiewicz University, 61-614 Poznan, Poland
2 Department of Animal Physiology and Developmental Biology, Laboratory of Electron and Confocal Microscopy, Faculty of Biology, Adam Mickiewicz University, 61-614 Poznan, Poland
* Correspondence: wojciech.niedbala@amu.edu.pl
† urn:lsid:zoobank.org:act:3DD36354-8814-42AB-8729-EDFA7D077778.

Simple Summary: In this work, a new species of ptyctimous mite ("box mite")—*Protophthiracarus afthonos*—is described and illustrated. It is exceptional by its very rich body setation—bigger than in any other known member of the ptyctimous oribatids. The new species was found in a forest soil sample from the Andes in Peru. The discovery confirms the uniqueness of the montane arthropod fauna of South America. The biogeographic distribution of ptyctimous oribatids in Peru is summarized and supplied with a key to the species of Peruvian fauna.

Abstract: *Protophthiracarus afthonos* sp. nov. is described and illustrated using line drawings, transmitted light and SEM imaging. It is characterized by an extraordinary richness of notogastral setae (ca. 166 pairs) that has been previously unseen among phthiracaroid mites. The species originates from the material collected from the litter of primary forest in the Peruvian Andes. The genus *Protophthiracarus* is well represented in the Neotropical Region. Many species of ptyctimous mites have been found in Peru, representing both widespread and endemic biogeographic elements. Among a total of 37 species, 20 from Peru have been described for the first time. Currently, the ptyctimous fauna consists of 12 endemite, 11 neotropical, 4 semicosmopolitan and 9 pantropical biogeographic elements.

Keywords: ptyctimous mites; *Protophthiracarus*; neotropical realm; neotrichy; new species

Citation: Niedbała, W.; Adamski, Z.; Laniecki, R.; Magowski, W.L. Ptyctimous Mites (Acari, Oribatida) of Peru with the Description of an Extraordinary New Phthiracaroid Mite from the Peruvian Andes. *Animals* **2023**, *13*, 2403. https://doi.org/10.3390/ani13152403

Academic Editor: Xiaolei Huang

Received: 19 May 2023
Revised: 11 July 2023
Accepted: 13 July 2023
Published: 25 July 2023

1. Introduction

Ptyctimous mites (Acari: Oribatida) are doubtlessly one of the best-known taxonomic groups of oribatid mites worldwide. Despite so many new species being described from all geographic regions of the world, new habitats are being explored, yielding yet more new taxa.

Peru is covered by a diverse range of habitats, from the Amazon rainforest in the east, to the high Andes mountains in the west. These habitats support a remarkable array of plant and animal life, including mites. Currently, there are three land domains of faunas: Amazon (with Amazon, Yungas, Pacific, Equatorial and Páramo provinces), Chaco and Andean-Patagonian (with Puno, Deserts and Mountain Steppes provinces). Knowledge of the distribution of most of invertebrate groups among those biogeographic units is fragmentary at best, if not nonexistent.

Peru is occupied by various vegetation units, namely mixed zones with evergreen wet tropical forests, mountainous tropical and subtropical wet forests, dry equatorial climate forests, savannahs and semideserts, trees and shrubs, grasslands, cushionplants and shrubs, and moss and lichens [1].

In this paper, we report a new, endemic species of the genus *Protophthiracarus* (Phthiracaroidea) from the Peruvian Andes that has a high, previously unseen number of notogastral setae, with the main purpose being to supply a description and images of this new, unusual taxon. In addition, our objective is to showcase the current picture of the ptyctimous mite fauna of Peru.

2. Materials and Methods

Mites were extracted from soil samples into 75% ethanol using Berlese's funnels with electric lamps in the laboratory. Thereafter, the specimens were mounted temporarily in lactic acid in cavity slides for making measurements and illustrations. The identification and illustration of the specimens were performed under a phase-contrast microscope Olympus BX50, equipped with a drawing attachment.

For microscopic imaging, the holotype female was photographed in transmitted light using Canon 5D or Olympus E5 SLR cameras attached to the Olympus BX51 microscope with differential interference contrast (DIC). Obtained frames were stacked and processed using PICOLAY software [2]. For SEM imaging, the specimens were prepared as follows: the mites were air-dried, attached to stubs with double-sided sticky tape, coated in gold and observed in a Zeiss Evo 40 Scanning Electron Microscope.

The morphological terminology of the description follows that of Grandjean (referenced by [3]; overview by [4]). Moreover, some terms and formulas follow [5]. The body measurements are given in micrometers. The prodorsum was measured in lateral view from the tip of the rostrum to the posterior edge; the notogaster was measured as a maximum length in lateral view. The width of the body was expressed as the maximum measurement in the dorsal aspect. The subcapitulum, genitoaggenital and anoadanal plates were measured on the ventral side. Similarly, the lengths of the body setae were measured in the lateral view.

The following abbreviations are used in the description: *ro*, *le*, *in*, *ex*—rostral, lamellar, interlamellar and exobothridial setae; *tr*—trichobothrium; *c*, *d*, *f*, *h*, *ps*—notogastral setae; *ia*, *im*—notogastral lyrifissures; *h*—subcapitular seta; *g*, *ag*, *an*, *ad*—genital, aggenital, anal and adanal setae; *iad*—adanal lyrifissure; and *d*, *l*, *v*—leg setae.

The holotype and paratype were deposited in the Department of Animal Taxonomy and Ecology, the Adam Mickiewicz University in Poznan, Poland.

3. Results

Systematics

Protophthiracarus afthonos Niedbała sp. nov. (Figures 1–5)

Description. Measurements of the holotype: prodorsum: length 414, width 293, height 151; setae: *tr* 88, *in* 38, *le* 51, *ro* 63, *ex* 56; notogaster: length 808, width 586, height 545, seta c_1 268, left genitoaggenital plate 186 × 136, right genitoaggenital plate 202 × 141, anoadanal plate 303 × 146; length of plates measured from the side: genitoaggenital 182, anoadanal 308. Paratype: prodorsum: length 429, height 162; notogaster: length 909, height 616.

Rather large species with strong neotrichy of notogaster (Figure 1), weaker adanal setae. Color: light brown. The surface of the body is punctated (spaced regular mounds visible under high magnification—Figure 4C). Setae finely serrate.

Prodorsum (Figure 2A,C and Figure 4A) with weak lateral carinae. Sigillar fields narrow and well-marked. Posterior furrows absent. Trichobothria (Figures 2B and 4B) long, filiform but rigid, with a clear inner core, distinctly serrate and pointed distally. Interlamellar, lamellar and rostral setae short, spiniform rough: lamellar and interlamellar procumbent; rostral semierect; *tr* > *ro* > *ex* > *le* > *in*.

Notogaster (Figure 3) with ca., 166 pairs of very long ($c_1 > c_1$–d_1) serrated setae, generally rigid but more flexible towards tips. Due to the huge number of setae, it is extremely difficult to distinguish the setae of appropriate rows. Vestigial setae invisible. Three pairs of lyrifissures (Figure 3), *ia*, *im* and *ip* present.

Venter (Figure 4D). Setae *h* of subcapitulum considerably longer than the distance between them (Figure 2D). Genital setae formula of right genitoaggenital plate (Figure 2F): 5 + 4:1; formula of left plate (Figure 2E): 4 + 4:1; setae $g_{1\text{–}5}$ shorter than setae $g_{6\text{–}9}$. In paratype, the formula of both genitoaggenital plates: 4 + 4:1. Anoadanal plates (Figure 2G) with three pairs of setae at their paraxial border (in a position of anal setae) and seven pairs of setae remote from the border (in a position of adanal setae). Setae long, filiform, weakly serrate, slightly more than notogastral setae.

Legs (Figures 2H, 4E,F and 5). Chaetome complete. Setae *d*, *l″* and *v′* slightly remote from the anterior end of its segment and situated almost at the same level; seta *v″* situated posteriorly.

Type material. Holotype and paratype: South America, Peru, Central Peru, Andes, 09°42′58″ S 75°05′33″ W, Huánuco Department, Huánuco Province, Chinchao District, NW Tunel de Carpish, 2770 m a.s.l.; upper soil and leaf litter in primary mountain forest, Winkler extraction, 14 April 2016; leg. S. Friedrich, F. Wachtel and D. Hauth.

Etymology. The specific epithet is derived from Greek *afthonos* meaning "abundant" and refers to the polytrichy of the notogastral setae.

Comparison and diagnosis. This species is distinguishable by the huge neotrichy of the notogastral setae. Multiplication of a number of setae applies to all rows on notogaster, namely: *c*, *d*, *e*, *h* and *ps*. In other species revealing notogastral neotrichy, the enlarged number (multiplication) of setae applies typically to the setae of rear notogastral rows, usually *h* and *ps*.

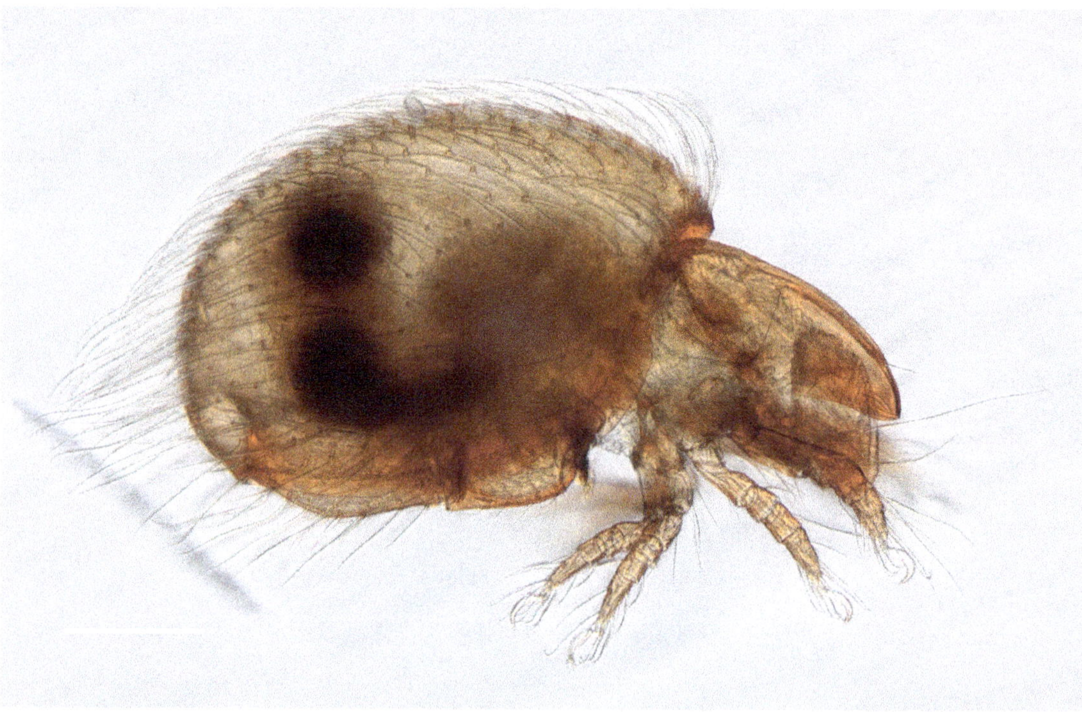

Figure 1. *Protophthiracarus afthonos* Niedbała sp. nov. holotype (DIC), habitus—side view. Scale bar 200 μm.

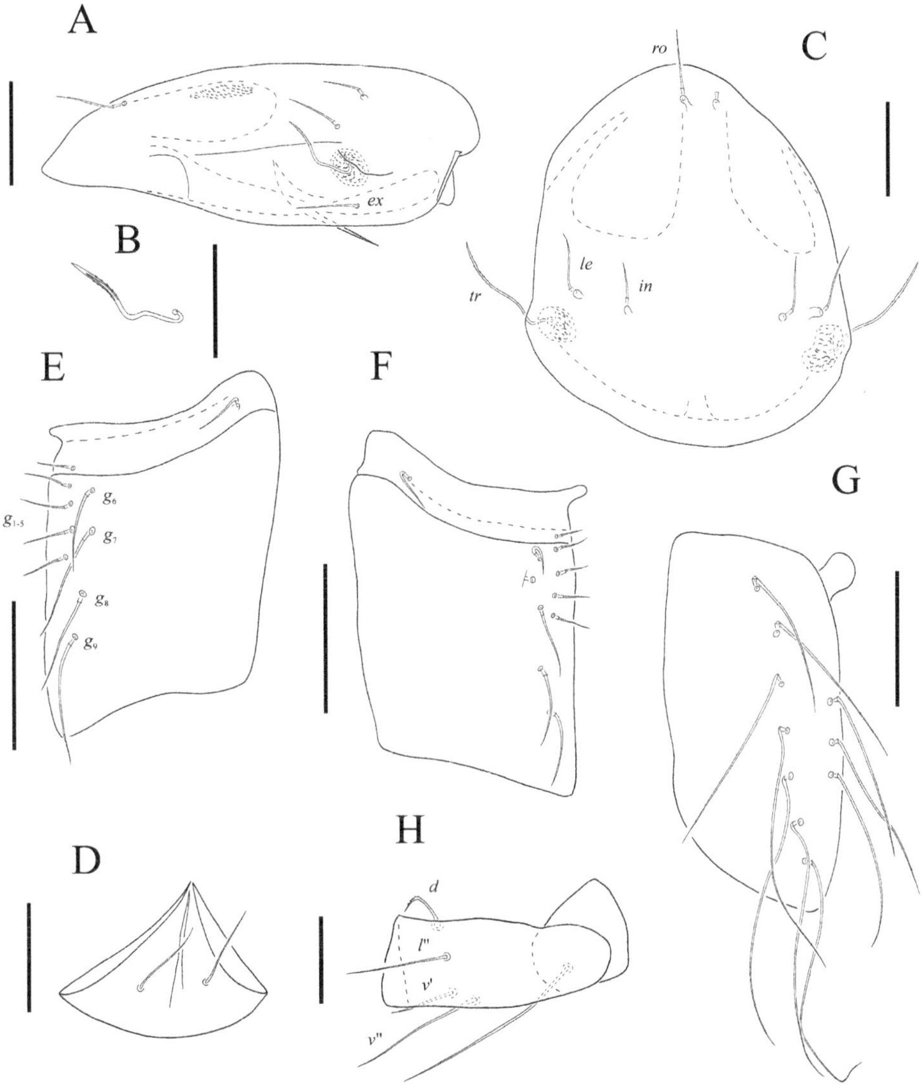

Figure 2. *Protophthiracarus afthonos* Niedbała sp. nov. holotype, (**A**) prodorsum, lateral view; (**B**) trichobothrium, lateral view; (**C**) prodorsum, dorsal view; (**D**) mentum of subcapitulum; (**E**) left genitoaggenital plate; (**F**) right genitoaggenital plate; (**G**) right anoadanal plate; (**H**) trochanter and femur of leg I. Scale bars 100 μm (**A**,**C**,**E**–**G**), 25 μm (**B**,**D**,**H**).

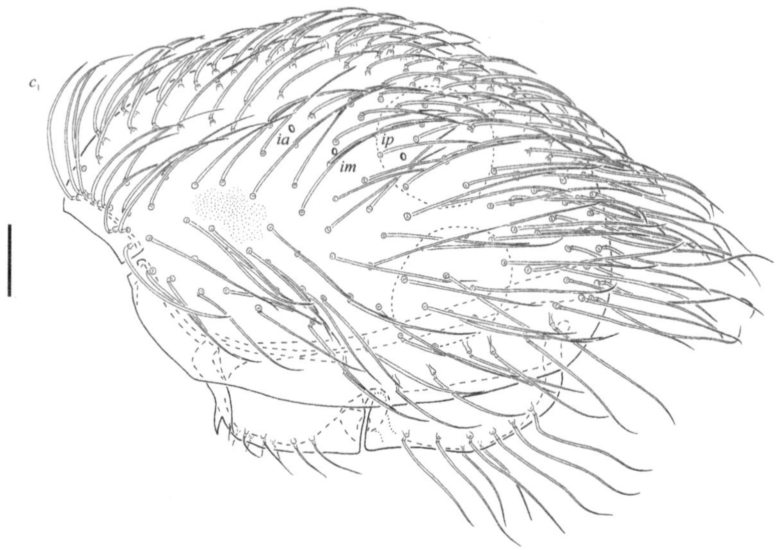

Figure 3. *Protophthiracarus afthonos* Niedbała sp. nov. holotype, opisthosoma, lateral view. Scale bar 100 μm.

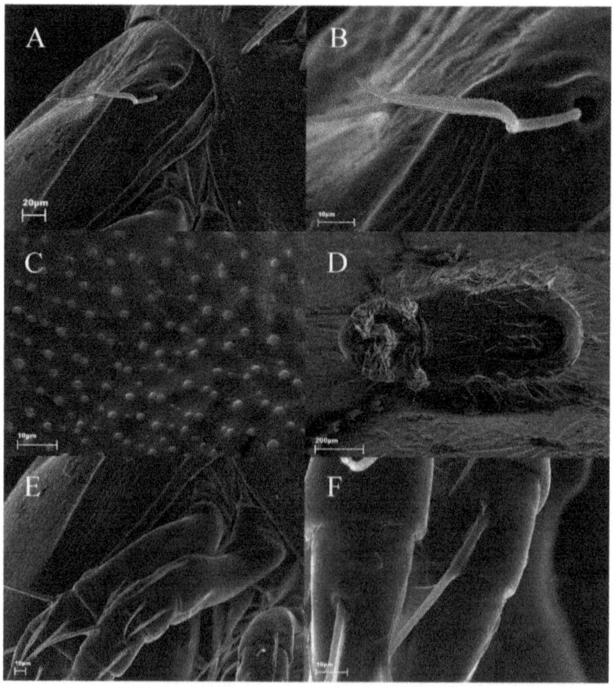

Figure 4. *Protophthiracarus afthonos* Niedbała sp. nov. paratype (SEM); (**A**) part of prodorsum and anterior part of notogaster, lateral view; (**B**) trichobothrium, lateral view; (**C**) notogaster surface texture, lateral view; (**D**) habitus, ventral side; (**E**) parts of prodorsum and legs I and II, lateral view; (**F**) tibia of leg IV, dorsal view.

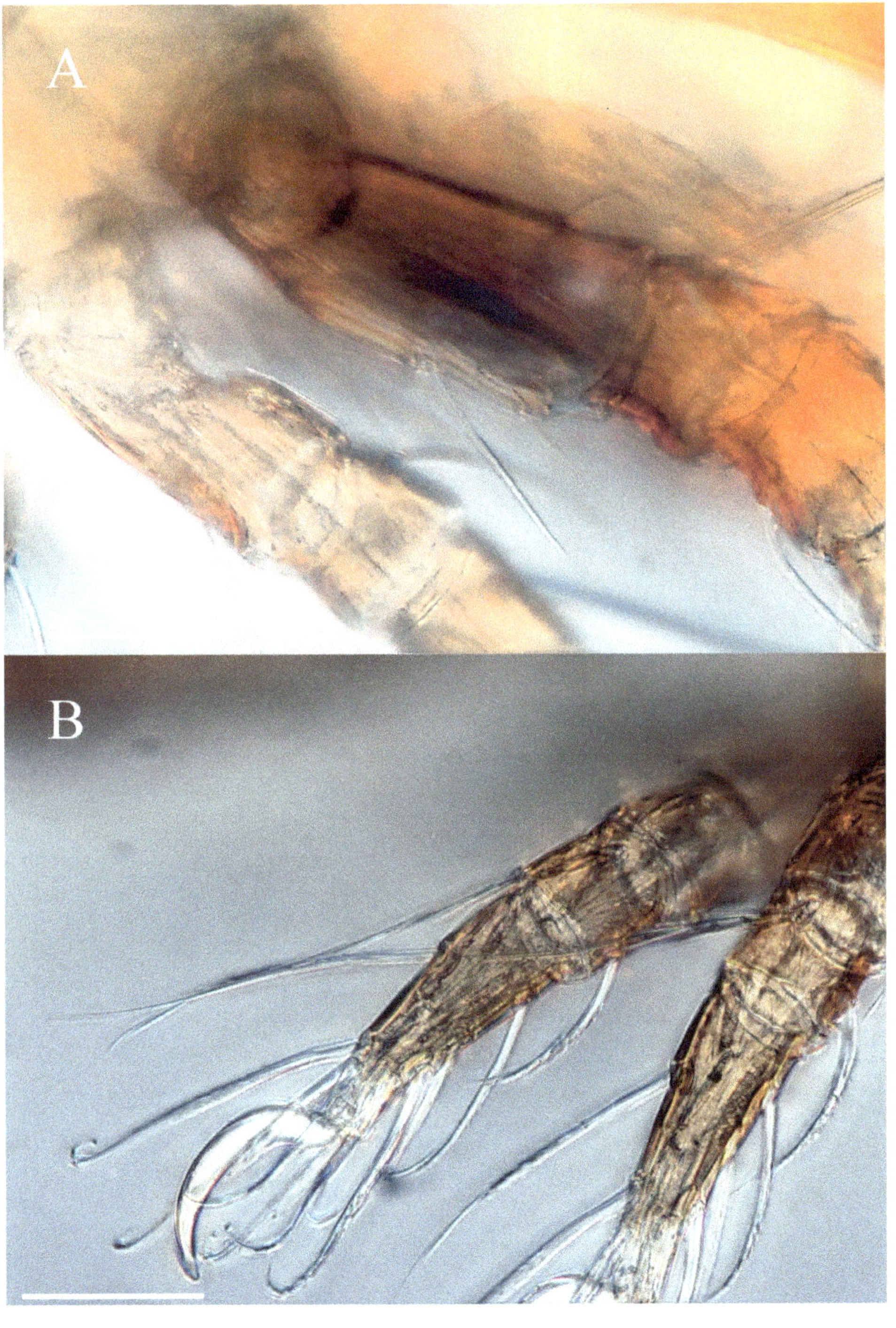

Figure 5. *Protophthiracarus afthonos* Niedbała sp. nov. holotype (DIC); (**A**) parts of legs I and II lateral view; (**B**) parts of leg III and IV. Scale bars 100 μm.

A systematic list of ptyctimous mite species (Oribatida) of Peru

Synonymes are only supplied for species included in the literature cited in this paper; for a complete list of the synonymies of species recorded in Peru, see Niedbała 1992 and 2004. Species names are supplied with a biogeographic element epithet. Species described from Peru as new are marked with asterisk (*). Spellings and collocations of taxonomic names follow the recent catalogue by Niedbała and Liu [6].

Enarthronota Grandjean, 1947
Hypochthonioidea Berlese, 1910

Mesoplophoridae Ewing, 1917
Mesoplophora Berlese, 1904
Mesoplophora subgen. nom.
- *bacilla* Niedbała, 2004; neotropical
- *hauseri* Mahunka, 1982; neotropical
-* *quasigaveae* Niedbała, 2016; endemic
-* *sparsa* Niedbała, 2004; endemic
Parplophora Niedbała, 1985
-* *subtilis* Niedbała, 1981; pantropical

Mixonomata Grandjean, 1969
Euphthiracaroidea Jacot, 1930
Oribotritiidae Balogh, 1943
Oribotritia Jacot, 1924
Oribotritia subgen. nom.
- *didyma* Niedbała et Schatz, 1996; neotropical

Mesotritia Forsslund, 1963 (=*Perutritia* Märkel, 1964)
-* *amazonensis* (Märkel, 1964); endemic
-* *curviseta* (Hammer, 1961); neotropical

Indotritia Jacot, 1929
Indotritia subgen. nom.
- *bellingeri* Niedbała et Schatz, 1996; pantropical
- *krakatauensis* (Sellnick, 1923) (=*Indotritia acanthophora* Märkel, 1964); pantropical

Euphthiracaridae Jacot, 1930
Acrotritia Jacot, 1923
-* *clavata* (Märkel, 1964); nearctic and neotropical
- *dikra* (Niedbała et Schatz, 1996); nearctic and neotropical
-* *peruensis* (Hammer, 1961); neotropical
- *refracta* (Niedbała, 1998); pantropical
- *vestita* (Berlese, 1913) (=*Rhysotritia comteae* Mahunka, 1983); pantropical

Microtritia Märkel, 1964
-* *tropica* Märkel, 1964; pantropical

Phthiracaroidea Perty, 1841
Phthiracaridae Perty, 1841
Phthiracarus Berlese, 1920
- *anonymus* Grandjean, 1933; semicosmopolitan
- *boresetosus* Jacot, 1930; semicosmopolitan
-* *helluonis* (Niedbała, 1982); endemic
- *nitens* (Nicolet, 1855); palaearctic (likely introduced)
-* *octosetosus* Niedbała, 2004; endemic

Steganacaridae Niedbała, 1986
Hoplophthiracarus Jacot, 1933
-* *incredibilis* (Niedbała, 1982); endemic

Steganacarus Ewing, 1917
Rhacaplacarus Niedbała, 1986
-* *stenodes* Niedbała, 2004; neotropical

Protophthiracarus Balogh, 1972
-* *ventosus* (Hammer, 1961); endemic
-* *afthonos* sp. nov.; endemic

Notophthiracarus Ramsay, 1966
-* *fornicarius* (Niedbała, 1982); neotropical
-* *improvisus* (Niedbała, 1982); endemic
-* *inauditus* (Niedbała, 1982); neotropical

Austrophthiracarus Balogh et Mahunka, 1978
-* *excellens* (Niedbała, 1982); endemic

Arphthicarus Niedbała, 1994
- *inelegans* (Niedbała, 1986); pantropical
-* *simplex* Niedbała, 2017; endemic

Atropacarus Ewing, 1917
Hoplophorella Berlese, 1923
- *andrei* (Balogh, 1958); pantropical
- *hamatus* (Ewing, 1909); semicosmopolitan
- *lanceosetus* (Balogh et Mahunka, 1981) (=*Hoplophorella neglecta* Niedbała, 1984, *H. neglectus* Niedbała, 1992); neotropical
-* *stilifer* (Hammer, 1961); pantropical
- *vitrinus* (Berlese, 1913) (=*Hoplophorella scapellata* Aoki, 1965, *H. africana* Wallwork, 1967); semicosmopolitan.

List of localities and species of ptyctimous Oribatida found in Peru.

The list contains standardized data, with corrections of apparent mistakes in primary sources, modern transliterations of geographic names and the numbers of collected individuals of each species recorded. In addition, it follows the original format as much as possible. Biogeographic distribution is depicted in Figure 6, with locations marked by numbers in braces below.

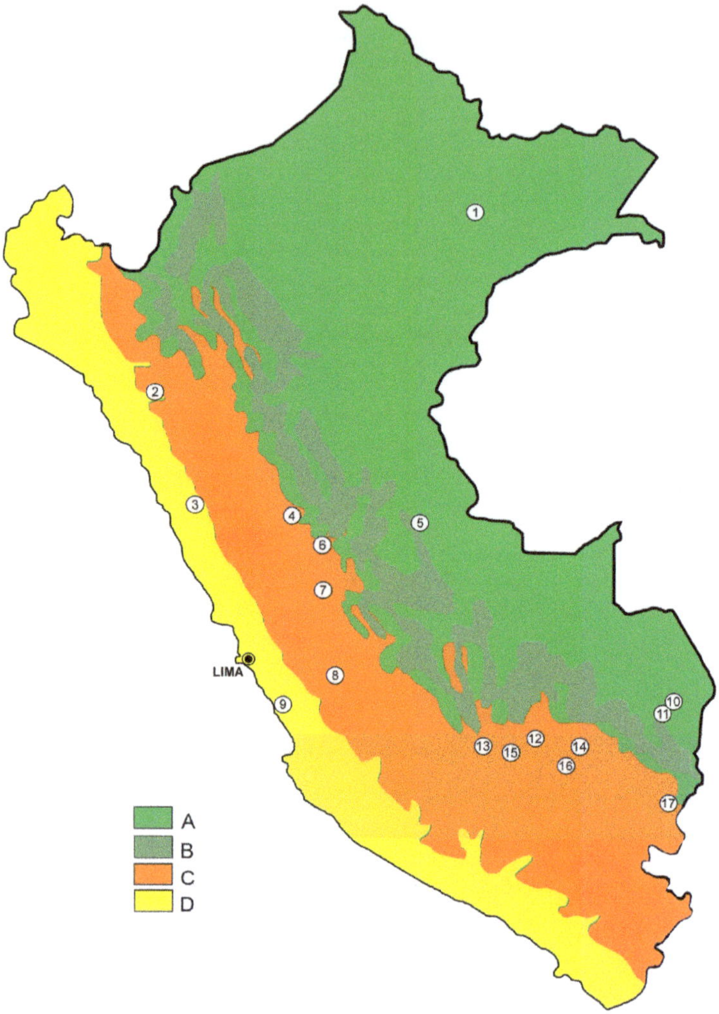

Figure 6. Map of distribution of Peruvian species of ptyctimous oribatid mites plotted against main ecoregions: A—tropical rain forest, B—mountain rain forest, C—mountain grass, scrub and alpine wastes, D—deserts (after [1]; changed). Numbers in circles represent major localities corresponding to those in braces in the "List of localities and species of ptyctimous Oribatida found in Peru".

Loreto Region

{1}

Amazon Basin, Muyuy Island near Iquitos, 105 m a.s.l., lowland rainforest area, jungle, litter, 1956/57, leg. F. Schaller. *I. (I.) krakatauensis*—1 [7].

Cajamarca Region

{2}

At Cajamarca, ca. 7° S, 3000 m a.s.l., in low cushion with stiff leaves on dry moldering soil, below an agave hedge, 5–6 October 1957, leg. M. Hammer. *A. (H.) stilifer*—2 [8].

At Cajamarca, ca. 7° S, 3000 m a.s.l., in almost dry moldering soil with grass, white clover and *Equisetum* below agave and bramble, near the river, 5–6 October 1957, leg. M. Hammer. *P. ventosus*—1 [8].

At Cajamarca, ca. 7° S, 3000 m a.s.l., in dry moss on sun-dried moldering soil below a hedge of agave, 5–6 October 1957, leg. M. Hammer. *A. peruensis*—1 [8].

Ancash Region

{3}

Cordillera Blanca, Llanganuco valley, Huascaran National Park, 4000 m a.s.l., litter from bushes and tree *Schinus* sp., 24 August 1976, leg. J. Michejda. *N. fornicarius*—18 [9].

Cordillera Blanca, Llanganuco valley, Huascaran National Park, 3000 m a.s.l., *Schinus* sp., exp. S, 24 August 1976, leg. J. Michejda. *M. curviseta*—2; *A. vestita*—1 [10].

Cordillera Blanca, Llanganuco valley, Huascaran National Park, near Huapi, *Eucalyptus* forest, very dry, 26 August 1976, leg. J. Michejda. *A. vestita*—6 [10].

Cordillera Blanca, Llanganuco valley, Huascaran National Park, near Huapi, *Eucalyptus* forest, wet sample, 26 August 1976, leg. J. Michejda. *A. vestita*—1 [10].

Huanuco Region

{4}

Mountain forest area near Tingo Maria, 780 m a.s.l., leaf litter from rubber tree plantation; 1956/57, leg. F. Schaller. *A. clavata*—No. of specimens not given [7].

Mountain forest area near Tingo Maria, 780 m a.s.l., jungle litter; 1956/57, leg. F. Schaller. *M. tropica*—1 [7].

Environments of Tingo Maria, humus on the rocks, decayed wood, and soil (originally three separate samples); 1956/57, leg. F. Schaller. *M. tropica*—No. of specimens not given [7].

{5}

Puerto Inca Province, Yuyapichis District, Área de Conservación Privada Panguana (biological field station), near Rio Yuyapichis (river), 230 m. a.s.l., primary evergreen lowland rainforest, upper soil and leaf litter, Winkler extraction, 09°37′ S, 74°56′ W, 1–21 May 2015, leg. S. Friedrich and F. Wachtel. *A. simplex*—2; *M. (M.) quasigaveae*—86; *M. curviseta*—5; *I. (I.) bellingeri*—2; *A. vestita*—1; *A. (H.) andrei*—2 [11].

Puerto Inca Province, Yuyapichis District, Área de Conservatión Privada, Panguana (biological field station), near Rio Yuyapichis (river), 230–260 m a.s.l., primary evergreen lowland rainforest, upper soil and leaf litter, Winkler extraction, 09°37′ S, 74°56′ W, 20 September–7 October 2013, leg. S. Friedrich and F. Wachtel. *M. (M.) quasigaveae*—41; *M. curviseta*—8; *O. (O.) didyma*—5; *A. clavata*—1; *A. dikra*—19; *A. refracta*—4; *M. tropica*—1; *A. inelegans*—49; *A. (H.) andrei*—7; *A. (H.) hamatus*—7 [12].

Puerto Inca Province, Yuyapichis District, Área de Conservación Privada, Panguana (biological field station), near Rio Yuyapichis (river), 230–260 m a.s.l., primary evergreen lowland rainforest, upper soil and leaf litter, 09°37′ S, 74°56′ W, 23 April—9 May 2016, leg. S. Friedrich, F. Wachtel and D. Hauth. *M. (M.) quasigaveae*—51; *A. clavata*—2; *A. refracta*—5; *A. inelegans*—5; *A. (H.) andrei*—9 [13].

{6}

Huánuco Province, Chinchao District, NW Tunel de Carpish, 2770 m a.s.l., 09°42′58″ S 75°05′33″ W, upper soil and leaf litter in primary mountain forest, Winkler extraction, 14 April 2016; leg. S. Friedrich, F. Wachtel and D. Hauth. *P. afthonos* n. sp.—2.

Oxapampa Region

{7}

Near Oxapampa, Rio Esperanza, 2000 m a.s.l., foggy forest area, leaf litter (two separate samples are listed with same collection data in original text); 1956/57; leg. F. Schaller. *M. tropica*—No. of specimens not given; *A. clavata*—8 [7].

Rio Esperanza, 2150 m a.s.l., moss from tree plantation at height of 2 m; 1956/57; leg. F. Schaller. *A. clavata*—No. of specimens not given [7].

Rio Esperanza, 2150 m a.s.l., tree plantation, leaf litter; 1956/57; leg. F. Schaller. *A. clavata*—No. of specimens not given [7].

Rio Esperanza, 2200 m a.s.l., primary forest on steep slope, rotten wood; 1956/57; leg. F. Schaller. *A. clavata*—No. of specimens not given [7].

Junin Region

{8}

At Huancayo, ca. 12° S, 3550 m a.s.l., in moist–wet low moss on a vertical slope, always shaded, 2 November 1957, leg. M. Hammer. *A. peruensis*—1 [8].

At Huancayo, ca. 12° S, 3550 m a.s.l., in wet liverworts and moss, shaded, 2 November 1957, leg. M. Hammer. *A. peruensis*—2 [8].

{9}

La Huerta (habitat and substrate unknown), 24–28 November 1955, leg. L. Peña. *A. peruensis*—3; *N. fornicarius*—3 [10].

Madre de Dios Region

{10}

Tambopata National Reserve, ex litter along river, 25 October 1982, L. E. Watrous and G. Mazurek. *A. peruensis*—11; *M. tropica*—10; *A. (H.) vitrinus*—1 [10].

Tambopata National Reserve, ex rotten palm flowers, 28 October 1982, leg. L. E. Watrous and G. Mazurek. *A. (H.) vitrinus*—2 [10].

Tambopata National Reserve, ex bamboo litter, 28 October 1982, leg. L. E. Watrous and G. Mazurek. *M. (M.) sparsa*—1; *M. curviseta*—3; *I. (I.) bellingeri*—3 [10].

{11}

Rio Madre de Dios basin, Puerto Maldonado, 220 m a.s.l., rainforest, 1956/57, leg. F. Schaller. *M. amazonensis*—2 [7].

Rio Madre de Dios, near Puerto Maldonado ("Maldano" in original text, very likely typo), 250 m a.s.l., rain forest (two separate samples but same data); 1956/57, leg. F. Schaller. *M. tropica*—No. of specimens not given [7].

Rio Madre de Dios, Puerto Maldonado, litter under logs at farm pen, 5 September 1976, leg. J. Michejda. *A. (H.) vitrinus*—1 [5].

Rio Madre de Dios, Puerto Maldonado, forest near airport, 5 September 1976, leg. J. Michejda. *A. (H.) lanceosetus*—1 [5].

Rio Madre de Dios, Puerto Maldonado, 500 m a.s.l., wood dust from log laying at a farm pen, 3 September 1976, leg. J. Michejda. *M. subtilis*—6 [14],—8 [10]; *A. (H.) lanceosetus*—3 [15]; *A. (H.) vitrinus*—6 [5].

Cusco Region

{12}

Pillahuata, Manu road, 128 km, ex litter in dry streambed, 18 September 1982, leg. L. E. Watrous and G. Mazurek. *M. (M.) bacilla*—1; *M. curviseta*—4; *A. vestita*—11 [10].

Pillahuata, Manu road, 128 km, ex leaf litter, 27 September 1982, leg. L. E. Watrous and G. Mazurek. *M. (M.) bacilla*—3; *M. curviseta*—8; *A. vestita*—6 [10].

Pillahuata, Manu road, 128 km, ex damp leaf litter; 26 September 1982, leg. L. E. Watrous and G. Mazurek. *M. (M.) bacilla*—2; *M. curviseta*—12; *A. vestita*—7; *P. helluonis*—1; *P. octosetosus*—1 [10].

Pillahuata, Manu road, 128 km, ex litter under ferns, 16 September 1982, leg. L. E. Watrous and G. Mazurek. *M. curviseta*—3; *A. vestita*—11 [10].

Pillahuata, Manu road, 128 km, ex litter at seepage area 17 September 1982, leg. L. E. Watrous and G. Mazurek. *M. curviseta*—7; *A. vestita*—1 [10].

Pillahuata, Manu road, 128 km, ex moss and litter on xeric slope, 26 September 1982, leg. L. E. Watrous and G. Mazurek. *M. (M.) bacilla*—1; *A. vestita*—2 [10].

Pillahuata, Manu road, 128 km, ex litter under grass clumps, 16 September 1982, leg. L. E. Watrous and G. Mazurek. *M. curviseta*—1; *A. peruensis*—2 [10].

Pillahuata, Manu road, 128 km, ex rotten logs, 26 September 1982, leg. L. E. Watrous and G. Mazurek. *M. curviseta*—7; *A. peruensis*—3; *S. (R.) stenodes*—1 [10].

Pillahuata, Manu road, 128 km, ex leaf litter after rain, 17 September 1982, leg. L. E. Watrous and G. Mazurek. *M. curviseta*—19; *A. vestita*—4; *P. helluonis*—1 [10].

Pillahuata, Manu road, 128 km, ex litter along stream, 26 September 1982, leg. L. E. Watrous and G. Mazurek. *M. (M.) bacilla*—1; *M. curviseta*—6; *A. vestita*—2 [10].

Consuelo, Manu road, 165 km, ex rotten palm, 5 October 1982, leg. L. E. Watrous and G. Mazurek. *M. hauseri*—2 [10].

{13}

At foot of Machu Picchu, valley of Urubamba River, 2700 m a.s.l., in forest, 2 September 1976, leg. J. Michejda. *H. incredibilis*—1 [9]; *P. boresetosus*—18; *N. fornicarius*—1; *N. inauditus*—10 [5].

Machu Picchu, valley of Urubamba river, near track, 2700 m a.s.l., forest, 2 September 1976, leg. J. Michejda. *A. vestita*—50; *M. curviseta*—4; *P. anonymus*—4 [10].

Urubamba valley, 2700 m a.s.l., dry litter, 3 September 1976, leg. J. Michejda. *A. vestita*—3 [10].

Urubamba valley, Cuscichaca river, 2485 m a.s.l., cloud forest, decayed stump, 30 September 1982, leg. J. Sale. *A. dikra*—19; *P. boresetosus*—3; *P. nitens*—1 [10].

{14}

Rio Madre de Dios basin, near Quince Mil, 650 m a.s.l., mountain forest area, epiphytes, 1956/57, leg. F. Schaller. *A. clavata*—No. of specimens not given [6].

Rio Madre de Dios, near Quince Mil, 650 m a.s.l., in mountain forest, 1956/57, leg. F. Schaller. *M. tropica*—No. of specimens not given [7].

{15}

At the pass Cusco—Pisac, ca. 13°30′ S, 3750 m a.s.l., in moist 2–3 cm high moss between ankle-deep heathery shrubs, 5 February 1955, leg. M. Hammer. *A. peruensis*—1 [8].

Foot of Machu Picchu, ca. 13° S, 2200 m a.s.l., in wet moss on a vertical cliff wall, 1 February 1955, leg. M. Hammer. *M. curviseta*—2; *A. peruensis*—2 [8].

Foot of Machu Picchu, ca. 13° S, 2200 m a.s.l., in wet *Selaginella* sp. on the ground below meter-high vegetation, 1 February 1955, leg. M. Hammer. *M. curviseta*—1; *A. peruensis*—1 [8].

Machu Picchu, exp. N, 3400 m a.s.l., bamboo forest, wet, 1 September 1976, leg. J. Michejda. *P. boresetosus*—3; *A. excellens*—1; *M. curviseta*—1; *A. vestita*—4 [10].

Machu Picchu, 3400 m a.s.l., litter under dense shrubs among ruins, 1 September 1976, leg. J. Michejda. *A. excellens*—1; *P. boresetosus*—3; *N. inauditus*—7 [5].

Machu Picchu, 3600 m a.s.l., tropical rain forest, 1 September 1976, leg. J. Michejda. *M. curviseta*—3; *A. vestita*—10 [10].

Near tourist trail from Wiñay Wayna towards Machu Picchu, 3600 m a.s.l., litter in tropical forest, 1 September 1976, leg. J. Michejda. *A. excellens*—8; *P. helluonis*—4; *N. improvisus*—1; *N. inauditus*—6 [16].

Near tourist trail from Wiñay Wayna towards Machu Picchu, 3400 m a.s.l., litter in tropical forest, E exposure, 1 September 1976, leg. J. Michejda. *A. excellens*—2; *N. inauditus*—3 [5].

Near tourist trail from Wiñay Wayna towards Machu Picchu, 3600 m a.s.l., in forest near railroad, 1 September 1976, leg. J, Michejda. *P. boresetosus*—1; *N. inauditus*—1; *P. anonymus*—1 [5].

Machu Picchu, forest near track, 2 September 1976, leg. J. Michejda. *P. anonymus*—1; *P. boresetosus*—1 [10].

Machu Picchu, rain forest, exp. E, 3400 m a.s.l., 2 September 1976, leg. J. Michejda. *A. vestita*—8 [10].

{16}

Marcapata, road to Puerto Maldonado, km 175, ex leaf litter, 21 October 1982, leg. L. E. Watrous. *M. curviseta*—2; *P. helluonis*—1 [10].

Puno Region

{17}

At Sillustani, north of Puno, ca. 15° S, 3900 m a.s.l., almost dry moss on a vertical slope below shrubs, shaded, 16 November 1957, leg. M. Hammer. *M. curviseta*—1 [8].

The mite material was collected from soil (below an agave hedge, dry moldering soil with grass, humus on the rocks), litter (under ferns, at seepage areas, under grass clumps, from bushes and trees, in a bamboo forest, in a tropical rain forest, under trunks, in dry streambed, damp leaf litter, stiff leaves on dry moldering), moss, liverworts and clubmoss (between ankle-deep heathery shrubs, wet and dry moss on a vertical cliff-wall, wet moss on a vertical cliff, wet liverworts and moss, wet *Selaginella* on the ground), decomposing plant matter (rotten palm flowers, decayed wood, wood dust from a fallen tree trunk, rotten logs, rotten palms, decayed stump in a cloud forest) and other living plant matter (white clover and *Equisetum* below agave and bramble, epiphytes from mountain forest).

Key to ptyctimous Oribatida of Peru

1. Genital and anal openings well separated .... Enarthronota 2
-. Genital and anal openings joined .... Mixonomata 6
2. Three pairs of anal setae .... *Mesoplophora (Parplophora)* .... *M. (P.) subtilis* (Figures 1–24 in [14])
-. Two pairs of anal setae .... *Mesoplophora (Mesoplophora)* 3
3. "Notogastral" setae smooth, formula of genital setae 5:2 .... *M. (M.) bacilla* (Figure 3F–I in [10])
-. "Notogastral" setae spinose or rough, formula of genital setae 6:1 4
4. Trichobothria with slightly fusiform head covered with nine small setae .... *M. (M.) hauseri* (Figures 1–3 in [17]; Figure 4H–K in [10])
-. Trichobothria setiform covered with more than ten small setae 5
5. Anal setae an_1 located in posterior half of plates; trichobothria covered with 11–12 pairs of small setae .... *M. (M.) quasigavae* (Figures 12–22 in [12])
-. Anal setae an_1 located in anterior half of plates; trichobothria covered with 15 pairs of small setae .... *M. (M.) sparsa* (Figure 6E–G in [10])
6. Body considerably compressed laterally, anogenital region narrow, V-shaped .... Euphthiracaroidea 7
-. Body less compressed laterally, anogenital region relatively wide, U-shaped .... Phthiracaroidea 17
7. Ventral plates not completely fused, at least anal plates separated by suture, longitudinal suture of ano-genital region without interlocking triangle .... Oribotritiidae 8
-. Ventral plates completely fused, at least one triangle in longitudinal suture of ano-genital region present .... Euphthiracaridae 12
8. Genitoaggenital suture incomplete, two plates well delineated from each other only posteriorly .... *Indotritia (Indotritia)* 9
-. Genitoaggenital suture complete 10
9. Interlamellar setae almost as the half of height of prodorsum, not bent distally, exobothridial setae well developed .... *I. (I.) bellingeri* (Figures 34–49 in [18])
-. Interlamellar setae fine, no longer than one fourth of height of prodorsum, bent distally, exobothridial setae vestigial .... *I. (I.) krakatauensis* (Figure 18M–O in [10])
10. Bothridial scale situated above bothridium, scisure between genital and anal plates present .... *Oribotritia (Oribotritia)* .... *O. (O.) didyma* (Figures 7–14 in [18])
-. Bothridial scale situated below bothridium, scisure between genital and anal plates absent .... *Mesotritia* 11
11. Rostral setae situated distinctly anteriorly of lamellar setae, adanal setae ad_1 and ad_2 similar in length .... *M. amazonensis* (Figure 6A–G in [7]; Figure 13A–H in [10])
-. Rostral setae situated at the level with lamellar setae, adanal setae ad_1 considerably longer than setae ad_2 .... *M. curviseta* (Figures 134–134C in [8]; Figure 15A–I in [10])
12. Genitoaggenital plates with 4–6 genital setae .... *Microtritia* .... *M. tropica* (Figure 11 in [7]; Figure 35E–H in [10])
-. Genitoaggenital plates with 7–9 genital setae .... *Acrotritia* 13
13. Two pairs of lateral prodorsal carinae present on each side 14
-. One single pair of lateral prodorsal carinae present on each side 15
14. Nine pairs of genital setae, one pair in progenital position; tarsi of legs monodactylous .... *A. refracta* (Figures 155–159 in [19])

-. Nine pairs of genital setae, all in genital position; tarsi I bi-, tarsi II-IV tridactylous .... *A. peruensis* (Figures 133, 133A in [8]; Figure 32M–O in [10])
15. Lateral carinae of prodorsum not forked distally .... *A. clavata* (Figures 16A–C in [7]; Figures 29E–I in [10])
-. Lateral carinae of prodorsum forked distally 16
16. Trichobothria setiform .... *A. dikra* (Figures 111–117 in [18])
-. Trichobothria with distinct fusiform head .... *A. vestita* (= *Rhysotritia comteae*: Figures 4–7 in [20]; Figure 29J–L in [10])
17. Setae smooth, fine, attenuate, tapering to distal end .... Phthiracaridae .... *Phthiracarus* 18
-. Setae (except exobothridial) rough or covered with small spines of different shapes but not smooth and attenuate .... Steganacaridae 22
18. Trichobothria long and narrow, their length more than 10 times of width 19
-. Trichobothria short and wide, length not more than 10 times of width 20
19. Neotrichy of setae present, notogaster with 16 pairs of setae, anoadanal plates with six pairs of adanal setae .... *P. octosetosus* (Figure 39A–F in [10])
-. Neotrichy of setae absent, notogaster with 15 pairs of setae, anoadanal plates with three pairs of adanal setae .... *P. boresetosus* (Figures 15–17 in [21]; Figures 517–522 in [22]; Figure 36I-O in [10])
20. Neotrichy present, 21–28 pairs of notogastral setae, three to four pairs anal and six to nine pairs of adanal setae present .... *P. helluonis* (Figures 22–43 in [16])
-. Neotrichy absent, always 15 pairs of notogastral setae, two pairs of anal and three pairs of adanal setae present 21
21. Four pairs of lyrifissures, *ia*, *im*, *ip*, *ips* present, adanal setae ad_1 and ad_2 vestigial .... *P. nitens* (Figures 1–6 in [23]; Figures 38M–T in [10])
-. Two pairs of lyrifissures, *ia* and *im* present, adanal setae well developed .... *P. anonymus* (Figures 1A,B; 2A–C in [24]; Figure 36E–H in [10])
22. Three setae (ad_1, an_1, an_2) in a row near paraxial margin of anoadanal plate 23
-. Two setae (an_1 and an_2) near paraxial margin of anoadanal plate 30
23. Setae *d* on tibiae IV long, independent of solenidia .... *Steganacarus* (*Rhacaplacarus*) .... *S.* (*R.*) *stenodes* (Figure 52F–J in [10])
-. Setae *d* on tibiae IV short, coupled with solenidia 24
24. Genital setae $g_{7\text{-}9}$ displaced towards paraxial margin of genitoaggenital plates and arranged in a row with setae $g_{1\text{-}5}$, setae g_6 not displaced .... *Protophthiracarus* 25
-. All genital setae located in a row along paraxial margin .... *Atropacarus (Hoplophorella)* 26
25. Notogaster with 17 pairs of short ($c_1 < c_1\text{–}d_1$) setae .... *P. ventosus* (Figures 131, 131A in [8]; Figure 88P–U in [10])
-. Notogaster with ca. 166 pairs of long setae; c_1 setae considerably longer than distance $c_1\text{-}d_1$ .... *P. afthonos* sp. nov.
26. Notogastral setae wide, phylliform 27
-. Notogastral setae longer, slightly lanceolate 29
27. Rostral setae directed forwards .... *A. (H.) hamatus* (Figure 99D–J in [10])
-. Rostral setae directed inwards 28
28. Notogaster with median band .... *A. (H.) andrei* (Figures 11–23 in [25])
-. Notogaster without median band .... *A. (H.) vitrinus* (Figure 102J–O in [10])
29. Notogastral setae longer ($c_1 > 1/2c_1\text{–}d_1$), setae v' of femora absent .... *A. (H.) lanceosetus* (Figure 100G–T in [10])
-. Notogastral setae shorter ($c_1 < 1/2c_1\text{–}d_1$), setae v' of femora present .... *A. (H.) stilifer* (Figures 132–132C in [8]; Figure 101O–T in [10])
30. Setae *d* on tibiae of legs IV long, independent of solenidia .... *Hoplophthiracarus* .... *H. incredibilis* (Figures 1–16 in [9])
-. Setae *d* on tibiae IV short, coupled with solenidia 31

31. Genital setae arranged in a row near paraxial margin of plates .... *Notophthiracarus* 32
-. Genital setae arranged in two rows; setae g_6 and g_7 remote from paraxial margin or only setae g_6 remote from margin 34
32. Majority of notogastral setae hooked distally; two pairs of lyrifissures *ia* and *im* present .... *N. improvisus* (Figures 44–61 in [16])
-. Notogastral setae not hooked distally; three or four pairs of lyrifissures present 33
33. Surface of notogaster punctated; three pairs of lyrifissures *ia, im, ip* present; formula of genital setae 6:3 .... *N. fornicarius* (Figures 17–35 in [16])
-. Surface of notogaster covered with distinct concavities; four pairs of lyrifissures *ia, im, ip, ips* present; formula of genital setae 5:4 .... *N. inauditus* (Figures 62–79 in [16])
34. Genital setae arranged in two rows; setae g_6 and g_7 remote from paraxial margin .... *Austrophthiracarus* .... *A. excellens* (Figures 1–21 in [16])
-. Only setae g_6 remote from margin, other genital setae forming one row near paraxial margin .... *Arphthicarus* 35
35. Prodorsum with long line in extension of sinus; notogaster with setae c_3 the smallest, spiniform, rough, other setae longer, ciliate, obtuse distally; three pairs of lyrifissures *ia, im, ip* present .... *A. simplex* (Figure 2A–I in [18])
-. No line in extension of sinus on prodorsum; notogaster with setae c_3 ad *cp* the smallest and different shape than other notogastral setae; four pairs of lyrifissures *ia, im, ip, ips* present .... *A. inelegans* (Figures 1–7 in [26]).

4. Discussion

The phenomenon of setal multiplication on the particular body area of mites is known under the name neotrichy: "Néotrichie est la terme générale [...] est la formation secondaire de poils, dans un territoire, par la multiplication de poils préexistant dans ce territoire" [27,28]. Neotrichy is well known among phthiracaroid mites, where extra setae appear usually on the notogaster and anoadanal plates. It also sometimes results in hyper-trichy, as exemplified by *P. afthonos* n. sp.

We report a new species of the genus *Protophthiracarus* (Phthiracaroidea) from the Peruvian Andes, which has a high, previously unseen number of notogastral setae. To our best knowledge, this is the second case of such rich neotrichy of notogastral setae ever known. Until now, the most hairy ptyctimous mite species known and probably the most neotrichous oribatid mite ever was *Atropacarus (A.) niedbalai* from New Zealand [29]. This species shows extreme neotrichy on the prodorsum, notogaster, genitoaggenital and anoadanal plates, whereas the neotrichy of the newly described *Protophthiracarus afthonos* is seen on the notogaster and anoadanal plates only. However, the number of setae on the notogaster being ca. 166 pairs is fairly higher than that of the New Zealand species (109–115 pairs). The neotrichy on the notogaster has a form of plethotrichy because numerous setae are displaced unevenly and asymmetrically arranged.

Neotrichy occurs independently in different phylogenetic linages of phthiracaroid mites [5] and, in the case reported herewith, concerns the taxon belonging to the genus *Protophthiracarus*, which is generally poor in species. It is more frequent in species from the Southern Hemisphere, especially Neotropical and Australasian regions [27–30].

It is also worth mentioning that the neotrichy itself cannot be considered an argument strong enough to create upper-level taxa [5,27,30].

The genus *Protophthiracarus* was proposed by Balogh [31], with a type species *Notophthiracarus chilensis* Balogh et Mahunka, 1967 [32]. It is well represented in the fauna of the southern hemisphere, except the Australasian region. The genus *Protophthiracarus* comprises 47 described species and 2/3 of them originate from the Neotropical region.

Approximately half of the 36 species (19) known in Peru have been described as new for science. They belong to very different genera: Mesoplophoridae: *Mesoplophora*—3 spp.; Euphthiracaroidea: Oribotriidae: *Oribotritia*—1 sp., *Mesotritia*—1 sp., *Acrotritia*—2 spp., *Microtritia*—1 sp., Phthiracaroidea: Phthiracaridae: *Phthiracarus*—2 spp.; Steganacaridae:

Hoplophthiracarus—1 sp., *Steganacarus*—3 spp., *Austrophthiracarus*—1 sp., *Arphthicarus*—1 sp., *Notophthiracarus*—3 spp. (see Systematic list). Thus, the species described from Peru represent ptyctimous mites of both main orders of Oribatida, namely Enarthronota and Mixonomata. Out of the abovementioned 36 species, 19 have been described by the senior author (including three with H. Schatz [18]).

All species reveal similar proportions in biogeographic distribution, those being endemites (11 species), neotropical (11 species) and widely distributed ones (semicosmopolitan—4 species, and pantropical—9 species), each group sharing ca. 1/3rd of the pool [6]. Endemic species are generally scarce, possibly except for *Mesoplophora (Mesoplophora) quasigaveae*. Two endemic species are numerously represented in few samples, but each from one region only: *A. excellens* at Machu Picchu, and *M. quasigavae* in the Puerto Inca Province.

The majority of more broadly distributed species occur in a larger number of various localities, e.g., pantropical *A. vestita* in 17, neotropical *M. curviseta* in 20 and *A. peruensis* in 10 localities. The number of individuals is not distribution-dependent, and even though the most numerous are neotropical species, pantropical and endemic species are also numerous, and two pantropical and endemic species are richest in numbers.

Three neotropical species, *A. clavata*, *N. fornicarius* and *N. inauditus*, reveal that Guyanan distribution occurs only in the northern part of the Neotropical region (Antilles, Venezuela, Ecuador, Peru, Bolivia). This may prove that quite specific climatic and environmental factors shape the distribution of ptyctimous mites in Peru.

The reported resulting individual counts were collected from a variety of substrates, ranging from soil and litter through to decayed organic plant matter and to lower vascular and epiphytic plants.

The most numerous ptycimous mites (total number of individuals in samples/share in total number of individuals of all spp.) are as follows: *Acrotritia vestita*—pantropical, 138/18.1%; *Mesoplophora (Mesoplophora) quasigaveae*—endemic, 137/18%; *Mesotritia curviseta* —neotropical, 99/13%; *Notophthiracarus inauditus*—neotropical, 54/7.1%; *Arphthicarus inelegans*—pantropical, 49/6.4%; *Notophthiracarus fornicarius*—neotropical, 41/5.4%; *Acrotritia dikra*—nearctic and neotropical, 38/5%; *Phthiracarus boresetosus*—semicosmopolitan, 37/4.9%; *Acrotritia peruensis*—neotropical, 25/3.3%; *Austrophthiracarus excellens*—endemic, 19/2.5%; *Atropacarus (Hoplophorella) vitrinus*—semicosmopolitan, 17/2.2%. The most numerous species belong to various genera of Mesoplophoridae (one species), Euphthiracaroidea (four species) and Phthiracaroidea (six species).

Most of the geographic localities from where the mite material is reported herewith are within the Andes range and its highland vicinities. This apparently reflects more the attitude of the collectors than the true geographic distribution of the species. Thus, our conclusion is that ptyctimous Oribatida needs more extensive sampling from mountain and highland (and to some extend lowland tropical) areas in order to evaluate its real abundance and density.

Author Contributions: Conceptualization, W.N.; visualization, W.N.; line drawings, R.L.; light microscopy imaging, W.L.M.; SEM imaging, Z.A.; methodology, W.N., Z.A., R.L., W.L.M.; writing original draft, W.N.; writing: review and editing, W.L.M. and R.L. All authors have read and agreed to the published version of the manuscript.

Funding: This research received no external funding.

Informed Consent Statement: Not applicable.

Data Availability Statement: Data sharing not applicable.

Acknowledgments: We wish to express our thanks to S. G. Ermilov (Tyumen State University, Russia), who let us study the material of mites from Peru and all other sample collectors. We are also indebted to J. Bloszyk (Department of General Zoology, A. Mickiewicz Univeristy) for his invaluable help in mapping fauna of ptyctimous mites in Peru.

Conflicts of Interest: The authors declare no conflict of interest.

References

1. Fittkau, E.J. The Fauna of South America. In *Biogeography and ecology in South America*; Fittkau, E., Illies, J.J., Klinge, H., Schwabe, G.H., Sioli, H., Eds.; Junk: The Hague, The Netherlands, 1969; Volume 2, pp. 624–650. [CrossRef]
2. Cypionka, H. PICOLAY Version 2022-07-15. Available online: http://www.picolay.icbm.de/index.html (accessed on 1 November 2022).
3. Travé, J.; Vachon, M. François Grandjean 1882–1975 (Notice biographique et bibliographique). *Acarologia* **1975**, *17*, 1–19. [PubMed]
4. Norton, R.A.; Behan-Pelletier, V.M. Suborder Oribatida, Chapter 15. In *A Manual of Acarology*; Krantz, G.W., Walter, D.E., Eds.; Texas Tech University Press: Lubbock, TX, USA, 2009; pp. 430–564.
5. Niedbała, W. *Phthiracaroidea (Acari, Oribatida). Systematic Studies*; Elsevier: Amsterdam, The Netherlands; Oxford, UK; New York, NY, USA; Tokyo, Japan; Państwowe Wydawnictwo Naukowe: Warsaw, Poland, 1992; pp. 1–612.
6. Niedbała, W.; Liu, D. Systematic, synonymic and biogeographical list of ptyctimous mites (Acari, Oribatida) in the world (1799–2022). *Zootaxa* **2023**, *5265*, 1–442. [CrossRef]
7. Märkel, K. Die Euphthiracaridae Jacot, 1930, und ihre Gattungen (Acari, Oribatei). *Zool. Verh.* **1964**, *67*, 3–178.
8. Hammer, M. Investigations on the oribatid fauna of the Andes Mountains. II. Peru. *Biol. Skr.* **1961**, *13*, 1–159 (+43 plates).
9. Niedbała, W. Nouveaux Phthiracaridae tropicaux (Acari, Oribatida). *Pol. J. Entomol.* **1982**, *52*, 189–229.
10. Niedbała, W. Ptyctimous mites (Acari, Oribatida) of the Neotropical Region. *Ann. Zool.* **2004**, *54*, 1–288.
11. Niedbała, W.; Ermilov, S.G. New data on ptyctimous mites (Acari, Oribatida) from Mexico and Peru with descriptions of two new species. *Syst. Appl. Acarol.* **2017**, *22*, 759–765. [CrossRef]
12. Ermilov, S.G.; Niedbała, W.; Friedrich, S. Additions to the Peruvian oribatid mite fauna including new records and descriptions of three new species. *Spixiana* **2016**, *39*, 61–74.
13. Niedbała, W.; Ermilov, S.G. New species and records of ptyctimous mites (Acarina, Oribatida) from the Neotropical region. *Syst. Appl. Acarol.* **2022**, *27*, 1566–1573. [CrossRef]
14. Niedbała, W. *Mesoplophora subtilis* sp. n. du Pérou (Acari, Oribatida, Mesoplophoridae). *Pol. J. Entomol.* **1981**, *51*, 511–517.
15. Niedbała, W. *Hoplophorella neglecta* sp. nov. (Acari, Oribaida, Phthiracaridae) du Pérou. *Bull. Soc. Am. Sci. Lett. Poznan* **1984**, *24*, 133–137.
16. Niedbała, W. Phthiracaridae (Acari, Oribatida) nouveaux de Pérou. *Ann. Zool.* **1982**, *36*, 449–464.
17. Mahunka, S. Neue und interessante Milben aus dem Genfer Museum XLIV Oribatida Americana 5: Costa Rica (Acari). *Arch. Sci.* **1982**, *35*, 179–193.
18. Niedbała, W.; Schatz, H. Euptyctimous mites from the Galapagos Islands, Cocos Island, and Central America (Acari: Oribatida). *Genus* **1996**, *7*, 239–317.
19. Niedbała, W. Ptyctimous mites of the Pacific Islands. Recent knowledge, origin, descriptions, redescriptions, diagnoses and zoogeography (Acari, Oribatida). *Genus* **1998**, *9*, 431–558.
20. Mahunka, S. Neue und interessante Milben aus dem Genfer Museum XLV. Oribatida Americana 6: Mexico II (Acari). *Rev. Suisse Zool.* **1983**, *90*, 269–298.
21. Jacot, A.P. More box-mites of the northeastern United States. *J. N. Y. Entomol. Soc.* **1938**, *46*, 109–145.
22. Niedbała, W. Ptyctimous mites (Acari, Oribatida) of the Nearctic Region. *Monogr. Upp. Siles. Mus.* **2002**, *4*, 1–261.
23. Hammen, L. van der. The Oribatid family Phthiracaridae. III. Redescription of *Phthiracarus nitens* (Nicolet). *Acarologia* **1964**, *6*, 400–411.
24. Grandjean, F. *Phthiracarus anonymum* n. sp. *Rev. Française Entomol.* **1934**, *1*, 51–58.
25. Liu, D.; Oconnor, B.M. Ptyctimous mites (Acari, Oribatida) from Colombia, with description of a new species and some remarks on the validity of *Atropacarus* (*Hoplophorella*) *andrei* (Balogh, 1958). *Syst. Appl. Acarol.* **2015**, *20*, 61–70.
26. Niedbała, W. *Hoplophthiracarus inelegans* sp. n. (Acari, Oribatida, Phthiracaridae) de Costa Rica. *Bull. Soc. Am. Sci. Lett. Poznan* **1986**, *25*, 115–117.
27. Grandjean, F. Complément a mon travail de 1953 sur la classification des Oribates. *Acarologia* **1965**, *7*, 713–734.
28. Travé, J. La néotrichie chez les Oribates (Acariens). *Acarologia* **1978**, *20*, 590–602.
29. Liu, D.; Zhang, Z.Q. *Atropacarus* (*Atropacarus*) *niedbalai* sp. nov., an extreme case of neotrichy in oribatid mites (Acari: Oribatida: Phthiracaridae). *Int. J. Acarol.* **2013**, *39*, 507–512. [CrossRef]
30. Niedbała, W. Unusual case of neotrichy in a new species of oribatid mite (Acari, Oribatida, Steganacaridae). *Aust. J. Entomol.* **2009**, *48*, 317–320. [CrossRef]

31. Balogh, J. *The Oribatid Genera of the World*; Akadémiai Kiadó: Budapest, Hungary, 1972; 188p.
32. Balogh, J.; Mahunka, S. The Scientific Results of the Hungarian Soil Expedition to South America. 2. *Notophthiracarus chilensis n. gen., n. sp. (Acari). Opusc. Zool. Inst. Zoosyst. Oecol. Univ. Bp.* **1967**, *7*, 43–45.

animals

MDPI

Article

A Review of the Feather Mite Genus *Lopharalichus* Gaud & Atyeo, 1996 (Acariformes: Pterolichidae), with Descriptions of Three New Species from Brazilian Parrots (Psittaciformes: Psittacidae) †

Fabio Akashi Hernandes

Departamento de Ecologia e Zoologia (ECZ), Centro de Ciências Biológicas, Federal University of Santa Catarina, Florianópolis 88040-970, Brazil; abakashi@gmail.com

† ZooBank link: urn:lsid:zoobank.org:pub:EAD0A1AF-70CB-4060-B9D3-C58C91F0973C.

Simple Summary: Understanding the current biodiversity of our planet is an ongoing challenge, as natural habitats are being destroyed at a faster rate than species are described. This is especially true for South America, which harbors over one-third of the parrot species in the world. A diverse yet poorly studied group of mites associated with birds are feather mites, which currently include about 2500 known species, and estimates range from 10,000 to 20,000 species. Herein, three new species of feather mites of the genus *Lopharalichus* are described from parrots in Brazil.

Abstract: Feather mites of the genus *Lopharalichus* Gaud & Atyeo, 1996 (Pterolichidae: Pterolichinae), formerly containing three described species, are associated with New World parrots (Psittaciformes: Psittacidae) of the subfamily Arinae. Three new species of this genus are described: *Lopharalichus tuim* **sp. nov.** from *Forpus xanthopterygius* (Spix, 1824), *L. spinosus* **sp. nov.** from *Ara ararauna* (Linnaeus, 1758), and *L. chiriri* **sp. nov.** from *Brotogeris chiriri* (Vieillot, 1818). Type specimens of the previously described *Lopharalichus* species were examined, and a key to the known species is provided.

Keywords: avian mites; diversity; taxonomy; systematics; Psoroptidia; Pterolichoidea

Citation: Hernandes, F.A. A Review of the Feather Mite Genus *Lopharalichus* Gaud & Atyeo, 1996 (Acariformes: Pterolichidae), with Descriptions of Three New Species from Brazilian Parrots (Psittaciformes: Psittacidae). *Animals* **2023**, *13*, 2360. https://doi.org/10.3390/ani13142360

Academic Editors: Maciej Skoracki and Monika Fajfer

Received: 19 June 2023
Revised: 8 July 2023
Accepted: 14 July 2023
Published: 19 July 2023

1. Introduction

Three groups of feather mite genera from the subfamily Pterolichinae (Acariformes: Pterolichidae) are found on parrots (Psittaciformes): *Protolichus*, *Psittophagus*, and *Rhytidelasma* groups [1–3]. The *Protolichus* generic group, incorporating nearly 100 described species in 24 genera, is the most diverse of these groups, with 11 genera found on parrots of the New World [4]. The genus *Lopharalichus* Gaud & Atyeo, 1996 belongs to this group and has included, to date, three species [5,6]: *Lopharalichus denticulatus* (Mégnin & Trouessart, 1884) from *Pyrrhura cruentata* (Wied-Neuwied, 1820) from Brazil, *L. cribiformis* (Mégnin & Trouessart, 1884) from *Forpus passerinus* (Linnaeus, 1758) from Guyana, and *L. beckeri* Mironov, Dabert & Ehrnsberger, 2005 from *Conuropsis carolinensis* (Linnaeus, 1758), an extinct parrot of North America. Gaud & Atyeo [5] presented illustrations of two undescribed species from two other New World parrots, *Thectocercus acuticaudatus* (Vieillot, 1818) (formerly *Aratinga acuticaudata*) and *Forpus modestus sclateri* (Gray, 1859) (formerly *Forpus sclateri*). An undetermined *Lopharalichus* species was reported from *Brotogeris chiriri* (Vieillot, 1818) [7] (not confirmed whether it corresponds to the new species described herein form the same parrot species). Pedroso and Hernandes [8] reported three undescribed species of *Lopharalichus* from Brazil, and these mites are described below.

The most distinctive feature of the genus *Lopharalichus* is the presence of prominent spiny crests on the femora and genua of legs I and II of both males and females, after which the genus was named (Gr. *lophos* = crest, mane). Other noticeable features are as follows: in both sexes, the lateral regions of hysterosoma have small cuticular spines, setae *c*2 are bifid, scapular setae *si* are subequal to or longer than setae *se*, setae *se* are very short (at maximum $\frac{1}{2}$ the distance *si:se*), the prodorsal shield is entire (unlike species of the genera *Aralichus* Gaud 1966, *Chelomatolichus* Gaud & Atyeo 1996, and *Pararalichus* Atyeo 1989, in which the shield is divided by a transverse band of weakly sclerotized area at level of scapular setae *si*, *se*), and setae *h*1 are absent. Additionally, in males, setae *h*2 and *h*3 are flatly expanded with a filamentous tip, setae *e*2 are bifid with a short basal spine (except in *L. denticulatus*), setae *f*2 are expanded (leaf-like), setae *ps*1 are broad, and in females, setae *e*2 and *ps*1 are short, expanded with minute spines.

In this paper, three new species of *Lopharalichus* are described from parrots of Brazil, and a key to the known species of this genus is presented.

2. Materials and Methods

The new mites studied herein were collected from either wild bird specimens found dead in the field or from taxidermied bird specimens (see below). In the former case, the birds were collected and frozen for a later study; in laboratory, they were washed in a plastic tray with water and detergent to remove the ectoparasites [9], and the water was filtered through a paper filter. The mites were collected from the filters with a fine brush under a dissecting microscope. A few specimens from *Ara ararauna* (Linnaeus) were also retrieved from dry museum skins deposited at the Museu de História Natural Capão da Imbuia (MHNCI), Curitiba, following the ruffling technique described in Gaud & Atyeo [5]. The mites obtained with both methods were cleared and distended in 30% lactic acid at 50 °C for 24 h, mounted on microscopic slides using Hoyer's medium [10], and heated and dried at 50 °C for 5 days. Finally, the edges of the coverslips were sealed with transparent varnish and the slides were labeled. The specimens were studied under an Olympus CX31 microscope, and illustrations were prepared from pictures of the mites taken with a digital camera (Omax A35140U 14mpx, Chengdu, China) attached to the ocular lenses and produced on Adobe Illustrator CS5 using a Wacom Bamboo Create tablet. The chaetotaxies of idiosoma and legs follow Griffiths et al. [11] and Atyeo & Gaud [12], respectively, with further corrections for coxal setae [13]. The nomenclature of birds is according to Gill et al. [14].

The species descriptions are given according to the formats proposed by Mironov et al. [6] and Hernandes [4]. Type specimens of the new species are deposited at the Acari Collection of the Department of Ecology and Zoology of the Universidade Federal de Santa Catarina, Florianópolis (ECZ–UFSC). Additional material examined consisted of types and other specimens of *Lopharalichus cribriformis* and *L. denticulatus* determined by W.T. Atyeo and are deposited at the Trouessart collection of the Muséum National d'Histoire Naturelle (MNHN), Paris, France. Photos of non-type specimens of *L. beckeri* deposited at the Zoology Institute, Russian Academy of Sciences (ZISP), St. Petersburg, were also examined.

3. Results

Systematics

Pterolichidae Trouessart & Mégnin, 1884

Lopharalichus Gaud & Atyeo, 1996

(*Lopholichus*, Gaud & Atyeo 1996:121, *sic*)

Type species: *Pterolichus* (*Pterolichus*) *denticulatus* Mégnin & Trouessart, 1884, by original designation.

Lopharalichus denticulatus **(Mégnin & Trouessart, 1884)**

(Figures 1A, 2A, 3A and 4)

Pterolichus (*Pterolichus*) *denticulatus* Mégnin & Trouessart, 1884 [15]: 211.

Pterolichus (*Eupterolichus*) *denticulatus*; Canestrini & Kramer, 1899 [16]: 37.

Pterolichus denticulatus; Radford, 1953 [17]: 201; Gaud & Atyeo, 1996 [5]: 128.

Type material examined: Lectotype male ex *Pyrrhura cruentata* (Wied-Neuwied, 1820) (Psittaciformes: Psittacidae) from BRAZIL, no further data, MNHN#969.236.3 (slide 35-I-6) (the remounted slide also contains a paralectotype male of *Neorhytidelasma tritiventris* (Trouessart, 1884)).

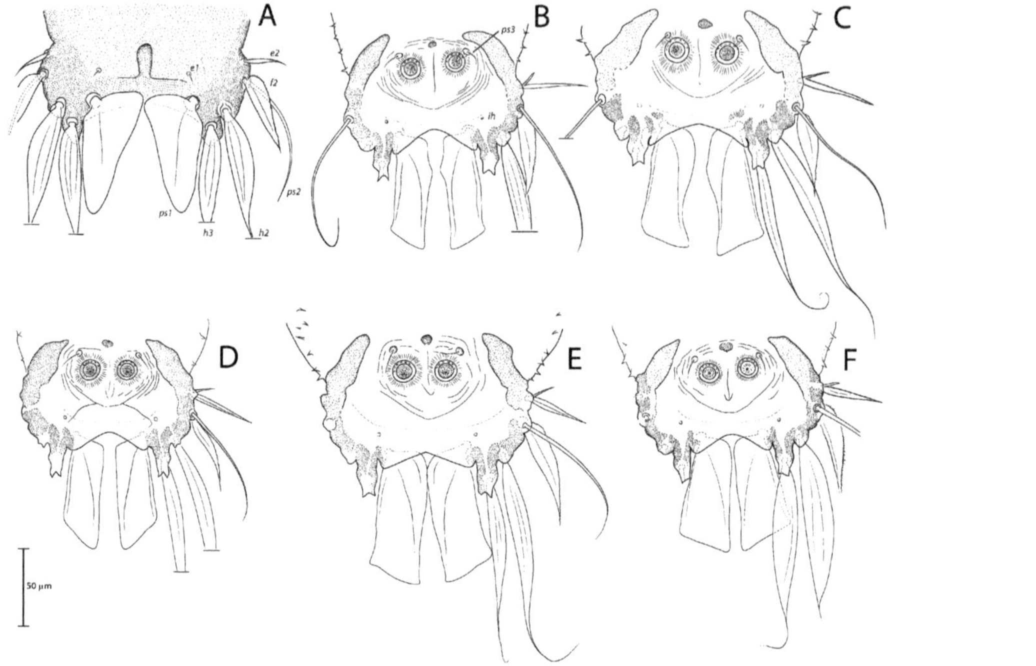

Figure 1. Opisthosoma of *Lopharalichus* spp. males (A = dorsal; B–F = ventral): *L. denticulatus* (**A**); *L. cribriformis* (**B**); *L. beckeri* (**C**,); *L. tuim* sp. nov. (**D**); *L. spinosus* sp. nov. (**E**); *L. chiriri* sp. nov. (**F**).

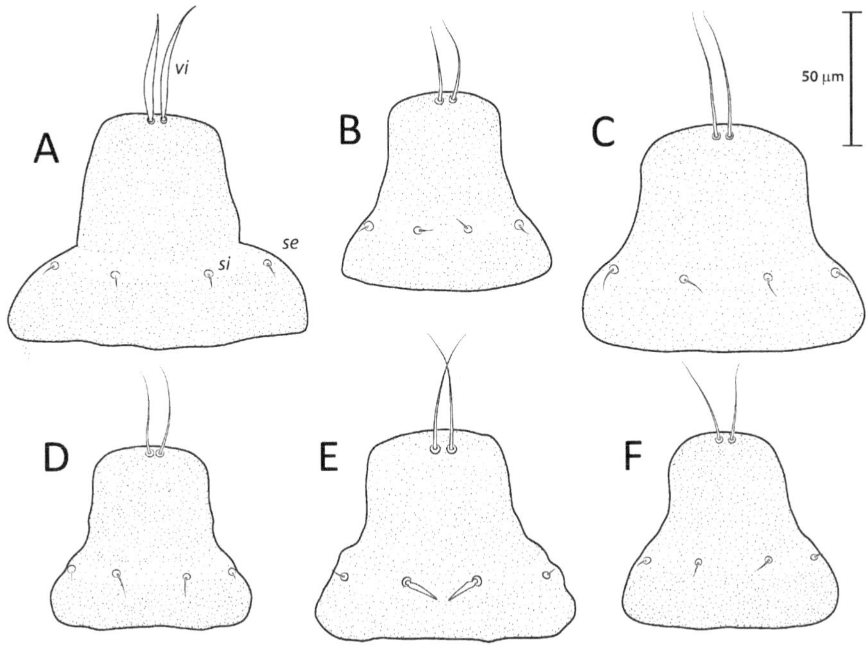

Figure 2. Prodorsal shield of *Lopharalichus* spp. males: *L. denticulatus* (**A**); *L. cribriformis* (**B**); *L. beckeri* (**C**); *L. tuim* sp. nov. (**D**); *L. spinosus* sp. nov. (**E**); *L. chiriri* sp. nov. (**F**).

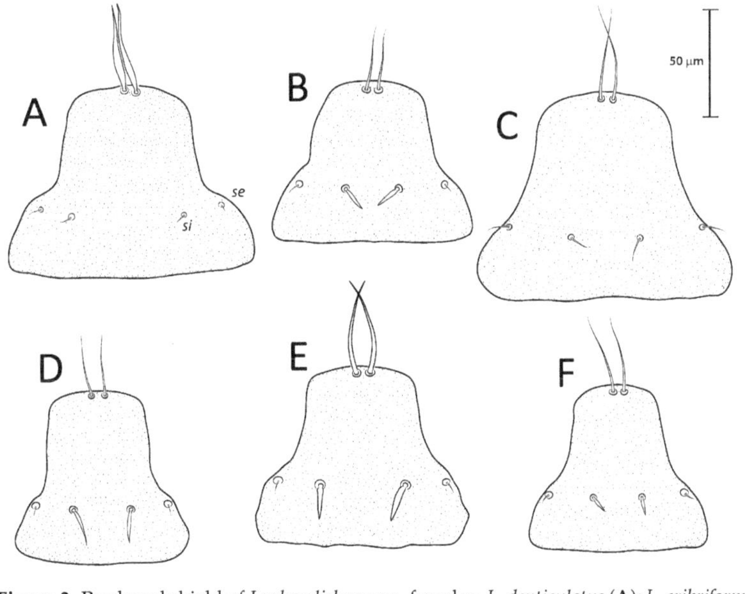

Figure 3. Prodorsal shield of *Lopharalichus* spp. females: *L. denticulatus* (**A**); *L. cribriformis* (**B**); *L. beckeri* (**C**); *L. tuim* sp. nov. (**D**); *L. spinosus* sp. nov. (**E**); *L. chiriri* sp. nov. (**F**).

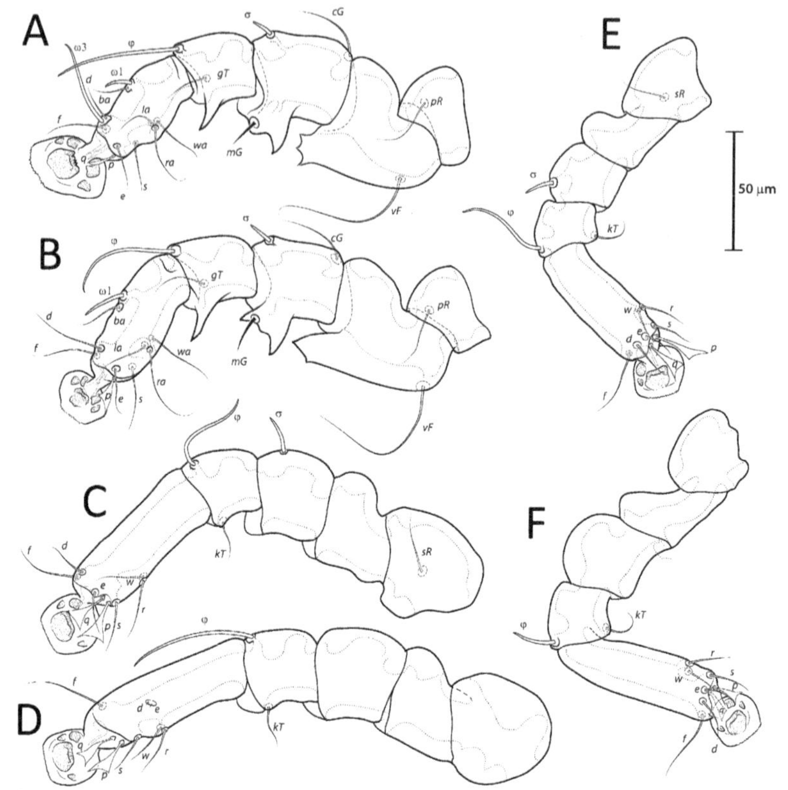

Figure 4. *Lopharalichus denticulatus*, legs I–IV (**A–D**) of male; legs III–IV (**E**,**F**) of female.

Additional material examined: One male ex *P. cruentata*, BRAZIL, Bahia state, Boa Nova, 5 June 1928, E. Kaempfer (AMNH241747, UGA10,450), MNHN#1060.31.2 (slide 65-D-6) (W.T. Atyeo det. 1993); one female ex *P. cruentata*, BRAZIL, Espírito Santo state, Lagoa Juparanã, 11 November 1929, E. Kaempfer (AMNH317283, UGA10,452), MNHN#1060.31.1 (slide 65-D-5) (W.T. Atyeo det. 1993).

Remarks: *Lopharalichus denticulatus* stands out from other species in having, in males, setae *ps*1 roughly triangular, setae *e*2 simple and not bifurcate basally; in females, the prodorsal setal pair *si* is well spaced by about three-times the distance *si:se* (Figure 3A); in both sexes, vertical setae *vi* are slightly expanded (Figures 2A and 3A), genua I, II have prominent, thick antiaxial crests, and the hysteronotal shield is usually devoid of lacunae; in one non-type male examined, there are a few small, sparse circular lacunae in the center of the shield, about 1–3 μm in diameter. The only examined female is broken, with legs, epigynum, and other structures displaced from their original position.

***Lopharalichus cribriformis* (Mégnin & Trouessart, 1884)**

(Figures 1B, 2B, 3B and 5A–F)

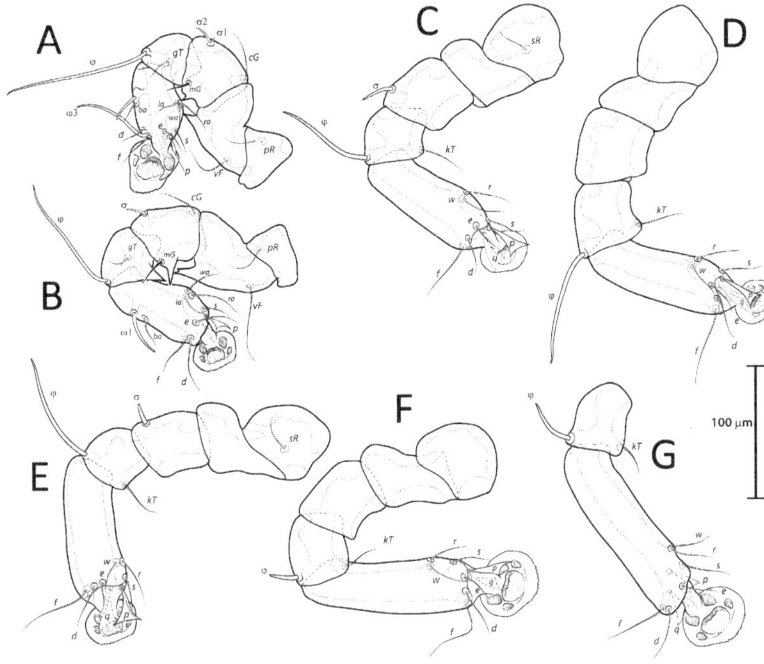

Figure 5. *Lopharalichus cribriformis*, legs I–IV (**A–D**) of male; legs III–IV (**E**,**F**) of female. Lopharalichus beckeri, tibia, and tarsus IV of female (**G**).

Pterolichus (*Pterolichus*) *denticulatus* var. *cribriformis* Mégnin & Trouessart, 1884 [15]: 213.
Pterolichus (*Eupterolichus*) *cribriformis*; Canestrini & Kramer, 1899 [16]: 38.
Pterolichus denticulatus; Radford, 1953 [17]: 201.

Type material examined: Syntypes 15 males and 19 females (in the same original slide, not remounted) ex *Forpus passerinus* (Linnaeus, 1758) (=*Psittaculus passerinus*), from GUYANA, MNHN#969.237.1 (slide 35-I-8).

Additional material examined: one male and one female ex *Forpus passerinus cyanochlorus* (Schlegel, 1864), BRAZIL, Amazonas state, Frechal, Rio Surumu, 6 September 1929, T.D. Carter col. (AMNH236355, UGA12,742) MNHN#1060.30 (slide 65-D-4) (W.T. Atyeo det. 1993).

Remarks: *Lopharalichus cribriformis* is very similar to *L. beckeri* Mironov et al. (2005), differing from that species in having, in males, the terminal cleft angular and the paragenital apodemes indistinctly developed, and in females, setae *si* distinctly longer and more robust than *se* (at least twice longer and twice thicker) (Figure 3B), and the solenidion on tibia IV as long as half the width of this segment (Figure 5F). In males of *L. beckeri*, the lobar cleft is nearly semicircular (Figure 1C), and the paragenital apodemes are distinctly formed; and in females, setae *se* and *si* are both piliform and similar in structure (Figure 3C), and solenidion φ on tibia IV is about the same length as the width of tibia (Figure 5G).

Mironov et al. [6] stated that, in males of *L. cribriformis*, setae *e*2 are twice as long as *f*2, and in females, setae *f*2 are "large and foliform, almost circular, and with a vein". However, in the examined specimens of this species, setae *e*2 and *f*2 have about the same length in males, and setae *f*2 of females are roughly triangular, like in *L. beckeri*. The type series of *L. cribriformis* consists of a single slide containing 34 poorly clarified syntypes, still with the original label by E.L. Trouessart. The illustrations presented here are based on non-type material collected from the type host species and determined by W.T. Atyeo.

Lopharalichus beckeri **Mironov, Dabert & Ehrnsberger, 2005**

Figures 1C, 2C, 3C and 5G)

Lopharalichus beckeri Mironov, Dabert & Ehrnsberger, 2005 [5]: 2259

Material examined: Photos of 1 male and 1 female (ZISP 6760, 6767) ex *Conuropsis carolinensis* (MCZ 209911, UNAM 110), USA, Florida, Tampa, no date, coll. W. Brewster.

Remarks: *Lopharalichus beckeri* was described from *Conuropsis carolinensis* (Linnaeus, 1758), an extinct parrot from North America. This species is very similar to *L. cribriformis* (see differential characters in the remarks of the previous species).

Lopharalichus tuim **sp. nov.**

(Figures 1D, 2D, 3D and 6, Figures 7 and 8)

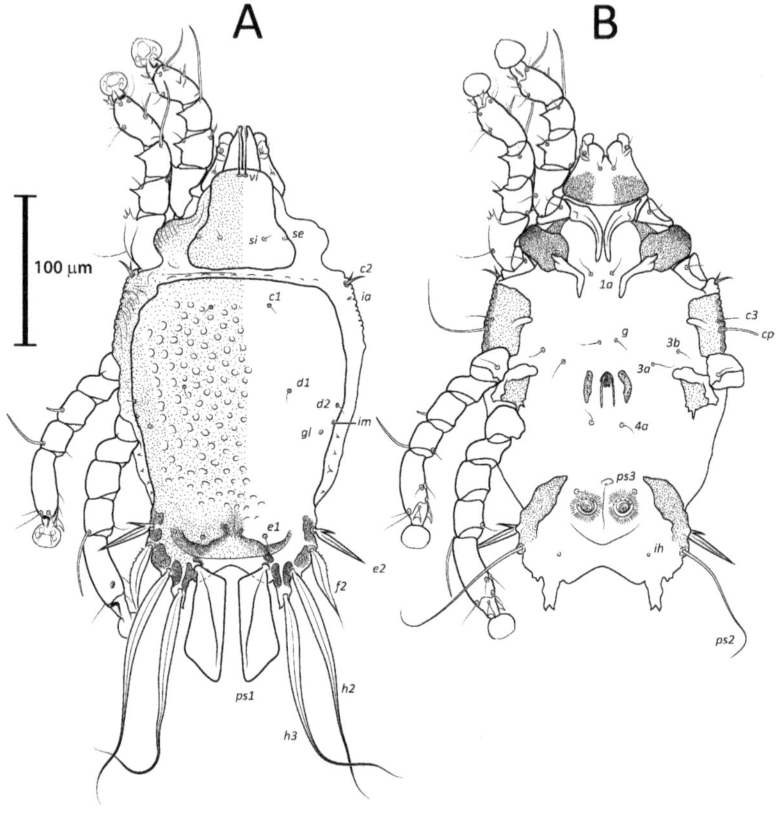

Figure 6. *Lopharalichus tuim* sp. nov. male: dorsal (**A**) and ventral (**B**) views.

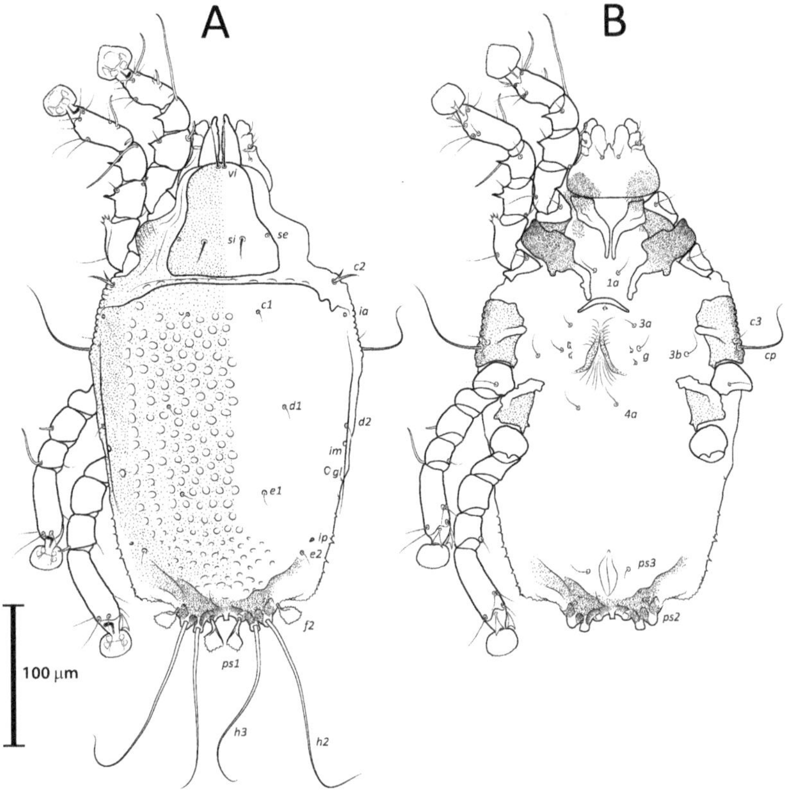

Figure 7. *Lopharalichus tuim* sp. nov. female: dorsal (**A**) and ventral (**B**) views.

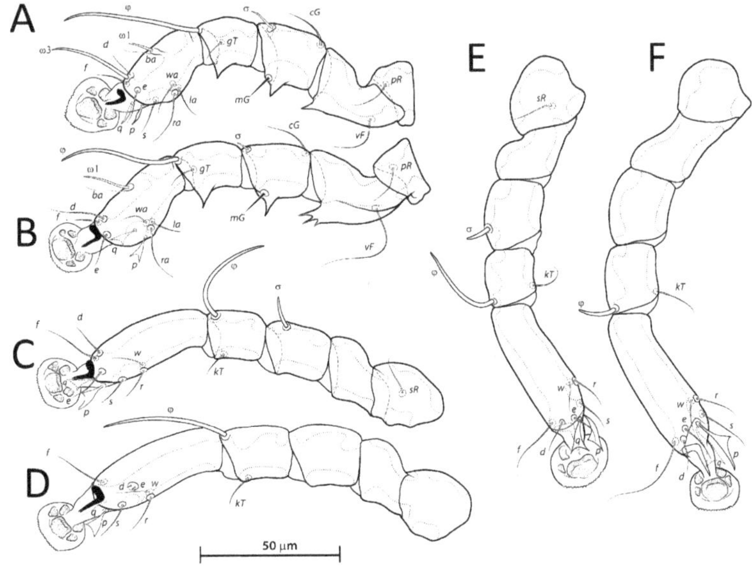

Figure 8. *Lopharalichus tuim* sp. nov., legs I–IV (**A–D**) of male; legs III–IV (**E**,**F**) of female.

Zoobank registration: urn:lsid:zoobank.org:act:7BAEC958-A174-42FC-8C28-0198480DC854

Type material. Holotype male, paratypes 10 males, 31 females, and 1 nymph ex *Forpus xanthopterygius* (Spix, 1824) (Psittaciformes: Psittacidae), BRAZIL, São Paulo State, Pedreira, 22°44′ S, 46°54′ W, June 2012, D.V. Boas-Filho col. (#1074).

Male (holotype, range for five paratypes in parentheses). Idiosoma length (from the level of setae *vi* to the base of setae *h*3) 284 (284–297), greatest width at level of humeral shields 162 (163–176). Prodorsal shield shaped as an Erlenmeyer flask (elongated trapezoid), with rounded edges, posterior margin slightly sinuous, surface without ornamentation, 64 (64–72) in length from the level of setae *vi* to the posterior margin, 69 (69–78) in width at the widest part. Scapular setae *si* thin spiculiform, 9 (8–11) long, setae *se* piliform, reduced, distance between bases of scapular setae *si:si* 29 (25–27), *se:se* 57 (56–61). Hysterosoma 212 (211–226) in length from sejugal area to the bases of setae *h*3. Hysteronotal shield: anterior margin straight, length from anterior margin to bases of setae *h*3 207 (212–227), greatest width at the level of setae *d*2 138 (144–158), surface with numerous circular lacunae posterior to level of setae *c*1 (Figure 6A), supranal concavity poorly distinct, anterior to level of setae *e*1. A bow-shaped transverse fold between levels of setae *e*1 and *ps*1. Membranous margin of terminal cleft (=contour of free margin of interlobar membrane) blunt-angular, 28 (30–34) long, opisthosomal lobes with prominent tubercles at bases of setae *h*3, narrow interlobar membrane between bases of setae *ps*1. Setae *c*2 bifid, 12 (12–15) long; setae *e*2 lanceolate with short basal bifurcation, 45 (42–49) long; setae *f*2 lanceolate with outer edge minutely serrate, 54 (56–65); setae *ps*1 roughly parallelogram-shaped, 78 (77–84) long. Distances between hysteronotal setae: *c2:d2* 72 (84–90), *d2:e2* 84 (76–82), *e2:h3* 40 (45–50), *d1:d2* 10 (8–13), *e1:e2* 4 (5–13), *ps1:ps1* 45 (43–51), *h3:h3* 69 (66–76), *h2:h2* 82 (82–92), and *ps2:ps2* 106 (106–117).

Bases of epimerites I and II with inflations and dark sclerotized (Figure 6B). Humeral shields developed ventrally and bearing setae *c*3, *cp*. Setae *c*3 thin piliform, 14 (12–16) long; coxal fields I–II without sclerotized areas. Genital apparatus situated between levels of trochanters III and IV, 24 (24–28) long, 11 (10–13) wide; paragenital apodemes as a pair of longitudinal sclerites lateral to the genital apparatus and bearing the genital acetabula. Distances between setae: *g:4a* 55 (51–58), *g:g* 8 (6–9). Cupules *ih* ventrally at the level of setae *ps*2. Adanal suckers 13 (13–15) in diameter, distance between centers of suckers 24 (22–26), corolla with 5–7 teeth on anterior half, posterior half without teeth (Figures 1D and 6B).

Femora I with 1–3 apicoventral spines or crests, femur II with 2–6 apicoventral spines. Acute apicoventral spines on genua, tibiae I, II. Length of tarsi excluding ambulacra: tarsus I 37 (35–38), tarsus II 44 (46–50), tarsus III 49 (49–54), tarsus IV 55 (55–57). Seta *kT* present on tibia IV. Setae *d*, *e* minute spiculiform, inserted close together (Figure 8D). Setae *p*, *q* on tarsi I thinner and apically less expanded than on tarsi of other legs. Solenidion $\sigma 2$ of genu I apparently absent. Length of solenidia: $\sigma 2I$ 10 (9–12), σIII 9 (7–9), φI 50 (50–55), φII 45 (43–47), φIII 33 (30–33), φIV 40 (35–43), $\omega 1I$ 10 (10–12), $\omega 3I$ 29 (29–32), and $\omega 1II$ 19 (17–18).

Female (range for 6 paratypes). Idiosoma length 309–334, greatest width 173–185. Prodorsal shield-shaped as in the male, 73–79 long, 76–79 wide (Figure 7A); scapular setae *si* spiculiform, 14–17 long, setae *se* piliform, reduced; distances between scapular setae *si:si* 21–28, *se:se* 59–63. Hysteronotal shield 239–247 in length, 162–171 in width at the widest part; surface with numerous circular lacunae posterior to level of setae *c*1. Setae *c*2 bifid, setae *f*2, *ps*1 flat, spiky leaf-like, setae *c*1, *d*1, *d*2, *e*1, *e*2 piliform. Terminal region of opisthosoma shaped as a semicircular concavity between a pair of tubercles bearing setae *h*2, *h*3, and with a small external copulatory tube in the center about 5–7 long located between setae *ps*1. Posterolateral margins of opisthosoma with small spines. Length of setae: *c*2 12–15, *c*3 16–31, *e*2 11–16, and *f*2 16–22. Distances between dorsal setae: *c2:d2* 95–101, *d2:e2* 89–93, *d1:d2* 11–20, *e1:e2* 40–46, *ps1:ps1* 17–21, *h3:h3* 37–40, *h2:h2* 52–56.

Epimerites I free. Bases of epimerites I, II inflated, dark-sclerotized (Figure 7B). Epigynum as a low arch, 9–13 in length, 24–35 in width. Distance between ventral setae *1a:3a* 36–42, *3a:g* 20–30. Legs I, II as in the male, except for a shorter apicoventral spine on genu and tibia I. Length of tarsi excluding ambulacra: tarsus I 35–42, tarsus II 44–50, tarsus III 53–56, tarsus IV 59–65. Length of solenidia: $\sigma 2I$ 10–13, σIII 9–10, φI 61–62, φII 50–55, φIII 35–38, φIV 12–16, $\omega 1I$ 11–13, $\omega 3I$ 30–34, and $\omega 1II$ 18–25.

Differential diagnosis: The new species, *L. tuim* sp. nov., is very close to *L. cribriformis* (Mégnin & Trouessart, 1884) described from *Forpus passerinus* in having a blunt-angular terminal cleft in males. In males of *L. denticulatus* and *L. beckeri,* the lobar cleft is concave and semi-circular. The new species most clearly differs from *L. cribriformis* in the relative length and arrangement of prodorsal setae *si*: in females of *L. tuim sp.* nov., *si* reaches the base of *se* of the same side (Figure 3D), and in males, *si* reaches at least halfway to the base of corresponding setae *se*, *si* being about twice longer than *se* (Figure 2D). Also, in males of the new species, setae *si* are inserted slightly closer to the corresponding *se* than to the other member of the pair *si* (distance *si:si* is about $1\frac{1}{2}$ the distance *si:se*). In *L. cribriformis* females, setae *si* only reach about halfway to the bases of corresponding *se* (Figure 3B), and in males of that species, these setae reach one-third of that distance (*si* is about the same length as *se*) (Figure 2B), and in both sexes, the scapular setae *si* and *se* are uniformly spaced (distance *si:se* = *si:si*).

Etymology: The name of the new species is based on the Brazilian common name of the host (tuim) and is a noun in apposition.

***Lopharalichus spinosus* sp. nov.**

(Figures 1E, 2E, 3E and 9, Figures 10 and 11)

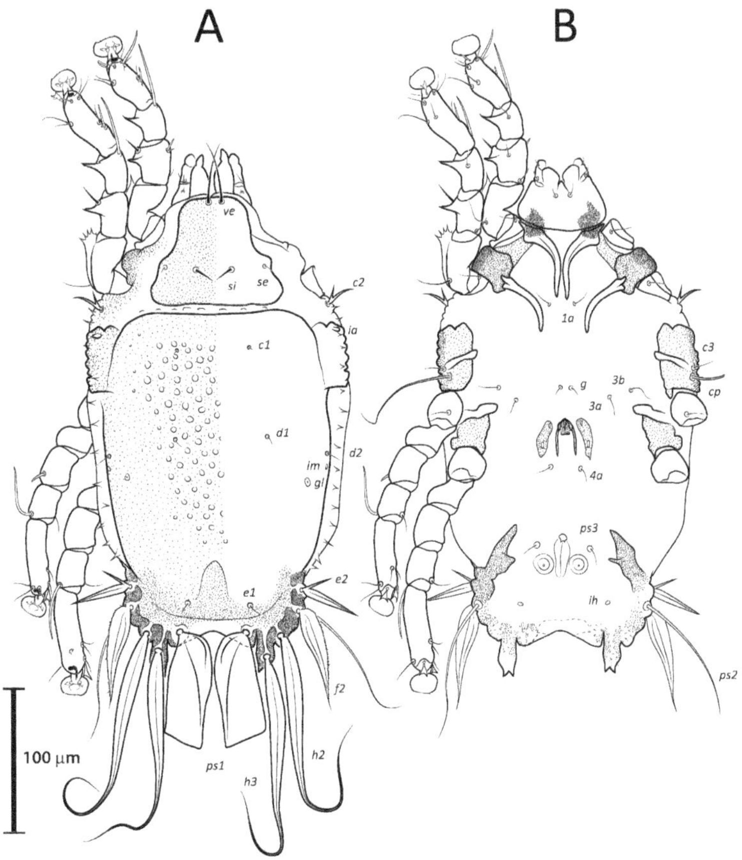

Figure 9. *Lopharalichus spinosus* sp. nov. male: dorsal (**A**) and ventral (**B**) views.

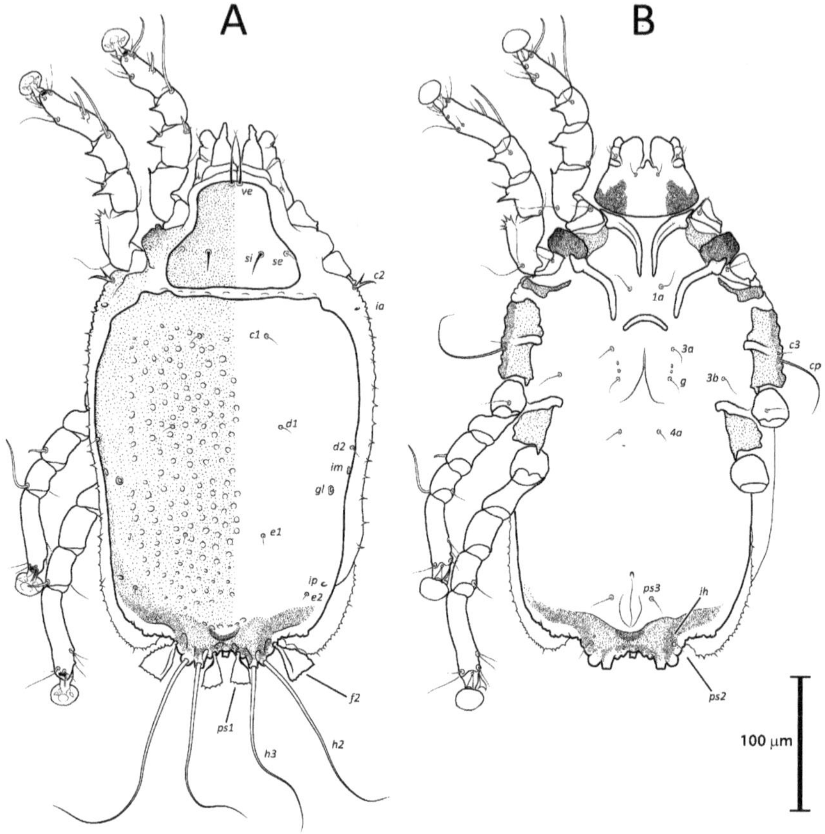

Figure 10. *Lopharalichus spinosus* sp. nov. female: dorsal (**A**) and ventral (**B**) views.

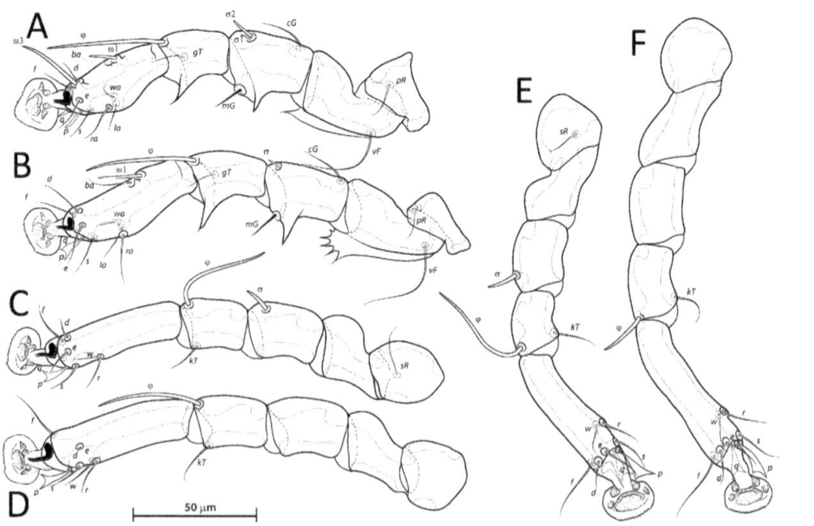

Figure 11. *Lopharalichus spinosus* sp. nov., legs I–IV (**A–D**) of male; legs III–IV (**E**,**F**) of female.

Zoobank registration: urn:lsid:zoobank.org:act:12DC2DDA-DE5D-4356-AF19-53F77-CB37A96

Type material: holotype male, seven male and six female paratypes ex *Ara ararauna* (Linnaeus, 1758) (Psittaciformes: Psittacidae), BRAZIL, São Paulo State, Itatiba, 23°00′ S, 46°50′ W, 24 March 2007, U. Kawazoe col. (#152). Paratypes from the same host species: five males and four females, Pernambuco State, 29 September 1953 (MHNCI#1557), mites collected from the bird skin by FAH in November 2016.

Male (holotype, range for two paratypes in parentheses). Idiosoma length from the level of setae *vi* to the base of setae *h*3 345 (338–350), greatest width at level of humeral shields 190 (190–193). Prodorsal shield shaped roughly as an isosceles trapezoid, with sinuous lateral margins and rounded edges, 78 (76–78) in length from the level of setae *vi* to the posterior margin, 96 (93–96) in width at the posterior margin. Scapular setae *si* as a short spike, about as long as the distance between their bases, 13 (11–13) long, distance between scapular setae *si:si* 25 (24–25), *se:se* 75 (73–77). Hysterosomal region 263 (256–265) in length from sejugal area to the bases of setae *h3*. Hysteronotal shield: anterior margin straight, length from anterior margin to bases of setae *h*3 260 (254–256), greatest width around the level of setae *d*2 165 (167–176), surface with numerous circular lacunae from the level of setae *c*1 to genua IV (Figure 9A). A bow-shaped transverse fold between levels of setae *e*1 and *ps*1. Membranous margin of terminal cleft blunt-angular, 30 (30–30) long, opisthosomal lobes with prominent tubercles at bases of setae *h*3, narrow interlobar membrane between bases of setae *ps*1. Setae *c*2 bifid, 21 (17–21) long; setae *e*2 lanceolate with short basal bifurcation, greatest length 43 (40–43); setae *f*2 lanceolate with external margin minutely serrate, 66 (64–66); setae *ps*1 roughly parallelogram-shaped with sharp posterior edges, 89 (88–90) long. Distances between hysteronotal setae: *c2:d*2 111 (106–111), *d2:e*2 101 (95–98), *e2:h*3 52 (50–52), *d*1*:d*2 11 (13–14), *e*1*:e*2 11 (9–11), *ps*1*:ps*1 49 (45–52), *h3:h*3 78 (75–78), *h2:h*2 96 (92–96), and *ps2:ps*2 126 (116–126).

Bases of epimerites I, II inflated, dark-sclerotized (Figure 9B). Humeral shields bearing setae *c*3, *cp* ventrally. Setae *c*3 thin piliform, 19 (17–19) long, coxal fields I, II without sclerotized areas. Genital apparatus situated between levels of trochanters III, IV, 30 (27–30) long, 12 (12–14) wide; paragenital apodemes as a pair of longitudinal sclerites roughly parallel to the arms of genital arch and bearing genital acetabula. Distances between setae: *g:4a* 66 (61–67), *g:g* 9 (7–9). Cupules *ih* ventrally at the level of setae *ps*2. Adanal suckers 15 (14–17) in diameter, distance between centers of suckers 27 (25–27), corolla with 5–7 teeth on anterior half, posterior half without teeth.

Femora I, II with 3–5 apicoventral spines or crests. Acute apicoventral spines on genua, tibiae I, II. Length of tarsi excluding ambulacra: tarsus I 44 (40–42), tarsus II 55 (53–55), tarsus III 57 (58–60), tarsus IV 67 (62–65). Seta *kT* present on tibia IV. Setae *d*, *e* minute spiculiform inserted together (Figure 11D). Genual solenidion *σ*1 on genu I present, minute, about 5 in length. Length of solenidia: *σ2I* 10 (10–11), *σIII* 10 (8–10), *φI* 49 (46–50), *φII* 43 (42–44), *φIII* 47 (39–42), *φIV* 45 (37–42), *ω*1*I* 11 (10–11), *ω3I* 30 (26–30), and *ω*1*II* 20 (17–20).

Female (range for six paratypes). Idiosoma length 367–399, greatest width 197–212. Prodorsal shield shaped as in the male, 82–90 long, 94–106 wide (Figure 10A); scapular setae *si* spiculiform, 17–20 long, setae *se* piliform; distances between scapular setae *si:si* 29–39, *se:se* 78–88. Hysteronotal shield 278–307 in length, 179–190 in width at the widest part at the level of setae *d*1; surface with numerous circular lacunae from the level between setae *c*1 to *e*2. Lateral hysterosomal setae *c*2 bifid, *c*1, *d*1, *d*2, *e*1, *e*2 thin piliform, setae *f*2, *ps*1 flat, spiky leaf-like. Terminal region of opisthosoma shaped as a semicircular concavity flanked by a pair of tubercles bearing setae *h*2 and *h*3. Posterolateral margins of opisthosoma with small spines. Terminal margin of opisthosoma between setae *ps*1 with small copulatory extension about 5–7 long. Length of setae: *c*2 14–20, *e*2 10–14, *c*3 20–29, and *f*2 23–32. Distances between dorsal setae: *c2:d*2 113–124, *d2:e*2 110–121, *d*1*:d*2 7–13, *e*1*:e*2 39–48, *ps*1*:ps*1 21–26, *h3:h*3 42–49, and *h2:h*2 59–65.

Epimerites I free, bases of epimerites I, II inflated, dark-sclerotized (Figure 10B). Epigynum as a low arch, 12–18 in length, 34–46 in width. Distance between ventral setae

1*a*:3*a* 46–55, 3*a*:*g* 14–24. Legs I, II as in the male. Length of tarsi excluding ambulacra: tarsus I 39–47, tarsus II 56–61, tarsus III 61–66, tarsus IV 70–76. Solenidion $\sigma 1I$ present, minute, about 5 in length. Length of solenidia: $\sigma 2I$ 10–13, σIII 11–13, φI 56–64, φII 51–60, φIII 41–58, φIV 13–18, $\omega 1I$ 10–14, $\omega 3I$ 27–34, $\omega 1II$ 19–21.

Differential diagnosis: *Lopharalichus spinosus* **sp. nov.** is close to *L. beckeri* and *L. cribriformis* in having, in males, well-formed cuticular spines in the lateral part of idiosomal anterior to setae *e*2. In both sexes of *L. spinosus*, however, those spines are much more numerous and occupy a larger area, from the level of setae *cp* to that of setae *e2*; in addition, in males of the new species, scapular setae *si* are spiculiform, noticeably more robust than *se* (Figure 2E). In both sexes of *L. beckeri* and *L. cribriformis*, the lateral spines are present only from the level of trochanter IV to the level of setae *e*2. In males of *L. cribriformis* and *L. beckeri*, and in females of the latter species, both scapular setae *si* and *se* are thin piliform (Figure 2B,C and Figure 3C); in females of *L. cribriformis*, setae *si* are more robust than *se*, but they only reach halfway to the distance between those setae (Figure 3B), whereas in *L. spinosus* sp. nov. females, *si* reaches the bases of corresponding setae *se* (Figure 3E).

Etymology: the specific name is an adjective (masculine) referring to the numerous cuticular spines on the lateral margins of hysterosoma, more pronounced and numerous than in other known species.

Lopharalichus chiriri **sp. nov.**

(Figures 1F, 2F, 3F and 12, Figures 13 and 14)

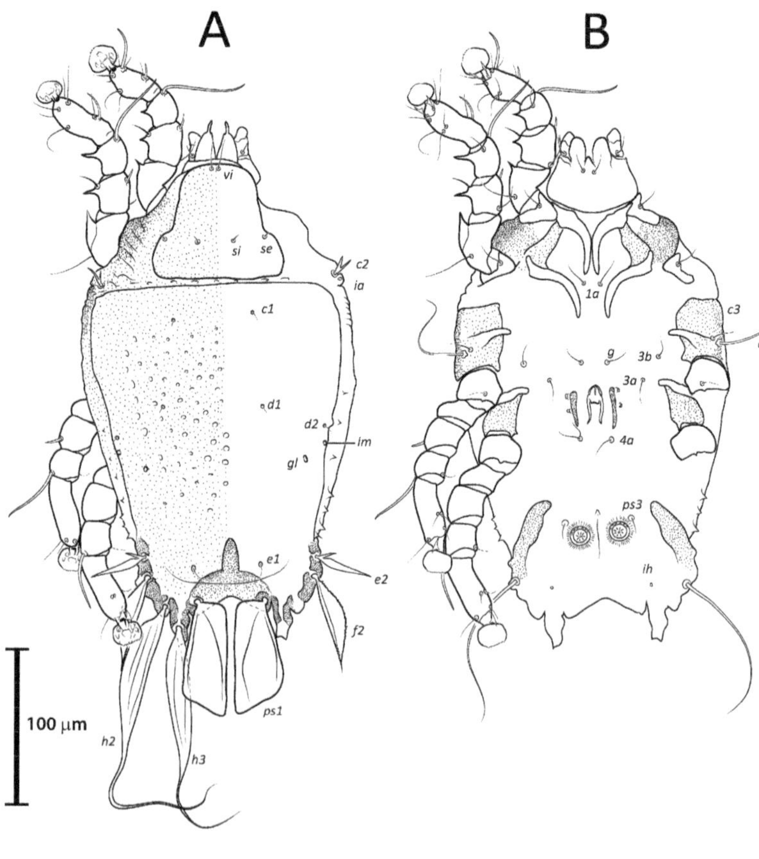

Figure 12. *Lopharalichus chiriri* sp. nov. male: dorsal (**A**) and ventral (**B**) views.

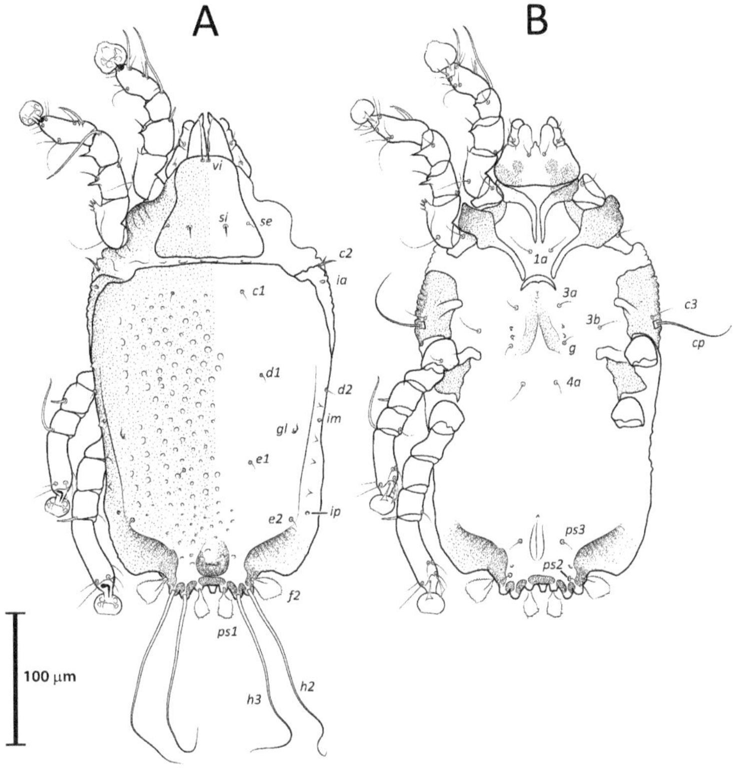

Figure 13. *Lopharalichus chiriri* sp. nov. female: dorsal (**A**) and ventral (**B**) views.

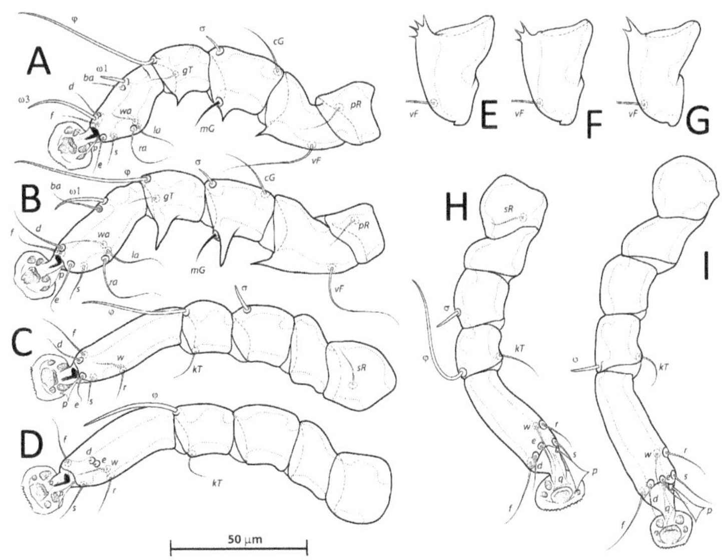

Figure 14. *Lopharalichus chiriri* sp. nov., legs I–IV (**A–D**) of male; variation in femora I in males (**E–G**); legs III–IV (**H,I**) of female.

Zoobank registration: urn:lsid:zoobank.org:act:6DD91CE5-499F-43AD-8210-7B84DF879959

Type material: holotype male, 15 male and 8 female paratypes ex *Brotogeris chiriri* (Vieillot, 1818) (Psittaciformes: Psittacidae), BRAZIL, São Paulo State, Pedreira, 22°44′ S, 46°54′ W, October 2013, D.V. Boas Filho col. (#1113); paratypes 4 females and 1 nymph, same host species, Pará State, Santana do Araguaia, Fazenda Fartura, 09°40′ S/50°23′ W, 07 September 2011, D.V. Boas-Filho coll. (#1006).

Male (holotype, range for six paratypes in parentheses). Idiosoma length from the level of setae *vi* to the base of setae *h*3 285 (294–308), greatest width at level of humeral shields 160 (160–167). Prodorsal shield roughly as an isosceles trapezoid with rounded posterior corners, 76 (67–74) in length from the level of setae *vi* to the posterior margin, 74 (76–79) in width at the widest part. Scapular setae *si* piliform, 7 (6–7) long, distance between *si:si* 23 (23–26), *se:se* 59 (57–62), *si:se* 17 (17–19). Hysterosomal region 224 (213–219) in length from sejugal area to the bases of setae *h*3. Hysteronotal shield: anterior margin straight, length from anterior margin to bases of setae *h*3 207 (213–219), greatest width around the level of setae *d*2 150 (140–155), surface with sparse circular lacunae from the level of setae *c*1 to genua IV (Figure 12A). A bow-shaped transverse fold between levels of setae *e*1 and *ps*1. Membranous margins of terminal cleft blunt-angular, 30 (28–31) long, opisthosomal lobes with prominent tubercles at bases of setae *h*3, and narrow interlobar membranes between bases of setae *ps*1. Setae *c*2 bifid, 12 (11–15) long; setae *e*2 lanceolate with short basal bifurcation, greatest length 35 (35–44); setae *f*2 lanceolate with outer margin minutely serrate, 54 (59–68); setae *ps*1 roughly parallelogram-shaped, 73 (73–82) long. Distances between hysteronotal setae: *c2:d2* 95 (94–100), *d2:e2* 81 (71–81), *e2:h3* 42 (41–50), *d1:d2* 16 (9–16), *e1:e2* 9 (8–14), *ps1:ps1* 42 (42–47), *h3:h3* 66 (67–72), *h2:h2* 82 (85–92), *ps2:ps2* 106 (107–115).

Bases of epimerites I, II inflated, dark-sclerotized (Figure 12B). Humeral shields bearing setae *c*3, *cp* ventrally. Setae *c*3 thin piliform, 17 (12–15) long, coxal fields I–II without sclerotized areas. Genital apparatus situated between levels of trochanters III and IV, 13 (10–13) long, 11 (10–11) wide; paragenital apodemes as a pair of thin longitudinal sclerites roughly parallel to the arms of genital arch and bearing genital acetabula. Distances between setae: *g:4a* 51 (47–53), *g:g* 9 (7–13). Cupules *ih* ventrally at the level of setae *ps*2. Adanal suckers 13 (13–15) in diameter, distance between centers of suckers 24 (22–27), corolla with 5–7 teeth on anterior half, posterior half without teeth.

Femora I, II with 1–4 apical spines on. Acute apicoventral spines on genua, tibiae I, II (slightly more developed on legs II than in legs I). Length of tarsi excluding ambulacra: tarsus I 33 (31–36), tarsus II 43 (40–45), tarsus III 46 (42–49), tarsus IV 48 (48–50). Seta *kT* present on tibia IV. Setae *d*, *e* minute spiculiform inserted close together. Solenidion $\sigma 2$ of genu I apparently absent. Length of solenidia: $\sigma 2I$ 7 (7–9), σIII 9 (7–10), φI 50 (48–55), φII 43 (39–48), φIII 41 (34–45), φIV 40 (33–40), $\omega 1I$ 12 (10–11), $\omega 3I$ 26 (25–28), $\omega 1II$ 19 (18–20).

Female (range for six paratypes). Idiosoma length 296–338, greatest width 171–189. Prodorsal shield shaped as an Erlenmeyer flask (elongated trapezoid), 68–80 long, 74–83 wide (Figure 13A); scapular setae *si* short spiculiform, 9–11 long, setae *se* piliform; distances between scapular setae *si:si* 21–28, *se:se* 59–65, *si:se* 16:21. Hysteronotal shield 233–254 in length, 163–174 in width at the widest part around level of setae *d*1; surface with numerous circular lacunae from the level between setae *c*1 to supranal concavity. Lateral hysterosomal setae *c*2 bifid, *c*1, *d*1, *d*2, *e*1, *e*2 thin piliform, setae *f*2, *ps*1 flat, spiky leaf-like. Terminal region of opisthosoma shaped as a semicircular concavity flanked by a pair of tubercles bearing setae *h*2, *h*3, and a small external copulatory tube around 5–7 in length between bases of setae *ps*1. Lateral margins of opisthosoma with few small spines. Length of setae: *c*2 9–13, *e*2 8–12, *c*3 14–17, *f*2 22–25. Distances between dorsal setae: *c2:d2* 99–112, *d2:e2* 85–101, *d1:d2* 10–20, *e1:e2* 31–46, *ps1:ps1* 16–20, *h3:h3* 35–41, *h2:h2* 53–57.

Epimerites I free, bases of epimerites I, II inflated, dark-sclerotized (Figure 13B). Epigynum as a low arch, 9–12 in length, 27–29 in width. Distance between ventral setae *1a:3a* 37–54, *3a:g* 17–21. Length of tarsi excluding ambulacra: tarsus I 30–37, tarsus II 40–46, tarsus III 42–48, tarsus IV 51–58. Length of solenidia: $\sigma 2I$ 8–11, σIII 7–11, φI 54–64, φII 48–58, φIII 38–47, φIV 10–14, $\omega 1I$ 10–13, $\omega 3I$ 24–29, $\omega 1II$ 18–24.

Differential diagnosis: *Lopharalichus chiriri* sp. nov. is very similar to *L. cribriformis* due to the blunt-angular shape of terminal cleft in males but can be distinguished by the relatively longer distance between prodorsal setae *si-si*. In males of the new species, this distance is about 3.5-times the length of setae *si*, against 2.5-times that length in *L. cribriformis*. Also, the new species has smaller dorsal lacunae and relatively shorter solenidion on tibia IV in males, reaching only about half of the length of tarsus (it reaches at least 3/4 of tarsus length in *L. cribriformis*). The new species is also distinguished from all previously known species in having, in both sexes, considerably longer solenidion on tibia III, roughly longer than the length of genu and tibia III combined. In females of *L. chiriri*, setae *si* are relatively shorter, their tips not touching each other (Figure 3F), while in *L. cribriformis* females, these setae do touch each other. Additionally, in both sexes of *L. chiriri*, tibial solenidion φIII is equal to the length of genu + tibia III (Figure 14C,H), while in other known species of *Lopharalichus*, solenidion φIII is shorter than the length of corresponding genu and tibia.

Etymology: the specific name is a noun in apposition referring to the species name of the type host.

Key to species of *Lopharalichus* Gaud & Atyeo, 1996

1. Both sexes: wide apicoventral spines on genua I, II around base of seta *mG*, much wider than spines on corresponding tibiae I, II; setae *vi* dilated; cuticular spines absent; males with setae *ps*1 roughly triangular with rounded edges (gradually narrowed toward distal end, width of basal part about 4-times wider than distal part); setae *e*2 not bifid basally . . . . *L. denticulatus* (Mégnin & Trouessart, 1884)

1′. Both sexes: spines on genua I, II about as wide as those on tibiae I, II; setae *vi* not dilated; cuticular spines present on posterolateral margins of opisthosoma; males with setae *ps*1 parallelogram-shaped (width of base subequal to that of distal end); setae *e*2 bifid basally 2

2. In both sexes, setae *si* and *se* piliform, subequal in length (Figures 2C and 3C) . . . . .. *L. beckeri* Mironov et al., 2005

2′ In females, setae *si* always spiculiform; in males, setae *si* either spiculiform or piliform . . . 3

3. In both sexes, lateral margins of hysterosoma with pronounced spines from level of setae *cp* to *e*2 (Figures 9A and 10A); in males, setae *si* spiculiform, noticeably more robust than *se* (Figure 2E) . . . *L. spinosus* sp. nov.

3′: In both sexes, spines on the lateral margins of hysterosoma limited to the levels between setae *d*1 to *e*2 (in males), and *d*1 to *f*2 (in females) 4

4. In both sexes, solenidion φIII longer or equal to the length of genu + tibia III (Figure 14C,H); in females, tips setae *si* not reaching each other . . . . *L. chiriri* sp. nov.

4′. In both sexes, solenidion φIII shorter than the length of genu + tibia III; in females, setae *si* relatively longer, their tips touching each other . . . 5

5. In both sexes, distance *si:si* about 1.5 longer than distances between *si:se* (Figures 2D and 3D); in males, *si* about twice longer than *se*; in females, setae *si* equal to distance *si:se* . . . .. *L. tuim* sp. nov.

5′ In both sexes, distance *si:si* approximately equal to the distance *si:se* (Figures 2B and 3B); in males, *si* and *se* subequal in length; in females, setae *si* shorter than the distance between setae *si* and *se* . . . .. *L. cribriformis* (Mégnin & Trouessart, 1884)

4. Discussion

By the time Gaud & Atyeo [5] established the genus *Lopharalichus*, they mentioned that it occurred solely on parrots of the subfamily Aratinginae (sensu Wolters [18]). However, they also referred to this genus as having two undescribed species [5] from parrots then considered in the subfamily Forpinae (sensu Wolters): *Forpus passerinus* and *F. sclateri* (the latter is currently regarded as a subspecies of *Forpus modestus* (Cabanis, 1849)). Herein, a new species is described from the genus *Brotogeris*, previosuly considered in yet another subfamily of Wolters, Brotogerinae. In the current classification of parrots [19], the hosts of *Lopharalichus* are parrots of the family Psittacidae, subfamily Arinae, tribes Arini, Forpini, and Androglossini—it remains to be discovered whether *Lopharalichus* is also present on the tribe Amoropsittacini. Those three tribes account for nearly 140 parrot species (~93% of the arine species), and *Lopharalichus* spp. has been reported from only eight of those hosts so far, including two undescribed species illustrated by Gaud & Atyeo [5].

According to Wright et al. [20], the Arinae—the New World parrots—diverged from the African Psittacinae around the K-T boundary (~66 mya) and diversified approximately 55 mya. *Lopharalichus*, being found only in New World parrots, probably originating between those dates, and given its seemingly uneven distribution on three out of four arine tribes (see above), it probably independently colonized those hosts horizontally rather than vertically. Recent studies have demonstrated that horizontal transfer is an important means of colonizing new hosts e.g., [21,22]. An alternative but less likely scenario would be *Lopharalichus* being present on the arine ancestor and having independently become extinct from several hosts of the tribe Arini (e.g., *Anodorhynchus* Spix, *Cyanopsitta* Bonaparte, *Deroptyus* Wagler, *Diopsittaca* Ridgway, *Enicognathus* Gray, *Leptosittaca* Berlepsch & Stolzmann, *Pionites* Heine, and *Pyrrhura* Bonaparte) and Androglossini (most genera excepting *Brotogeris*, see [19]). In a series of papers, W.T. Atyeo and co-workers investigated the pterolichine feather mites from several of those Arini hosts and did not retrieve any mites that would be later classified in the genus *Lopharalichus* [23–28]. Valdebenito et al. [29] examined feather mites from the two species of *Enicognathus* from Chile (also belonging to the Arini) and did not retrieve *Lopharalichus*. As for the tribe Androglossini, only one *Lopharalichus* is known, *L. chiriri* sp. nov. from *Brotogeris chiriri*; the latter tribe contains 10 genera and at least 66 species [14]. Since many of those hosts have not been thoroughly investigated for feather mites, it is reasonable to anticipate that other *Lopharalichus* species may be present in some of those hosts. In the past decade, only a few studies have examined feather mites associated with psittaciform birds in Brazil e.g., [4,30–34]. It is clear, however, that several species remain to be discovered, as nearly 90 parrot species (Psittacidae: Arinae) are found in the country [35].

As in other genera of the *Protolichus* group, the solenidion $\sigma 1$ of genu I in *Lopharalichus* is highly reduced, vestigial, and depending on the position of the specimen on the slide, barely visible. Although the presence of this solenidion was confirmed for some *Lopharalichus* species (e.g., *L. cribriformis*, *L. beckeri*, and *L. spinosus* sp. nov.), it was not possible to confirm its presence in the remaining species studied.

Despite the presence of cuticular spines in the adults, the two examined immature specimens belonging to the species described herein lack such spines. The retention of small cuticular spines on the posterolateral margins of opisthosoma in most adults of *Lopharalichus* species (except in *L. denticulatus*) is not unique to this genus. In other pterolichines belonging to the *Protolichus* generic group, like *Aralichus* Gaud, 1966 and *Distigmesikya* Atyeo, Gaud et Pérez, 1984, the immatures have numerous such spines—in *Aralichus*, they are mostly located caudally, and in *Distigmesikya*, they abundantly cover most of the dorsum) [5,25]. As these mites undergo their final moult to adulthood, those spines disappear in most species. In some of them, however, spines are present in adults, like in both sexes of *Aralichus glaucogularis* Atyeo et Pérez, 1990, and in females of *Scolaralichus vazquezae* Pérez et Atyeo, 1986, *Aralichus menchacai* Pérez et Atyeo, 1989, and *Tanyaralichus elongatus* Pérez et Atyeo, 1989. However, the immatures of the latter two species were not illustrated with cuticular spines [23,24,28].

5. Conclusions

With the description of three new species, *Lopharalichus* has effectively doubled its known species count, now encompassing six species: *L. denticulatus* (Mégnin & Trouessart, 1884) (type species), *L. cribriformis* (Mégnin & Trouessart, 1884), *L. beckeri* Mironov et al. 2005, *L. tuim* **sp. nov.**, *L. spinosus* **sp. nov.**, and *L. chiriri* **sp. nov.** However, since most neotropical parrots remain uninvestigated for their feather mites, it is safe to assume that many other *Lopharalichus* species may exist and will eventually be discovered.

Funding: This research received no external funding.

Institutional Review Board Statement: Not applicable.

Informed Consent Statement: Not applicable.

Data Availability Statement: Data are available upon request to the corresponding author.

Acknowledgments: The author thanks David Vilas Boas-Filho for collecting and donating some of the bird specimens studied here, Antenor Silva Junior and Patrícia Weckerlin for allowing me to collect the feather mites from the ornithological skins at MHNCI, Sergey V. Mironov who kindly sent pictures of *Lopharalichus beckeri*, and Mark Judson (MNHN) who kindly allowed me to examine specimens from the Trouessart collection.

Conflicts of Interest: The author declares no conflict of interest.

References

1. Mironov, S.V.; Pérez, T.M. A review of feather mites of the *Rhytidelasma* generic group (Pterolichoidea Pterolichidae), specific parasites of parrots (Aves: Psittaciformes). *Bull. Inst. R. Sci. Nat. Belg. Entomol.* **2003**, *73*, 135–176.
2. Mironov, S.V.; Dabert, J.; Ehrnsberger, R. A review of feather mites of the *Psittophagus* generic group (Astigmata, Pterolichidae) with descriptions of new taxa from parrots (Aves, Psittaciformes) of the Old World. *Acta Parasitol.* **2003**, *48*, 280–293.
3. Mironov, S.V.; Dabert, J. Systematic revision of the feather mite genus *Protolichus* Trouessart, 1884 (Astigmata, Pterolichidae). *Zootaxa* **2010**, *2526*, 1–36. [CrossRef]
4. Hernandes, F.A. *Genoprotolichus tilae* sp. nov. (Acariformes: Pterolichidae), a new feather mite species from *Psittacara leucophthalmus* (Müller PLS) (Psittacidae) in Brazil. *Syst. Appl. Acarol.* **2022**, *27*, 1618–1628. [CrossRef]
5. Gaud, J.; Atyeo, W.T. Feather mites of the World (Acarina, Astigmata): The Supraspecific Taxa. *Ann. Mus. R. Afr. Centr. Sci. Zool.* **1996**, *277*, 1–193, (Pt. 1, text), 1–436 (Pt. 2, illustrations)..
6. Mironov, S.V.; Dabert, J.; Ehrnsberger, R. Six new feather mite species (Acari: Astigmata) from the carolina parakeet *Conuropsis carolinensis* (Psittaciformes: Psittacidae), an extinct parrot of North America. *J. Nat. Hist.* **2005**, *39*, 2257–2278. [CrossRef]
7. Enout, A.M.J.; Lobato, D.N.C.; Diniz, F.C.; Antonini, Y. Chewing lice (Insecta, Phthiraptera) and feather mites (Acari, Astigmata) associated with birds of the Cerrado in central Brazil. *Parasitol. Res.* **2012**, *111*, 1731–1742. [CrossRef]
8. Pedroso, L.G.A.; Hernandes, F.A. New records of feather mites (Acariformes: Astigmata) from nonpasserine birds (Aves) in Brazil. *Check List* **2016**, *12*, 1–25. [CrossRef]
9. Clayton, D.H.; Walther, B.A. Collection and quantification of arthropod parasites of birds. In *Host-Parasite Evolution: General Principles and Avian Models*; Clayton, D.H., Moore, J., Eds.; Oxford University Press: Oxford, UK, 1997; pp. 419–440.
10. Krantz, J.; Walter, D.E. *A Manual of Acarology*, 3rd ed.; Texas Tech University Press: Lubbock, TX, USA, 2009; 807p.
11. Griffiths, D.A.; Atyeo, W.T.; Norton, R.A.; Lynch, C.A. The idiosomal chaetotaxy of astigmatid mites. *J. Zool.* **1990**, *220*, 1–32. [CrossRef]
12. Atyeo, W.T.; Gaud, J. The chaetotaxy of sarcoptiform feather mites (Acarina: Analgoidea). *J. Kansas Entomol. Soc.* **1966**, *39*, 337–346.
13. Norton, R. Morphological evidence for the evolutionary origin of Astigmata (Acari: Acariformes). *Exp. Appl. Acarol.* **1998**, *22*, 559–594. [CrossRef]
14. Gill, F.; Donsker, D.; Rasmussen, P. (Eds.) IOC World Bird List (v13.1). 2023. Available online: https://www.worldbirdnames.org/new/ (accessed on 24 June 2023). [CrossRef]
15. Mégnin, P.; Trouessart, E.L. Les Sarcoptides plumicoles. *J. Microgr.* **1884**, *8*, 257–266.
16. Canestrini, G.; Kramer, P. Demodicidae und Sarcoptidae. *Tierreich* **1899**, *7*, 1–193. [CrossRef]
17. Radford, C.D. The mites (Acarina: Analgesidae) living on or in the feathers of birds. *Parasitology* **1953**, *42*, 199–230. [CrossRef] [PubMed]
18. Wolters, H.E. *Die Vogelarten der Erde: Eine Systematische Liste Mit Verbreitungsangaben Sowie Deutschen und Englischen Namen*; Paul Parey: Hamburg, Germany, 1975–1982; part I–XX; pp. 1–748.
19. Schodde, R.; Remsen, J.V., Jr.; Schirtzinger, E.E.; Joseph, L.; Wright, T.F. Higher classification of New World parrots (Psittaciformes; Arinae), with diagnoses of tribes. *Zootaxa* **2013**, *3691*, 591–596. [CrossRef] [PubMed]
20. Wright, T.F.; Schirtzinger, E.E.; Matsumoto, T.; Eberhard, J.R.; Graves, G.R.; Sanchez, J.J.; Capelli, S.; Müller, H.; Scharpegge, J.; Chambers, G.K.; et al. A multilocus molecular phylogeny of the parrots (Psittaciformes): Support for a Gondwanan origin during the cretaceous. *Mol. Biol. Evol.* **2008**, *25*, 2141–2156. [CrossRef]
21. Doña, J.; Proctor, H.; Mironov, S.; Serrano, D.; Jovani, R. Host specificity, infrequent major host switching and the diversification of highly host-specific symbionts: The case of vane-dwelling feather mites. *Glob. Ecol. Biogeogr.* **2017**, *27*, 1–11. [CrossRef]
22. Matthews, M.E.; Wijeratne, A.J.; Sweet, A.D.; Hernandes, F.A.; Toews, D.P.L.; Boves, T.J. Dispersal-Limited Symbionts Exhibit Unexpectedly Wide Variation in Host Specificity. *Syst. Biol.* **2023**, *72*, syad014. [CrossRef]
23. Pérez, T.M.; Atyeo, W.T. Una especie nueva de *Aralichus* Gaud (Acarida: Pterolichidae, Pterolichinae), representante de un complejo de especies nuevo. *Anales Inst. Biol. Univ. Nac. Autón. México Zool.* **1986**, *56*, 31–38.

24. Pérez, T.M.; Atyeo, W.T. New species of *Aralichus* Gaud (Acarina, Pterolichidae) from the White-capped Parrot, *Pionus senilis* (Spix). *J. Parasitol.* **1989**, *75*, 11–20. [CrossRef]
25. Atyeo, W.T. Feather mites of the *Aralichus canestrinii* (Trouessart) complex (Acarina, Pterolichidae) from New World parrots (Psittacidae). I. From the genera *Ara* Lacépède and *Anodorhynchus* Spix. *Fieldiana Zool.* **1988**, *47*, 1–26. [CrossRef]
26. Atyeo, W.T. *Pararalichus* gen.n. (Acarina, Pterolichidae) from New World parrots (Aves, Psittacidae). *Zool. Scr.* **1989**, *18*, 331–346. [CrossRef]
27. Atyeo, W.T. *Aralichus porrectus* (Mégnin & Trouessart) and related feather mite species (Acarina, Pterolichide) from parrots of the genus *Brotogeris* Vigors (Aves, Psittacidae). *Syst. Parasitol.* **1989**, *14*, 101–111. [CrossRef]
28. Atyeo, W.T.; Pérez, T.M. Feather mites of the *Aralichus canestrinii* (Trouessart) complex (Acarina, Pterolichidae) from New World parrots (Psittacidae). II. From the genera *Aratinga* Spix, *Deroptyus* Wagler, *Leptopsittaca* Berlepsch & Stolzmann, *Ognorhynchus* Bonaparte, *Pionites* Heine, and *Pyrrhura* Bonaparte, and conclusions to the study. *Fieldiana Zool.* **1990**, *62*, 1–30. [CrossRef]
29. Valdebenito, J.O.; Moreno, L.; Landaeta-Aqueveque, C.; Kinsella, J.M.; Mironov, S.; Cicchino, A.; Troncoso, I.; González-Acuña, D. Gastrointestinal and external parasites of *Enicognathus ferrugineus* and *Enicognathus leptorhynchus* (Aves, Psittacidae) in Chile. *Rev. Bras. Parasitol. Vet.* **2015**, *24*, 422–431. [CrossRef]
30. Hernandes, F.A.; Pedroso, L.G.A. Two new feather mites of the genus *Protonyssus* Trouessart, 1916 (Acariformes: Xolalgidae) from Brazilian parakeets (Psittacidae), with a key to species. *Int. J. Acarol.* **2017**, *43*, 204–211. [CrossRef]
31. Pereira, D.M.; Hernandes, F.A.; Santos, A.C.G.; Nogueira, R.M.S. Feather mites (Acari: Astigmata) of captive Psittaciformes in Brazil. Arq. Bras. Med. *Vet. Zootec.* **2018**, *70*, 843–849. [CrossRef]
32. Jardim, C.C.G.; Cunha, L.M.; Rezende, L.C.; Teixeira, C.M.; Martins, N.R.S.; Oliveira, P.R.; Faccini, J.L.H.; Leite, R.C. Quill mites in Brazilian psittacine birds (Aves: Psittaciformes). *J. Zoo Wildl. Med.* **2012**, *43*, 511–516. [CrossRef] [PubMed]
33. Hernandes, F.A. *Psittophagus hollandicus* n. sp., a new feather mite species (Acariformes: Pterolichidae) from the cockatiel *Nymphicus hollandicus* (Kerr, 1792) (Psittaciformes: Cacatuidae) in Brazil. *Acarologia* **2017**, *57*, 893–900. [CrossRef]
34. Albuquerque, D.A.D.; Brener, B.; Menna-Barreto, R.F.S.; Brun, S.F. The first identification of *Nymphicilichus perezae* Mironov and Galloway, 2002 in cockatiels in Brazil and the first record of *Psittophagus* sp. Gaud and Atyeo, 1996 and cf Dubininia sp. Vassilev, 1958 in cockatiels (Nymphicus hollandicus Kerr, 1792). *Parasitol. Int.* **2012**, *61*, 572–578. [CrossRef]
35. Pacheco, J.F.; Silveira, L.F.; Aleixo, A.; Agne, C.E.; Bencke, G.A.; Bravo, G.A.; Brito, G.R.R.; Cohn-Haft, M.; Maurício, G.N.; Naka, L.N.; et al. Annotated checklist of the birds of Brazil by the Brazilian Ornithological Records Committee—Second edition. *Ornithol. Res.* **2021**, *29*, 1–123. [CrossRef]

animals

Article

New Definition of *Neoprotereunetes* Fain et Camerik, Its Distribution and Description of the New Genus in Eupodidae (Acariformes: Prostigmata: Eupodoidea) †

Ronald Laniecki and Wojciech L. Magowski *

Department of Animal Taxonomy and Ecology, Adam Mickiewicz University, Uniwersytetu Poznańskiego 6, 61-614 Poznań, Poland; ronald.laniecki@amu.edu.pl or ronlaniecki@gmail.com
* Correspondence: magowski@amu.edu.pl
† urn:lsid:zoobank.org:act:515FA061-406E-46E6-A605-91912AEF714E.

Simple Summary: The mite genus *Neoprotereunetes*, long neglected in the literature, is revised according to modern taxonomic standards. Six species from both Arctic and Antarctic locations, previously placed in the genera *Protereunetes* or *Eupodes*, are transferred to *Neoprotereunetes*. The new genus *Antarcteupodes* is created to accommodate one Antarctic species *A. maudae* comb. nov, originally described in *Protereunetes*. An identification key to *Neoprotereunetes* is provided.

Abstract: The genus *Neoprotereunetes* Fain et Camerik, 1994 is revised and its definition is extended in order to incorporate some species of the invalid genus *Protereunetes* Berlese, 1923. The former type species *Neoprotereunetes—Ereunetes lapidarius* Oudemans, 1906 is redescribed and transferred to *Filieupodes* Jesionowska, 2010 (Cocceupodidae); *Proterunetes boerneri* is redescribed and designated the new type species. Two species groups are proposed to embrace Arctic and Antarctic species, respectively. *Protereunetes paulinae* Gless, 1972 is redescribed, whereas *Protereunetes maudae* Strandtmann, 1967 is redescribed and designated the type species of the new genus *Antarcteupodes* gen. nov. A key to the species of *Neopretereunetes* is provided.

Keywords: Acari; *Protereunetes*; taxonomy; biogeography; polar regions

Citation: Laniecki, R.; Magowski, W.L. New Definition of *Neoprotereunetes* Fain et Camerik, Its Distribution and Description of the New Genus in Eupodidae (Acariformes: Prostigmata: Eupodoidea). *Animals* **2023**, *13*, 2213. https://doi.org/10.3390/ani13132213

Academic Editor: Alexis Ribas

Received: 20 May 2023
Revised: 22 June 2023
Accepted: 29 June 2023
Published: 5 July 2023
Corrected: 6 December 2023

1. Introduction

Superfamily Eupodoidea C.L. Koch, 1842 gathers mostly cosmopolitan, terrestrial, soft-bodied and often-colorful mites. Most of them are mycophagous, but there are also predacious (Rhagidiidae) and phytophagous groups (Penthaleidae and Penthalodidae). Some of them, like *Penthaleus major* (Dugès, 1834) (Penthaleidae) and *Halotydeus destructor* (Tucker, 1925) (Penthalodidae), are significant crop pests, whereas *Linopodes* sp. (Cocceupodidae) is considered an economic pest in mushroom houses [1]. Eupodoidea is divided into nine families: Eupodidae C.L. Koch, 1842; Rhagidiidae Oudemans, 1922; Penthaleidae Oudemans, 1931; Penthalodidae Thor, 1933; Strandtmanniidae Zacharda, 1979; Eriorhynchidae Qin et Halliday, 1997; Pentapalpidae Olivier et Theron, 2000; Dendrochaetidae Olivier, 2008 and Cocceupodidae Jesionowska, 2010 [2]. However, internal relationships among families within Eupodoidea remain uncertain [3].

Family Eupodidae C.L. Koch, 1842 currently includes 11 genera: *Eupodes* C.L. Koch, 1835; *Benoinyssus* Fain, 1958; *Claveupodes* Strandtmann et Prasse, 1976; *Caleupodes* Baker, 1987; *Niveupodes* Barillo, 1991; *Neoprotereunetes* Fain et Camerik, 1994; *Aethosolenia* Baker et Lindquist, 2002; *Xerophiles* Jesionowska, 2003; *Pseudoeupodes* Khaustov, 2014; *Pseudopenthaleus* Khaustov, 2015 and *Echinoeupodes* Khaustov, 2017. Genera *Linopodes* Koch, 1835 and *Cocceupodes* Thor, 1934 (previously in Eupodidae) along with one new genus *Filieupodes* (Jesionowska, 2010), were placed by Jesionowska [4] in the separate family Cocceupodidae Jesionowska, 2010, still within Eupodoidea.

Representatives of the genus *Neoprotereunetes* (as diagnosed herewith) are small, inconspicuous mites. Their bodies are pale white often with dark- to light-green-colored idiosoma, divided by a white medial longitudinal stripe, that apparently being an intestine, showing through the lucent integument. Two pigment eye spots occur on the prodorsum but do not preserve in permanent microscopic slides. These fast-moving mites inhabit soil, mosses, lichens, grasses and mammal nests, and have been observed feeding on algae [5].

The history of the genus is long and complex. The subgenus *Protereunetes* was erected within the genus *Micrereunetes* (Tydeoidea: Ereynetidae) by Berlese [6], with the type species *M.* (*P.*) *agilis* and another species, *M.* (*P.*) *brevipes*. Thor [7] raised *Protereunetes* to the generic rank and placed it in the family Eupodidae. Next, Thor and Willmann [8] included five species in *Protereunetes*: *P. striatellus* (C.L. Koch, 1838), *P. lapidarius* (Oudemans, 1906), *P. agilis* (Berlese, 1923), *P. brevipes* (Berlese, 1923) and *P. börneri* Thor, 1934. Subsequently, Fain [9] redescribed *P. agilis* and *P. brevipes*, showing them actually belonging in the genus *Ereynetes* Berlese, 1883 (Tydeoidea: Ereynetidae), meaning that *Protereunetes* is a junior synonym of *Ereynetes*. Regardless of that, in the next twenty years, some new eupodid species were still described in *Protereunetes*: *P. minutus* Strandtmann, 1967; *P. maudae* Strandtmann, 1967; *P. crozeti* Strandtmann et Davies, 1972 and *P. paulinae* Gless, 1972. Although Strandtmann [10], noticing the results of Fain's study [9], transferred *P. minutus* to the genus *Eupodes*, he did not sustain his own view in subsequent papers [11,12]. This new combination, however, was widely accepted by subsequent authors, e.g., Goddard [13], Booth et al. [14], Baker [15]. Lastly, Fain and Camerik [16] created a new genus—*Neoprotereunetes* with the type species *Neoprotereunetes lapidarius* (Oudemans, 1906)—to bracket those species, which were described in *Protereunetes* until then and, unlike *P. agilis* and *P. brevipes*, belonged to the family Eupodidae. However, Fain and Camerik did not present any firm diagnosis for the new genus and only pointed at inaccurate and outdated definition of *Protereunetes* of Thor and Willmann [8]. Moreover, thanks to the present study, their designated type species, *Neoprotereunetes lapidarius* (Oudemans, 1906) appears to be a senior synonym of *Filieupodes filistellatus* Jesionowska, 2010, from the family Cocceupodidae (Eupodoidea). *Neoprotereunetes* as a genus-level taxon appeared in the literature only once more, in the revision of the family Eupodidae by Khaustov [17]. Thus, the aims of the present study are: (1) to redescribe *Ereunetes lapidarius* and correct its systematic position; (2) to provide new definition for the genus *Neoprotereunetes*; (3) to list species in this genus according to the revised diagnosis; (4) to designate its new type species; and (5) to create an identification key for the species within the genus.

2. Material and Methods

The material of *Neoprotereunetes boerneri* was extracted from soil samples using a Berlese–Tullgren funnel (photo-eclector) for one day and stored in 75% ethanol. Thereafter, specimens were cleared in lactic acid, mounted in Hoyer's medium on glass slides and heated for 10–15 days at the temperature of 55 °C. The type material of *Eupodes minutus*, *Protereunetes maudae* and *Protereunetes paulinae* was loaned from collection at Bishop Museum in Honolulu (Hawaii) and the type material of *Neoprotereunetes lapidarius* was loaned from collection at Naturalis Biodiversity Center in Leiden (the Netherlands) (Figure 1). Mites were studied with a phase contrast (PC) (Olympus BX41, BX50) and differential interference contrast (DIC) (BX51) microscopes and identified using keys of Booth et al. [14], Jesionowska [4] and Khaustov [17], as well as original descriptions. Measurements were obtained from the specimens with the aid of an ocular micrometer and are given in micrometers (μm). The drawings were performed using a drawing tube (camera lucida) and processed in the Corel PHOTO-PAINT X5 program. Micrographs were taken using a Canon D5 Mk. II DSLR camera; pictures were assembled and processed with PICOLAY stacking software [18] or manually in Corel PHOTO-PAINT X5.

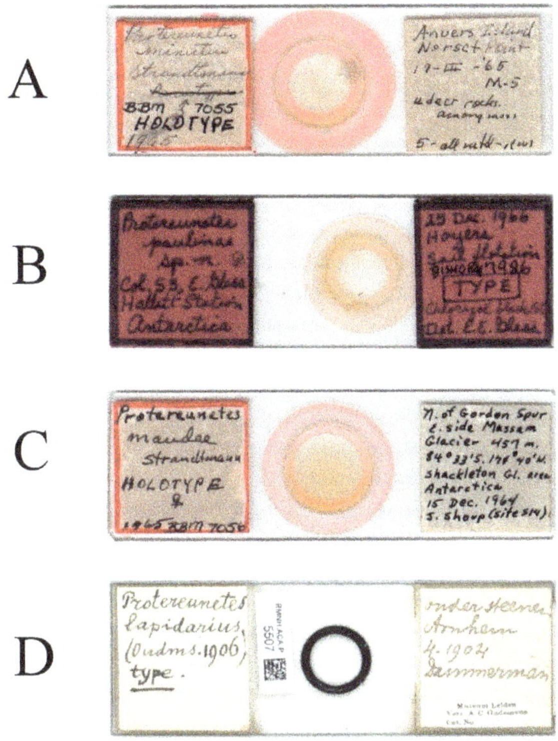

Figure 1. Microscopic slides. (**A**) *Neoprotereunetes minutus* (Strandtmann, 1967), holotype male; (**B**) *Neoprotereunetes paulinae* (Gless, 1972), holotype female; (**C**) *Antarcteupodes maudae* (Strandtmann, 1967), holotype female; (**D**) *Filieupodes lapidarius* (Oudemans, 1906), holotype female.

Morphological nomenclature for idiosoma and gnathosoma follows Baker and Lindquist [3]; for leg chaetotaxy, a universal Grandjean's notation system, reviewed by Norton [19] and applied for eupodoids by Lindquist and Zacharda [20] and Baker [21], is used. The spine-like seta on tibia I is treated herein as a famulus, and thus designated by the Greek letter kappa (κ) rather than the Latin letter *k*, analogically to the famuli on tarsi I and II designated by the Greek letter epsilon (ε). Palpal and leg setal formulae are given from trochanters to tarsi with solenidia and famuli indicated in parentheses. The setae for basi- and telofemora are given separately, even when segment is not divided. The terms "long" and "short" related to dorsal hysterosomal setae mean values equal to or longer than the distance between members of a pair of setae and shorter than this distance, respectively. This excludes lateral hysterosomal setae, i.e., c_2, f_2 and h_2, and also setae f_1 and h_1, which are more tightly clustered at rear part of hysterosoma (caudal bent). Those are longer than remaining hysterosomal setae, and the latter in some eupodid genera (e.g., *Benoinyssus*, *Aethosolenia*) differentiated into trichobothria. Eupathidia are treated herein as setae (1) completely hollow, and (2) with a widely open base and designated by the Greek letter zeta (ζ), subtending the name of a seta. When a seta does not fulfill both conditions (e.g., it is partially hollow) it is then designated by "ζ?". Abbreviations used: *ap*—subcapitular apodema, *cpc*—podocephalic canal, *LL*—lateral lip, *LS*—labrum, *OE*—esophagus, *tr*?—trachea?. Diagnoses and descriptions of taxa refer to adult females if not stated otherwise.

3. Results

3.1. Systematics

Superfamily Eupodoidea C.L. Koch, 1842
Family Eupodidae C.L. Koch, 1842

***Neoprotereunetes* Fain et Camerik, 1994**

Type species: *Protereunetes boerneri* Thor, 1934 by new designation.

Diagnosis. Sejugal furrow present. Idiosomal integument with striate-spiculate ornamentation. Internal vertical setae (v_1) inserted in common areolae, on well-delimited naso. Prodorsal trichobothria (sc_1) filiform and pilose. Hysterosomal setae short and pilose. No hysterosomal trichobothria. Coxisternal setal formula: 3–1–4–3. Six (or exceptionally five) pairs of genital setae in single row and none more lateral than others. Three pairs of pseudanal setae. Adanal setae absent. Four pairs of lyrifissures. Palpal setal formula: 0–2–3–9(ω). All legs shorter than body. Femora IV not enlarged. Leg integument with spiculate ornamentation. Tibiae I and II each with two rhagidial organs.

Description. Idiosomal dorsum. Sejugal furrow present. Integument with striate-spiculate ornamentation. Prodorsum bearing four pairs of setae: internal verticals (v_1), external verticals (v_2), internal scapulars (sc_1) and external scapulars (sc_2). Naso basally delimited from prodorsal shield and bearing setae v_1. Setae sc_1 trichobothrial, short, not reaching the posterior edge of naso, pilose. Remaining prodorsal setae short, pilose and inserted in typical areolae. Hysterosoma bearing eight pairs of dorsal setae: internal humerals (c_1), external humerals (c_2), first dorsals (d_1), second dorsals (e_1), internal lumbars (f_1), external lumbars (f_2), internal sacrals (h_1) and external sacrals (h_2). All hysterosomal setae short, pilose, inserted in typical areolae and none of them trichobothrial. Three pairs of dorsal lyrifissures (*ia*, *im*, *ip*) present.

Idiosomal venter. Coxisternal fields integument with weakly striate-spiculate ornamentation. Coxisternal setal formula: 3–1–4–3. Small cavities near outer margin of coxae I-III present. Genital aperture posteroventral, flanked by four or five pairs of aggenital setae ($ag_{1\text{-}4 \text{ or } \text{-}5}$). Genital valves bearing six (or exceptionally five) pairs of genital setae ($g_{1\text{-}6\ (\text{-}5)}$) of which the anterior first is longer than the second and the second is longer than the remaining ones. All setae g are always in single row and none more lateral than others. Internal genital structures consist of two pairs of genital papillae and four or six pairs of eugenital setae ($eu_{1\text{-}4 \text{ or } \text{-}6}$) set on protuberances. Anal opening terminal, flanked by three pairs of pseudanal setae: $ps_{1,2}$ posteriorly (sometimes located terminally or dorsally) and shorter ps_3 anteriorly. No anal setae (*an*) on anal valves. One pair of ventral lyrifissures (*ih*) present.

Gnathosoma. Subcapitulum roughly triangular, bearing four pairs of setae: two pairs of pilose subcapitular setae ($sbc_{1,2}$) and two pairs of minute smooth adoral setae ($or_{1,2}$). Setae sbc_1 usually thinner and shorter than sbc_2, both located along the base of each lateral lip, at antiaxial and paraxial end of subcapitular apodema, respectively. Setae or_1 and or_2 closely clustered at the tip of each lateral lip and often hard to discern. Apex of labrum acuminate. Chelicerae slender, bearing short, smooth dorsal seta *cha*. Palps four-segmented with weakly barbed supracoxal seta *ep*. Palpal setal formula: 0–2–3–9(ω). Tarsus laterally flattened, bearing nine setae: dorsal (*d*), two laterals (l', l''), sublateral (sl''), anteroculminal (*acm*), two prorals (p', p''), ventral (v), basal (*ba*) and small rhagidial organ ω.

Legs. Legs I and IV longer than II and III, but all shorter than body. Femora I subdivided ventrally, II undivided, III and IV divided. All apoteles with ambulacra, consisting of pad-like empodium with dense setulae arranged in bands on lateral margins and pair of hooked claws with short outgrows on its ventral surface. Integument with spiculate ornamentation. All setae densely pilose except for sparsely pilose v' on trochanters I and II and weakly barbed supracoxal seta *el*. Solenidia and famuli. Leg I. Genu with one dorsomedial erect solenidion σ. Tibia with one anterior complex of rhagidial organ φ_1 and spiniform famulus κ, and one medial rhagidial organ φ_2, tandemly or obliquely in separated depressions. Tibial rhagidial organs long and T-shaped or L-shaped, or either short and ellipsoid to almost spherical. Tarsus with two rhagidial organs $\omega_{1,2}$ and one stellate famulus ε, in tandem in confluent or separate depressions. Posterior one two to three times longer than anterior one. Leg II. Genu with or without medial, erect solenidion σ. Tibia with two rhagidial organs $\varphi_{1,2}$ (anterior and medial), tandemly in separate depressions. Tarsus with two or three rhagidial organs and with or without spiniform famulus ε, variously arranged. Mostly three rhagidial organs present, in confluent depression arranged alternately, i.e., anterior and posterior rhagidial organs situated antiaxially, whereas the medial one is situated paraxially. However, only two rhagidial organs can be present and situated obliquely in separate depressions or in tandem in confluent depression. Leg III. Genu without solenidion. Tibia with or without proximal rhagidial organ. Tarsus without rhagidial organs. Leg IV without solenidia.

Differential diagnosis. The genus resembles *Caleupodes* Baker, 1987 in having short dorsal setae, all legs shorter than the body, femora IV not enlarged and two rhagidial organs on both tibiae I and II. It differs from *Caleupodes* in having integument with striate-spiculate ornamentation (reticulate in *Caleupodes*), pilose dorsal setae (weakly serrate in *Caleupodes*), six or five genital setae (seven in *Caleupodes*) and three pairs of pseudanal setae (two in *Caleupodes*). *Neoprotereunetes* also shares some similarities with the genus *Pseudoeupodes* Khaustov, 2014, i.e., short dorsal setae, legs shorter than body and femora IV not enlarged. It differs from *Pseudoeupodes* in having five or six genital setae (six in *Pseudoeupodes*), three pairs of pseudanal setae (two in *Pseudoeupodes*), two rhagidial organs on both tibiae I and II (one rhagidial organ and one erect solenidion in *Pseudoeupodes*).

Species belonging to the genus *Neoprotereunetes*:

1. *Protereunetes boerneri* Thor, 1934
2. *Protereunetes crozeti* Strandtmann et Davies, 1972
3. *Eupodes exiguus* Booth, Edwards et Usher, 1985
4. *Eupodes minutus* (Strandtmann, 1967)
5. *Eupodes parvus* Booth, Edwards et Usher, 1985
6. *Protereunetes paulinae* Gless, 1972

Neoprotereunetes boerneri species group

Diagnosis. Genital region with five aggenital, six genital and six eugenital setae. Tarsus I with 21 setae (additional ventro-lateral antiaxial seta on tarsus I between setae pv'' and v_1''). Tarsus IV with 13 setae. Tibia I with five setae. Genua III and IV each with four setae. Femur I with 13 setae. Arctic and sub-Arctic distribution. Currently the group contains only one species (*N. boerneri*).

***Neoprotereunetes boerneri* (Thor, 1934)** comb. nov. (Figures 2–7)

Protereunetes börneri [7,8]
Protereunetes boerneri [11,22]
Protereynetes boerneri (sic!) [23]
Neoprotereunetes borneri (sic!) [24]

Diagnosis. Genital region with five pairs of *ag* and six pairs of *g* setae. An extra ventro-lateral, antiaxial seta on tarsus I, located between setae pv'' and v_1''. Trochanter IV with one seta. Two rhagidial organs on tarsus II, slantwise in separated depressions. Proximal rhagidial organ on tibia I and II long, at most four times shorter than its segment.

Redescription. Female. Idiosoma 220 long, 113 wide.

Idiosomal dorsum (Figures 2 and 7A). Prodorsal shield 57 long, 82 wide, triangular. Prodorsal integument with weakly striate-spiculate ornamentation, but course of striae hard to retrace. Naso (Figures 2 and 7B) 9 long, 13 wide, rounded. Lengths of prodorsal setae: v_1 9, v_2 17, sc_1 ca. 40, sc_2 16; distances: v_1–v_1 3, v_2–v_2 44, sc_1–sc_1 29, sc_1–sc_1 64. Hysterosoma tapering caudally, its frontal corners protruding laterally over prodorsum. Hysterosomal integument with striate-spiculate ornamentation. Lengths of hysterosomal setae: c_1 11, c_2 19, d_1 11, e_1 10, f_1 12, f_2 19, h_1 17, h_2 17; distances: c_1–c_1 22, c_1–c_2 44, d_1–d_1 37, e_1–e_1 24, f_1–f_1 20, f_1–f_2 20, h_1–h_1 12, h_1–h_2 ca. 14.

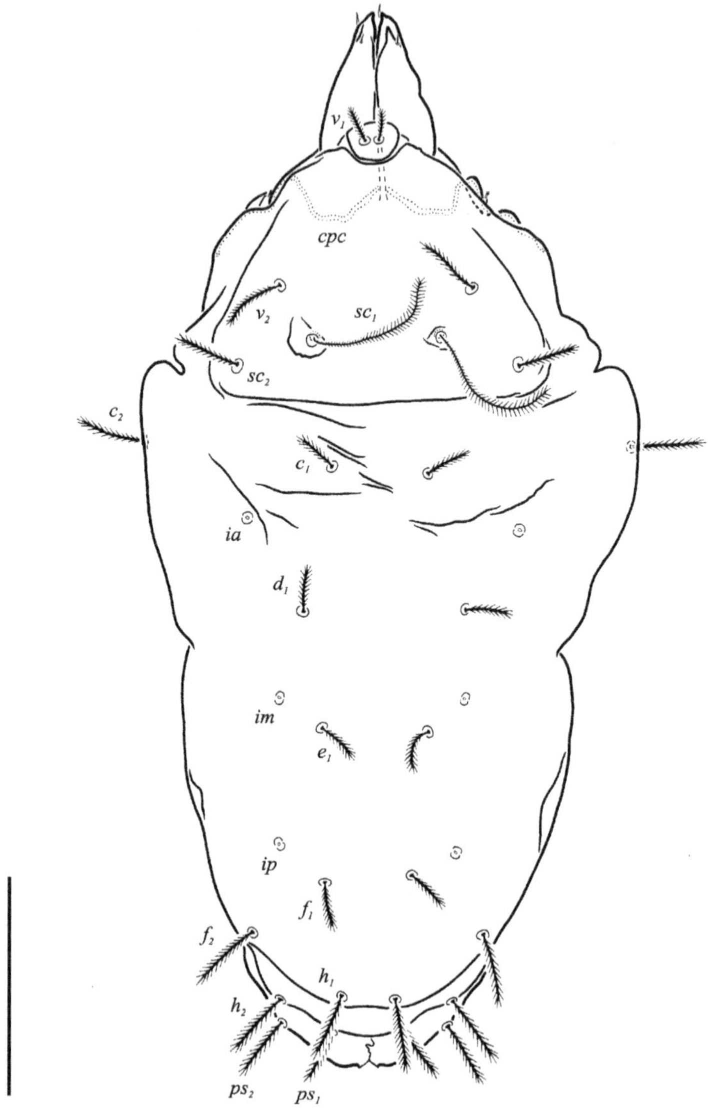

Figure 2. *Neoprotereunetes boerneri* (Thor, 1934), female. Body, dorsal view. Scale bar: 50 μm.

Idiosomal venter (Figures 3 and 7C). Coxisternal fields outlined, with weakly striate-spiculate ornamentation, separated medially by striate-spiculate ornamentation of longitudinal course. Lengths of coxisternal setae: *1a* 12, *1b* 14, *1c* 9, *2b* 11, *3a* 10, *3b* 11, *3c* 11, *3d* 11, *4a* 9, *4b* 10, *4c* 10; distances: *1a–1a* 19, *1b–1b* 40, *1c–1c* 65, *2b–2b* 66, *3a–3a* 15, *3b–3b* 80, *3c–3c* 90, *3d–3d* 61, *4a–4a* 20, *4b–4b* 81, *4c–4c* 57. Coxal cavities well defined. Genital region (Figure 4A) with five pairs of aggenital setae: ag_1 8 long, $ag_{2\text{-}5}$ 7 long, and six pairs and genital setae: g_1 8 long, g_2 7 long, $g_{3\text{-}6}$ 6 long. Six pairs of eugenital setae, ca. 6 long, on protuberances (Figure 4B). Sternal (*1a*, *3a*, *4a*), genital, aggenital and ps_3 slightly expanded distally. Lengths of pseudanal setae: ps_1 14, ps_2 16, ps_3 10.

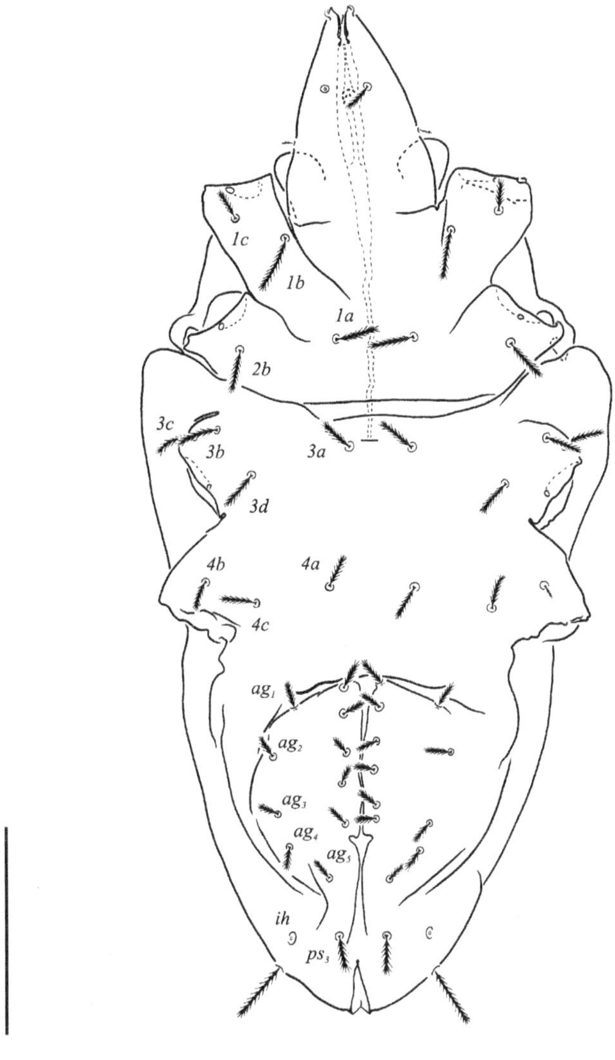

Figure 3. *Neoprotereunetes boerneri* (Thor, 1934), female. Body, ventral view. Scale bar: 50 µm.

Gnathosoma (Figures 4C–G and 7D,E). Subcapitulum (Figure 4C) 66 long, 31 wide, slender, roughly triangular, with spiculate ornamentation. Subcapitular apodema not visible. Setae sbc_2 8 long, densely pilose, thicker than sparsely pilose sbc_1, 3 long. Chelicerae (Figure 4D) 52 long, 15 wide, with spiculate ornamentation, bearing small smooth dorsal seta *cha*. Fixed digit with two pointed tips, ventral pointing forward and dorsal slightly curved backward; movable digit sharp, clawlike. Palps (Figures 4E–G and 7D,E) with spiculate ornamentation, spiculate-cuspidate on femorogenu. Palpal femorogenu 30 long, with indication of division dorsolaterally (Figure 4E). Palp tibial seta l' nearly twice as long as l''. Palpal tarsus 18 long, oval in lateral aspect, triangular in dorsoventral aspect. Setae d, l', l'', v and ba pilose; setae sl'', acm, p' and p'' smooth; rhagidial organ ω ellipsoid with proximal stock.

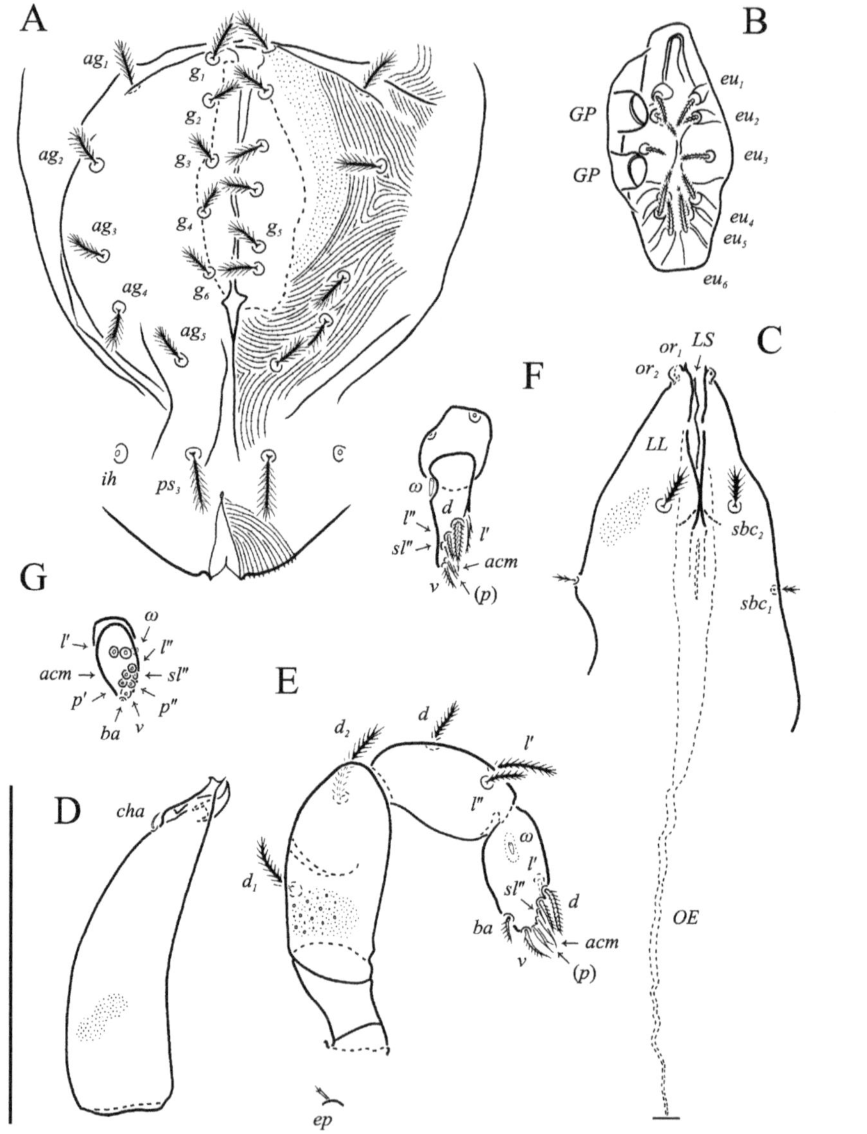

Figure 4. *Neoprotereunetes boerneri* (Thor, 1934), female. (**A**) Genital and anal region; (**B**) progenital chamber; (**C**) subcapitulum, ventral view; (**D**) right chelicera, lateral view; (**E**) left palp, lateral view; (**F**) tarsus of left palp, dorsal view; (**G**) tarsus of right palp, apical view. Scale bar: 50 μm.

Legs (Figures 5, 6 and 7F–H). Lengths of legs: I 164, II 94, III 108, IV 140. Lengths of leg segments: I: Ts: 47, Tb 30, G 30, F 64, Tr 27; II: Ts 33, Tb 22, G 18, F 40, Tr 22; III: Ts 34, Tb 23, G 19, TF 15, BF 31, Tr 22; IV: Ts 36, Tb 29, G 24, TF 16, BF 39, Tr 29. Integument with spiculate ornamentation, spiculate-cuspidate on basifemur III and from tibia to trochanter of leg IV. Leg setal formulae: I: 1–8+5–6(σ)–5(2φ, κ)–21(2ω, ε); II: 1–5+5–4(σ)–5(2φ)–13(2ω, ε); III: 1–4+4–4–5(φ)–12; IV: 1–3+3–4–5–13. Leg eupathidial setae: I: Tb: all except v'; Ts: all except (v_3) II: Tb: d, l''; Ts: all except tc'' and it''; III: Tb: d, v'; Ts: it', (p); IV: G: l'; Tb: d, l', v'; Ts: tc. Solenidia and famuli. Leg I. Genu with one dorsomedial erect solenidion

σ. Tibia with one anterior complex of L-shaped rhagidial organ φ_1, 6 long, plunge into integument basally and spiniform famulus κ, and one medial T-shaped rhagidial organ φ_2, 11 long, obliquely in separated depressions. Tarsus with two rhagidial organs and one stellate famulus ε, obliquely in confluent depression. Posterior one (ω_1) T-shaped, 10 long and anterior one (ω_2) L-shaped, 4 long. Leg II. Genu with one dorsomedial erect solenidion σ. Tibia with two rhagidial organs (L-shaped φ_1, 4 long and T-shaped φ_2, 6 long), tandemly in separated depressions. Tarsus with two rhagidial organs (T-shaped ω_1, 10 long and L-shaped ω_2, 4 long), obliquely in separated depressions. Posterior one (ω_1) subtended by spiniform famulus ε. Leg III. Tibia with proximal T-shaped rhagidial organ φ, 6 long.

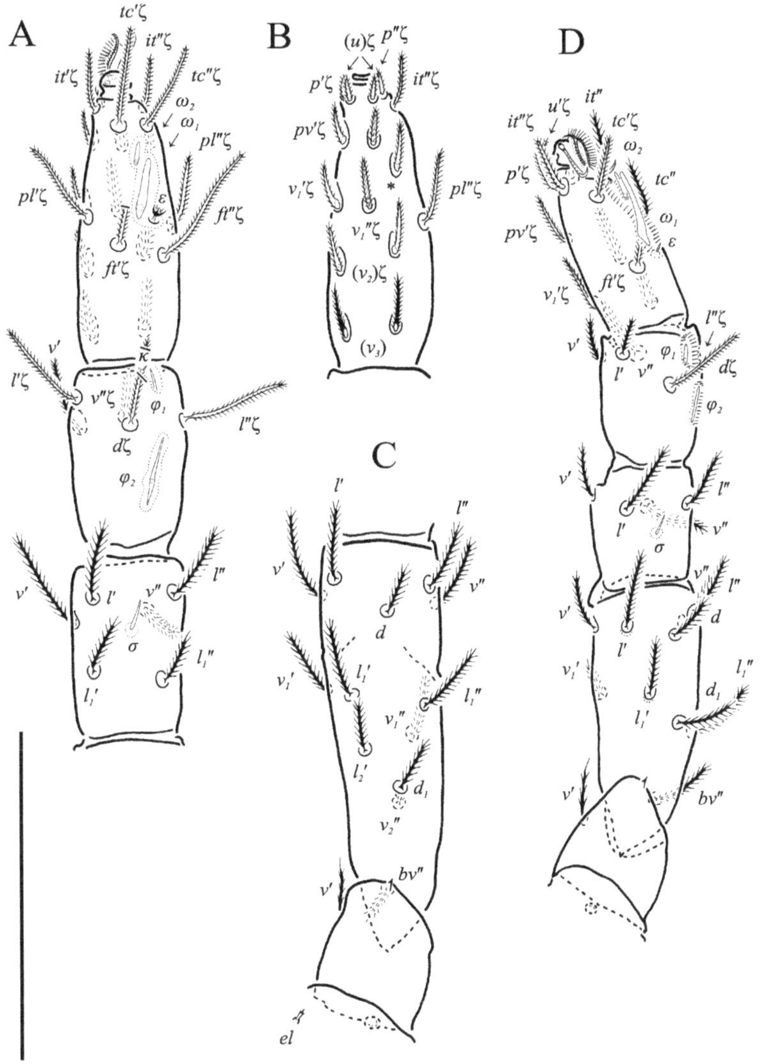

Figure 5. *Neoprotereunetes boerneri* (Thor, 1934), female. (**A**) Tarsus, tibia and genu of right leg I, dorsal view; (**B**) tarsus of right leg I, ventral view (apotele omitted); (**C**) femur and trochanter of right leg I, dorsal view; (**D**) right leg II, dorsolateral view. Asterisk denotes unpaired tarsal seta. Scale bar: 50 μm.

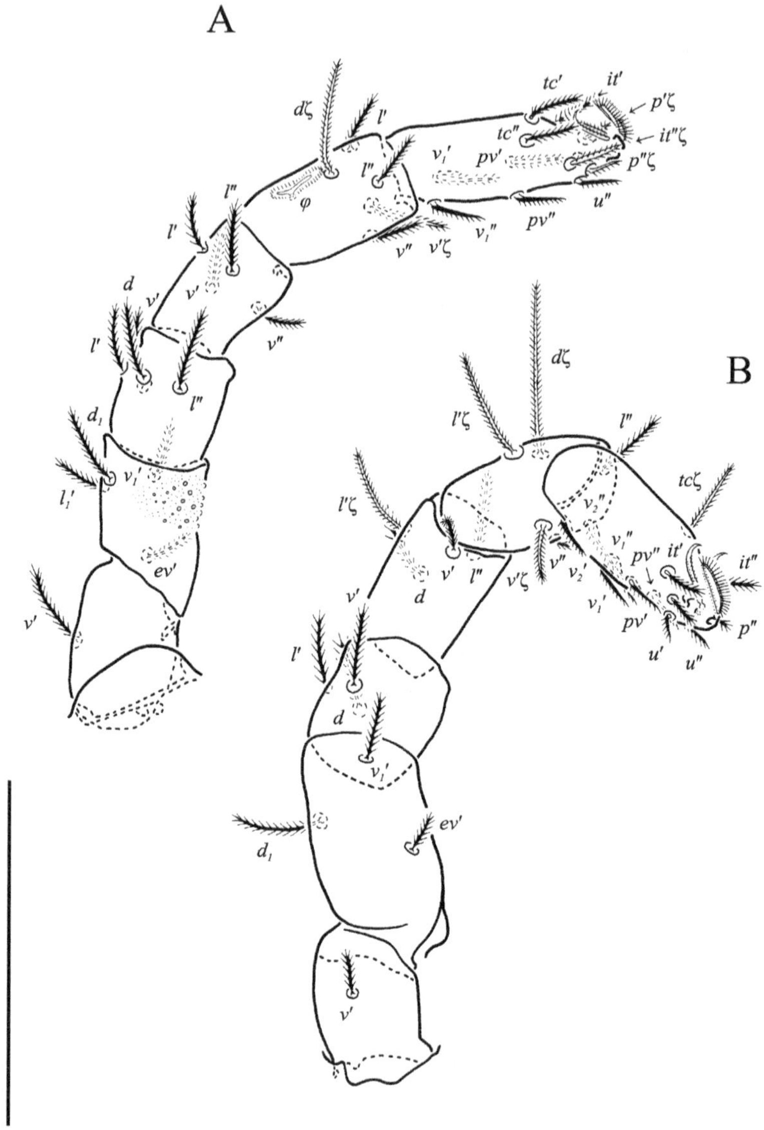

Figure 6. *Neoprotereunetes boerneri* (Thor, 1934), female. (**A**) Right leg III, dorsolateral view; (**B**) right leg IV, lateral view. Scale bar: 50 μm.

Tritonymph. Body length 214. Four and three pairs of *ag* and *g* setae, respectively; *eu* setae absent. Leg setal formulae: I: 1–6+5–6(σ)–5(2φ, κ)–18(2ω, ε); II: 1–5+5–4(σ)–5(2φ)–12(2ω, ε); III: 1–4+4–4–5(φ)–10; IV: 1–3+3–4–5–11. Other characters as in adults.

Deutonymph. Body length 198. Coxisternal setal formula: 3–1–3–2. Two pairs of both *ag* and *g* setae; *eu* setae absent. Leg setal formulae: I: 1–5+5–4(σ)–5(2φ, κ)–17(2ω, ε); II: 1–3+5–4(σ)–5(2φ)–12(2ω, ε); III: 1–2+4–4–4(φ)–10/11; IV: 0–1+3–4–4–11. Other characters as in adults.

For male, protonymph and larva see [11].

Differential diagnosis. *N. boerneri* resembles *N. parvus* by presence of two rhagidial organs on tarsus II. It differs from *N. parvus* in having five pairs of *ag* setae (four in

N. parvus), one seta on trochanter IV (lacking in *N. parvus*) and long proximal rhagidial organs on tibiae I-III (short on tibia I and II, and lacking on tibia III in *N. parvus*).

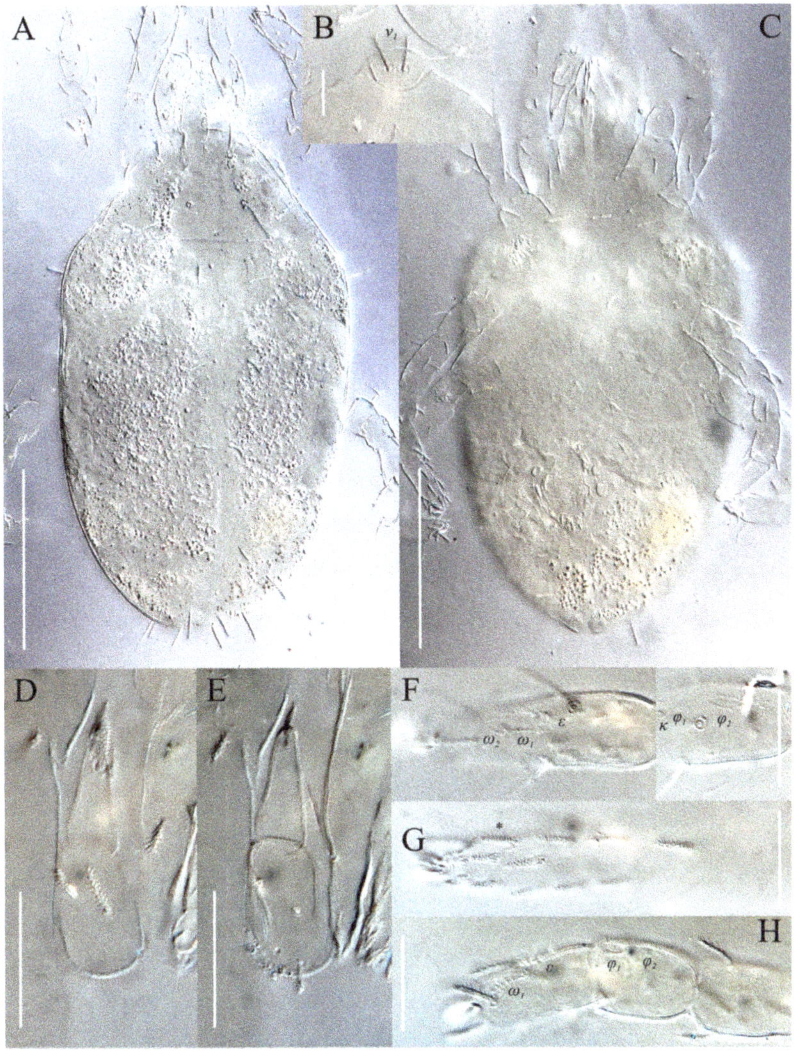

Figure 7. *Neoprotereunetes boerneri* (Thor, 1934), female. (**A**) Body, dorsal view; (**B**) naso; (**C**) body, ventral view; (**D**) tarsus and tibia of left palp, dorsal view; (**E**) tarsus and tibia of left palp, ventral view; (**F**) tarsus and tibia of right leg I, dorsal view; (**G**) tarsus of right leg I, ventral view; H—tarsus and tibia of right leg II, dorsolateral view. Asterisk denotes unpaired tarsal seta. Scale bar: (**A**,**C**) 100 μm; (**B**) 10 μm; (**D**–**H**) 20 μm.

Distribution. Temple Bay, "Grosser Trichter", Magdalena Bay, Spitsbergen, Svalbard, Norway [7]; Utqiaġvik (formerly Barrow), Anaktuvuk Pass, Wainwright, Alaska, USA [11]; Bolshevik Island, Severnaya Zemlya, Russia [22].

Material examined. Four females, one tritonymph and one deutonymph: Svalbard, Spitsbergen, mountain slope, NW exposition, 150 m a.s.l., 78°14′08″ N 15°20′05″ E, soil in vicinity of little auk (*Alle alle*) rookery, 17 July 2022, leg. K. Zawierucha, M. Zacharyasiewicz, M. Jastrzębski.

Remarks. The original description lacks some valid diagnostic characters and thus the species is redescribed herewith. The type material of *N. boerneri* does not exist ([25], p. 408; correspondence with Dr. Vladimir Gusarov, Natural History Museum, University of Oslo), but the specimens collected from Spitsbergen fully fit the original description and figures by Thor [7].

The species was redescribed by Strandtmann [11] on the basis of specimens collected from tundra and from the nests of brown lemming (*Lemmus trimucronatus*) in Alaska. The Strandtmann's material was not available for this study, but no significant differences between the specimens from Alaska and those from Svalbard were found.

N. boerneri possesses a unique character, i.e., one extra ventro-lateral, antiaxial seta on tarsus I, located between setae pv'' and v_1''. An additional tarsal seta is present in yet another eupodid species, *Echinoeupodes echinus* Khaustov, 2017. In that species additional seta (designted as "*vs*" by Khaustov [26]) is situated ventrally, between the pair of *pv* setae, and occurs on tarsi of all four legs. As it is hard to determine whether these two cases deal with homologous setae, the extra seta is marked only with an asterisk (*) in our study (Figures 5B and 7G).

Neoprotereunetes minutus species group

Diagnosis. Genital region with four aggenital, six (or exceptionally five) genital and four eugenital setae. Tarsus I with 20 setae. Tarsus IV with 11 setae. Tibia I with four setae. Genu III with two or three setae. Genu IV with three setae. Femur I with 12 setae. Antarctic and sub-Antarctic distribution.

***Neoprotereunetes crozeti* (Strandtmann et Davies, 1972)** comb. nov.

Protereunetes crozeti [12,22,27]

Diagnosis. Genital region with four pairs of *ag* and six pairs of *g* setae. Trochanter IV with one seta. Tarsus II with three rhagidial organs and without spiniform famulus. Both tarsal rhagidial organs in separate depressions. Proximal rhagidial organs on tibia I and II long, at most four times shorter than its segment. No rhagidial organs on tibia III.

Differential diagnosis. *N. crozeti* resembles *N. minutus* by long proximal rhagidial organs on tibiae and lack of famulus on tarsus II. It differs from *N. minutus* in lacking proximal rhagidial organ on tibia III (present in *N. minutus*) and in arrangement of tarsal rhagidial organs. On tarsus I, in *N. crozeti* tip of antiaxial ω_1 and base of paraxial ω_2 overlap, whereas in *N. minutus* both are situated medially in tandem. On tarsus II, in *N. crozeti* ω_2 and ω_3 lie side by side and in *N. minutus* ω_3 is displaced anteriorly in relation to ω_2.

Distribution. Possession Island, Crozet Islands, ATF [12].

Material examined. None.

Remarks. The original description lacks some valid diagnostic characters, but as the type- or any other material was not available for this study, only standardized diagnosis is given. There is no information on type material deposition in the original paper. It is not deposited in Bishop Museum (courtesy of Dr. Jeremy Frank, Entomology Collections Manager at Bishop Museum).

***Neoprotereunetes exiguus* (Booth, Edwards et Usher, 1985)** comb. nov.

Eupodes exiguus [14,22,27]

Diagnosis. Genital region with four pairs of *ag* and six pairs of *g* setae. Trochanter IV with one seta. Tarsus II with three rhagidial organs and spiniform famulus. Both tarsal rhagidial organs in confluent depressions. Proximal rhagidial organs on tibiae I–III short, at least seven times shorter than their segment.

Differential diagnosis. *N. exiguus* resembles *N. parvus* by very short, globular proximal rhagidial organs on tibiae, and T-shaped rhagidial organs on tarsi I and II. It differs from *N. parvus* in number of rhagidial organs on tarsus II (two instead of three) as well as in presence of rhagidial organ on tibia III and seta on trochanter IV (both absent in *N. parvus*).

Distribution. Signy Island, South Orkney Islands [14]; South Shetland Islands [28].

Material examined. One female and one male: King George Island, South Shetland Islands, 62°05′00″ S, 58°23′28″ W, Grasses near the Comandante Ferraz Antarctic Station, 8 February 2016, leg. D.J. Gwiazdowicz.

Remarks. The original description contains all valid diagnostic characters and thus only standardized diagnosis is given.

***Neoprotereunetes minutus* (Strandtmann, 1967)** comb. nov.

Protereunetes minutus [29]

Eupodes minutus [10,13,14,22,27,30–32]

Diagnosis. Genital region with four pairs of *ag* and six pairs of *g* setae. Trochanter IV with one seta. Tarsus II with three rhagidial organs and without spiniform famulus. Both tarsal rhagidial organs in confluent depressions. Proximal rhagidial organs on tibiae I and II at most four times shorter than their segment.

Differential diagnosis. *N. minutus* closely resembles *N. crozeti*, by long proximal rhagidial organs on tibiae and lack of famulus on tarsus II. *N. minutus*, however, possess a proximal rhagidial organ on tibia III (lacking in *N. crozeti*). Additionally, the arrangement and shape of rhagidial organs is different in these two species. On tarsus I, in *N. minutus* ω_1 and ω_2 lie parallel, while in *N. crozeti* they lie in tandem. On tarsus II, in *N. minutus* ω_3 is displaced anteriorly in relation to ω_2 and in *N. crozeti* ω_2 and ω_3 lie side by side.

Distribution. Anvers Island, Palmer Archipelago [29]; Signy Island, South Orkney Islands [14]; Dunedin, New Zeland [30]; Marion Island, Prince Edward Islands, South Africa [31]; King George Island, Halfmoon Island, Deception Island, South Shetland Islands [28].

Material examined. Holotype male (Bishop Museum, slide labeled "BBM 7055"): Antarctic Peninsula, Anvers Island, Norsel Point, 64°30′ S 63°30′ W, under stones and mosses, March 17. 1965, Coll. D. Strong; one female: King George Island, South Shetland Islands, 62°05′00″ S, 58°23′28″ W, Grasses near the Comandante Ferraz Antarctic Station, 8 February 2016, leg. D.J. Gwiazdowicz.

Remarks. The redescription by Booth et al. [14] contains all valid diagnostic characters and thus only a standardized diagnosis is given here.

Except the type locality, records published before 1985 are not included as suggested in [14].

Mites collected from subalpine grasslands of Mt. Aso and Mt. Kamegamori in Japan were identified by Shiba [33] as *P. minutus*. However, the depicted specimen does not fully agree with the original description and figures as well as the holotype of *P. minutus*. It has shorter rhagidial organs on tibiae I and II and shows rather unusual solenidiotaxy of tibia II (two rhagidial organs and one erect solenidion; see [33], Figure 7e), which does not occur in any other eupodoid species. As the solenidiotaxy of tibiae is not commented in the description and thus cannot be confronted with that figure, this record remains dubious.

Luxton [30] recorded *N. minutus* from Dunedin, New Zeland and refered this species to *Eupodes antipodus* (Womersley, 1937). As no nomenclatorial act was established or synonymy commented it is not included here.

***Neoprotereunetes parvus* (Booth, Edwards et Usher, 1985)** comb. nov.

Eupodes parvus [14,22,27]

Diagnosis. Genital region with four pairs of *ag* and six pairs of *g* setae. Tarsus II with two rhagidial organs and spiniform famulus. Both tarsal rhagidial organs in confluent depressions. Proximal rhagidial organs on tibia I and II short, at least seven times shorter than its segment. No rhagidial organs on tibia III. Trochanter IV without setae.

Differential diagnosis. *N. parvus* resembles *N. exiguus* by short proximal rhagidial organs on tibiae and T-shaped rhagidial organs on tarsi I and II. It differs from *N. exiguus* in number of rhagidial organs on tarsus II (two instead of three) as well as in absence of rhagidial organ on tibia III and seta on trochanter IV (both present in *N. exiguus*).

Distribution. Signy Island, South Orkney Islands, South Shetland Islands, Antarctic Peninsula [14]; King George Island and Ardley Island [28].

Material examined. One female and one male: King George Island, South Shetlands, the Antarctic, 62°09′49″ S 58°27′57″ W, nest of the south polar skua (*S. maccormicki*), 27, 28, 31 January 2016, leg. D.J. Gwiazdowicz.

Remarks. The original description contains all valid diagnostic characters, and therefore only a standardized diagnosis is given here.

Two subspecies of *N. parvus* were proposed by Booth et al. [14]: *N. parvus parvus* from South Orkney Islands and *N. parvus grahamensis* from South Shetland Islands and Antarctic Peninsula, which differs from nominative subspecies only in body length and lengths of idiosomal setae (see [14]).

Figure 8. Neoprotereunetes paulinae (Gless, 1972), holotype female. Body, dorsal view. Scale bar: 100 µm.

Figure 9. *Neoprotereunetes paulinae* (Gless, 1972), holotype female. Body, ventral view. Scale bar: 100 μm.

***Neoprotereunetes paulinae* (Gless, 1972)** comb. nov. (Figures 8–13)

Protereunetes paulinae [5,22,27]

Diagnosis. Genital region with four pairs of *ag* and five pairs of *g* setae. Trochanter IV without setae. Tarsus II with three rhagidial organs and spiniform famulus. Rhagidial organs on tarsi I and II in confluent depression. Proximal rhagidial organs on tibia I and II short, at least seven times shorter than its segment.

Redescription. Holotype female. Idiosoma 268 long, 178 wide.

Idiosomal dorsum (Figures 8 and 13A). Prodorsal shield (Figure 13B) 64 long, 80 wide, triangular. Prodorsal integument with weakly striate-spiculate ornamentation, but course of striae hard to retrace. Naso 10 long, 20 wide, rounded. Lengths of prodorsal setae: v_1 11, v_2 20, sc_1 ca. 47, sc_2 19; distances: v_1–v_1 3, v_2–v_2 49, sc_1–sc_1 34, sc_2–sc_2 73. Hysterosoma tapering caudally, its frontal corners protruding laterally over prodorsum. Lengths of hysterosomal setae: c_1 16, c_2 21, d_1 16, e_1 15, f_1 18, f_2 22, h_1 24, h_2 23; distances: c_1–c_1 26, c_1–c_2 65, d_1–d_1 42, e_1–e_1 41, f_1–f_1 30, f_1–f_2 22, h_1–h_1 17, h_1–h_2 19. Prodorsal integument with weakly striate-spiculate ornamentation, hysterosomal ornamentation striate-spiculate.

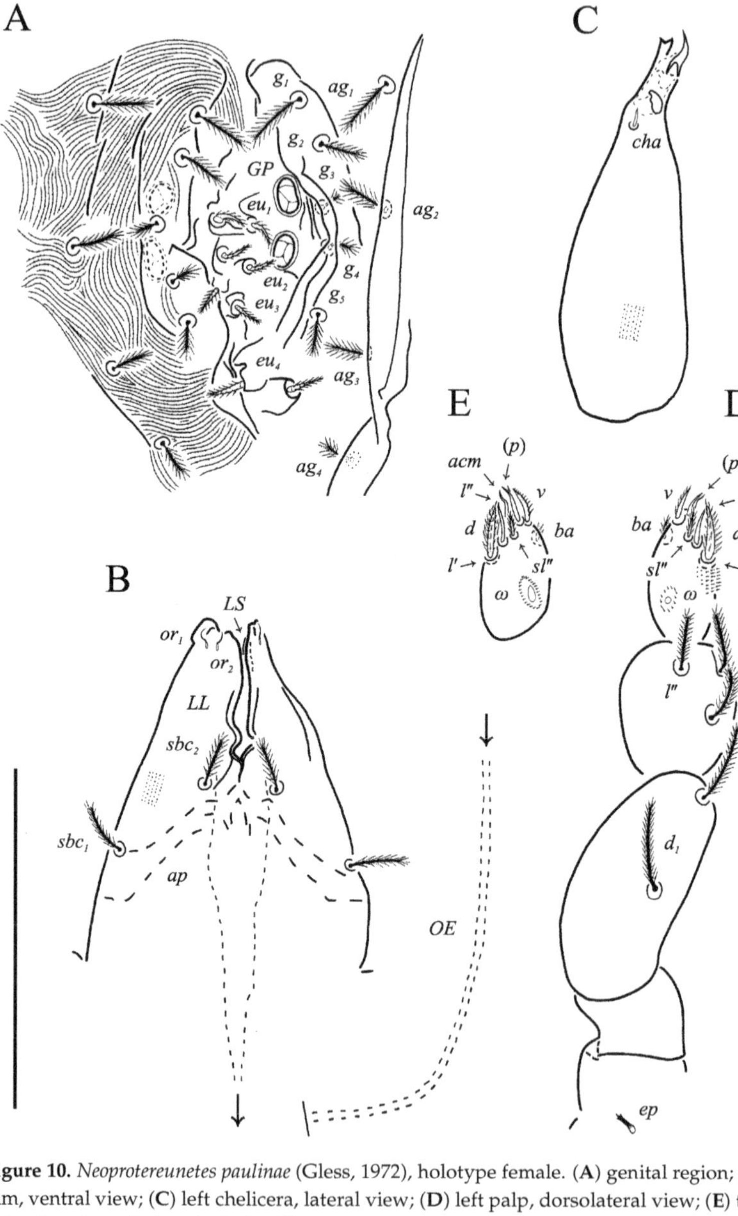

Figure 10. *Neoprotereunetes paulinae* (Gless, 1972), holotype female. (**A**) genital region; (**B**) subcapitulum, ventral view; (**C**) left chelicera, lateral view; (**D**) left palp, dorsolateral view; (**E**) tarsus of right palp, dorsolateral view. Scale bar 50 μm.

Idiosomal venter (Figures 9 and 13A). Coxisternal fields outlined with weakly striate-spiculate ornamentation, separated medially by striate-spiculate ornamentation of longitudinal course. Lengths of coxisternal setae: *1a* 12, *1b* 17, *1c* 10, *2b* 15, *3a* 11, *3b* 15, *3c* 16, *3d* 17, *4a* 11, *4b* 15, *4c* 17; distances: *1a*–*1a* 12, *1b*–*1b* 46, *1c*–*1c* 70, *2b*–*2b* 75, *3a*–*3a* 26, *3b*–*3b* 77, *3c*–*3c* 99, *3d*–*3d* 114, *4a*–*4a* 27, *4b*–*4b* 67, *4c*–*4c* 95. Genital region (Figures 10A and 13C) with four pairs of aggenital setae: ag_1 10 long, ag_2 9 long, $ag_{3\text{-}4}$ 6 long, and five genital setae: g_1 13 long, g_2 8 long, $g_{3\text{-}4}$ 5 long, g_3 6 long. Four pairs of *eu* setae, 6 long, on protuberances. Sternal setae (*1a*, *3a*, *4a*), genital, aggenital and ps_3 setae slightly expanded distally. Lengths of pseudanal setae: ps_1 19, ps_2 19, ps_3 14.

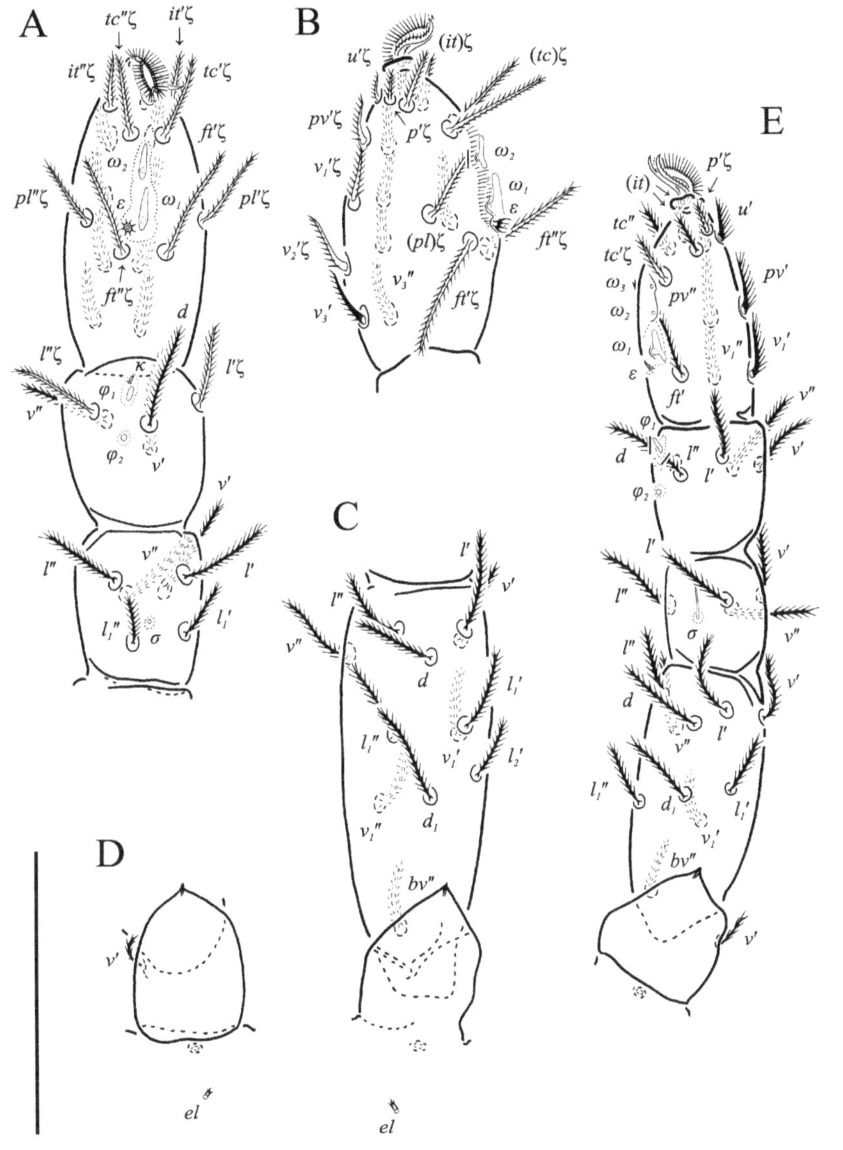

Figure 11. *Neoprotereunetes paulinae* (Gless, 1972), holotype female. (**A**) Tarsus, tibia and genu of left leg I, dorsal view; (**B**) tarsus of left leg I, lateral view; (**C**) femur and trochanter of left leg I, dorsal view; (**D**) trochanter of right leg I, dorsal view; (**E**) left leg II, dorsolateral view. Scale bar: 50 μm.

Gnathosoma (Figures 10B–E and 13D,E). Subcapitulum (Figure 10B) 51 long, 40 wide, slender, roughly triangular, with spiculate ornamentation. Subcapitular apodema visible under integument. Setae sbc_1 10 long, sbc_2 9 long, densely pilose, subequal. Chelicerae (Figure 10C) 60 long, 20 wide, with spiculate ornamentation, bearing small smooth dorsal seta *cha*; fixed digit with two pointed tips directed forward; movable digit sharp, clawlike. Palps (Figure 10D,E) 92 long, with spiculate ornamentation, spiculate-cuspidate on femorogenu. Palpal femorogenu 36 long. Palpal tibia 20 long, seta l'' 2/3 length of l'. Palpal tarsus 19 long, oval in lateral aspect, triangular in dorsoventral aspect. Setae d, l', l'', sl'', v and *ba* pilose; setae *acm*, p' and p'' smooth; rhagidial organ ω ellipsoid with proximal stock.

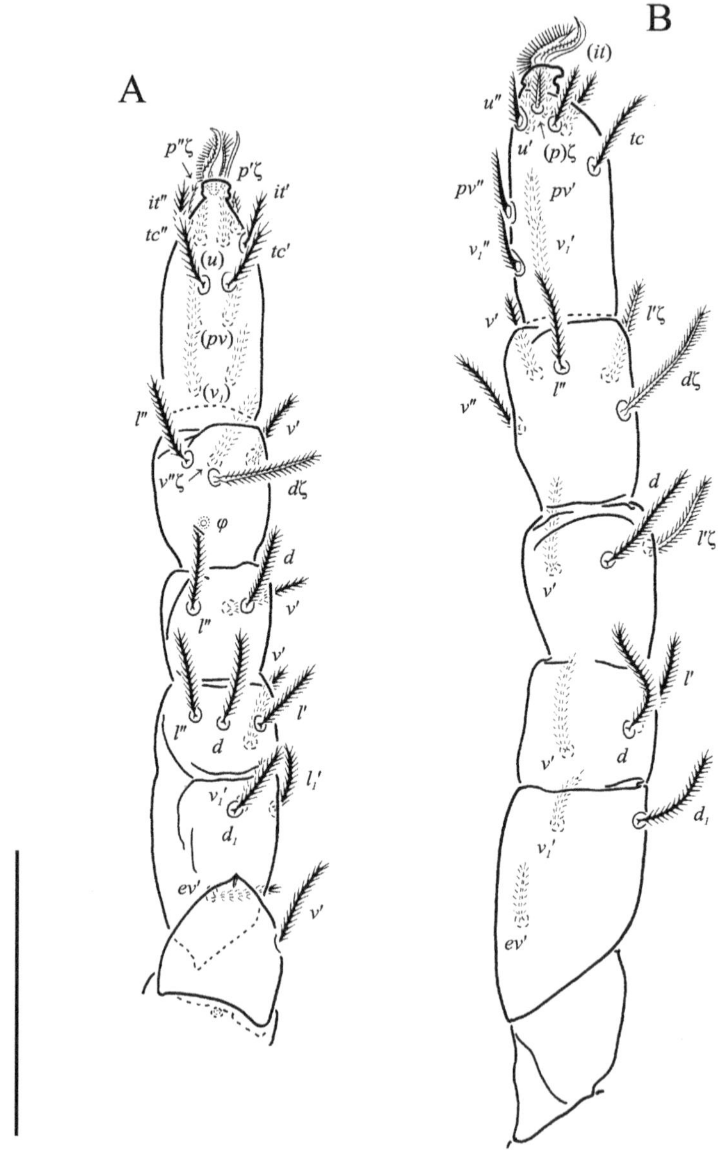

Figure 12. *Neoprotereunetes paulinae* (Gless, 1972), holotype female. (**A**) Left leg III, dorsal view; (**B**) left leg IV, lateral view. Scale bar: 50 µm.

Legs (Figures 11, 12 and 13F,G). Lengths of legs: I 193, II 147, III 149, IV 190. Lengths of leg segments: I: Ts 68, Tb 32, G 29, F 71, Tr 28; II: Ts 43, Tb 25, G 23, F 50, Tr 23; III: Ts 46, Tb 26, G 22, TF 188, BF 35, Tr 25; IV: Ts 48, Tb 37, G 29, TF 25, BF 47, Tr 30. Integument with spiculate ornamentation, spiculate-cuspidate on basifemur III and from tibia to trochanter of leg IV. Leg setal formulae: I: 1–7+5–6(σ)–5(2φ, κ)–20(2ω, ε); II: 1–5+5–4(σ)–5(2φ)–13(3ω, ε); III: 1–4+4–3–4(φ)–12; IV: 0–3+3–3–5–11. Leg eupathidial setae: I: Tb: (l) Ts: all except (v_3); II: Ts: tc', (p); III: Tb: d, v''; Ts: (p); IV: G: l'; Tb: d, l'; Ts: (p).

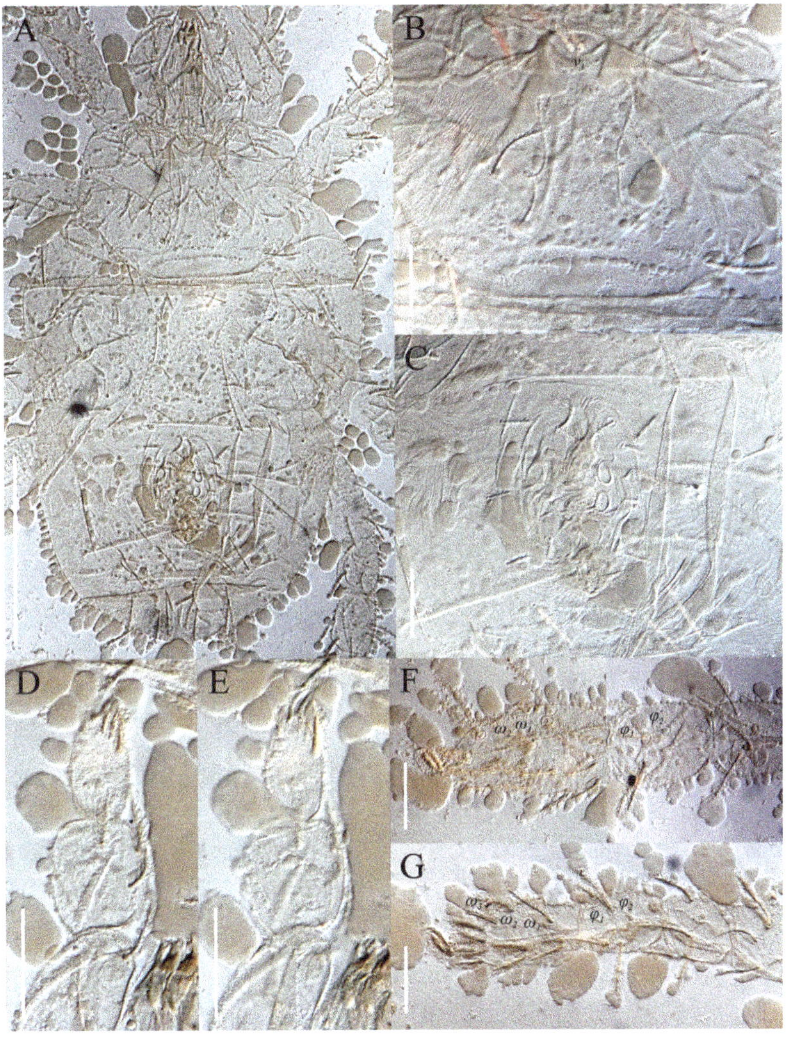

Figure 13. *Neoprotereunetes paulinae* (Gless, 1972), holotype female. (**A**) Body, dorsal and ventral view; (**B**) prodorsum; (**C**) genital region (**D**) tarsus and tibia of left palp, lateral view, antiaxial facesurface; (**E**) tarsus and tibia of left palp, lateral view, paraxial facesurface; (**F**) tarsus and tibia of right leg I, lateral view; (**G**) tarsus, tibia and genu of right leg II, ventrolateral view. Scale bar: (**A**) 100 µm; (**B–G**) 20 µm.

Solenidia and famuli. Leg I. Genu with dorsomedial erect solenidion σ. Tibia with one anterior rhagidial organ φ_1 3 long, associated with spiniform famulus κ and one medial, globular rhagidial organ φ_2 1 long, in separate depressions. Tarsus with two T-shaped rhagidial organs: ω_1 9 long, ω_2 6 long, and stellate famulus ε, tandemly in confluent depression. Leg II. Genu with dorsomedial erect solenidion σ. Tibia with one short dorsodistal rhagidial organ φ_1 3 long and dorsomedial globular rhagidial organ φ_2 1 long. Tarsus with three T-shaped rhagidial organs ω_1 7 long, $\omega_{2,3}$ 5 long, tandemly in confluent depression, subtended by spiniform famulus ε. Leg III. Tibia with globular rhagidial organ φ 1 long.

Differential diagnosis. *N. paulinae* resembles *N. parvus* by lack of seta on trochanter IV and short, globular proximal rhagidial organs on tibiae I and II. It differs from *N. parvus* in number of genital setae (five instead of six) and number of rhagidial organs on tarsus II (three instead of two).

Distribution. Hallett Peninsula, Antarctica [5].

Material examined. Holotype female (Bishop Museum, slide labeled "Bishop 7986"): Hallett Peninsula, Cape Hallett, about 1000 m southeast of Hallett Station on a talus slope, 72°20′ S 170°10′ E, loose soil in north shadow of rock, 25 December 1966, leg. E. Gless.

Remarks. Chaetotaxy of holotype differs significantly from that in original description by Gless [5]. The most apparent seems to be the discrepancy in genital chaetotaxy, i.e., six genital setae in original description and undoubtedly five in holotype female (Figures 10A and 13C). On one hand, this can be attributable to misfortunate arrangement of the last pair of eugenital setae (eu_4) which is everted outward the progenital chamber and supplants, (evidently lacking) last pair of genital setae (g_6). On the other hand, the depicted body ventral side of a female and genital region of a male in the original paper (Figures 32 and 33 in [5]) clearly show six pairs of genital setae in both sexes. As no other specimens of *N. paulinae* are available for this study, it is impossible to decide if it is a both-sided anomaly in holotype or typical state of the species, and thus this character is excluded from the couplet No. 4 of the key.

Species formerly listed as, but not belonging to the genus *Neoprotereunetes* according to the newly proposed diagnosis:

1. *Ereunetes lapidarius* Oudemans, 1906: a senior synonym of *Filieupodes filistellatus* Jesionowska, 2010 (Cocceupodidae).
2. *Protereuntes maudae* Strandtmann, 1967: transferred herewith to the new genus *Antarcteupodes* gen. nov.
3. *Protereunetes turgidus* Shiba, 1978: transferred by Khaustov (2017) to the genus *Echinoeupodes* Khaustov, 2017.
4. *Protereunetes villosus Shiba, 1978: probably belongs to the genus* Benoinyssus *Fain, 1958.*
5. *Protereunetes perforatus* Shiba, 1978: resembles *Caleupodes reticulatus* Baker, 1987, but it differs in body size and form of solenidia.

Species Inquirenda

Protereunetes striatellus (C.L. Koch, 1838): the species description is not sufficient to determine its generic affiliation and the type material most probably does not exist.

***Antarcteupodes* Laniecki** gen. nov.

Type species: *Protereunetes maudae* Strandtmann, 1967; monobasic.

Diagnosis. Sejugal furrow present. Idiosomal integument with lightly striate-spiculate ornamentation. Internal vertical setae (v_1) inserted in bothridia, on well-delimited naso. Prodorsal trichobothria (sc_1) filiform and pilose. Hysterosomal setae short, thin and setose. No hysterosomal trichobothria. Coxisternal setal formula: 3–1–3–2. Six pairs of genital setae in single row and none more lateral than others. Three pairs of pseudanal setae. Adanal setae absent. Four pairs of lyrifissures. Palpal setal formula: 0–2–3–8(ω). All legs shorter than body. Femora IV not enlarged. Leg integument with striate-spiculate ornamentation. Tibiae I and II each with one distal rhagidial organ and one proximal erect solenidion.

Description. Idiosomal dorsum. Sejugal furrow present. Integument with lightly striate-spiculate ornamentation. Prodorsum bearing four pairs of setae: v_1, v_2, sc_1 and sc_2. Naso basally delimited from prodorsal shield and bearing short setae v_1 inserted in bothridia. Setae sc_1 trichobothrial, short, not reaching the posterior edge of naso. Remaining prodorsal setae short, setose, inserted in typical areolae and none of them trichobothrial. Hysterosoma bearing eight pairs of dorsal setae: $c_1, c_2, d_1, e_1, f_1, f_2, h_1$ and h_2. All hysterosomal setae short, thin and setose. Three pairs of dorsal lyrifissures (*ia*, *im*, *ip*) present.

Idiosomal venter. Coxisternal fields integument with weakly striate-spiculate ornamentation. Coxisternal formula: 3–1–3–2; setae *3d* and *4c* not present. Small cavities near outer margin of coxae I-III present. Genital aperture postero-ventral, flanked by five pairs of aggenital setae ($ag_{1\text{-}5}$). Genital valves bearing six pairs of genital setae ($g_{1\text{-}6}$) of which anterior first is longer than second and second is longer than remaining ones. All setae g in single row and none more lateral than others. Internal genital structures consisting of two pairs of genital papillae and six pairs of eugenital setae ($eu_{1\text{-}6}$) set on protuberances. Anal opening terminal, flanked by three pairs of pseudanal setae: $ps_{1,\ 2}$, posteriorly and shorter ps_3, anteriorly. No anal setae (*an*) on anal valves. One pair of ventral lyrifissures (*ih*) present.

Gnathosoma. Subcapitulum roughly triangular, squat bearing four pairs of setae: two pairs of setose subcapitular setae ($sbc_{1,2}$) and two pairs of minute smooth adoral setae ($or_{1,\ 2}$). Setae sbc_1 thinner and shorter than sbc_2, both located along the base of each lateral lip, at antiaxial and paraxial end of subcapitular apodema, respectively. Setae or_1 and or_2 closely clustered at the tip of each lateral lip and hard to discern. Apex of labrum acuminate. Chelicerae thick, bearing long, nude dorsal seta *cha*. Palps four-segmented with weakly barbed supracoxal seta *ep*. Palpal setal formula: 0–2–3–8(ω). Palparsus laterally flattened, bearing eight setae: $d, l', l'', acm, p', p'', v, ba$ and small rhagidial organ ω; seta sl'' not present. Cheliceral and palpal ornamentation spiculate (spiculate-cuspidate on palpal femorogenu).

Legs. Legs I and IV longer than II and III, but all shorter than body. Femora of I and II leg undivided. Femora III and IV divided. All apoteles consist of pad-like empodium with dense setulae arranged in bands on lateral margins and pair of hooked claws with short outgrows on its ventral surface. Integument with striate-spiculate ornamentation. All setae setose except weakly barbed supracoxal seta *el*. Solenidia and famuli. Leg I. Genu with one dorsomedial erect solenidion σ. Tibia with one anterior complex of short ellipsoid rhagidial organ φ and weakly furcate famulus κ, and one proximal short erect solenidion. Tarsus with two L-shaped rhagidial organs (ω) and one small weakly stellate famulus ε, tandemly in separated depressions. Leg II. Genu without solenidion. Tibia with one ellipsoid rhagidial organ φ and one proximal erect solenidion. Tarsus with three L-shaped rhagidial organs and with weakly furcate famulus ε, arranged alternately, i.e., anterior and posterior rhagidial organs situated antiaxially, whereas medial one—paraxially, each in separated depression. Leg III. Genu without solenidion. Tibia with proximal erect solenidion φ. Tarsus without rhagidial organs. Leg IV. Genu without solenidion. Tibia with proximal erect solenidion φ. Tarsus without rhagidial organs.

Differential diagnosis. The new genus is similar to *Pseudoeupodes* Khaustov, 2014 because of legs shorter than body, femora IV not enlarged, short dorsal setae, and number and location of genital setae. It differs from *Pseudoeupodes* by internal vertical setae located in bothridia (in common areolae in *Pseudoeupodes*), coxisternal formula: 3–1–3–2 (3–1–4–2 in *Pseudoeupodes*) and three pairs of pseudanal setae (two in *Pseudoeupodes*). It resembles also *Neoprotereunetes* Fain et Camerik, 1994 in having short dorsal setae, same number and location of genital setae, all legs shorter than body, and not enlarged femora IV. It differs from *Neoprotereunetes* in internal vertical setae located in bothridia (in common areolae in *Neoprotereunetes*), coxisternal formula: 3–1–3–2 (3–1–4–3 in *Neoprotereunetes*) and in presence of one rhagidial organ and one erect solenidion on both tibiae I and II (two rhagidial organs in *Neoprotereunetes*).

Antarcteupodes maudae **(Strandtmann, 1967)** comb. nov. (Figures 14–20)

Protereunetes maudae [22,27,29]

Redescription. Holotype female. Idiosoma flattened and ruptured along its right margin, 360 long, ca. 220 wide.

Idiosomal dorsum (Figures 14 and 20A). Prodorsal shield 74 long, 100 wide, triangular. Prodorsal integument with weakly striate-spiculate ornamentation, but course of striae hard to retrace. Naso (Figure 20B) 15 long, 28 wide, rounded. A pair of canals, probably representing tracheae (*tr*?), extending from anterior end of idiosoma to posterior corners of prodorsum (Figure 14). Lengths of prodorsal setae: v_1 18, v_2 14, sc_1 ca. 40, sc_2 18; distances: v_1–v_1 8, v_2–v_2 60, sc_1–sc_1 37, sc_2–sc_2 92. Hysterosoma roughly rectangular, slightly rounded caudally. Lengths of hysterosomal setae: c_1 14, c_2 24, d_1 14, e_1 14, f_1 19, f_2 20, h_1 24, h_2 23;

distances: c_1–c_1 50, c_1–c_2 78, d_1–d_1 58, e_1–e_1 63, f_1–f_1 52, f_1–f_2 33, h_1–h_1 22, h_1–h_2 27. Prodorsal and hysterosomal integument with lightly striate-spiculate ornamentation.

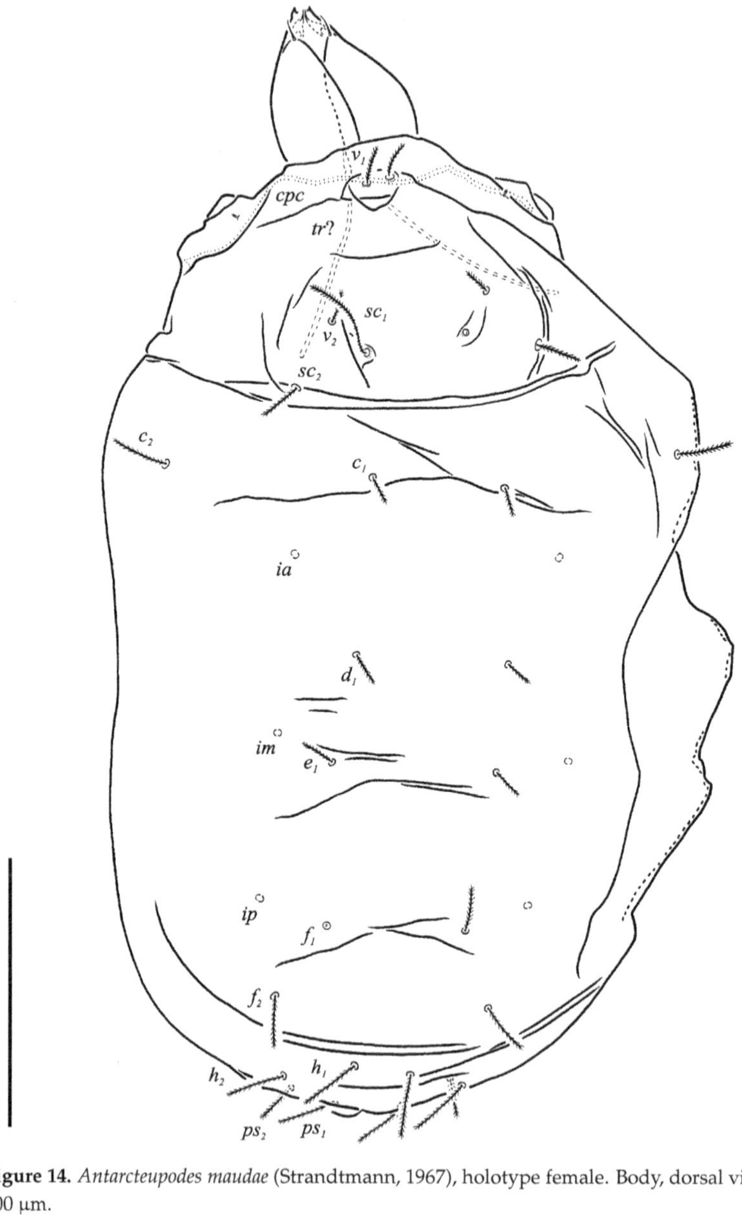

Figure 14. *Antarcteupodes maudae* (Strandtmann, 1967), holotype female. Body, dorsal view. Scale bar: 100 μm.

Idiosomal venter (Figures 15 and 20C). Coxisternal fields outlined, with weakly striate-spiculate ornamentation, separated medially by striate-spiculate ornamentation of longitudinal course. Lengths of coxisternal setae: *1a* 14, *1b* 16, *1c* 10, *2b* 13, *3a* 12, *3b* 15, *3c* 15, *4a* 10, *4b* 15; distances: *1a*–*1a* 27, *1b*–*1b* 69, *1c*–*1c* 97, *2b*–*2b* 82, *3a*–*3a* 39, *3b*–*3b* 92, *3c*–*3c* 122, *4a*–*4a* 42, *4b*–*4b* 104. Coxal cavities weakly defined, half-open. Genital region (Figure 16A) with five pairs of aggenital setae: ag_1 10 long, ag_2 9 long, ag_{3-4} 8 long, ag_5 7 long, and six pairs of genital setae: g_1 10 long, g_2 9 long, g_{3-6} 7 long. Six pairs of eugenital setae, ca. 10 long, on protuberances (Figure 16B). Genital, aggenital and ps_3 setae slightly expanded distally. Lengths of pseudanal setae: ps_1 24, ps_2 19, ps_3 9.

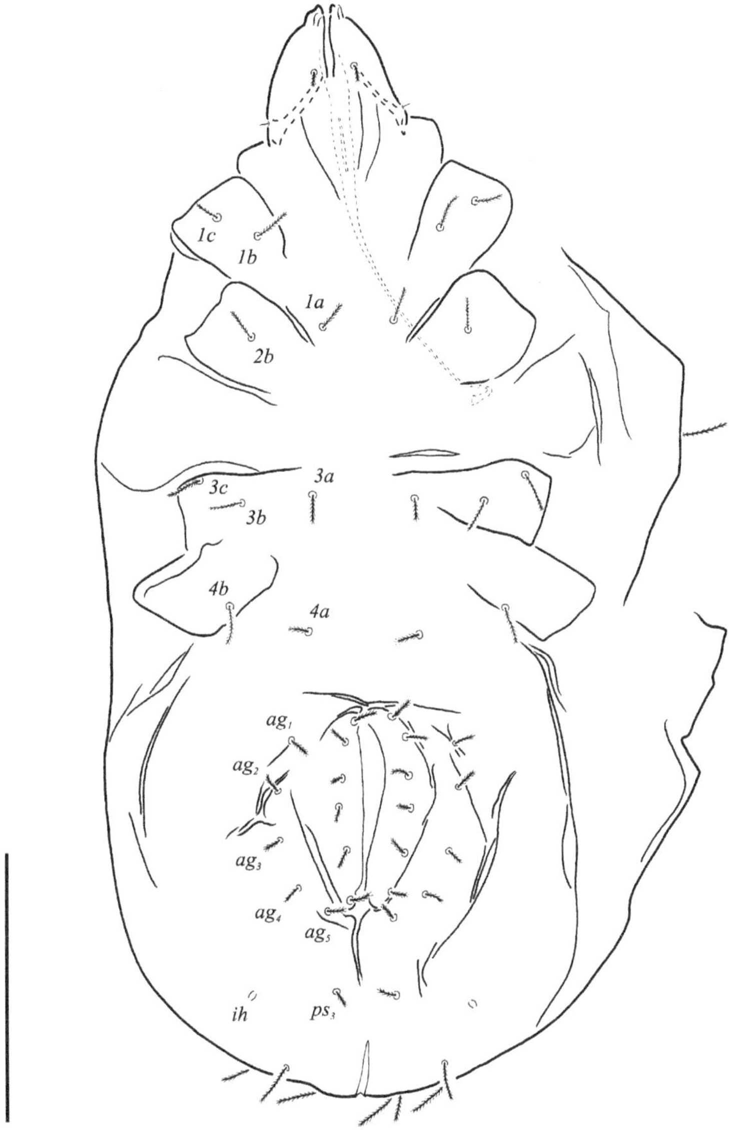

Figure 15. *Antarcteupodes maudae* (Strandtmann, 1967), holotype female. Body, ventral view. Scale bar: 100 µm.

Gnathosoma (Figures 16C, 17A–C and 20D,E). Subcapitulum (Figure 17A) 47 long, 50 wide. Border between lateral lips and subcapitular base visible under integument. Setae sbc_2 7 long, setose, thicker than sparsely setose sbc_1 5 long. Chelicerae (Figure 16C) 70 long, 29 wide, with small smooth dorsal seta *cha*; fixed digit with two pointed tips directed forward; movable digit sharp, clawlike. Palps (Figures 17B,C and 20D,E) 118 long, with striate-spiculate ornamentation, spiculate-cuspidate on femorogenu. Palpal femorogenu 50 long. Palpal tibia 25 long, seta l'' 2/3 length of two times thicker l'. Palpal tarsus 37 long, ellipsoid in dorsoventral aspect. Setae d, l', l'', v and ba pilose; setae acm, p' and p'' smooth; rhagidial organ ω minute, protruding on both tarsi in dorsoventral view. Subcapitular, cheliceral and palpal integument with striate-spiculate ornamentation (spiculate-cuspidate on palpal femorogenu).

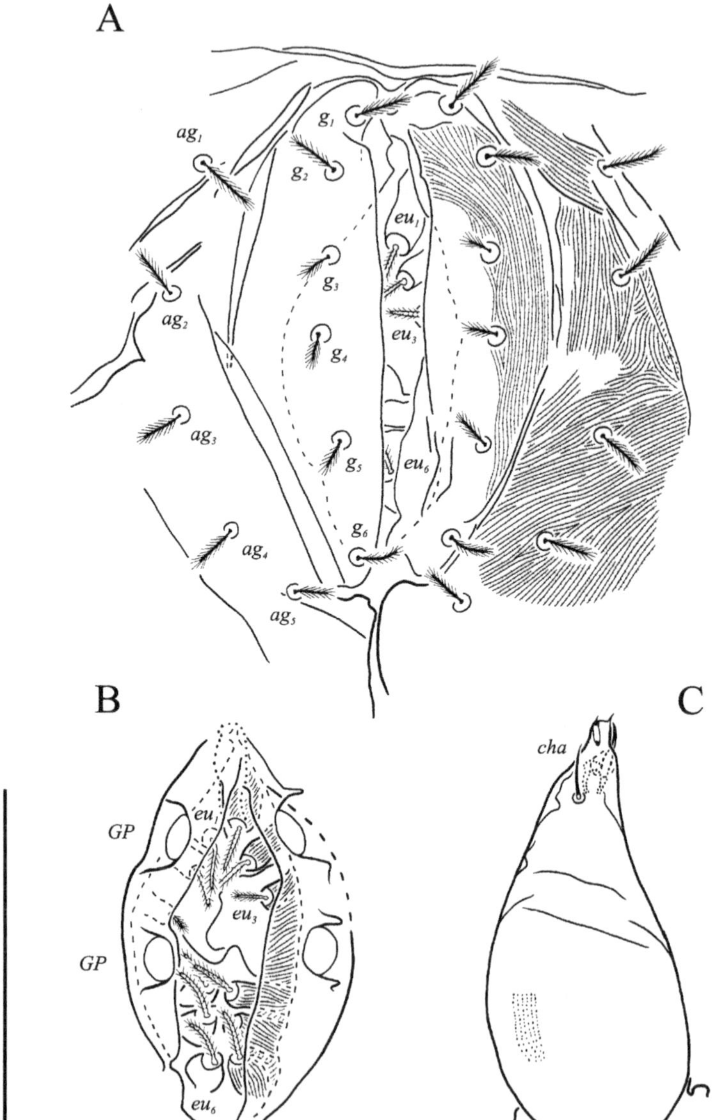

Figure 16. *Antarcteupodes maudae* (Strandtmann, 1967), holotype female. (**A**) Genital region; (**B**) progenital chamber; (**C**) left chelicera, dorsal view. Scale bar: 50 μm.

Legs (Figures 18, 19 and 20F,G). Lengths of legs: I 247, II 197, III 200, IV 255. Lengths of leg segments: I: Ts 64, Tb 45, G 35, F 88, Tr 35; II: Ts 53, Tb 34, G 30, F 68, Tr 28; III: Ts 53, Tb 37, G 27, TF 30, BF 38, Tr 32; IV: Ts 63, Tb 43, G 35, TF 35, BF 60, Tr 43. Integument with striate-spiculate ornamentation, spiculate-cuspidate on basifemur of leg III and from genu to basifemur of leg IV. Leg setal formulae: I: 1–3+5–4(σ)–5(2φ, κ)–17(2ω, ε); II: 1–2+5–

4–5(2φ)–11(3ω, ε); III: 1–2+3–3–3(φ)–11; IV: 1–2+3–3–4(φ)–9. Eupathidial setae: I: Tb: all; Ts: all and v_2''?; II: Tb: *d* (only on left leg), *l''*, *v''*; Ts: all; III: Tb: *d*?, *v'*?; Ts: all and (*tc*)? (*it*)?; IV: Tb: *d*?; Ts: all and *tc*?, (*it*)?. Solenidia and famuli as in generic description. Lengths of rhagidial organs: leg I: φ_1 4, ω_1 8, ω_2 5; leg II: φ_1 4, ω_1 8, ω_2 8, ω_3 8.

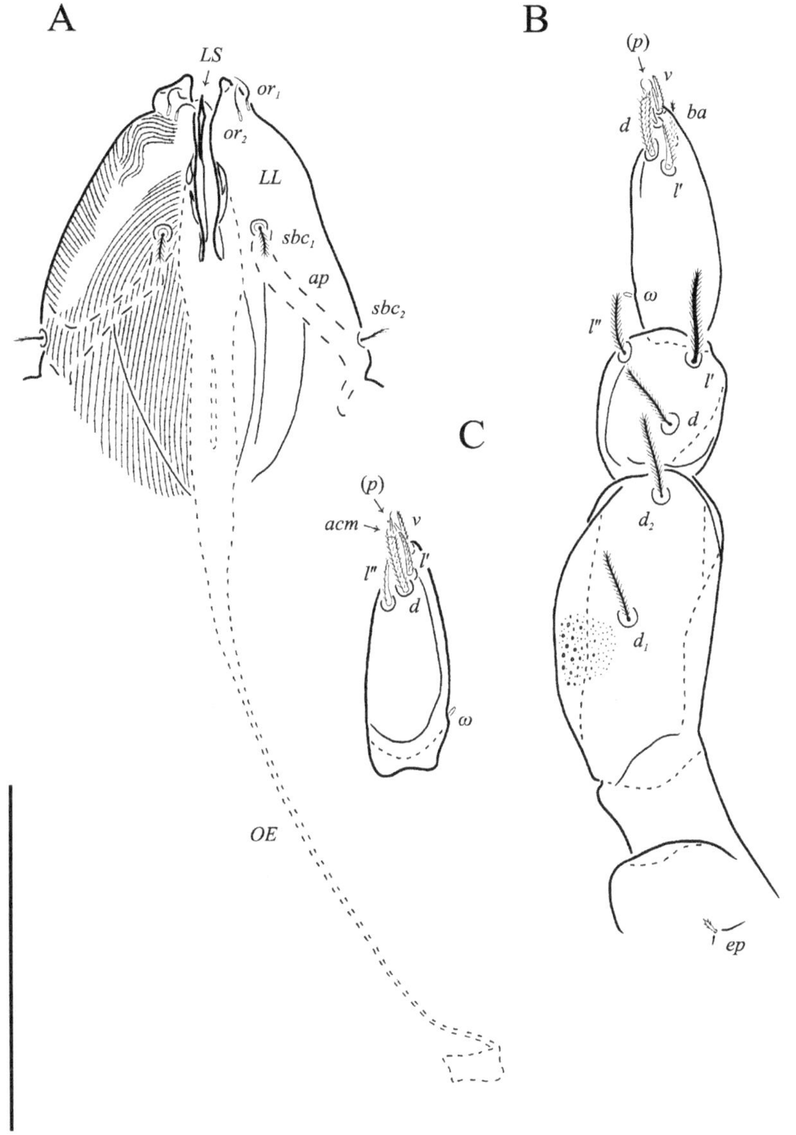

Figure 17. *Antarcteupodes maudae* (Strandtmann, 1967), holotype female. (**A**) Subcapitulum, ventral view; (**B**) left palp, dorsal view; (**C**) tarsus of right palp, dorsal view. Scale bar: 50 µm.

Distribution. Victoria Land, Antarctica [29].

Material examined. Holotype female (Bishop Museum, slide labeled "BBM 7056"): Shackleton Glacier area, north of Garden Spur, east side of Massam Glacier, 457 m elevation, 84°33′ S 174°40′ E, 15 December 1964, leg. J. Shoup.

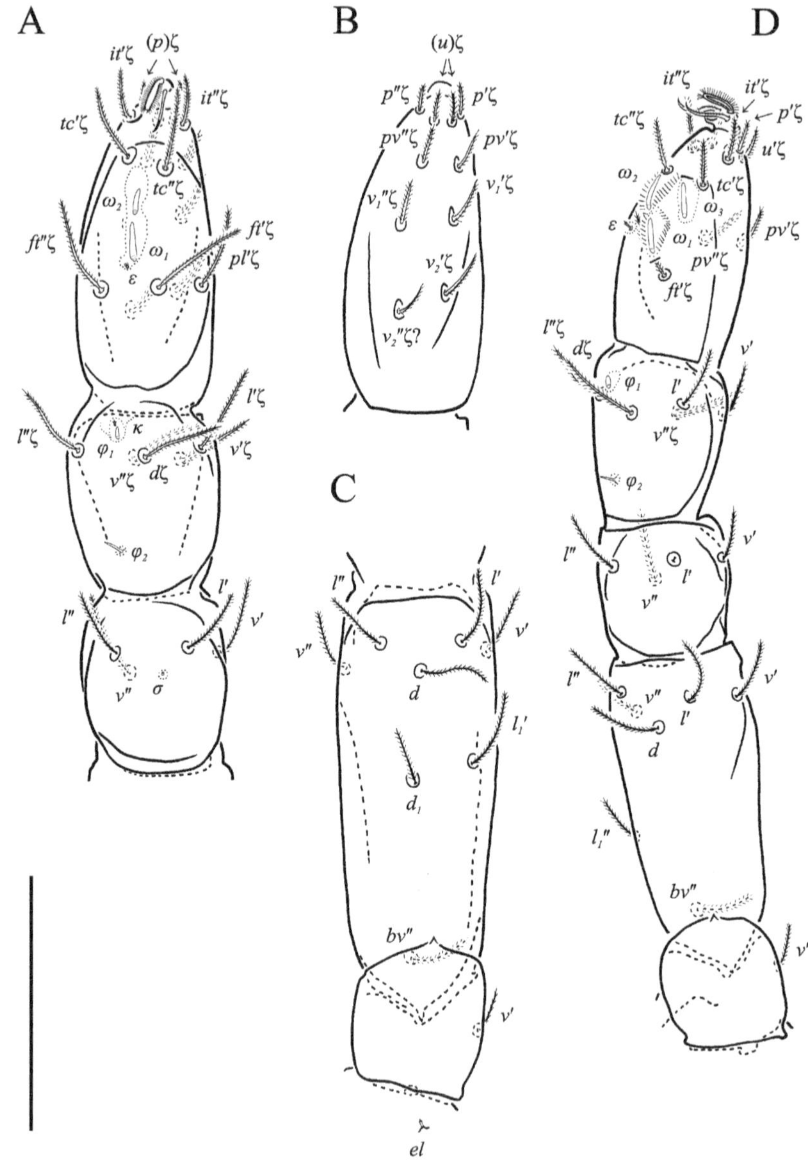

Figure 18. *Antarcteupodes maudae* (Strandtmann, 1967), holotype female. (**A**) Tarsus, tibia and genu of left leg I, dorsal view; (**B**) tarsus of left leg I, ventral view; (**C**) femur and trochanter of left leg I, dorsal view; (**D**) left leg II, dorsolateral view. Scale bar: 50 μm.

Remarks. The species is characterized by the unique combination of character states, not present in any hitherto described eupodid genus, including the most reduced coxisternal and leg chaetotaxy among the family Eupodidae, and sufficient to represent a separate genus.

Family: Cocceupodidae Jesionowska, 2010

Filieupodes Jesionowska, 2010

Type species: *Filieupodes filiformis* Jesionowska, 2010 by original designation.

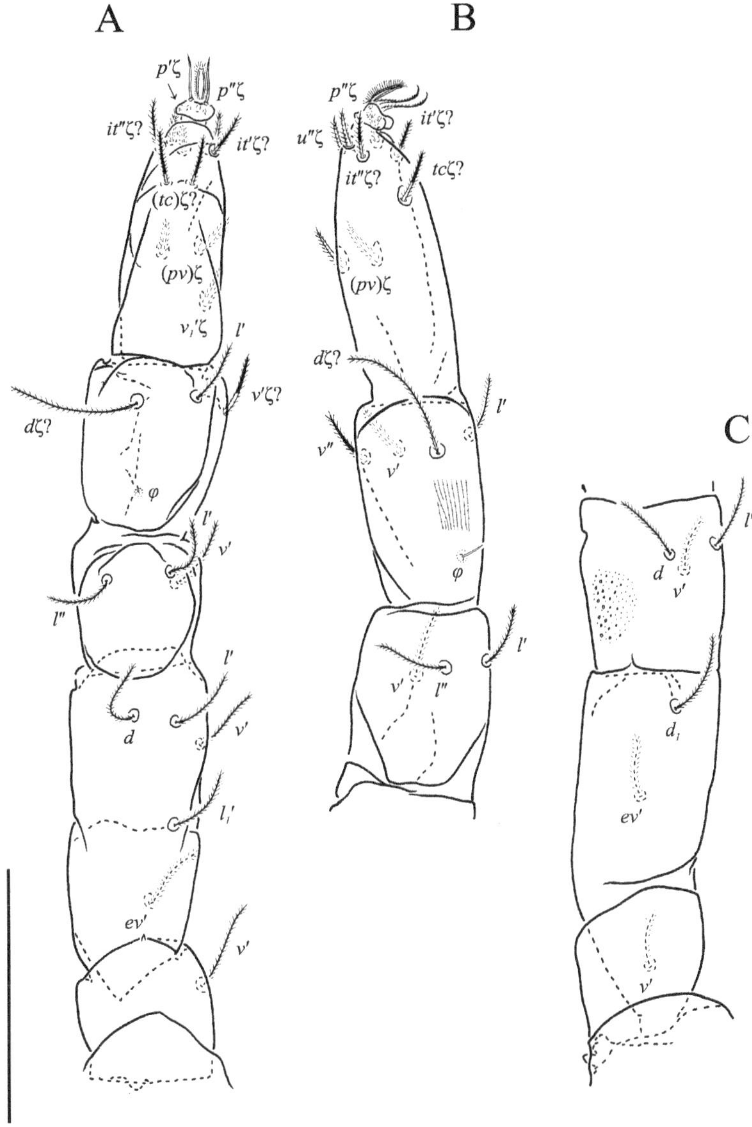

Figure 19. *Antarcteupodes maudae* (Strandtmann, 1967), holotype female. (**A**) Left leg III, dorsal view; (**B**) tarsus, tibia and genu of left leg IV, dorsolateral view; (**C**) telo-, basifemur and trochanter of left leg IV, dorsolateral view. Scale bar: 50 μm.

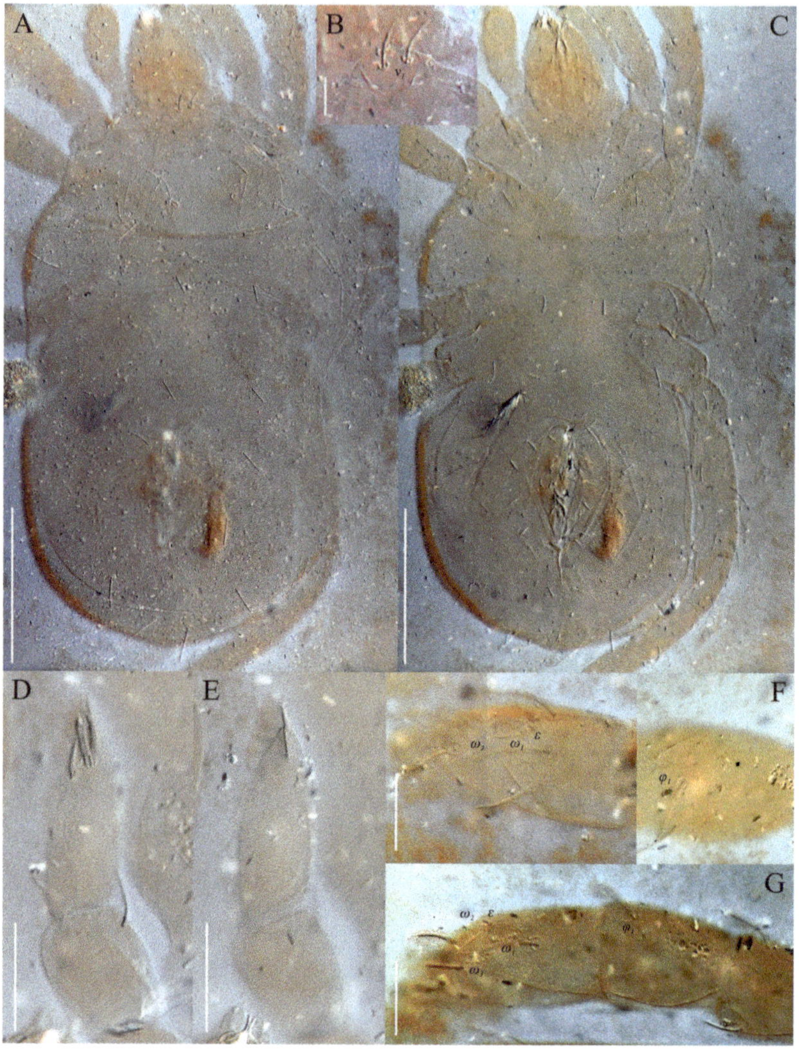

Figure 20. *Antarcteupodes maudae* (Strandtmann, 1967), holotype female. (**A**) Body, dorsal view; (**B**) naso; (**C**) body, ventral view; (**D**) tarsus and tibia of right palp, dorsal view; (**E**) tarsus and tibia of right palp, ventral view; (**F**) tarsus and tibia of right leg I, dorsolateral view; (**G**) tarsus and tibia of right leg II, dorsolateral view. Scale bar: (**A**,**C**) 100 μm; (**B**) 10 μm; (**D**–**G**) 20 μm.

***Filieupodes lapidarius* (Oudemans, 1906)** comb. nov. (Figures 21–26)

Ereunetes lapidarius [34]
Ereynetes lapidarius [35,36]
Micrereunetes (Protereunetes) lapidarius [6]
Protereunetes lapidarius [7,8,11]
Neoprotereunetes lapidarius [16,22]
Filieupodes filistellatus Jesionowska, 2010 syn. nov.

Diagnosis. Naso well delimited. Dorsal hysterosomal setae short. Tarsus I with two parallel rhagidial organs in separate depressions, of which proximal one posterolaterad of distal one. Stellate famulus well removed proximo-laterally from proximal rhagidial organ. Tarsus II with three parallel rhagidial organs in separate depressions, of which medial one posterolaterad of proximal and distal ones. Spiniform famulus well removed laterally from proximal rhagidial organ.

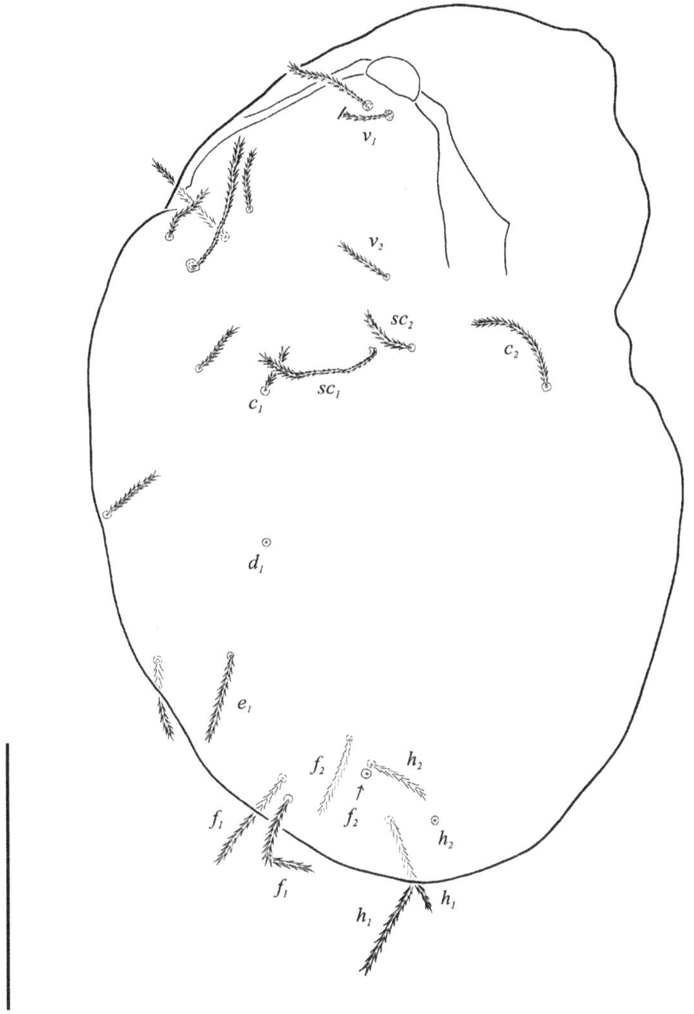

Figure 21. *Filieupodes lapidarius* (Oudemans, 1906), holotype female. Idiosoma, dorsolateral view. Scale bar: 100 μm.

Redescription. Holotype female. Idiosoma 330 long, 200 wide.

Idiosomal dorsum (Figures 21 and 26A). Prodorsal shield (Figure 26B) 74 long, 100 wide, triangular. Prodorsal integument with weakly striate-spiculate ornamentation, but course of striae hard to retrace. Naso 15 long, 28 wide, rounded. Lengths of prodorsal setae: v_1 34, v_2 22, sc_1 ca. 50, sc_2 25. Hysterosoma oval. Hysterosomal integument with striate-spiculate ornamentation. Lengths of hysterosomal setae: c_1 24, c_2 40, d_1 27, e_1 30, f_1 40, f_2 33, h_1 41, h_2 27.

Idiosomal venter (Figure 22). Coxisternal fields poorly outlined, with weakly striate-spiculate ornamentation, separated medially by striate-spiculate ornamentation of longitudinal course. Lengths of coxisternal setae: *1a* 18, *1b* 20, *1c* 13, *2b* 26, *3a* 18, *3b* 18, *3c* 20, *3d* 18, *4a* 13, *4b* 16, *4c* 15. Genital region (Figures 23A and 26C) with four pairs of aggenital setae, *ag* 9 long, *g* 10 long and six pairs of genital setae, all ca. 10 long. Two pairs of genital papillae and five pairs of eugenital setae, 8 long, on protuberances. Lengths of pseudanal setae: ps_1 32, ps_3 15. Lyrifissures *ih* not visible.

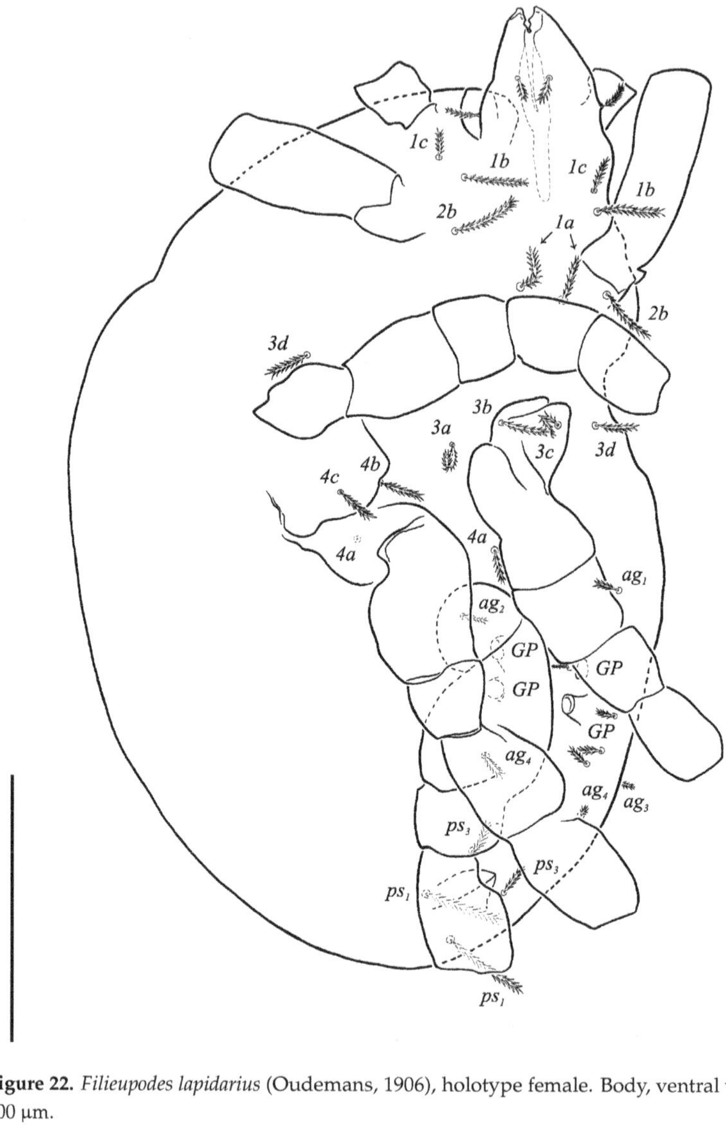

Figure 22. *Filieupodes lapidarius* (Oudemans, 1906), holotype female. Body, ventral view. Scale bar: 100 μm.

Gnathosoma (Figures 23B–D and 26D,E). Subcapitulum (Figure 23B) 52 long, 46 wide roughly triangular, with striate-spiculate ornamentation. Subcapitular apodema visible under integument. Setae sbc_2, 9 long, subequal to sbc_1, 12 long, both pilose. Chelicerae (Figure 23C) 60 long, with spiculate ornamentation, bearing pilose dorsal seta *cha*; fixed digit with blunt tip; movable digit sharp, clawlike. Palps (Figures 23D and 26D,E) with spiculate ornamentation. Palpal femorogenu 29 long. Palpal tibia 33 long, setae l'' and l' subequal in length. Palpal tarsus 20 long, ellipsoid in dorsoventral aspect. All setae except *acm* smooth; rhagidial organ *ω* small, protruding on both tarsi in dorsoventral view.

Subcapitular, cheliceral and palpal integument with spiculate ornamentation (spiculate-cuspidate on palpal femorogenu).

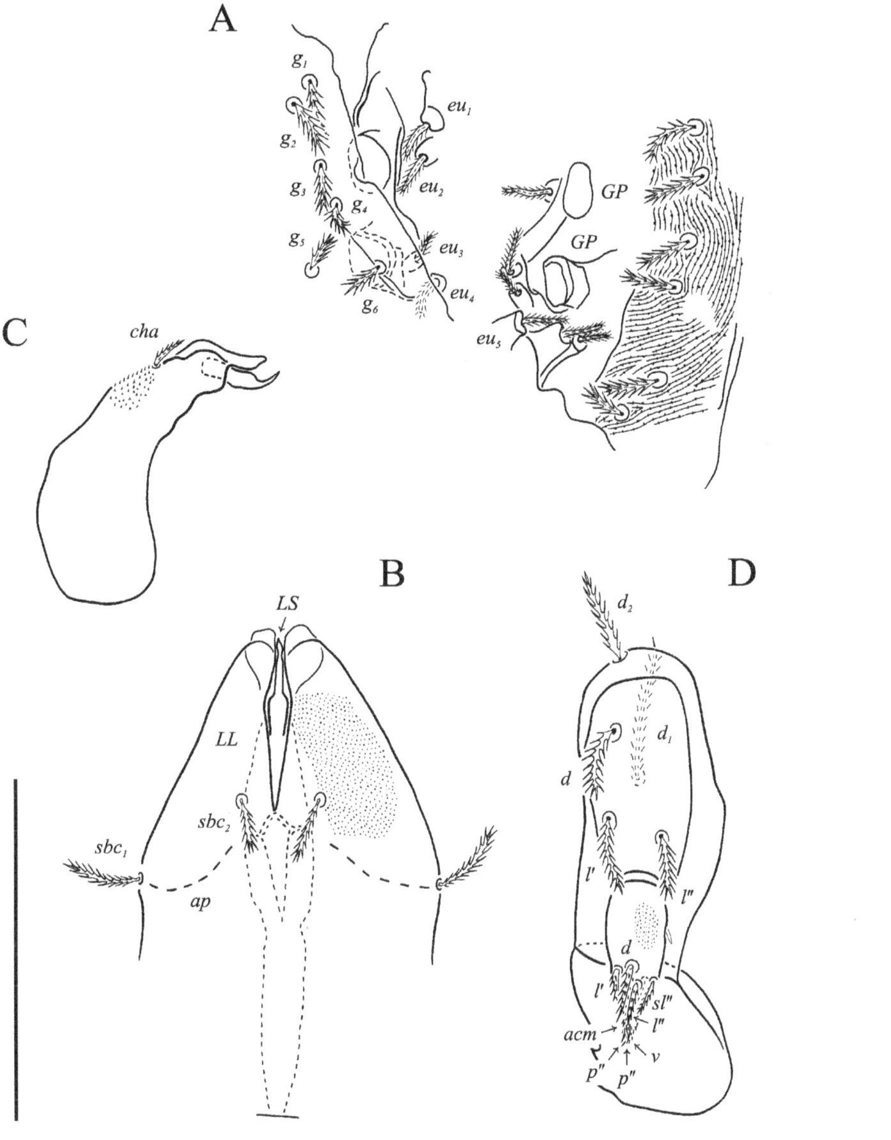

Figure 23. *Filieupodes lapidarius* (Oudemans, 1906), holotype female. (**A**) Genital region; (**B**) subcapitulum, ventral view; (**C**) left chelicera, lateral view; (**D**) left palp, dorsal view. Scale bar: 50 μm.

Legs (Figures 24, 25 and 26F,G). Lengths of legs: I 315, II 198, III 221, IV 255. Lengths of leg segments: I: Ts 77, Tb 64, G 51, F 112, Tr 27; II: Ts 58, Tb 37, G 24, F 74, Tr 20; III: Ts 62, Tb 37, G 32, TF 29, BF 41, Tr 20; IV: Ts 72, Tb 46, G 51, TF 23, BF 64, Tr 23. Integument with spiculate ornamentation. Leg setal formulae: I: 1–6+5–8–13(2φ)–21(2ω, ε); II: 1–5+5–4–5(2φ)–12(3ω, ε); III: 1–4–4–4–5–12; IV: 1–3–3–4–5–12. Eupathidial setae: I: G: (*l*); Tb: all except (l_{1-2}); Ts: all; II: Tb: *d*, *v'*; Ts: all except *ft'*; III: G: *l'*; Tb: *d*, *l'*; Ts: all; IV: BF *d*; G: *l"*; Tb: *d*, *l'*; Ts: all.

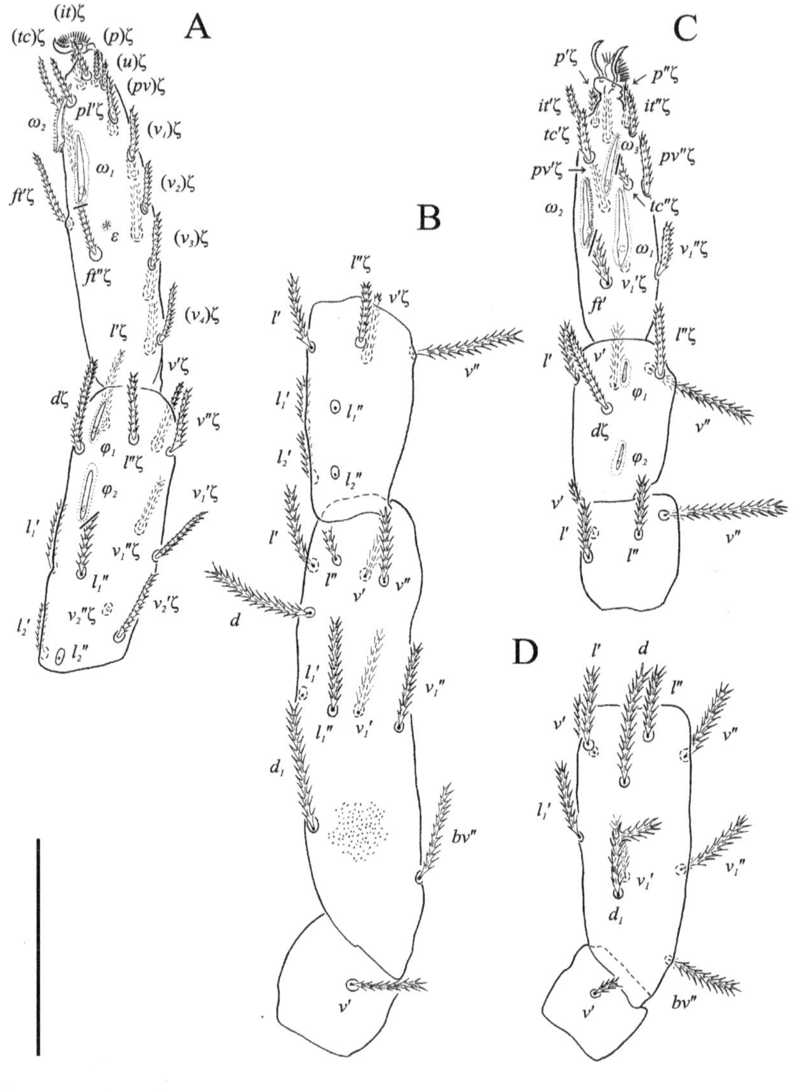

Figure 24. *Filieupodes lapidarius* (Oudemans, 1906), holotype female. (**A**) Tarsus and tibia of right leg I, dorsolateral view; (**B**) genu, femur and trochanter of right leg I, dorsolateral view; (**C**) tarsus, tibia and genu of right leg II, dorsal view; (**D**) femur and trochanter of right leg II, dorsal view. Scale bar: 50 μm.

Solenidia and famuli. Leg I. Tarsus with two parallel T-shaped rhagidial organs in confluent depression and one stellate famulus ε well moved antiaxially to the lateral side. Posterior, dorsolateral rhagidial organ (ω_1) reaching half of the length of anterior, dorsal one (ω_2). Tibia with one distal (φ_1) and one medial rhagidial organ φ_2, both T-shaped, tandemly in separated depressions. Leg II. Tarsus with three parallel T-shaped rhagidial organs in separated depressions, of which the smallest anterior one (ω_3) oblique antiaxially and flanked by two bigger posterior ones (ω_1 and ω_2). Spiniform famulus ε not visible. Tibia with two T-shaped rhagidial organs (anterior φ_1 and medial φ_2), in separated depressions. Leg III and IV without solenidia.

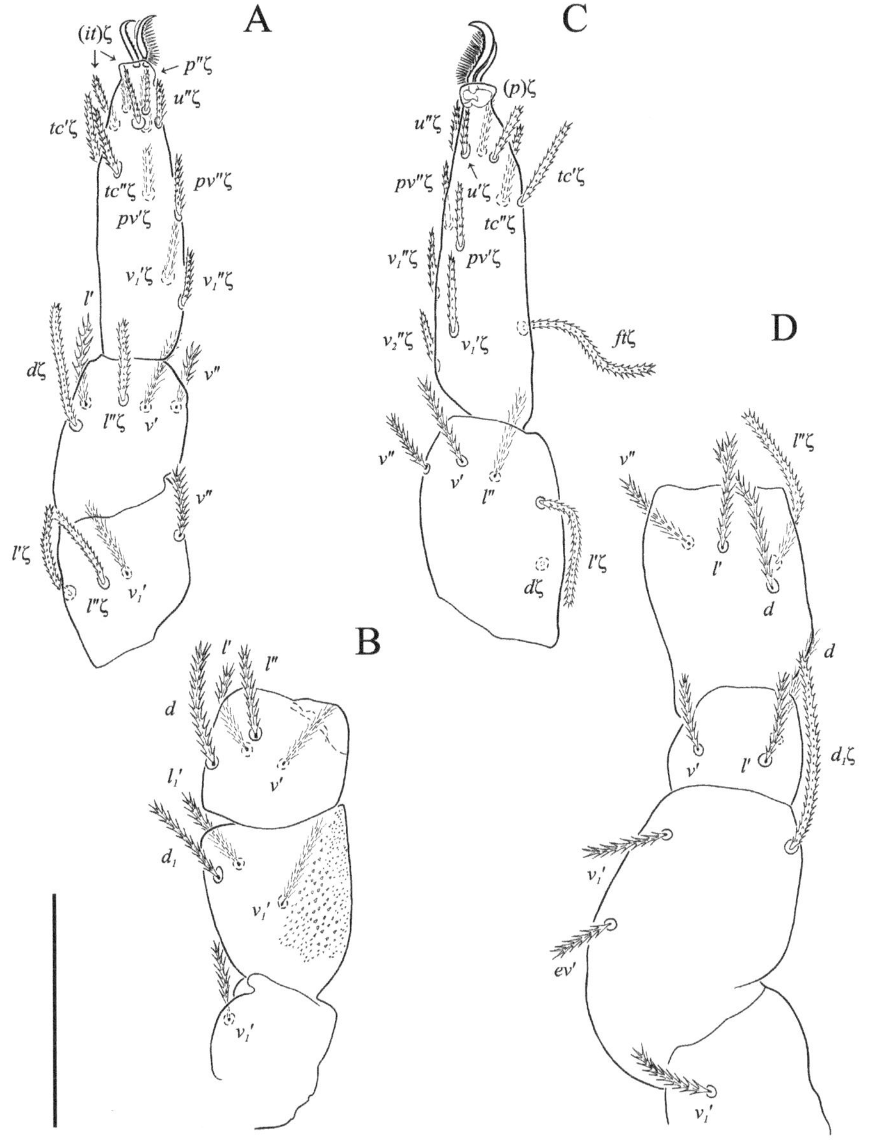

Figure 25. *Filieupodes lapidarius* (Oudemans, 1906), holotype female. (**A**) Tarsus, tibia and genu of right leg III, lateral view; (**B**) femur and trochanter of right leg III, lateral view; (**C**) tarsus and tibia of left leg IV, lateral view; (**D**) genu, femur and trochanter of left leg IV, lateral view. Scale bar: 50 μm.

Differential diagnosis. *F. lapidarius* is similar to *F. shepardi* Strandtmann, 1971 because of naso delimited dorsally and the same number of aggenital and genital setae. It differs from *F. shepardi* in short dorsal hysterosomal setae (long in *F. shepardi*) and parallel arrangement of rhagidial organs on tarsi I and II (tandem in *F. shepardi*).

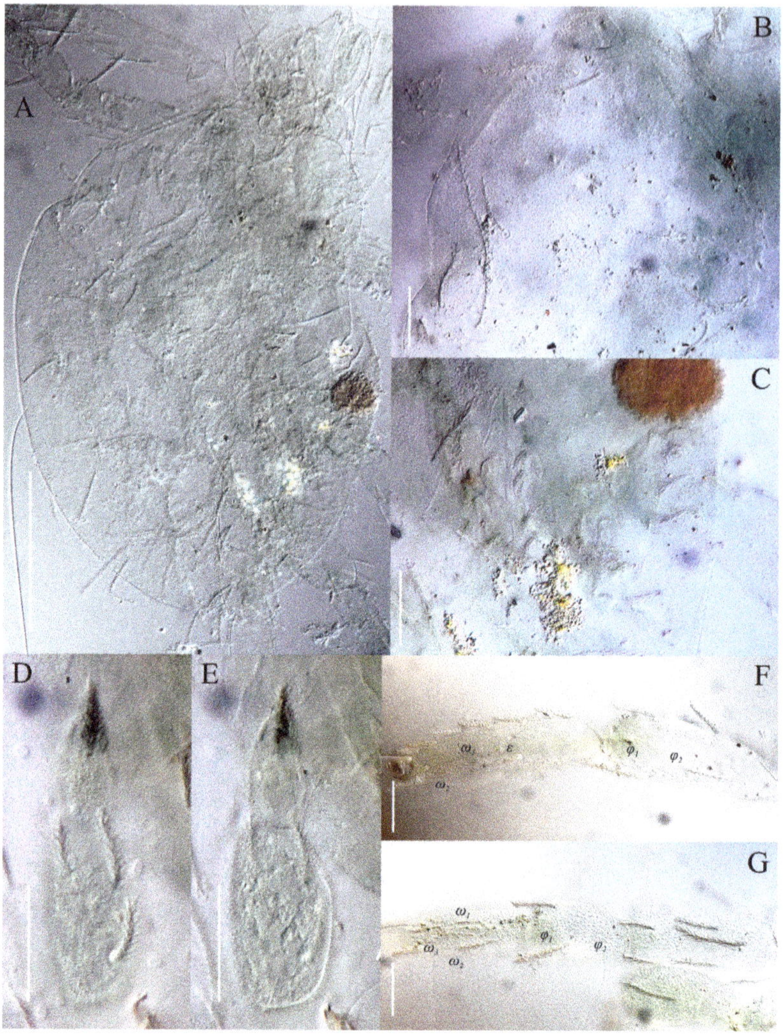

Figure 26. *Filieupodes lapidarius* (Oudemans, 1906), holotype female. (**A**) Body, dorsal view; (**B**) prodorsum; (**C**) genital region; (**D**) tarsus and tibia of right palp, dorsal view; (**E**) tarsus and tibia of right palp, ventral view; (**F**) tarsus and tibia of right leg I, lateral view; (**G**) tarsus, tibia and genu of right leg II, dorsal view. Scale bar: (**A**) 100 μm; (**B–G**) 20 μm.

Distribution. Arnhem, Netherlands [34]; nature reserve of halophitic vegetation, Ciechocinek near Toruń, Kujawsko-Pomorskie District; "Zielona Góra" nature reserve, vicinity of Częstochowa, Śląskie District (both [4]); Morasko Campus, Poznań, Wielkopolskie District [37], all latter localities in Poland.

Material examined. Holotype female (Naturalis Biodiversity Centre, slide labeled "RMNH.ACA.P 5507"): Netherlands, Arnhem, under stones, 1903, leg. Dammermann; four females and two males: Poland, Wielkopolskie district, Poznań, Morasko University Campus, 52°27′58″ N 16°55′21″ E, Fresh meadow with often reaped *Arrhenatheretum elatioris*, 19 February 2019, leg. R. Laniecki.

Remarks. The original description [34], as well as subsequent redescriptions [35,36], lacks some valid diagnostic characters and thus the species is redescribed herewith. Despite high similarity of the holotype of *Filieupodes lapidarius* and specimens previously identified as *F. filistellatus* collected in Poland, some differences can be observed. In the original description of *F. filistellatus* proximal rhagidial organs are present on all tibiae, whereas they were not found on tibiae III and IV in holotype of *F. lapidarius*. This, however, could be a result of age and condition of the original material (117 years). Because of this, some structures (e.g., lyrifissures, supracoxal setae) are not visible. Moreover, due to the position of the specimen on the slide, some structures could not be entirely distinguished, e.g., some coxisternal and aggenital setae. Those, however, which are visible, fit the setal patterns of *F. filistellatus*.

The type material of *F. filistellatus* is lost (courtesy of Prof. Andrzej J. Zawal, former superior of Dr. Katarzyna Jesionowska), and thus only newly collected material along with the original description by Jesionowska [4] were used to compare the species with holotype of *N. lapidarius*.

3.2. Key to the Species of Neoprotereunetes (Adults)

1. Five pairs of genital setae, tarsus I with 21 setae, femur I with 13 setae, tibia II with five setae, genu III and IV each with four setae, arctic species .... *boerneri* species group .... *boerneri* (Thor, 1934)

– Five or six pairs of genital setae, tarsus I with 20 setae, femur I with 12 setae, tibia II with four setae, genu III and IV each with three setae, Antarctic or sub-Antarctic species .... *minutus* species group 2

2. Tibial proximal rhagidial organs long, at most four times shorter than its segment, famulus ε on tarsus II absent 3

– Tibial proximal rhagidial organs short, at least seven times shorter than its segment, famulus ε on tarsus II present 4

3. Tibia III with proximal rhagidial organs, two anterior rhagidial organs ($\omega_{2,\ 3}$) on tarsus II parallel and arranged side by side, both tarsal rhagidial organs in confluent depressions..*minutus* (Strandtmann, 1967)

– Tibia III without proximal rhagidial organ, two anterior rhagidial organs ($\omega_{2,\ 3}$) on tarsus II parallel, but ω_3 displaced anteriorly in relation to ω_2, both tarsal rhagidial organs in separated depressions .... *crozeti* (Strandtmann et Davies, 1972)

4. Tarsus II with three rhagidial organs of unequal size, tibia III with short ellipsoid rhagidial organ, trochanter IV with one seta .... *exiguus* (Booth, Edward et Usher, 1985)

– Tarsus II with three rhagidial organs of equal size, tibia III with small spherical rhagidial organ, trochanter IV without setae .... *paulinae* (Gless, 1972)

– Tarsus II with two rhagidial organs, tibia III without rhagidial organ, trochanter IV without setae .... *parvus* (Booth, Edward et Usher, 1985)

4. Discussion

The family Eupodidae is composed mostly of monotypic genera, e.g., *Claveupodes*, *Caleupodes*, *Aethosolenia*. Two non-monotypic genera, i.e., *Pseudopenthaleus* and *Echinoeupodes*, have two species each, but in both cases, only one of them is accurately described. The remaining two non-monotypic genera, i.e., *Eupodes* and *Benoinyssus* are highly heterogenous. It is, therefore, hard to establish diagnostic characters at the generic level. We decided not to base generic diagnoses on leg setal patterns until more data on intra-generic variability in this respect will be collected. Body dimensions and shape are also excluded from diagnoses as these characters are contingent on age and condition of an individual as well as specimen treatment and preparation technique and may even change with an age of the slide.

In the present study, six species were classified within the genus *Neoprotereunetes*. Though the Arctic species differ slightly from Antarctic and sub-Antarctic congeners (mostly in genital and leg chaetotaxy), we decided not to divide them into separate genera or subgenera until the intrageneric variability in eupodid genera is better understood. However, to express these differences, two species groups were proposed: one, *boerneri*, containing *N. boerneri*, and another one, *minutus*, containing *N. crozeti*, *N. exiguus*, *N. minutus*,

N. parvus and *N. paulinae*, based on type or newly collected material as well as original descriptions and redescriptions. Additionally, seven species that were described in or transferred to *Protereunetes* are not included in *Neoprotereunetes*. The first one was originally described by Oudemans [34] as *Ereunetes lapidarius* (Ereynetidae). In [35], the description of this species, with the (then) corrected generic name (*Ereynetes*), was extended and supplied with pictures by the same author. Next Oudemans [36] moved *E. lapidarius* to the family Eupodidae, without reference to its generic rank. Subsequently, Thor [7] transferred *E. lapidarius* to the genus *Protereunetes*. Finally, this species was designated as a type species of *Neoprotereunetes* Fain et Camerik, 1994. However, after the present examination of the holotype, it turned out that *Neoprotereunetes lapidarius* is the senior synonym of *Filieupodes filistellatus* Jesionowska, 2010 (Cocceupodidae). The second one, *P. maudae*, described as a congener of *N. minutus* by Strandtmann [29], is designated as a type species of the new genus *Antarcteupodes* on the basis of its unique combination of character states, not present in any hitherto described eupodid genus, including the most reduced coxisternal and leg chaetotaxy among the family Eupodidae. The third one, *Protereunetes turgidus* Shiba, 1978, was transferred by Khaustov [26] to the genus *Echinoeupodes* Khaustov, 2017. The fourth, *Protereunetes villosus* Shiba, 1978, possesses long and slender setae f_1 (presumably trichobothrial) and characteristic solenidiotaxy of tarsi I and II (each with two rhagidial organs, of which distal one is much smaller than proximal one), suggesting its affiliation to *Benoinyssus* Fain, 1958. The fifth species, *Protereunetes perforatus* Shiba, 1978, resembles *Caleupodes reticulatus* Baker, 1987 with respect to its reticulated body ornamentation and almost smooth setae. These characters are extremely rare in the family Eupodidae and might suggest a close relationship between these two species. Even if so, *P. perforatus* slightly differs from *C. reticulatus* in the solenidiotaxy of tarsus II (three rhagidial organs, instead of two) and the tibiae (one rhagidial organ, instead of two), and also in terms of its much larger body. The last species, *Protereunetes striatellus* (C.L. Koch, 1838) was originally described in *Eupodes* and then transferred by Thor and Willmann [8] to *Protereunetes*. As the description of this species is insufficient to determine its generic affiliation, and the type material probably does not exist, it is considered a *species inquirenda*. To confirm the above proposals, the type material should be examined, if (or when) available.

In reply to the transfer of *Protereuntes* (junior synonym of *Ereynetes*) back to Ereynetidae by Fain [9], Strandtmann [10] moved one of his species (*P. minutus*) to the genus *Eupodes* without reference to the second one (*P. maudae*). Although in subsequent papers Strandtmann was still using the name *Protereunetes* in relation to eupodoid mites, the usage of *Eupodes* sensu [10] was widely accepted by other authors [13–15]. However, in our opinion, the six species assigned herein to *Neoprotereunetes* possess a set of characters sufficient to constitute a separate genus. They have short and plumose dorsal body setae (long and lightly plumose in *Eupodes*); normal setae f_1 (sometimes trichobothrial in *Eupodes*); all legs shorter than body (legs I and II longer than body in *Eupodes*); femur IV slender (usually swollen in *Eupodes*) short and plumose leg setae (long and pilose in *Eupodes*); two or three L-shaped or T-shaped rhagidial organs on tarsi I and II (always two L-shaped rhagidial organs in *Eupodes*); two rhagidial organs on tibiae I and II (one rhagidial organ and one erect solenidion in *Eupodes*). *Eupodes* is still a highly heterogeneous taxon demanding a major revision. Nevertheless, the abovementioned characters enable the separation of *Neoprotereunetes* from *Eupodes*.

According to the Principle of Priority [38], the synonymy of original type species of *Neoprotereunetes*, namely *Ereunetes lapidarius* Oudemans, 1906 with *Filieupodes filistellatus* Jesionowska, 2010 implies that *Neoprotereunetes* is the valid genus-level name and should replace *Filieupodes* as its senior synonym.

This, however, does not resolve the problem of the lack of a replacement for the genus-level name *Protereunetes*—the primary aim of creating *Neoprotereunetes* by Fain and Camerik [16]. As the descriptions, redescriptions and original figures of *E. lapidarius* [33–35] do not imply that this species belongs to the family Cocceupodidae, only the present examination of the type could demonstrate that. As the type species fixation of the genus

Neoprotereunetes was based on a misidentification (even at the time of its inception it did not meet its own diagnosis) for the sake of nomenclatural stability, we suggest retaining the name *Filieupodes* for the genus in the family Cocceupodidae (in line with the original proposal by Jesionowska [4]) and *Neoprotereunetes* for the genus in the family Eupodidae (as used by Khaustov [17]).

Thus, because the designation of new type species for *Neoprotereunetes* becomes necessary, we propose establishing *Protereunetes boerneri* Thor, 1934 (the oldest known species after *E. lapidarius* listed by Thor and Willamnn [8]) as the type species of the newly diagnosed genus.

Such practices are justified and encouraged by ICZN [38], as expressed in its initial chapter "Introduction. Development of underlying principles" (p. 14) by the following statement: "Also when individual zoologists discover that the type species had been misidentified when a genus or subgenus was established, they are given the power to fix as the type species either the species actually nominated by the original author or the nominal species in conformity with the name in use".

The representatives of *Neoprotereunetes* thus far have been found only in the high latitudes of either hemisphere. The *boerneri* species group is restricted to the Arctic (Svalbard, Severnaya Zemlya, Arctic Alaska) and sub-Arctic (sub-Arctic Alaska) locations, whereas the *minutus* species group is restricted to the Antarctic (e.g., Antarctic Peninsula, South Orkney Islands, South Shetland Islands) and sub-Antarctic (Crozet Islands, Prince Edward Islands) locations. Additionally, *N. minutus* has also been recorded in Dunedin, New Zealand. Apart from the latter, all the locations are characterized by harsh climate and low yearly temperatures that seem to be favorable to eupodoid mites. Among terrestrial Prostigmata, Eupodoidea dominate in the Antarctic (36 species described) are one of the dominating groups in the Arctic.

5. Conclusions

Establishing *Neoprotereunetes* as a replacement for *Protereunetes* constitutes an important step in dividing the large and highly heterogeneous eupodid genus *Eupodes* and contributes to increasing the stability within Eupodidae. Even though *Neoprotereunetes* displays no unique characters specific only to this genus, it can be easily defined by a combination of characters. This might be one of the reasons that it remained so poorly defined for such a long time.

Author Contributions: Conceptualization, both authors; visualization, R.L. (line drawings), W.L.M. (light microscopy imaging); methodology, both authors; writing: original draft, all authors; writing: review and editing, all authors. All authors have read and agreed to the published version of the manuscript.

Funding: This research received no external funding.

Institutional Review Board Statement: Not applicable.

Informed Consent Statement: Not applicable.

Data Availability Statement: The data presented in this study are available on request from the corresponding author.

Acknowledgments: We wish to express our thanks to staff of the Bishop Museum in Honolulu (Hawaii) for loan of the type material of *N. minutus*, *N. paulinae* and *Antarcteupodes maudae* and to staff of the Naturalis Biodiversity Center in Leiden (the Netherlands) for loan of the type material of *Neoprotereunetes lapidarius*. We are also grateful to Dariusz J. Gwiazdowicz from Poznań University of Life Sciences for providing the samples from King George Island. The samples from Spitsbergen were obtained during the summer school for students of Faculty of Biology at Adam Mickiewicz University, financed by the Dean of Faculty, Doctoral School of Environmental Sciences and Rector of AMU.

Conflicts of Interest: The authors declare no conflict of interest.

References

1. Krantz, G.W.; Walter, D.E. *A Manual of Acarology*, 3rd ed.; Texas Tech University Press: Lubbock, TX, USA, 2009; p. 807.
2. Szudarek-Trepto, N.; Kaźmierski, A.; Dabert, M.; Dabert, J. Molecular phylogeny of Eupodidae reveals that the family Cocceupodidae (Actinotrichida; Eupodoidea) and its genus *Filieupodes* are valid taxa. *Exp. Appl. Acarol.* **2020**, *80*, 43–57. [CrossRef] [PubMed]
3. Baker, A.S.; Lindquist, E.E. *Aethosolenia laselvensis* gen. n., sp. n., a new eupodoid mite from Costa Rica (Acari: Prostigmata). *Syst. Appl. Acarol. Spec. Publ.* **2002**, *11*, 1–11. Available online: https://www.nhm.ac.uk/hosted-sites/acarology/saas/saasp/2002/saasp11.pdf (accessed on 1 November 2022). [CrossRef]
4. Jesionowska, K. Cocceupodidae, a new family of eupodoid mites, with description of a new genus and two new species from Poland. Part I. (Acari: Prostigmata: Eupodoidea). *Genus* **2010**, *21*, 637–658.
5. Gless, E.E. Life cycle studies of some Antarctic mites and description of a new species, *Protereunetes paulinae* (Acari: Eupodidae). *Antarct. Res. Ser. Wash.* **1972**, *20*, 289–306.
6. Berlese, A. Centuria sesta di Acari nuovi. I. Prostigmata. *Redia* **1923**, *15*, 237–262.
7. Thor, S. Neue Beiträge zur Kenntnis der invertebraten Fauna von Svalbard. *Zool. Anz.* **1934**, *107*, 201–215.
8. Thor, S.; Willmann, C. Eupodidae, Penthalodidae, Penthaleidae, Rhagidiidae, Pachygnathidae, Cunaxidae. *Das Tierreich* **1941**, *71a*, 1–186.
9. Fain, A. Les Ereynetidae de la Collection Berlese à Florence. Désignation d'une espèce type pour le genre *Ereynetes* Berlese. *Redia* **1964**, *49*, 87–111.
10. Strandtmann, R.W. Acarina: Eupodiform Prostigmata of South Georgia. *Pacif. Insects Monogr.* **1970**, *23*, 89–106.
11. Strandtmann, R.W. The eupodoid mites of Alaska. *Pacif. Insects* **1971**, *13*, 75–118.
12. Strandtmann, R.W.; Davies, L. Eupodiform mites from Possession Island, Crozet Islands, with a key to the species of *Eupodes* (Acarina: Prostigmata). *Pacif. Insects* **1972**, *14*, 39–56.
13. Goddard, D.G. Biological observations on the free-living mites of Signy Island in the maritime Antarctic. *Bull. Br. Antarct. Surv.* **1980**, *49*, 181–205.
14. Booth, R.G.; Edwards, M.; Usher, M.B. Mites of the genus *Eupodes* (Acari, Prostigmata) from maritime Antarctica: A biometrical and taxonomic study. *J. Zool.* **1985**, *207*, 381–406. [CrossRef]
15. Baker, A.S. *Caleupodes*, a new genus of eupodoid mite (Acari: Acariformes) showing primary opisthosomal segmentation. *Bull. Br. Mus. Nat. Hist.* **1987**, *53*, 103–113. [CrossRef]
16. Fain, A.; Camerik, A.M. Notes on the mites of the genus Ereynetes Berlese (Acari: Ereynetinae), with description of five new species from South Africa. *Bull. Inst. R. Sci. Nat. Belg. Entomol.* **1994**, *64*, 145–164.
17. Khaustov, A.A. A new genus and species in the mite family Eupodidae (Acari, Eupodoidea) from Crimea. *ZooKeys* **2014**, *422*, 11–22. [CrossRef] [PubMed]
18. Cypionka, H. PICOLAY 2022, Version 2022-07-15. Available online: http://www.picolay.icbm.de/index.html (accessed on 1 November 2022).
19. Norton, R.A. A review of F. Grandjean's system of leg chaetotaxy in the Oribatei (Acari) and its application to the family Damaeidae. In *Biology of Oribatid Mites*; Dindal, D.L., Ed.; SUNY College of Environmental Science and Forestry: Syracuse, NY, USA, 1977; pp. 33–61.
20. Lindquist, E.E.; Zacharda, M. A new genus and species of Rhagidiidae (Acari: Prostigmata) from Chihuahuan Desert litter in New Mexico. *Can. J. Zool.* **1987**, *65*, 2149–2158. [CrossRef]
21. Baker, A.S. A redescription of *Halotydeus destructor* (Tucker) (Prostigmata: Penthaleidae) with a survey of ontogenetic setal development in the superfamily Eupodoidea. *Int. J. Acarol.* **1995**, *21*, 261–282. [CrossRef]
22. Beron, P. Acarorum catalogus X. Trombidiformes, Prostigmata, Superfamilia Labidostommatoidea (Labidostommatidae), Superfamilia Eupodoidea, (Eupodidae, Dendrochaetidae, Rhagidiidae, Eriorhynchidae, Pentapalpidae, Penthalodidae, Penthaleidae, Proterorhagiidae, Strandtmanniidae), Superfamilia Tydeoidea, Ereynetidae, Superfamily Paratydeoidea, Paratydeidae, Superfamilia Anystoidea, (Anystidae, Erythracaridae, Teneriffiidae, Pseudocheylidae, Stigmocheylidae), Superfamilia Caeculoidea (Caeculidae), Superfamilia Adamystoidea (Adamystidae), Superfamilia Pomerantzioidea (Pomerantziidae). In *Advanced Books*; Sophia: Moscow, Russia, 2022; p. 424. [CrossRef]
23. Makarova, O.L. Acarocenoses (Acariformes, Parasitiformes) in Polar Deserts: 2. Cenotic Relations, Structure of Communities, and the Proportion of Suborders. *Entomol. Rev.* **2002**, *82*, 857–875.
24. Seniczak, A.; Seniczak, S.; Schwarzfeld, M.D.; Coulson, S.J.; Gwiazdowicz, D.J. Diversity and distribution of mites (Acari: Ixodida, Mesostigmata, Trombidiformes, Sarcoptiformes) in the Svalbard Archipelago. *Diversity* **2020**, *12*, 323. [CrossRef]
25. Winston, J.E. *Describing Species: A Practical Taxonomic Procedure for Biologists*; Cambridge University Press: Cambridge, UK, 1999; p. 518.
26. Khaustov, A.A. A new genus and species of Eupodidae (Acari: Eupodoidea) from mosses in Crimea. *Acarina* **2017**, *25*, 29–44. [CrossRef]
27. Pugh, P.J.A. A synonymic catalogue of the Acari from Antarctica, the sub-Antarctic Islands and the Southern Ocean. *J. Nat. Hist.* **1993**, *27*, 323–421. [CrossRef]

28. Russell, D.J.; Hohberg, K.; Potapov, M.; Brückner, A.; Otte, V.; Christian, A. Native terrestrial invertebrate fauna from the northern Antarctic Peninsula: New records, state of current knowledge and ecological preferences. Summary of a German federal study. *Soil Org.* **2014**, *86*, 1–58.
29. Strandtmann, R.W. Terrestrial Prostigmata (*Trombidiform mites*). *Antarct. Res. Ser. Wash.* **1967**, *10*, 51–80.
30. Luxton, M. The marine littoral mites of the New Zealand region. *J. R. Soc. N. Z.* **1990**, *20*, 367–418. [CrossRef]
31. Mercer, R.D.; Chown, S.L.; Marshall, D.J. Mite and insect zonation on a Marion Island rocky shore: A quantitative approach. *Polar Biol.* **2000**, *23*, 775–784. [CrossRef]
32. Mortimer, E.; Jansen van Vuuren, B. Phylogeography of *Eupodes minutus* (Acari: Prostigmata) on sub-Antarctic Marion Island reflects the impact of historical events. *Polar Biol.* **2007**, *30*, 471–476. [CrossRef]
33. Shiba, M. On some eupodiform mites from Japan (Acarina: Prostigmata). *Rep. Res. Matsuyama Shinonome Jun. Coll.* **1978**, *19*, 133–152.
34. Oudemans, A.C. Acarologische Aanteekeningen XXI. *Entomol. Ber.* **1906**, *27*, 37–43.
35. Oudemans, A.C. Notizen uber Acari. *Arch. Naturgesch.* **1915**, *5*, 1–78.
36. Oudemans, A.C. Acarologische Aanteekeningen XCIV. *Entomol. Ber.* **1928**, *164*, 374–382.
37. Laniecki, R.; Kaźmierski, A.; Mąkol, J.; Laniecka, I.; Magowski, W. Know your campus: Salient research potential of prostigmatic soil mite fauna (Acariformes: Prostigmata, Endeostigmata) within university campus area. *Acarologia* **2021**, *61*, 650–663. [CrossRef]
38. ICZN. *International Code of Zoological Nomenclature*, 4th ed.; International Trust for Zoological Nomenclature: London, UK, 1999; p. 306.

animals MDPI

Article

Molecular Identification and Phylogenetic Analysis of Laelapidae Mites (Acari: Mesostigmata)

Evelina Kaminskienė [1], Jana Radzijevskaja [1], Loreta Griciuvienė [1], Michal Stanko [2], Justina Snegiriovaitė [1], Dalytė Mardosaitė-Busaitienė [1] and Algimantas Paulauskas [1,*]

[1] Faculty of Natural Sciences, Vytautas Magnus University, Donelaičio Str. 58, LT-44248 Kaunas, Lithuania; evelina.kaminskiene@vdu.lt (E.K.); jana.radzijevskaja@vdu.lt (J.R.); loreta.griciuviene@vdu.lt (L.G.); justina.snegiriovaite@vdu.lt (J.S.); dalyte.mardosaite-busaitiene@vdu.lt (D.M.-B.)

[2] Department of Vector-Borne Diseases, Institute of Parasitology, Slovak Academy of Sciences, Hlinkova 3, 04001 Košice, Slovakia; stankom@saske.sk

* Correspondence: algimantas.paulauskas@vdu.lt; Tel.: +421-370-6146-1805

Simple Summary: Mites from the family Laelapidae are frequently associated with small mammals, mainly rodents, and can be found on their body surface or in their nests. Classification of the Laelapidae is complicated because of high levels of their morphological and ecological variability. This study aimed to undertake molecular characterization and to assess the phylogenetic relationship among eight Laelapidae mite species collected from different rodent hosts in Lithuania, Norway, Slovakia, and the Czech Republic using the nuclear and mitochondrial molecular markers. Our study provides new molecular data on *Laelaps agilis*, *Laelaps hilaris*, *Laelaps jettmari*, *Haemogamasus nidi*, *Eulaelaps stabularis*, *Hyperlaelaps microti*, *Myonyssus gigas*, and *Hirstionyssus* sp. mites collected from seven different rodent hosts and three geographical regions in Europe. This study, for the first time, registered sequences of four mite species: *H. microti*, *Hirstionyssus* sp., *M. gigas*, and *E. stabularis*.

Abstract: The family Laelapidae (Dermanyssoidea) is morphologically and ecologically the most diverse group of Mesostigmata mites. Although molecular genetic data are widely used in taxonomic identification and phylogenetic analysis, most classifications in Mesostigmata mites are based solely on morphological characteristics. In the present study, eight species of mites from the Laelapidae (Dermanyssoidea) family collected from different species of small rodents in Lithuania, Norway, Slovakia, and the Czech Republic were molecularly characterized using the nuclear (28S ribosomal RNA) and mitochondrial (cytochrome oxidase subunit I gene) markers. Obtained molecular data from 113 specimens of mites were used to discriminate between species and investigate the phylogenetic relationships and genetic diversity among Laelapidae mites from six genera. This study provides new molecular data on *Laelaps agilis*, *Laelaps hilaris*, *Laelaps jettmari*, *Haemogamasus nidi*, *Eulaelaps stabularis*, *Hyperlaelaps microti*, *Myonyssus gigas*, and *Hirstionyssus* sp. mites collected from different rodent hosts and geographical regions in Europe.

Keywords: Laelapidae mites; 28S ribosomal RNA; cytochrome oxidase subunit I gene; phylogenetic analysis; genetic diversity

Citation: Kaminskienė, E.; Radzijevskaja, J.; Griciuvienė, L.; Stanko, M.; Snegiriovaitė, J.; Mardosaitė-Busaitienė, D.; Paulauskas, A. Molecular Identification and Phylogenetic Analysis of Laelapidae Mites (Acari: Mesostigmata). *Animals* **2023**, *13*, 2185. https://doi.org/10.3390/ani13132185

Academic Editors: Monika Fajfer and Maciej Skoracki

Received: 14 June 2023
Revised: 30 June 2023
Accepted: 1 July 2023
Published: 3 July 2023

1. Introduction

Mesostigmata mites represent the most taxon-rich group of Parasitiformes and comprise approximately 11,000 described species [1]. Numerous species of mesostigmatic mites can occasionally infest humans and cause dermatitis and severe allergic reactions. These mites can be potential vectors of the human pathogenic tick-borne encephalitis virus (TBEV) [2] and various rickettsial agents [3–6]. The superfamily Dermanyssoidea is the largest subdivision of mesostigmatid mites. It consists of 15 families [7], including Laelapidae, which is morphologically and ecologically the most diverse group of

Mesostigmata mites [8,9]. Laelapidae currently includes 92 known genera with more than 1300 described species [10–13]. This family was divided into nine subfamilies: Hypoaspidinae Vitzthum, 1940; Melittiphidinae Evans and Till, 1966; Haemogamasinae Oudemans, 1926; Myonyssinae Bregetova, 1956; Hirstionyssinae Evans and Till, 1960; Mesolaelapinae Tenori and Radovsky, 1974; Alphalaelapinae Tipton, 1960; Laelapinae Berlese, 1892; and Acanthochelinae Radovsky and Gettinger, 1999 [14]. Laelapid mites are frequently associated with small mammals, mainly rodents, and can be found on their body surface or in their nests [15]. Classification of the Laelapidae is complicated. High levels of morphological variability in these mites are causing difficulties. Therefore, molecular evidence is needed to identify mites' taxonomy at the species level. The phylogenetic analysis provides important information on biodiversity and taxonomy. Most modern taxonomic studies have a total evidence approach incorporating both morphology and DNA sequencing [16–20].

The large (28S) and small (18S) subunit ribosomal RNA (rRNA) genes are most frequently used in taxonomic studies of arthropods [21]. The 18S rRNA gene is generally considered more appropriate for resolving relationships among phyla and superphyla, with the 28S rRNA gene providing more signals at slightly lower taxonomic levels [22,23]. Nuclear rRNA genes have great advantages: they are generally easy to amplify and appear to contain more signals than other genes used for higher-order questions in animal phylogeny [24]. In a previous study, Dowling and OConnor [7] reported the first large-scale phylogenetic relationships within Dermanyssoidea and the evolution of parasitic lineages within the superfamily using the 28S region (domains 1–3) of the nuclear rDNA. With the aim of screening DNA barcodes for mites, in recent studies, Zhao et al. [25] evaluated the universality of the divergent domains with high identification efficiency in Acari. Researchers showed that domains D5, D6, and D8 of 28S rDNA are universal DNA barcodes for molecular classification and identification of mites [25].

The mitochondrial cytochrome c oxidase subunit I (COI) gene was used for the taxonomical identification of mesostigmatic mites and the determination of their intra- and interspecific variation [26–29]. Recent genetic studies investigated the population genetic structure of *Laelaps agilis* mites across Europe and revealed their phylogenetic relationships [28]. Yang et al. [29] used COI sequence data and morphological characters to assess the phylogenetic relationships of Laelapidae mites from China. However, the phylogenetic relationships and genetic diversity of laelapid mites from Europe are still insufficiently described.

This study aimed to undertake molecular characterization and to assess the phylogenetic relationship among eight Laelapidae mite species collected from different rodent hosts in Lithuania, Norway, Slovakia, and the Czech Republic using the nuclear (28S ribosomal RNA) and mitochondrial (cytochrome oxidase subunit I gene) molecular markers.

2. Materials and Methods

2.1. Sample Collection

Small rodents were captured with live traps at six locations in Lithuania (Trakai (54°39′24.94″ N, 24°49′29.48″ E), Guodžiai peatland (55°58′56.97″ N, 24°36′50.86″ E), Curonian Spit (55°33′06.0″ N, 21°07′31.5″ E), Rusnė (55°19′26.23″ N, 21°20′24.15″ E), Beištrakiai (54°54′22.3″ N, 24°20′28.6″ E) and Nemunas Loops (54°35′19.04″ N, 23°59′49.56″ E)); three locations in Slovakia (Ptičie (48°54′07.3″ N, 21°55′50.8″ E), Svetlice (48°34′56.8″ N, 20°46′37.9″ E), Hrhov (48°34′53.9″ N, 20°46′44.4″ E)); one location in the Czech Republic (České Budějovice (48°59′56.1438″ N, 14°27′20.217″ E)); and one location in Norway (Mjävatn (58°32′19.32″ N, 8°29′22.92″ E)) during 2014–2018.

All trapped rodents were marked and identified by species level and sex. Ectoparasites were collected using soft tweezers, placed into 1.5 mL tubes with 70% ethanol solution, and then stored at 4 °C until processed. The collected mites were determined using morphological identification keys by Mašán, Fend'a [15], Bregetova [30], Baker [31], and Kaminskienė et al. [32].

2.2. DNA Extraction

Ammonium hydroxide solution (2.5%) was used for DNA extraction from mites. The laelapid mites were taken from the ethanol solution, dried (3–5 min) on the paper towel at room temperature, and then put in a 0.5 mL microcentrifuge tube. A quantity of 40 μL of 2.5% NH_4OH solution was added for each adult mite. In the solution the mites were crushed with a sterile plastic pestle and stored at room temperature for 30 min until incubated at 100 °C for 30 min, allowing for maximal DNA recovery. Subsequently, the tubes were centrifuged at 13,000/min for 1 min to collect condensate from the cap and sides of the tube. All opened tubes with the solution were placed back in the heating block and incubated at 100 °C for 20 min to evaporate the ammonia. After incubation, the tubes were closed and placed on the ice for 2–3 min. Then tubes were centrifuged at 13,000/min for 30 s. Extracted DNA was stored at −20 °C until further usage.

2.3. PCR Amplification and Sequencing

Domains 1–3 from the 28S nuclear ribosomal RNA gene region and the COI gene of mitochondrial DNA were used for molecular characterization and phylogenetic reconstruction within the family Laelapidae [33].

Conventional PCR was performed to amplify 856 bp fragment of mites 28S rRNA using 43F 5′- GCT GCG AGT GAA CTG GAA TCA AGC CT3′ and 929R 5′-AGG TCA CCA TCT TTC GGG TC-3′ primers [7]. Each 25 μL reaction contained a mixture of 13.7 μL ddH_2O, 2.5 μL 10× PCR buffer (KCl-(NH_4) SO_4) (Thermo Fisher Scientific Baltics, Vilnius, Lithuania), 20 μL 25 mM $MgCl_2$, 0.5 μL 25 mM dNTP, 1 μL of each 10 mM primer, 0.3 μL Taq DNA Polymerase (Thermo Fisher Scientific Baltics, Vilnius, Lithuania) (5 U/μL), and 4 μL DNA. The PCR reaction conditions were as follows: initial denaturation at 94° for 2 min; followed by 35 cycles of denaturation at 94° for 25 s, annealing at 53° for 20 s, and extension at 72° for 1min; with a final extension at 72° for 7 min after completion of all cycles.

To amplify a 709 bp fragment of the COI gene, conventional PCR with primers LCO1490 (5′-GGT CAA CAA ATC ATA AAG ATA TTG G-3′) and HCO2198 (5′-TAA ACT TCA GGG TGA CCA AAA AAT CA-3′) was performed [33]. Each 25 μL reaction contained a mixture of 16.5 μL ddH_2O, 5 μL 5× MyTaq reaction buffer (Thermo Fisher Scientific Baltics, Vilnius, Lithuania), 0.5 μL of each 10 mM primer, 0.5 μL MyTaq DNA polymerase (Thermo Fisher Scientific Baltics, Vilnius, Lithuania) (5 U/μL), and 2 μL DNA. The PCR reaction conditions were as follows: initial denaturation at 94° for 3 min; followed by 40 cycles of denaturation at 94° for 45 s, annealing at 45° for 45 s, and extension at 72° for 1 min; with a final extension at 72° for 5 min after completion of all cycles.

PCR products were subjected to electrophoresis on 1.5% agarose gel and analyzed by UV transilluminator. The DNA fragment was excised from agarose gel and purified using a GenJET PCR purification kit (Thermo Fisher Scientific Baltics, Vilnius, Lithuania) according to the manufacturer's protocol. All purified PCR products were sent for DNA sequencing to a sequencing service (Macrogen, Amsterdam, The Netherlands).

2.4. Sequence Analysis

The sequences obtained in this study were analyzed using the BLAST program to confirm the morphological identification of mite species and were aligned with the corresponding sequences of other laelapid mites available in GenBank using ClustalW [34] multiple alignments implemented in MegaX [35]. The partial 28S rRNA and COI gene sequences were aligned in two independent datasets. The intraspecific and interspecific pairwise genetic distances, variable sites, conserved sites, and parsimony-informative sites were computed by Mega X. The non-synonymous mutation rate (Ka) and synonymous mutation rate (Ks), haplotype diversity (Hd), nucleotide diversity (Π), and polymorphic sites (S) were calculated using DnaSP v5.10.01 [36]. The representative sequences of 28S rRNA and COI gene were deposited to GenBank.

2.5. Phylogenetic Analysis

Phylogenetic trees were constructed using maximum likelihood (ML) and Bayesian inference (BI) methods. The best-fitting nucleotide substitution model (GTR + I + G) was determined by the Bayesian Information Criterion (BIC) yielded using jModelTest v2.1.10 [37]. The ML trees were generated using the Tamura–Nei parameter model in MEGA X, with each node supported by 1000 bootstraps. Bayesian inference (BI) analyses were run with MrBayes v.3.2.7 [38]. The Markov chain was run with 40,000,000 generations, and trees were sampled every 1000th generation. The first 25% of samples were discarded as burn-in, and the remaining saved samples were used to estimate the posterior probabilities (PP) of each bipartition. The phylogenetic tree was visualized using FigTree v1.4.4 [39].

To estimate the phylogenetic relationships among the COI gene haplotypes of *L. agilis* derived from different rodent hosts and geographical regions, median-joining (MJ) networks were constructed using Network 10.2.0.0 [40].

3. Results

3.1. 28S rRNA Region

A total of 53 sequences of partial 28S rRNA gene were obtained from eight species of Laelapidae mites (*Laelaps agilis, Laelaps jettmari, Laelaps hilaris, Haemogamasus nidi, Eulaelaps stabularis, Hyperlaelaps microti, Myonyssus gigas*, and *Hirstionyssus* sp.) collected from seven small rodent species (*Apodemus flavicollis, Apodemus agrarius, Apodemus sylvaticus, Clethrionomys glareolus, Microtus arvalis, Micromys minutus*, and *Microtus oeconomus*) in Lithuania, Slovakia, the Czech Republic, and Norway (Table 1). The lengths of the analyzed 28S rRNA sequences varied between 527 and 821 bp; the AT content ranged from 55.5 to 57.2% (Table 1). Sequence comparison showed 174 variable sites among 28S rRNA gene sequences of all examined Laelapidae mites and 51 variable sites among mites from the *Laelaps* genus (Table 2).

Partial 28S rRNA sequences obtained from *L. agilis* (MZ043837–MZ043844), *L. hilaris* (MZ043845, MZ043846), *L. jettmari* (MZ043833, MZ043834, ON763742), *M. gigas* (MZ043831, MZ043832), and *Hirstionyssus* sp. (ON775520, ON775521) showed no intraspecific variability (Table 1). However, two genotypes of *H. microti* (MZ043835; MZ043836), *E. stabularis* (MZ043828, MZ043829, MZ043830), and *Hg. nidi* (MZ061928, MZ061929, MZ061930, MZ061931) were identified. *H. microti* sequences differed at one nucleotide position showing ambiguous nucleotide Y (C/T—transition). Two genotypes of *E. stabularis* detected in Lithuania (MZ043828; $n = 2$) and Norway (MZ043829; MZ043830) were specific to their respective locations (Table 1) and differed at two nucleotide positions. Two genotypes representing six 28S rRNA sequences derived from *Hg. nidi* differed at three nucleotide positions (three sequences (MZ061928-MZ061930) had one ambiguous nucleotide W (A/T transversion).

The overall mean genetic distance between laelapid mite sequences obtained in this study was 0.0820. The intra- and interspecific genetic distances of Laelapidae species are shown in Table 3. The highest interspecific distances were detected between *H. microti* and the other Laelapidae mite species.

Table 1. Hosts and GenBank nucleotide accession numbers of the 28S rRNA gene region sequences of Laelapidae mites.

Taxonomic Status of Species	Host Species	Country	Length (bp)	GenBank Accession No.	AT%	No of Representative Samples
Genus *Laelaps*						
L. agilis	*A. agr*	Lithuania	527	MZ043838	56.4	3
	A. fla	Lithuania	760	MZ043837	56.8	3
	C. gla	Lithuania	527	MZ043839	56.4	1
	M. min	Lithuania	818	MZ043840	56.6	1
	A. agr	Slovakia	818	MZ043842	56.7	1
	A. fla	Slovakia	805	MZ043841	56.8	8
	A. syl	Czech Republic	818	MZ043843	56.7	8
	C. gla	Czech Republic	805	MZ043844	56.8	1
L. jettmari	*A. fla*	Lithuania	802	MZ043834	57.2	1
	A. fla	Slovakia	803	MZ043833	57.2	4
	A. agr	Lithuania	821	ON763742	57.1	1
L. hilaris	*A. agr*	Lithuania	814	MZ043845	56.6	1
	M. arv	Lithuania	805	MZ043846	56.8	1
Genus *Eulaelaps*						
E. stabularis (1 gen.)	*C. gla*	Lithuania	820	MZ043828	57.0	2
E. stabularis (2 gen.)	*A. fla*	Norway	743	MZ043830	55.9	1
	C. gla	Norway	770	MZ043829	56.1	1
Genus *Haemogamasus*						
Hg. nidi (1 gen.)	*A. agr*	Lithuania	768	MZ061928	55.9	1
	A. fla	Lithuania	768	MZ061929	55.9	1
	C. gla	Lithuania	768	MZ061930	55.9	1
Hg. nidi (2 gen.)	*A. fla*	Lithuania	768	MZ061931	55.8	3
Genus *Hyperlaelaps*						
H. microti (1 gen.)	*M. arv*	Lithuania	760	MZ043835	55.8	1
H. microti (2 gen.)	*M. arv*	Lithuania	760	MZ043836	55.9	1
Genus *Myonyssus*						
M. gigas	*A. fla*	Lithuania	771	MZ043832	55.9	3
	C. gla	Lithuania	811	MZ043831	55.5	2
Genus *Hirstionyssus*						
Hirstionyssus sp.	*A. agr*	Lithuania	818	ON775520	55.9	1
	A. agr	Lithuania	818	ON775521	55.9	1

Abbreviations: *A. agr—Apodemus agrarius, A. fla—Apodemus flavicollis, A. syl—Apodemus sylvaticus, M. arv—Microtus arvalis, C. gla—Clethrionomys glareolus, M. min—Micromys minutus.*

Table 2. Comparison of the 28S rRNA gene sequences of *Laelaps* genus mites in this study.

Nucleotide Position / Samples						1	1	1	2	2	2	3	3	3	3	3	3	3	3	3	3	3	3	3	4	
	1	2	3	8	8	9	0	0	1	8	9	9	1	3	4	6	6	6	7	7	7	8	8	8	9	0
	7	8	3	3	6	0	7	9	5	9	5	9	7	5	8	3	6	7	3	4	9	1	3	5	6	5
MZ043837 *L. agilis A. flavicollis* Lithuania	G	G	T	A	T	G	A	G	T	G	T	G	C	T	A	G	T	G	A	A	G	C	G	T	T	C
MZ043838 *L. agilis A. agrarius* Lithuania	.	.	.	.	.	.	.	.	.	.	.	.	.	.	.	.	.	.	.	.	.	.	.	.	.	.
MZ043839 *L. agilis C. glareolus* Lithuania	.	.	.	.	.	.	.	.	.	.	.	.	.	.	.	.	.	.	.	.	.	.	.	.	.	.
MZ043840 *L. agilis M. minutus* Lithuania	.	.	.	.	.	.	.	.	.	.	.	.	.	.	.	.	.	.	.	.	.	.	.	.	.	.
MZ043841 *L. agilis A. flavicollis* Slovakia	.	.	.	.	.	.	.	.	.	.	.	.	.	.	.	.	.	.	.	.	.	.	.	.	.	.
MZ043842 *L. agilis A. agrarius* Slovakia	.	.	.	.	.	.	.	.	.	.	.	.	.	.	.	.	.	.	.	.	.	.	.	.	.	.
MZ043843 *L. agilis A. sylvaticus* Czech Republic	.	.	.	.	.	.	.	.	.	.	.	.	.	.	.	.	.	.	.	.	.	.	.	.	.	.
MZ043844 *L. agilis C. glareolus* Czech Republic	.	.	.	.	.	.	.	.	.	.	.	.	.	.	.	.	.	.	.	.	.	.	.	.	.	.
MZ043846 *L. hilaris M. arvalis* Lithuania	C	A	.	T	C	A	G	A	A	T	C	A	T	C	T	T	C	.	G	G	.	.	A	G	A	A
MZ043833 *L. jettmari A. flavicollis* Slovakia	T	.	A	T	.	.	.	A	G	.	.	.	T	.	T	.	C	A	G	G	A	T	.	G	A	A
MZ043834 *L. jettmari A. flavicollis* Lithuania	T	.	A	T	.	.	.	A	G	.	.	.	T	.	T	.	C	A	G	G	A	T	.	G	A	A
ON763742 *L. jettmari A. agrarius* Lithuania	T	.	A	T	.	.	.	A	G	.	.	.	T	.	T	.	C	A	G	G	A	T	.	G	A	A

Nucleotide Position / Samples	4	4	4	4	4	4	4	4	4	4	4	4	5	5	5	5	5	5	5	5	5	5	5	6	7
	0	1	1	2	4	5	7	7	8	8	9	9	1	1	1	2	4	7	8	8	9	9	9	0	2
	9	0	8	9	5	9	4	6	4	7	2	5	2	3	4	6	4	4	4	8	0	6	7	7	9
MZ043837 *L. agilis A. flavicollis* Lithuania	G	G	A	A	G	T	A	A	C	C	T	T	G	A	G	T	A	T	A	T	T	A	C	C	C
MZ043838 *L. agilis A. agrarius* Lithuania	.	.	.	.	.	.	.	.	.	.	.	.	.	.	.	.	.	.	.	.	.	.	.	.	.
MZ043839 *L. agilis C. glareolus* Lithuania	.	.	.	.	.	.	.	.	.	.	.	.	.	.	.	.	.	.	.	.	.	.	.	.	.
MZ043840 *L. agilis M. minutus* Lithuania	.	.	.	.	.	.	.	.	.	.	.	.	.	.	.	.	.	.	.	.	.	.	.	.	.
MZ043841 *L. agilis A. flavicollis* Slovakia	.	.	.	.	.	.	.	.	.	.	.	.	.	.	.	.	.	.	.	.	.	.	.	.	.
MZ043842 *L. agilis A. agrarius* Slovakia	.	.	.	.	.	.	.	.	.	.	.	.	.	.	.	.	.	.	.	.	.	.	.	.	.
MZ043843 *L. agilis A. sylvaticus* Czech Republic	.	.	.	.	.	.	.	.	.	.	.	.	.	.	.	.	.	.	.	.	.	.	.	.	.
MZ043844 *L. agilis C. glareolus* Czech Republic	.	.	.	.	.	.	.	.	.	.	.	.	.	.	.	.	.	.	.	.	.	.	.	.	.
MZ043846 *L. hilaris M. arvalis* Lithuania	.	A	G	G	T	C	T	G	.	G	C	A	T	T	.	C	T	.	G	A	.	T	.	T	T
MZ043833 *L. jettmari A. flavicollis* Slovakia	A	A	.	G	A	.	T	G	T	G	C	.	.	.	A	C	C	C	.	.	C	T	T	T	T
MZ043834 *L. jettmari A. flavicollis* Lithuania	A	A	.	G	A	.	T	G	T	G	C	.	.	.	A	C	C	C	.	.	C	T	T	T	T
ON763742 *L. jettmari A. agrarius* Lithuania	A	A	.	G	A	.	T	G	T	G	C	.	.	.	A	C	C	C	.	.	C	T	T	T	T

Table 3. Genetic distances within and between Laelapidae species.

Species	Genetic Distance			
	28S rRNA		COI	
	Within Species	Between Species [a]	Within Species	Between Species [a]
L. a	0	0–0.046312 (vs. *L. j*) 0–0.056138 (vs. *L. h*) 0–0.116236 (vs. *E. s*) 0–0.113036 (vs. *Hg. n*) 0–0.137949 (vs. *H. m*) 0–0.113032 (vs. *M. g*) 0–0.112793 (vs. *Hirst.* sp.)	0–0.033718	0–0.125923 (vs. *L. j*) 0–0.146954 (vs. *L. h*) - (vs. *E. s*) 0–0.302688 (vs. *Hg. n*) 0–0.211204 (vs. *H. m*) 0–0.308024 (vs. *M. g*) - (vs. *Hirst.* sp.)
L. j	0	0–0.042192 (vs. *L. h*) 0–0.099372 (vs. *E. s*) 0–0.100931 (vs. *Hg. n*) 0–0.125555 (vs. *H. m*) 0–0.099202 (vs. *M. g*) 0–0.105226 (vs. *Hirst.* sp.)	0–0.019213	0–0.155239 (vs. *L. h*) - (vs. *E. s*) 0–0.318748 (vs. *Hg. n*) 0–0.206667 (vs. *H. m*) 0–0.337406 (vs. *M. g*) - (vs. *Hirst.* sp.)
L. h	0	0–0.114634 (vs. *E. s*) 0–0.113034 (vs. *Hg. n*) 0–0.128336 (vs. *H. m*) 0–0.105137 (vs. *M. g*) 0–0.118999 (vs. *Hirst.* sp.)	0	- (vs. *E. s*) 0–0.304940 (vs. *Hg. n*) 0–0.211069 (vs. *H. m*) 0–0.320741 (vs. *M. g*) - (vs. *Hirst.* sp.)
E. s	0–0.002626	0–0.022748 (vs. *Hg. n*) 0–0.169238 (vs. *H. m*) 0–0.024086 (vs. *M. g*) 0–0.043391 (vs. *Hirst.* sp.)	-	- (vs. *Hg. n*) - (vs. *H. m*) - (vs. *M. g*) - (vs. *Hirst.* sp.)
Hg. n	0–0.002632	0–0.162658 (vs. *H. m*) 0–0.026818 (vs. *M. g*) 0–0.046214 (vs. *Hirst.* sp.)	0	0–0.337478 (vs. *H. m*) 0–0.344651 (vs. *M. g*) - (vs. *Hirst.* sp.)
H. m	0	0–0.165662 (vs. *M. g*) 0–0.179227 (vs. *Hirst.* sp.)	0.088065	0–0.332020 (vs. *M. g*) - (vs. *Hirst.* sp.)
M. g	0	0–0.040487 (vs. *Hirst.* sp.)	0	- (vs. *Hirst.* sp.)
Hirst. sp.	0	-	-	-

Abbreviations: *L. a*—*Laelaps agilis*, *L. j*—*Laelaps jettmari*, *L. h*—*Laelaps hilaris*, *E. s*—*Eulaelaps stabularis*, *Hg. n*—*Haemogamasus nidi*, *H. m*—*Hyperlaelaps microti*, *M. g*—*Myonyssus gigas*, *Hirst.* sp.—*Hirstionyssus* sp., - no data available. [a] Mean distances are shown between species.

The phylogenetic analysis based on 28S rRNA gene included sequences of other dermanysoid mite species available in GenBank: *L. jettmari* (*pavlovskyi*) (GU440635), *L. hilaris* (GU440637), *Laelaps stupkai* (GU440596), *Laelaps clethrionomydis* (GU440636), *Laelaps kochi* (GU440626), *Laelaps muris* (GU440638), *Ondatralaelaps multispinosus* (FJ911778), *Laelaps vansomereni* (GU440619), *Laelaps zumpti* (GU440623), *Laelaps spinigera* (GU440613), *Laelaps mazzai* (GU440590), *Haemogamasus reidi* (GU440583), *Brevisterna morlani* (FJ911773), *Haemogamasus* sp. (FJ911772), and *Dermanyssus gallinae* (FJ911771).

The phylogenetic tree of 28S rRNA gene sequences constructed using the ML method is divided into two main clusters: one cluster groups sequences of twelve *Laelaps* genus species and *H. microti*, while the other cluster consists of six species of *Hirstionyssus*, *Haemogamasus*, *Myonyssus*, *Brevisterna*, and *Eulaelaps* genera. The members of each species form individual subclusters on the phylogenetic tree (Figure 1).

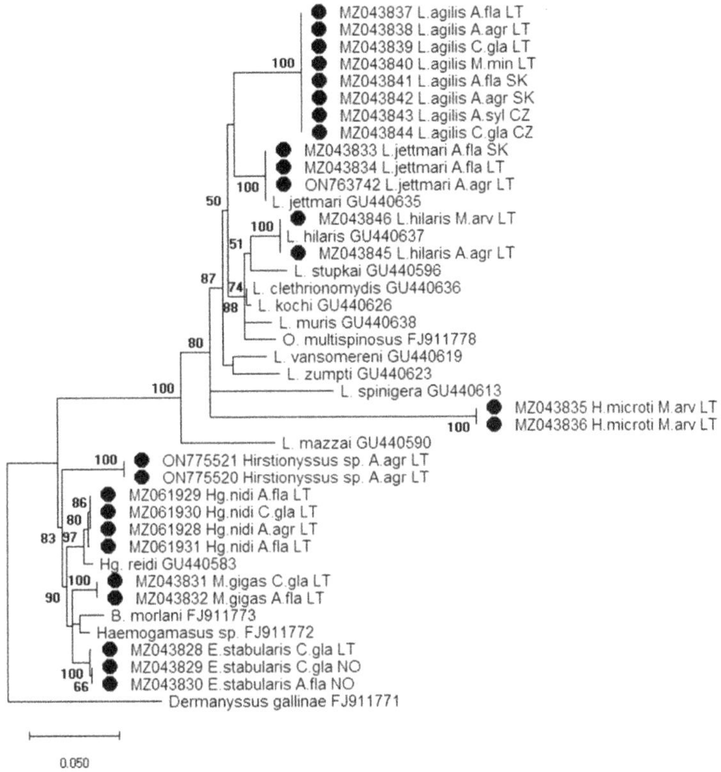

Figure 1. Phylogenetic tree of *28S rRNA* gene sequences of Laelapidae mites generated using the maximum likelihood method and Tamura–Nei model and bootstrap analysis of 1000 replicates. Samples sequenced in the present study are marked (•). Abbreviations: A. agr—*Apodemus agrarius*, A. fla—*Apodemus flavicollis*, A. syl—*Apodemus sylvaticus*, M. arv—*Microtus arvalis*, C. gla—*Clethrionomys glareolus*, M. min—*Micromys minutus*, LT—Lithuania, SK—Slovakia, CZ—Czech Republic, NO—Norway.

The 28S rRNA gene sequences of *L. jettmari* and *L. hilaris* obtained in the present study were 100% identical to corresponding sequences derived from GenBank: GU440635 and GU440637, respectively (Figure 1). Sequences of *Hg. nidi* (MZ061928, MZ061929, MZ061931, MZ061930) collected in Lithuania shared 98.95–99.08% similarity to *Hg. reidi* (synonym *Hg. nidi*) sequences from GenBank: GU440583.

3.2. COI Gene

The partial sequences of the COI gene were successfully obtained from six species of Laelapidae mites (*L. agilis*, *L. jettmari*, *L. hilaris*, *Hg. nidi*, *H. microti*, and *M. gigas*) collected from six species of small rodents (*A. flavicollis*, *A. agrarius*, *A. sylvaticus*, *C. glareolus*, *M. arvalis* and *M. oeconomus*). A total of 60 good-quality COI sequences were analyzed (among them 47 sequences of *L. agilis*, four sequences of *L. jettmari*, three sequences of *L. hilaris*, two sequences of *Hg. nidi*, two sequences of *M. gigas*, and two sequences of *H. microti*). COI sequences of Laelapidae mites ranged from 582 to 699 bp in length and from 64.9 to 74.6% in AT content (Table 4); there were 253 variable sites, 330 conserved sites, and 245 parsimony-informative sites. A total of 23 nucleotide variable sites were detected among *L. agilis* species (Table 5). The mean value of Ka/Ks of COI gene sequences obtained in this study was 2.31.

Table 4. Hosts and GenBank nucleotide accession numbers of the COI gene sequences of Laelapidae mites.

Taxonomic Status of Species	Host Species	Country	Length (bp)	GenBank Accession No.	AT%	No of Representative Samples
Genus *Laelaps*						
L. agilis (1 hap.)	*A. fla*	Slovakia	699	MZ315167	74.1	1
	A. fla	Slovakia	695	MZ315168	74.4	3
	A. fla	Lithuania	650	ON754956	73.8	4
	C. gla	Lithuania	651	ON754955	73.7	2
	A. syl	Czech Republic	685	MZ315172	74.5	1
	C. gla	Czech Republic	651	OP199248	73.7	1
L. agilis (2 hap.)	*A. syl*	Czech Republic	684	MZ315169	74.3	1
	A. syl	Czech Republic	688	MZ315171	74.2	1
L. agilis (3 hap.)	*A. syl*	Czech Republic	684	MZ315170	73.5	1
L. agilis (4 hap.)	*A. agr*	Lithuania	582	MZ048460	74.6	1
	A. agr	Lithuania	582	MZ048461	74.6	1
	C. gla	Lithuania	582	MZ048462	74.6	2
	A. fla	Lithuania	646	ON754957	74.0	7
L. agilis (5 hap.)	*C. gla*	Lithuania	582	MZ048463	74.2	1
L. agilis (6 hap.)	*M. oec*	Lithuania	582	MZ048464	74.4	1
	A. fla	Lithuania	649	ON754958	73.7	1
L. agilis (7 hap.)	*C. gla*	Lithuania	647	ON754963	73.7	2
	A. fla	Lithuania	650	ON754962	73.7	2
	A. fla	Lithuania	650	ON754961	73.7	5
	M. min	Lithuania	649	ON754960	73.7	1
	C. gla	Lithuania	652	ON754959	73.5	3
L. agilis (8 hap.)	*A. fla*	Lithuania	650	ON754964	73.8	1
L. agilis (9 hap.)	*A. fla*	Norway	650	ON754965	73.1	1
L. jettmari (1 hap.)	*A. agr*	Lithuania	582	MZ048465	73.9	1
L. jettmari (2 hap.)	*A. agr*	Lithuania	582	MZ048466	73.7	1
L. jettmari (3 hap.)	*A. agr*	Lithuania	657	OP199246	72.0	1
	A. agr	Lithuania	645	OP199245	72.2	1
L. hilaris	*A. agr*	Lithuania	582	MZ048455	72.0	1
	M. arv	Lithuania	582	MZ048456	72.0	1
	M. arv	Lithuania	582	MZ048457	72.0	1
Genus *Haemogamasus*						
Hg. nidi	*A. fla*	Lithuania	582	MZ049956	64.9	1
	C. gla	Lithuania	582	MZ049957	64.9	1
Genus *Hyperlaelaps*						
H. microti (1 hap.)	*M. arv*	Lithuania	582	MZ048467	73.7	1
H. microti (2 hap.)	*M. oec*	Lithuania	582	MZ048468	72.3	1
Genus *Myonyssus*						
M. gigas	*C. gla*	Lithuania	582	MZ048469	70.1	1
	C. gla	Lithuania	582	MZ048470	70.1	1

Abbreviations: *A. agr*—*Apodemus agrarius*, *A. fla*—*Apodemus flavicollis*, *A. syl*—*Apodemus sylvaticus*, *M. arv*—*Microtus arvalis*, *M. oec*—*Microtus oeconomus*, *C. gla*—*Clethrionomys glareolus*.

Table 5. Comparison of the COI gene sequences of *L. agilis* mites in this study.

Samples	Nucleotide Position	No of Representative Samples	15	33	48	75	90	138	156	162	171	172	201	216	288	309	321	372	396	450	483	510	516	522	549
1 HAP	MZ315167 *A. fla* SK	1	A	T	A	T	T	A	G	A	G	G	A	A	G	A	A	A	T	T	G	A	G	A	A
	MZ315168 *A. fla* SK	3	.	.	.	.	.	.	.	.	.	.	.	.	.	.	.	.	.	.	.	.	.	.	.
	ON754956 *A. fla* LT	4	.	.	.	.	.	.	.	.	.	.	.	.	.	.	.	.	.	.	.	.	.	.	.
	ON754955 *C. gla* LT	2	.	.	.	.	.	.	.	.	.	.	.	.	.	.	.	.	.	.	.	.	.	.	.
	MZ315172 *A. syl* CZ	1	.	.	.	.	.	.	.	.	.	.	.	.	.	.	.	.	.	.	.	.	.	.	.
	OP199248 *C. gla* CZ	1	.	.	.	.	.	.	.	.	.	.	.	.	.	.	.	.	.	.	.	.	.	.	.
2 HAP	MZ315171 *A. syl* CZ	1	.	.	.	.	C	.	.	.	.	.	.	.	.	.	.	.	.	.	.	.	.	.	.
	MZ315169 *A. syl* CZ	1	.	.	.	.	C	.	.	.	.	.	.	.	.	.	.	.	.	.	.	.	.	.	.
3 HAP	MZ315170 *A. syl* CZ	1	T	.	G	.	.	G	A	G	.	.	.	G	A	T	G	T	A	A	A	G	A	G	.
4 HAP	MZ048462 *C. gla* LT	2	.	.	.	.	.	.	.	.	A	.	.	.	.	.	.	.	.	.	.	.	.	.	G
	MZ048461 *A. agr* LT	1	.	.	.	.	.	.	.	.	A	.	.	.	.	.	.	.	.	.	.	.	.	.	G
	MZ048460 *A. agr* LT	1	.	.	.	.	.	.	.	.	A	.	.	.	.	.	.	.	.	.	.	.	.	.	G
	ON754957 *A. fla* LT	7	.	.	.	.	.	.	.	.	A	.	.	.	.	.	.	.	.	.	.	.	.	.	G
5 HAP	MZ048463 *C. gla* LT	1	.	.	.	C	.	.	.	.	.	.	.	.	.	.	.	.	.	.	.	.	.	.	G
6 HAP	MZ048464 *M. oec* LT	1	.	.	.	.	.	.	.	.	.	.	.	.	.	.	.	.	.	.	.	.	.	.	G
	ON754958 *A. fla* LT	1	.	.	.	.	.	.	.	.	.	.	.	.	.	.	.	.	.	.	.	.	.	.	G
7 HAP	ON754963 *C. gla* LT	2	.	.	.	C	.	.	.	.	.	.	.	.	.	.	.	.	.	.	.	.	.	.	.
	ON754962 *A. fla* LT	2	.	.	.	C	.	.	.	.	.	.	.	.	.	.	.	.	.	.	.	.	.	.	.
	ON754961 *A. fla* LT	5	.	.	.	C	.	.	.	.	.	.	.	.	.	.	.	.	.	.	.	.	.	.	.
	ON754960 *M. min* LT	1	.	.	.	C	.	.	.	.	.	.	.	.	.	.	.	.	.	.	.	.	.	.	.
	ON754959 *C. gla* LT	3	.	.	.	C	.	.	.	.	.	.	.	.	.	.	.	.	.	.	.	.	.	.	.
8 HAP	ON754964 *A. fla* LT	1	.	.	.	C	.	.	.	.	.	A	.	.	.	.	.	.	.	.	.	.	.	.	.
9 HAP	ON754965 *A. fla* NO	1	T	C	.	.	.	G	A	G	.	.	G	G	A	T	G	T	A	A	A	G	A	G	.

Abbreviations: *L. a—Laelaps agilis, A. agr—Apodemus agrarius, A. fla—Apodemus flavicollis, A. syl—Apodemus sylvaticus, C. gla—Clethrionomys glareolus, M. oec—Microtus oeconomus; M. min—Micromys minutus*, LT—Lithuania, SK—Slovakia, CZ—Czech Republic, NO—Norway.

Nine COI haplotypes (h = 9) between 23 *L. agilis* sequences were detected with estimated haplotype diversity of Hd = 0.870, nucleotide diversity $\Pi = 0.00720$, and a total number of polymorphic sites S = 23. In total, 559 conserved sites, one singleton site, and 19 parsimony-informative sites were detected. Haplotype H_1 of *L. agilis* was the most frequent. It was found in three out of four different locations (Lithuania, Slovakia, and the Czech Republic) (Table 4, Figure 2). Haplotypes H_2 and H_3 (the Czech Republic), H_4-H_8 (Lithuania), and H_9 (Norway) of *L. agilis* were specific for their respective sampling locations.

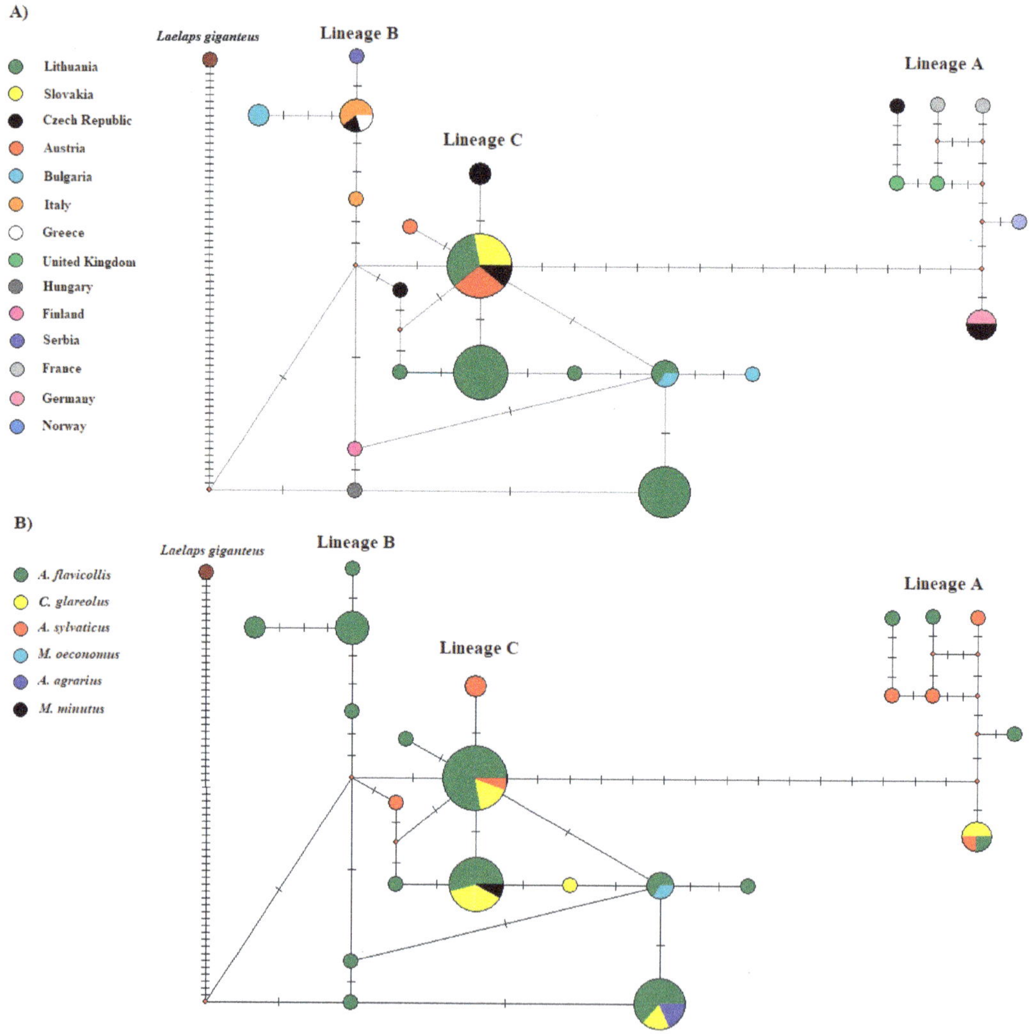

Figure 2. A median-joining network of haplotypes based on COI sequences of *L. agilis* from different European regions. The circles represent different haplotypes with size proportional to relative frequencies. (**A**): different colors represent geographic distribution; (**B**): different colors represent host species. The network branches linking the cycles indicate one mutation step; two or more mutations are represented by slashes crossed with the network branches. The red points indicate undetected intermediate haplotypes.

In this study, six haplotypes of *L. agilis* were detected in Lithuania. From these sequences, four haplotypes of *L. agilis* detected in Lithuania (H_4, H_5, H_7, and H_8) were unique and differed from the most similar sequences in GenBank (Figure 2A). The distribution of *L. agilis* haplotypes in different areas of Lithuania showed that the highest haplotype diversity was detected in the Lithuanian coastal area—the Curonian Spit where five of six haplotypes (H_1, H_5-H_8; n = 21) were found. In the continental part of the country (northern and south-eastern parts), three haplotypes were detected (H_1, H_4, H_6; n = 13) (Figure 3). Distribution of different *L. agilis* haplotypes did not reveal specificity to host species. Five haplotypes were detected in *A. flavicollis*, four haplotypes in *C. glareolus*, and *A. agrarius*, *M. oeconomus*, and *M. minutus* each harbored one haplotype H_4, H_6, and H_7, respectively. This study detected three COI haplotypes of *L. jettmari* (n = 4) and two COI haplotypes of *H. microti* (n = 2). In contrast, only one haplotype was found among *L. hilaris*, *Hg. Nidi*, and *M. gigas* sequences (Table 4).

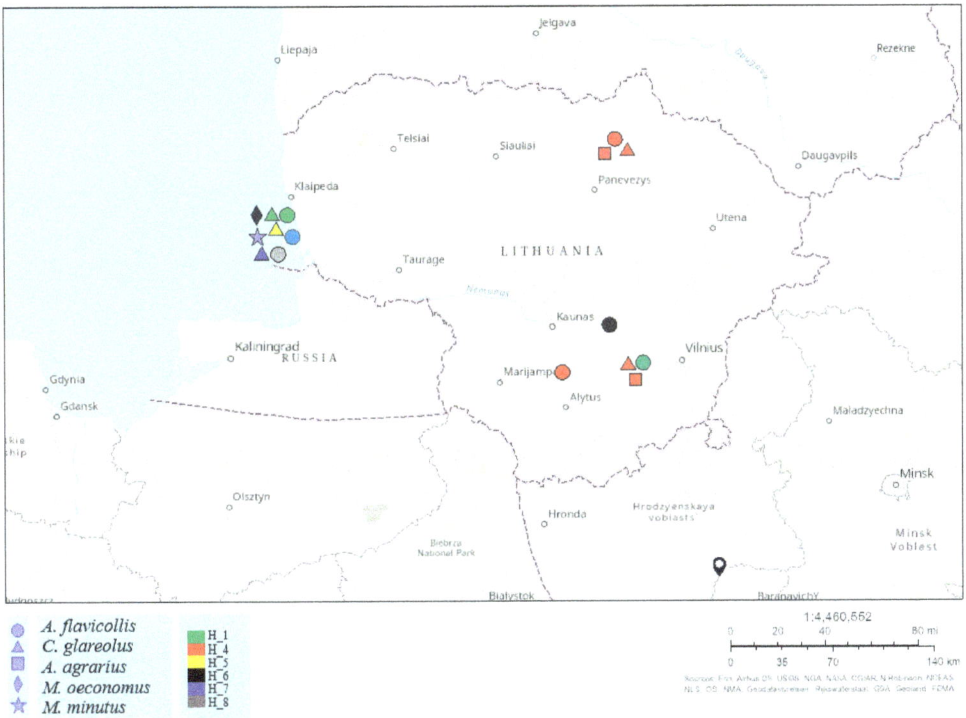

Figure 3. The distribution of COI haplotypes of *L. agilis* collected from different host species and sampling sites in Lithuania. Different host species and haplotypes are shown by various shapes and colors. The number of samples varied from 1 to 7.

The overall mean genetic distance between laelapid mites' COI gene sequences obtained in this study was 0.1215. The inter- and intraspecific genetic distances based on the COI gene are shown in Table 3. The highest interspecific distances were detected between *M. gigas* and the other Laelapidae mite species. The intraspecific genetic distance among *L. agilis* sequences was 0.0074.

The phylogenetic analysis based on the COI gene included sequences of other dermanysoid mite species available in GenBank: *Laelaps muricola* (KU166735; KU166676; KU166784; KU166789), *Laelaps giganteus* (KU166660; KU166413; KU166425), *L. kochi* (MF914881; MG413303), *Haemogamasus ambulans* (KM831963), *Gaeolaelaps debilis* (MW367907), *E. stabularis* (OP960202), and *Dermanyssus hirundinis* (MN355089). The phylogenetic tree of COI gene sequences constructed using the ML method showed a clear separation of different species of Laelapidae mites into different clusters. *L. agilis* sequences were heterogenic and, together with *L. jettmari* and *L. hilaris*, formed a separate cluster on the phylogenetic tree (Figure 4).

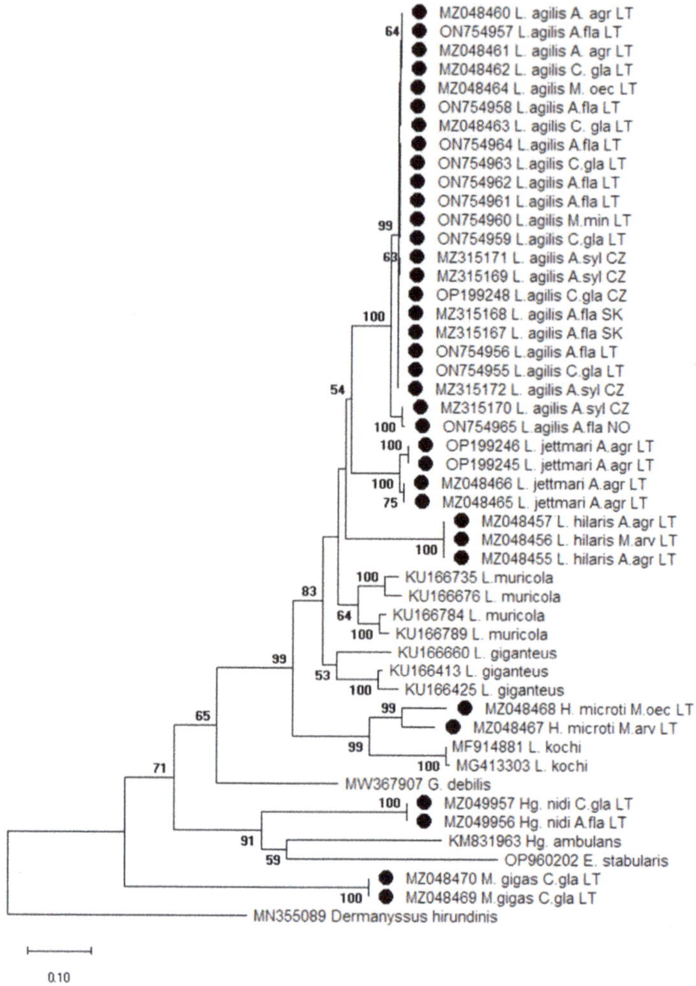

Figure 4. Phylogenetic tree of COI gene sequences of Laelapidae mites generated using the maximum likelihood method and General Time Reversible model (Gamma Distributed with Invariant Sites (G + I)) model and bootstrap analysis of 1000 replicates. Samples sequenced in the present study are marked (•). Abbreviations: *A. agr– Apodemus agrarius*, *A. fla—Apodemus flavicollis*, *A. syl—Apodemus sylvaticus*, *M. arv—Microtus arvalis*, *M. oec—Microtus oeconomus*, *C. gla—Clethrionomys glareolus*, *M. min—Micromys minutus*, LT—Lithuania, SK—Slovakia, CZ—Czech Republic, NO—Norway.

Another phylogenetic tree of Laelapidae mites was constructed using the BI method (Figure 3). ML and BI phylogenetic trees differed slightly in topology and branching structures (Figures 4 and 5). The Bayesian tree (Figure 5) exhibited higher posterior probabilities (PPs) values (52–100%) than the bootstrap values (38–100%) of the ML (Figure 4).

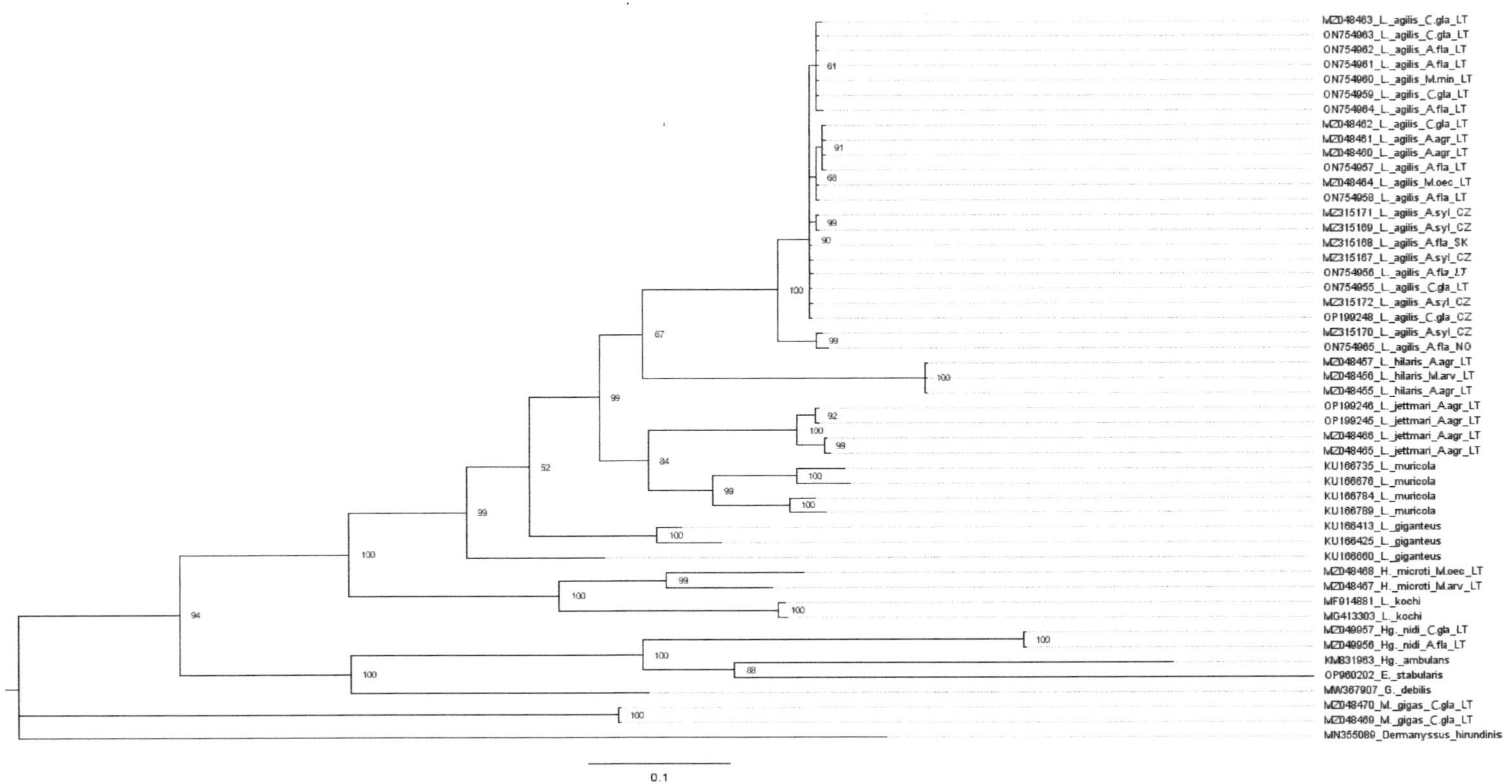

Figure 5. A Bayesian phylogenetic tree of the Laelapidae mites, based on the COI gene sequences. Numbers above the branches show the Bayesian posterior probabilities. Samples sequenced in the present study are marked (•). Abbreviations: *A. agr—Apodemus agrarius, A. fla—Apodemus flavicollis, A. syl—Apodemus sylvaticus, M. arv—Microtus arvalis, M. oec—Microtus oeconomus, C. gla—Clethrionomys glareolus, M. min—Micromys minutus*, LT—Lithuania, SK—Slovakia, CZ—the Czech Republic, NO—Norway. The geographical structure of *L. agilis* was supported by the median-joining network, which showed at least three major Lineages A, B, and C (Figure 2A,B). In the median-joining (MJ) network were included the COI gene

Haplotypes assigned to Lineage A were found in Norway, Finland, the United Kingdom, France, and the Czech Republic. Most of the haplotypes were not shared between different geographic areas, except for one haplotype identified in Germany and the Czech Republic (Figure 2A). This lineage (A) showed haplotype sharing between three host species: *A. flavicollis*, *A. sylvaticus*, *C. glareolus* (Figure 2B). A single haplotype was assigned to Lineage B, shared between the *L. agilis* from Italy, Greece, and the Czech Republic, and three unique to the individuals from Italy, Bulgaria, and Serbia. This lineage (B) only included mites collected from *A. flavicollis*. All sequences of *L. agilis* from Lithuania belonged to Lineage C. They clustered together with samples from Slovakia, Bulgaria, Austria, the Czech Republic, Hungary, and Finland. Additionally, the network showed that four haplotypes have so far been found only in Lithuania (Figure 2A). This lineage (C) was composed of *L. agilis* found in different host species (*A. flavicollis*, *A. sylvaticus*, *A. agrarius*, *C. glareolus*, *M. minutus*, and *M. oeconomus*) (Tables 2B and 3).

4. Discussion

In the present study, eight species of Laelapidae mites collected from different rodent hosts and geographical regions in Europe were molecularly characterized based on both nuclear 28S rRNA and mitochondrial COI gene regions. Our findings confirm that these molecular markers could be successfully used for molecular identification of Laelapidae mite species and inference of their phylogenetic relationships [7,27–29]. On the other hand, mitochondrial DNA evolves much faster and is more evolutionarily variable than the ribosomal DNA of the nuclear genome [41]. Thus, the COI gene sequences are more appropriate for analyzing intraspecific phylogenetic relationships [26,42]. In this study, our results based on the COI gene indicated a high intraspecific variation (9 haplotypes out of 23 obtained sequences) in *L. agilis* species. Intraspecific variations on the COI gene were also detected in *L. jettmari* (three haplotypes identified among four obtained sequences) and *H. microti* (two haplotypes among two obtained sequences).

Our findings provide new data on the intra- and interspecific phylogenetic relationships of Laelapidae mites belonging to six genera. This study, for the first time, registered sequences of four mite species: *H. microti*, *Hirstionyssus* sp., *M. gigas*, and *E. stabularis*.

Phylogenetic relationships based on 28S rRNA exhibited polyphyly of the different species from the family Laelapidae. The previous study also determined a polytomy structure in the phylogenetic relationships [7]. In contrast, Li et al. [43] and Yang et al. [44] showed that based on mitochondrial barcoding region, the family Laelapidae is a monophyletic group.

The results of the phylogenetic analysis based on 28S rRNA revealed the separation of Laelapidae mites into two different groups. The first group consists of sequences belonging to obligate parasitic mites from two genera, *Laelaps* and *Hyperlaelaps*. The second group contains two clusters—one cluster consists of sequences belonging to facultative parasitic mites *Eulaelaps*, *Haemogamasus*, and *Myonyssus*, whereas sequences of obligate parasitic *Hirstionyssus* sp. formed a separate cluster (Figure 1).

It should be noticed that phylogenetic analysis based on both genes (28S rRNA and COI) indicated the clustering of *H. microti* with the species of the genus *Laelaps* and did not show separation into distinct clades. The differences between the molecular and morphological taxonomy of this species were also observed in recent studies [29,44].

In line with a previous study [28], our results of the phylogenetic analysis based on mt DNA also corroborated three lineages (Lineages A, B, and C) within *L. agilis* (Figure 2). The results did not indicate clear specificity according to geographical locations. Lineages A and C comprised specimens from diverse geographical regions of Europe (North, Central-Eastern, and West) (Figure 2A), which was also revealed in a recent study [28]. However, our results supplemented Lineage A with one specimen from Norway and Lineage C with sequences from Lithuania (Figure 2A). Moreover, our findings showed no clear host species specificity and confirmed the results previously obtained by Nazarizadeh et al. [28]. However, the number of host species in these lineages (A and C) was supplemented by three additional species (*A. agrarius*, *M. minutus*, and *M. oeconomus*) in this study. Only one *L. agilis* lineage (B) showed clear specificity according to host species (*A. flavicollis*) (Figure 2B), and it is consistent with the results of the Nazarizadeh et al. study [28].

Considering several species of rodents as important hosts of the parasitic mites analyzed in this study, it should be mentioned that populations of rodents of the genera *Apodemus* and *Clethrionomys* in Europe are genetically heterogeneous. During the glaciation in the Quaternary, they survived in various refugia in southern Europe [45,46] and had complex recolonization routes in Europe. A specific species in this regard is *Apodemus agrarius*, which only relatively recently colonized Europe from Asia [47].

Based on published data, at least 21 parasitic mite species belonging to the Laelapidae family have been morphologically identified in Lithuania [32,48–51]. This study provides the first molecular characterization of eight species of laelapid mites collected from different rodent hosts in Lithuania. Therefore, the more comprehensive phylogenetic analysis of Laelapidae mites in Lithuania must be further investigated.

5. Conclusions

Our study provides new molecular data on *Laelaps agilis*, *Laelaps hilaris*, *Laelaps jettmari*, *Haemogamasus nidi*, *Eulaelaps stabularis*, *Hyperlaelaps microti*, *Myonyssus gigas*, and *Hirstionyssus* sp. mites collected from seven different rodent hosts and three geographical regions in Europe. This study is the first molecular characterization of eight Laelapidae mite species in Baltic countries. Specifically, 28S rRNA and COI sequences of four mite species were, for the first time, registered in the NCBI database (2021–2022).

Author Contributions: Conceptualization, J.R., E.K. and A.P.; methodology, E.K. and J.R.; software, L.G.; formal analysis, E.K., J.S. and D.M.-B.; investigation, E.K., L.G., J.R., M.S. and A.P.; resources, A.P. and M.S.; writing—original draft: E.K., L.G. and J.S.; writing—review and editing: J.R., A.P., M.S. and D.M.-B.; supervision: J.R. and A.P. All authors have read and agreed to the published version of the manuscript.

Funding: This research received no external funding.

Institutional Review Board Statement: Rodent sampling was conducted with permission from the Environmental Protection Agency (EPA) and approved by the Ministry of Environment of the Republic of Lithuania, licenses No. 15 (31 March 2014), No. 22 (10 April 2015), No. 12 (30 March 2016, No. 13 (22 March 2017) and No. 1 (2 February 2018) in accordance with Lithuanian (the Republic of Lithuania Law on Welfare and Protection of Animals No. XI-2271) and European legislation (Directive 2010/63/EU) on the protection of animals.

Informed Consent Statement: Not applicable.

Data Availability Statement: The data presented in this study are available within the article.

Acknowledgments: The authors are grateful to Linas Balčiauskas and Vaclovas Gedminas for the collection of small rodents in Lithuania and Olav Rosef for providing mite samples from rodents collected in Norway.

Conflicts of Interest: The authors declare no conflict of interest.

References

1. Zhang, Z.-Q.; Fan, Q.-H.; Pesic, V.; Smit, H.; Bochkov, A.V.; Khaustov, A.A.; Baker, A.; Wohltmann, A.; Wen, T.H.; Amrine, J.W.; et al. *Animal Biodiversity: An Outline of Higher-Level Classification and Survey of Taxonomic Richness*; Zhang, Z.-Q., Ed.; Magnolia Press: Auckland, New Zealand, 2011; pp. 1–237.
2. Kocianova, E.; Kozuch, O.; Bakoss, P.; Rehacek, J.; Kovacova, E. The prevalence of small terrestrial mammals infected with tick-borne encephalitis virus and leptospirae in the foothills of the southern Bavarian forest, Germany. *Adv. Parasitol.* **1993**, *34*, 283–290.
3. Reeves, W.K.; Dowling, A.P.G.; Dasch, A.G.A. Rickettsial agents from parasitic Dermanyssoidea (Acari: Mesostigmata). *Exp. Appl. Acarol.* **2006**, *38*, 181–188. [CrossRef]
4. Merhej, V.; Angelakis, E.; Socolovschi, C.; Raoult, D. Genotyping, evolution and epidemiological findings of *Rickettsia* species. *Infect. Genet. Evol.* **2014**, *25*, 122–137. [CrossRef]
5. Mit'ková, K.; Berthová, L.; Kalúz, S.; Kazimírová, M.; Burdová, L.; Kocianová, E. First detections of *Rickettsia helvetica* and *R. monacensis* in ectoparasitic mites (Laelapidae and Trombiculidae) infesting rodents in south-western Slovakia. *Parasitol. Res.* **2015**, *114*, 2465–2472. [CrossRef]
6. Radzijevskaja, J.; Kaminskienė, E.; Lipatova, I.; Mardosaitė-Busaitienė, D.; Balčiauskas, L.; Stanko, M.; Paulauskas, A. Prevalence and diversity of *Rickettsia* species in ectoparasites collected from small rodents in Lithuania. *Parasites Vectors* **2018**, *11*, 375. [CrossRef] [PubMed]
7. Dowling, A.P.G.; OConnor, B.M. Phylogeny of Dermanyssoidea (Acari: Parasitiformes) suggests multiple origins of parasitism. *Acarologia* **2010**, *50*, 113–129. [CrossRef]
8. Faraji, F.; Halliday, B. Five new species of mites (Acari: Laelapidae) associated with large Australian cockroaches (Blattodea: Blaberidae). *Int. J. Acarol.* **2009**, *35*, 245–264. [CrossRef]
9. Kavianpour, M.; Nemati, A.; Gwiazdowicz, D.J.; Kocheili, F. A new species of the genus *Gaeolaelaps* (Acari, Mesostigmata, Laelapidae) from Iran. *Zookeys* **2013**, *277*, 1–11. [CrossRef]
10. Beaulieu, F.; Dowling, A.P.G.; Klompen, H.; de Moraes, G.J.; Walter, D.E. Superorder Parasitiformes Reuter, 1909. *In*: Zhang, Z.-Q. (Ed.) Animal biodiversity. *Zootaxa* **2011**, *3148*, 123–128. [CrossRef]
11. Kazemi, S.; Rajaei, A.; Beaulieu, F. Two new species of *Gaeolaelaps* (Acari: Mesostigmata: Laelapidae) from Iran, with a revised generic concept and notes on significant morphological characters in the genus. *Zootaxa* **2014**, *3861*, 501–530. [CrossRef]
12. Nemati, A.; Gwiazdowicz, D.J. A new genus and species of Laelapidae from Iran with notes on *Gymnolaelaps* Berlese and *Laelaspisella* Marais & Loots (Acari, Mesostigmata). *Zookeys* **2016**, *549*, 23–49. [CrossRef]

13. Kazemi, S.; Beaulieu, F. A new genus and species of Laelapidae (Acari: Mesostigmata) from Iran. *Zootaxa* **2016**, *4200*, 487–500. [CrossRef]
14. Kazemi, S.; Rajaei, A.; Paktinat Saeej, S. The Laelapidae mites (Acari: Mesostigmata) from Mazandaran Province, North Iran. In Proceedings of the 2nd International Persian Congress of Acarology, The Graduate University of Advanced Technology (GUAT), Kerman, Iran, 29–31 August 2013.
15. Mašán, P.; Fenda, P. *A Review of the Laelapid Mites Associated with Terrestrial Mammals in Slovakia, with a Key to the European Species*; Institute of Zoology, Slovak Academy of Sciences: Bratislava, Slovakia, 2010.
16. Bickford, D.; Lohman, D.J.; Sodhi, N.S.; Ng, P.K.L.; Meier, R.; Winker, K.; Ingram, K.K.; Das, I. Cryptic species as a window on diversity and conservation. *Trends Ecol. Evol.* **2007**, *22*, 148–155. [CrossRef]
17. Morelli, M.; Spicer, G.S. Cospeciation between the nasal mite *Ptilonyssus sairae* (Acari: Rhinonyssidae) and its bird hosts. *Syst. Appl. Acarol.* **2007**, *12*, 179–188. [CrossRef]
18. Skoracka, A.; Dabert, M. The cereal rust mite *Abacarus hystrix* (Acari: Eriophyoidea) is a complex of species: Evidence from mitochondrial and nuclear DNA sequences. *Bull. Entomol. Res.* **2010**, *100*, 263–272. [CrossRef] [PubMed]
19. Apanaskevich, D.A.; Horak, I.G.; Matthee, C.A.; Matthee, S. A new species of *Ixodes* (Acari: Ixodidae) from South African mammals. *J. Parasitol.* **2011**, *97*, 389–398. [CrossRef] [PubMed]
20. Knee, W.; Beaulieu, F.; Skevington, J.H.; Kelso, S.; Forbes, M.R. Cryptic species of mites (Uropodoidea: *Uroobovella* spp.) associated with burying beetles (Silphidae: *Nicrophorus*): The collapse of a host generalist revealed by molecular and morphological analyses. *Mol. Phylogenet. Evol.* **2012**, *65*, 276–286. [CrossRef] [PubMed]
21. Cruickshank, R.H. Molecular markers for the phylogenetics of mites and ticks. *Syst. Appl. Acarol.* **2002**, *7*, 3–14. [CrossRef]
22. Mallatt, J.M.; Garey, J.R.; Shultz, J.W. Ecdysozoan phylogeny and Bayesian inference: Wrst use of nearly complete 28S and 18S rRNA gene sequences to classify the arthropods and their kin. *Mol. Phylogenet. Evol.* **2004**, *31*, 178–191. [CrossRef]
23. Mallatt, J.; Giribet, G. Further use of nearly complete 28S and 18S rRNA genes to classify Ecdysozoa: 37 more arthropods and a kinorhynch. *Mol. Phylogenet. Evol.* **2006**, *40*, 772–794. [CrossRef]
24. Giribet, G. Current advances in the phylogenetic reconstruction of metazoan evolution. A new paradigm for the Cambrian explosion? *Mol. Phylogenet. Evol.* **2002**, *24*, 345–357. [CrossRef]
25. Zhao, Y.; Zhang, W.Y.; Wang, R.L.; Niu, D.L. Divergent domains of 28S ribosomal RNA gene: DNA barcodes for molecular classification and identification of mites. *Parasites Vectors* **2020**, *13*, 251. [CrossRef]
26. Yang, B.; Cai, J.; Cheng, X. Identification of astigmatid mites using ITS2 and COI regions. *Parasitol. Res.* **2011**, *108*, 497–503. [CrossRef]
27. Savchenko, E.; Lareschi, M. A new species of *Laelaps* Koch, 1836 (Mesostigmata: Laelapidae) parasitic of the sigmodontine rodent *Oligoryzomys flavescens* Waterhouse, 1837 (Rodentia: Cricetidae): Molecular and morphological characterization. *Acta Trop.* **2019**, *199*, 105146. [CrossRef]
28. Nazarizadeh, M.; Martinů, J.; Nováková, M.; Stanko, M.; Štefka, J. Phylogeography of the parasitic mite *Laelaps agilis* in Western Palearctic shows lineages lacking host specificity but possessing different demographic histories. *BMC Zool.* **2022**, *7*, 15. [CrossRef] [PubMed]
29. Yang, H.; Yang, Z.; Dong, W. Morphological Identification and Phylogenetic Analysis of Laelapin Mite Species (Acari: Mesostigmata: Laelapidae) from China. *Korean J. Parasitol.* **2022**, *60*, 273–279. [CrossRef]
30. Bregetova, N.G. Gamasid mites (Gamasoidea). The short keys. In Брегетова Н.Г. *Гамазовые клещи (Gamasoidea).* Краткий Определитель; Izdatelstvo Akademii Nauk SSSR: Moskow, Russia, 1956. (In Russian)
31. Baker, E.W.; Evans, T.M.; Gould, D.J.; Hull, W.B.; Keegan, H.L. *A Manual of Parasitic Mites of Medical or Economic Importance*; National Pest Control Association: New York, NY, USA, 1956; pp. 1–170.
32. Folmer, O.; Black, M.; Hoeh, W.; Lutz, R.; Vrijenhoek, R. DNA primers for amplification of mitochondrial cytochrome c oxidase subunit I from diverse metazoan invertebrates. *Mol. Marine Biol. Biotechnol.* **1994**, *3*, 294–297.
33. Kaminskienė, E.; Radzijevskaja, J.; Balčiauskas, L.; Gedminas, V.; Paulauskas, A. Laelapidae mites (Acari: Mesostigmata) infesting small rodents in the Curonian Spit, Lithuania. *Biologija* **2017**, *63*, 169–176. [CrossRef]
34. Thompson, J.D.; Higgins, D.G.; Gibson, T.J. CLUSTAL W: Improving the sensitivity of progressive multiple sequence alignment through sequence weighting, position-specific gap penalties and weight matrix choice. *Nucleic Acids Res.* **1994**, *22*, 4673–4680. [CrossRef] [PubMed]
35. Kumar, S.; Stecher, G.; Li, M.; Knyaz, C.; Tamura, K. MEGA X: Molecular Evolutionary Genetics Analysis across computing platforms. *Mol. Biol. Evol.* **2018**, *35*, 1547–1549. [CrossRef] [PubMed]
36. Librado, P.; Rozas, J. DnaSP v5: A software for comprehensive analysis of DNA polymorphism data. *Bioinformatics* **2009**, *25*, 1451–1452. [CrossRef]
37. Darriba, D.; Taboada, G.L.; Doallo, R.; Posada, D. jModelTest 2: More models, new heuristics and parallel computing. *Nat. Methods* **2012**, *9*, 772. [CrossRef]
38. Ronquist, F.; Teslenko, M.; van der Mark, P.; Ayres, D.; Darling, A.; Hohna, S.; Larget, B.; Liu, L.; Suchard, M.A.; Huelsenbeck, J.P. MrBayes 3.2: Efficient Bayesian phylogenetic inference and model choice across a large model space. *Syst. Biol.* **2012**, *61*, 539–542. [CrossRef]
39. Rambaut, A. FigTree v.1.4.4. 2018. Available online: http://tree.bio.ed.ac.uk/software/figtree (accessed on 25 October 2022).

40. Bandelt, H.J.; Forster, P.; Rohl, A. Median-joining networks for inferring intraspecific phylogenies. *Mol. Biol. Evol.* **1999**, *16*, 37–48. [CrossRef]
41. Hwang, U.W.; Kim, W. General properties and phylogenetic utilities of nuclear ribosomal DNA and mitochondrial DNA commonly used in molecular systematics. *Korean J. Parasitol.* **1999**, *37*, 215–228. [CrossRef]
42. Liu, Y.; Nie, Y.; Chen, J.; Lu, T.; Niu, L.; Jia, J.; Ye, Z.; Fu, Y. Genetic diversity of three major spider mites damaging rubber trees. *Syst. Appl. Acarol.* **2022**, *27*, 2025–2037. [CrossRef]
43. Li, W.-N.; Shao, R.; Zhang, Q.; Deng, W.; Xue, X.-F. Mitochondrial genome reorganization characterizes various lineages of mesostigmatid mites (Acari: Parasitiformes). *Zool. Scr.* **2019**, *48*, 679–689. [CrossRef]
44. Yang, H.; Chen, T.; Dong, W. Divergence time of mites of the family Laelapidae based on mitochondrial barcoding region. *PLoS ONE* **2023**, *18*, e0279598. [CrossRef]
45. Michaux, J.R.; Libois, R.; Filippucc, M.-G. So close and so different: Comparative phylogeography of two small mammal species, the Yellow-necked fieldmouse (*Apodemus flavicollis*) and the Woodmouse (*Apodemus sylvaticus*) in the Western Palearctic region. *Heredity* **2005**, *94*, 52–63. [CrossRef]
46. Kotlík, P.; Marková, S.; Horníková, M.; Escalante, M.A.; Searle, J.B. The bank vole (*Clethrionomys glareolus*) as a model system for adaptive phylogeography in the European theater. *Front. Ecol. Evol.* **2022**, *10*, 866605. [CrossRef]
47. Kozyra, K.; Zając, T.M.; Ansorge, H.; Wierzbicki, H.; Moska, M.; Stanko, M.; Stopka, P. Late Pleistocene expansion of small murid rodents across the Palearctic in relation to the past environmental changes. *Genes* **2021**, *12*, 642. [CrossRef]
48. Jezerskienė, E. Ectoparasites of the bank-vole *Clethrionomydis glareolus* Schreb. under Lithuania conditions. *Acta Zool. Litu.* **1974**, *12*, 193–197. (In Russian)
49. Podenaite, V.I. Ticks and fleas fauna of small mammals in Lithuania SSR. Natural foci of infections and invasions. In Proceedings of the Scientific Conference, Vilnius, Lithuania, 6–7 September 1979. (In Russian).
50. Kaminskienė, E.; Radzijevskaja, J.; Stanko, M.; Balčiauskas, L.; Paulauskas, A. Associations between different Laelapidae (Mesostigmata: Dermanyssoidea) mites and small rodents from Lithuania. *Exp. Appl. Acarol.* **2020**, *81*, 149–162. [CrossRef] [PubMed]
51. Kitrytė, N.; Križanauskienė, A.; Baltrūnaitė, L. Ecological indices and factors influencing communities of ectoparasitic laelapid mites (Acari, Mesostigmata, Laelapidae) of small mammals in Lithuania. *J. Vector Ecol.* **2022**, *47*, 99–108. [CrossRef]

animals

MDPI

Article

Species Composition of Parasitic Mites of the Subfamily Picobiinae (Acariformes: Syringophilidae) Associated with African Barbets (Piciformes: Lybiidae) †

Bozena Sikora [1], Mathieu Mahamoud-Issa [2], Markus Unsoeld [3], Martin Hromada [4] and Maciej Skoracki [1,*]

1 Department of Animal Morphology, Faculty of Biology, Adam Mickiewicz University, Uniwersytetu Poznańskiego 6, 61-614 Poznań, Poland; bozena.sikora@amu.edu.pl
2 Department of Behavioural Ecology, Faculty of Biology, Adam Mickiewicz University, Uniwersytetu Poznańskiego 6, 61-614 Poznań, Poland; mathieumahamoudissa@hotmail.fr
3 SNSB-Bavarian State Collection of Zoology, Section Ornithology, Münchhausenstr. 21, 81247 Munich, Germany; unsoeld@snsb.de
4 Laboratory and Museum of Evolutionary Ecology, Department of Ecology, Faculty of Humanities and Natural Sciences, University of Prešov, 080 01 Prešov, Slovakia; hromada.martin@gmail.com
* Correspondence: skoracki@amu.edu.pl

† urn:lsid:zoobank.org:act:151432A7-2536-4343-8FD8-12375D075856; urn:lsid:zoobank.org:act:3DAE24CB-CD57-4C2D-901B-AC6DE769D789.

Simple Summary: This study investigated the picobiine mites parasitising African barbets. The results showed that this bird family is more widely infested by feather mites than previously thought, with three species of the genus *Tanopicobia* found on ten African barbet species. Birds belonging to the family Lybiidae have a unique parasite fauna consisting exclusively of mites from the genus *Tanopicobia*, which is restricted solely to African barbets. Based on the distribution of the genus *Tanopicobia* on the studied barbet hosts, our results also provide indirect cues that the host genus *Trachyphonus*, attributed to such different avian families, e.g., Capitonidae, Ramphastidae, is part of the family Lybiidae, whereas other/related bird families have their own distinct quill mite fauna.

Abstract: In this study, we conducted a parasitological investigation of the quill mite fauna of the subfamily Picobiinae (Acariformes: Prostigmata: Syringophilidae) associated with African barbets (Aves: Piciformes: Lybiidae). We examined twenty-seven host species, representing 57% of the forty-seven known host species in the family Lybiidae, belonging to seven genera (70% of the ten genera in the family). Our research revealed that ten host species were infested by three species of picobiine mites belonging to the genus *Tanopicobia*: (1) *Tanopicobia hallae* Sikora and Skoracki, sp. n., from three species of the genus *Lybius* and two species of the genus *Tricholaema*, (2) *Tanopicobia stactolaema* Sikora and Skoracki, sp. n., from two species of the genus *Stactolaema*, and (3) *Tanopicobia trachyphoni* Skoracki et al., 2020, collected from three host species of the genus *Trachyphonus*. Our findings demonstrate that birds belonging to the family Lybiidae have a specific parasite fauna consisting exclusively of mites of the genus *Tanopicobia*; this mite genus is apparently restricted to African barbets.

Keywords: acari; birds; diversity; ectoparasites; quill mites

Citation: Sikora, B.; Mahamoud-Issa, M.; Unsoeld, M.; Hromada, M.; Skoracki, M. Species Composition of Parasitic Mites of the Subfamily Picobiinae (Acariformes: Syringophilidae) Associated with African Barbets (Piciformes: Lybiidae). *Animals* **2023**, *13*, 2007. https://doi.org/10.3390/ani13122007

Academic Editor: Alexis Ribas

Received: 18 May 2023
Revised: 12 June 2023
Accepted: 14 June 2023
Published: 16 June 2023

1. Introduction

Quill mites of the subfamily Picobiinae (Acariformes: Prostigmata: Syringophilidae) are permanent and host-specific ectoparasites of birds. Their whole life cycle takes place inside quills of the contour feathers where they live, feed, and reproduce [1–4]. The exception is an ambiguous species *Calamincola lobatus* Casto, which inhabits quills of secondaries [5]. Currently, the subfamily comprises c.a. 80 described species grouped in 12 genera and recorded from more than 200 host species belonging to neognathous birds (Aves: Neognathae) [4,6]. The quill mite fauna of picobiines, known from birds of the order Piciformes, are still understudied. Although the first record of picobiine mites from Piciformes was presented more than a hundred years ago when Haller, in 1878, described the first species *Picobia heeri* [7], it is only relatively recently that intensive research on this group of hosts has begun. In several papers, the picobiine fauna have been presented for birds of the families Lybiidae [8], Picidae [1,3,4,9–14], Semniornithidae [15], and Ramphastidae [16]. To this time, we have no data on the presence of picobiine mites from birds of the families Megalaimidae, Capitonidae, and Indicatoridae.

In this paper, we present the results of our study on quill mites of the subfamily Picobiinae parasitising African barbets (Lybiidae). This avian family comprises about 41 species grouped in seven genera distributed mainly in sub-Saharan Africa [17–19]. The African barbets are medium-sized birds, ranging from 15 to 30 cm in length and are found in a variety of habitats, including forests, woodlands, savannas, and gardens. Some species are more specialised, inhabiting specific habitats such as montane forests or riverine woodlands. Several species of African barbets are considered threatened or endangered due to habitat loss and fragmentation, particularly in West and Central Africa. Overall, the African barbets are a fascinating group of birds that play important ecological roles in African ecosystems [18–20]. The Lybiidae family consists of bird species with different social organizations: some species live in single pairs, while other species live in family groups and even small colonies and are considered as group breeding species [21]. It is thus an excellent family to study the evolution of the host–parasite speciation and transmission according to the degree of complexity of the social organization, as well as possible inter-species contamination. Moreover, African barbets are nesting in tree cavities, which also could modify the probability of infestation by ectoparasites, and even more interestingly, they are brood-parasitised by other piciform birds, e.g., Indicatoridae [19].

Because to date, only one species, *Tanopicobia trachyphoni* Skoracki et al., has been recorded from one host species, i.e., the red and yellow barbet, *Trachyphonus erythrocephalus* Cabanis [8], we conducted a parasitological investigation of the picobiine fauna associated with the birds of the family Lybiidae. Our research revealed that ten host species were infested by three species of picobiine mites, including two species described herein as new to science. Our findings demonstrate that birds belonging to the family Lybiidae have a unique parasite fauna consisting exclusively of mites from the genus *Tanopicobia* and that the distribution of this mite genus is restricted solely to African barbets.

2. Materials and Methods

The mite material used in this study was collected from dry bird skins housed in the ornithological collection of the Bavarian State Collection of Zoology (Munich, Germany) (by M.S and M.U.), according to the technique described by Skoracki [3] (Figure 1A–D). Additional mite material was collected from the yellow-breasted barbet *Trachyphonus margaritatus* (Cretzschmar) captured during a field expedition in Djibouti (by B.S. and M.M-I) (permit no. 438/DEDD/2020 to M.M-I) (Figure 1E).

We examined the quills of approximately ten contour feathers in the proximity of the cloaca region for each bird. Before mounting, mites were treated in Nesbitt's solution at room temperature for three days, following the procedure outlined by Walter and Krantz [22] and Skoracki [3] to soften and clear them. Subsequently, the mites were mounted on slides in Hoyer's medium and examined under a light microscope (ZEISS Axioscope, Oberkochen, Germany) with differential interference contrast (DIC) illumination. To illustrate the findings, we created drawings using a camera lucida drawing attachment. Finally, drawings were made using a camera lucida drawing attachment.

All measurements provided in the description are in micrometers. The paratypes' dimensional ranges are indicated in parentheses alongside the holotype data. Idiosomal setation adheres to Grandjean's [23] classification, adapted for Prostigmata by Kethley [24]. Leg chaetotaxy follows the proposal made by Grandjean [25]. Finally, the morphological terminology follows Skoracki [3,4].

Figure 1. Representatives of the host genera of the family Lybiidae infested by picobiine mites. (**A**)—the banded barbet *Lybius undatus* (Rüppell); (**B**)—the spot-flanked barbet *Tricholaema lacrymosa* Cabanis; (**C**)—the white-eared barbet *Stactolaema leucotis* (Sundevall); (**D**)—the red-and-yellow barbet *Trachyphonus erythrocephalus* Cabanis; (**E**)—the yellow-breasted barbet, *Trachyphonus margaritatus* (Cretzschmar). Photos: (**A–D**) M.U.; (**E**) M.M-I.

Specimen depositories are cited using the following abbreviations: AMU—Adam Mickiewicz University, Department of Animal Morphology, Poznan, Poland; SNSB—ZSM—Bavarian Natural History Collections—Bavarian State Collection of Zoology, Munich, Germany.

3. Results

Systematics

Family Syringophilidae Lavoipierre
Subfamily Picobiinae Johnston and Kethley
Genus *Tanopicobia* Skoracki, Sikora, Jerzak and Hromada, 2020

Descriptions

Tanopicobia hallae Sikora and Skoracki sp. n. (Figure 2)

Female. Total body length of holotype 450 (440–530 in seven female paratypes). *Gnathosoma*. Infracapitulum apunctate. Stylophore 100 (100–110) long; exposed portion of stylophore (stylophoral shield) apunctate, 70 (70–80) long. Each medial branch of peritremes with six chambers, each lateral branch with weakly marked borders between chambers. Movable cheliceral digit edentate on proximal end. *Idiosoma*. Setae *vi*, *ve*, *si*, *se*, *c1*, *c2*, *d1*, *d2*, *e2*, *3b*, *4b*, *3c*, *4c*, and *3a* strongly ornamented. Setae *1a* and *ag1–3* smooth. Propodonotal shield divided into three sclerites: two lateral shields bearing bases of setae *si* and *se* narrowly separated from large medial shield bearing bases of setae *vi*, *ve*, and *c1*; all propodonotal sclerites punctate. Length ratio of setae *vi:ve:si* 1:1.6:2.2–2.4. Hysteronotal shield reduced to two well developed and punctate sclerites surrounding bases of setae *d1*. Hysteronotal setae *d1*, *d2*, and *e2* subequal in length. Pygidial shield present, well sclerotised and punctate, 95 (90–95) long. Setae *f2* 3.5–4 times longer than *f1*. Genital plate present, punctate. Pseudanal setae as microsetae. Setae *ag1* 3.3–4 times longer than *ag2*. Coxal fields I–II apunctate, III and IV punctate. Setae *3c* 1.6 times longer than *3b*, *4c* about twice as long as *4b*. *Legs*. Setae *dFI*, *dGI*, *dTI*, *l'GI–IV*, *l'TI–IV*, and *l'RIII–IV* strongly knobbed, other leg setae slightly ornamented or smooth.

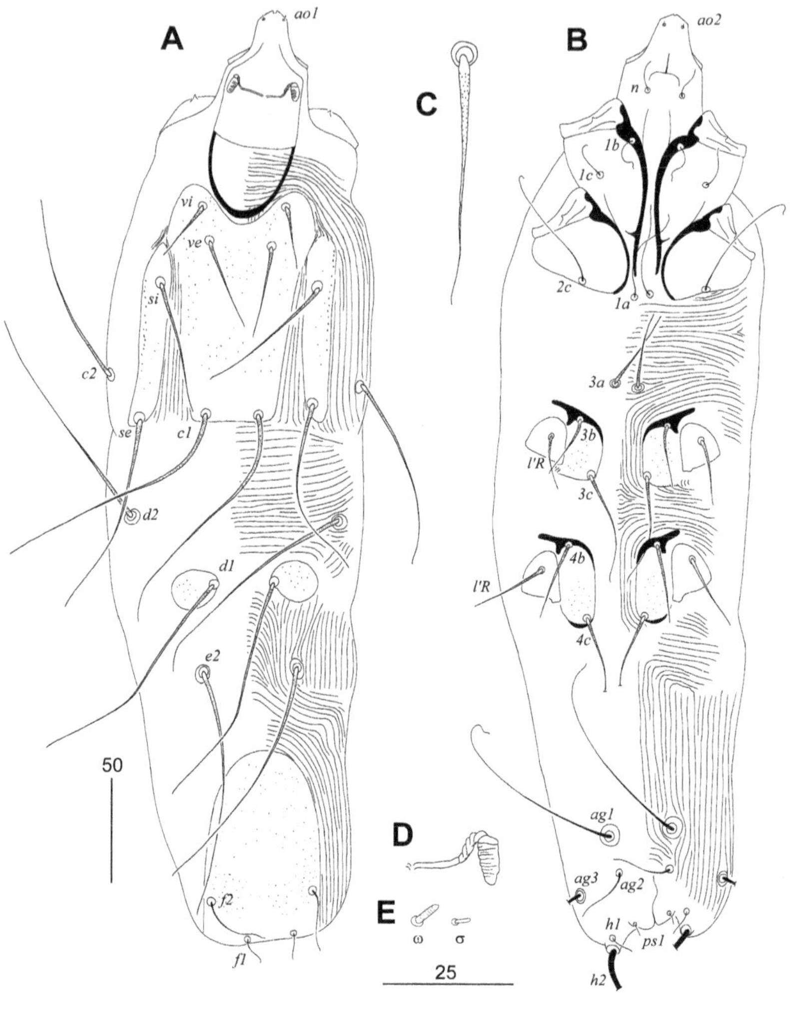

Figure 2. *Tanopicobia hallae* Sikora and Skoracki sp. n., female. (**A**)—dorsal view; (**B**)—ventral view; (**C**)—propodonotal seta *ve*; (**D**)—peritreme; (**E**)—solenidia of leg I. Scale bars: (**A**,**B**) = 50 μm, (**C**–**E**) = 25 μm.

Lengths of setae: *vi* 30 (30–35), *ve* 45 (45), *si* 70 (50–70), *se* 95, *c1* 125 (120–130), *c2* 95 (95–100), *d1* 115 (105–115), *d2* 115 (110–115), *e2* 125 (100–125), *f1* (10), *f2* 30 (30–40), *h1* (10), *h2* (225), *ag1* 105 (100–105), *ag2* 25 (25–30), *ps1* (5), *3a* 40 (40), *3b* 30 (30–40), *3c* 50 (50–65), *4b* 40 (40), *4c* (75–80), *l'RIII* 30 (30–35), *l'RIV* 35 (30–35), *tc'III–IV* (25), *tc''III–IV* (45).

Male. Not found.

Type Material

Female holotype and seven female paratypes from quill of contour feathers of the banded barbet *Lybius undatus* (Rüppell) (host reg. no. SNSB-ZSM 66.200, female); ETHIOPIA: Benishangul-Gumuz Region, Lekamti, 15–17 November 1965, coll. K.E. Linsenmair.

Type Material Deposition

Holotype and paratypes are deposited in the AMU (reg. no. MS 22-1022-053), except three female paratypes in the SNSB-ZSM (reg. no. SNSB-ZSM A20112197).

Additional material

Ex quill of contour feather of the black-collared barbet *Lybius torquatus* (Dumont) (host reg. no. SNSB-ZSM 1287; male); TANZANIA: Morogoro Region, Morogoro District, 27 November 1952, coll. Th. Andersen—four females deposited in AMU (reg. no. AMU MS 22-1022-052a) and two females deposited in SNSB-ZSM (reg. no. SNSB-ZSM A20112198).

Ex quill of contour feather of the brown-breasted barbet *Lybius melanopterus* (Peters) (host reg. no. SNSB-ZSM 60.1784; female); TANZANIA: Arusha Region, Arusha District, Usa-River, 22 February 1960, coll. unknown—five females deposited in AMU (reg no. AMU MS 22-1022-054) and two females in SNSB-ZSM (reg. no. SNSB-ZSM A20112198).

Ex quill of contour feather of the spot-flanked barbet *Tricholaema lacrymosa* Cabanis (host reg. no. SNSB-ZSM 64.894; female); TANZANIA: Lindi Region, Kilwa District, 18 July 1952, coll. Th. Andersen—4 females deposited in AMU (reg. no. AMU MS 22-1022-119) and 4 females in SNSB-ZSM (reg. no. SNSB-ZSM A20112205).

Ex quill of contour feather of the red-fronted barbet *Tricholaema diademata* (Heuglin) (host reg. no. SNSB-ZSM 60.1774; male); TANZANIA: Arusha Region, Arusha District, near Arusha, 14 March 1960, coll. von Nagy—three females and one male deposited in AMU (reg. no. AMU MS 22-1022-122) and three females in the SNSB-ZSM (reg. no. SNSB-ZSM A20112204).

Differential Diagnosis

This new species differs from *T. trachyphoni* as follows: in females of *T. hallae*, the lengths of setae *ve* and *si* are 45 and 50–70, respectively; and the hysteronotal shields are well-developed and punctate. In females of *T. trachyphoni*, the lengths of setae *ve* and *si* are 70–80 and 95–110, respectively; and the hysteronotal shields are absent.

Etymology

This species is named in honour of the British ornithologist Dr. Beryl Patricia Hall (1917–2010), an expert in the distribution and speciation of African birds.

Tanopicobia stactolaema Sikora and Skoracki, sp. n. (Figures 3 and 4)

Female (Figure 3). Total body length 430 in holotype (460–525 in six paratypes). *Gnathosoma*. Stylophore 115 (115) long; exposed portion of stylophore (stylophoral shield) apunctate, 80 (80) long. Each medial branch of peritremes with six or seven chambers, each lateral branch with weakly marked borders between chambers. Movable cheliceral digit edentate on proximal end. *Idiosoma*. Setae *vi, ve, si, se, c1, c2, d1, d2, e2, 3b, 4b, 3c, 4c*, and *3a* strongly ornamented. Setae *1a* and *ag1–3* smooth. Propodonotal shield divided into three sclerites: two lateral shields bearing bases of setae *si* and *se* narrowly separated from large medial shield bearing bases of setae *vi, ve*, and *c1*; all propodonotal sclerites apunctate. Length ratio of setae *vi:ve:si* 1:1.6–1.9:2–2.6. Two hysteronotal shields well developed and apunctate, posterior margin of each shield reaching bases of setae *e2*. Hysteronotal setae *d1, d2*, and *e2* subequal in length. Pygidial shield present, well sclerotised and apunctate. Setae *f2* slightly (1.3 times) longer than *f1*. Setae *ag1* 2.8–3.2 times longer than *ag2*. Coxal fields I–IV apunctate. Setae *3c* and *4c* 1.3–1.6 times longer than *3b* and *4b*. *Legs*. Setae *dFI, dGI, dTI, l'GI–IV, l'TI–IV*, and *l'RIII–IV* strongly knobbed, other leg setae slightly ornamented or smooth.

Lengths of setae: *vi* 35 (35–45), *ve* 65 (60–70), *si* 90 (80–90), *se* 110 (110–120), *c1* 115 (115), *c2* 110 (95–115), *d1* (70–80), *d2* 95 (80–95), *e2* 95 (90–95), *f1* 20 (15–20), *f2* 25 (25), *h1* 10 (10–15), *h2* (205–235), *ag1* (85–95), *ag2* 30 (20–30), *ag3* 70 (65–70), *ps1* 5 (5), *3a* 40 (40), *3b* 35 (35), *3c* 45 (45–55), *4b* 35 (35), *4c* 45 (45–55), *l'RIII* 35 (30–35), *l'RIV* 35 (30–35), *tc'III–IV* 30 (30), *tc"III–IV* 45 (45).

Male (Figure 4). Total body length 275 in one paratype. *Gnathosoma*. Infracapitulum apunctate. Stylophore 80 long; exposed portion of stylophore apunctate, 65 long. Each medial branch of peritremes with six chambers, each lateral branch with weakly marked borders between chambers. *Idiosoma*. Propodonotal setae *vi, ve, si, se*, and *c1* strongly ornamented. Setae *1a* and *ag1–2* smooth. Propodonotal shield divided into three apunctate sclerites: two lateral shields bearing bases of setae *si* and *se*, both shields narrowly separated from large medial shield bearing bases of setae *vi* and *ve*. Setae *ve* and *si* twice as long as *vi*. Hysteronotal shield apunctate, bearing bases of setae *d1* and *e2*. Pygidial shield apunctate. Hysteronotal setae *d1*, *d2*, *e2*, and *f2* short and smooth. Aggenital plate entire and apunctate, bearing bases of setae *ag1* on posterior margin of this plate. Setae *ag1* distinctly longer than *ag2*. Coxal fields I–IV apunctate. Setae *3b*, *4b*, *3c*, *4c*, and *3a* strongly ornamented.

Lengths of setae: *vi* 20, *ve* 40, *si* 40, *se* 30, *c1* 45, *c2* 15, *d1* 8, *d2* 8, *e2* 5, *f2* 4, *ag1* 15, *ag2* 4.

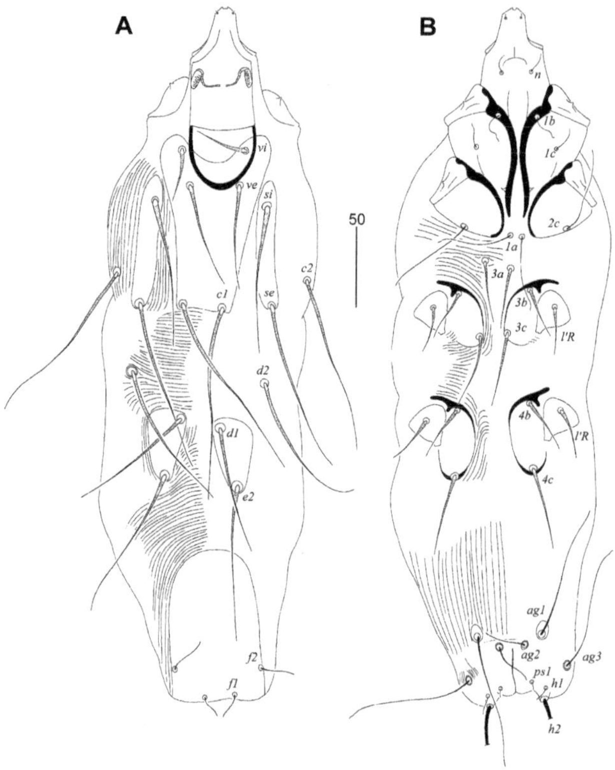

Figure 3. *Tanopicobia stactolaema* Sikora and Skoracki sp. n., female. (**A**)—dorsal view; (**B**)—ventral view.

Type Material

Female holotype, six female paratypes from quill of contour feather of the white-eared barbet *Stactolaema leucotis* (Sundevall) (host reg. no. SNSB-ZSM 59.257; female); TANZANIA: Arusha Region, Arusha District, Arusha, 10 February 1958, coll. von Nagy. One female paratype from same host species (host reg. no. SNSB-ZSM uncatalogued; female); TANZANIA: Kilimanjaro Region, Same District, 9 February 1954, coll. Th. Andersen. Two female and one male paratypes from same host species (host reg. no. SNSB-ZSM 60.31; female); TANZANIA: Arusha Region, Arusha District, Usa-River, 6 December 1959, coll. unknown.

Type Material Deposition

Holotype and paratypes are deposited in the AMU (reg. no. AMU MS 22-1022-114/115/116), except two female paratypes in SNSB-ZSM (reg. no. SNSB-ZSM A20112200).

Additional Material

Ex quill of contour feather of the green barbet *Stactolaema olivacea* (Shelley) (host reg. no. SNSB-ZSM uncatalogued; male); TANZANIA: Morogoro Region, Morogoro District,

5 March 1955, coll. Th. Andersen—three females in AMU (reg. no. AMU MS 22-1022-110) and four females in SNSB-ZSM (reg. no. SNSB-ZSM A20112201).

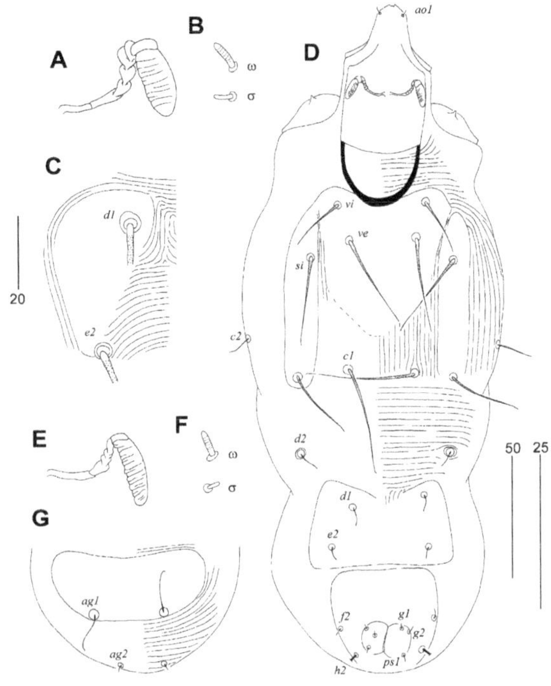

Figure 4. *Tanopicobia stactolaema* Sikora and Skoracki sp. n. Female (**A–C**). (**A**)—peritreme; (**B**)—solenidia of leg I; (**C**)—left hysteronotal shield. Male (**D–G**). (**D**)—dorsal view; (**E**)—peritreme; (**F**)—solenidia of leg I; (**G**)—opisthosoma in ventral view. Scale bars: (**A**,**B**,**E**,**F**) = 25 µm; (**C**) = 20 µm; (**D**, **G**) = 50 µm.

Differential Diagnosis

Tanopicobia stactolaema sp. n. is morphologically similar to the above-described species, *T. hallae* sp. n., by the presence of the well-visible hysteronotal shields and is distinguishable by the following features: in females of *T. stactolaema*, the propodonotal, hysteronotal, and pygidial shields are apunctate; the posterior margins of the hysteronotal shields reach bases of setae *e2*; the lengths of propodonotal setae *ve* and *si* are 60–70 and 80–90, respectively; the lengths of hysteronotal setae *d1* and *d2* are 70–80 and 80–95, respectively; setae *f2* are slightly (1.3 times) longer than *f1*; coxal fields III and IV are apunctate, and setae *4c* are 45–55 long. In females of *T. hallae*, the propodonotal, hysteronotal, and pygidial shields are punctate; the posterior margins of the hysteronotal shields not reach bases of setae *e2*; the lengths of propodonotal setae *ve* and *si* are 45 and 50–70, respectively; the lengths of hysteronotal setae *d1* and *d2* are 105–115 and 110–115, respectively; setae *f2* are 3.5–4 times longer than *f1*; coxal fields III and IV are punctate, and setae *4c* are 75–80 long.

Etymology

The specific name "*stactolaema*" is taken from the generic name of the host.

Tanopicobia trachyphoni Skoracki, Sikora, Jerzak and Hromada, 2020 (Figure 5)

This species was recently described from the red-and-yellow barbet *Trachyphonus erythrocephalus* Cabanis in Tanzania [8], and there have been no other records since the first description. Herein, we report two new hosts species for this quill mite: the yellow-breasted barbet, *Trachyphonus margaritatus* (Cretzschmar), from Djibouti and Eritrea and the d'Arnaud's barbet *Trachyphonus darnaudii* (Prévost and des Murs), from Tanzania.

Because this species was described based only on the holotype and two female paratypes, herein, we give additional data for its description based on the material collected from the type of host species, *T. erythrocephalus*, from Tanzania.

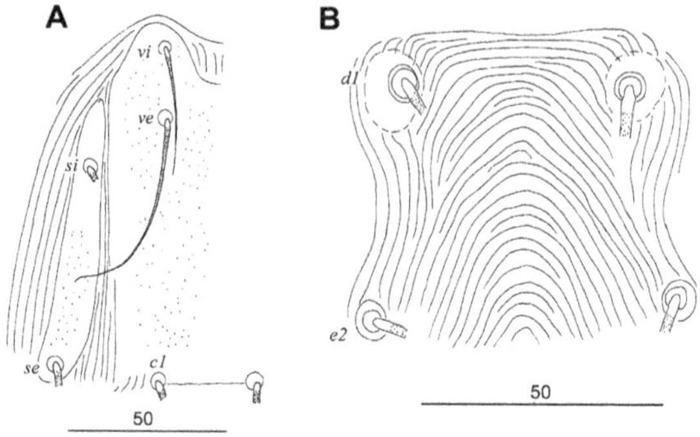

Figure 5. *Tanopicobia trachyphoni* Skoracki et al., 2020, female. (**A**) Propodonotal shield; (**B**) hysteronotum with hysteronotal shields around bases of setae *d1*.

Complementary Description

Female (14 specimens from *T. erythrocephalus* from Tanzania; measurements of type material [8] are in square brackets). Propodonotal shield entire or lateral sclerites bearing bases of setae *si* and se narrowly separated from large medial shield bearing bases of setae *vi*, *ve*, and *c1*. Length of stylophore and stylophoral shield 110–115 and 80–85, respectively. Lengths of setae: *vi* 35–40 [40], *ve* 75–80 [70–80], *si* 95–110 [95], *se* 95–115 [100–110], *c1* 140–155 [140–145], *c2* 110–130 [115], *d1* 110–130 [110–120], *d2* 110–130 [115], *e2* 145 [115–120], *f1* 5–7 [7], *f2* 15–20 [25–35], *h1* 5–7 [10], *h2* 220–265 [260], *ag1* 90 [120–125], *ag2* 25–35 [35], *ag3* 125 [130–140], *ps1* 5 [5], *3a* 30, *3b* 35–40 [40], *3c* 50–55 [50–55], *4b* 35–40 [40], *4c* 50 [50–55], *l'RIII* 35 [35], *l'RIV* 35, *tc'III–IV* 25–30 [30], *tc"III–IV* 60–70 [70].

New Material Examined

Ex the red-and-yellow barbet *Trachyphonus erythrocephalus* Cabanis; TANZANIA: Arusha Region, Arusha District, S. Arusha, 25 March 1960, coll. von Nagy—four females and one male deposited in the AMU (reg. no. AMU MS 22-1022-033) and five females in the SNSB-ZSM (SNSB-ZSM A20112200). From same host species; TANZANIA: Kilimanjaro Region., Same Distr., Lembani, 25 January 1954, coll. Th. Andersen—two females and two males deposited in the AMU (reg. no. AMU MS 22-1022-034) and three females in the SNSB-ZSM (SNSB-ZSM A20112201).

Ex the yellow-breasted barbet, *Trachyphonus margaritatus* (Cretzschmar) (new host species); ERITREA: Northern Red Sea Region, Massawa, 18–26 December 1965, coll. K. E. Linsenmair—six females deposited in the AMU (reg. no. AMU MS 22-1022-(035-036)) and three females in the SNSB-ZSM (reg. no. SNSB-ZSM A20112195). From same host species; DJIBOUTI: Assamo, 2–6 February 2020, coll. B. Sikora and M. Mahamoud-Issa—thirteen females and two males deposited in the AMU (reg. no. AMU MS 22-1114-(001-008))

Ex the d'Arnaud's barbet *Trachyphonus darnaudii* (Prévost and des Murs) (new host species); TANZANIA: Manyara–Arusha Region, 13–30 November 1959, coll. von Nagy—ten females and two males (reg. no. AMU MS 22-1022-(037-039, 045)). Ex same host species; TANZANIA: Manyara Region, Babati District, Magugu, 25 July–6 August 1960, coll. von Nagy—four females and one male deposited in the AMU (reg. no. AMU MS 22-1022-(040-042, 046) and three females in the SNSB-ZSM (reg. no. SNSB-ZSM A20112196). Ex same host species; TANZANIA: Dodoma Region, Bahi District 15 June 1953, coll. Th. Andersen—one female deposited in the AMU (reg. no. AMU MS 22-1022-043). Ex same host species; TANZANIA: Dodoma Region, Kondoa District, Busi, 22 August 1960, coll. von Nagy—three females deposited in the AMU (reg. no. AMU MS 22-1022-047).

4. Discussion

To date, only one species, *Tanopicobia trachyphoni* Skoracki et al., has been recorded from one host species, the red and yellow barbet, *Trachyphonus erythrocephalus* Cabanis [8]. Herein, we conducted a parasitological investigation of the picobiine quill mite fauna associated with birds of the family Lybiidae. Above, we have demonstrated that this bird family is much more widely infested by quill mites than previously thought [8]. Our research has identified the presence of three species of picobiine mites of the genus *Tanopicobia* on ten African barbets belonging to the four genera, i.e., *Trachyphonus* (three species), *Tricholaema* (two species), *Lybius* (three species), and *Stactolaema* (two species). Our findings presented in this study, demonstrate that birds belonging to the family Lybiidae have a unique parasite fauna consisting exclusively of mites from the genus *Tanopicobia* (lack members of the other picobiine genera) and that the distribution of this mite genus is restricted solely to the African barbets. Unfortunately, the small sample size of individuals examined from the genera *Gymnobucco, Pogoniulus*, and *Buccanodon* has not allowed us to confirm the presence of mites on these birds, but we are rather confident that future studies will demonstrate the occurrence of mites on these bird genera. Additionally, given that all examined bird genera of the African barbets were infested by members of the genus *Tanopicobia*, it is expected that species (supposedly new to science) of this genus will also be present on these birds.

Systematic position of Trachyphonus birds vs. *quill mites.* The genus *Trachyphonus* Ranzani, comprises five species distributed exclusively in sub-Saharan Africa [17,20]. Although *Trachyphonus* is currently classified in the family Lybiidae, the relationship of this genus seems to be one of the most intricate problems in the barbet phylogeny. Swierczewski and Raikow [26], using variations in the hind limb muscle, and Bellman [27], analysing fossil records, proposed that *Trachyphonus* is the sister group to the rest of the species in the family Capitonidae. Later, Prum [28] used a cladistic analysis of morphological characters and placed the representatives of *Trachyphonus* in the newly erected subfamily Trachymphoninae in the family Ramphastidae. In 2000, Barker and Lanyon used mitochondrial DNA sequence data [29] and placed it as the sister taxon to the Neotropical radiation. In contrast, in 2004, Moyle, based on the combined gene analyses, placed *Trachyphonus* as the basal taxon of the African radiation and indicated that African barbets (Lybiidae) are monophyletic [30]. Moreover, *Trachyphonus* is considered to be an old and early diverging lineage [26,28,30–32], which may not even be closely related to other African barbets [30]. Our parasitological investigation of the picobiine mites associated with African barbets provides indirect but rather supportive evidence that birds of the genus *Trachyphonus* are indeed part of the family Lybiidae, as they host the same quill mites of the genus *Tanopicobia* as other members of this bird family. In contrast, the other families of the order Piciformes have their own distinct quill mite fauna, e.g., mites of the genera *Picobia* and *Neopicobia* infest birds in the family Picidae, *Pseudopicobia* infests birds in the family Bucconidae, and *Rafapicobia* infests birds in the families Semniornithidae and Ramphastidae [1,3,4,8–16].

5. Conclusions

This study investigated the quill mites on birds of the family Lybiidae (African barbets). The results showed that this bird family is more widely infested by feather mites than previously thought, with three *Tanopicobia* species found on ten African barbet species. Birds belonging to the family Lybiidae have a unique parasite fauna consisting exclusively of mites from the genus *Tanopicobia*, which is restricted solely to African barbets. The study also provides indirect evidence that birds of the genus *Trachyphonus* are indeed part of the family Lybiidae based on their hosting of *Tanopicobia* quill mites, while other bird families have their own distinct quill mite fauna.

Author Contributions: Conceptualisation, B.S., M.U. and M.S.; methodology, M.U. and M.S.; investigation, B.S., M.U., M.M.-I. and M.S.; data curation, M.U. and M.S.; formal analysis, M.U. and M.S.; writing—original draft preparation, B.S., M.U., M.M.-I., M.H. and M.S.; writing—review and editing, B.S., M.U., M.M.-I., M.H. and M.S.; supervision, B.S., M.M.-I., M.U., M.H. and M.S.; project administration, B.S., M.M.-I., M.H. and M.S.; funding acquisition, B.S., M.M.-I., M.H. and M.S. All authors have read and agreed to the published version of the manuscript.

Funding: This research was funded by The National Scholarship Programme, SAIA, n. o. (2022/2023) to B.S., the AMU grant (Excellence Initiative—Research University, grant no. 037/02/POB1/0007) to M.S., the Slovak Research and Development Agency (grant no. APVV-16-0411) and the Agency of the Ministry of Education, Research and Sport of the Slovak Republic and Slovak Academy of Sciences (grant no. 1/0876/21) to M.H., and the Polish National Science Centre's "PRELUDIUM" (grant no. UMO-2017/27/N/NZ8/01554) to M. M-I and B.S.

Institutional Review Board Statement: Not applicable.

Informed Consent Statement: Not applicable.

Data Availability Statement: Data are available upon request from the corresponding author.

Acknowledgments: We thank the reviewers for their valuable comments and suggestions to the manuscript.

Conflicts of Interest: The authors declare no conflict of interest. The funders had no role in the design of the study; in the collection, analyses, or interpretation of data; in the writing of the manuscript; or in the decision to publish the results.

References

1. Kethley, J.B. A revision of the family Syringophilidae (Prostigmata: Acarina). *Contrib. Am. Entomol. Inst.* **1970**, *6*, 1–76.
2. Kethley, J.B. Population regulation in quill mites (Acari: Syringophilidae). *Ecology* **1971**, *52*, 1113–1118. [CrossRef]
3. Skoracki, M. Quill mites (Acari: Syringophilidae) of the Palaearctic region. *Zootaxa* **2011**, *2840*, 1–414. [CrossRef]
4. Skoracki, M.; Sikora, B.; Spicer, G.S. A review of the subfamily Picobiinae Johnston and Kethley, 1973 (Acariformes: Prostigmata: Syringophilidae). *Zootaxa* **2016**, *4113*, 1–95. [CrossRef] [PubMed]
5. Casto, S.D. *Cuculiphilus lobatus* gen. n., sp. n. Representing a new subfamily of quill mites (Acarina: Syringophilidae) from the groove-billed ani, *Crotophaga sulcirostris* (Cuculiformes: Cuculidae). *Southwest Nat.* **1977**, *22*, 169–175. [CrossRef]
6. Zmudzinski, M.; Skoracki, M.; Sikora, B. An Updated Checklist of Quill Mites of the Family Syringophilidae (Acariformes: Prostigmata). 2022. Available online: https://sites.google.com/site/syringophilidae/v2022 (accessed on 1 May 2023).
7. Haller, G. Freyana und Picobia. *Zeitschr. Wissensch. Zool.* **1878**, *30*, 81–98.
8. Skoracki, M.; Sikora, B.; Jerzak, L.; Hromada, M. *Tanopicobia* gen. nov., a new genus of quill mites, its phylogenetic placement in the subfamily Picobiinae (Acariformes: Syringophilidae) and picobiine relationships with avian hosts. *PLoS ONE* **2020**, *15*, e0225982. [CrossRef]
9. Skoracki, M.; Bochkov, A.V.; Wauthy, G. Revision of the quill mites of the genus *Picobia* Haller, 1878 (Acari: Syringophilidae) with notes on their host-parasite relationships. *Insect Syst. Evol.* **2004**, *35*, 155–176. [CrossRef]
10. Skoracki, M.; Spicer, G.S.; OConnor, B.M. A review of mites of the subfamily Picobiinae Johnston & Kethley, 1973 (Prostigmata: Syringophilidae) from North American birds. *Syst. Parasitol.* **2014**, *87*, 99–110. [CrossRef]
11. Skoracki, M.; Hendricks, S.; Spicer, G.S. New species of parasitic quill mites of the genus *Picobia* (Acari: Syringophilidae: Picobiinae) from North American birds. *J. Med. Entomol.* **2010**, *47*, 727–742. [CrossRef]
12. Skoracki, M.; Unsoeld, M.; Kavetska, K.; Kaszewska, K. Quill mites of the subfamily Picobiinae (Acari: Syringophilidae) associated with woodpeckers (Aves: Piciformes: Picidae). *Acta Parasitol.* **2014**, *59*, 68–79. [CrossRef] [PubMed]
13. Glowska, E.; Laniecka, I. A new quill mite species *Neopicobia hepburni* sp. nov. (Cheyletoidea: Syringophilidae) parasitising picid birds (Piciformes: Picidae) in Peru. *Acta Parasitol.* **2014**, *59*, 635–637. [CrossRef]
14. Klimovicova, M.; Skoracki, M.; Wamiti, W.; Hromada, M. New data on the systematics of quill mites of the subfamily Picobiinae (Acari: Syringophilidae) parasitising African birds with description of two new species. *Folia Parasitol.* **2014**, *61*, 394–400. [CrossRef]
15. Skoracki, M.; Zmudzinski, M.; Sikora, B. *Rafapicobia olszanowskii*, a new species of syringophilid mite (Acariformes: Syringophilidae) from *Semnornis ramphastinus* (Piciformes: Semnornithidae. *Annales Zool.* **2020**, *70*, 449–452. [CrossRef]
16. Fain, A.; Bochkov, A.V.; Mironov, S.V. New genera and species of quill mites of the family Syringophilidae (Acari: Prostigmata). *Bull. Inst. Roy. Sci. Nat. Belgique* **2000**, *70*, 33–70.
17. Clements, J.F.; Schulenberg, T.S.; Iliff, M.J.; Roberson, D.; Fredericks, T.A.; Sullivan, B.L.; Wood, C.L. The eBird/Clements Checklist of Birds of the World. 2022. Available online: http://www.birds.cornell.edu/clementschecklist/download/v2022 (accessed on 1 May 2023).
18. del Hoyo, J. (Ed.) *All the Birds of the World*; Lynx Edicions: Barcelona, Spain, 2020; pp. 325–329.

19. Winkler, D.W.; Billerman, S.M.; Lovette, I.J. African Barbets (*Lybiidae*), ver. 1.0. In *Birds of the World*; Billerman, S.M., Keeney, B.K., Rodewald, P.G., Schulenberg, T.S., Eds.; Cornell Lab of Ornithology: Ithaca, NY, USA, 2020. [CrossRef]
20. Fry, C.H.; Keith, S.; Urban, E.K. (Eds.) *The Birds of Africa*; Academic Press: Cambridge, MA, USA, 1988; Volume III, 612p.
21. Soma, M.; Brumm, H. Group living facilitates the evolution of duets in barbets. *Biol. Let.* **2020**, *16*, 20200399. [CrossRef]
22. Walter, D.E.; Krantz, G.W. Collecting, rearing, and preparing specimens. In *A Manual of Acarology*; Krantz, G.W., Walter, D.E., Eds.; Texas Tech University Press: Lubbock, TX, USA, 2009; pp. 83–96.
23. Grandjean, F. Les segments post-larvaires de l'hystérosoma chez les Oribates (Acariens). *Bull. Soc. Zool. Fr.* **1939**, *64*, 273–284.
24. Kethley, J.B. Acarina: Prostigmata (Actinedida). In *Soil Biology Guide*; Dindal, D.L., Ed.; John Wiley & Sons: New York, NY, USA, 1990; pp. 667–756.
25. Grandjean, F. Observations sur les Acariens de la famille des Stigmaeidae. *Arch. Sci. Phys. Nat.* **1944**, *26*, 103–131.
26. Swierczewski, E.V.; Raikov, R.J. Hind limb musculature, phylogeny and classification of the Piciformes. *Auk* **1981**, *98*, 466–480.
27. Ballmann, P. A new species of fossil barbet (Aves: Piciformes) from the Middle Miocene of the Nordlinger Ries (Southern Germany). *J. Vertebr. Paleontol.* **1983**, *3*, 43–48. [CrossRef]
28. Prum, R.O. Phylogenetic interrelationships of the barbets (Aves: Capitonidae) and toucans (Aves: Ramphastidae) based on morphology with comparisons to DNA-DNA hybridisation. *Zool. J. Linn. Soc.* **1988**, *92*, 313–343. [CrossRef]
29. Barker, F.K.; Lanyon, S.M. The impact of parsimony weighting schemes on inferred relationships among toucans and neotropical barbets (Aves: Piciformes). *Mol. Phylogenet. Evol.* **2000**, *15*, 215–234. [CrossRef] [PubMed]
30. Moyle, R.G. Phylogenetics of barbets (Aves: Piciformes) based on nuclear and mitochondrial DNA sequence data. *Mol. Phylogenet. Evol.* **2004**, *30*, 187–200. [CrossRef] [PubMed]
31. Ripley, S.D. The barbets. *Auk* **1945**, *62*, 542–563. [CrossRef]
32. Goodwin, D. Some aspects of taxonomy and relationships of barbets (Capitonidae). *Ibis* **1964**, *106*, 198–220. [CrossRef]

animals

MDPI

Article

A New Species of *Ultratenuipalpus* (Acari: Tenuipalpidae) from Brazil and Re-Description of *Ultratenuipalpus meekeri* (De Leon), the Type Species of the Genus, with DNA Barcodes †

Elizeu B. Castro [1,*], Jennifer J. Beard [2], Ronald Ochoa [3], Gary R. Bauchan [4,‡], Gabriel Otero-Colina [5], Ashley P. G. Dowling [6], Antonio C. Lofego [7] and Reinaldo J. F. Feres [7]

1 Graduate Program in Biodiversity, Department of Biological Sciences, São Paulo State University (UNESP), São José do Rio Preto 15054-000, SP, Brazil
2 Queensland Museum, P.O. Box 3300, South Brisbane, QLD 4101, Australia
3 Systematic Entomology Laboratory (SEL), Agricultural Research Service (ARS), United States Department of Agriculture (USDA), Beltsville Agricultural Research Centre (BARC), Beltsville, MD 20705, USA
4 Electron and Confocal Microscopy Unit (ECMU), Agricultural Research Service (ARS), United States Department of Agriculture (USDA), Beltsville Agricultural Research Centre (BARC), Beltsville, MD 20705, USA
5 Posgrado en Entomología y Acarología, Colegio de Postgraduados, Texcoco 56264, Estado de Mexico, Mexico
6 Department of Entomology and Plant Pathology, University of Arkansas, Fayetteville, AR 72704, USA
7 Department of Biological Sciences, Institute of Biosciences, Humanities and Exact Sciences (IBILCE), São Paulo State University (UNESP), São José do Rio Preto 15054-000, SP, Brazil
* Correspondence: elizeu_unesp@yahoo.com.br
† urn:lsid:zoobank.org:pub:14131833-2594-4279-8F3D-8D9D184311B1.
‡ Deceased 12 January 2021.

Citation: Castro, E.B.; Beard, J.J.; Ochoa, R.; Bauchan, G.R.; Otero-Colina, G.; Dowling, A.P.G.; Lofego, A.C.; Feres, R.J.F. A New Species of *Ultratenuipalpus* (Acari: Tenuipalpidae) from Brazil and Re-Description of *Ultratenuipalpus meekeri* (De Leon), the Type Species of the Genus, with DNA Barcodes. *Animals* **2023**, *13*, 1838. https://doi.org/10.3390/ani13111838

Academic Editors: Monika Fajfer and Maciej Skoracki

Received: 7 March 2023
Revised: 23 May 2023
Accepted: 24 May 2023
Published: 1 June 2023

Simple Summary: The flat mite family Tenuipalpidae includes 41 genera and more than 1100 species worldwide, and is considered one of the most important families of phytophagous mites. The *Ultratenuipalpus* is a small genus with 25 known species present in almost all zoogeographic regions. Here, a new species *Ultratenuipalpus parameekeri* Castro, Ochoa & Feres sp. nov. is described from specimens collected on ferns from Brazil. It represents the first species of the genus described from the country. The type species of the genus *Ultratenuipalpus meekeri* (De Leon) is redescribed based on types and newly collected material from Mexico. Highly detailed low-temperature scanning electron image (LT-SEM) micrographs and DNA barcodes are provided for both species. The taxonomy of the genus *Ultratenuipalpus* and the ontogenetic additions of leg setae are discussed.

Abstract: Species of the genus *Ultratenuipalpus* bear a broad subquadrate propodosoma with many large, flattened, lanceolate to ovate dorsal setae. They also bear some plesiomorphic character states, such as the presence of three pairs of ventral *ps* setae. Here, we describe *Ultratenuipalpus parameekeri* Castro, Ochoa & Feres sp. nov. based on adult females, males, and immatures, collected on ferns from Brazil. We also re-describe *Ultratenuipalpus meekeri* (De Leon), the type species of the genus, based on types and newly collected material from Mexico, and include additional novel data (e.g., dorsal and ventral ornamentation, leg chaetotaxy, and setal measurements) in a standardized form. We include highly detailed images obtained using LT-SEM, accompanied by DNA barcodes, for both species. The ontogenetic additions of leg chaetotaxy are presented and discussed.

Keywords: flat mites; new species; ferns; integrative taxonomy; LT-SEM; COI; ontogeny

1. Introduction

The *Ultratenuipalpus* Mitrofanov is a small genus of the family Tenuipalpidae (Trombidiformes: Tetranychoidea), with 25 known species to date [1–3]. Most species are described from three countries: New Zealand (10 species), Australia (3), and Chile (3) [3,4]. Species of the genus bear a broad subquadrate propodosoma with many large, flattened, lanceolate, and/or obovate to ovate dorsal setae [2]. They also bear some potentially plesiomorphic character states, such as the presence of three pairs of pseudanal setae (*ps1–ps3*) and the absence of genital plates [1,2], which are both shared across the Tetranychoidea [5].

The *Ultratenuipalpus* shares many character states with the genera *Extenuipalpus* and *Tenuipalpus*, such as having the prodorsum broader than the opisthosoma [2,6]. The *Ultratenuipalpus* and *Extenuipalpus* also share the character of dorsal opisthosomal setae *h2* not being flagellate; the *Ultratenuipalpus* and *Tenuipalpus sensu stricto* share the presence of lateral body projections associated with prodorsal setae *sc2* [2,6].

The genus *Extenuipalpus* was recently reinstated and includes only three species described from South Africa [2,3], while *Tenuipalpus* is the largest genus in the flat mite family, with over 300 described species worldwide [3]. According to Beard et al. [2], the *Ultratenuipalpus* and these two genera are closely allied, and the *Extenuipalpus* may occupy a position intermediate between the *Ultratenuipalpus* and *Tenuipalpus*.

Species of the *Ultratenuipalpus* occur on different families of ferns (e.g., Dennstaedtiaceae, Thelypteridaceae), conifers (e.g., Araucariaceae, Podocarpaceae), monocots (e.g., Arecaceae, Asteliaceae), and dicots plants (e.g., Proteaceae, Rubiaceae) [3,4]. While most species have only ever been recorded from the original host plant, there are some species (e.g., *U. aberrans*, *U. coprosmae*) that have been found on multiple plants of different families.

Here, we describe a new species of the *Ultratenuipalpus* collected on ferns from Brazil and re-describe the type species of the genus, *U. meekeri* (De Leon), in a standardized form. As these two species are morphologically similar and share several character states, including the shape of dorsal setae and chaetotaxy of the legs, molecular analyses were undertaken to confirm their separation using material freshly collected from Brazil and Mexico.

2. Materials and Methods

A portion of the samples collected of each species was maintained in 70% ethanol for subsequent use in low temperature SEM (LT-SEM) studies. Mites for LT-SEM were studied using the previously described methodology [7]. Another portion of the samples of each species was maintained in 100% alcohol for a subsequent molecular analysis.

DNA was extracted from individual mites using the QIAGEN DNeasy Blood & Tissue kits following standard protocols with the following exceptions: (1) mite specimens were carefully pierced with a sterilized minutin pin and then incubated overnight in a solution of buffer ATL and Proteinase K as per instructions and (2) a final elution was performed with 100 μL to increase the total DNA concentration. A portion of cytochrome oxidase I was amplified by PCR with previously published primers [8,9]. The amplification reactions were performed in 25 μL volumes containing 2.5 μL of manufacturer supplied buffer, 0.2 μL (5 units) of Platinum Taq polymerase (Invitrogen), 2.5 μL dNTP (0.25 mM of each base), 1 μL of each primer (10 mM), 2.5 μL of $MgCl_2$ (25 mM), 11.3 μL of ddH_2O, and 4 μL of the template DNA. The samples were denatured at 94 °C for 2 min, followed by 30 cycles of 1 min denaturation at 92 °C, 1 min annealing at 50 °C, and 1.5 min extension at 72 °C, with a final elongation of 10 min after the completion of all cycles. PCR products were visualized on a 1% agarose gel saturated with GelRed (Biotium, Hayward, CA, USA). DNA was then purified with a QIAquick PCR Purification kit (Qiagen, Germantown, MD, USA). The amplified fragments were sent to Macrogen USA for sequencing. No additional primers were used for sequencing. COI sequences have been deposited in GenBank (https://www.ncbi.nlm.nih.gov/, accessed on 15 January 2023).

All measurements are given in micrometers (μm). Measurements are presented for the holotype followed by the range for all types in parentheses. The number of leg setae is written as the total number of setae followed by the number of solenidia in parentheses. Terminology of leg and body setation is adapted from [10–12]. Photographs of slide-mounted specimens were obtained using a Zeiss Axioscope™ microscope (Carl Zeiss

Inc., Thornwood, NY, USA) with a differential interference contrast (DIC) 100× Plan Apochromatic objective with an NA 1.4.

Type specimens and vouchers of non-type specimens are deposited in the Collection of Acari, Departamento de Zoologia e Botânica, UNESP, São José do Rio Preto, State of São Paulo, Brazil (DZSJRP, http://www.splink.cria.org.br, accessed on 10 December 2022) and in the National Insect and Mite Collection, National Museum of Natural History, Smithsonian Institution, located at the Systematic Entomology Laboratory (SEL), USDA, Beltsville, MD, USA (NMNH). The holotype of *U. meekeri* is deposited in the Museum of Comparative Zoology (MCZ), Harvard University, Cambridge, MA, USA.

3. Results

3.1. Description of Ultratenuipalpus parameekeri Castro, Ochoa & Feres sp. nov.

Family Tenuipalpidae Berlese, 1913
Genus *Ultratenuipalpus* Mitrofanov, 1973
Type species: *Ultratenuipalpus meekeri* (De Leon), 1957

Diagnosis of the genus (Based on [2]). "Body shape from elongate-ovate to broadly rounded; broad propodosoma differentiated from narrower opisthosoma (although anterior opisthosoma is broad at junction with propodosoma). Anterior margin of prodorsum usually with median forked projection forming a short notch". Prodorsum with one pair of lateral body projections anterior to setae *sc2* present or absent. Posterior margin of opisthosoma with a broad rounded projection between setae *h1* usually present. "Prodorsal shield divided by two oblique folds running from vicinity of the eyes angled medially to posterior margin of shield, superficially dividing the shield into three smaller plate-like regions; a small plate is indicated between setae *c3–d3* on dorsal opisthosomal margin; posteroventral body margin often with band of pustulate cuticle. Dorsal opisthosoma with setae *c1, c3, d1, d3, e1, e3, f3, h1, h2* present (except setae *f3* absent in *U. aberrans*); setae *f2* present or absent, when present then inserted on lateral margin aligned with lateral setae *e3, f3, h1, h2*; setae *c2, d2, e2* absent. Setae *h2* not flagellate, similar in form to *h1*; setae *sc2, e3, f2, f3, h1, h2* flattened, lanceolate, oblanceolate, obovate to ovate, with *sc2* often falcate; form of other dorsal setae variable (e.g., *sc2* and *f3* flagellate in *U. bunyai*). Three pairs of *ps* (pseudanal) setae present; female with *ps3* positioned anteriorly on anal valves and much shorter than *ps1–2*, which are closely associated with each other and positioned posterolaterad anal valves; setae *ps2* usually much longer than *ps1*; male with *ps3* modified into accessory genital stylets and inserted on elongate genitoanal valves, with *ps1–2* positioned as in female. Ventral, genital and anal regions membranous, without defined sclerotized plates; flap of ovipore and anus surrounded by strongly plicate and wrinkled membranous cuticle. Genital setae *g2* inserted slightly anterior to *g1* on reduced genital flap; *g1–2* often aligned longitudinally with setae *ag*. Intercoxal setae (*3a, 4a*) not multiplied. Palps four segmented; palp tibiae with 1–2 setae; palp tarsi with 1–3 phaneres, with solenidion always present, sometimes curved, often inserted basally on palp tarsus segment at junction with palp tibia. Dorsal setae on legs inserted in lateral position. Femora of legs I–II with four setae (*d, l′, v′, bv″*); genua I–II with three setae (*d, l′, l″*) (except some species variously described with two setae—*U. acharis* (genua I–II with 3–2 setae; possibly *d* absent), *U. pterophilus* (genua I–II with 2–2)); tibiae I–II usually with five setae (except two species described with four setae, *U. avarua* (*v″* absent) and *U. lacorpuzrarosae* (possibly *d* absent)). Tarsal claws pad-like. Immature stages with setae *c1* inserted distinctly anterior to level of setae *c3*. See also diagnosis of [1].

Description

Diagnosis. Female: As per genus, in addition to: prodorsal setae *v2, sc1* minute to short, and *sc2* large, flattened obovate to ovate; dorsal opisthosoma with 10 pairs of setae (*f2* present); most of the dorsal opisthosomal setae large, flattened, obovate to ovate, except setae *d3* is distinctly short and *c3* almost orbicular; pair lateral body projections anterior to setae *sc2* and projection between opisthosomal setae *h1* both present; palp four segmented, setal formula 0, 0, 2, 2. Male: Opisthosoma narrower than that of the female, with distinct transverse constriction (waist) between setae *d1* and *e1*; many dorsal setae similar in form to those of females, except *c1*, *d1*, and *e1* short to minute, *d3* longer, and setae along posterior margin of opisthosoma (especially *e3*) narrower and more elongated than those of the female. Tarsi I–II each with two solenidia (ω' paraxial and ventrolateral; ω'' antiaxial); tarsus III with one solenidion ω' paraxial and ventrolateral. Immatures: with lateral body projection anterior to setae *sc2* present (except absent in larvae); posterior projection between setae *h1* absent; dorsal setae similar in general form to those of the female, except setae *c1*, *d1*, and *e1* short to minute. Larvae with central prodorsum and pygidial region of posterior opisthosoma with finely colliculated integument.

Female (n = 10) (Figures 1–10)

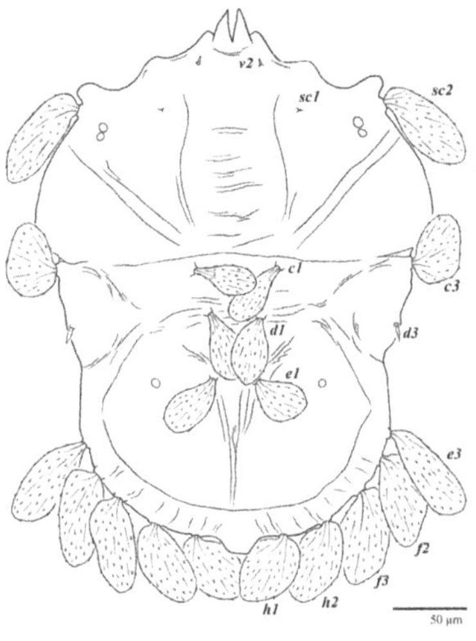

Figure 1. *Ultratenuipalpus parameekeri* Castro, Ochoa & Feres sp. nov. (Female): view of dorsum.

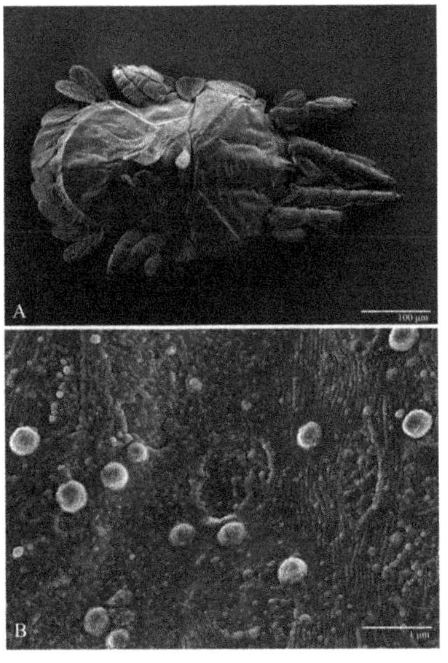

Figure 2. *Ultratenuipalpus parameekeri* Castro, Ochoa & Feres sp. nov. (Female): (**A**) dorsal view; (**B**) view of cuticular microplates on the dorsum.

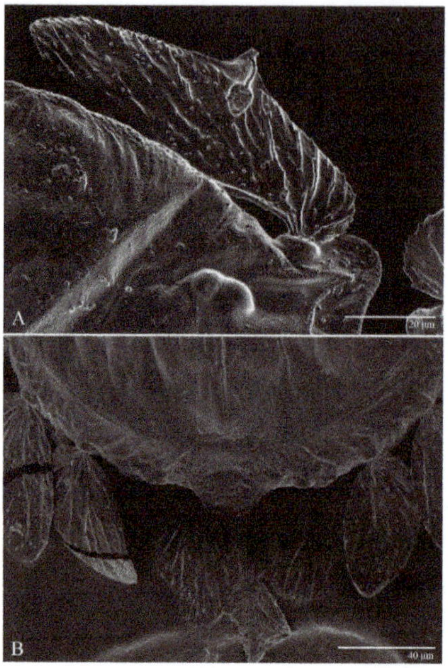

Figure 3. *Ultratenuipalpus parameekeri* Castro, Ochoa & Feres sp. nov. (Female): (**A**) detail of the lateral region of prodorsum, with setae *sc2*. Note the presence of body projection anterior to *sc2*; (**B**) posterior region of opisthosoma, indicating the body projection on posterior margin.

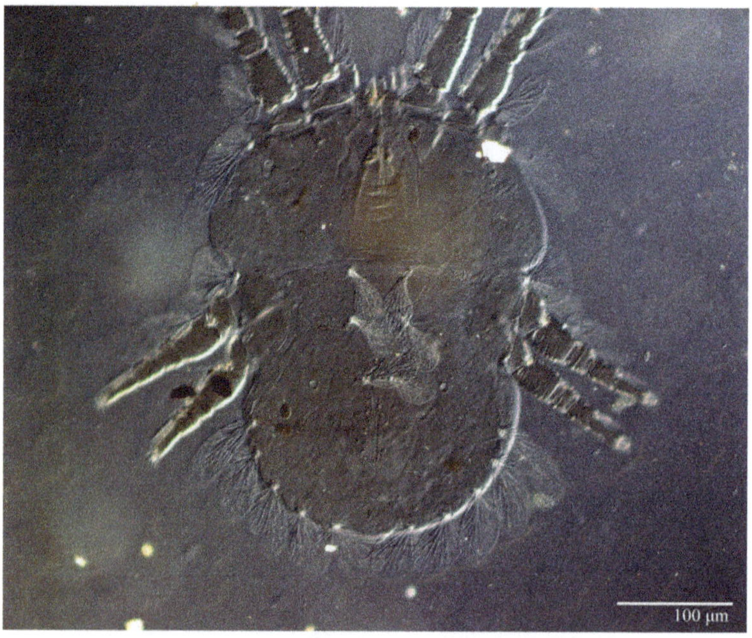

Figure 4. *Ultratenuipalpus parameekeri* Castro, Ochoa & Feres sp. nov. (Female): view of dorsum.

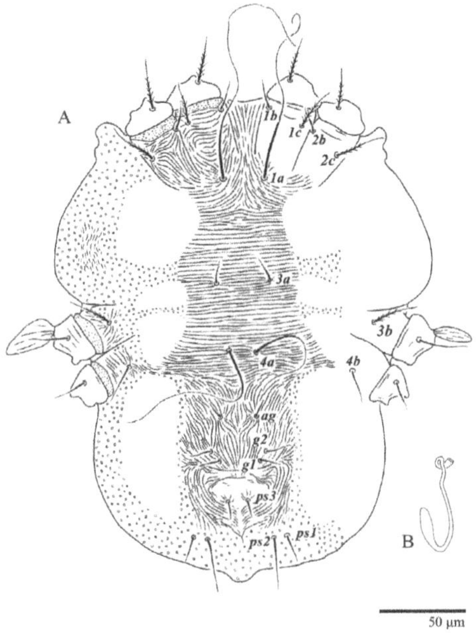

Figure 5. *Ultratenuipalpus parameekeri* Castro, Ochoa & Feres sp. nov. (Female): (**A**) view of venter; (**B**) spermatheca.

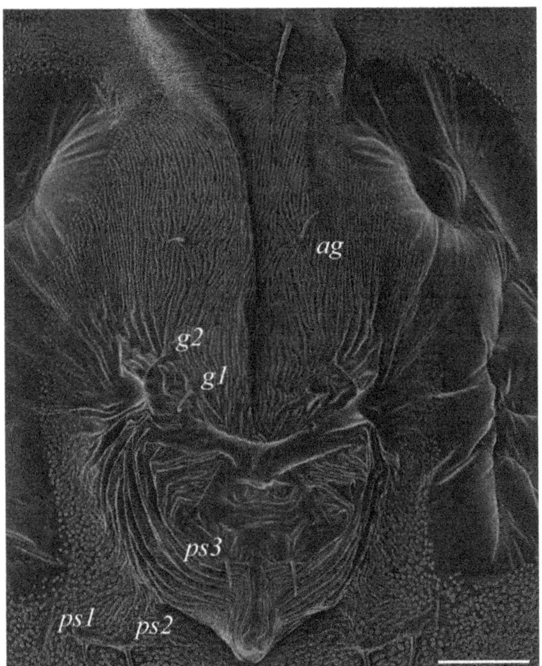

Figure 6. *Ultratenuipalpus parameekeri* Castro, Ochoa & Feres sp. nov. (Female): posterior ventral opisthosoma.

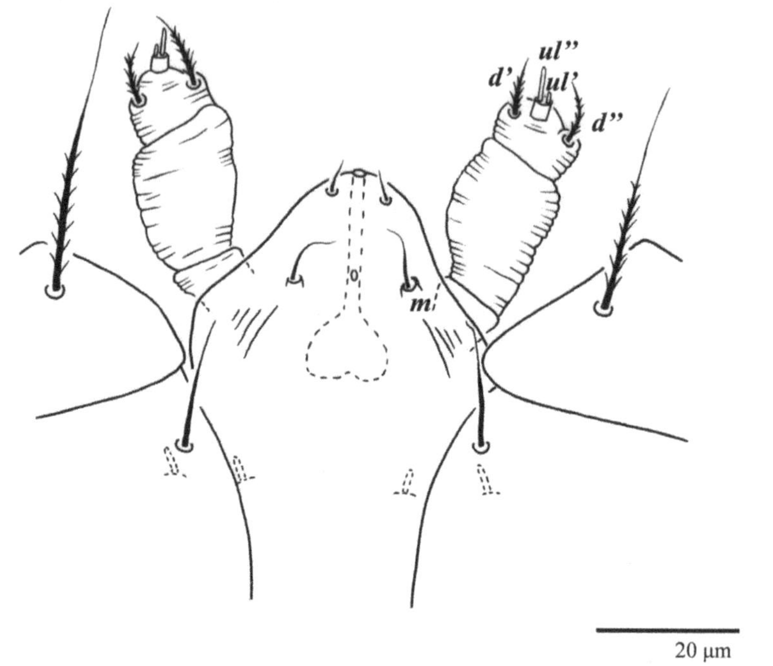

Figure 7. *Ultratenuipalpus parameekeri* Castro, Ochoa & Feres sp. nov. (Female): ventral infracapitulum.

Figure 8. *Ultratenuipalpus parameekeri* Castro, Ochoa & Feres sp. nov. (Female): (**A**) view of ventral infracapitulum; (**B**) detail of palp.

Figure 9. *Ultratenuipalpus parameekeri* Castro, Ochoa & Feres sp. nov. (Female): detail of stylet tip with lateral serrations.

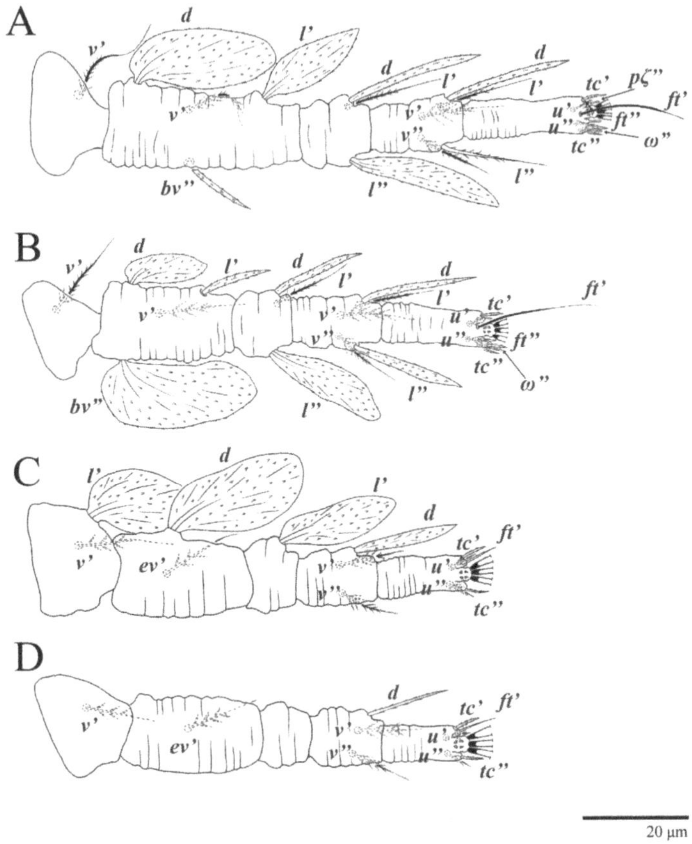

Figure 10. *Ultratenuipalpus parameekeri* Castro, Ochoa & Feres sp. nov. (Female): (**A**) leg I; (**B**) leg II; (**C**) leg III; (**D**) leg IV. (Right legs).

Body measurements: distance between setae *v2–h1* 350 (340–375), *sc2–sc2* 220 (215–230); other measurements: *v2–v2* 45 (37–45), *sc1–sc1* 93 (90–110), *c1–c1* 60 (55–63), *c3–c3* 260 (245–265), *d1–d1* 40 (37–45), *d3–d3* 240 (230–240), *e1–e1* 32 (27–33), *e3–e3* 220 (215–225), *f2–f2* 205 (205–215), *f3–f3* 170 (170–185), *h1–h1* 58 (58–68), *h2–h2* 120 (115–135).

Dorsum (Figures 1–4). Anterior margin of prodorsum with a short median forked projection forming a short notch 27 (20–27). Dorsum smooth, with pair lateral projections anterior to setae *sc2* and single projection between opisthosomal setae *h1* (Figure 3B). A pair of converging folds from the eyes to near the sejugal furrow on the prodorsum posterior margin. Prodorsal setae *v2* and *sc1* short to minute; *sc2* large, flattened elongate obovate (Figures 1 and 3A); most opisthosomal setae similar to prodorsal setae *sc2*, except *d3* short. Setal measurements: *v2* 5 (4–7), *sc1* 3 (3–5), *sc2* 74 (74–82), *c1* 52 (52–58), *c3* 36 (36–45), *d1* 55 (54–55), *d3* 10 (8–10), *e1* 50 (48–55), *e3* 70 (70–81), *f2* 65 (65–70), *f3* 61 (60–68), *h1* 52 (52–60), *h2* 58 (58–65).

Venter (Figures 5A and 6). Ventral integument weakly striate along central region and densely colliculated around lateral body margin; ventral, genital, and anal plates not developed, and entire region membranous and distinctly plicate; ventral setae filiform, with coxal setae *1c*, *2c*, and *3b* barbed; setae *ps2* distinctly longer than *ps1*. Setal measurements: *1a* 105 (100–135), *1b* 13 (13–16), *1c* 30 (25–30), *2b* 22 (22–26), *2c* 38 (30–39), *3a* 18 (15–18), *3b* 32 (31–36), *4a* 145 (115–145), *4b* 22 (18–22), *ag* 10 (10–11), *g1* 16 (14–16), *g2* 16 (15–18), *ps1* 28 (22–28), *ps2* 53 (50–55), *ps3* 12 (10–13).

Gnathosoma (Figures 7–9). Palps four segmented, setal formula: 0, 0, 2, 2; tibia with two setae, d' 7 (6–8), d'' 6 (5–6), tarsus with one eupathidium 5 (3–5) and one solenidion 1 (1–2). Ventral setae *m* 8 (6–8); distance between setae *m–m* 14 (13–16). Tips of cheliceral stylets with a few bluntly rounded lateral projections (Figure 9).

Spermatheca (Figure 5B). Duct length ca. 75–85, terminating in smooth rounded bulb.

Legs (Figure 10). Setation (from coxae to tarsi): I 3–1–4–3–5–8(1), II 2–1–4–3–5–8(1), III 1–2–2–1–3–5, IV 1–1–1–0–3–5. Tarsi I–II each with one solenidion ω'' 7 (6–8) (for both tarsi I and tarsi II) and two eupathidia $p\zeta'$–$p\zeta''$ (5–6, 5–6; 5, 5–6, respectively); femur I with setae *d* obovate and l' broadly lanceolate; femur II with setae *d* narrowly obovate, l' lanceolate, and bv'' obovate to broadly falcate. Femora, genua, and tibiae with setae *d* inserted in lateral position. Detail of the development of leg chaetotaxy in Table 1.

Table 1. Additions of leg setae during ontogeny in both *Ultratenuipalpus parameekeri* Castro, Ochoa & Feres and *Ultratenuipalpus meekeri* (De Leon). The stage in which each seta first appears is indicated. Setae in parentheses represent pairs.

	Coxa	Trochanter	Femur	Genu	Tibia	Tarsi
Leg I						
Larva	*1b*	-	d, v', bv''	l'	d, (v), (l)	(u), ($p\zeta$), (ft), ω''
Protonymph	*1c*	-	-	-	-	-
Deutonymph	-	v'	l'	d, l''	-	(tc)
Female/male	-	-	-	-	-	ω' [1]
Leg II						
Larva	-	-	d, v', bv''	l'	d, (v), (l)	(u), ($p\zeta$), (ft), ω''
Protonymph	*2c*	-	-	-	-	-
Deutonymph	*2b*	v'	l'	d, l''	-	(tc)
Female/male	-	-	-	-	-	ω' [1]
Leg III						
Larva	-	-	d, ev'	l'	d, (v)	(u), ft'
Protonymph	*3b*	l'	-	-	-	-
Deutonymph	-	v'	-	-	-	-
Female/male	-	-	-	-	-	(tc), ω' [1]
Leg IV						
Protonymph	-	-	ev'	-	d, (v)	(u), ft'
Deutonymph	*4b*	-	-	-	-	-
Female/male	-	v'	-	-	-	(tc)

[1] Solenidion ω' added only on tarsi I, II, and III in the males.

Microplates (Figure 2B). The microplate layer forms a reticulate network of thick ridges covered in small, single, irregularly-shaped wax-like crystals or masses.

Color. The body is mostly orange with the margin of prodorsum and opisthosoma with dark spots, eyes red, and legs orange. The dorsal body setae and leg setae white to translucent.

Male (n = 5) (Figures 11–15)

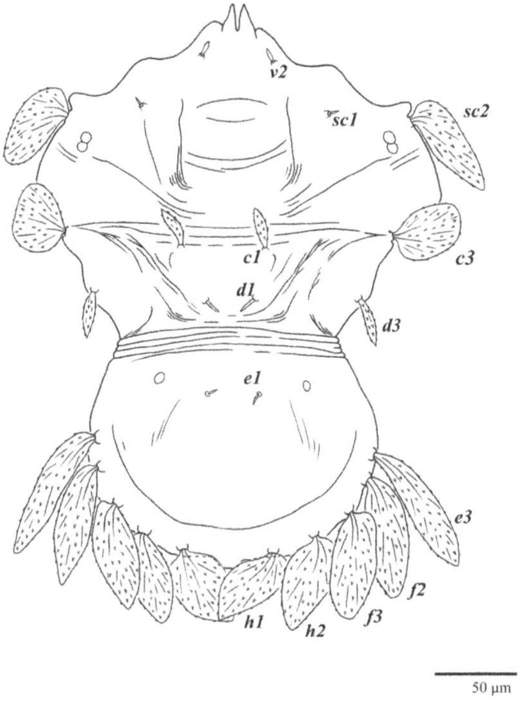

Figure 11. *Ultratenuipalpus parameekeri* Castro, Ochoa & Feres sp. nov. (Male): view of dorsum.

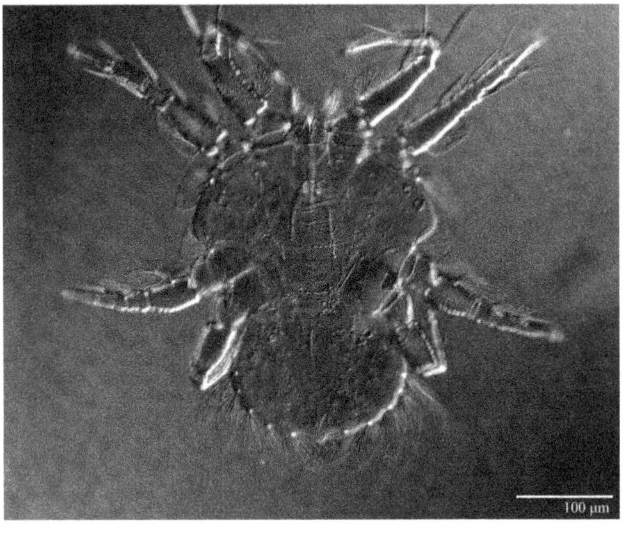

Figure 12. *Ultratenuipalpus parameekeri* Castro, Ochoa & Feres sp. nov. (Male): view of dorsum.

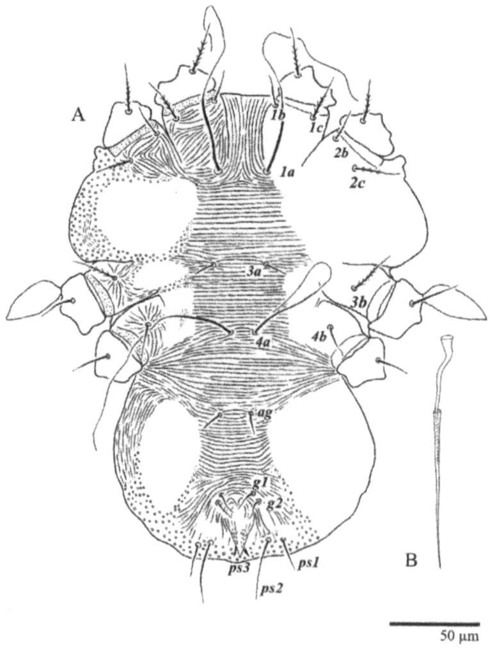

Figure 13. *Ultratenuipalpus parameekeri* Castro, Ochoa & Feres sp. nov. (Male): (**A**) view of venter; (**B**) aedeagus.

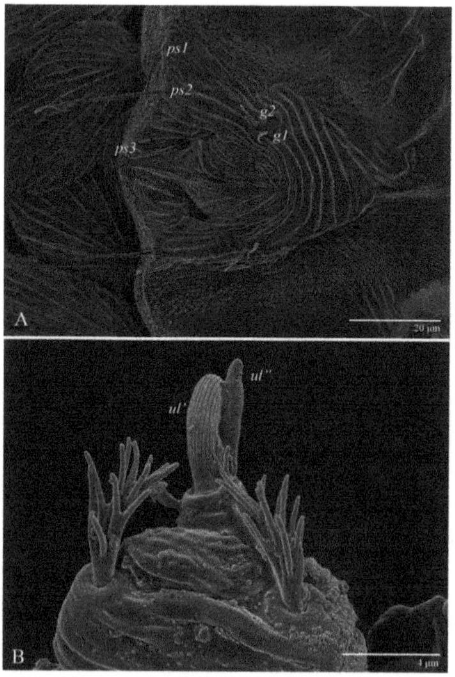

Figure 14. *Ultratenuipalpus parameekeri* Castro, Ochoa & Feres sp. nov. (Male): (**A**) posterior ventral opisthosoma; (**B**) detail of palp. Note the well-developed solenidion.

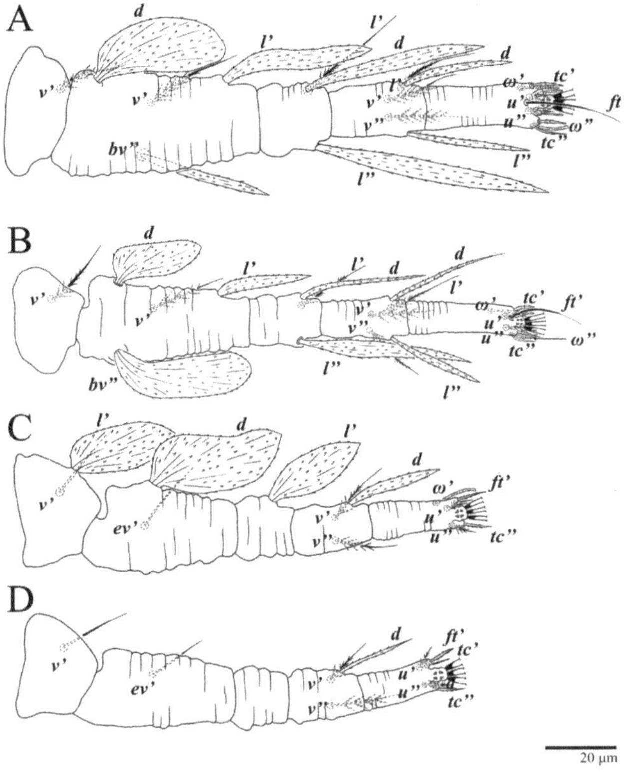

Figure 15. *Ultratenuipalpus parameekeri* Castro, Ochoa & Feres sp. nov. (Male): (**A**) leg I; (**B**) leg II; (**C**) leg III; (**D**) leg IV. (Right legs).

Body measurements: distance between setae, *v2–h1* 285–310, *sc2–sc2* 200–220; other measurements: *v2–v2* 45–50, *sc1–sc1* 105–120, *c1–c1* 50–58, *c3–c3* 190–215, *d1–d1* 27–30, *d3–d3* 155–175, *e1–e1* 27–33, *e3–e3* 165–175, *f2–f2* 160–170, *f3–f3* 140–145, *h1–h1* 55–58, *h2–h2* 105–115.

Dorsum (Figures 11 and 12). Anterior margin of prodorsum with a short median forked projection forming a short notch. The dorsum is smooth, with a pair of lateral projections anterior to setae *sc2* and a single projection between opisthosomal setae *h1*. Many dorsal setae similar in general form to those of female, except *c1*, *d1*, and *e1* short to minute, *d3* longer, and setae along posterior margin of opisthosoma (especially *e3*) narrower and more elongated than those of the female. Setal measurements: *v2* 5–7, *sc1* 4–6, *sc2* 60–67, *c1* 29–30, *c3* 40–45, *d1* 8–10, *d3* 19–27, *e1* 5–7, *e3* 77–80, *f2* 60–72, *f3* 60–63, *h1* 49–50, *h2* 53–55.

Venter (Figures 13 and 14A). Ventral integument weakly striate along central region and densely colliculated along lateral body margin; ventral setae filiform, with coxal setae *1c*, *2c*, and *3b* barbed; setae *ps2* distinctly longer than *ps1*; setae *ps3* thickened and inserted ventrally on the elongate tapered anal valves. Setal measurements: *1a* 100–105, *1b* 18–21, *1c* 27–30, *2b* 23–29, *2c* 34–35, *3a* 17–23, *3b* 40–42, *4a* 120–130, *4b* 23–30, *ag* 14–15, *g1* 12–13, *g2* 14–17, *ps1* 21–23, *ps2* 42–55, *ps3* 13–14.

Gnathosoma (Figure 14B). Palps four segmented, setal formula: 0, 0, 2, 2; tibia with two setae, d' 7–8, d'' 7–8, tarsus with one eupathidium 5–6 and one solenidion 6. Ventral setae *m* 6–7; distance between setae *m–m* 13–14.

Legs (Figure 15). Setation (from coxae to tarsi): I 3–1–4–3–5–8(2), II 2–1–4–3–5–8(2), III 1–2–2–1–3–5(1), IV 1–1–1–0–3–5. Tarsi I–II each with two solenidia (one abaxial, one adaxial), tarsi I ω'' 10–11, ω' 16–17, tarsi II ω'' 11–12, ω' 14–15 and two eupathidia $p\zeta'$–$p\zeta''$

(6–7, 7; 5–6, 5–6), and tarsus III with one solenidion (paraxial and ventrolateral) ω' 13–15. Leg setae similar to that of the female; seta l'' on genu I distinctly elongated. Detail of the development of leg chaetotaxy in Table 1.

Aedeagus (Figure 13B). As figured; ca. 130 long.

Deutonymph (n = 3) (Figure 16)

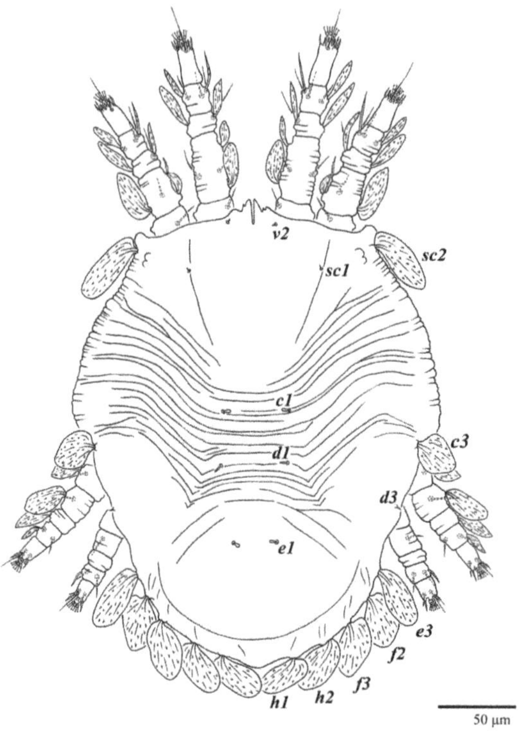

Figure 16. *Ultratenuipalpus parameekeri* Castro, Ochoa & Feres sp. nov. (Deutonymph): dorsum, with detail of legs (unguinal setae u'–u'' on tarsus I and II are not included in the drawing).

Body measurements: distance between setae *v2–h1* 335–365, *sc2–sc2* 165–180; other measurements: *v2–v2* 37–40, *sc1–sc1* 93–105, *c1–c1* 42–55, *c3–c3* 235–265, *d1–d1* 40–50, *d3–d3* 200–215, *e1–e1* 18–28, *e3–e3* 155–175, *f2–f2* 145–160, *f3–f3* 120–135, *h1–h1* 45–50, *h2–h2* 87–95.

Dorsum (Figure 16). Anterior margin of prodorsum with a short median forked projection forming a short notch; pair of lateral projections anterior and adjacent to setae *sc2* present; projection not formed (or rudimentary) between setae *h1*. Prodorsal region smooth; region between setae *sc2–c3* with transverse plicae and folds; region posterior to setae *d1–d3* smooth. Dorsal setae similar in general form to that of females, except setae *c1*, *d1* and *e1* short to minute. Setal measurements: *v2* 3–4, *sc1* 2–3, *sc2* 60–64, *c1* 3–5, *c3* 32–35, *d1* 2–3, *d3* 3–4, *e1* 3–4, *e3* 45–54, *f2* 40–42, *f3* 41–42, *h1* 35–36, *h2* 38–45.

Gnathosoma. Palps similar to those of female, setal formula: 0, 0, 2, 2; tibia with two setae, d' 4–5, d'' 4–5, tarsus with one eupathidium 3–4 and one minute solenidion, 1 long. Ventral setae *m* 4–5; distance between setae *m–m* 10–11.

Venter. Cuticle covered with fine and mostly transverse striae. Coxal, genital, and anal setae fine. Setal lengths: *1a* 80–100, *1b* 10–15, *1c* 10–12, *2b* 10–14, *2c* 18–20, *3a* 12–13, *3b* 12–17, *4a* 50–80, *4b* 12–21, *ag* 7–8, *g1* 8–9, *ps1* 12–14, *ps2* 25–27, *ps3* 9–10. Setae *g2* absent.

Legs (Figure 16). Setation (from coxae to tarsi): I 3–1–4–3–5–8(1), II 2–1–4–3–5–8(1), III 1–2–2–1–3–5, IV 1–0–1–0–3–5. Leg chaetotaxy similar to that of the female, except by trochanter IV nude; tarsi I–II each with one solenidion ω'' (tarsi I 4–5 and tarsi II 5), and two eupathidia $p\zeta'$–$p\zeta''$ (4–5, 5; 4–5, 4–5, respectively). Detail of the development of leg chaetotaxy in Table 1.

Protonymph (n = 3) (Figure 17)

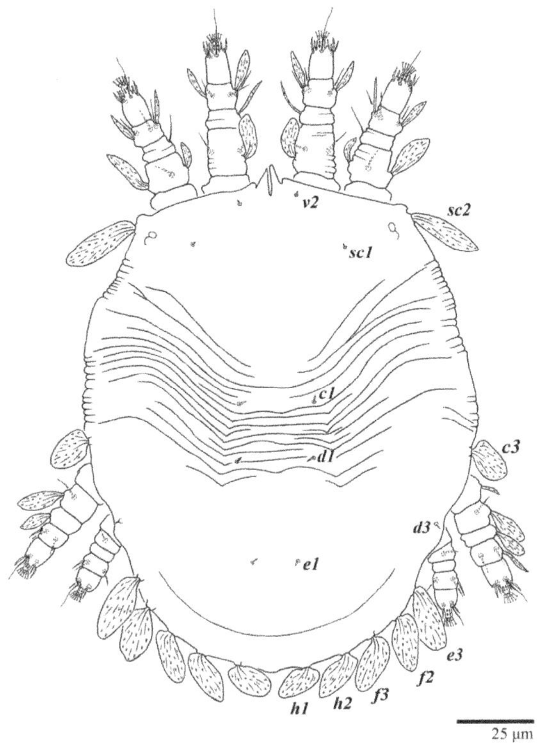

Figure 17. *Ultratenuipalpus parameekeri* Castro, Ochoa & Feres sp. nov. (Protonymph): dorsum, with detail of legs (unguinal setae u'–u'' on tarsus I and II are not included in the drawing).

Body measurements: distance between setae *v2–h1* 275–290, *sc2–sc2* 135–145; other measurements: *v2–v2* 32–35, *sc1–sc1* 80–83, *c1–c1* 40–43, *c3–c3* 185–195, *d1–d1* 35–38, *d3–d3* 150–155, *e1–e1* 22–25, *e3–e3* 120–130, *f2–f2* 110–115, *f3–f3* 90–95, *h1–h1* 30–33, *h2–h2* 62–65.

Dorsum (Figure 17). Anterior margin of prodorsum with a short median forked projection forming a short notch; pair of lateral body projections anterior and adjacent to setae *sc2* present. Prodorsal region smooth; region between setae *sc2–c3* with transverse striations and region posterior to setae *c3* smooth; dorsal setae similar to that of the female, except setae *c1*, *d1*, and *e1* short. Setal measurements: *v2* 2–3, *sc1* 2–3, *sc2* 40–44, *c1* 3–4, *c3* 24–25, *d1* 3–4, *d3* 3–4, *e1* 2–3, *e3* 30–32, *f2* 27–28, *f3* 24–26, *h1* 22–25, *h2* 24–25.

Gnathosoma. Palps similar to those of the female, setal formula: 0, 0, 2, 2; tibia with two setae, d' 4–5, d'' 3–4, tarsus with one eupathidium 3–4 and one solenidion, 1 long. Ventral setae *m* 4–5; distance between setae *m–m* 10–12.

Venter. Cuticle covered with fine and mostly transverse striae. Coxal, genital and anal setae fine. Setal measurements: *1a* 65–67, *1b* 10–11, *1c* 9–12, *2c* 12–13, *3a* 10–13, *3b* 14–17, *ag* 6–7, *ps1* 7–9, *ps2* 13–15, *ps3* 7–8. Setae *2b*, *4a*, *4b*, *g1* and *g2* absent.

Legs (Figure 17). Setation (from coxae to tarsi): I 3–0–3–1–5–6(1), II 1–0–3–1–5–6(1), III 1–0–2–0–3–5, IV 0–0–1–0–3–3. Tarsi I–II each with one solenidion ω'' 4–5 (for both tarsi I and tarsi II) and two eupathidia $p\zeta'$–$p\zeta''$ (all 3–4). Detail of the development of leg chaetotaxy in Table 1.

Larva (n = 3) (Figure 18)

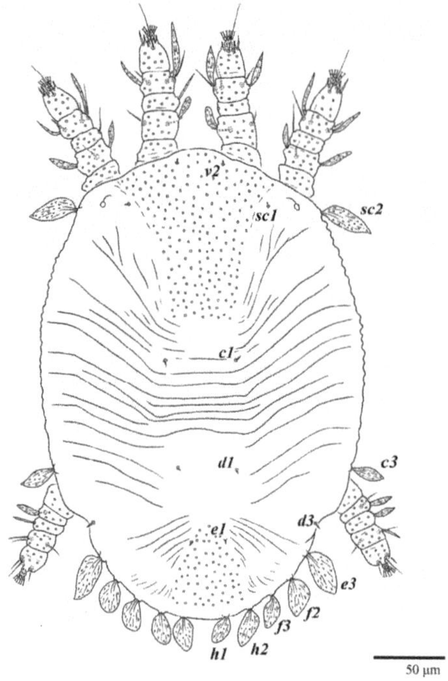

Figure 18. *Ultratenuipalpus parameekeri* Castro, Ochoa & Feres sp. nov. (Larva): dorsum, with detail of legs (unguinal setae *u′–u″* on tarsus I and II are not included in the drawing).

Body measurements: distance between setae *v2–h1* 220–230, *sc2–sc2* 115–125; other measurements: *v2–v2* 22–25, *sc1–sc1* 70–73, *c1–c1* 32–38, *c3–c3* 140–150, *d1–d1* 30–38, *d3–d3* 110–115, *e1–e1* 16–18, *e3–e3* 100–105, *f2–f2* 86–88, *f3–f3* 67–70, *h1–h1* 20–23, *h2–h2* 40–45.

Dorsum (Figure 18). Prodorsal region with colliculated integument anteromedially; region between setae *sc2–c3* with oblique and transverse folds; pygidial region posterior to setae *e1* with colliculated integument; dorsal setae similar in general form to those of females except much smaller and setae *c1*, *d1*, and *e1* minute. Setal measurements: *v2* 2–3, *sc1* 1–2, *sc2* 25–26, *c1* 2–4, *c3* 16–18, *d1* 2–3, *d3* 2–3, *e1* 2–3, *e3* 22–23, *f2* 16–20, *f3* 15–18, *h1* 15–17, *h2* 16–17.

Gnathosoma. Palps similar to those of female, setal formula: 0, 0, 2, 2; tibia with two setae, *d′* 3–4, *d″* 5, tarsus with one eupathidium 3–4 and one minute solenidion, 1 long. Setae *m* absent.

Venter. Cuticle covered with fine and mostly transverse striae. Coxal, genital, and anal setae fine. Setal measurements: *1a* 55–65, *1b* 7–8, *3a* 10–11, *ps1* 5–7, *ps2* 10–11, *ps3* 5–6. Setae *1c*, *2b*, *2c*, *3b*, *4a*, *4b*, *ag*, *g1*, and *g2* absent.

Legs (Figure 18). Setation (from coxae to tarsi): I 2–0–3–1–5–6(1), II 0–0–3–1–5–6(1), III 0–0–2–1–3–3. Tarsi I–II each with one solenidion ω'' 3–4 (for both tarsi I and II) and two eupathidia $p\zeta'$–$p\zeta''$ (3–4, 3–4; 3–4, 3–4, respectively). Cuticle of all legs covered with colliculated sculpturing. Detail of the development of leg chaetotaxy in Table 1.

Etymology. The specific name *parameekeri* refers to the morphological similarity of this species and *U. meekeri* (De Leon), the type species of the genus.

Differential diagnosis. This new species resembles *Ultratenuipalpus meekeri* (De Leon) (herein redescribed) as they both have dorsal setae of a similar shape and length and the same leg and palp chaetotaxy in all developmental stages. These two species also share several other characteristics, such as the pair of lateral projections anterior to setae *sc2* and a single posterior projection between opisthosomal setae *h1*. However, the two species can be separated: the prodorsum is distinctly broader in adult females and males (measured at the widest point between setae *sc1* and *c1*) in *U. meekeri* (325–345) than *U. parameekeri* (290–315) (in females); notch in anterior forked projection is shorter in *U. meekeri* (8–13) than in *U. parameekeri* (20–27) (in females); *e3* is narrower and more lanceolate on male (and to a lesser extent on females) *U. meekeri* than in *U. parameekeri*; *l″* on ti I on *U. meekeri* is thicker than on *U. parameekeri*; *d* on fe II is longer and more falcate on female *U. meekeri* than on *U. parameekeri*; *c3* in larvae is narrower and more lanceolate in *U. meekeri* than *U. parameekeri*. In addition to the morphological differences, the molecular analyses confirmed that *U. parameekeri* and *U. meekeri* represent distinct species, with a 15.7% difference between their COI sequences.

DNA Barcoding. DNA was successfully amplified and the mitochondrial cytochrome C oxidase subunit I gene (COI) sequenced from one specimen of *U. parameekeri* collected on *Cyclosorus interruptus* (Thelypteridaceae) from Pindorama, São Paulo, Brazil; sequence data have been deposited in GenBank (https://www.ncbi.nlm.nih.gov/, accessed on 15 January 2023), with the following accession code: female, 398 base pairs (GenBank: OQ533138).

Type material examined. Holotype: female collected on ferns *Rumohra adiantiforme* (Dryopteridaceae) from Ilha do Cardoso, São Paulo, Brazil, 22 March 2017, coll. G.C.O. Piccoli (DZSJRP). Paratypes: 3 females, 1 protonymph, and 2 larvae, with the same data as the holotype (DZSJRP); 4 females, 3 males, 4 deutonymphs, 5 protonymphs, and 2 larvae collected on *Psychotria nuda* (Rubiaceae) from Ilha do Cardoso, São Paulo, Brazil, 22 March 2017, coll. G.C.O. Piccoli (DZSJRP); 2 females and 2 males collected on *P. nuda* from Ilha do Cardoso, São Paulo, Brazil, 22 March 2017, coll. G.C.O. Piccoli (NMNH); 4 females and 1 deutonymph collected on ferns *C. interruptus* from Pindorama, São Paulo, Brazil, 15 December 2002, coll. R. Kishimoto (DZSJRP).

Other material examined. 1 female and 1 larva collected on ferns *C. interruptus* from Pindorama, São Paulo, Brazil, 15 March 2003, coll. P. Demite (DZSJRP); 2 females, 2 deutonymphs, 3 protonymphs, and 1 larva collected on ferns *C. interruptus* from Pindorama, São Paulo, Brazil, 15 December 2002, coll. R. Kishimoto (DZSJRP); 2 females, 2 deutonymphs, 3 protonymphs, and 1 larva collected on ferns *C. interruptus* from Pindorama, São Paulo, Brazil, 15 March 2005, coll. P. Demite (DZSJRP, USNM).

3.2. Redescription of Ultratenuipalpus meekeri (De Leon, 1957)

Tenuipalpus meekeri De Leon: De Leon [13]—original designation
Ultratenuipalpus meekeri (De Leon): Mitrofanov [14]

Redescriptions [4,15–19].

Diagnosis. Female: As per genus, in addition to: prodorsal setae *v2*, *sc1* minute to short, and *sc2* large, flattened, obovate to ovate; dorsal opisthosoma with 10 pairs of setae (*f2* present); most of the dorsal opisthosomal setae large, flattened, obovate to ovate, except setae *d3* is distinctly short and *c3* is almost orbicular; pair lateral projections anterior to setae *sc2* and single posterior projection between opisthosomal setae *h1* present; palp four segmented, setal formula 0, 0, 2, 2. Male: Opisthosoma narrower than that of females, with a distinct transverse constriction (waist) between setae *d1* and *e1*; many dorsal setae similar to those of the female, except *c1* much smaller, *d1* and *e1* short to minute, and *v2* and *d3* longer. Tarsi I–II each with two solenidia (ω' paraxial and ventrolateral; ω'' antiaxial); tarsus III with one solenidion ω' paraxial and ventrolateral. Immatures: with lateral body projections anterior to setae *sc2* present (except absent in larvae); single posterior projection between setae *h1* absent; dorsal setae similar in general form to those of the female, except *c1*, *d1*, and *e1* minute. Larvae with anterior margin colliculated and central prodorsum smooth; pygidial region of posterior opisthosoma with colliculated integument.

Female (n = 3) (Figures 19–24)

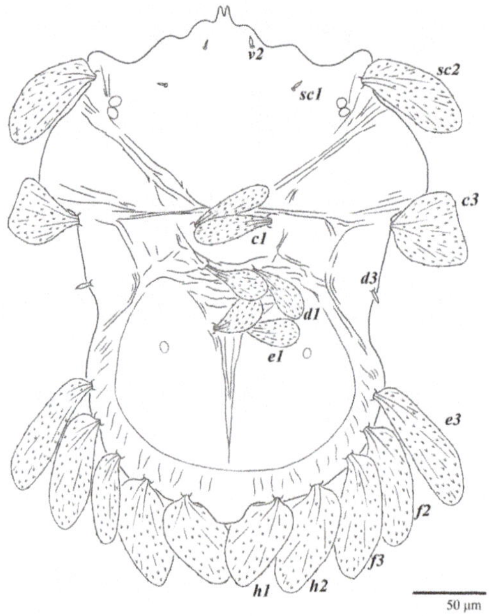

Figure 19. *Ultratenuipalpus meekeri* (De Leon). (Female, paratype): view of dorsum.

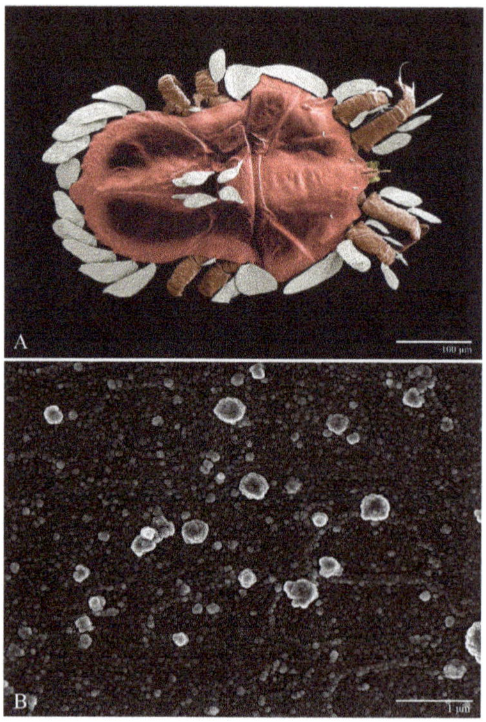

Figure 20. *Ultratenuipalpus meekeri* (De Leon). (Female): (**A**) dorsal view; (**B**) view of cuticular microplates on the dorsum.

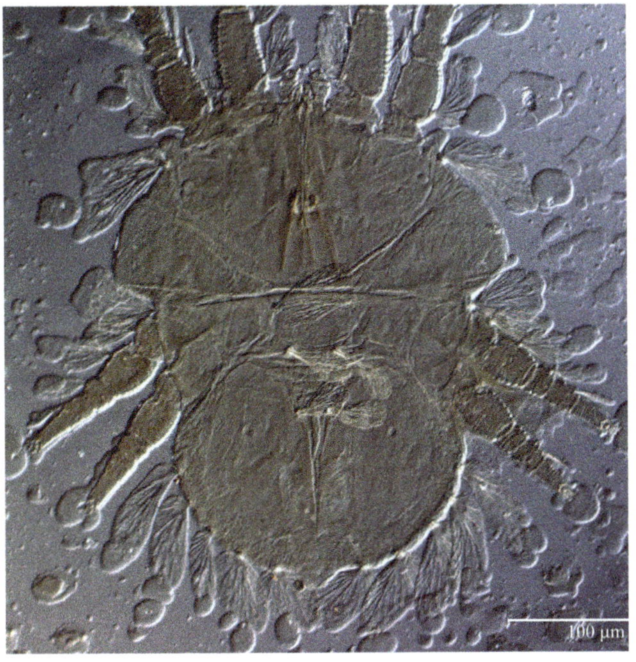

Figure 21. *Ultratenuipalpus meekeri* (De Leon). (Female, paratype): view of dorsum.

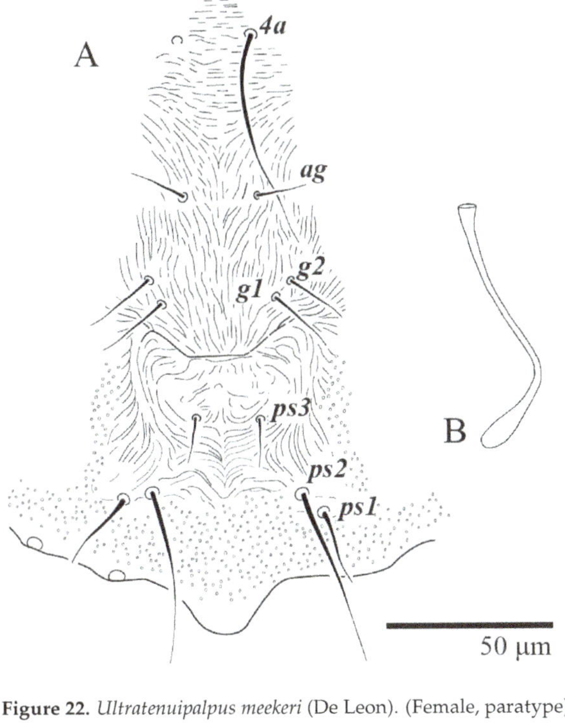

Figure 22. *Ultratenuipalpus meekeri* (De Leon). (Female, paratype): (**A**) posterior ventral opisthosoma; (**B**) spermatheca.

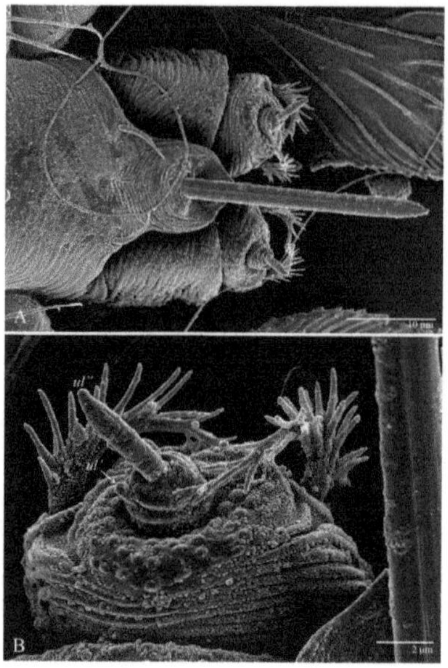

Figure 23. *Ultratenuipalpus meekeri* (De Leon). (Female): (**A**) view of ventral infracapitulum; (**B**) detail of palp; note the basal insertion of solenidion.

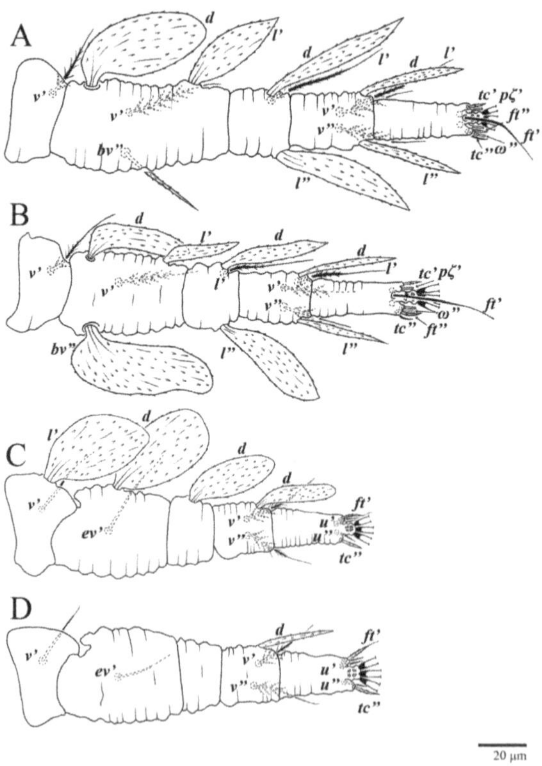

Figure 24. *Ultratenuipalpus meekeri* (De Leon). (Female, paratype): (**A**) leg I; (**B**) leg II; (**C**) leg III; (**D**) leg IV. (Right legs).

Body measurements: distance between setae *v2–h1* 390 (375–390), *sc2–sc2* 230 (230–235); other measurements: *v2–v2* 42 (42–48), *sc1–sc1* 115 (115), *c1–c1* 65 (65–68), *c3–c3* 290 (260–290), *d1–d1* 43 (37–43), *d3–d3* 250 (235–250), *e1–e1* 28 (25–30), *e3–e3* 235 (230–240), *f2–f2* 225 (215–225), *f3–f3* 190 (185–195), *h1–h1* 72 (72–78), *h2–h2* 135 (130–140).

Dorsum (Figures 19–21). Anterior margin of prodorsum with a short median forked projection forming a short notch 8 (8–13). Dorsum smooth, with pair of lateral projections anterior to setae *sc2* and a single projection between opisthosomal setae *h1* present. A pair of converging folds from the eyes to near the sejugal furrow on the prodorsum posterior margin. Prodorsal setae *v2* and *sc1* short to minute; *sc2* large, flattened elongated obovate (Figures 19 and 20A); most opisthosomal setae similar to prodorsal setae *sc2*, except *d3* is short. Setal measurements: *v2* 8 (4–8), *sc1* 4 (4–10), *sc2* 83 (83–94), *c1* 65 (65–69), *c3* 57 (54–57), *d1* 55 (55–58), *d3* 14 (14–15), *e1* 48 (39–48), *e3* 95 (92–95), *f2* 80 (80–84), *f3* 77 (77–83), *h1* 63 (62–66), *h2* 70 (70–73).

Venter (Figure 22A). Ventral integument weakly striate along central region and densely colliculated around lateral body margin; ventral, genital, and anal plates not developed, entire region membranous and distinctly plicate; ventral setae filiform, with coxal setae *1c*, *2c*, and *3b* barbed; setae *ps2* distinctly longer than *ps1*. Setal measurements: *1a* 105 (105–115), *1b* 19 (12–19), *1c* 29 (26–29), *2b* 27 (27–28), *2c* 47 (41–47), *3a* 20 (20–21), *3b* 43 (37–43), *4a* 105 (95–115), *4b* 26 (23–26), *ag* 15 (15–17), *g1* 19 (15–19), *g2* 17 (17–20), *ps1* 15 (12–15), *ps2* 48 (48–60), *ps3* 31 (23–31).

Gnathosoma (Figure 23). Palps four segmented, setal formula: 0, 0, 2, 2; tibia with two setae, *d′* 7 (7–11), *d″* 8 (7–8), tarsus with one eupathidium 5 (5) and one solenidion, 1 (1) long. Ventral setae *m* 7 (7–8); distance between setae *m–m* 15 (13–15).

Spermatheca (Figure 22B). Duct length ca. 70–85, terminating in smooth rounded bulb.

Legs (Figure 24). Setation (from coxae to tarsi): I 3–1–4–3–5–8(1), II 2–1–4–3–5–8(1), III 1–2–2–1–3–5, IV 1–1–1–0–3–5. Tarsi I–II each with one solenidion ω'' 9 (8–9) (for both tarsi I and tarsi II) and two eupathidia $p\zeta'$–$p\zeta''$ (7, 7; 7, 6–7, respectively); femur I with setae *d* obovate and *l′* broadly lanceolate; femur II with setae *d* elongate obovate to weakly falcate, *l′* lanceolate and *bv″* obovate to broadly falcate. Femora, genua, and tibiae with setae *d* inserted in lateral position. Detail of the development of leg chaetotaxy in Table 1.

Color (Figure 20A). The body is reddish with the central region becoming darker, eyes red, and legs orange. Dorsal body setae and legs setae are white.

Male (n = 1) (Figures 25–29)

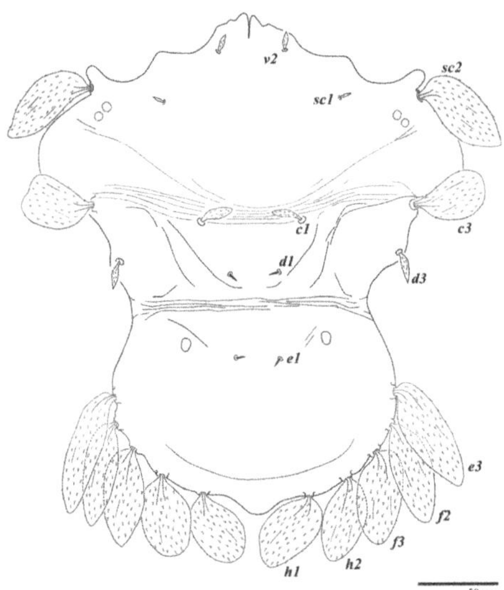

Figure 25. *Ultratenuipalpus meekeri* (De Leon). (Male, paratype): view of dorsum.

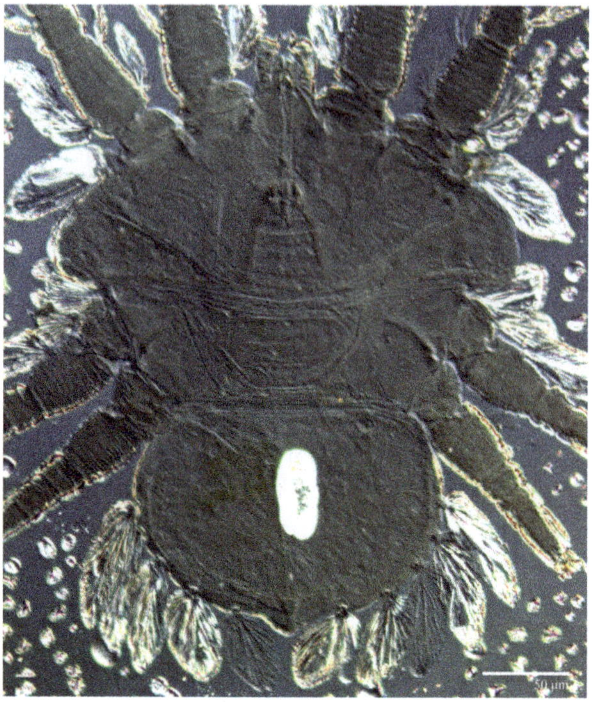

Figure 26. *Ultratenuipalpus meekeri* (De Leon). (Male, paratype): view of dorsum.

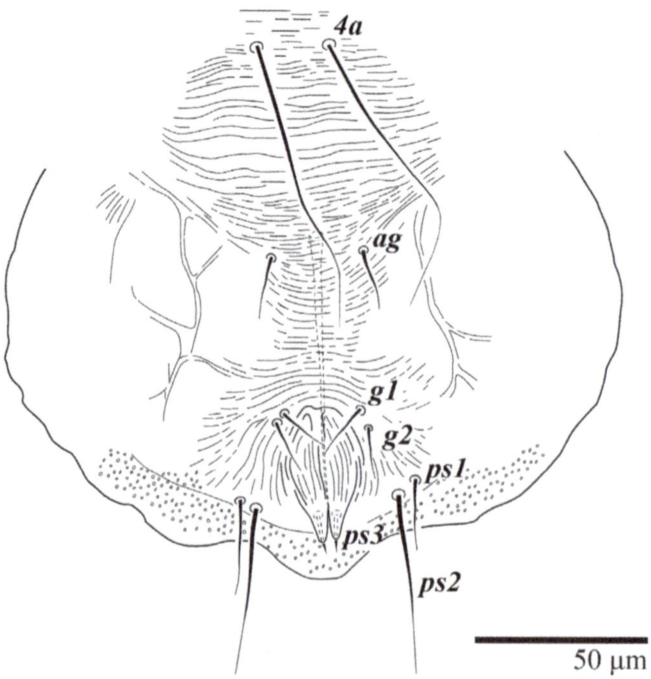

Figure 27. *Ultratenuipalpus meekeri* (De Leon). (Male, paratype): posterior ventral opisthosoma.

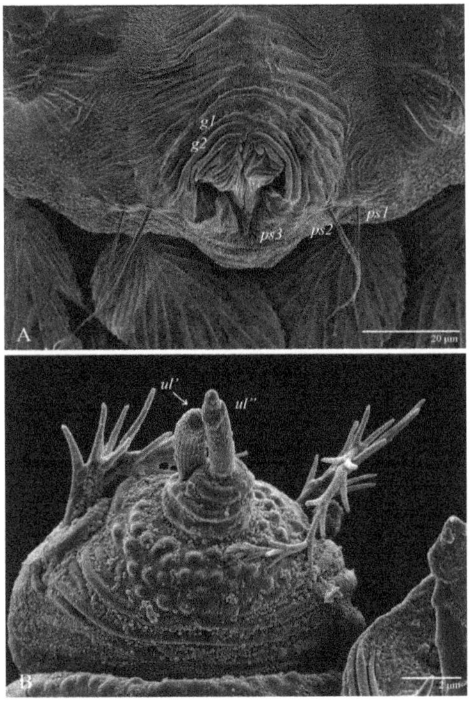

Figure 28. *Ultratenuipalpus meekeri* (De Leon). (Male): (**A**) posterior ventral opisthosoma; (**B**) detail of palp; note the basal insertion of solenidion.

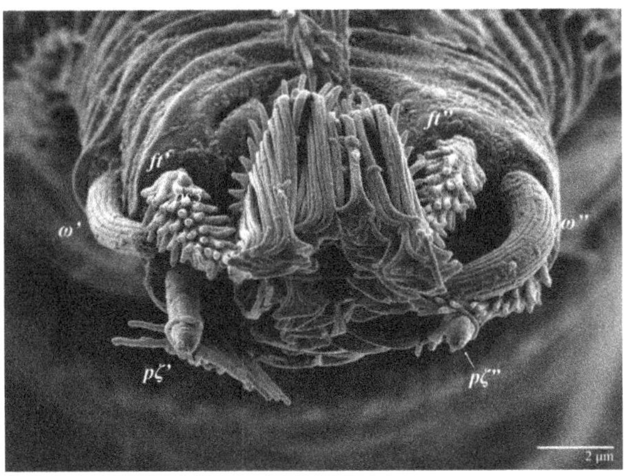

Figure 29. *Ultratenuipalpus meekeri* (De Leon). (Male): detail of tarsus II. Note the presence of solenidion ω' paraxial and ventrolateral.

Body measurements: distance between setae *v2–h1* 280, *sc2–sc2* 210; other measurements: *v2–v2* 43, *sc1–sc1* 110, *c1–c1* 65, *c3–c3* 205, *d1–d1* 30, *d3–d3* 165, *e1–e1* 28, *e3–e3* 175, *f2–f2* 170, *f3–f3* 150, *h1–h1* 63, *h2–h2* 110.

Dorsum (Figures 25 and 26). Anterior margin of prodorsum with a short median forked projection forming a short notch. Dorsum smooth, with pair lateral projections anterior to setae *sc2* and a single projection between opisthosomal setae *h1* present. Prodorsum with a pair of converging folds from the eyes to near the sejugal furrow on the posterior margin. Dorsal setae similar in general form to those of the female, except *c1*, *d1*, and *e1* small to minute, and *d3* longer. Setal measurements: *v2* 10, *sc1* 8, *sc2* 65, *c1* 23, *c3* 49, *d1* 6, *d3* 18, *e1* 5, *e3* 74, *f2* 70, *f3* 66, *h1* 53, *h2* 57.

Venter (Figures 27 and 28A). Ventral integument weakly striated along central region and densely colliculated around lateral margin of body; ventral setae filiform; coxal setae *1c*, *2c*, and *3b* barbed; setae *ps2* distinctly longer than *ps1*; setae *ps3* thickened and inserted ventrodistally on elongated, tapered anal valves. Setal measurements: *1a* 85, *1b* 22, *1c* 28, *2b* 26, *2c* 35, *3a* 21, *3b* 35, *4a* 90, *4b* 23, *ag* 20, *g1* 19, *g2* 16, *ps1* 24, *ps2* 60, *ps3* 16.

Gnathosoma (Figure 28B). Palps four segmented, setal formula: 0, 0, 2, 2; tibia with two setae, *d′* 8, *d″* 7, tarsus with one eupathidium 5 and one solenidion 6. Ventral setae *m* 8; distance between setae *m–m* 14.

Legs. Setation (from coxae to tarsi): I 3–1–4–3–5–8(2), II 2–1–4–3–5–8(2), III 1–2–2–1–3–5(1), IV 1–1–1–0–3–5. Tarsi I–II (Figure 29) each with two solenidia (one abaxial, one adaxial), tarsi I ω' 12, ω'' 9, tarsi II ω' 13, ω'' 9, and two eupathidia $p\zeta'$–$p\zeta''$ (all 6–7), and tarsus III with one solenidion (paraxial and ventrolateral) ω' 12. Leg setae similar to that of the female. Detail of the development of leg chaetotaxy in Table 1.

Deutonymph (n = 3) (Figures 30 and 31)

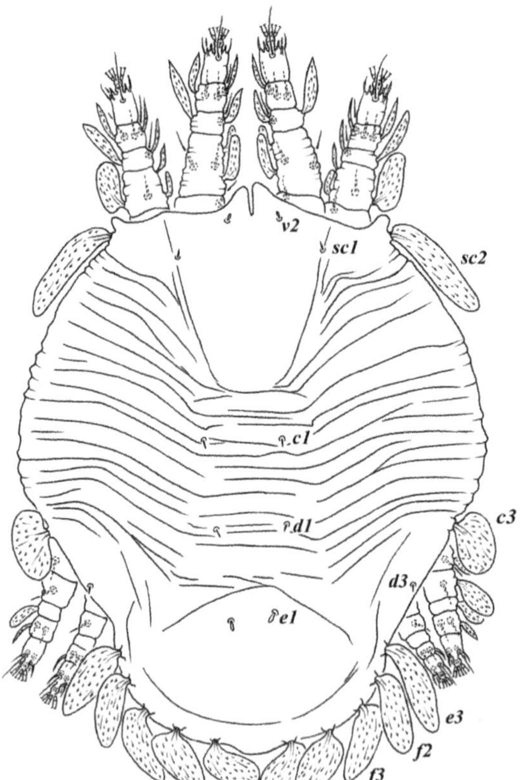

Figure 30. *Ultratenuipalpus meekeri* (De Leon). (Deutonymph, paratype): dorsum, with detail of legs (unguinal setae *u′*–*u″* on tarsus I and II are not included in the drawing).

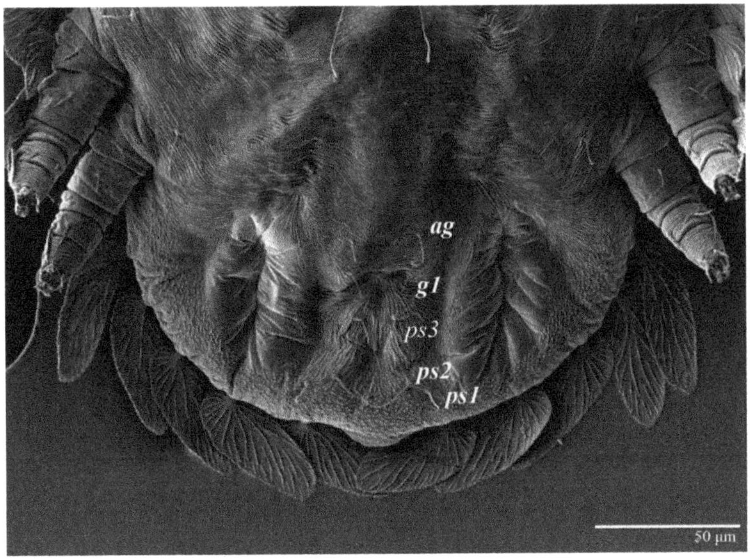

Figure 31. *Ultratenuipalpus meekeri* (De Leon). (Deutonymph): posterior ventral opisthosoma.

Body size measurements: distance between setae *v2–h1* 310–350, *sc2–sc2* 170–190; other measurements: *v2–v2* 37–40, *sc1–sc1* 95–105, *c1–c1* 43–53, *c3–c3* 220–270, *d1–d1* 45–55, *d3–d3* 200–225, *e1–e1* 25–30, *e3–e3* 160–180, *f2–f2* 145–170, *f3–f3* 120–145, *h1–h1* 40–55, *h2–h2* 83–105.

Dorsum (Figure 30). Anterior margin of prodorsum with a short median forked projection forming a short notch; a pair of body projections anterior and adjacent to setae *sc2* present; posterior projection between setae *h1* absent. Prodorsum with central region smooth; region between setae *sc2–c3* with transverse folds and plicae; region posterior to setae *e1* smooth. Dorsal setae similar in general form to those of the female, except setae *c1*, *d1*, and *e1* are short to minute. Setal measurements: *v2* 3–5, *sc1* 4–5, *sc2* 64–78, *c1* 6–12, *c3* 36–41, *d1* 5–7, *d3* 5–7, *e1* 4–8, *e3* 52–64, *f2* 51–55, *f3* 47–55, *h1* 36–42, *h2* 43–51.

Gnathosoma. Palps similar to those of female, setal formula: 0, 0, 2, 2; tibia with two setae, *d′* 6–7, *d″* 5–6, tarsus with one eupathidium 4–5 and one minute solenidion 1. Ventral setae *m* 5–7; distance between setae *m–m* 12–13.

Venter (Figure 31). Cuticle covered with fine and mostly transverse striae; with band of a colliculated cuticle around posterior body margin. Coxal, genital, and anal setae fine. Setal lengths: *1a* 75–90, *1b* 9–12, *1c* 11–15, *2b* 10–18, *2c* 13–16, *3a* 10–15, *3b* 16–17, *4a* 60–80, *4b* 11–12, *ag* 11–15, *g1* 8–11, *ps1* 9–10, *ps2* 27–33, *ps3* 15–17. Setae *g2* absent.

Legs (Figure 30). Setation (from coxae to tarsi): I 3–1–4–3–5–8(1), II 2–1–4–3–5–8(1), III 1–2–2–1–3–5, IV 1–0–1–0–3–5. Leg chaetotaxy similar to that of the female, except by trochanter IV nude; tarsi I–II each with one solenidion ω'' (tarsi I 5–6 and tarsi II 5, 6) and two eupathidia $p\zeta'$–$p\zeta''$ (5–6, 5–6; 5, 5 respectively). Detail of the development of leg chaetotaxy in Table 1.

Protonymph ($n = 1$) (Figure 32)

Body size measurements: distance between setae *v2–h1* 230, *sc2–sc2* 150; other measurements: *v2–v2* 28, *sc1–sc1* 85, *c1–c1* 35, *c3–c3* 190, *d1–d1* 25, *d3–d3* 155, *e1–e1* 23, *e3–e3* 130, *f2–f2* 120, *f3–f3* 100, *h1–h1* 38, *h2–h2* 73.

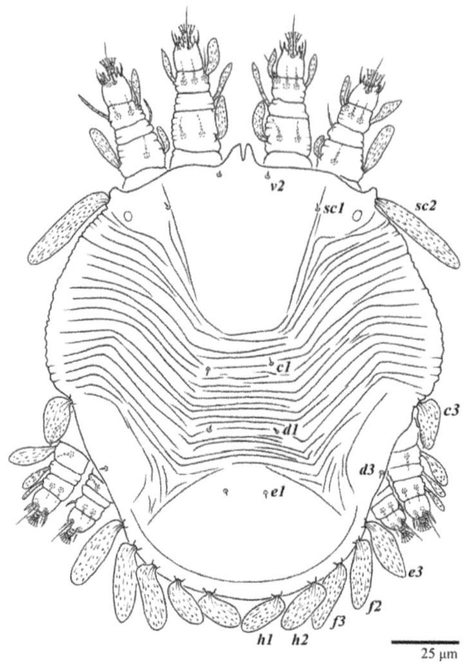

Figure 32. *Ultratenuipalpus meekeri* (De Leon). (Protonymph, paratype): dorsum, with detail of legs (unguinal setae *u′*–*u″* on tarsus I and II are not included in the drawing).

Dorsum (Figure 32). Anterior margin of prodorsum with a short median forked projection forming a short notch; a pair of lateral body projections anterior and adjacent to setae *sc2* present; posterior projection between setae *h1* absent. Prodorsum with central region smooth; region between setae *sc2*–*c3* with transverse folds and plicae; region posterior to setae *e1* smooth; dorsal setae similar in general form to those of the female, except setae *c1*, *d1*, and *e1* short to minute. Setal measurements: *v2* 3, *sc1* 2, *sc2* 54, *c1* 4, *c3* 27, *d1* 4, *d3* 4, *e1* 4, *e3* 45, *f2* 38, *f3* 35, *h1* 30, *h2* 35.

Gnathosoma. Palps similar to those of the female, setal formula: 0, 0, 2, 2; tibia with two setae, *d′* 4, *d″* 4, tarsus with one eupathidium 3 and one minute solenidion, 1 long. Ventral setae *m* 5; distance between setae *m*–*m* 12.

Venter. Cuticle covered with fine and mostly transverse striae. Coxal, genital, and anal setae fine. Setal measurements: *1a* 70, *1b* 9, *1c* 8, *2c* 13, *3a* 14, *3b* 11, *ag* 10, *ps1* 8, *ps2* 16, *ps3* 10. Setae *2b*, *4a*, *4b*, *g1*, and *g2* absent.

Legs (Figure 32). Setation (from coxae to tarsi): I 3–0–3–1–5–6(1), II 1–0–3–1–5–6(1), III 1–0–2–1–3–5, IV 0–0–1–0–3–3. Tarsi I–II each with one solenidion ω'' 4 (for both tarsi I and tarsi II) and two eupathidia $p\zeta'$–$p\zeta''$ (all 5). Detail of the development of leg chaetotaxy in Table 1.

Larva (n = 1) (Figure 33)

Body size measurements: distance between setae *v2*–*h1* 225, *sc2*–*sc2* 120; other measurements: *v2*–*v2* 33, *sc1*–*sc1* 68, *c1*–*c1* 33, *c3*–*c3* 160, *d1*–*d1* 30, *d3*–*d3* 120, *e1*–*e1* 18, *e3*–*e3* 115, *f2*–*f2* 105, *f3*–*f3* 88, *h1*–*h1* 23, *h2*–*h2* 58.

Dorsum (Figure 33). Prodorsal region with broad band of a colliculated integument anteromedially between setae *sc1*; region between setae *sc2*–*c3* with oblique and transverse folds and plicae; pygidial region posterior to setae *e1* with a small region of colliculated integuments; dorsal setae similar in general form to those of the female, except setae *c1*, *d1*, and *e1* are short to minute. Setal measurements: *v2* 3, *sc1* 3, *sc2* 28, *c1* 4, *c3* 23, *d1* 2, *d3* 3, *e1* 3, *e3* missing, *f2* 30, *f3* 25, *h1* 21, *h2* 21.

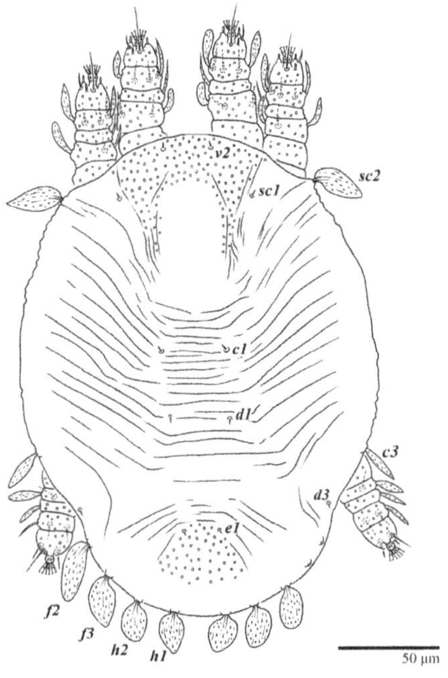

Figure 33. *Ultratenuipalpus meekeri* (De Leon). (Larva, paratype): dorsum, with detail of legs (unguinal setae u'–u'' on tarsus I and II are not included in the drawing).

Gnathosoma. Palps similar to those of female, setal formula: 0, 0, 2, 2; tibia with two setae, d' 3, d'' 5, tarsus with one eupathidium 3 and one minute solenidion, 1 long. Setae *m* absent.

Venter. Cuticle covered with fine and mostly transverse striae. Coxal, genital, and anal setae fine. Setal measurements: *1a* 50, *1b* 7, *3a* 10, *ps1* 7, *ps2* 11, *ps3* 6. Setae *1c*, *2b*, *2c*, *3b*, *4a*, *4b*, *ag*, *g1*, and *g2* absent.

Legs (Figure 33). Setation (from coxae to tarsi): I 2–0–3–1–5–6(1), II 0–0–3–1–5–6(1), III 0–0–2–1–3–3. Tarsi I–II each with one solenidion ω'' 3 (for both tarsi I and II) and two eupathidia $p\zeta'$–$p\zeta''$ (5, 5; 4, 4, respectively). Cuticles of all legs covered with colliculated cuticles. Detail of the development of leg chaetotaxy in Table 1.

Remarks. The new specimens examined in this study have similar body and setal measurements to those of the type specimens. In addition, the palp and leg chaetotaxy of those specimens match those of the type specimens.

DNA Barcoding. DNA was successfully amplified and the mitochondrial cytochrome C oxidase subunit I gene (COI) sequenced from one specimen of *U. meekeri* collected on *Acrostichum danaeifolium* (Pteridaceae) from Tecpan de Galeana, Guerrero State, Mexico; sequence data have been deposited in GenBank (https://www.ncbi.nlm.nih.gov/, accessed on 15 January 2023), with the following accession code: female, 446 base pairs (GenBank: OQ533137).

Type material examined: Holotype: the female collected on a fern in a mangrove swamp, from San Blas, Nayarit State, Mexico, 21 March 1957, coll. D. De Leon, was deposited in the Museum of Comparative Zoology (MCZ), Harvard University. Paratypes: 2 females, 1 male, 3 deutonymphs, 1 protonymph, and 1 larva, with the same data as the holotype, were deposited in the National Insect and Mite Collection, National Museum of Natural History (NMNH), Smithsonian Institution.

Other material examined: Non-type material: 5 females collected on ferns *A. danaeifolium* in a mangrove swamp, from Tecpan de Galeana, Guerrero State, **Mexico**, 5 September 2017, coll. G. Otero-Colina (USNM, DZJSRP).

4. Discussion

4.1. Ontogeny

Studies on possible patterns of ontogenetic development of chaetotaxy provide information potentially useful for understanding mite taxonomy, phylogeny, and biology [20]. The family Tenuipalpidae has the highest number of ontogenetic studies among all Trombidiformes mites [21], with ontogenetic data available for 60 species in 20 genera [22]. However, ontogenetic development is known only for one species of *Ultratenuipalpus*, *U. jubatus* Otley, Beard & Seeman [2,22]. Here, we discuss the ontogeny of the two species, *U. parameekeri* and *U. meekeri*, which share the same pattern of additions of leg setae (Table 1).

Trochanters. Setae *v′* are added to trochanters I, II, and III in the deutonymph and on trochanter IV in the adults. This is the standard pattern for other flat mites [22–24], and also for Tetranychidae [10]. Setae *l′* are added to trochanters III in the protonymph, and this addition also occurs in *U. jubatus*; although the expression of setae *l′* and *v′* varies within the family, this pattern has been commonly reported [2,22,23,25,26].

Femora. Setae *l′* on femora I and II are added in the deutonymph. The expression of setae *l′* on the deutonymph also occurs in *U. jubatus* and in many species of *Tenuipalpus* [2,22,23]. Setae *d* and *ev′* are present on femora III in the larva. This pattern is common in the Tenuipalpidae [22], but in *U. jubatus*, the addition of setae *ev′* is delayed until the protonymph. Setae *ev′* are added on the femora IV in the protonymph of *U. parameekeri* and *U. meekeri* (n.b., femora IV are not nude as described for *U. meekeri* in [17] and in the keys of [1,2]).

Genua. There is great variation in the chaetotaxy of genua I and II in the Tenuipalpidae [22,23]. Here, setae *l′* is present on genua I and II in the larva, and setae *d* and *l″* are added on genua I and II in the deutonymph. This pattern also occurs in *U. jubatus* and is common in the *Tenuipalpus* [22,23]. Setae *l′* is present on the genu III in the larva of the new species, and the same pattern occurs in *U. jubatus*; while many species of *Tenuipalpus* add setae *l′* or *d* on genu II in the deutonymph [22,23].

Tibiae. Although the number of tibial setae varies across the family, there are no post-larval additions made to the tibiae in the Tenuipalpidae [10]. Here, the number of tibial setae for both species is 5–5–3–3, as is also seen on *U. jubatus*, whereas setae *l′* are suppressed on tibiae III and IV on *U. avarua* Xu, Fan & Zhang [1,2,22].

Tarsi. *Ultratenuipalpus parameekeri* and *U. meekeri* have a pair of tectal setae added to tarsi I–III in the protonymph, as occurs in *U. jubatus*. However, many species of the *Tenuipalpus* added these setae in the deutonymph. As is the case for many additions to leg IV, the addition of the tectal setae on tarsi IV is delayed to the deutonymph. This same pattern also occurs in *U. jubatus*, while many *Tenuipalpus* add tectal setae to tarsi IV in the adults [22,23].

The solenidion ω' is added on each tarsi I–III in males of *U. parameekeri* and *U. meekeri*. Similarly, the male of *U. jubatus* has this solenidion added to tarsus I and II, but not to tarsus III [2]. This characteristic also occurs in other tenuipalpid genera, such as *Tenuipalpus*, *Prolixus*, and *Acaricis* [26–28], with some species of *Acaricis* also bearing one solenidion ω' on tarsus IV.

In response to the detailed work of Lindquist [10] on the patterns of setal additions to the legs in the family Tetranychidae, ontogenetic studies regarding the family Tenuipalpidae have received increasing attention in recent years. For example, the genus *Raoiella* has ontogenetic data available for 13 of the 22 known species [22]. However, despite this increase in attention, there are still many tenuipalpid genera that have received little or no such focused research [22], such as the *Ultratenuipalpus*. Only three of the 26 known species of the *Ultratenuipalpus* have so far been studied ontogenetically, and it is one of the genera that should receive priority in future studies. Filling these gaps may allow an adequate comparison of ontogenetic data between species and genera of the flat mite family, and as Lindquist [10] suggests, may contribute to our further understanding of the superfamily as a whole.

4.2. Distribution, Taxonomy and Systematic

The genus *Ultratenuipalpus* is known from all zoogeographic regions of the world with the exception of the Nearctic and Western Palearctic regions [2–4]. To date, six species of the *Ultratenuipalpus* have been described from the Neotropical region: the new species herein described from Brazil (which represents the first record of the genus for the country), in addition to two species from Mexico (*U. meekeri* and *U. younguisti* Baker & Tuttle) and three species from Chile (*U. acharis* (Gonzalez), *U. canelae* (Gonzalez), and *U. charlini* (Gonzalez)) [3,29].

According to Beard et al. [2], the presence or absence of opisthosomal setae *f2* may indicate a biogeographic pattern within the genus. Those species that lack *f2* show a putative Gondwanan distribution, being found in Chile, Australia, New Zealand, and the Cook Islands. Those species with setae *f2* present are found in China, the Philippines, Mexico, and now with the new species herein described, in Brazil. The unique exception for this pattern is *U. younguisti,* which lacks the setae *f2* and was described from Mexico (based on specimens intercepted in the USA).

The presence of a pair of lateral body projections anterior to setae *sc2* in some species of the *Ultratenuipalpus* (e.g., *U. meekeri*, *U. parameekeri*, and *U. avarua*) and the *Tenuipalpus sensu stricto* group could indicate a strong relationship between these two genera. Within the *Ultratenuipalpus,* the presence of a single posterior body projection between the setae *h1* may be an important character for separating a subgrouping, since it is shared by at least six species of the genus: *U. avarua*, *U. hainanensis* (Wang), *U. lacorpuzrarosae* Rimando, *U. meekeri*, *U. parameekeri*, and *U. umtataensis* Meyer.

5. Conclusions

The study of body morphology, spermathecae, geographic distribution, and plant associations will allow a broader and deeper understanding of the internal relationships within the genus *Ultratenuipalpus*, as well its relationships with other related genera (e.g., *Tenuipalpus*, *Extenuipalpus*, *Acaricis*, and *Prolixus*). In addition, we believe that the study of possible patterns of ontogenetic additions of leg setae can provide important insights into the systematics and origin of these taxa.

Author Contributions: Conceptualization, E.B.C., J.J.B., R.O., G.R.B., G.O.-C., R.J.F.F. and A.C.L.; methodology, E.B.C., J.J.B., R.O., G.R.B., G.O.-C., R.J.F.F. and A.C.L.; molecular analysis, A.P.G.D.; data curation, R.J.F.F., A.C.L. and R.O.; writing—original draft preparation, E.B.C.; writing—review and editing, E.B.C., J.J.B., R.O., G.R.B., G.O.-C., R.J.F.F. and A.C.L.; supervision, R.J.F.F., A.C.L. and R.O.; project administration, R.J.F.F. and A.C.L.; funding acquisition, R.J.F.F., A.C.L. and E.B.C. All authors have read and agreed to the published version of the manuscript.

Funding: E.B.C. was supported by a FAPESP (Fundação de Amparo à Pesquisa do Estado de São Paulo) Post-doctoral Scholarship, grant numbers 2016/01193-5 and 2017/00458-8; and by a UNESP Post-doctoral Scholarship, with project number 4324 (Edital PROPe 13/2022). Part of the studied specimens of the new species was collected with financial support of the FAPESP, grant numbers 2004/04820-3, 2006/55725-6 and 2006/57868-9.

Institutional Review Board Statement: Not applicable.

Informed Consent Statement: Not applicable.

Data Availability Statement: All data are available in this paper.

Acknowledgments: The authors would like to thank the late Debra Creel and Andrew Ulsamer (SEL-USDA) for their help with references and technical support. We also thank Laura Leibensperger, Department of Invertebrate Zoology, Museum of Comparative Zoology (MCZ), Harvard University, for lending specimens for study; and the Smithsonian Natural History Museum (NMNH) and National Agricultural Library (NAL-USDA), SEL-USDA for support and assistance with specimens and references. The mention of trade names or commercial products in this publication is solely for the purpose of providing specific information and does not imply recommendation or endorsement by the USDA. USDA is an equal opportunity provider and employer.

Conflicts of Interest: The authors declare no conflict of interest.

References

1. Xu, Y.; Fan, Q.-H.; Zhang, Z.-Q. A new species of *Ultratenuipalpus* (*Acari*: *Tenuipalpidae*) from Cook Islands, with a key to the known species. *Zootaxa* **2013**, *3731*, 223–233. [CrossRef]
2. Beard, J.J.; Otley, J.; Seeman, O.D. A review of *Ultratenuipalpus* (Trombidiformes: *Tenuipalpidae*) and related genera, with a new species from forest oak *Allocasuarina torulosa* (Aiton) (Casuarinaceae). *Int. J. Acarol.* **2016**, *42*, 285–302. [CrossRef]
3. Tenuipalpidae Database. Available online: http://www.tenuipalpidae.ibilce.unesp.br (accessed on 12 February 2023).
4. Mesa, N.C.; Ochoa, R.; Welbourn, W.C.; Evans, G.A.; de Moraes, G.J. A catalog of the *Tenuipalpidae* (*Acari*) of the World with a key to genera. *Zootaxa* **2009**, *2098*, 1–185. [CrossRef]
5. Lindquist, E.E. Anatomy, phylogeny and systematics, Chapter 1.1.3. Diagnosis and phylogenetic relationships. In *Spider Mites: Their Biology, Natural Enemies and Control*; Helle, W., Sabelis, M.W., Eds.; Elsevier Science Publishers: Amsterdam, The Netherlands, 1985; Volume 1A, pp. 63–74.
6. Castro, E.B.; Kane, E.C.; Feres, R.J.F.; Ochoa, R.; Bauchan, G.R. Definition of *Tenuipalpus* sensu stricto (*Acari, Tenuipalpidae*), with redescription of *Tenuipalpus caudatus* (Dugès) and description of a new species from Costa Rica. *Int. J. Acarol.* **2016**, *42*, 106–126. [CrossRef]
7. Castro, E.B.; Ochoa, R.; Feres, R.J.F.; Beard, J.J.; Bauchan, G.R. Reinstatement of the genus *Colopalpus* Pritchard and Baker (1958) and re-description of *Colopalpus matthyssei* Pritchard and Baker (1958), the type species of the genus (*Acari, Tenuipalpidae*). *Int. J. Acarol.* **2015**, *41*, 310–328. [CrossRef]
8. Navajas, M.; Lagnel, J.; Gutierrez, J.; Boursot, P. Species-wide homogeneity of nuclear ribosomal ITS2 sequences in the spider mite *Tetranychus urticae* contrasts with extensive mitochondrial COI polymorphism. *Heredity* **1998**, *80*, 742–752. [CrossRef]
9. Rodrigues, J.C.V.; Childers, C.C.; Gallo-Meagher, M.; Ochoa, R.; Adams, B.J. Mitochondrial DNA and RAPD polymorphisms in the haploid mite *Brevipalpus phoenicis* (*Acari*: *Tenuipalpidae*). *Exp. Appl. Acarol.* **2004**, *34*, 274–290. [CrossRef] [PubMed]
10. Lindquist, E.E. Anatomy, phylogeny and systematics, Chapter 1.1.1. External anatomy. In *Spider Mites: Their Biology, Natural Enemies and Control*; Helle, W., Sabelis, M.W., Eds.; Elsevier Science Publishers: Amsterdam, The Netherlands, 1985; Volume 1A, pp. 3–28.
11. Xu, Y.; Fan, Q.-H. *Tenuipalpus orilloi* Rimando, a new record to the Chinese fauna (*Acari*: *Tenuipalpidae*). *Syst. Appl. Acarol.* **2010**, *15*, 135–138. [CrossRef]
12. Seeman, O.D.; Beard, J.J. A new species of *Aegyptobia* (*Acari*: *Tenuipalpidae*) from Myrtaceae in Australia. *Syst. Appl. Acarol.* **2011**, *16*, 73–89. [CrossRef]
13. De Leon, D. The genus *Tenuipalpus* in Mexico (*Acari*na: *Tenuipalpidae*). *Fla. Entomol.* **1957**, *40*, 81–93. [CrossRef]
14. Mitrofanov, V.I. Revision of the system of phytophagous mites of the subfamily Tenuipalpinae s. str. (Trombidiformes, *Tenuipalpidae*). *Zool. Zhurnal* **1973**, *52*, 1315–1320.
15. Corpuz-Raros, L.A. New Philippine Tetranychoidea (*Acari*na). *Kalikasan Philipp. J. Biol.* **1978**, *7*, 211–230.
16. Meyer, M.K.P. *The Tenuipalpidae (Acari) of Africa with Keys to the World Fauna*; Entomology Memoir; Department of Agricultural Technical Services: Pretoria, South Africa, 1979; Volume 50, pp. 1–135.
17. Baker, E.W.; Tuttle, D.M. *The False Spider Mites of Mexico (Tenuipalpidae: Acari)*; Technical Bulletin; United States Department of Agriculture, Agricultural Research Service: Washington, DC, USA, 1987; Volume 1706, pp. 1–236.
18. Meyer, M.K.P. The South African species of *Ultratenuipalpus* Mitrofanov (*Acari: Tenuipalpidae*), with a key to the species of the genus. *Int. J. Acarol.* **1993**, *19*, 39–43. [CrossRef]
19. Rimando, L.C. A review of the genus *Ultratenuipalpus* Mitrofanov (*Acari: Tenuipalpidae*) with descriptions of two new species from the Philippines. *Philipp. Entomol.* **2001**, *15*, 101–113.
20. Zhang, Z.-Q. Accelerating studies on the ontogeny and morphological diversity in immature mites. *Zootaxa* **2018**, *4540*, 5–6. [CrossRef]
21. Liu, J.F.; Zhang, Z.-Q. A survey of descriptions of immature instars of mites during the last three years. *Zootaxa* **2018**, *4540*, 211–224. [CrossRef] [PubMed]
22. Gu, X.-Y.; Zhang, Z.-Q. Ontogenetic development of chaetotaxy in *Tenuipalpidae*: A survey with special reference to sexual dimorphism. *Zootaxa* **2020**, *4900*, 154–200. [CrossRef] [PubMed]
23. Welbourn, W.C.; Beard, J.J.; Bauchan, G.R.; Ochoa, R. Description of a new species of *Tenuipalpus* (*Acari*: Trombidiformes) from succulent plants in Florida, USA, and a redescription of *T. crassulus* Baker and Tuttle. *Int. J. Acarol.* **2017**, *43*, 112–136. [CrossRef]
24. Castro, E.B.; Beard, J.J.; Ochoa, R.; Bauchan, G.R.; Feres, R.J.F. Two new species of *Tenuipalpus* sensu stricto (*Acari*: *Tenuipalpidae*) from Brazil, with a discussion on the ontogeny of leg setae. *Zootaxa* **2018**, *4540*, 178–210. [CrossRef]
25. Khanjani, M.; Khanjani, M.; Seeman, O.D. The flat mites of the genus *Tenuipalpus* Donnadieu (*Acari*: *Tenuipalpidae*) from Iran. *Int. J. Acarol.* **2013**, *39*, 97–129. [CrossRef]
26. Castro, E.B.; Ramos, F.A.M.; Feres, R.J.F.; Ochoa, R. A new species of *Tenuipalpus* Donnadieu (*Acari*: *Tenuipalpidae*) from Brazil, with ontogeny of chaetotaxy. *Syst. Appl. Acarol.* **2015**, *20*, 339–356.
27. Beard, J.J.; Fan, Q.H.; Walter, D.E. A new genus and two new species of *Tenuipalpidae* (Prostigmata: Tetranychoidea) from an Australian sedge. *Acarologia* **2005**, *45*, 161–181.

28. Castro, E.B.; Beard, J.J.; Ochoa, R.; Feres, R.J.F. Two species of *Acaricis* (*Acari*: *Tenuipalpidae*) from New Zealand, moved from the genus *Tenuipalpus*, with a key to the known species. *Acarologia* **2018**, *58*, 855–867. [CrossRef]
29. Flat Mites of the World. Available online: https://idtools.org/tools/1074/index.cfm (accessed on 12 February 2023).

animals

MDPI

Article

The Genus *Neoseiulus* Hughes (Acari: Phytoseiidae) in Shanxi, China †

Yu Liu [1], Fang-Xu Ren [1], Qing-Hai Fan [2,*] and Min Ma [1,3,*]

1 College of Plant Protection, Shanxi Agriculture University, Taigu 030801, China
2 Plant Health & Environment Laboratory, Ministry for Primary Industries, Auckland 1072, New Zealand
3 State Key Laboratory for Biology of Plant Diseases and Insect Pests, Beijing 100193, China
* Correspondence: qinghai.fan@mpi.govt.nz (Q.-H.F.); acarimin@sxau.edu.cn (M.M.)
† ZooBank link: urn:lsid:zoobank.org:pub:9AEDDDFC-F810-43E5-8F59-BF8E53343101.

Simple Summary: Phytoseiid mites are widely distributed on plants and in soil; they can prey on many phytophagous mites and pests and play an important role in biological control programs. The genus *Neoseiulus* Hughes is one of the largest genera within the Phytoseiidae family, comprising 14.6% of the family's species worldwide. At present, there are few reports investigating *Neoseiulus* from Shanxi Province; however, the study on the species diversity of *Neoseiulus* is helpful to enrich the resource species bank of Phytoseiidae and provide more detailed basis for species identification. We report the discovery of five additional species in Shanxi, and redescribed four of them. *Neoseiulus paraki* (Ehara) is recorded for the first time in China, and *N. neoreticuloides* (Liang and Hu) is considered a new junior synonym of *N. bicaudus* (Wainstein). We provide a key to assist in the identification of the known species of *Neoseiulus* in Shanxi.

Abstract: The genus *Neoseiulus* in Shanxi Province is reviewed and seven species are recorded from the province. Four of these are redescribed and detailed taxonomic information are provided. *Neoseiulus paraki* (Ehara) is recorded for the first time in China and *Neoseiulus neoreticuloides* (Liang and Hu) is considered a new junior synonym of *Neoseiulus bicaudus* (Wainstein). Additionally, a diagnostic key to the known species of *Neoseiulus* in Shanxi is provided.

Keywords: Mesositgmata; taxonomy; predatory mite; new record

Citation: Liu, Y.; Ren, F.-X.; Fan, Q.-H.; Ma, M. The Genus *Neoseiulus* Hughes (Acari: Phytoseiidae) in Shanxi, China. *Animals* **2023**, *13*, 1478. https://doi.org/10.3390/ani13091478

Academic Editors: Maciej Skoracki and Monika Fajfer

Received: 23 March 2023
Revised: 24 April 2023
Accepted: 25 April 2023
Published: 26 April 2023

1. Introduction

The genus *Neoseiulus* Hughes [1] is one of the largest genera within the Phytoseiidae family, comprising 361 (14.6%) of the family's species worldwide [2–4]. In China, it is also a big genus, with 57 recorded species, accounting for 17.1% of the Phytoseiidae species in the country [3,5–7].

Shanxi is a province located in the northern part of China and covers a total area of 156,700 square kilometers. The majority of the province, over two-thirds, sits on the loess plateau. The climate is characterized as semiarid, with lower annual rainfall and longer dry seasons. In the south, the monthly 24 h average temperature ranges from −3 °C to 32 °C, while in the north, it ranges from −9.8 °C to 21.9 °C. The annual precipitation falls between 400 mm and 650 mm, with over 70% of the rainfall occurring between June and September. The most common natural vegetation consists of shrubs and grasses, and forested areas, mainly found on mountain slopes, only cover approximately 20% of the land area. Due to these factors, Shanxi has a lower biodiversity in comparison to areas located in southern China.

The knowledge of *Neoseiulus* species in Shanxi is limited. Currently, only two species have been recorded, namely *N. womersleyi* (Schicha) [8] (previously identified as *N. pseudolongispinosus*) [6,9–11] and *N. zwoelferi* Dosse [12]. The objective of this study is to report the discovery of five additional species in Shanxi, clarify the identity of *N. neoreticuloides* [13], and provide redescriptions of *N. bicaudus* (Wainstein) [14], *N. lushanensis* (Zhu and Chen) [15], *N. paraki* (Ehara) [16,17] and *N. tauricus* (Livshitz and Kuznetsov) [18]. Additionally, this paper aims to provide a key to assist in the identification of the known species of *Neoseiulus* in Shanxi.

2. Materials and Methods

The mites were collected using the beating method. The whole plant or branches of plants were beaten with a stick over a black rectangular plastic plate. Specimens were picked up with a fine soft hairbrush and kept in absolute ethanol before being taken to the laboratory. The mites were cleared and macerated in Nesbitt's fluid until they became translucent, then transferred to distilled water for 2–3 min to dissolve the Nesbitt's fluid before being mounted in Hoyer's medium on slides under a dissecting microscope (Zeiss DV4, Carl Zeiss Microscopy GmbH, Gottingen, Germany and Optec SZ650, Chongqing Optec Instrument Co., Ltd., Chongqing, China). Using a compound microscope (Nikon Eclipse 80i, Nikon Corporation, Tokyo, Japan) with differential interference contrast (DIC), the specimens were examined, measured, and photographed. Illustrations were created using a drawing tube (Nikon Y-IDT, Nikon Corporation, Tokyo, Japan) attached to the microscope and were edited with Photoshop CC2018. The voucher specimens are deposited in the Entomological Microscopy Laboratory of College of Plant Protection, Shanxi Agricultural University.

Measurements were taken along the midline from the anterior to posterior margins for the length of the dorsal shield, sternal shield, epigynal shield and ventrianal shield. The width of the dorsal shield was measured at the level of *s4*, the sternal shield (or sternogenital in males) at the level of *st2*, the genital shield at the level of *st5*, and the ventrianal shield at the level of *ZV2*. All measurements were presented in micrometers (μm) for the specimen used for illustration and other specimens in parentheses. The terminology of the idiosomal and leg chaetotaxy, and pore-like structures follow that of Rowell et al. (1978) [19] and Chant and McMurtry (2007) [2], Evans (1963) [20], and Athias-Henriot (1975) [21], respectively. The setal pattern system of idiosoma follows that of Chant and Yoshida-Shaul (1992) [22].

3. Results

***Neoseiulus* Hughes, 1948**

Type species: *Neoseiulus barkeri* Hughes, 1948: 141.

3.1. Redescriptions of Species

3.1.1. *Neoseiulus bicaudus* (Wainstein, 1962) (Figures 1–5)

Amblyseius bicaudus Wainstein, 1962: 146.
Typhlodromus bicaudus (Wainstein), Hirschmann 1962: 2 [23].
Amblyseius (Amblyseius) bicaudus (Wainstein), Ehara, 1966: 20 [24].
Neoseiulus bicaudus (Wainstein), Congdon, 2002: 23 [25]; Wang et al., 2015: 456 [26].
Amblyseius neoreticuloides Liang and Hu, 1988: 317 [13].
Amblyseius (Neoseiulus) neoreticuloides (Liang and Hu), Wu et al., 2009: 105 [6].
Neoseiulus neoreticuloides (Liang and Hu). **New synonym.**

1. **Diagnosis (female).** Dorsal shield elongate oval, reticulated throughout, bearing 17 pairs of setae, 16 pairs of lyrifissures and 7 pairs of solenostomes, *S4*, *S5*, *Z4* and *Z5* serrated, others smooth; *Z5* longer than others. Peritremes extending anteriorly to level of *j1*. Sternal shield reticulated, bearing three pairs of setae. Ventrianal shield approximately pentagonal, reticulated; solenostomes (*gv3*) posteromedian to *JV2*, circular. Calyx of spermathecal apparatus cup-shaped and basally stalked, and stalk approximately as long as width of atrium; atrium nodular at junction with minor duct, minor duct thread-like for a short distance and then expanded, forming a cylindrical tube. Fixed digit of chelicera with six teeth, movable digit with a tooth. Palpgenu

with genu setae *al1* and *al2* rod-like. Leg genu II with seven setae. Only basitarsus of leg IV with a macroseta.

2. **Redescription. Female** (n = 7). Dorsal idiosoma (Figures 1A and 2A). Idiosomal setal pattern 10A:9B/JV-3:ZV. Dorsal shield elongate oval, fully reticulate, has a waist located slightly below *R1*, 394 (380–412) long, 183 (181–189) wide; muscle marks visible between *j3* and *Z4*, a pair of muscle marks present in front of *J5*. Dorsum with 17 pairs of setae and 16 pairs of lyrifissures (*id1*, *id2*, *id4*, *id6*, *idx*, *idx1*, *idl2*, *idl3*, *idl4*, *idm1*, *idm2*, *idm3*, *idm4*, *idm5*, *idm6* and *is1*) and 7 pairs of solenostomes (*gd1*, *gd2*, *gd4*, *gd5*, *gd6*, *gd8* and *gd9*); lyrifissures *id3* and solenostomes *gd3* on peritremal shield. All dorsal setae smooth except for serrated *S4*, *S5*, *Z4* and *Z5*; *Z5* longer than others. Lateral setae *r3* and *R1* smooth, on interscutal membrane. Peritremes extending anteriorly close to *j1*, posterior part of peritremal shield (Figure 1B) curved and pointed, protuberance of exopodal shield at level of the stigmata. Lengths of setae: *j1* 22 (22–24), *j3* 28 (27–31), *j4* 13 (12–14), *j5* 13 (11–14), *j6* 16 (15–16), *J2* 18 (15–19), *J5* 12 (12–14), *r3* 27 (26–30), *R1* 25 (24–26), *s4* 31 (27–33), *S2* 29 (29–32), *S4* 34 (32–36), *S5* 41 (39–44), *z2* 22 (19–27), *z4* 20 (16–21), *z5* 11 (11–13), *Z1* 20 (19–22), *Z4* 35 (34–41), *Z5* 80 (80–95).

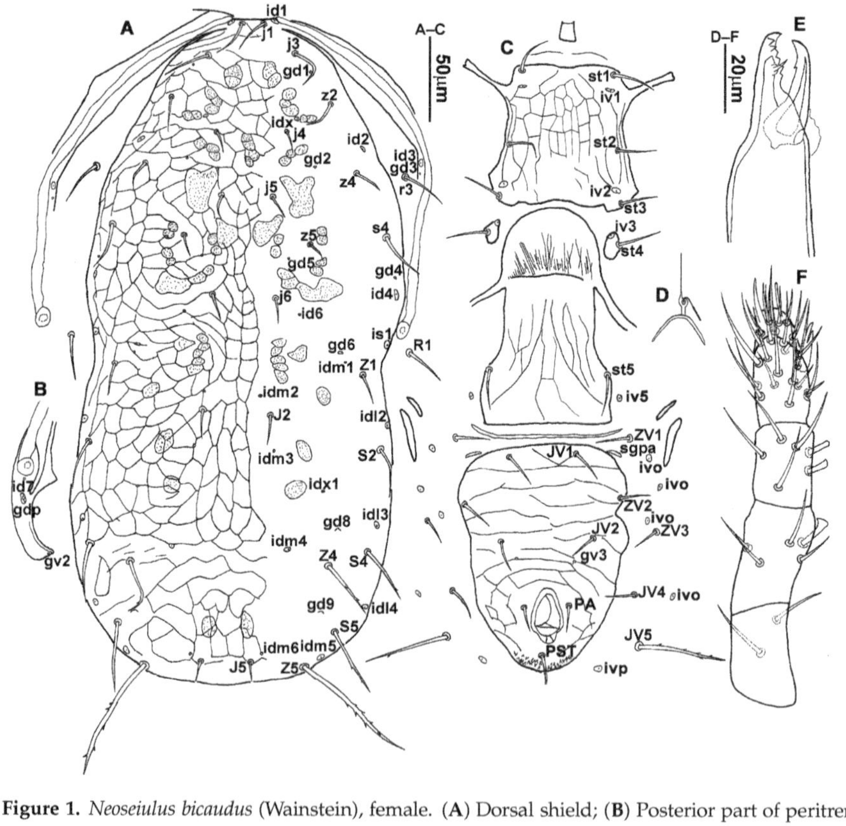

Figure 1. *Neoseiulus bicaudus* (Wainstein), female. (**A**) Dorsal shield; (**B**) Posterior part of peritremal shield and exopodal shield; (**C**) Ventral idiosoma; (**D**) Spermathecal apparatus; (**E**) Chelicera; (**F**) Palp.

3. Ventral idiosoma (Figures 1C and 2B). Sternal shield reticulated, 79 (77–83) long, 74 (74–77) wide; anterior margin laterally convex, forming a flat M-shaped projection, posterior margin concave, arched above the level of bases of *st3*; three pairs of setae (*st1*, *st2* and *st3*) and two pairs of lyrifissures (*iv1* and *iv2*) present on sternal shield, *iv1* positioned posteriad of *st1*, *iv2* positioned between *st2* and *st3*, and close to *st3*.

Metasternal platelets small, each bearing a seta *st4* and a lyrifissure *iv3*. Epigynal shield slightly striated, 132 (125–136) long, 76 (73–76) wide. Lengths of setae: *st1* 25 (24–29), *st2* 24 (21–24), *st3* 23 (21–25), *st4* 24 (22–27), *st5* 24 (20–25). Approximately four slender transverse sclerites present between epigynal and ventrianal shields. Ventrianal shield approximately pentagonal, transversally striated and a few oblique striae present between transverse striae, 136 (134–142) long, 100 (97–108) wide, bearing 3 pairs of preanal setae (*JV1*, *JV2* and *ZV2*), a pair of paranal setae (*PA*) and a postanal seta (*PST*), and a pair of circular solenostomes (*gv3*) posteromedial to *JV2*, distance *gv3–gv3* 36 (36–43); 4 pairs of setae (*JV4*, *JV5*, *ZV1* and *ZV3*) and 5 pairs of lyrifissures present on soft cuticle surrounding ventrianal shield, *JV5* serrate, others smooth. A pair of tiny platelets (*sgpa*) posteroparaxial to *ZV1* adjacent to anterior corners of ventrianal shield. Lengths of setae: *JV1* 18 (17–20), *JV2* 17 (16–18), *JV4* 17 (16–20), *JV5* 52 (52–60), *ZV1* 18 (18–20), *ZV2* 18 (17–19), *ZV3* 15 (15–17). Primary metapodal platelet 32 (30–35) long, 4 (4–6) wide; secondary platelet 15 (12–17) long, 2 (2–3) wide.

4. Spermatheca (Figures 1D and 3A,B). Calyx of spermathecal apparatus cup-shaped and basally stalked, 11 (10–12) long; stalk approximately as long as width of atrium, atrium nodular at junction with minor duct; minor duct thread-like for a short distance and then expanded, forming a cylindrical tube; major duct slender.
5. Gnathosoma. Chelicera (Figures 1E and 2C) with fixed digit 37 (33–37) long, bearing six teeth, pilus dentilis located at the level of fourth tooth, 6 (6–8) long; movable digit 32 (31–34) long, bearing single tooth. Palp (Figure 1F). Trochanter with two simple setae; femur with a spatulate and four simple setae; genu bearing two rod-like setae (*al1* and *al2*) and four simple setae; tarsal apotele two-tined.
6. Legs (Figure 4A–D). Leg I 357 (341–362) long, setal formula: coxa 0-0/1-0/1-0, trochanter 1-0/1-0/2-1, femur 2-3/1-2/2-2, genu 2-2/1-2/1-2, tibia 2-2/1-2/1-2, basitarsus 0-0/0-1/0-0. Apical sensorial setal cluster of tarsus I (Figure 4E) with nine modified setae. Leg II 300 (290–303) long, setal formula: coxa 0-0/1-0/1-0, trochanter 1-0/1-0/2-1, femur 2-3/1-2/1-1, genu 2-2/0-2/0-1, tibia 1-1/1-2/1-1, basitarsus 1-1/0-1/0-1. Leg III 307 (295–307) long, setal formula: coxa 0-0/1-0/1-0, trochanter 1-1/1-0/2-0, femur 1-2/1-1/0-1, genu 1-2/1-2/0-1, tibia 1-1/1-2/1-1, basitarsus 1-1/0-1/0-1. Leg IV 400 (386–404) long, setal formula: coxa 0-0/1-0/0-0, trochanter 1-1/1-0/2-0, femur 1-2/1-1/0-1, genu 1-2/1-2/0-1, tibia 1-1/1-2/0-1, basitarsus 1-1/0-1/0-1. Legs I–III without obvious macrosetae; leg IV with a smooth macroseta on basitarsus, 73 (73–78) long.
7. **Male** (n = 4). Dorsal idiosoma (Figure 5A). Dorsal shield elongate oval, presenting distinct reticulation throughout, 312 (301–312) long, 163 (153–166) wide; muscle marks visible between *j3* and *J5*; dorsum bearing 19 pairs of setae, 16 pairs of lyrifissures (*id1*, *id2*, *id4*, *id6*, *idx*, *idx1*, *idl2*, *idl3*, *idl4*, *idm1*, *idm2*, *idm3*, *idm4*, *idm5*, *idm6* and *is1*) and 7 pairs of solenostomes (*gd1*, *gd2*, *gd4*, *gd5*, *gd6*, *gd8* and *gd9*). Setae *S5*, *Z4* and *Z5* serrate, others smooth; *Z4* and *Z5* longer than others. Peritremes anteriorly ending at level between *j1* and *j3*. Lengths of setae: *j1* 21 (19–21), *j3* 24 (24–27), *j4* 14 (13–14), *j5* 13 (13–15), *j6* 16 (14–17), *J2* 19 (17–19), *J5* 12 (10–12), *r3* 27 (24–27), *R1* 23 (22–26), *s4* 29 (29–32), *S2* 31 (30–32), *S4* 34 (32–34), *S5* 36 (34–39), *z2* 20 (17–20), *z4* 15 (15–20), *z5* 14 (12–14), *Z1* 19 (19–21), *Z4* 43 (43–46), *Z5* 69 (67–69).
8. Ventral idiosoma (Figure 5B). Sternogenital shield sparsely striated between *st1* and *st4*, reticulate between *st3* and *st4*, 133 (132–137) long, 61 (60–63) wide; anterior margin prominently convex, posterior margin nearly straight, bearing five pairs of setae (*st1*, *st2*, *st3*, *st4* and *st5*) and three pairs of lyrifissures (*iv1*, *iv2* and *iv3*); lengths of setae: *st1* 23 (19–23), *st2* 18 (17–18), *st3* 17 (16–18), *st4* 18 (17–18), *st5* 19. Ventrianal shield subtriangular, reticulate throughout, 122 (114–122) long, 133 (127–133) wide; with three pairs of preanal setae (*JV1*, *JV2* and *ZV2*), a pair of paranal setae (*PA*) and a postanal seta (*PST*), three pairs of lyrifissures, a pair of circular solenostomes (*gv3*) posteromesad to *JV2*, distance *gv3–gv3* 28 (27–32), two pairs of marginal muscle marks situated anterolateral to anus. Setae *JV5* serrate, and a pair of lyrifissures on soft

cuticle surrounding ventrianal shield. Lengths of setae: *JV1* 15 (14–15), *JV2* 17 (16–17), *JV5* 44 (41–45), *ZV2* 19 (18–20).

9. Gnathosoma. Chelicera (Figure 5C) with fixed digit 24 (22–24) long, bearing four teeth, movable digit 23 (21–23) long, bearing a tooth. Spermatodactyl L-shaped, with acute toe and heel, shaft 16 (16–17) long, foot 7 (6–7) long. Palp and hypostome with same chaetotaxy as in female.
10. Legs. Leg chaetotaxy same as those in adult female.
11. **Materials examined.** A total of 10♀and 1♂, Dabaishi Village, Taigu County, Shanxi Province, 37°20′12″ N, 112°38′50″ E, 1340 m, e.g., *Setaria viridis* (L.) P. Beauv. (Poaceae), 31 August 2020, Y. Liu, M. Ma, B. Zhang and F.-X. Ren coll.; 13♀and 1♂, Shanxi Agriculture University, Taigu County, Shanxi Province, 37°25′15″ N, 112°34′37″ E, 794 m, *Hemerocallis fulva* (L.) L. (Asphodelaceae), 7 October 2013, Qing-Hai Fan coll. (accession no.: T13_0009); 8♀and 1♂, same locality and host as T13_0009, 2 July 2016, M. Ma coll.; 1♀, same locality and host as T13_0009, 5 July 2016, M. Ma coll. (T16_0002); 1♀and 1♂, locality and host as T13_0009, 27 June 2016, Y.-X. Li coll.; 9♀, same locality and host as T13_0009, but 28 June 2016, M. Ma coll.; 4♀and 1♂, same collection locality and host as T13_0009, 10 October 2013, M. Ma and Y.-N. Zhao coll. (T13_0017); 8♀and 3♂, same locality and host as T13_0009, 20 September 2014, M.-J. Yi and B.-Q. Su coll. (T14_0299).

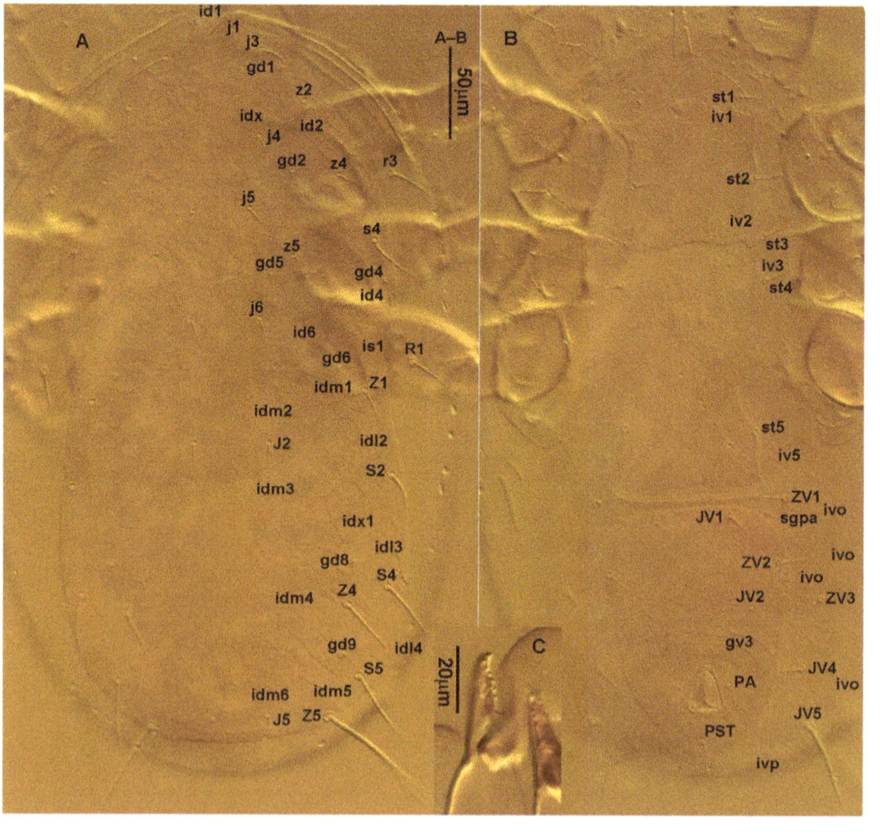

Figure 2. *Neoseiulus bicaudus* (Wainstein), female. (**A**) Dorsal shield; (**B**) Ventral idiosoma; (**C**) Chelicera.

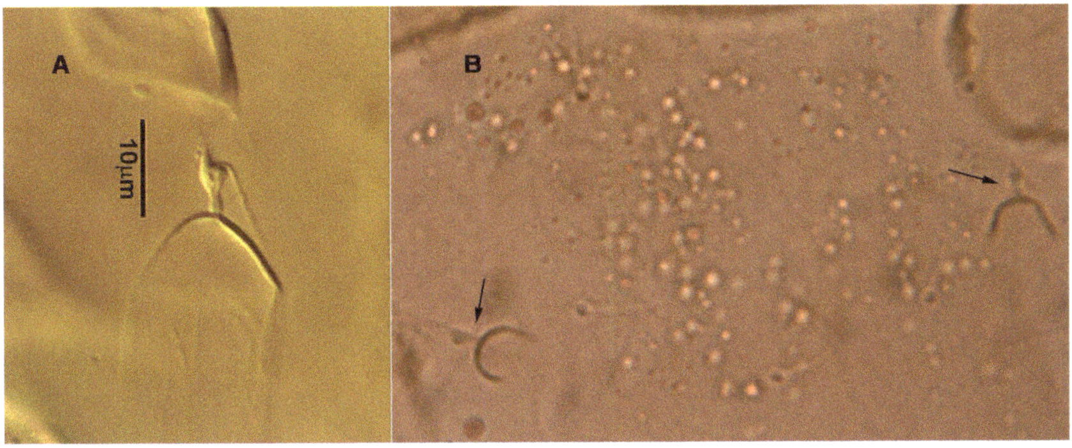

Figure 3. *Neoseiulus bicaudus* (Wainstein), female spermathecal apparatus: (**A**) Shanxi specimen; (**B**) Holotype of *N. neoreticuloides* (The spermathecal apparatuses are marked with arrows) (from Weinan Wu).

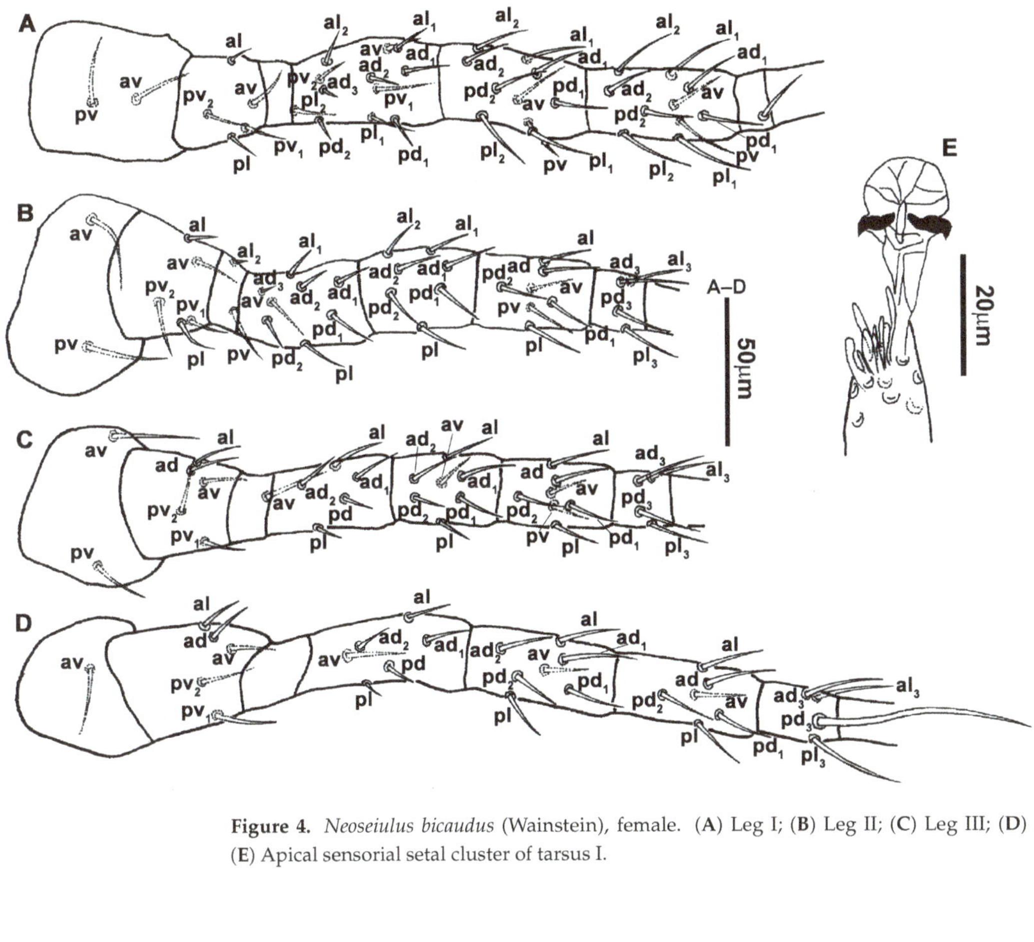

Figure 4. *Neoseiulus bicaudus* (Wainstein), female. (**A**) Leg I; (**B**) Leg II; (**C**) Leg III; (**D**) Leg IV; (**E**) Apical sensorial setal cluster of tarsus I.

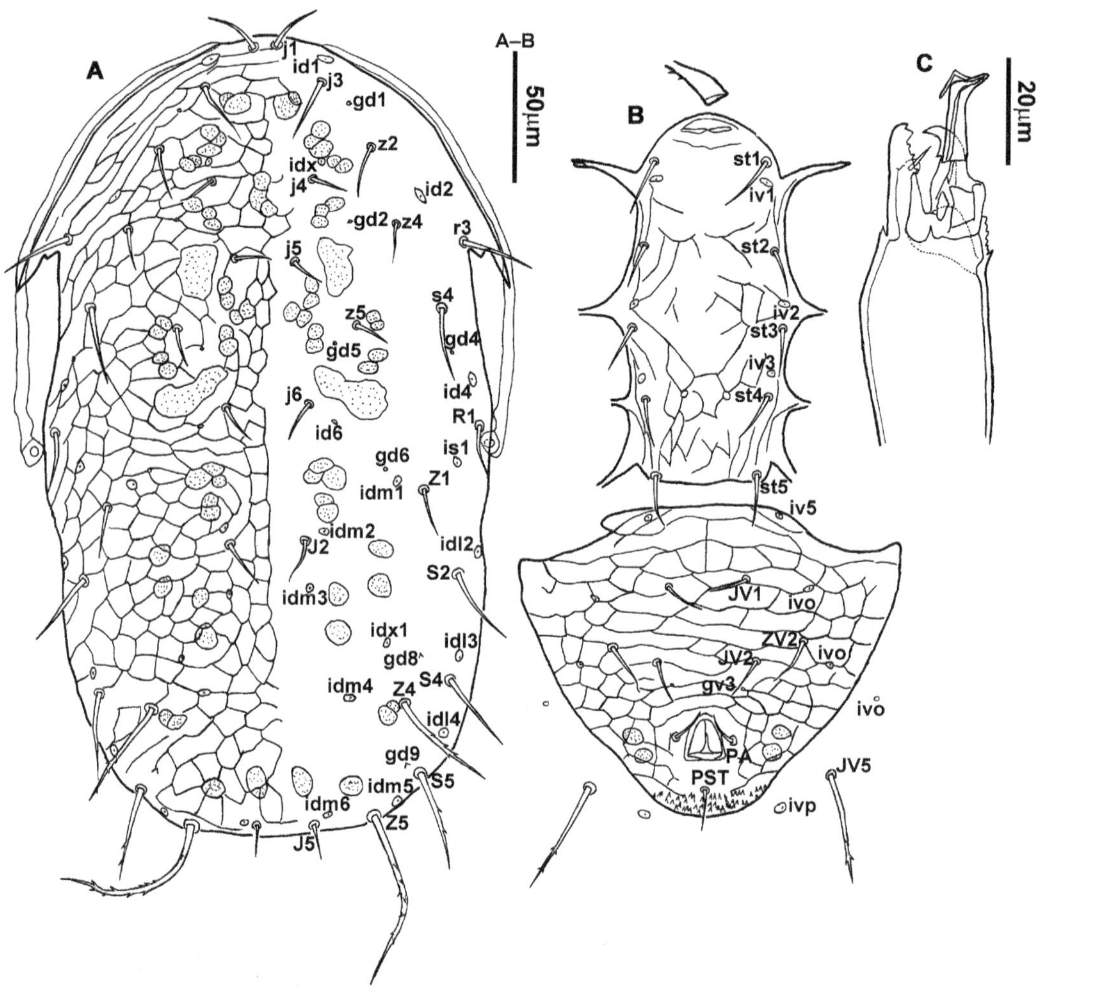

Figure 5. *Neoseiulus bicaudus* (Wainstein), male. (**A**) Dorsal shield; (**B**) Ventral idiosoma; (**C**) Chelicera.

12. **Remarks.** This species was originally described from grass found in Kazakhstan [14]. Subsequently, it has been collected from various hosts, including the following: almond trees [27], *Cichorium intybus* [28]; *Citrus* spp., *Cydonia oblonga*, *Fragaria ananassa*, *Malus communis*, *Prunus avium*, *Prunus domestica*, *Prunus persica* [29]; *Cupressus* sp. [30]; *Cynodon dactylon*, *Phoenix dactylifera* [31]; *Populus nigra* var. *thevestina* (Dode) bean [26]; pussy willow [32]; raspberry, low-growing vegetation [25]; *Setaria macrostachya*, *Phalaris minor*, *Cynodon dactylon*, *Pinus strobus*, *Aster spinosus*, *Haplopappus*, *Peyanum mexicanum*, *Rhynchelytrum repens*, *Asclepias curassavica*, *Distichilis stricta* [33]; *Setaria viridis* (present paper); *Tropaeolum majus* [34]; *Ulmus pumila* [13]; *Vitis vinifera* [35] in the Nearctic, Neotropical and Palearctic realms [3]. In China, it has been recorded as *Neoseiulus neoreticuloides* from Ningxia (Yinchuan) [13], and Xinjiang Uygur Autonomous Region [26].
13. The type specimen of *Neoseiulus neoreticuloides* was an adult female collected from *Ulmus pumila* (Siberian elm) in Ningxia. It should be noted that the host information was mistakenly attributed to *Platycladus orientalis* in the English abstract of the same publication, which actually pertains to a different species, *Typhlodromus* (*Anthoseius*)

yinchuanensis, by Liang and Hu. The specimens collected from Shanxi are consistent with the description provided by Liang and Hu (1988) [13] and Wu et al. (2009) [6]. Based on the presence of a stalk between the calyx and atrium of the spermatheca, *N. neoreticuloides* should be assigned to the *paraki* species subgroup. After comparing *N. neoreticuloides* with *N. bicaudus* (Table 1), we did not find any significant differences. Therefore, based on our observations, we consider *N. neoreticuloides* to be conspecific with *N. bicaudus*. As *N. bicaudus* has taxonomic priority, we conclude that *N. neoreticuloides* is a junior synonym of *N. bicaudus*.

Table 1. Comparison of morphological characteristics of *Neoseiulus neoreticuloides* and *N. bicaudus*.

	N. neoreticuloides [13]	*N. bicaudus* [36]	*N. bicaudus* [28]	*N. bicaudus* [31]	*N. bicaudus* [26]
Dorsal shield	reticulate, 415.7 long, 206.7 wide	thin net-like on posterior part, 400 long, 180 wide	reticulate, 402–410 long, 190–200 wide	reticulate, 388–401 long, 167–202 wide	reticulate, 375–425 long, 160–178 wide
Peritreme	extending to level between *j1* and *j3*	extending forward to level of bases of setae *j1*	extending to level between *j1* and *j3*	extending to level between *j1* and *j3*, close to *j3*	extending to level between *j1* and *j3*, close to *j1*
j1	25	25	22–25	23–25	22–26
j3	30	31	26–31	28–30	27–32
j4	15	15	14	14–16	12–15
j5	-	15	13–15	14–15	12–15
j6	16	17	16–20	16–19	15–18
J2	19.5	21	17–20	17–19	17–22
J5	14	14	14–16	12–15	12–15
r3	34	34	29–33	29–32	30–35
R1	30	30	25–31	28–32	30–32
s4	32	34	30–34	30–34	30–37
S2	33.5	35	32–38	32–36	33–40
S4	37.5	43	44–46	36–40	35–42
S5	39.5	45	53–61	42–46	40–49
z2	24	23	18–23	21–29	20–26
z4	20	20	18–19	13–18	16–20
z5	15.5	14	14–15	13–15	12–15
Z1	23	20	21–23	21–24	20–25
Z4	39	40	35–41	34–38	36–42
Z5	83	98	92–99	87–95	82–97
Sternal shield	scarcely striated	smooth	reticulated	scarcely striated	reticulated
Ventrianal shield	approximately triangle	135 long, 117 wide, sub-triangular	135–140 long, 110 wide, approximately triangle	133–141 long, 99–120 wide, approximately triangle	135–145 long, 102–114 wide, approximately triangle
Distance *gv3–gv3*	-	-	40–46	-	36–47
JV5	serrate; 60	serrate; 63	serrate; 64–78	serrate	serrate; 60 (55–65)
Fixed digit	6 teeth	6 teeth	6 teeth	6 teeth	7 teeth
Movable digit	1 tooth	1 tooth	1 tooth	1 tooth	2 teeth
Calyx of spermathecal apparatus	bowl-shaped	cup-shaped	bowl-shaped	bowl-shaped	bell-shaped
Macroseta *StIV*	74	73 (73–78)	64–74	70–73	67–85

Note: "-" indicates that the article does not describe this feature.

3.1.2. *Neoseiulus lushanensis* (Zhu and Chen, 1985) (Figures 6–11)

Amblyseius lushanensis Zhu and Chen, 1985: 273.

Amblyseius (*Neoseiulus*) *lushanensis* (Zhu and Chen), Wu et al. 2009: 152 [6].

Amblyseius (*Amblyseius*) *longisiphonulus* Wu and Lan, 1989: 248 [36]; synonym by Wu et al. 2009: 152.

Neoseiulus lushanensis (Zhu and Chen), Chant and McMurtry, 2003: 37 [37]; Moraes et al. 2004: 131 [38].

1. **Diagnosis (female).** Dorsal shield oval, mostly smooth but striated anterolaterally, bearing 17 pairs of setae, 16 pairs of lyrifissures and 7 pairs of solenostomes, all smooth but *Z4* and *Z5* serrate; *j3, S2, s4, z2, z4, Z4* and *Z5* longer than others. Peritremes extending anteriorly close to base of *j1*. Sternal shield obviously wider than long, sparsely striated, bearing three pairs of setae. Ventrianal shield approximately pentagonal, transversally striated and a few oblique striae present between transverse striae, with three pairs of preanal setae; solenostomes (*gv3*) posteromedian to *JV2*, crescent-shaped. Calyx of spermathecal apparatus elongate trumpet-shaped, arms distally thickened and flaring distally, atrium very large, bifurcate at junction with major duct, major duct membranous and thick-walled, as broad as atrium and then gradually reduced. Fixed digit of chelicera with five–six teeth and movable digit with one tooth. Palpgenual setae *al1* and *al2* rod-like. Leg genu II with 7 setae. Genu, tibia and basitarsus of leg IV each with a macroseta.

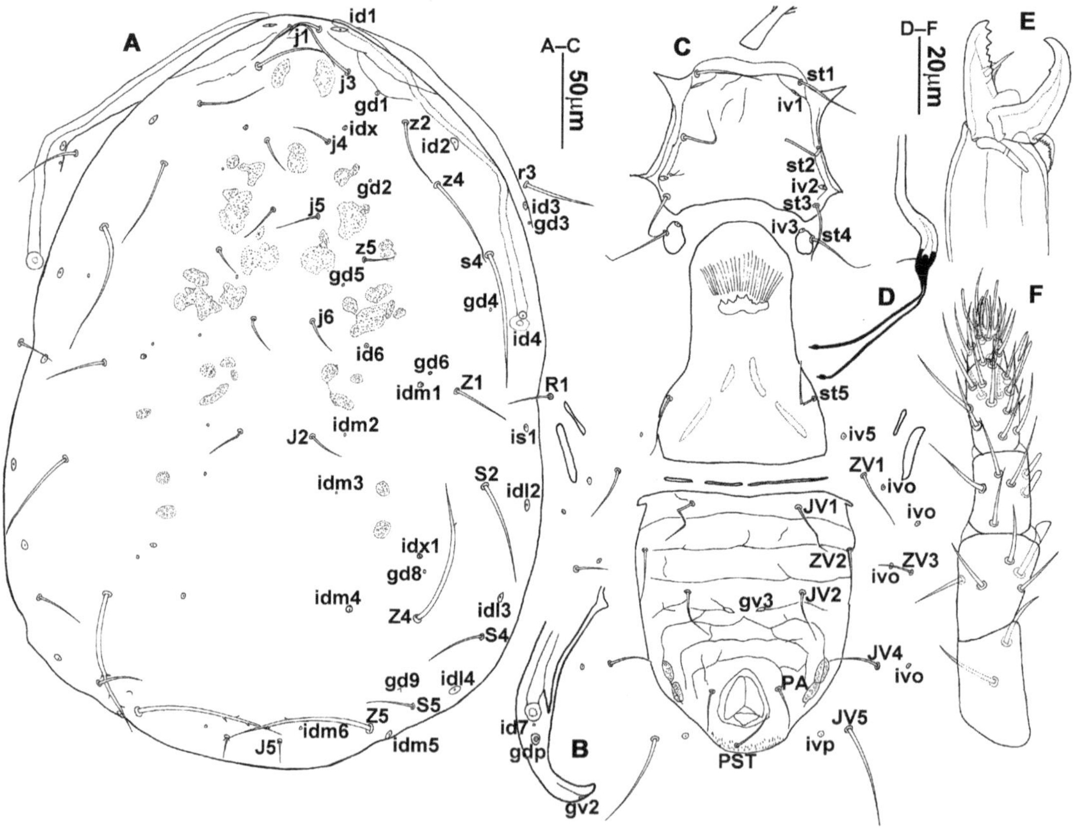

Figure 6. *Neoseiulus lushanensis* (Zhu and Chen), female. (**A**) Dorsal shield; (**B**) Posterior part of peritremal shield and exopodal shield; (**C**) Ventral idiosoma; (**D**) Spermathecal apparatus; (**E**) Chelicera; (**F**) Palp.

2. **Redescription. Female** (n = 7). Dorsal idiosoma (Figures 6A and 7A). Idiosomal setal pattern 10A:9B/JV-3:ZV. Dorsal shield oval, with a waist at level of *R1*; 399 (373–411) long, 262 (242–271) wide; shield mostly smooth but with a few striae at anterolateral margins between *j1* and *z2*; muscle marks visible between *j3* and *Z4*. Dorsum with 17 pairs of setae, 16 pairs of lyrifissures (*id1, id2, id4, id6, idx, idx1, idl2, idl3, idl4, idm1, idm2, idm3, idm4, idm5, idm6* and *is1*) and 7 pairs of solenostomes (*gd1, gd2, gd4,*

gd5, *gd6*, *gd8* and *gd9*); lyrifissures *id3* and solenostomes *gd3* on peritremal shield. Dorsal setae *Z4* and *Z5* serrated, others smooth, *j3*, *s4*, *S2*, *z2*, *z4*, *Z4* and *Z5* longer than others. Lateral setae *r3* and *R1* smooth, on interscutal membrane. Peritremes extending anteriorly close to bases of *j1*, posterior part of peritremal shield (Figure 6B) curved and pointed, protuberance of exopodal shield in front of stigmata. Lengths of setae: *j1* 30 (22–30), *j3* 50 (44–52), *j4* 19 (18–20), *j5* 20 (16–23), *j6* 23 (18–23), *J2* 21 (17–24), *J5* 10 (9–10), *r3* 36 (32–37), *R1* 21 (20–24), *s4* 77 (71–80), *S2* 58 (52–65), *S4* 34 (26–37), *S5* 23 (18–27), *z2* 39 (34–42), *z4* 48 (43–49), *z5* 14 (12–14), *Z1* 35 (28–36), *Z4* 82 (74–82), *Z5* 83 (70–83).

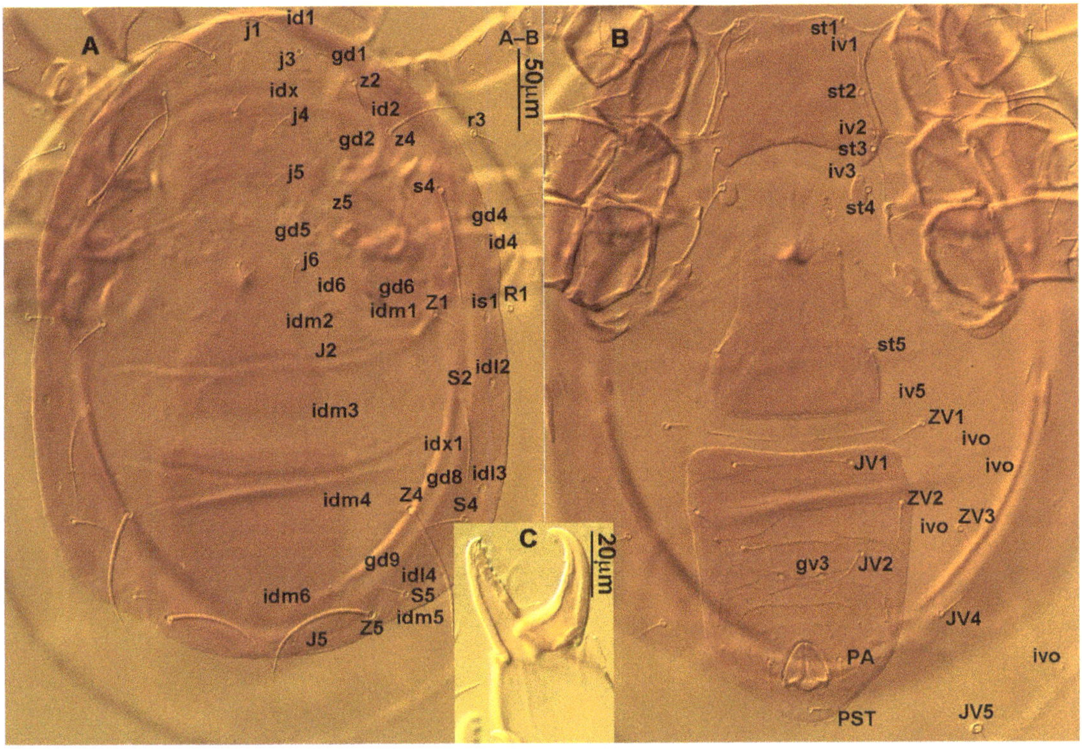

Figure 7. *Neoseiulus lushanensis* (Zhu and Chen), female. (**A**) Dorsal idiosoma; (**B**) Ventral idiosoma; (**C**) Chelicera.

3. Ventral idiosoma (Figures 6C and 7B). Sternal shield mostly smooth, striated laterally, 66 (65–69) long, 82 (82–88) wide; anterior margin weakly convex, forming a weak M-shaped median projection; posterior margin obviously concave, arched above the level of bases of *st3*; three pairs of setae (*st1*, *st2* and *st3*) and two pairs of lyrifissures (*iv1* and *iv2*) present, *iv1* positioned posteriad of *st1*, *iv2* positioned between *st2* and *st3*, and close to *st3*. Metasternal platelets small, each bearing a seta *st4* and a lyrifissure *iv3*. Epigynal shield smooth, 134 (127–139) long, 85 (78–85) wide, with two pairs of muscle scars between *st5*–*st5*. Lengths of setae: *st1* 29 (29–37), *st2* 31 (31–37), *st3* 31 (30–36), *st4* 29 (29–37), *st5* 25 (25–32). A slender transverse sclerite (sometimes broken into three or four parts) present between epigynal and ventrianal shields. Ventrianal shield approximately pentagonal, 134 (124–144) long, 119 (114–126) wide, transversally striated and a few oblique striae present between transverse striae, bearing three pairs of preanal setae (*JV1*, *JV2* and *ZV2*), a pair of paranal setae (*PA*) and a postanal seta (*PST*), and a pair of obviously crescent-shaped solenostomes (*gv3*)

posteromedial to *JV2*, distance *gv3–gv3* 21 (17–30); two pairs of marginal muscle marks situated anterolateral to anus; four pairs of setae (*JV4*, *JV5*, *ZV1* and *ZV3*) and five pairs of lyrifissures present on soft cuticle surrounding ventrianal shield. Lengths of setae: *JV1* 24 (23–30), *JV2* 28 (21–28), *JV4* 26 (20–26), *JV5* 59 (46–62), *ZV1* 29 (23–31), *ZV2* 25 (25–29), *ZV3* 21 (17–21). Primary metapodal platelet 29 (26–31) long, 5 wide; secondary platelet 12 (12–15) long, 2 wide.

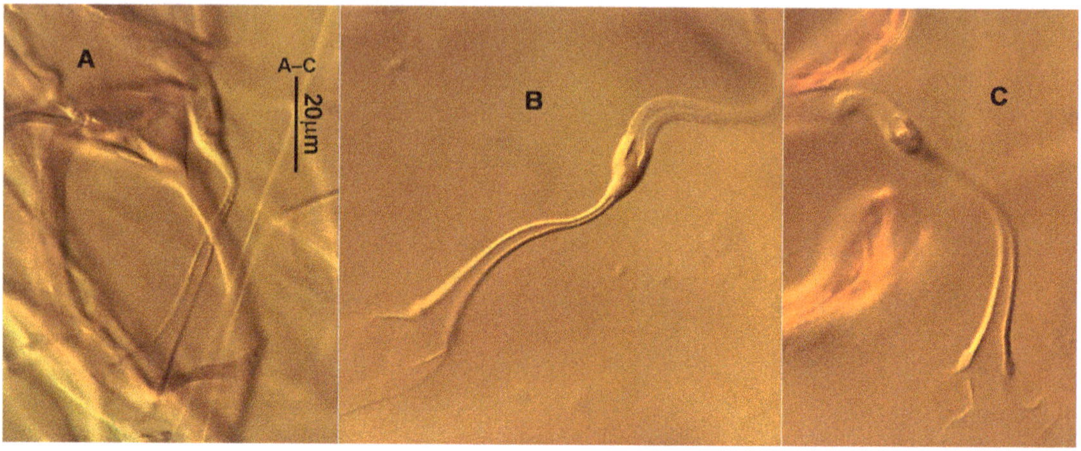

Figure 8. *Neoseiulus lushanensis* (Zhu and Chen), female. (**A–C**) Variation in spermathecal apparatus.

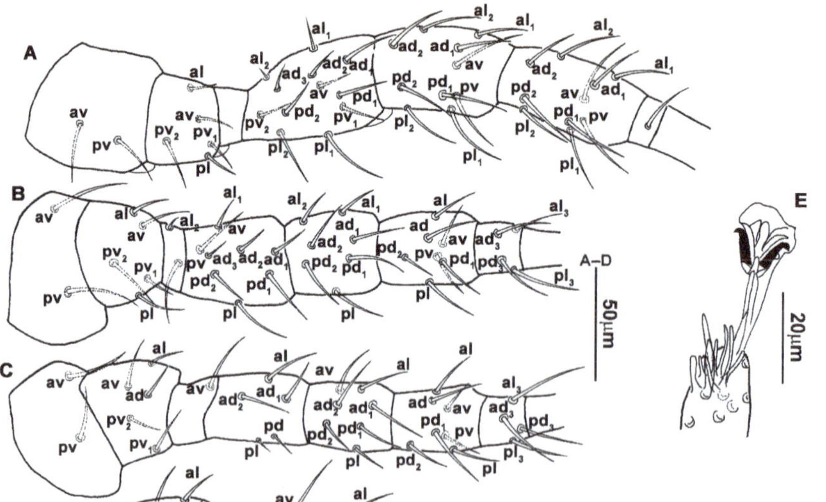

Figure 9. *Neoseiulus lushanensis* (Zhu and Chen), female. (**A**) Leg I; (**B**) Leg II; (**C**) Leg III; (**D**) Leg IV; (**E**) Apical sensorial setal cluster of tarsus I.

4. Spermatheca (Figures 6D and 8A–C). Calyx of spermathecal apparatus elongate trumpet-shaped, distally thickened and flaring, 50 (47–51) long; a large atrium nodular

at base of calyx and without a neck, bifurcate at junction with major duct; major duct membranous and thick-walled, as wide as atrium and then gradually reduced.

5. Gnathosoma. Chelicera (Figures 6E and 7C) with fixed digit 41 (38–41) long, bearing five–six teeth, pilus dentilis located at the level of tooth six, 7 (7–9) long, movable digit 37 (33–38) long, bearing a single tooth. Palp (Figure 6F). Trochanter with two setae; femur with a spatulate and four simple setae; genu bearing two rod-like setae (*al1* and *al2*) and four simple setae; tarsal apotele two-tined.
6. Legs (Figure 9A–D). Leg I 416 (413–433) long, setal formula: coxa 0-0/1-0/1-0, trochanter 1-0/1-0/2-1, femur 2-3/1-2/2-2, genu 2-2/1-2/1-2, tibia 2-2/1-2/1-2, basitarsus 0-0/0-1/0-0. Apical sensorial setal cluster (Figure 9E) with nine modified setae. Leg II 333 (323–334) long, setal formula: coxa 0-0/1-0/1-0, trochanter 1-0/1-0/2-1, femur 2-3/1-2/1-1, genu 2-2/0-2/0-1, tibia 1-1/1-2/1-1, basitarsus 1-1/0-1/0-1. Leg III 325 (318–329) long, setal formula: coxa 0-0/1-0/1-0, trochanter 1-1/1-0/2-0, femur 1-2/1-1/0-1, genu 1-2/1-2/0-1, tibia 1-1/1-2/1-1, basitarsus 1-1/0-1/0-1. Leg IV 445 (436–457) long, setal formula: coxa 0-0/1-0/0-0, trochanter 1-1/1-0/2-0, femur 1-2/1-1/0-1, genu 1-2/1-2/0-1, tibia 1-1/1-2/0-1, basitarsus 1-1/0-1/0-1. Legs I–III without macrosetae. Genu, tibia and basitarsus of leg IV each with a smooth macroseta, *SgeIV* 56 (50–59), *StiIV* 36 (33–38) and *StIV* 81 (76–86).

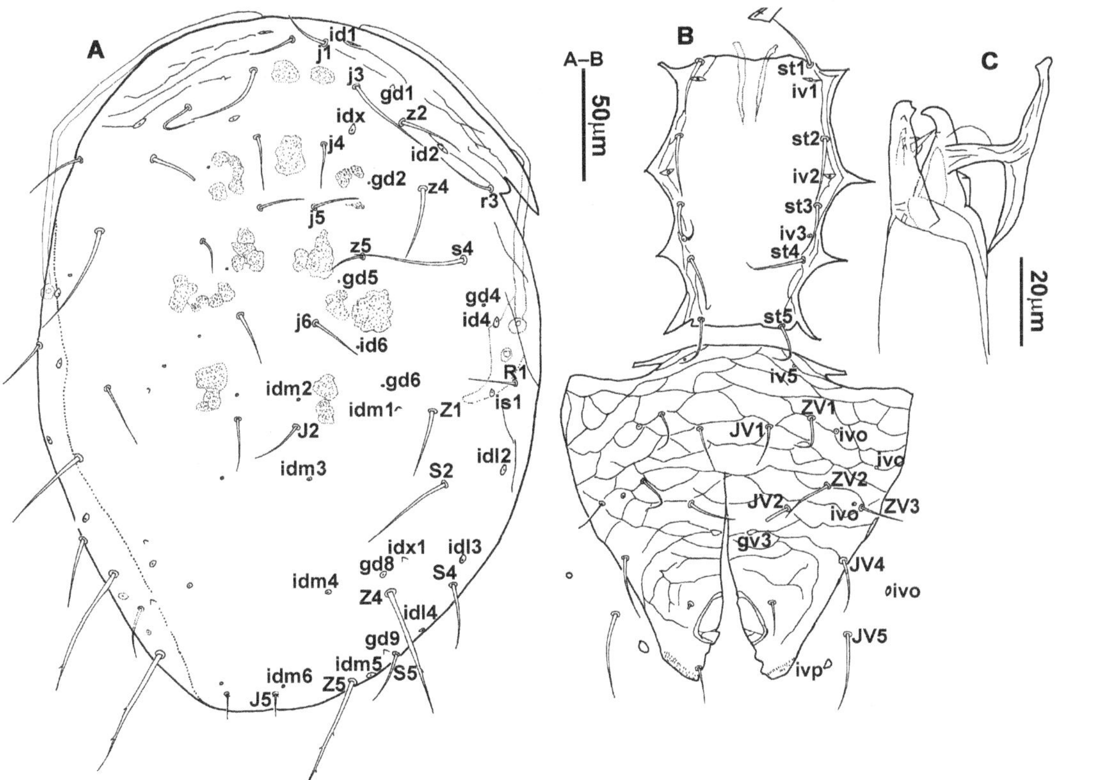

Figure 10. *Neoseiulus lushanensis* (Zhu and Chen), male. (**A**) Dorsal shield; (**B**) Ventral idiosoma; (**C**) Chelicera.

7. **Male** (n = 4). Dorsal idiosoma (Figures 10A and 11A). Dorsal shield oval, 308 (306–309) long, 216 (211–216) wide; mostly smooth but striated anterolaterally; muscle marks visible between *j3* and *J2*; dorsum bearing 19 pairs of setae, 16 pairs of lyrifissures (*id1*,

id2, *id4*, *id6*, *idx*, *idx1*, *idl2*, *idl3*, *idl4*, *idm1*, *idm2*, *idm3*, *idm4*, *idm5*, *idm6* and *is1*) and 7 pairs of solenostomes (*gd1*, *gd2*, *gd4*, *gd5*, *gd6*, *gd8* and *gd9*). *Z4* and *Z5* slightly serrated, other setae smooth; *j3*, *s4*, *S2*, *z2*, *z4*, *Z4* and *Z5* longer than others. Peritremes extending anteriorly close to *j1*. Lengths of setae: *j1* 21 (21–23), *j3* 41 (38–42), *j4* 19 (19–21), *j5* 19 (19–23), *j6* 23 (21–23), *J2* 21 (19–21), *J5* 9 (7–11), *r3* 30 (26–30), *R1* 22 (20–24), *s4* 56 (51–56), *S2* 48 (43–48), *S4* 32 (30–32), *S5* 28 (22–29), *z2* 33 (29–33), *z4* 28 (28–37), *z5* 14 (13–15), *Z1* 26 (26–32), *Z4* 58 (58–62), *Z5* 56 (56–63).

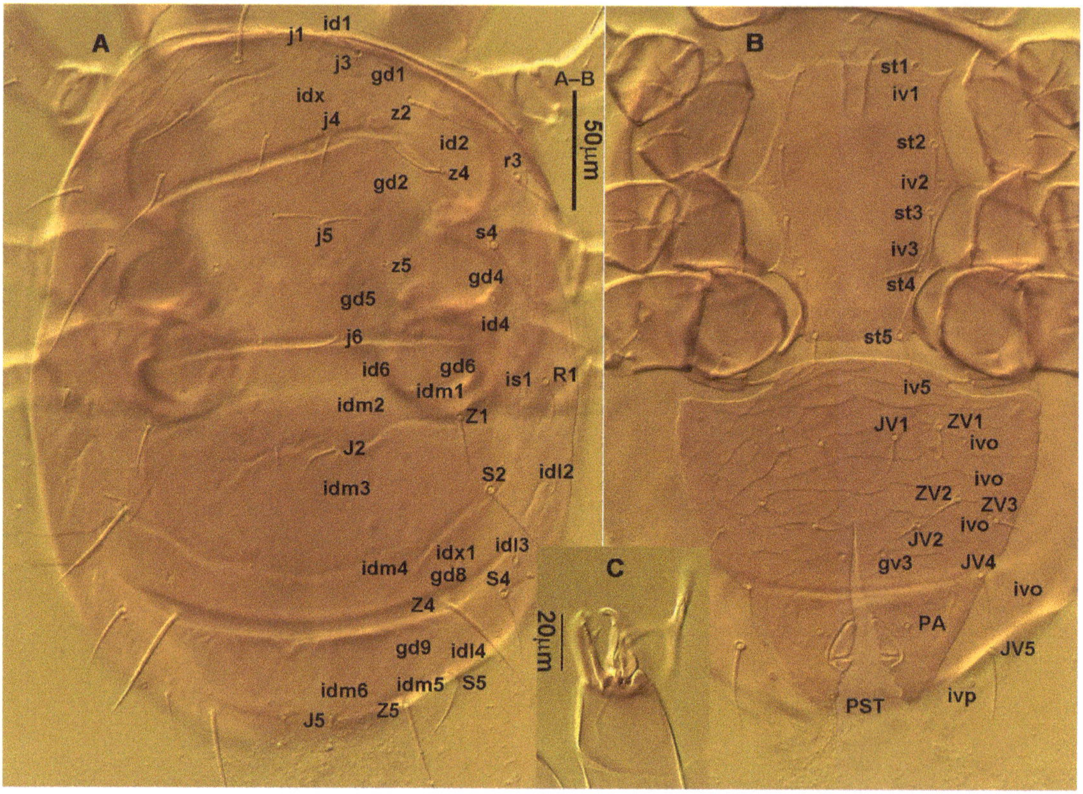

Figure 11. *Neoseiulus lushanensis* (Zhu and Chen), male. (**A**) Dorsal idiosoma; (**B**) Ventral idiosoma; (**C**) Chelicera.

8. Ventral idiosoma (Figures 10B and 11B). Sternogenital shield mostly smooth but a few striae present at lateral margins, 122 (120–124) long, 77 (72–77) wide; anterior margin and posterior margin nearly straight, bearing five pairs of setae (*st1*, *st2*, *st3*, *st4* and *st5*) and three pairs of lyrifissures (*iv1*, *iv2* and *iv3*); lengths of setae: *st1* 26 (26–31), *st2* 25 (25–30), *st3* 27 (27–30), *st4* 22 (22–29), *st5* 25 (23–25). Ventrianal shield subtriangular, reticulated throughout, anterior margin convex medially, 152 (134–152) long, 144 (141–144) wide, with six pairs of preanal setae (*JV1*, *JV2*, *JV4*, *ZV1*, *ZV2* and *ZV3*), a pair of paranal setae (*PA*) and a postanal seta (*PST*); four pairs of lyrifissures on ventrianal shield, a pair of crescent-shaped solenostomes (*gv3*) posteromedian to *JV2*, distance *gv3*–*gv3* 16 (16–22). Setae *JV5* and three pairs of lyrifissures on soft cuticle surrounding ventrianal shield. Lengths of setae: *JV1* 19 (19–25), *JV2* 20 (20–25), *JV4* 23 (21–23), *JV5* 33 (31–39), *ZV1* 18 (18–25), *ZV2* 22 (22–26), *ZV3* 17 (17–19).
9. Gnathosoma. Chelicera (Figures 10C and 11C) with fixed digit 28 (28–31) long, bearing five teeth; movable digit 26 (26–29) long, bearing a tooth. Spermatodactyl T-shaped,

with acute toe and heel, shaft 22 (20–22) long, foot 47 (46–49) long. Palp and hypostome with same chaetotaxy as in female.

10. Legs. Leg chaetotaxy same as that in adult female.
11. **Materials examined.** A total of 2♀and 2♂, Dabaishi Village, Taigu County, Shanxi Province, 37°20′12″ N, 112°38′50″ E, 1340 m asl, e.g., *Sanguisorba officinalis* L. (Rosaceae), 31 August 2020, M. Ma and B. Zhang coll.; 15♀and 3♂, same locality, *Setaria viridis* (Poaceae), 21 August 2021, Y. Liu, M. Ma, B. Zhang and F.-X. Ren coll.
12. **Remarks.** *Neoseiulus lushanensis* is classified in the *womersleyi* species subgroup within the *barkeri* species group [37]. Originally described from grass in Jiangxi Province, it has also been recorded in Guizhou, Henan, Hunan, Shandong, and Zhejiang provinces [6]. This paper reports the first record of it in Shanxi Province, where we observed a variation in the length of seta *Z1*. Compared to specimens from Jiangxi (23.4) [15] and Guizhou (26.5–27.5) [6], those from Shanxi have a longer *Z1* (35 (28–36)). In earlier descriptions, the striation on the anterolateral margins of the dorsal shield were often overlooked, as was the case in the original description provided by Zhu and Chen (1985) and the subsequent redescription presented by Wu et al. (2009). It is worth noting that the original description may have presented an incorrect number of preanal setae in males, which should be six pairs instead of three pairs.

3.1.3. *Neoseiulus makuwa* (Ehara, 1972)

Amblyseius (*Amblyseius*) *makuwa* Ehara, 1972: 154.

Amblyseius makuwa (Ehara), Wu et al., 1991 [39]: 147; Wu et al., 1997: 99 [40].

Amblyseius (*Neoseiulus*) *makuwa* (Ehara), Ehara and Amano, 1998: 37 [41]; Wu et al. 2009: 151 [6].

Neoseiulus makuwa (Ehara), Moraes et al., 1986: 87 [42]; Chant and McMurtry, 2003: 33 [37]; Moraes et al., 2004: 131 [38].

1. **Material examined.** A total of 1♀, Dabaishi Village, Taigu County, Shanxi Province, 37°20′12″ N, 112°38′50″ E, 1340 m, e.g., *Setaria viridis* (Poaceae), 31 August 2020, Y. Liu, M. Ma, B. Zhang and F.-X. Ren coll.
2. **Remarks.** The species was first described from *Cucumis melo* L. var. *makuwa* Makino in Kyushu, Japan, by Ehara (1972), and has since been reported in six other Asian countries including China, as well as one African country [3]. The current study documents the first record of this species in Shanxi Province.

3.1.4. *Neoseiulus paraki* (Ehara, 1967) (Figures 12–15)

Amblyseius (*Amblyseius*) *paraki* Ehara, 1967: 216; Ehara and Yokogawa, 1977: 52 [43]; Ehara et al., 1994: 126 [44].

Amblyseius (*Neoseiulus*) *paraki* (Ehara), Ehara and Amano, 1998: 35 [41].

Neoseiulus paraki (Ehara), Moraes et al., 1986: 92 [42]; Chant and McMurtry, 2003: 23 [37]; Moraes et al., 2004: 137 [38].

Typhlodromip paraki (Ehara), Ehara and Amano, 2004: 9 [45]; Ryu, 2013: 292 [46].

This is the first record for China.

1. **Diagnosis (female)**. Dorsal shield elongate oval, strongly reticulate, bearing 17 pairs of setae, 16 pairs of lyrifissures and 7 pairs of solenostomes, all smooth except *Z5* serrated; *s4*, *Z4* and *Z5* longer than others. Peritremes extending anteriorly to bases of *j1*. Sternal shield reticulated, bearing three pairs of setae. Ventrianal shield approximately pentagonal, loosely reticulated, solenostomes (*gv3*) small and rounded, posterior to *JV2*. Calyx of spermathecal apparatus bell-shaped and basally stalked, stalk approximately twice as long as width of atrium; atrium broadened at junction with minor duct, minor duct thread-like; major duct narrower than atrium. Fixed digit of chelicera with four–five teeth and movable digit edentate. Palpgenu with genu setae *al1* and *al2* rod-like. Genu II with eight setae. Genu, tibia and basitarsus of leg IV each with a macroseta.

2. **Redescription. Female** (n = 3). Dorsal idiosoma (Figures 12A and 13A). Idiosomal setal pattern 10A:9B/JV-3:ZV. Dorsal shield elongate oval, strongly reticulate, with a waist at level of *R1*, 404 (375–404) long, 198 (178–198) wide; muscle marks visible between *j3* and *Z4*. Dorsum with 17 pairs of setae, 16 pairs of lyrifissures (*id1, id2, id4, id6, idx, idx1, idl2, idl3, idl4, idm1, idm2, idm3, idm4, idm5, idm6* and *is1*) and 7 pairs of solenostomes (*gd1, gd2, gd4, gd5, gd6, gd8* and *gd9*)); lyrifissures *id3* and solenostomes *gd3* on peritremal shield. All dorsal setae smooth except *Z5* serrated; *Z4* and *Z5* longer than others. Lateral setae *r3* and *R1* smooth, on interscutal membrane. Peritremes extending anteriorly to bases of *j1*, posterior part of peritremal shield (Figure 12B) curved and bluntly pointed, protuberance of exopodal shield situated at level of stigmata. Lengths of setae: *j1* 26, *j3* 33 (32–35), *j4* 19 (18–19), *j5* 16 (16–18), *j6* 22 (21–25), *J2* 25 (24–26), *J5* 12 (12–14), *r3* 29 (29–32), *R1* 27 (27–35), *s4* 41 (41–44), *S2* 39 (38–42), *S4* 34 (33–38), *S5* 34 (33–37), *z2* 28 (28–29), *z4* 30 (30–33), *z5* 19 (19–21), *Z1* 26 (26–29), *Z4* 42 (42–49), *Z5* 65 (65–70).

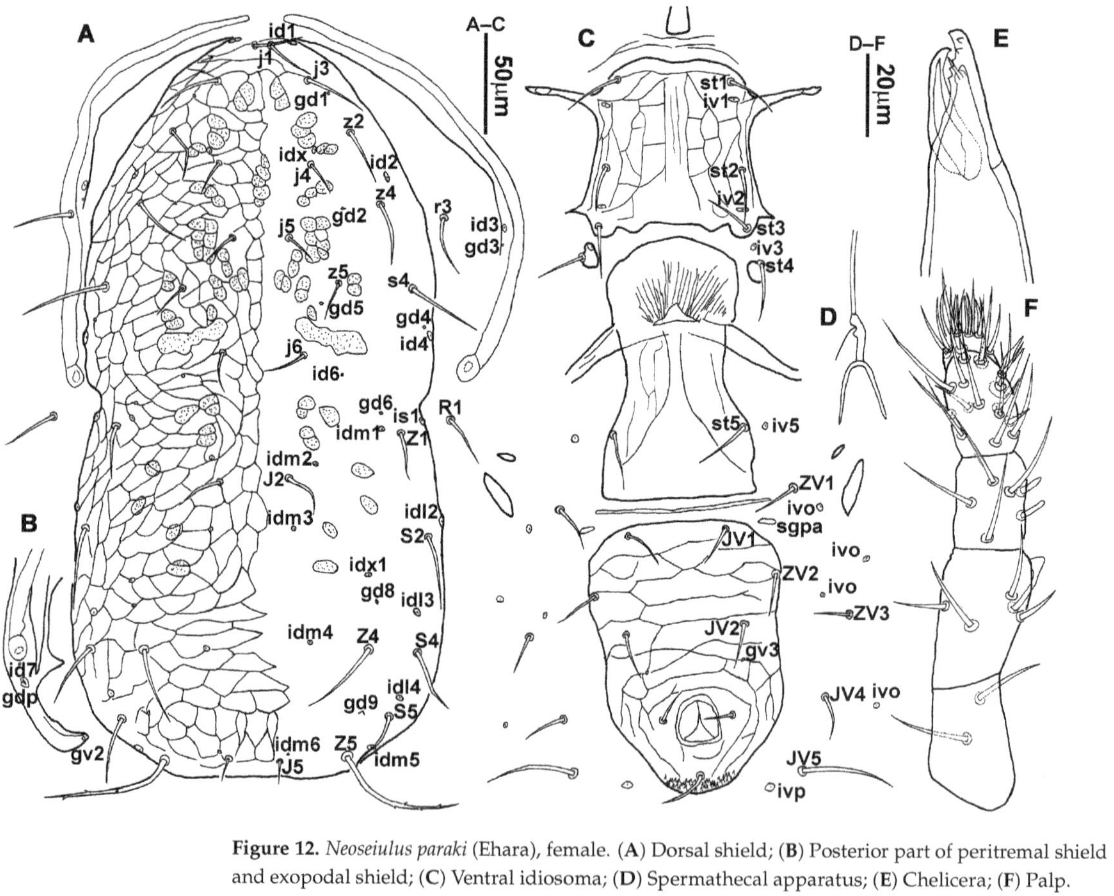

Figure 12. *Neoseiulus paraki* (Ehara), female. (**A**) Dorsal shield; (**B**) Posterior part of peritremal shield and exopodal shield; (**C**) Ventral idiosoma; (**D**) Spermathecal apparatus; (**E**) Chelicera; (**F**) Palp.

3. Ventral idiosoma (Figures 12C and 13B). Sternal shield reticulated, 81 (76–81) long, 85 (80–85) wide; anterior margin convex, forming a flat M-shaped projection, posterior margin weakly concave, with two small lateral projections; three pairs of setae (*st1, st2* and *st3*) and two pairs of lyrifissures (*iv1* and *iv2*) present on sternal shield, *iv1* positioned posteriad of *st1*, *iv2* positioned between *st2* and *st3*, and closer to *st3* than to *st2*. Metasternal platelets small, each bearing a seta *st4* and a lyrifissure *iv3*. Epigynal shield slightly striated, 136 (126–136) long, 85 (71–85) wide. Lengths of

setae: *st1* 29 (29–30), *st2* 25 (25–27), *st3* 28 (26–28), *st4* 30 (27–30), *st5* 25 (25–27). A slender transverse sclerite present between epigynal and ventrianal shields. Ventrianal shield (Figure 14A,B) approximately pentagonal, loosely reticulated throughout, 145 (135–145) long, 111 (97–111) wide, bearing three pairs of preanal setae (*JV1*, *JV2* and *ZV2*), a pair of paranal setae (*PA*) and a postanal seta (*PST*), and a pair of solenostomes (*gv3*) posterior to *JV2*, *gv3* small and round, distance *gv3*–*gv3* 53 (48–53); four pairs of setae (*JV4*, *JV5*, *ZV1* and *ZV3*) and five pairs of lyrifissures present on soft cuticle surrounding ventrianal shield. A pair of tiny platelets (*sgpa*) posteroparaxial to *ZV1* close to anterior corners of ventrianal shield. Lengths of setae: *JV1* 21 (18–21), *JV2* 25 (22–25), *JV4* 20 (19–22), *JV5* 45 (40–48), *ZV1* 22 (20–22), *ZV2* 22 (21–22), *ZV3* 19 (17–19). Primary metapodal platelet 25 (23–25) long, 5 (5–6) wide; secondary platelet 13 (11–13) long, 2 (2–3) wide.

4. Spermatheca (Figures 12D and 14C). Calyx of spermathecal apparatus 19 (18–20) long, bell-shaped and basally stalked, stalk approximately twice as long as width of atrium, atrium broadened at junction with minor duct, minor duct thread-like; major duct approximately half width of atrium.
5. Gnathosoma. Chelicera (Figures 12E and 13C) with fixed digit 34 long, bearing four–five teeth, pilus dentilis located at the level of fourth tooth, 5 (5–6) long; movable digit 33 (31–33) long, without teeth. Palp (Figure 12F). Trochanter with two setae; femur with a spatulate and four simple setae; genu bearing two rod-like setae (*al1* and *al2*) and four simple setae; tarsal apotele two-tined.
6. Legs (Figure 15A–D). Leg I 392 (374–406) long, setal formula: coxa 0-0/1-0/1-0, trochanter 1-0/1-0/2-1, femur 2-3/1-2/2-2, genu 2-2/1-2/1-2, tibia 2-2/1-2/1-2, basitarsus 0-0/0-1/0-0. Apical sensorial setal cluster of tarsus I (Figure 15E) with 10 modified setae. Leg II 323 (311–333) long, setal formula: coxa 0-0/1-0/1-0, trochanter 1-0/1-0/2-1, femur 2-3/1-2/1-1, genu 2-2/1-2/0-1, tibia 1-1/1-2/1-1, basitarsus 1-1/0-1/0-1. Leg III 324 (314–333) long, setal formula: coxa 0-0/1-0/1-0, trochanter 1-1/1-0/1-1, femur 1-2/1-1/0-1, genu 1-2/1-2/0-1, tibia 1-1/1-2/1-1, basitarsus 1-1/0-1/0-1. Leg IV 427 (419–451) long, setal formula: coxa 0-0/1-0/0-0, trochanter 1-1/1-1/1-0, femur 1-2/1-1/0-1, genu 1-2/1-2/0-1, tibia 1-1/1-2/0-1, basitarsus 1-1/0-1/0-1. Legs I–II without macrosetae, genu and tibia of leg III each with a smooth macroseta, *SgeIII* 31 (29–32), *StiIII* 27 (25–28); genu, tibia and basitarsus of leg IV each with a smooth macroseta, *SgeIV* 40 (39–40), *StiIV* 39 (39–41), *StIV* 78 (70–78).
7. **Males**. Not found in the current study.
8. **Materials examined.** A total of 3♀, Dabaishi Village, Taigu County, Shanxi Province, 37°20′12″ N, 112°38′50″ E, 1340 m, e.g., *Setaria viridis* (L.) P. Beauv. (Poaceae), 31 August 2020, Y. Liu, M. Ma, B. Zhang and F.-X. Ren coll.
9. **Remarks.** The female of this species was originally described from apple trees in Sapporo, Hokkaido, Japan [16]. Subsequently, Ehara and Yokogwa (1977) provided information on the male morphology. Both sexes of *N. paraki* have four–five teeth on the fixed digit, while the movable digit is edentate in females but has one tooth in males [43]. While this species has also been reported in South Korea, specimens from this country have seven teeth on the fixed digit and one tooth on the movable digit, and their epigynal shield is strongly reticulated [46]. In contrast, the epigynal shield of our specimens is slightly striated. One of our specimens exhibited a structural anomaly where *iv3* was positioned off the metasternal platelet on the right side.

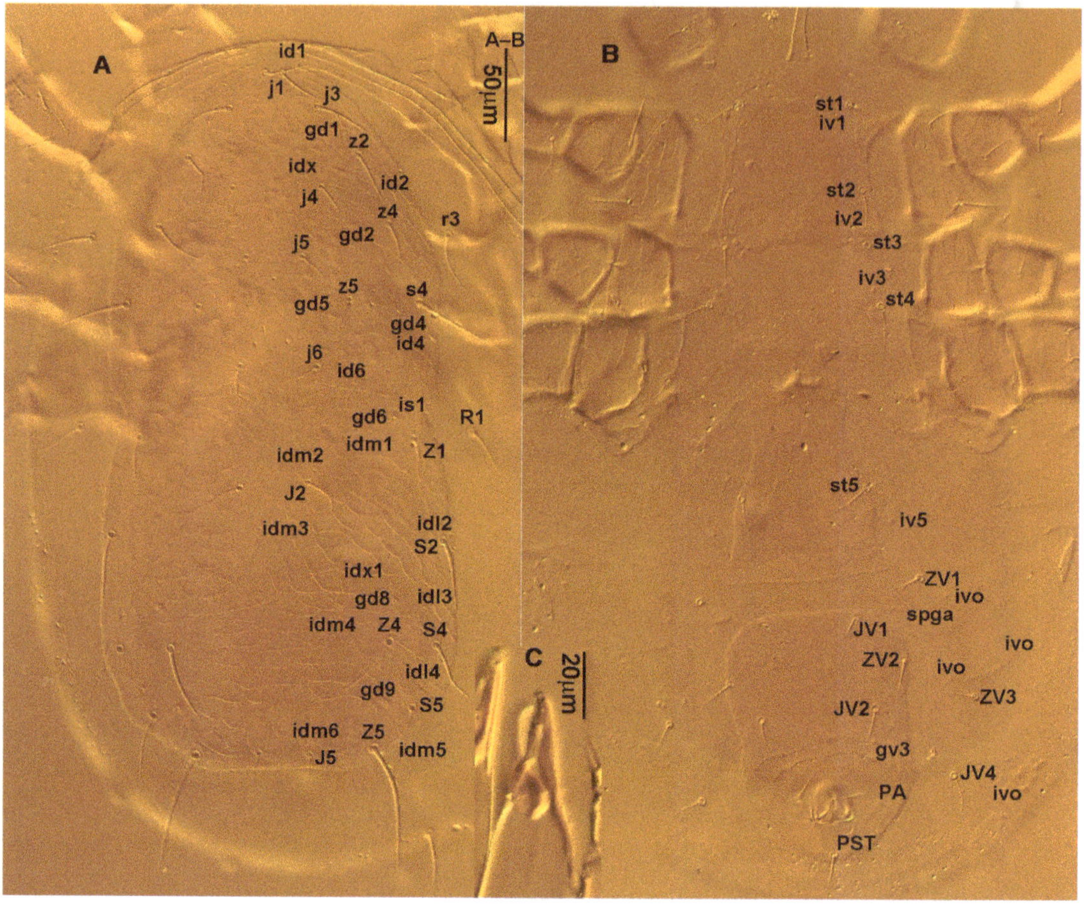

Figure 13. *Neoseiulus paraki* (Ehara), female. (**A**) Dorsal shield; (**B**) Ventral idiosoma; (**C**) Chelicera.

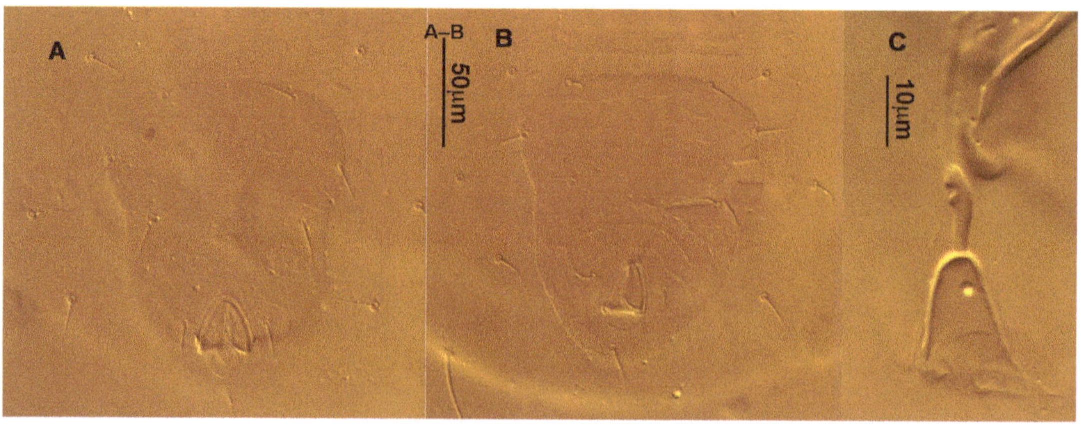

Figure 14. *Neoseiulus paraki* (Ehara), female. (**A**,**B**) Variation in ventrianal shield; (**C**) Spermathecal apparatus.

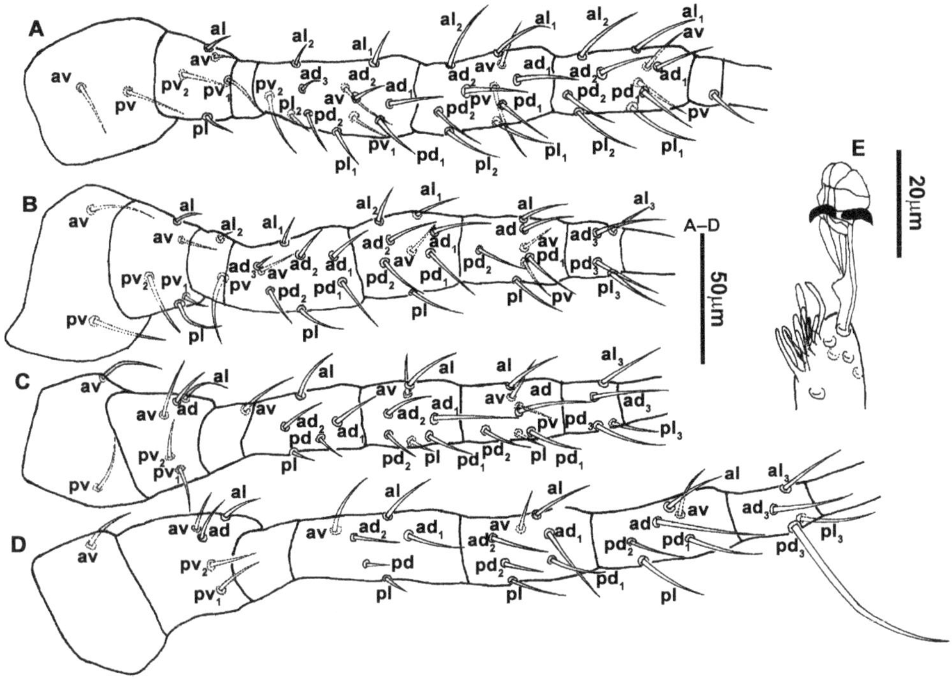

Figure 15. *Neoseiulus paraki* (Ehara), female. (**A**) Leg I; (**B**) Leg II; (**C**) Leg III; (**D**) Leg IV; (**E**) Apical sensorial setal cluster of tarsus I.

3.1.5. *Neoseiulus tauricus* (Livshitz and Kuznetsov, 1972) (Figures 16–19)

Amblyseius tauricus Livshitz and Kuznetsov, 1972: 24; Wu and Lan, 1991: 316 [47]; Wu and Ou, 1999: 105 [48].

Amblyseius (*Neoseiulus*) *tauricus* (Livshitz and Kuznetsov), Wu et al., 2009: 92 [6].

Neoseiulus tauricus (Livshitz and Kuznetsov), Moraes et al., 1986: 92 [42]; Chant and McMurtry, 2003: 23 [37]; Moraes et al., 2004: 147 [38].

1. **Diagnosis (female)**. Dorsal shield elongate oval, reticulated throughout, bearing 17 pairs of setae, 16 pairs of lyrifissures and 7 pairs of solenostomes, all smooth except *Z4* and *Z5*, which were serrated; *Z4* and *Z5* longer than others. Peritremes extending anteriorly to bases of *j1*. Sternal shield striated laterally, bearing three pairs of setae. Ventrianal shield approximately pentagonal, mostly transversally striated and a few oblique striae present between transverse striae; solenostomes (*gv3*) not discernible. Calyx of spermathecal apparatus funnel-shaped and constricted medially, arms apically flaring; atrium positioned right at base of calyx; major duct membranous, gradually expanded after a short distance. Fixed digit of chelicera with four teeth and movable digit with a tooth. Palpgenu with genu setae *al1* and *al2* tapered, spiniform. Genu II with seven setae. Genu, tibia and basitarsus of leg IV each with a macroseta.
2. **Redescription**. **Female** (n = 4). Dorsal idiosoma (Figures 16A and 17A). Idiosomal setal pattern 10A:9B/JV-3:ZV. Dorsal shield elongate oval, reticulated, with a waist at level of *R1*, 347 (328–347) long, 171 (160–171) wide; muscle marks visible between *j1* and *Z4*, 2 pairs of muscle marks present in front of *J5*. Dorsum with 17 pairs of setae and 16 pairs of lyrifissures (*id1, id2, id4, id6, idx, idx1, idl2, idl3, idl4, idm1, idm2, idm3, idm4, idm5, idm6* and *is1*) and 7 pairs of solenostomes (*gd1, gd2, gd4, gd5, gd6, gd8* and *gd9*); lyrifissures *id3* and solenostomes *gd3* on peritremal shield. All dorsal setae smooth except *Z4* and *Z5*, which were serrated; *Z4* and *Z5* longer than others. Lateral setae *r3* and *R1* smooth, on interscutal membrane. Peritremes extending anteriorly to bases of *j1*, posterior part

of peritremal shield (Figure 16B) nearly straight and with a blunt tip, protuberance of exopodal shield at level of stigmata. Lengths of setae: *j1* 19 (15–19), *j3* 18 (17–19), *j4* 11, *j5* 12 (11–12), *j6* 13 (12–13), *J2* 14 (12–14), *J5* 9 (9–10), *r3* 15 (15–18), *R1* 16 (14–18), *s4* 26 (24–26), *S2* 24 (24–26), *S4* 23 (21–23), *S5* 21 (20–22), *z2* 18 (14–18), *z4* 16 (15–17), *z5* 12 (11–13), *Z1* 19 (15–19), *Z4* 37 (36–37), *Z5* 48 (48–53).

3. Ventral idiosoma (Figures 16C and 17B). Sternal shield striated laterally, 68 (62–68) long, 73 (69–73) wide; anterior margin convex, posterior margin nearly plane; three pairs of setae (*st1*, *st2* and *st3*) and two pairs of lyrifissures (*iv1* and *iv2*) present on sternal shield, lyrifissure *iv1* positioned posteriad of *st1*, *iv2* positioned between *st2* and *st3*, and close to *st3*. Metasternal platelets small, each bearing a seta *st4* and a lyrifissure *iv3*. Epigynal shield with a few longitudinal and oblique striae and two pairs of muscle scars between *st5* and *st5*; 118 (110–120) long, 65 (63–68) wide. Lengths of setae: *st1* 29 (25–29), *st2* 24 (24–26), *st3* 28 (22–28), *st4* 23 (20–23), *st5* 22 (21–23). A slender transverse sclerite present between epigynal and ventrianal shields. Ventrianal shield approximately pentagonal, striated, 122 (110–122) long, 92 (89–92) wide, bearing three pairs of preanal setae (*JV1*, *JV2* and *ZV2*), a pair of paranal setae (*PA*) and a postanal seta (*PST*), solenostomes (*gv3*) not discernible; four pairs of setae (*JV4*, *JV5*, *ZV1* and *ZV3*) and five pairs of lyrifissures present on soft cuticle surrounding ventrianal shield. A pair of tiny platelets (*sgpa*) posteroparaxial to *ZV1* adjacent to anterior corners of ventrianal shield. Lengths of setae: *JV1* 20 (19–20), *JV2* 22 (17–22), *JV4* 21 (21–22), *JV5* 50 (50–56), *ZV1* 21 (17–21), *ZV2* 19 (18–20), *ZV3* 11 (11–13). Primary metapodal platelet 27 (26–32) long, 6 (5–6) wide; secondary platelet 13 (11–16) long, 3 (3–4) wide.
4. Spermatheca (Figures 16D and 18A,B). Calyx of spermathecal apparatus elongate, funnel-shaped and constricted medially, flaring distally, 24 (23–24) long; atrium incorporated within base of calyx; minor duct thread-like; major duct membranous, gradually broadened after a short distance.
5. Gnathosoma. Chelicera (Figures 16E and 17C) with fixed digit 27 (27–30) long, bearing four teeth, pilus dentilis located at the level of fourth tooth, 7 long; movable digit 26 (25–26) long, bearing a single tooth. Palp (Figure 16F). Trochanter with two setae; femur with a spatulate and four simple setae; genu bearing two tapered spiniform setae (*al1* and *al2*), and four simple setae; tarsal apotele two-tined.
6. Legs (Figure 19A–D). Leg I 339 (327–342) long, setal formula: coxa 0-0/1-0/1-0, trochanter 1-0/1-0/2-1, femur 2-3/1-2/2-2, genu 2-2/1-2/1-2, tibia 2-2/1-2/1-2, basitarsus 0-0/0-1/0-0. Apical sensorial setal cluster of tarsus I (Figure 19E) with 10 modified setae. Leg II 268 (253–268) long, setal formula: coxa 0-0/1-0/1-0, trochanter 1-0/1-0/2-1, femur 2-3/1-2/1-1, genu 2-2/0-2/0-1, tibia 1-1/1-2/1-1, basitarsus 1-1/0-1/0-1. Leg III 268 (245–268) long, setal formula: coxa 0-0/1-0/1-0, trochanter 1-1/1-0/1-1, femur 1-2/1-1/0-1, genu 1-2/1-2/0-1, tibia 1-1/1-2/1-1, basitarsus 1-1/0-1/0-1. Leg IV 352 (328–352) long, setal formula: coxa 0-0/1-0/0-0, trochanter 1-1/1-1/1-0, femur 1-2/1-1/0-1, genu 1-2/1-2/0-1, tibia 1-1/1-2/0-1, basitarsus 1-1/0-1/0-1. Legs I–III without macrosetae. Genu, tibia and basitarsus of leg IV each with a smooth macroseta, *SgeIV* 30 (30–33), *StiIV* 21 (20–23), *StIV* 49 (48–50).
7. **Males**. Not found in the present study.
8. **Materials examined.** A total of 4♀, Dabaishi Village, Taigu County, Shanxi Province, 37°20′12″ N, 112°38′50″ E, 1340 m, e.g., *Setaria viridis* (L.) P. Beauv. (Poaceae), 31 August 2020, Y. Liu, M. Ma, B. Zhang and F.-X. Ren coll.
9. **Remarks.** *Neoseiulus tauricus* was first described by Livshitz and Kuznetsov (1972). In China, it was previously recorded from weeds in Inner Mongolia [47]. We observed that seta *JV5* in our specimens is smooth, whereas it was originally described as serrated [18]. Additionally, we noted that the left lyrifissure *iv3* is off the metasternal platelet. The macroseta on tibia IV in our Shanxi specimens is indistinct, as reported in [6,40], whereas it is only slightly shorter than that on genu IV in Livshitz and Kuznetsov (1972).

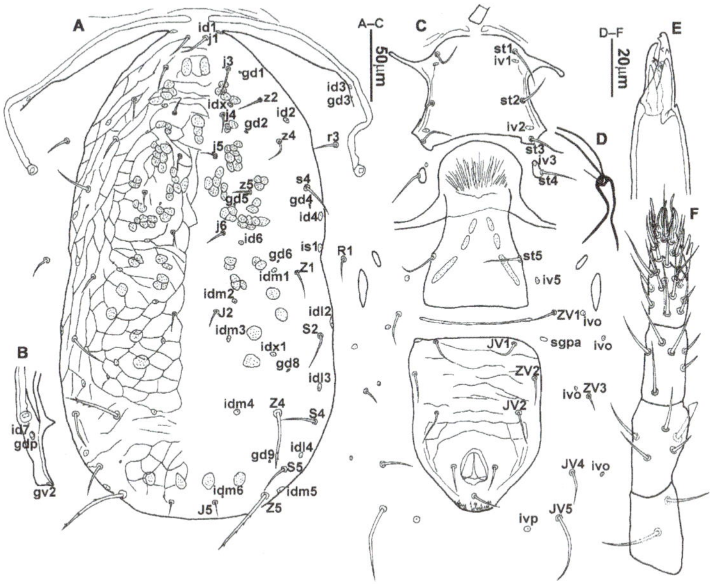

Figure 16. *Neoseiulus tauricus* (Livshitz and Kuznetsov), female. (**A**) Dorsal shield; (**B**) Posterior part of peritremal shield and exopodal shield; (**C**) Ventral idiosoma; (**D**) Spermathecal apparatus; (**E**) Chelicera; (**F**) Palp.

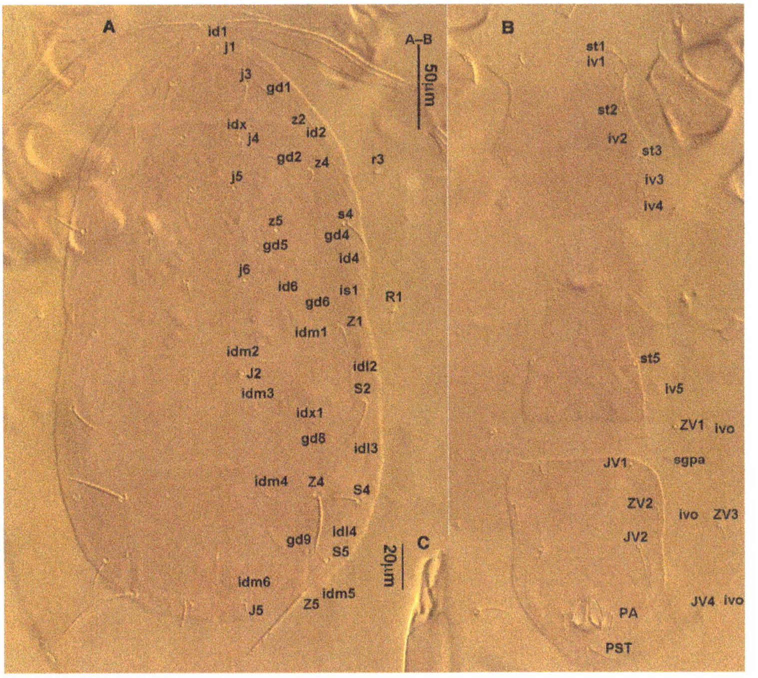

Figure 17. *Neoseiulus tauricus* (Livshitz and Kuznetsov), female. (**A**) Dorsal shield; (**B**) Ventral idiosoma; (**C**) Chelicera.

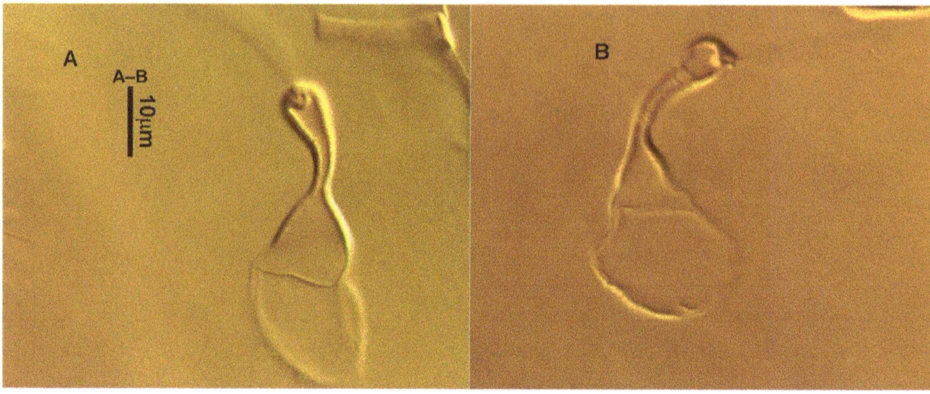

Figure 18. *Neoseiulus tauricus* (Livshitz and Kuznetsov), female. (**A**,**B**) Variation in spermathecal apparatus.

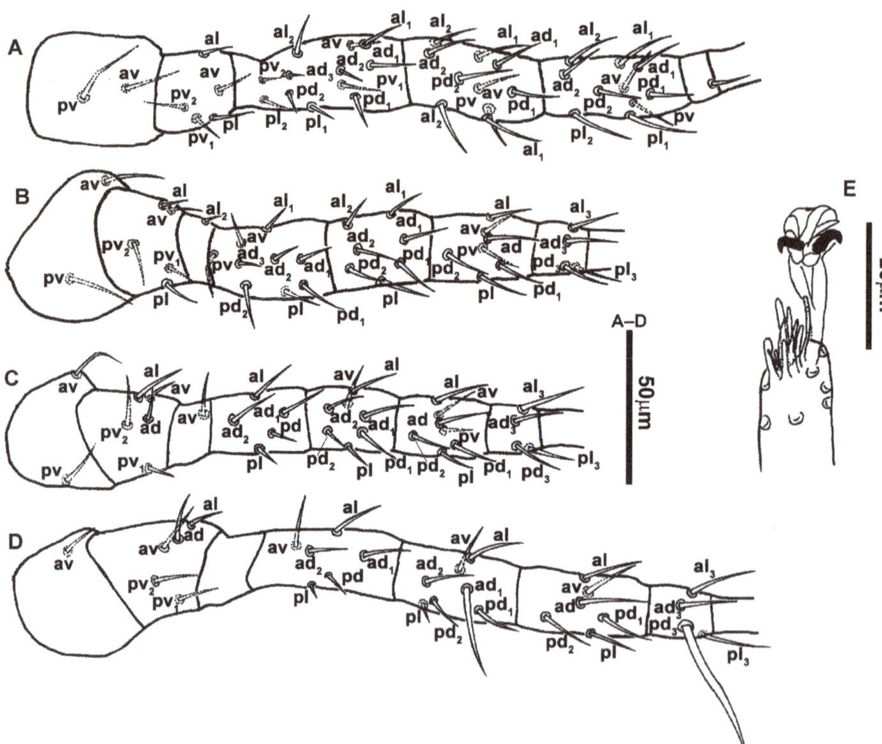

Figure 19. *Neoseiulus tauricus* (Livshitz and Kuznetsov), female. (**A**) Leg I; (**B**) Leg II; (**C**) Leg III; (**D**) Leg IV; (**E**) Apical sensorial setal cluster of tarsus I.

3.1.6. *Neoseiulus womersleyi* (Schicha, 1975)

Amblyseius womersleyi Schicha, 1975: 101; Schicha, 1987: 96 [49].

Amblyseius (*Amblyseius*) *womersleyi* (Schicha), Tseng, 1983: 54 [50]; Ehara et al., 1994: 123 [44].

Amblyseius (*Neoseiulus*) *womersleyi* (Schicha), Ehara and Amano, 1998: 30 [41]; Wu et al., 2009: 65 [6].

Neoseiulus womersleyi (Schicha), Moraes et al., 1986: 86 [42]; Beard, 2001: 84 [51]; Chant and McMurtry, 2003: 37 [37]; Moraes et al., 2004: 152 [38]; Liao et al., 2020: 288 [52].

Amblyseius pseudolongispinosus Xin, Liang and Ke, 1981: 75 [9]; Wu et al. 1997: 43 [40]; synonymy by Tseng, 1983: 57 [50].

Amblyseius (*Neoseiulus*) *pseudolongispinosus* (Xin, Liang and Ke), Wu et al., 2009: 65 [6].

Neoseiulus pseudolongispinosus (Xin, Liang and Ke), Ma et al., 2015: 15 [10].

1. **Materials examined.** A total of 2♀, Jincheng City, Shanxi Province, 35°29′31″ N, 112°49′47″ E, 624 m, e.g., weed, 19 August 2014, Y.-N. Zhao coll.; 1♂, Shanxi Agriculture University, Taigu County, Shanxi Province, 37°25′24″ N, 112°34′53″ E, 794 m, weed, 20 September 2014, M.-J. Yi and B.-Q. Su coll.
2. **Remarks.** This species was originally described from strawberry in Australia [8] and later recorded in China, Japan and South Korea [3]. In 1981, Xin et al. described a new species, *Amblyseius pseudolongispinosus*, based on specimens collected from various plant species in different regions of China. This species has been reported in Mainland China under the name *Amblyseius pseudolongispinosus*, although Tseng, from Taiwan, synonymized it with *Amblyseius womersleyi* in 1983. Our examination of specimens from different localities in China, as well as the original description and specimens from New Zealand, supports Tseng's decision to synonymize the two species.

3.1.7. *Neoseiulus zwoelferi* (Dosse, 1957)

Typhlodromus zwoelferi Dosse, 1957: 301 [53].

Cydnodromus zwoelferi (Dosse), Muma, 1961: 290 [54].

Amblyseius zwoelferi (Dosse), Schuster and Pritchard, 1963: 268 [55].

Amblyseius (*Amblyseius*) *zwoelferi* (Dosse), Wainstein, 1975: 920 [56].

Amblyseius subreticulatus Wu, 1987: 264 [57]; synonym by Zhang et al. 2021: 20 [12]. Third bullet.

1. **Materials examined.** A total of 17♀, 5♂, 8 deutonymph, 8 protonymphs, 6 larvae, e.g., laboratory culture in Shanxi Agriculture University, 8 June 2020–23 September 2020, B. Zhang, M. Ma and S. Jiao. coll.; 5♀, Taigu, e.g., *Hemerocallis fulva* L. (Asphodelaceae), 7 October 2013, Q.-H. Fan coll.; 1♀, Taigu, e.g., *Hemerocallis fulva* L. (Asphodelaceae), 6 August 2014, M. Ma coll.; 17♀, 2♂, 1 protonymph. Ningwu, Luyashan National Nature Reserve, e.g., weed, 6 August 2014, B. Zhang and M. Ma coll.; 1♀, Ningwu, Luyashan National Nature Reserve, e.g., *Hippophae rhamnoides* L. (Elaeagnaceae), B.-Q. Su and M.-J. Yin coll.; 1♀, 1♂, Taigu, e.g., weed, 20 August 2014, B.-Q. Su and M.-J. Yin coll.
2. **Remarks.** The original description of this species was based on specimens collected from apple leaves in Oldenburg, Germany [53]. Since then, it has been found in eighteen countries in the Palearctic realm, one country in the Indomalayan realm, and one country in the Nearctic realm [3]. In China, it has a broad geographical range, being present from the far northeast (Heilongjiang) to the far northwest (Xinjiang) and the southern coast (Guangdong).

3.2. *Key to Adult Females of Neoseiulus in Shanxi Province*

1. Most dorsal idiosomal setae (except j1 and J5) serrated; setae *j4–6*, *Z1*, *S2* and *S4* extending beyond bases of setae in next row. . . . . *N. womersleyi* (Schicha)

- Most dorsal idiosomal setae smooth, except *Z5*, which is barbed, *Z4*, S4 and S5, which are barbed or smooth; setae *j4–6*, *Z1*, *S2* and *S4* not extending beyond bases of setae in the next row. 2

2. Atrium of spermathecal apparatus bifurcated at junction with major duct; calyx elongate trumpet-shaped; spermatodactyl of male T-shaped. 3

- Atrium of spermathecal apparatus not bifurcated at junction with major duct; calyx does not elongate trumpet-shaped; spermatodactyl of male L-shaped. 4

3. Setae *z4*, *s4* and *Z4* long, extending beyond or nearly reaching to bases of setae in next row; macrosetae present on genu, tibia and basitarsus of leg IV. . . . . *N. lushanensis* (Zhu and Chen)

- Setae *z4*, *s4* and *Z4* short, far from bases of setae in next row; macrosetae only on genu and basitarsus of leg IV. . . . . *N. makuwa* (Ehara)

4. Calyx of spermathecal apparatus not stalked, immediately attached to calyx; major duct expanded, approximately as wide as medial part of calyx; palpgenu anterior lateral setae *al1* and *al2* tapered, spiniform; solenostomes (*gv3*) absent. . . . . *N. tauricus* (Livshitz and Kuznetsov)

- Calyx of spermathecal apparatus basally stalked; major duct slender; palpgenu setae *al1* and *al2* cylindrical, rod-like; solenostomes (*gv3*) present. 5

5. Genu II with 7 setae; genu IV without obvious macrosetae; *S4* and *S5* serrated. . . . . *N. bicaudus* (Wainstein)

- Genu II with 8 setae; genu IV with macroseta at least 1.5 times length of other setae; *S4* and *S5* smooth. 6

6. Calyx of spermathecal apparatus basally pointed, V-shaped; seta *S4* approximately half distance *S4–S5*; *s4* approximately half distance *s4–z5*. . . . . *N. zwoelferi* (Dosse)

- Calyx of spermathecal apparatus basally rounded, U-shaped; *S4* nearly as long as distance *S4–S5*; *s4* longer than distance *s4–z5*. . . . . *N. paraki* (Ehara)

4. Discussion

Neoseiulus is the third most abundant and most widely distributed genus across the globe after *Typhlodromus* and *Amblyseius* in Phytoseiidae. The species of this genus are predominantly found in plants on every continent, except Antarctica, and their largest diversity is concentrated in tropical and subtropical regions. The ancestral habitat of *Neoseiulus* is presumed to be amongst plants that grew close to the ground as well as the ground litter [37]. The species documented above are unexceptional inhabitants of plants, much as most other species in the genus *Neoseiulus*. The green foxtail (*Setaria viridis*), a species that is abundant in Shanxi and other areas in northern China, appears to be an ideal habitat for the *Neoseiulus* in Shanxi, hosting five (*N. bicaudus*, *N. lushanensis*, *N. makuwa*, *N. paraki* and *N. tauricus*) out of the seven species recorded; however, for some species, this is based on a small number of specimens.

Ten species groups are recognized in *Neoseiulus* and two of them contain four species subgroups each, while the others remain undivided [3,37,38]. The species found in Shanxi can be classified into two groups: *barkeri* species group (*N. lushanensis*, *N. makuwa* and *N. womersleyi*) and *cucumeris* species group (*N. bicaudus*, *N. paraki* and *N. zwoelferi* in *paraki* species subgroup and *N. tauricus* in *ceratoni* species subgroup).

The classification of *Neoseiulus* primarily relies on variations in the morphology of the spermatheca [2,6,33,37,49,51]. In addition, other characteristics utilized include the idiosomal setal pattern, dorsal idiosomal setal ratio, ornamentation of the dorsal shield, length-to-width ratio of the sternal shield, the position of sternal seta *st3*, the shape of the female ventrianal shield, the shape and position of solenostomes *gv3*, the number of teeth on the movable and fixed cheliceral digits of females, the number of macrosetae on leg IV, the number of setae on genu II, and the shape of the male spermatodactyl. Taxonomists have continuously searched for new morphological characteristics to improve the systematics of Phytoseiidae. Certain characteristics, such as the shape of the posterior section of the peritremal shield and exopodal shields, the location of gland pores and poroids/lyrifissures on the dorsal and ventral idiosomal shields, as well as on the peritremal shield, in addition to the shape of the specialized setae on femur (*al*) and genu (*al1*, *al2*), and leg chaetotaxy, are presented in a number of publications [2,33,37,51,52,58–63]. Further research is necessary to comprehensively evaluate these characteristics in a systematic way.

This is the first review of the genus *Neoseiulus* in Shanxi Province, recording seven species which account for 12.3% of the total known species of this genus in China. Among the findings, *N. paraki* (Ehara), recorded in the current work, was previously only known to be present in Japan and South Korea. Additionally, the study confirms *N. neoreticuloides* (Liang and Hu) as a new junior synonym of *N. bicaudus* (Wainstein). It should be noted that, as there have been no previous surveys of phytoseiid mites in the province, the data presented here may be limited, and it is likely that more species have yet to be discovered.

5. Conclusions

The study of *Neoseiulus* in Shanxi Province has added valuable information to the study of phytoseiid fauna in the area. The redescriptions of four species contribute to the enrichment and provide a more detailed basis for species identification. The recording of *N. paraki* in China for the first time expands the known distribution of this species, and synonymizing *N. neoreticuloides* with *N. bicaudus* clarifies the taxonomic status of these species. The diagnostic key provided will assist in the identification of the known species of *Neoseiulus* in Shanxi and will be useful for future studies on the genus. In brief, this study highlights the importance of conducting further investigations into the species diversity and distribution across different geographical locations.

Author Contributions: Conceptualization, Y.L. and M.M.; methodology, M.M. and Q.-H.F.; software, Y.L. and F.-X.R.; validation, Y.L., M.M. and Q.-H.F.; formal analysis, M.M. and Q.-H.F.; investigation, Y.L., M.M. and F.-X.R.; resources, M.M. and Q.-H.F.; data curation, Y.L. and F.-X.R.; writing—original draft preparation, Y.L.; writing—review and editing, M.M. and Q.-H.F.; visualization, Y.L., M.M. and F.-X.R.; supervision, M.M. and Q.-H.F.; project administration, Y.L. and F.-X.R.; funding acquisition, M.M. All authors have read and agreed to the published version of the manuscript.

Funding: This research was funded by the Shanxi Scholarship Council of China (HGKY2019044), the State Key Laboratory for Biology of Plant Diseases and Insect Pests (SKLOF202215), and the Plant Protection Pioneer Special Training Program of the Plant Protection College at Shanxi Agricultural University.

Institutional Review Board Statement: Not applicable.

Informed Consent Statement: Not applicable.

Data Availability Statement: The data presented in this study are available on request from the corresponding author.

Acknowledgments: We would like to express our appreciation to Xuenong Xu (Chinese Academy of Agricultural Sciences) for his valuable comments on the draft, Weinan Wu (Guangdong Institute of Applied Biological Resources) for sharing a photograph of the spermathecal apparatus of *N. neoreticuloides*, and Bin Zhang (Shanxi Agriculture University) for his assistance in collecting specimens.

Conflicts of Interest: The authors declare no conflict of interest.

References

1. Hughes, A.M. *The Mites Associated with Stored Food Products*; Ministry of Agriculture and Fisheries, Her Majesty's Stationery Office: London, UK, 1948; pp. 1–168.
2. Chant, D.A.; McMurtry, J.A. *Illustrated Keys and Diagnoses for the Genera and Subgenera of the Phytoseiidae of the World (Acari: Mesostigmata)*; Indira Publishing House: West Bloomfield, MI, USA, 2007; pp. 1–219.
3. Phytoseiidae Database. Available online: www.lea.esalq.usp.br/phytoseiidae (accessed on 14 January 2023).
4. Döker, İ.; Khaustov, V.A.; Joharchi, O. Redescriptions of two little known species of *Neoseiulus* Hughes (Acari: Phytoseiidae) with description of a new species from Russia. *Syst. Appl. Acarol.* **2021**, *26*, 672–683. [CrossRef]
5. Liao, J.-R.; Ho, C.-C.; Ko, C.-C. Survey of phytoseiid mites (Acari: Mesostigmata) in the Penghu Islands with two new records and descriptions of two new species. *Syst. Appl. Acarol.* **2021**, *26*, 641–671. [CrossRef]
6. Wu, W.N.; Ou, J.F.; Huang, J.L. *Fauna Sinica, Invertebrata Vol. 47, Arachnida, Acari, Phytoseiidae*; Science Press: Beijing, China, 2009; pp. 1–511.
7. Wu, W.N.; Fang, X.D. *Phytoseiidae Systematics and Management of Pests*; Guangdong Science and Technology Press: Guangzhou, China, 2021; pp. 1–411.
8. Schicha, E. A new predacious species of *Amblyseius* Berlese from strawberry in Australia, and *A. longispinosus* (Evans) redescribed (Acari: Phytoseiidae). *Aust. J. Entomol.* **1975**, *14*, 101–106. [CrossRef]
9. Xin, J.-L.; Liang, L.-R.; Ke, L.-S. A new species of the genus *Amblyseius* from China (Acarina: Phytoseiidae). *Int. J. Acarol.* **1981**, *7*, 75–80. [CrossRef]
10. Ma, M.; Li, S.-C.; Fan, Q.-H. Mites and ticks (Acari) in Shanxi Province, China: An annotated checklist. *Zootaxa* **2015**, *4006*, 1–39. [CrossRef]

11. Tang, Y.; Yang, R.Y.; Zeng, W. *Penthaleus major* (Dugès) damages young corns in Daxian. *China Plant Prot.* **2004**, *24*, 33.
12. Zhang, B.; Ma, M.; Fan, Q.-H. Morphological ontogeny of *Neoseiulus zwoelferi* (Acari: Phytoseiidae). *Zootaxa* **2021**, *5086*, 7–28. [CrossRef]
13. Liang, L.R.; Hu, C.Y. Two new species of the phytoseiid mites from Ningxia, China (Acarina: Phytoseiidae). *Entomotaxonomia* **1988**, *10*, 317–319.
14. Wainstein, B.A. Some new predatory mites of the family Phytoseiidae (Parasitiformes) of the USSR fauna. *Rev. Entomol. Urss* **1962**, *41*, 230–240.
15. Zhu, Z.-M.; Chen, X.-W. A new species of *Amblyseius* Berlese from Mt. Lushan of Jiangxi (Acarina: Phytoseiidae). *Acta Zootaxon. Sin.* **1985**, *10*, 273–275.
16. Ehara, S. Phytoseiid Mites from Hokkaido (Acarina: Mesostigmata)(With 71 Text-figures). *J. Fac.Sci. Hokkaido Univ.* **1967**, *16*, 212–233.
17. Ehara, S. Some phytoseiid mites from Japan, with descriptions of thirteen new species (Acarina: Mesostigmata). *Mushi* **1972**, *46*, 137–173.
18. Livschitz, I.Z. Phytoseiid mites from Crimea (Parasitiformes: Phytoseiidae). *Proc. State Nikita Bot. Gard.* **1972**, *61*, 13–64.
19. Rowell, H.J.; Chant, D.A.; Hansell, R.I.C. The determination of setal homologies and setal patterns on the dorsal shield in the family Phytoseiidae (Acarina: Mesostigmata). *Can. Entomol.* **1978**, *110*, 859–876. [CrossRef]
20. Evans, G.O. Observations on the Chaetotaxy of the Legs in the Freeliving Gamasina (Acari: Mesostigmata). *Bull. Br. Mus. Nat. Hist.* **1963**, *10*, 277–303. [CrossRef]
21. Athias-Henriot, C. Nouvelles notes sur les Amblyseiini. II. Le relevé organotaxique de la face dorsale adulte (Gamasides protoadéniques, Phytoseiidae). *Acarologia* **1975**, *17*, 20–29.
22. Chant, D.A.; Yoshida-Shaul, E. Adult idiosomal setal patterns in the family Phytoseiidae (Acari: Gamasina). *Int. J. Acarol.* **1992**, *18*, 177–193. [CrossRef]
23. Hirschmann, W. Gangystematik der Parasitiformes. Acarologie Schriftenreihe fur Vergleichende Milbenkunde. *Hirschmann-Verl. Furth. Bay* **1962**, *5*, 80.
24. Ehara, S. A tentative catalogue of predatory mites of Phytoseiidae known from Asia, with descriptions of five new species from Japan. *Mushi* **1966**, *39*, 9–30.
25. Congdon, B.D. The family Phytoseiidae (Acari) in western Washington State with descriptions of three new species. *Int. J. Acarol.* **2002**, *28*, 3–27. [CrossRef]
26. Wang, B.; Wang, Z.; Jiang, X.; Zhang, J.; Xu, X. Re-description of *Neoseiulus bicaudus* (Acari: Phytoseiidae) newly recorded from Xinjiang, China. *Syst. Appl. Acarol.* **2015**, *20*, 455–461. [CrossRef]
27. Asali Fayaz, B.; Khanjani, M.; Molavi, F.; Ueckermann, E.A. Phytoseiid mites (Acari: Phytoseiidae) of apple and almond trees in regions of western and southwestern Iran. *Acarologia* **2011**, *51*, 371–379. [CrossRef]
28. Asali Fayaz, B.; Khanjani, M.; Uckermann, E. Description of immature stages and redescription of female and male of *Neoseiulus bicaudus* (Wainstein, 1962)(Acari: Phytoseiidae) from west of Iran. *Acta Phytopathol. Entomol. Hung.* **2011**, *46*, 329–338. [CrossRef]
29. Faraji, F.; Çobanoğlu, S.; Çakmak, I. A checklist and a key for the Phytoseiidae species of Turkey with two new species records (Acari: Mesostigmata). *Int. J. Acarol.* **2011**, *37*, 221–243. [CrossRef]
30. Sahraoui, H.; Grissa, K.L.; Kreiter, S.; Douin, M.; Tixier, M.-S. Phytoseiid mites (Acari: Mesostigmata) of Tunisian citrus orchards: Catalogue, biogeography and key for identification. *Acarologia* **2012**, *52*, 433–452. [CrossRef]
31. Palevsky, E.; Gal, S.; Ueckermann, E.A. Phytoseiidae from date palms in Israel with descriptions of two new taxa and a key to the species found on date palms worldwide (Acari: Mesostigmata). *J. Nat. Hist.* **2009**, *43*, 1715–1747. [CrossRef]
32. Rahmani, H.; Kamali, K.; Faraji, F. Predatory mite fauna of Phytoseiidae of northwest Iran (Acari: Mesostigmata). *Turk. J. Zool.* **2010**, *34*, 497–508. [CrossRef]
33. Denmark, H.A.; Evans, G.A. *Phytoseiidae of North America and Hawaii: (Acari: Mesostigmata)*; Indira Publishing House: West Bloomfield, MI, USA, 2011; pp. 1–415.
34. Negm, M.W.; Alatawi, F.J.; Aldryhim, Y.N. A new species of *Neoseiulus* Hughes, with records of seven species of predatory mites associated with date palm in Saudi Arabia (Acari: Phytoseiidae). *Zootaxa* **2012**, *3356*, 57–64. [CrossRef]
35. Liao, J.-R.; Ho, C.-C.; Ko, C.-C. Predatory mites (Acari: Mesostigmata: Phytoseiidae) intercepted from samples imported to Taiwan, with description of a new species. *Zootaxa* **2021**, *4927*, 301–330. [CrossRef]
36. Wu, W.N.; Lan, W.M. Two new species of *Amblyseius* Berlese from Kuizhou Province (Acarina: Phytoseiidae). *Acta Entomol. Sin.* **1989**, *32*, 248–250.
37. Chant, D.A.; McMurtry, J.A. A review of the subfamily Amblyseiinae Muma (Acari: Phytoseiidae): Part I. Neoseiulini new tribe. *Int. J. Acarol.* **2003**, *29*, 3–46. [CrossRef]
38. De Moraes, G.J.; McMurtry, J.A.; Denmark, H.A.; Campos, C.B. A revised catalog of the mite family Phytoseiidae. *Zootaxa* **2004**, *434*, 1–494. [CrossRef]
39. Wu, W.N.; Lan, W.M.; Liu, Y.H. Phytoseiid mites of lychee from China, and their value for application. *Nat. Enemies Insects* **1991**, *13*, 82–91.
40. Wu, W.N.; Liang, L.R.; Lan, W.M. *Acari: Phytoseiidae*; Economic Insect Fauna of China, Science Press: Beijing, China, 1997; pp. 1–227.

41. Ehara, S.; Amano, H. A revision of the mite family Phytoseiidae in Japan (Acari, Gamasina), with remarks on its biology. *Species Divers.* **1998**, *3*, 25–73. [CrossRef]
42. de Moraes, G.J.; McMurtry, J.A.; Denmark, H.A. *A Catalog of the Mite Family Phytoseiidae: References to Taxonomy, Synonymy, Distribution and Habitat*; EMBRAPA—DDT: Brasilia, Brazil, 1986; pp. 1–353.
43. Ehara, S.; Yokogawa, M. Two new *Amblyseius* from Japan with notes on three other species (Acarina: Phytoseiidae). *Proc. Jpn. Soc. Syst. Zool.* **1977**, *13*, 50–58.
44. Ehara, S. Contribution to the knowledge of the mite family Phytoseiidae in Japan (Acari: Gamasina). *J. Fac. Educ. Tottori Univ. Nat. Sci.* **1994**, *42*, 119–160.
45. Ehara, S.; Amano, H. Checklist and keys to Japanese Amblyseiinae (Acari: Gamasina: Phytoseiidae). *J. Acarol. Soc. Jpn.* **2004**, *13*, 1–30. [CrossRef]
46. Ryu, M.O. Key to the species of the genus *Typhlodromips* (Acari: Phytoseiidae) with a new and a newly recorded species in Korea. *Korean J. Appl. Entomol.* **2013**, *52*, 291–294. [CrossRef]
47. Wu, W.N.; Lan, W.M. Two new species and one new record of the genus *Typhlodromus* from Northwest China (Acari: Phytoseiidae). *Acta Zootaxonomica Sin.* **1991**, *16*, 328–332.
48. Wu, W.-N.; Ou, J.-F. A new species group of the genus *Amblyseius*, with descriptions of two new species (Acari: Phytoseiidae) from China. *Syst. Appl. Acarol.* **1999**, *4*, 103–110. [CrossRef]
49. Schicha, E. *Phytoseiidae of Australia and Neighboring Areas*; Indira Publishing House: West Bloomfield, MI, USA, 1987; pp. 1–187.
50. Tseng, Y.-H. Further study on phytoseiid mites from Taiwan (Acarina: Mesostigmata). *Chin. J. Entomol.* **1983**, *3*, 33–74.
51. Beard, J.J. A review of Australian *Neoseiulus* Hughes and *Typhlodromips* De Leon (Acari: Phytoseiidae: Amblyseiinae). *Invertebr. Syst.* **2001**, *15*, 73–158. [CrossRef]
52. Liao, J.-R.; Ho, C.-C.; Lee, H.-C.; Ko, C.-C. *Phytoseiidae of Taiwan (Acari: Mesostigmata)*; National Taiwan University Press: Taipei, Taiwan, 2020; pp. 1–538.
53. Dosse, G. Morphologie und biologie von *Typhlodromus zwölferi* n. sp. (Acar., Phytoseiidae). *Z. Angew. Entomol.* **1957**, *41*, 301–311. [CrossRef]
54. Muma, M.H. Subfamilies, genera, and species of Phytoseiidae (Acarina: Mesostigmata). *Bull. Fla. State Mus. Biol. Sci.* **1961**, *5*, 267–302.
55. Schuster, R.; Pritchard, A. Phytoseiid mites of California. *Hilgardia* **1963**, *34*, 191–285. [CrossRef]
56. Wainstein, B.A. Predatory mites of the family Phytoseiidae (Parasitiformes) of Yaroslavl Province. *Entomol. Obozr.* **1975**, *54*, 914–922.
57. Wu, W.N. New species and new records of phytoseiid mites from Northeast China. II. Amblyseius Berlese (Acarina: Phytoseiidae). *Acta Zootaxon. Sin.* **1987**, *12*, 260–270.
58. Athias-Henriot, C. Phytoseiidae et Aceosejidae (Acarina, Gamasina) d'Algérie. I. Genres Blattisocius Keegan, Iphiseius Berlese, Amblyseius Berlese, Phytoseius Ribaga, Phytoseiulus Evans. *Bull. Soc. Hist. Nat. Afr. Nord* **1957**, *48*, 319–352.
59. Denmark, H.A.; Muma, M.H. A revision of the genus *Amblyseius* Berlese, 1914 (Acari: Phytoseiidae). *Occas. Pap. Fla. State Collect. Arthropods* **1989**, *4*, 1–149.
60. Kolodochka, L.A. Four new species of phytoseiid mites of the fauna of the USSR (Parasitiformes, Phytoseiidae). *Vestn. Zool.* **1979**, *5*, 32–40.
61. Kolodochka, L.A. New species of phytoseiid mites from the fauna of the USSR (Parasitiformes: Phytoseiidae). *Vestn. Zool.* **1980**, *2*, 64–70.
62. Kolodochka, L.A. Two new species of the genus *Neoseiulus* (Parasitiformes, Phytoseiidae) with redescriptions of *N. bicaudus* and *N. micmac* based on holotypes. *Vestn. Zool.* **2018**, *52*, 295–306. [CrossRef]
63. Ragusa, S. Observation on the genus *Neoseiulus* Hughes (Parasitiformes, Phytoseiidae). Redefinition. Composition. Geography. Description of two new species. *Rev. Suisse Zool.* **1983**, *90*, 657–678.

animals

MDPI

Article

The Unique Cauda-Liked Structure Represents a New Subfamily of Cunaxidae: Description of New Taxa and Discussion on Functional Morphology †

Jianxin Chen [1,2,3,4], Maoyuan Yao [1,2,3,4], Jianjun Guo [1,2,3], Tianci Yi [1,2,3] and Daochao Jin [1,2,3,*]

1 Institute of Entomology, Guizhou University, Guiyang 550025, China; jianxinchen000@163.com (J.C.)
2 The Guizhou Provincial Key Laboratory for Plant Pest Management of Mountainous Region, Guiyang 550025, China
3 The Scientific Observing and Experimental Station of Crop Pest in Guiyang, Ministry of Agriculture, Guiyang 550025, China
4 College of Agriculture, Anshun University, Anshun 561000, China
* Correspondence: dcjin@gzu.edu.cn
† urn:lsid:zoobank.org:pub:39937F54-7C30-4A8C-AB02-F25D3343C7C5; urn:lsid:zoobank.org:act:A2281D2A-EA1D-4B78-82BA-3FC4BA57BD57; urn:lsid:zoobank.org:act:34C94A2D-F1FD-47DE-BD63-044BB3CB4CA8; urn:lsid:zoobank.org:act:6B6635C3-284C-4390-BDB0-164CECA9C37A.

Simple Summary: In this study, a cauda-like structure found in Cunaxidae is defined, and with it the new taxa, Cunaxicaudinae Chen & Jin subfam. nov., and its two new genera, *Cunaxicaudus* Chen & Jin gen. nov. (type genus) and *Brevicaudus* Chen & Jin gen. nov., are erected. It is proposed that the specialized cauda may be the result of the evolution of the sperm transfer mode.

Abstract: A cauda-like structure was found, firstly in Cunaxidae, and with it the new taxa Cunaxicaudinae Chen & Jin subfam. nov., and its two new genera, *Cunaxicaudus* Chen & Jin gen. nov. (type genus) and *Brevicaudus* Chen & Jin gen. nov., were erected. Cunaxicaudinae Chen & Jin subfam. nov. differs from the known members of the family Cunaxidae by the unique conspicuous cauda derived from the posterior end of the hysterosoma. The generic features of *Cunaxicaudus* Chen & Jin gen. nov. are as follows: the posterior of the hysterosoma elongated as a much longer cauda; palp between genu and tibiotarsus without apophysis; *e1* closer to *d1* than *f1*; and *e1* closer to mid-line than *c1* and *d1*. The generic features of *Brevicaudus* Chen & Jin gen. nov. are as follows: the posterior of hysterosoma elongated as a short cauda; palp between genu and tibiotarsus with one apophysis; distance between setae *e1* and *d1* approximately equal to *e1*; and *f1*, *e1* as close to mid-line as *c1* and *d1* to mid-line. It is proposed that the specialized cauda may be the result of the evolution of the sperm transfer mode.

Keywords: Acariformes; Bdelloidea; taxonomy; predator; China

Citation: Chen, J.; Yao, M.; Guo, J.; Yi, T.; Jin, D. The Unique Cauda-Liked Structure Represents a New Subfamily of Cunaxidae: Description of New Taxa and Discussion on Functional Morphology. *Animals* **2023**, *13*, 1363. https://doi.org/10.3390/ani13081363

Academic Editors: Maciej Skoracki, Monika Fajfer and Alexis Ribas

Received: 26 February 2023
Revised: 24 March 2023
Accepted: 6 April 2023
Published: 15 April 2023

1. Introduction

Cunaxidae (Prostigmata: Bdelloidea) erected by Thor [1] is a predatory mite group that can prey on phytophagous mites, other small arthropods and nematodes, etc. [2–7]. They commonly inhabit various terrestrial habitats, including forest leaf litter and soil, tree holes, moss, etc., and being important predators, they play a crucial role in agricultural ecosystems [7–9].

According to the recent literature, there are six subfamilies, 30 genera and more than 450 species described as Cunaxidae in the world [10–21]. In this work, with unique hysterosoma, a new subfamily with two new genera and three new species is described in Mohan Port in Mengla county, Xishuangbanna Dai Autonomous Prefecture, Yunnan province, China. Skvarla et al. [10] reviewed the Cunaxidae with keys to the world subfamilies, genera and species. Here, we provide an updated key to the subfamilies of the Cunaxidae to include the new subfamily.

The following abbreviations are used: prodorsum: anterior trichobothria (*at*), posterior trichobothria (*pt*), lateral proterosomal (*lps*), median proterosomal (*mps*); hysterosoma: internal humerals (*c1*), external humerals (*c2*), internal dorsals (*d1*), internal lumbals (*e1*), internal sacrals (*f1*), external sacrals (*f2*), internal clunals (*h1*), external clunals (*h2*); venter: propodogastral seta (*ppgs*), hysterogastral seta (*hgs*); anal region: pseudanal (*ps*); genital region: aggenitals (*ag*), genitals (*g1–4*); gnathosoma: hypognathals (*hg1–4*); leg: attenuate (sharply) solenidion (*asl*), blunt-pointed rod-like solenidion (*bsl*), famulus (*fam*), trichobothria (*T*), simple tactile seta (*sts*), microseta (*mst*), dorsoterminal solenidion (*dtsl*). Duplex and triplex setae are indicated in brackets ({}).

2. Materials and Methods

2.1. Sampling Area

Samples of fallen leaves were collected from the woodland at Mohan Port (21°11′22.66″ N, 101°41′51.80″ E, elevation 893 m) in Mengla county, Xishuangbanna Dai Autonomous Prefecture, Yunnan province, China.

Mohan Port, bordering the Botan port of Laos, is the only national port between China and Laos. Mohan Port, bordering the Botan port of Laos, is the only national port between China and Laos, where the climate is pleasant: there is no chilly winter and hot summer, four seasons is not clear but raining season with the distinction of the dry quarter, and the average annual temperature is 21.2 °C. The rainfall here averages 1615 mm.

2.2. Laboratory Activities

Fallen leaves were placed in a modified Berlese-Tullgren funnel for at least eight hours to isolate mite specimens. The specimens were preserved in 75% alcohol and then mounted on slides in Hoyer's medium [22]. Coordinates and altitudes were obtained by smartphone with GPS. Line drawings were produced with the aid of a drawing tube attached to a phase contrast Nikon Ni E microscope with DIC optics, and photographs were taken using a camera (Nikon DS-Ri 2) attached to a Nikon Ni E microscope with DIC optics. All figures were edited with Adobe Photoshop CC 2019. All measurements were taken with the software Nikon NIS Elements AR 4.50 and provided in μm for the holotype and paratypes in parentheses. The nomenclature and abbreviations of idiosoma follow Den Heyer and Castro [23] and Skvarla et al. [10], except for propodosomal setae, which follows Fisher et al. [24], and legs setal notation follows Den Heyer [25].

3. Results

Family Cunaxidae Thor, 1902; Subfamily Cunaxicaudinae Chen & Jin subfam. nov.; Type genus: *Cunaxicaudus* Chen & Jin gen. nov.

The new subfamily was established by diagnostic caudal structure, which consists of three sections: caudal base, caudal petiole and caudal xiphoid (Figures 1, 2, 3B and 4).

Caudal base: the extended posterior end of the hysterosoma, with the genital region, and with no suture separating the hysterosoma from the caudal base, was present but clearly less sclerotized than the main hysterosoma.

Caudal petiole: a tubelike extension from the caudal base, weakly sclerotized, translucent or transparent, with a distinct suture line present between it and the caudal base.

Caudal xiphoid: the extension from the caudal petiole, sword shaped and transparent, with a clear suture line demarcating it from the caudal petiole.

The subfamily Cunaxicaudinae Chen & Jin subfam. nov. can be easily distinguished from other members of the family Cunaxidae by the unique caudal structure. According to the literature, in most cunaxids sperm transfer is indirect, but in some species it is direct by mating with a visible aedeagus. Therefore, we infer that the caudal structure may be more conducive to the cunaxid's mating.

Etymology. The new subfamily is named from the stem of Cunaxidae (Cunaxi-), and the posterior of hysterosoma being noticeably elongated as a cauda (-caudinae), which means the tail or tail-like structure of an animal, bird, fish, or other creature in Latin.

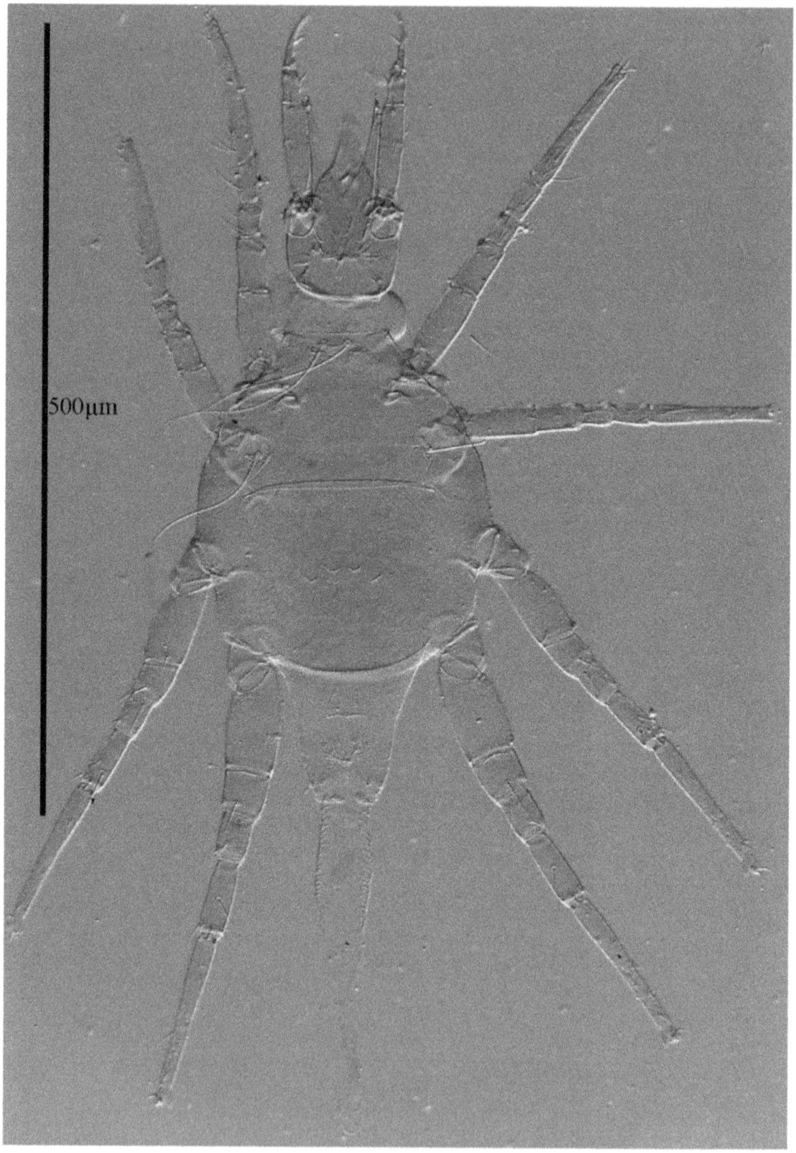

Figure 1. *Cunaxicaudus mohanensis* sp. nov. (male). Entire specimen (photo). Scale bar = 500 μm.

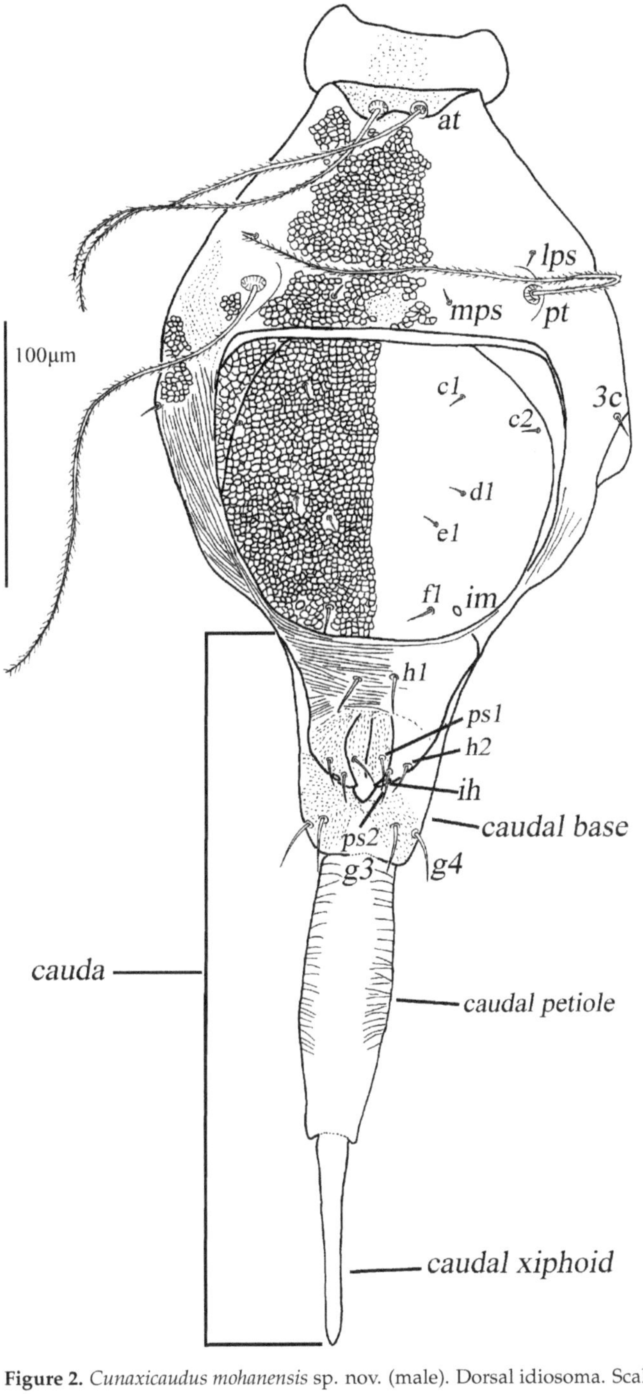

Figure 2. *Cunaxicaudus mohanensis* sp. nov. (male). Dorsal idiosoma. Scale bar = 100 μm.

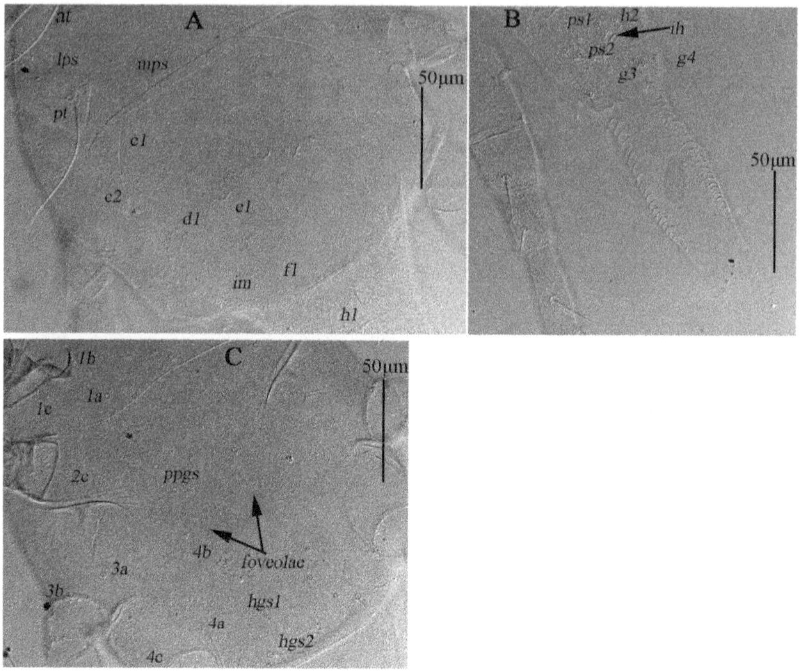

Figure 3. *Cunaxicaudus mohanensis* sp. nov. (male). (**A**,**B**)—Dorsal idiosoma (photo). (**C**)—Ventral idiosoma (photo). Scale bar = 50 μm.

3.1. *Cunaxicaudus Chen & Jin gen. nov.*

Species Type: *Cunaxicaudus mohanensis* Chen & Jin sp. nov.

Generic features (male): posterior of hysterosoma is much elongated as a conspicuous cauda; the caudal petiole and caudal xiphoid are long; palp between genu and tibiotarsus without apophysis; distance between setae *e1* and *d1* about 1/3 of that between *e1* and *f1*, and *e1* closer to mid-line than *c1* and *d1*; lyrifissures (*im*) close to and at the same level as *f1*; leg IV longest and leg II shortest, and tarsal lobes well-developed.

3.1.1. *Cunaxicaudus mohanensis* Chen & Jin sp. nov.

Diagnosis. The *h1* was longer than other dorsal setae (*lps*, *mps*, *c1*, *c2*, *d1*, *e1*, *f1*); two pairs of hysterogastral setae (*hgs1–hgs2*); basifemora I–IV: 4-4-3-0 *sts*.

Description (Figures 1–8); male (n = 17).

The idiosoma length was 324 (305–364) from the base of subcapitulum to the posterior edge of median shield, and the width was 188 (170–219); the posterior end of the hysterosoma was elongated as a very long cauda (Figures 1, 2 and 4).

Dorsum (Figures 2 and 3A,B). Propodosomal and hysterosomal shields entirely complemented by reticulations, integument with striae. The propodosomal shield was 101 (101–128) long and 155 (135–160) wide, sclerotized and with a reticulated pattern, and was bearing one pair of anterior (*at*) and one pair of posterior (*pt*) setose trichobothria and two pairs of tactile setae (*lps* and *mps*); *at* was shorter than the length of *pt*, *lps* near *pt* base; the area was anterior to *at* papillary. The lengths of setae and the distances between the bases of setae were *at* 151 (145–170), *pt* 178 (160–220), *lps* 7 (5–7), *mps* 6 (5–6); *at-at* 18 (12–19), *pt-pt* 110 (83–128), *lps-lps* 109 (84–118), *mps-mps* 46 (44–57), *lps-mps* 39 (27–40), *at-lps* 69 (59–74), *pt-mps* 35 (24–38), *pt-lps* 16 (14–24), *at-mps* 77 (67–83) and *at-pt* 86 (74–93).

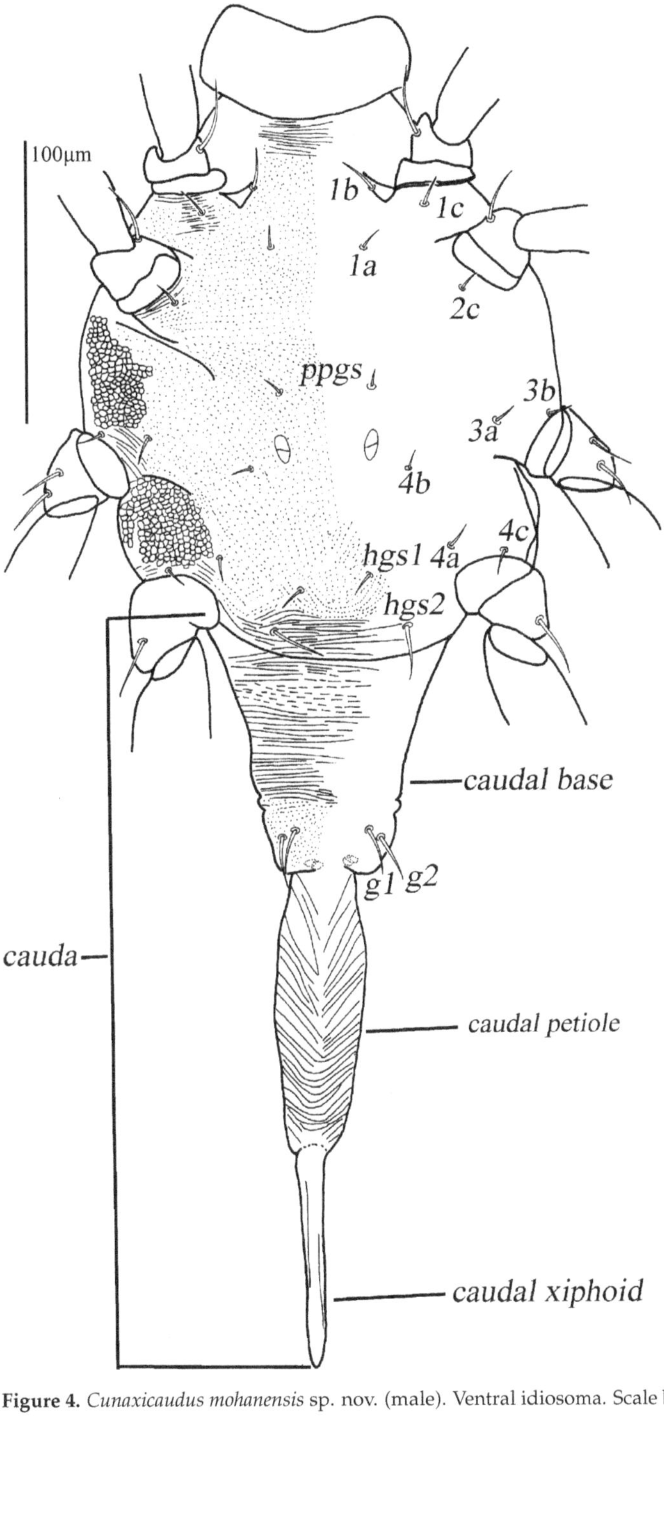

Figure 4. *Cunaxicaudus mohanensis* sp. nov. (male). Ventral idiosoma. Scale bar = 100 µm.

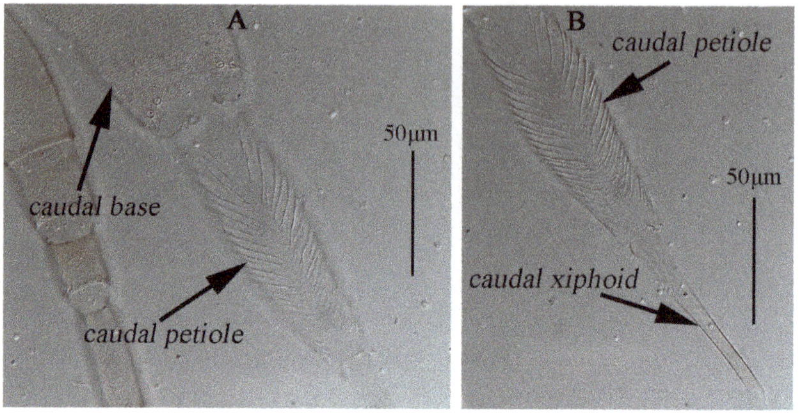

Figure 5. *Cunaxicaudus mohanensis* sp. nov. (male). (**A**)—Ventral caudal base and petiole (photo). (**B**)—Ventral caudal petiole and xiphoid (photo). Scale bar = 50 μm.

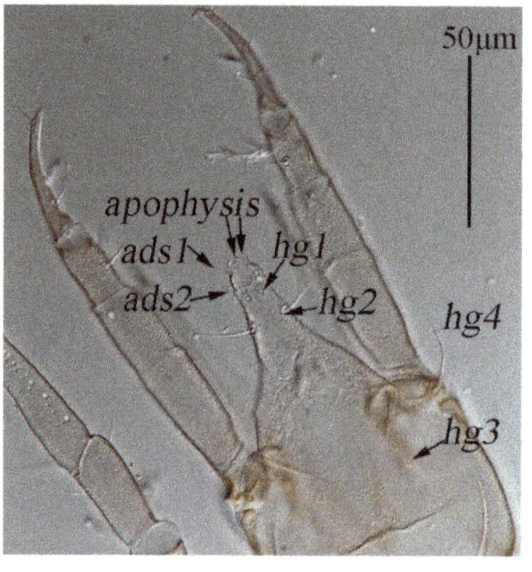

Figure 6. *Cunaxicaudus mohanensis* sp. nov. (male). Gnathosoma (photo). Scale bar = 50 μm.

The hysterosomal median shield was 123 (99–135) long, 134 (110–140) wide, with five pairs of dorsal setae (*c1*, *c2*, *d1*, *e1*, *f1*) and one pair of lyrifissures (*im*) close to and at same level with *f1*. The distance between setae *e1* and *d1* was about one-third of that between *e1* and *f1*; *e1* was closer to the mid-line of the median shield than *c1* and *d1*. Setae *h1* was situated on the striated integument of the caudal base and was longer than other dorsal setae. The lengths of six pairs of dorsal setae were *c1* 6 (5–8), *c2* 5 (5–8), *d1* 7 (5–7), *e1* 7 (5–9), *f1* 8 (8–10), *h1* 15 (11–16). Distances of setae: *c1-c1* 63 (66–77), *c2-c2* 117 (113–129), *d1-d1* 66 (53–68), *e1-e1* 43 (39–45), *f1-f1* 40 (37–42), *h1-h1* 16 (11–19), *c1-c2* 30 (26–36), *c1-d1* 39 (30–46), *c2-d1* 38 (36–44), *d1-e1* 14 (12–18), *e1-f1* 37 (33–38) and *f1-h1* 30 (26–38).

The *Cauda dorsum* is shown in Figures 1, 2 and 3B. From the dorsal view, the posterior end of the idiodoma was elongated significantly as a long cauda, clearly defined and gradually narrowed, consisting of a caudal base with light striae, a caudal petiole with transverse plicated striae on edge, and a smooth caudal xiphoid. The cauda was 290 (255–303) long, measured from the posterior edge of the median shield to the end; the caudal base was 80 (78–108), the caudal petiole was 125 (80–132) and the caudal xiphoid was 85 (75–90). The anal region was dorsally located on the caudal base with dotted fine papillae, bearing two pairs of pseudanal setae (*ps1–ps2*), 4 (8–15) and 10 (8–14) in length, respectively; one pair of *h2* was 10 (8–11) in length and one pair of lyrifissures (*ih*), and the rest of the caudal base was also covered with fine papillae, bearing genital setae *g3*–*g4* representing the genital region.

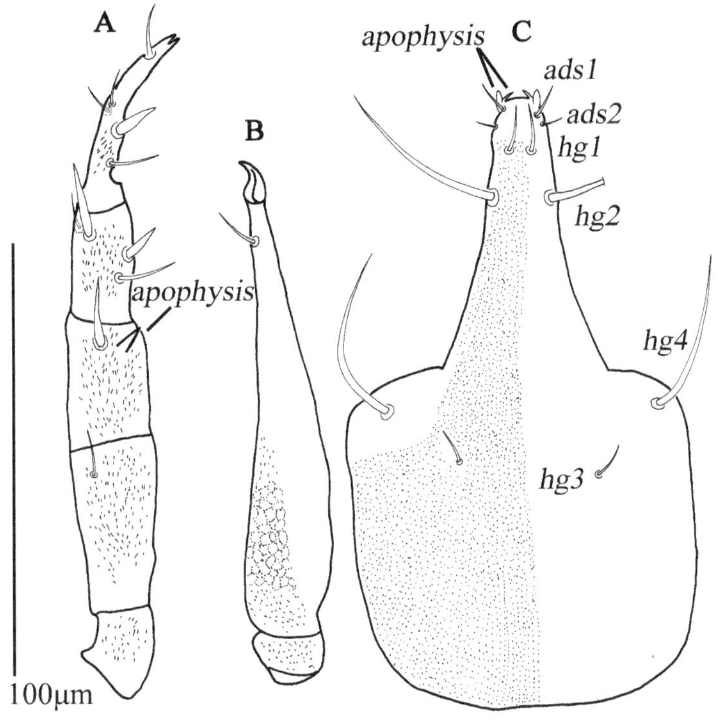

Figure 7. *Cunaxicaudus mohanensis* sp. nov. (male). (**A**)—Palp. (**B**)—Chelicerae. (**C**)—Subcapitulum. Scale bar = 100 μm.

Venter (Figures 3C, 4 and 5). Coxae I–IV fused, resulting in a clearly complete ventral shield with dotted papillae; coxae III–IV with visible reticulated pattern. There was one pair of propodogastral setae (*ppgs*) 6 (5–7) close to coxae II and two pairs of hysterogastral setae (*hgs1*–*hgs2*), and the lengths of setae *hgs1*–*hgs2* were 7 (6–8) and 21 (15–21); one pair of clear foveolae was medially located between the coxae III groups. The area between hysterogastral setae *hgs2* with transverse striation. The setal formula of coxal plates I–IV was 3-1-3-3 *sts*.

Cauda venter (Figures 4 and 5). The cauda was 290 (267–283) long: the caudal base was 81 (89–105), including the genital area; the caudal petiole with plicated striae was 124 (82–129) long; the caudal xiphoid was smooth and 85 (79–90) long. The cauda was clearly defined, and gradually narrowed as in its dorsal view. The caudal base had horizontal striation except for the genital region (genital shield); the genital region had dotted papillae, genital suckers (two pairs) visible, and four pairs of genital setae (*g1*–*g4*), of which *g3*–*g4* were dorsally located (Figures 2 and 3B). Lengths of setae *g1*–*g4* were 15 (12–15), 16 (11–18), 20 (12–21) and 17 (13–20), respectively.

Gnathosoma. Palp (Figures 6 and 7A). Five-segmented, 136 (117–144) long, with granulated papillae and terminated with a bifid claw. Palp chaetotaxy was as follows: trochanter without setae; basifemur with one dorsal simple seta; telofemur with one dorsal stout seta and one short and tapering apophysis; genu with two stout setae and two simple setae; and tibiotarsus with one stout seta, three simple setae and one acute solenidion.

Chelicerae (Figure 7B): 107 (98–132) long, with fine papillae and a reticulated pattern; one cheliceral seta, 8 (7–10) in length; terminating in a well developed chela.

Subcapitulum (Figures 6 and 7C) had dotted papillae, 128 (109–132) long and 69 (67–84) wide; two pairs of apophyses, of which one pair was smaller and claw-like and another pair was blunt rod-like; two pairs of adoral setae (*ads1*, *ads2*), of which *ads1* 6 (6–9) and *ads2* were 3 (3–5) in length; four pairs of hypostomal setae (*hg1–hg4*), where *hg2* and *hg4* were subequal in length and both four times longer than *hg1* and *hg3*; and lengths of *hg1–hg4* of 11 (10–13), 36 (33–53), 6 (6–9) and 37 (28–43), respectively. The distances of paired *hg*-setae were *hg1-hg1* 5 (5–7), *hg2-hg2* 14 (12–15), *hg3-hg3* 28 (25–31), *hg4-hg4* 55 (52–65), *hg1-hg2* 10 (9–12), *hg2-hg3* 53 (46–58) and *hg3-hg4* 19 (15–23).

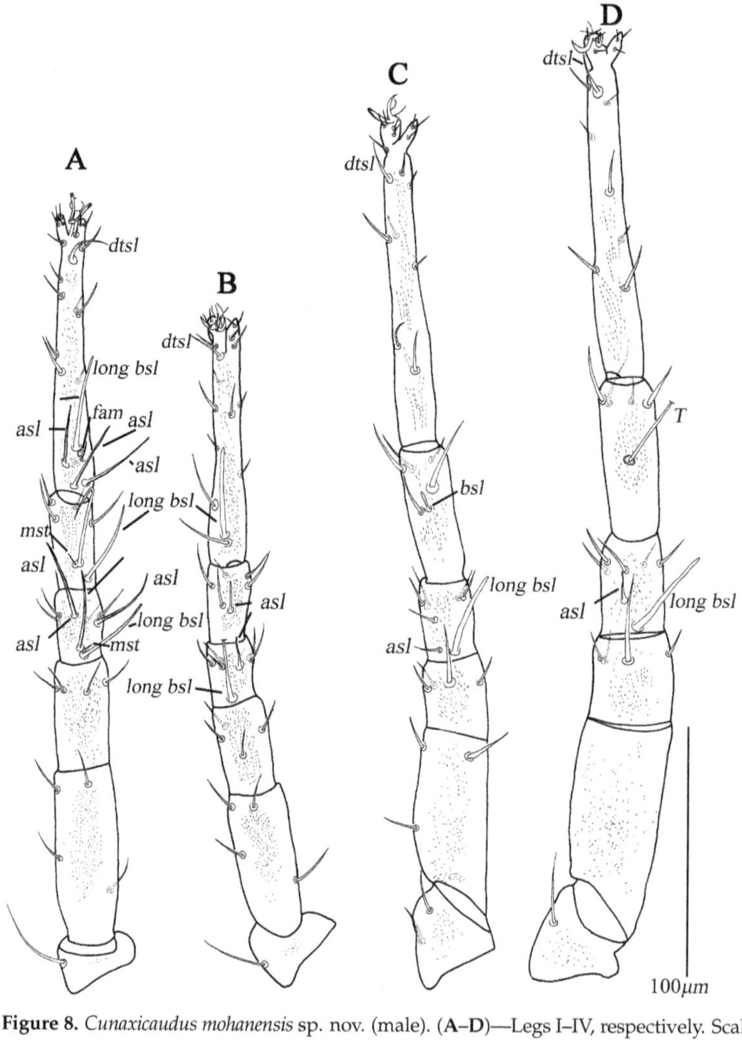

Figure 8. *Cunaxicaudus mohanensis* sp. nov. (male). (**A–D**)—Legs I–IV, respectively. Scale bar = 100 μm.

Legs (Figure 8). Leg IV was the longest and leg II was the shortest, with tarsal lobes fairly well-developed; the dorsum of all the leg segments had finely granulated papillae. Lengths of legs I–IV: 246 (217–259), 210 (198–224), 280 (238–280), 300 (261–309). Lengths of tarsi I–IV: 93 (80–99), 77 (70–84), 108 (91–113) and 116 (93–122). Leg I–IV chaetotaxy (including coxae) was as follows: coxae I–IV 3-1-3-3 *sts*; trochanters I–IV 1-1-2-1 *sts*; basifemora I–IV 4-4-3-0 *sts*; telofemora I–IV 4-4-4-4 *sts*; genua I–IV 2 *asl*, {1 *asl*, 1 long *bsl*, 1 *mst*}, 4 *sts*-1 *asl*, 1 long *bsl*, 5 *sts*-1 *asl*, 1 long *bsl*, 5 *sts*-1 *asl*, 1 long *bsl*, 5 *sts*; tibiae I–IV 1 long *bsl*, {1 *asl*, 1 *mst*}, 4 *sts*-1 *bsl*, 5 *sts*-1 *bsl*, 5 *sts*-1 *T*, 4 *sts*; and tarsi I–IV 3 *asl*, 1 long *bsl*, 1 *fam*, 1 *dtsl*, 17 *sts*-1 long *bsl*, 1 *dtsl*, 17 *sts*-1 *dtsl*, 15 *sts*-1 *dtsl* and 14 *sts*.

Female and other developmental stages: unknown.

Remarks: The new species is distinguished from other known species by its unique cauda-like structure.

Material examined: Holotype, male, collected from fallen leaves, Mohan Port (21°11′22.66″ N, 101°41′51.80″ E, elevation 893 m), Mengla County, Xishuangbanna Dai Autonomous Prefecture, Yunnan Province, China, on 20 November 2018, collector, Jianxin Chen and Xuesong Zhang, slide No. YN-CU-201811201001. Three paratype males, with the same data as the holotype, slides No. YN-CU-201811201002, YN-CU-201811202001 and YN-CU-201811201301. The paratypes, six males, were collected from fallen leaves in Mohan Port (21°11′22.66″ N, 101°41′51.80″ E, elevation 893 m), Mengla County, Xishuangbanna Dai Autonomous Prefecture, Yunnan Province, China, on 6 June, 2019, collector, Jian-Xin Chen, slides No. YN-CU-201906060301–YN-CU-201906060306. The paratype, one male, was collected from fallen leaves in Mohan Port (21°11′22.66″ N, 101°41′51.80″ E, elevation 893 m), Mengla County, Xishuangbanna Dai Autonomous Prefecture, Yunnan Province, China, on 7 June 2019, collector, Jianxin Chen, slide No. YN-CU-201906070201. Seven paratypes, male, collected from fallen leaves in Mohan Port (21°11′22.66″ N, 101°41′51.80″ E, elevation 893 m), Mengla County, Xishuangbanna Dai Autonomous Prefecture, Yunnan Province, China, on 9 June 2019 by Jian-Xin Chen, slide No. YN-CU-201906090201–YN-CU-201906090207. All types of materials were deposited at the Institute of Entomology, Guizhou University, Guiyang, P. R. China (GUGC).

Etymology. The new genus name is derived from the subfamily name Cunaxicaudinae subfam. nov., as above; the new species name refers to Mohan Port where the types were originated.

3.1.2. *Cunaxicaudus neomohanensis* Chen & Jin sp. nov.

Diagnosis. Setae *h1* were longer than other dorsal setae (*lps*, *mps*, *c1*, *c2*, *d1*, *e1*, *f1*) in length, with two pairs of hysterogastral setae (*hgs1*–*hgs2*); basifemora I–IV: 4-4-3-0 *sts*. The cauda was short, and the caudal petiole was wider and the caudal xiphoid was shorter (compared to *C. mohanensis* Chen & Jin sp. nov.).

Description (Figures 9–13); male ($n = 1$).

The idiosoma length was 276 from the base of the subcapitulum to the posterior edge of the median shield and the width was 189; the posterior end of the hysterosoma was elongated, forming a long cauda (Figures 9 and 10).

Dorsum (Figure 9). The propodosomal and hysterosomal shields were entirely complemented by reticulations, surrounding integument striate. The propodosomal shield was 81 long, 130 wide, sclerotized and with a reticulated pattern, and had one pair of anterior (*at*) and one pair of posterior (*pt*) setose trichobothria and two pairs of tactile setae (*lps* and *mps*); *at* was shorter than *pt*, and *lps* was near the *pt* base; the area anterior to *at* was covered with papillae. Lengths of the setae and distances between the bases of setae were *at* 150, *pt* 182, *lps* 7, *mps* 6; *at*-*at* 23, *pt*-*pt* 103, *lps*-*lps* 98, *mps*-*mps* 45, *lps*-*mps* 34, *at*-*lps* 56, *pt*-*mps* 28, *pt*-*lps* 22, *at*-*mps* 65 and *at*-*pt* 79.

The hysterosomal median shield had five pairs of simple setae (*c1*, *c2*, *d1*, *e1*, *f1*) and one pair of lyrifissures (*im*) closer to and at same level as *f1*. The distance between setae *e1* and *d1* was about one-third of that between *e1* and *f1*; *e1* was closer to the mid-line of the median shield than *c1* and *d1*. Setae *h1* was situated on the striated integument of the caudal base and was longer than other dorsal setae. The lengths of six pairs of dorsal setae were *c1* 8, *c2* 7, *d1* 6, *e1* 6, *f1* 8 and *h1* 8. The distances of setae were *c1*-*c1* 66, *c2*-*c2* 115, *d1*-*d1* 44, *e1*-*e1* 34, *f1*-*f1* 32, *h1*-*h1* 18, *c1*-*c2* 29, *c1*-*d1* 35, *c2*-*d1* 42, *d1*-*e1* 13, *e1*-*f1* 34 and *f1*-*h1* 31.

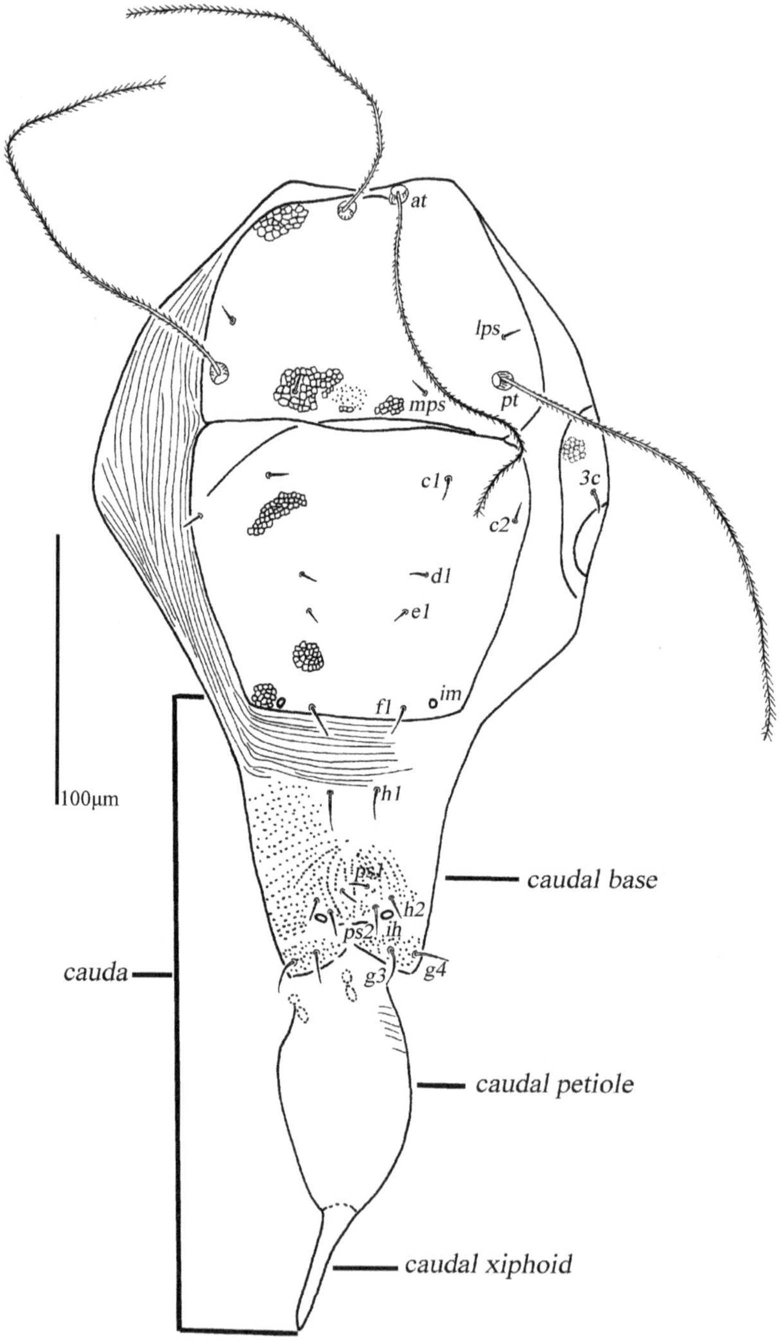

Figure 9. *Cunaxicaudus neomohanensis* sp. nov. (male). Dorsal idiosoma. Scale bar = 100 µm.

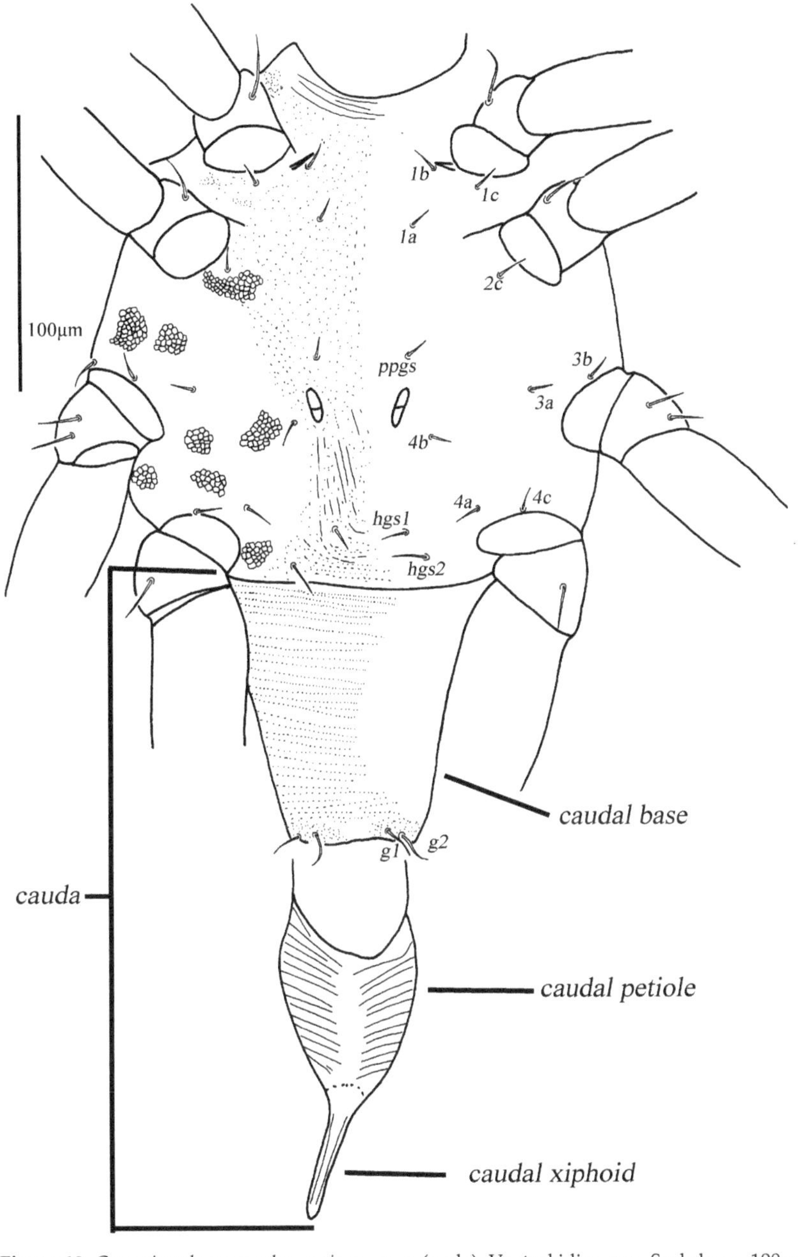

Figure 10. *Cunaxicaudus neomohanensis* sp. nov. (male). Ventral idiosoma. Scale bar = 100 µm.

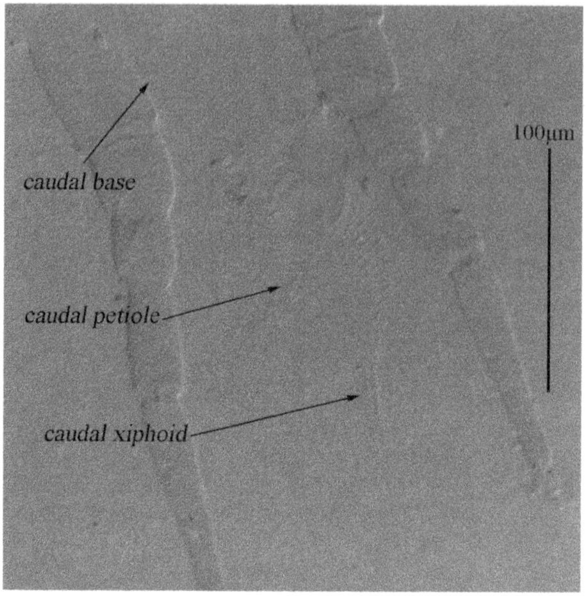

Figure 11. *Cunaxicaudus neomohanensis* sp. nov. (male). Cauda (photo). Scale bar = 100 μm.

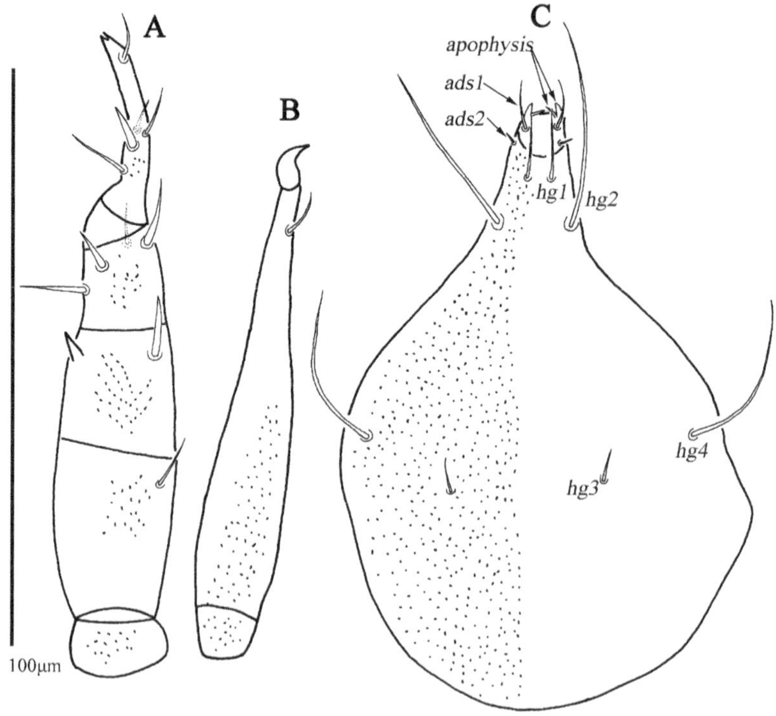

Figure 12. *Cunaxicaudus neomohanensis* sp. nov. (male). (**A**)—Palp. (**B**)—Chelicerae. (**C**)—Subcapitulum. Scale bar = 100 μm.

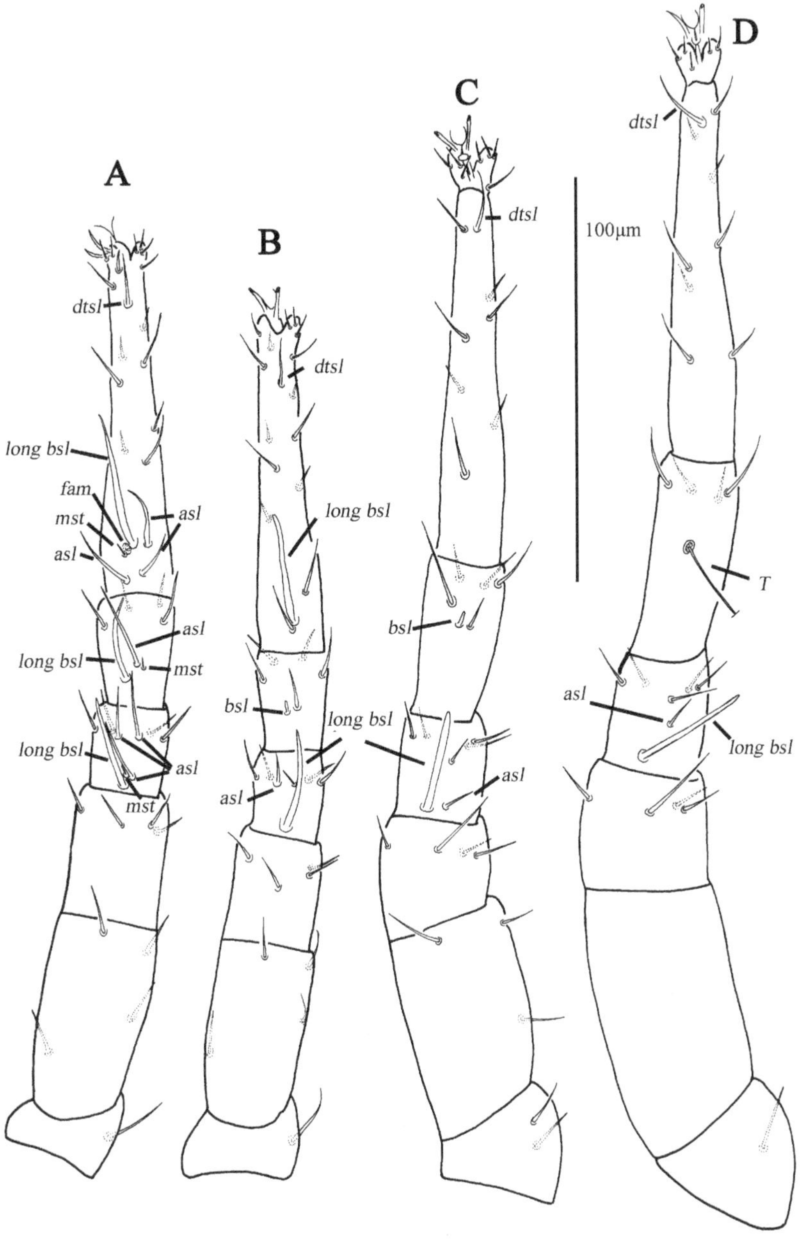

Figure 13. *Cunaxicaudus neomohanensis* sp. nov. (male). (**A–D**)—Legs I–IV, respectively. Scale bar = 100 μm.

Cauda dorsum (Figure 9). From the dorsal view, the posterior end of the idiodoma elongated significantly as a clearly defined long cauda, and gradually narrowed. It consisted of a caudal base with light striae, a caudal petiole with broad transverse plicated striae on the edge, and a smooth caudal xiphoid. The cauda was 226 long, measured from the posterior edge of the median shield to the end: the caudal base was 101, the caudal petiole was 74, and the caudal xiphoid was 51 in length. The anal region was dorsally located on the caudal base with fine dotted papillae, and the dorsal area of the genital region with *g3*–*g4* also had dotted papillae; the anal region had two pairs of pseudanal setae (*ps1–ps2*), 6 and 10 long, respectively; the length of the setae were *h2* 7, and lyrifissures (*ih*) were present.

Venter (Figures 10 and 11). Coxae I–IV were fused, resulting in a clearly whole ventral shield completely covered with dotted papillae; coxae III–IV had an obviously reticulated pattern. One pair of propodogastral setae (*ppgs*) was close to coxae II, there were two pairs of hysterogastral setae (*hgs1–hgs2*), 6 and 14 in length, and there was one pair of clear foveolae medially located between the coxae III groups. The area between hysterogastral setae *hgs2* had transverse striation. The setal formula of coxal plates I–IV was 3-1-3-3 *sts*.

Cauda venter (Figures 10 and 11). The cauda was 224 long: the caudal base was 99, including the genital area; the caudal petiole was broad with a plicated striae, 75 long; the caudal xiphoid was smooth and 50 long. The cauda was clearly defined and gradually narrowed, as in its dorsal view. The caudal base had horizontal striation except for the genital region (genital shield), with dotted papillae; the genital region had four pairs of genital setae (*g1–g4*), of which *g3–g4* were dorsally located (Figure 9), and two pairs of visible genital suckers. The lengths of setae *g1–g4* were 13, 14, 15 and 14, respectively.

Gnathosoma. Palp (Figure 12A) were five-segmented, 112 long, with granulated papillae and terminating with a bifid claw. Palp chaetotaxy was as follows: trochanter without setae; basifemur with one dorsal simple seta; telofemur with one dorsal stout seta and one short and tapering apophysis; genu with two stout setae and two simple setae; and tibiotarsus with one stout seta, three simple setae and one acute solenidion.

Chelicerae (Figure 12B): 92 long, with fine papillae; one cheliceral seta, 9 in length; it developed chela terminally.

Subcapitulum (Figure 12C): it had dotted papillae, 105 long, 70 wide, and two pairs of apophyses, of which one pair was smaller and claw-like and the other pair was blunt rod-like. It had two pairs of adoral setae (*ads1*, *ads2*), of which *ads1* was 6 and *ads2* was 4 in length, four pairs of hypostomal setae (*hg1–hg4*), where *hg2* and *hg4* were subequal in length and both were three times longer than *hg1* and *hg3*, and the lengths of *hg*-setae were *hg1* 12, *hg2* 29, *hg3* 10 and *hg4* 30. The distances between the bases of *hg*-setae were *hg1-hg1* 5, *hg2-hg2* 12, *hg3-hg3* 26, *hg4-hg4* 55, *hg1-hg2* 9, *hg2-hg3* 48 and *hg3-hg4* 19.

Legs (Figure 13). Leg IV was the longest and leg II was the shortest, with the tarsal lobes fairly well-developed; the dorsum of all leg segments had finely granulated papillae. The lengths of legs I–IV were 234, 220, 259 and 289. The lengths of tarsi I–IV were 84, 83, 100 and 101. Legs I–IV's chaetotaxy was as follows: coxae I–IV 3-1-3-3 *sts*; trochanters I–IV 1-1-2-1 *sts*; basifemora I–IV: 4-4-3-0 *sts*; telofemora I–IV 4-4-4-4 *sts*; genua I–IV 2 *asl*, {1 *asl*, 1 long *bsl*, 1 *mst*}, 4 *sts*-1 *asl*, 1 long *bsl*, 5 *sts*-1 *asl*, 1 long *bsl*, 5 *sts*-1 *asl*, 1 long *bsl*, 5 *sts*; tibiae I–IV 1 long *bsl*, {1 *asl*, 1 *mst*}, 4 *sts*-1 *bsl*, 5 *sts*-1 *bsl*, 5 *sts*-1 *T*, 4 *sts*; and tarsi I–IV 3 *asl*, 1 long *bsl*, 1 *fam*, 1 *dtsl*, 16 *sts*-1 long *bsl*, 1 *dtsl*, 15 *sts*-1 *dtsl*, 14 *sts*-1 *dtsl* and 13 *sts*.

Female and other developmental stages: unknown.

Remarks. This new species is similar to *C. mohanensis* Chen & Jin sp. nov., but differs from it in the following features: (1) *hg2* and *hg4* were subequal in length and both three times *hg1* and *hg3* (vs. *hg2* and *hg4* subequal in length and both four times *hg1* and *hg3* in *C. mohanensis* Chen & Jin sp. Nov.); (2) the chelicera reticulated pattern was absent (vs. present in *C. mohanensis* Chen & Jin sp. nov.); (3) the cauda was short, the caudal petiole was wider and the caudal xiphoid was shorter (vs. long and slender in *C. mohanensis* Chen & Jin sp. nov.).

Material examined. Holotype, male, collected from fallen leaves in Mohan Port (21°11′22.66″ N, 101°41′51.80″ E, elevation 893 m), Mengla County, Xishuangbanna Dai Autonomous Prefecture, Yunnan Province, China, on 20 November 2018, collector, Jianxin Chen and Xuesong Zhang. Slide No. YN-CU-201811201003. The slide is deposited in Institute of Entomology, Guizhou University, Guiyang, P. R. China (GUGC).

Etymology. The new specific epithet was formed by adding *neo-* (meaning new) to the name *C. mohanensis* Chen & Jin sp.nov., indicating its similarity to the latter species.

3.2. Brevicaudus Chen & Jin gen. nov.

Type species: *Brevicaudus trapezoides* Chen & Jin sp. nov.

Generic features (male): posterior of hysterosoma elongated with short conspicuous cauda, caudal petiole and caudal xiphoid short (as opposed to long in *Cunaxicaudus* Chen & Jin gen. nov.); palp between genu and tibiotarsus had one apophysis; the distance between setae *e1* and *d1* was approximately equal to that between *e1* and *f1; e1, c1* and *d1* were in a longitudinal line; lyrifissures (*im*) was situated lateral to *e1* and *f1*.

Brevicaudus Trapezoides Chen & Jin sp. nov.

Diagnosis. Setae *e1, f1* and *h1* were subequal and longer than the other dorsal setae (*lps, mps, c1, c2, d1*), with three pairs of hysterogastral setae (*hgs1–hgs3*); basifemora I–IV: 5-5-3-0 *sts*.

Description (Figures 14–21); male (n = 4).

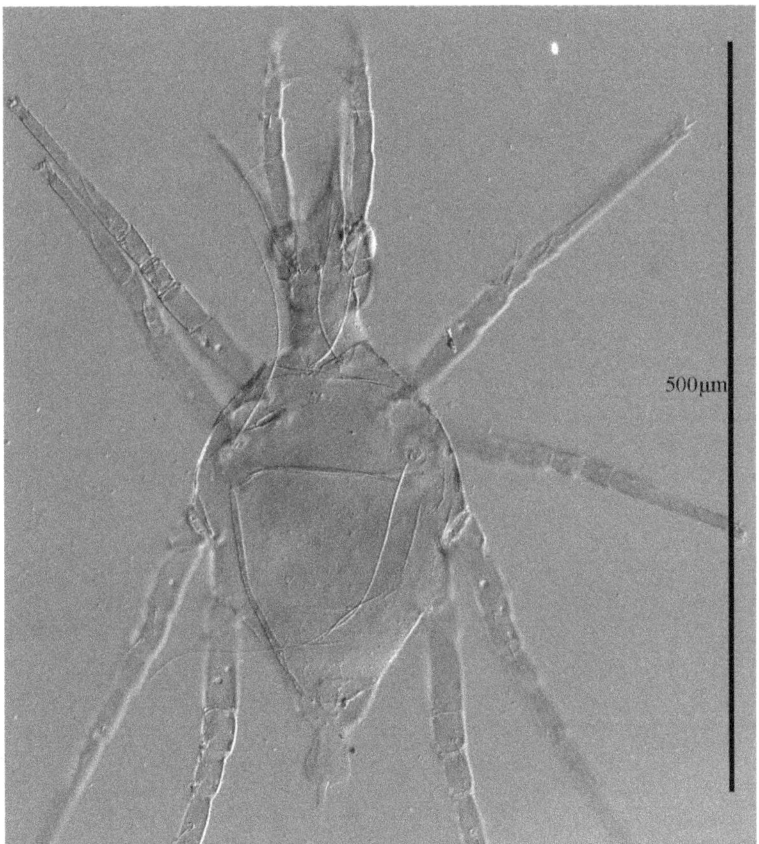

Figure 14. *Brevicaudus trapezoides* sp. nov. (male). Entire specimen (photo). Scale bar = 500 μm.

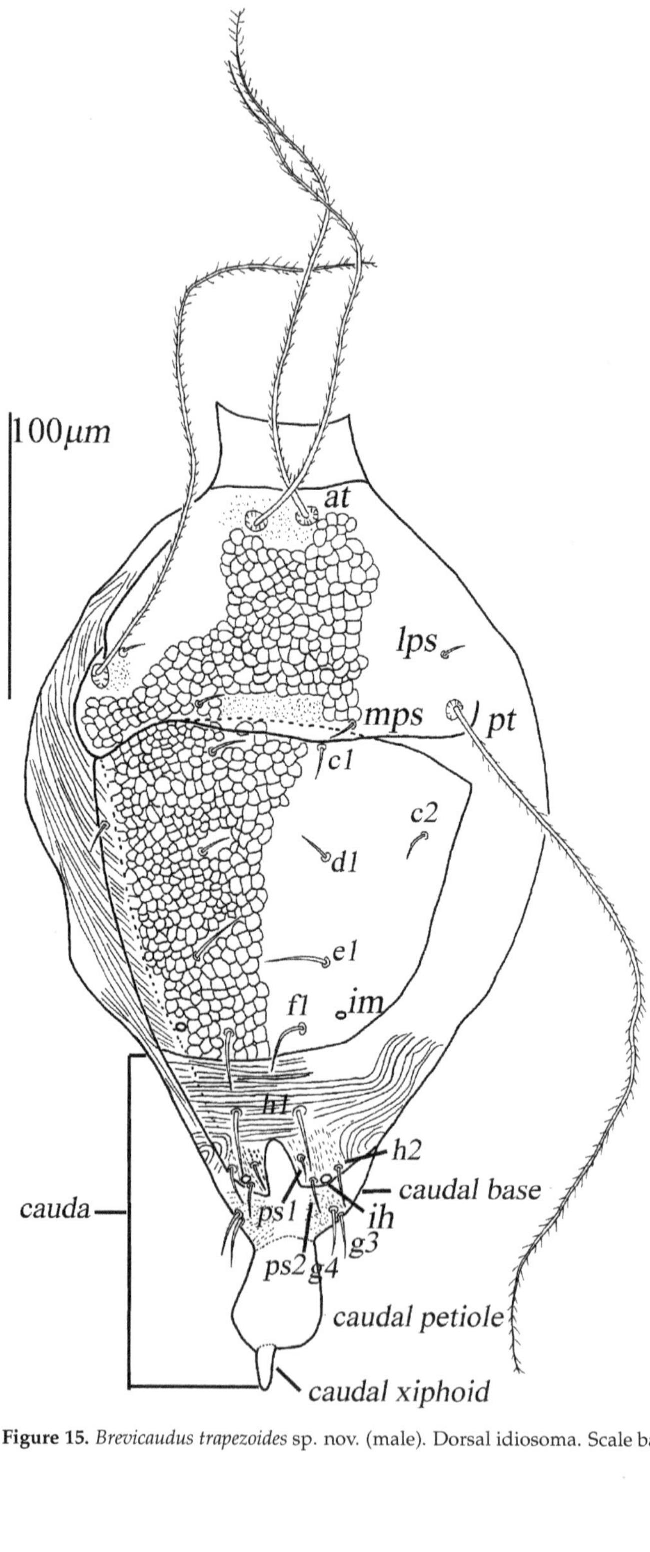

Figure 15. *Brevicaudus trapezoides* sp. nov. (male). Dorsal idiosoma. Scale bar = 100 μm.

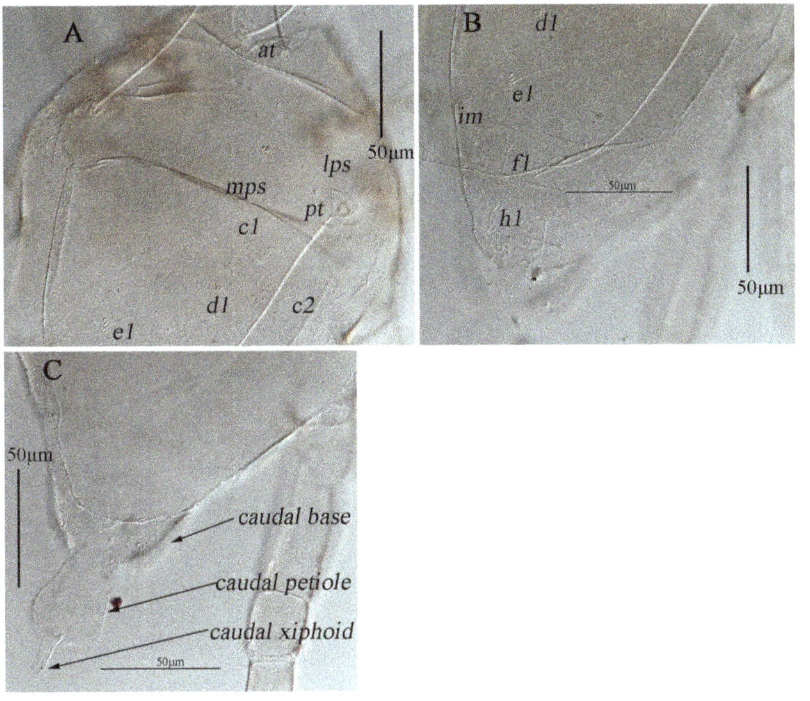

Figure 16. *Brevicaudus trapezoides* sp. nov. (male). (**A**,**B**)—Dorsal idiosoma (photo). (**C**)—Cauda (photo). Scale bar = 50 μm.

The idiosoma length was 284 (283–315) from the base of the subcapitulum to the posterior edge of the hysterosomal shield, the width was 195 (190–220), and the posterior end of the hysterosoma was elongated as a short cauda (Figures 14, 15 and 16C).

Dorsum (Figures 14, 15 and 16A,B). The propodosomal and hysterosomal shields were entirely complemented by reticulations, integument with striae. The propodosomal shield was 70 (70–85) long and 118 (95–130) wide, sclerotized and with a reticulated pattern, bearing one pair of anterior (*at*) and one pair of posterior (*pt*) setose trichobothria and two pairs of tactile setae (*lps* and *mps*); *at* was shorter than *pt*, *lps* was nearer to *pt* than to *at*; the area was anterior to setae *at* papillary. The setal lengths and distances of setae were *at* 175 (190–200), *pt* 259 (242–265), *lps* 10 (7–10), *mps* 12 (9–10); *at-at* 20 (20–23), *pt-pt* 125 (109–133), *lps-lps* 110 (96–124), *mps-mps* 53 (45–60), *lps-mps* 43 (31–42), *at-lps* 68 (69–75), *pt-mps* 37 (30–36), *pt-lps* 20 (10–22), *at-mps* 74 (75–88) and *at-pt* 82 (84–93).

The hysterosomal median shield was 121 (118–126) long and 126 (117–142) wide, with five pairs of dorsal simple setae (*c1*, *c2*, *d1*, *e1*, *f1*) and one pair of lyrifissures (*im*), which was situated laterally to *e1* and *f1*. The distance between setae *e1* and *d1* was about equal to that between *e1* and *f1*; *e1*, *c1* and *d1* were in a longitudinal line. *e1*, *f1* and *h1* were subequal and longer than *lps*, *mps*, *c1*, *c2* and *d1*. Setae *h1* was situated on the striated integument of the caudal base and was equal to *e1*, *f1* in length. The lengths of six pairs of dorsal setae were *c1* 12 (9–11), *c2* 10 (8–11), *d1* 12 (8–11), *e1* 23 (13–19), *f1* 18 (13–19), *h1* 21 (16–20). Distances of setae: *c1-c1* 39 (35–45), *c2-c2* 111 (103–114), *d1-d1* 42 (31–42), *e1-e1* 44 (24–46), *f1-f1* 26 (24–33), *h1-h1* 22 (16–26), *c1-c2* 47 (41–46), *c1-d1* 36 (32–37), *c2-d1* 36 (31–37), *d1-e1* 37 (34–36), *e1-f1* 29 (27–31) and *f1-h1* 30 (24–39).

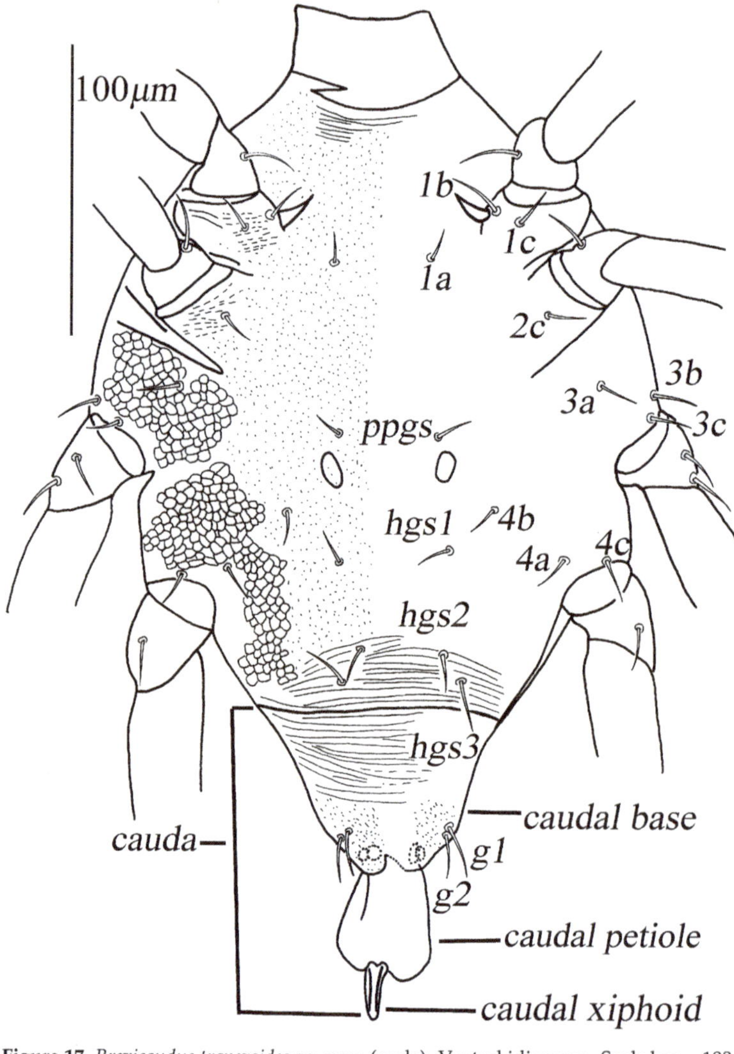

Figure 17. *Brevicaudus trapezoides* sp. nov. (male). Ventral idiosoma. Scale bar = 100 µm.

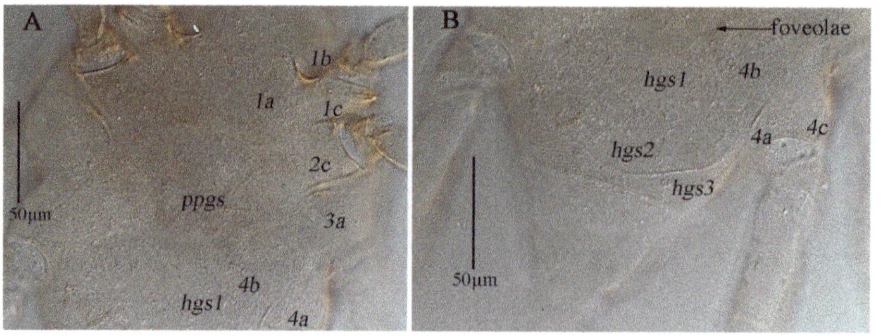

Figure 18. *Brevicaudus trapezoides* sp. nov. (male). (**A**,**B**)—Ventral idiosoma (photo). Scale bar = 50 µm.

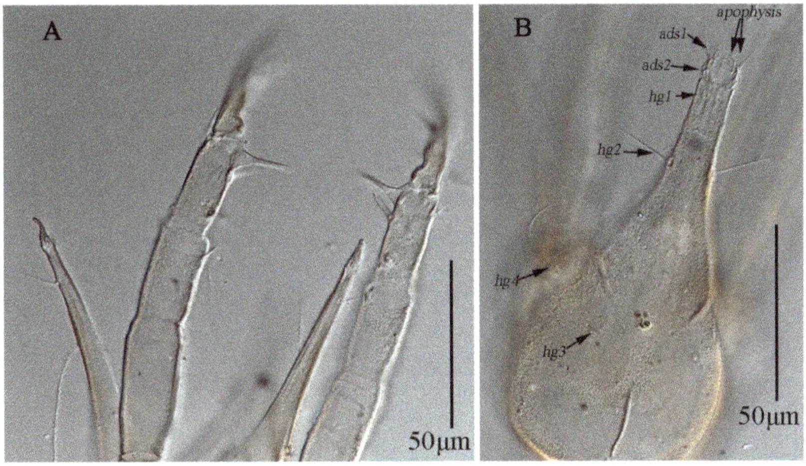

Figure 19. *Brevicaudus trapezoides* sp. nov. (male). (**A**)—Palp and chelicerae (photo). (**B**)—Subcapitulum (photo). Scale bar = 50 μm.

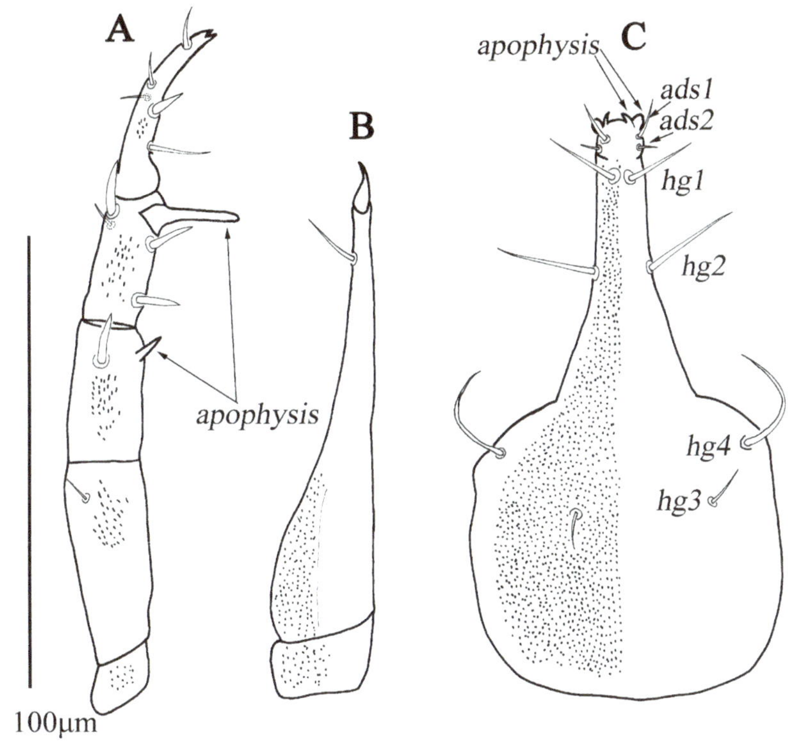

Figure 20. *Brevicaudus trapezoides* sp. nov. (male). (**A**)—Palp. (**B**)—Chelicerae. (**C**)—Subcapitulum. Scale bar = 100 μm.

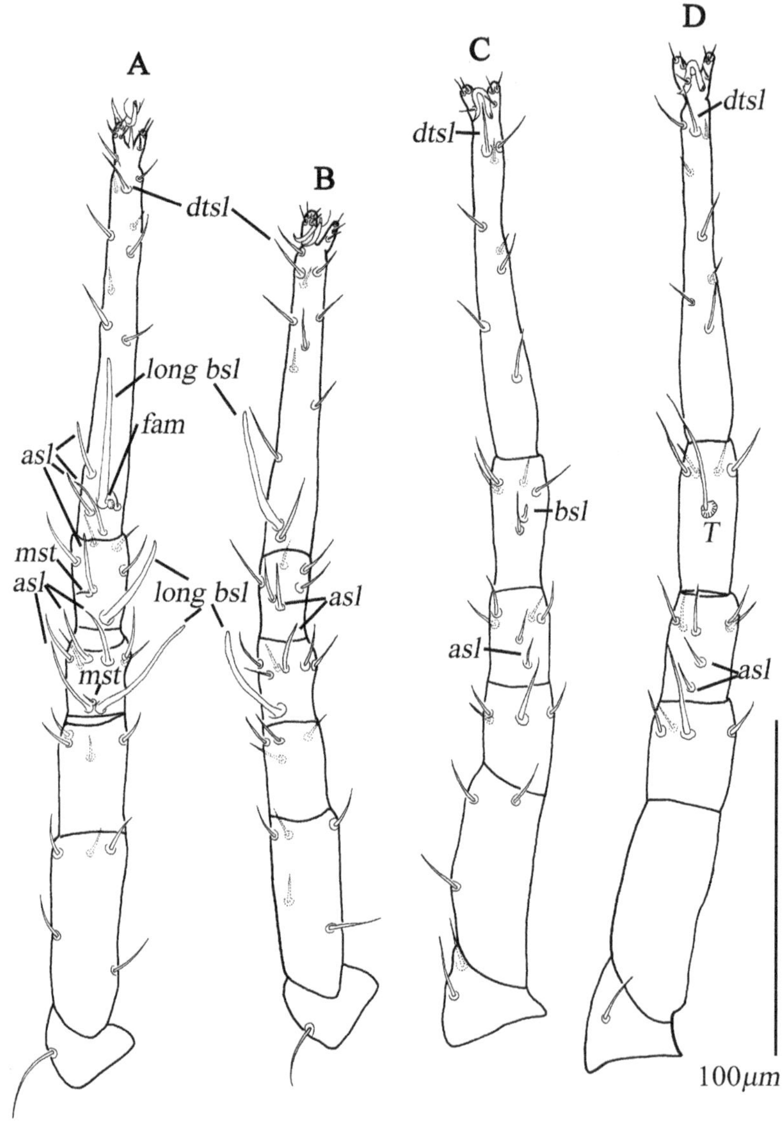

Figure 21. *Brevicaudus trapezoides* sp. nov. (Male). (**A–D**)—Legs I–IV, respectively. Scale bar = 100 μm.

Cauda dorsum (Figures 14, 15 and 16C). The posterior end of the idiodoma terminated into a distinctly developed cauda, which consisted of a caudal base with light striae and an approximately trapezoid (posteriorly widened) caudal petiole and caudal xiphoid, together almost as long as the caudal base, sharply reduced from the end of the caudal petiole. The cauda length was 100 (85–117): the caudal base was 50 (40–65), the caudal petiole was 30 (25–35) and the caudal xiphoid was 20 (18–22). The anal-genital region on the caudal base had fine dotted papillae; the genital region was *g3–g4*; the anal region had two pairs of pseudanal setae (*ps1–ps2*), 10 (12–14) and 11 (11–12) in length, respectively, and one pair of *h2* was 10 (8–11) in length and one pair was lyrifissures (*ih*).

Venter (Figures 17 and 18A,B). Coxae I–IV fused, forming a clearly whole ventral shield completely covered with dotted papillae; coxae III–IV had an obvious reticulated pattern. One pair of propodogastral setae (*ppgs*), 10 (10–12), were close to coxae II; three pairs of hysterogastral setae (*hgs1–hgs3*) had lengths of setae *hgs1–hgs3*: 10 (10–16), 13 (10–11), 17 (11–17); one pair of clear foveolae were medially located between the coxae III groups. Arae between *hgs2* and the posterior edge of the median shield had horizontal striation. The setal formula of coxal plates I–IV was 3-1-3-3 *sts*.

Cauda venter (Figure 17). The cauda was clearly defined as in its dorsal view, clearly separated from the ventral shield. The cauda was 110 (85–117) long: the caudal base was 55 (40–65) long, including the genital area; the caudal petiole was smooth and 35 (25–30) long; the caudal xiphoid was smooth and 20 (18–22) long. The caudal base had horizontal striation, except the genital area (genital shield); the genital shield had dotted papillae, two visible pairs of genital suckers, and four pairs of genital setae (*g1–g4*), of which *g3–g4* were dorsally located (Figure 15). The lengths of setae *g1– g4* were 14 (15–16), 14 (15–19), 18 (15–20) and 15 (15–17), respectively.

Gnathosoma. Palp (Figures 19A and 20A): five-segmented, 155 (145–169) long, with granulated papillae and terminating with a claw. Palp chaetotaxy was as follows: trochanter without setae; basifemur with one dorsal simple seta; telofemur with one dorsal stout seta and one short and tapering apophysis; genu with three stout setae, one simple seta and an apophysis close to base of tibiotarsus; and tibiotarsus with one stout seta, three simple setae and one terminating solenidion.

Chelicerae (Figures 19A and 20B): 125 (115–126) long, with fine papillae; the length of the cheliceral seta was 14 (13–14), which was located approximately 103 (95–106) from the base; one developed chela terminally.

Subcapitulum (Figures 19B and 20C): dotted papillae 132 (113–137) long and 63 (70–90) wide, with two pairs of apophyses, of which one pair was smaller and claw-like and the other was blunt and rod-like. There were two pairs of adoral setae (*ads1*, *ads2*), of which *ads1* was 11 (8–9) and *ads2* was 5 (4–5) in length, and four pairs of hypostomal setae (*hg1–hg4*). The lengths of *hg*-setae were *hg1* 16 (14–17), *hg2* 21 (23–24), *hg3* 8 (7–8) and *hg4* 31 (30–37). Distances between bases of *hg*-setae: *hg1-hg1* 6 (5–6), *hg2-hg2* 14 (13–17), *hg3-hg3* 30 (25–30), *hg4-hg4* 58 (59–69), *hg1-hg2* 22 (19–25), *hg2-hg3* 53 (47–65) and *hg3-hg4* 25 (20–24).

Legs (Figure 21). Leg IV was the longest and leg II was the shortest, and the tarsal lobes were well-developed. The lengths of legs I–IV were 270 (237–289), 230 (207–256), 265 (239–286), 288 (260–293). Lengths of tarsi I–IV: 110 (97–120), 99 (78–100), 97 (94–118) and 98 (93–118). Leg I–IV's chaetotaxy: coxae I–IV 3-1-3-3 *sts*; trochanters I–IV 1-1-2-1 *sts*; basifemora I–IV: 5-5-3-0 *sts*; telofemora I–IV 4-4-4-4 *sts*; genua I–IV 2 *asl*, {1 *asl*, 1 long *bsl*, 1 *mst*}, 4 *sts*-1 long *bsl*, 1 *asl*, 5 *sts*-1 *asl*, 5 *sts*-2 *asl*, 5 *sts*; tibiae I–IV 1 long *bsl*, {1 *asl*, 1 *mst*}, 4 *sts*-1 *asl*, 5 *sts*-1 *bsl*, 5 *sts*-1 *T*, 4 *sts*; and tarsi I–IV 3 *asl*, 1 long *bsl*, 1 *fam*, 1 *dtsl*, 17 *sts*-1 long *bsl*, 1 *dtsl*, 17 *sts*-1 *dtsl*, 13 *sts*-1 *dtsl* and 13 *sts*.

Female and other developmental stages. unknown.

Remarks. the new species resembles *C. mohanensis* Chen & Jin sp. nov. and *C. neomohanensis* Chen & Jin sp. nov., but significantly differs from the latter two by generic features.

Material examined. Holotype, male, collected from fallen leaves in Mohan Port (21°11′22.66″ N, 101°41′51.80″ E, elevation 893 m), Mengla County, Xishuangbanna Dai Autonomous Prefecture, Yunnan Province, China, on 9 June 2019, collector, Jianxin Chen, slide No. YN-CU-201906090208. Paratypes, two males, were collected from fallen leaves in Mohan Port (21°11′22.66″ N, 101°41′51.80″ E, elevation 893 m), Mengla County, Xishuangbanna Dai Autonomous Prefecture, Yunnan Province, China, on 6 June 2019, by Jianxin Chen; slides No. YN-CU-201906060101, YN-CU-201906060207. Paratype, one male, was collected from fallen leaves in Mohan Port (21°11′22.66″ N, 101°41′51.80″ E, elevation 893 m), Mengla County, Xishuangbanna Dai Autonomous Prefecture, Yunnan Province, China, on 8 June 2019, collector, Jianxin Chen; slide No. YN-CU-201906080201. All types of materials were deposited at the Institute of Entomology, Guizhou University, Guiyang, P. R. China (GUGC).

Etymology. Latin word 'brevis' means short in general. This refers to the short cauda of the new genus, which is much longer in *Cunaxicaudus* Chen & Jin gen. nov.; the new species is named for the caudal lobe, which is approximately trapezoid.

4. Discussion

According to Simely [5] and Skvarla et al. [10], the number of segments and length of palp, absence or presence of multi-branched seta on palp III (telofemur), cheliceral seta absent or present and *T* absent or present on tibiae IV were used as the main diagnostic features for the subfamilies in Cunaxidae. For example: palp with three segments = Counaxoidinae; palp with four segments = Scirulinae; palp with five segments and multi-branched seta present on palp III (telofemur) = Bonzinae; palp with five segments and multi-branched seta absent on palp III (telofemur), and palp extending beyond the subcapitulum by at least the distal half of palp IV(telofemur) = Cunaxinae; palp with five segments, multi-branched seta absent on palp III (telofemur), and palp not extending beyond the subcapitulum by more than the distal half of palp IV (genua), *T* present on tibiae IV, cheliceral seta usually present = Coleoscirinae; palp with five segments, multi-branched seta absent on palp III (telofemur), and palp not extending beyond the subcapitulum by more than the distal half of palp IV(genua), *T* absent on tibiae IV, and cheliceral seta absent, seta *hg1* geniculate = Orangescirulinae.

Our specimens of the three proposed new species have the following characteristics: palp with five segments, multi-branched seta absent on palp III (telofemur), and palp extending beyond the subcapitulum by at least the distal half of the palp IV (genua), cheliceral seta present and *T* present on tibiae IV. Moreover, we think that the feature of posterior configuration of hysterosoma can also be used as a main diagnostic feature for the subfamilies in the family Cunaxidae. The posterior end of the idiosoma elongated remarkably to form a cauda-like structure, which is an unusual new trait unknown in the family to date.

There was a tendency for the posterior area to the hysterosomal shield, including the genital region and anal plate (region) in the ventral view, to have the idiosoma elongated rearward in some members of Cunaxidae, in which case the anal pore was located on the end of the idiosoma and the anal plate (region) covered both ventral and dorsal around the pore in both males and females. What is more significant than that is that the end of the idiosoma obviously extended with the exposed aedeagus in males of some species, such as *Dactyloscirus humuli* (illustrated and mentioned) [26] and *Armascirus hastus* (illustrated but not mentioned) [27]. However, the completed and fully equipped cauda-like device is primarily defined in the present study.

A similar structure, with cauda and petiole as mating apparatuses [28], presents in some males of the water mite family Arrenuridae, especially those of the subfamily Arrenurinae, which makes the arrenurid mites sexual dimorphic [29]. The Arrenurid cauda is formed by an elongation of the hysterosoma posteriorly to a variable extent, and the petiole, either simple or complicated, and is a device derived from the integument of the idiosoma either caudally or directly from the posterior edge of dorsal shield [30]. The genital field or anal area may be or not on the cauda. Jin et al. [31] hypothesized that the formation and evolution of the arrenurid cauda and petiole was driven by the behavioral evolution of the reproductive mechanism from non-mating (indirect sperm transfer) to mating (direct sperm transfer). The cauda of Cunaxicaudinae Chen & Jin subfam. nov. is highly homologous, but remarkably different from that of the arrenurid mites. In cunaxicaud mites, both the genital area and the anal area were located on the male cauda, which implies that the traits (idiosoma posteriorly elongated with the accessory devices of cunaxicaud and arrenurid) evolved independently in two different branches. Interestingly, the aedeagus was not observed internally in all specimens of Cunaxicaudinae Chen & Jin subfam. nov., and therefore we propose that the caudal xiphoid might be the evaginated aedeagus with the idiosoma extending. Such an aedeagus was also observed in *Dactyloscirus condyles* [32].

Unfortunately, the females of Cunaxicaudinae Chen & Jin subfam. nov. were not collected. According to the literature, there are certain sexual differences, although they are not regarded as typical sexual dimorphism in the known members of Cunaxidae. These include the number of dorsal shields (females only have one, while males have two) [5,27,33,34], the setal formula of coxal plates (the setal formula of coxal plates I–IV was 3-2-3-3 *sts* in females, while 3-1-3-3 *sts* in males) [10], leg chaetotaxy (generally, males have longer *bsl* than females, especially on genua I–IV, tibia I and tarsi I–II) [35] and the number of genital setae (*g*) and eugenital setae (*eu*) (female genital setae were more than male, females had a pair of eugenital setae present, while in males this was absent) [13,16].

Therefore, we could put forward the hypothesis that the cauda of the new taxa may be the result of the evolution of the reproductive mode from indirect sperm transfer to direct sperm transfer; the new taxa may have sexual dimorphism, just like arrenurid mites, with the males having an exceptional body consisting of the mating device and females having a common oval body.

5. Conclusions

In the present work, a new subfamily, Cunaxicaudinae Chen & Jin subfam. nov., was described and illustrated based on a unique cauda-like structure in male Cunaxicaudinae. Additionally, we defined and discussed its morphology from the functional perspective. The findings highlight the species diversity and morphological evolution trends of the Cunaxidae.

6. Key to Subfamilies of Cunaxidae (Modified from Skvarla et al.) [10]

1. Hysterosoma cauda structure present in males (Figures 2, 3B, 4, 5A,B, 9, 10, 11, 14, 15, 16C and 17) **Cunaxicaudinae Chen & Jin subfam. nov.**

-Hysterosoma cauda structure absent in males 2

2. Palp telofemoral multi-branched seta present (except *Parabonzia*) **Bonziinae Den Heyer, 1978**

-Palp telofemoral multi-branched seta absent 3

3. Palp three-segmented **Cunaxoidinae Den Heyer, 1978**

-Palp four- or five-segmented 4

4. Palp four-segmented **Scirulinae Den Heyer, 1980**

-Palp five-segmented 5

5. Palp extended beyond the subcapitulum by at least the distal half of the genua **Cunaxinae Den Heyer, 1978**

-Palp not extended beyond the subcapitulum by more than the distal half of the genua 6

6. *T* on tibiae IV present; cheliceral seta usually present; seta *hg1* not geniculate **Coleoscirinae Den Heyer, 1978**

-*T* on tibiae IV absent; cheliceral seta absent; seta *hg1* geniculate **Orangescirulinae Bu & Li, 1987**

Author Contributions: Conceptualization, J.G., T.Y. and D.J.; methodology, all authors; project administration, D.J.; identification, J.C.; original draft preparation, J.C.; review and editing, D.J., M.Y., J.G. and T.Y. All authors have read and agreed to the published version of the manuscript.

Funding: This research was supported by the National Natural Science Foundation of China (31872275, 31272357), the Key Laboratory of Plants Protection Informatization for Featured and Efficient Agriculture in Central Guizhou Province (Qianjiaoji KY [2022] No. 052) and the PhD Foundation of Anshun University (asxybsjj202211 and asxybsjj202212).

Institutional Review Board Statement: Not applicable.

Informed Consent Statement: Not applicable.

Data Availability Statement: The data presented in this study are available on request from the corresponding author.

Acknowledgments: Thanks to Youfang Wu (Institute of Entomology, Guizhou University, Guiyang, China) and Xuesong Zhang (Forestry Bureau, Tongnan, Chongqing, China) for their great help in specimen collection.

Conflicts of Interest: The authors declare no conflict of interest.

References

1. Thor, S. Zur Systematik der Acarinenfamilien Bdellidae Koch, 1842, Grube 1859, Eupodidae Koch, 1842 und Cunaxidae Sig Thor, 1902. *Verh. Zool. Bot. Ges. Wien.* **1902**, *1902*, 159–165.
2. Zaher, M.A.; Soliman, Z.R.; El-Bishlawy, S.M. Feeding habits of the predacious mites, *Cunaxa capreola* (Acarina: Cunaxidae). *Entomophaga* **1975**, *20*, 209–212. [CrossRef]
3. Taha, H.A.; El-Naggar, M.E.E.; Abou, M.M.; Soliman, S.M. Effect of different prey species on the development and fecundity of the predaceous mite, *Neocunaxoides andrei* Baker and Hoff. (Acari: Cunaxidae). *Agric. Res. Rev.* **1988**, *66*, 129–135.
4. Walter, D.E.; Kaplan, D.T. Observation on *Colescirus simplex* (Acarina: Prostimata), a predatory mite colonizing greenhouse cultures of rootknot nematodes (Meloidogyne spp.) in the Cunaxidae. *Exp. Appl. Acarol.* **1991**, *12*, 47–59. [CrossRef]
5. Smiley, R.L. *The Predatory Mite Family Cunaxidae (Acari) of the World with a New Classification*; Indira Publishing House: West Bloomington, MN, USA, 1992; p. 356.
6. Walter, D.E.; Proctor, H.C. *Mites: Ecology, Evolution and Behavior. LIFE at a Microscale*, 2nd ed.; Springer: Dordrecht, The Netherlands; Heidelberg, Germany; New York, NY, USA; London, UK, 2013; p. 494. [CrossRef]
7. Hernandes, F.A.; Castro, T.M.M.G.; Venancio, R. Chapter 6 Prostigmata (Acari: Trombidiformes) as biological control agents. In *Prospects for Biological Control of Plant Feeding Mites and Other Harmful Organisms*; Carrillo, D., Moraes, G.J., Peña, J.E., Eds.; Springer: Berlin/Heidelberg, Germany, 2015; pp. 151–184. [CrossRef]
8. Skvarla, M.J.; Dowling, A.P.G. A Preliminary Phylogenetic Hypothesis for Cunaxidae (Acariformes: Trombidiformes: Prostigmata: Eupodina). In *Contemporary Acarology*; Skvarla, M., Ochoa, R., Verle Rodrigues, J., Hutcheson, H., Eds.; Springer: Cham, Switzerland, 2019; pp. 67–78. [CrossRef]
9. Wurlitzer, W.B.; Da-Costa, T.; De Azevedo, A.O.; Rocha, M.D.S.; Ferla, N.J. Edaphic diversity of cunaxid mites (Acari: Prostigmata) in different tree formations from northern region of Rio Grande Do Sul State, Brazil. *Syst. Appl. Acarol.* **2023**, *28*, 212–222. [CrossRef]
10. Skvarla, M.J.; Fisher, J.R.; Dowling, A.P.G. A review of Cunaxidae (Acariformes, Trombidiformes): Histories and diagnoses of subfamilies and genera, keys to world species, and some new locality records. *ZooKeys* **2014**, *418*, 1–103. [CrossRef]
11. Corpuz-Raros, L.A.; Naredo, J.C.B.; Garcia, R.C. Additional contributions to the knowledge of Philippine predatory mites mainly of the subfamily Cunaxinae and Cunaxoidinae (Acari: Prostigmata: Cunaxidae). *Acarologia* **2019**, *59*, 134–151. [CrossRef]
12. Kalúz, S.; Ermilov, S.G. A new genus of Pulaeini (Acari: Prostigmata: Cunaxidae) from South-East Asia. *Zootaxa* **2019**, *4619*, 382–390. [CrossRef] [PubMed]
13. Chen, J.X.; Guo, J.J.; Yi, T.C.; Jin, D.C. A new species of *Parabonzia* (Trombidiformes: Cunaxidae) based on adults and nymphs with a key to the world species. *Acarologia* **2020**, *60*, 806–824. [CrossRef]
14. Chen, J.X.; Yi, T.C.; Guo, J.J.; Jin, D.C. Two new species of *Armascirus* (Trombidiformes: Cunaxidae) from China. *Acarologia* **2021**, *61*, 453–467. [CrossRef]
15. Chen, J.X.; Yao, M.Y.; Guo, J.J.; Jin, D.C.; Yi, T.C. Two new species of *Neoscirula* (Trombidiformes: Cunaxidae) from China. *Syst. Appl. Acarol.* **2022**, *27*, 968–980. [CrossRef]
16. Khaustov, A.A. Contribution to systematics of the family Cunaxidae (Acari: Bdelloidea) of Western Siberia, Russia. *Syst. Appl. Acarol.* **2020**, *25*, 548–568. [CrossRef]
17. Laniecka, I.; Laniecki, R.; Kazmierski, A. New genus and four new species of the family Cunaxidae (Acariformes: Prostigmata: Bdelloidea) from Zambia. *Syst. Appl. Acarol.* **2021**, *26*, 981–1008. [CrossRef]
18. Wurlitzer, W.B.; Franklin, E.; Ferla, N.J.; Silva, G.L.; Rocha, M.S. *Pseudoscirus* gen. nov. of Coleoscirinae (Acari: Prostigmata: Cunaxidae) from the Amazon rainforest, Brazil, with a key to the genera of adult female Coleoscirinae. *J. Nat. Hist.* **2021**, *25–26*, 1639–1647. [CrossRef]
19. Wurlitzer, W.B.; Ferla, N.J.; Franklin, E.; Silva, G.L.; Rocha, M.D.S. First species of *Cunaxa* (Acari: Cunaxidae) from Brazil. *Syst. Appl. Acarol.* **2021**, *26*, 1508–1517. [CrossRef]
20. Wurlitzer, W.B.; Meira, A.A.; Vinhas, A.U.; Ferla, N.J. A new species and a new combination for the subfamily Cunaxinae (Acari: Cunaxidae). *Syst. Appl. Acarol.* **2022**, *27*, 141–148. [CrossRef]
21. Wurlitzer, W.B.; Ferla, N.J.; Franklin, E.; Rocha, M.D.S. A new species and a new report of Cunaxoidinae for the Brazil (Acari: Cunaxidae). *Syst. Appl. Acarol.* **2022**, *27*, 593–599. [CrossRef]
22. Walter, D.E.; Krantz, G.W. Collecting, rearing, and preparing specimens. In *a Manual of Acarology*, 3rd ed.; Krantz, G.W., Walter, D.E., Eds.; Texas Tech University Press: Lubbock, TX, USA, 2009; pp. 83–96.
23. Den Heyer, J.; Castro, T.M.M.G. A new cunaxid genus with descriptions of two new species from Brazil (Acari: Prostigmata: Bdelloidea: Cunaxidae). *Zootaxa* **2008**, *1731*, 42–50. [CrossRef]
24. Fisher, J.R.; Skvarla, M.J.; Bauchan, G.R.; Ochoa, R.; Dowling, A.P.G. *Trachymolgus purpureus* sp. n., an armored snout mite (Acari, Bdellidae) from the Ozark highlands: Morphology, development, and key to Trachymolgus Berlese. *ZooKeys* **2011**, *125*, 1–34. [CrossRef]
25. Den Heyer, J. Systematics of the family Cunaxidae Thor, 1902 (Actinedida: Acarida). *Public Univ. North Ser. A* **1981**, *24*, 1–19.
26. Liang, G.W. A new species and a new record of the genus *Dactyloscirus* from Shanghai, China (Acarina: Cunaxidae). *Entomotaxonomia* **1986**, *8*, 159–161. (In Chinese)
27. Paktinat-Saeij, S.; Bagheri, M.; Castro, T.M.D.; Saboori, A.; Gharekhani, G.; Ghobari, H. New species and records of Cunaxinae mites (Acari: Trombidiformes: Cunaxidae) from Iran. *Syst. Appl. Acarol.* **2017**, *22*, 1277–1294. [CrossRef]

28. Proctor, H.C.; Smith, B.P. Mating behaviour of the water mite *Arrenurus manubriator* (Acari: Arrenuridae). *Zool. Soc. Lond.* **1994**, *232*, 473–483. [CrossRef]
29. Cook, D.R. Water mite genera and subgenera. *Mem. Am. Entomol. Inst.* **1974**, *21*, 1–860.
30. Jin, D.C.; Wiles, P.R. New species of *Arrenurus* Duges 1834 (Acari: Hydrachnidia: Arrenuridae) from China and first records of watermites from Loas. *Acarologia* **1996**, *37*, 317–344.
31. Jin, D.C.; Li, L.S.; Wiles, P.R. The structure and evolution of male and petiole with a cladistic analysis of Chinese species of the Arrenurus (Acari: Arrenuridae). *Syst. Appl. Acarol.* **1997**, *2*, 195–209. [CrossRef]
32. Den Heyer, J. Notes on the cunaxid genus *Dactyloscirus* (Actinedida: Acarida) with descriptions of two new species from the Ethiopian Region. *Phytophylactica* **1979**, *11*, 87–98.
33. Den Heyer, J.; Castro, T.M.M.G. A new Neotropical genus of the family Cunaxidae (Acari: Prostigmata: Bdelloidea). *Zootaxa* **2008**, *1843*, 35–46. [CrossRef]
34. Den Heyer, J.; Ueckermann, E.A.; Khanjani, M. Iranian Cunaxidae (Acari: Prostigmata: Bdelloidea): Part 2. Subfamily Cunaxinae. *J. Nat. Hist.* **2011**, *45*, 1667–1678. [CrossRef]
35. Sergeyenko, A.L. Mites of the genera *Pulaeus* and *Lupaeus* (Acari: Prostigmata: Cunaxidae) of Crimea, Ukraine. *Zootaxa* **2011**, *2088*, 54–68. [CrossRef]

MDPI AG
Grosspeteranlage 5
4052 Basel
Switzerland
Tel.: +41 61 683 77 34

Animals Editorial Office
E-mail: animals@mdpi.com
www.mdpi.com/journal/animals

www.ingramcontent.com/pod-product-compliance
Lightning Source LLC
Chambersburg PA
CBHW041937260926
51850CB00018B/212
* 9 7 8 3 7 2 5 8 1 4 5 9 6 *